First European and Mediterranean Structural Engineering and Construction Conference (EURO-MED-SEC-1)

Interaction Between Theory and Practice in Civil Engineering and Construction

Istanbul, Turkey
24 - 29 May 2016

Editors:

R. Komurlu
A.P. Gurgun
A. Singh
S. Yazdani

ISBN: 978-1-7138-0002-6

First European and Mediterranean Structural Engineering and Construction Conference (EURO-MED-SEC-1)

Interaction Between Theory and Practice in Civil Engineering and Construction

Istanbul, Turkey
24 - 29 May 2016

Printed from e-media with permission by:

Curran Associates, Inc.
57 Morehouse Lane
Red Hook, NY 12571

Some format issues inherent in the e-media version may also appear in this print version.

ISEC PRESS
PO Box 1334
Fargo, North Dakota 58107
USA

isec.sec@gmail.com

Additional copies of this publication are available from:

Curran Associates, Inc.
57 Morehouse Lane
Red Hook, NY 12571 USA
Phone: 845-758-0400
Fax: 845-758-2633
Email: curran@proceedings.com
Web: www.proceedings.com

TABLE OF CONTENTS

KEYNOTE PAPERS

FIBER REINFORCED CEMENTITIOUS COMPOSITES

MECHANICAL PROPERTIES OF CONCRETE MATERIALS

BEHAVIOR OF CONCRETE AND CONCRETE-LIKE MATERIALS

CONCRETE STRUCTURES

STEEL STRUCTURES AND BRIDGES

STRUCTURAL ANALYSIS AND BEHAVIOR OF STRUCTURES

GEOTECHNICAL SOLUTION

SUSTAINABLE CONSTRUCTION PRACTICES

PROCUREMENT AND COST MANAGEMENT

ARCHITECTURAL SYSTEMS DESIGN

STRUCTURAL ENGINEERING AND CONCRETE

EURO-MED-SEC-1
The First European and Mediterranean Structural Engineering and Construction Conference
Istanbul, Turkey, May 24-29, 2016
Interaction between Theory and Practice in Civil Engineering and Construction

Welcome to EURO-MED-SEC-1

Dear Friends and Colleagues,

Welcome to the first European Mediterranean Conference 1 (EURO-MED-SEC-1) on Structural Engineering and Construction on May 24-29, 2016, in Istanbul and hosted by Yildiz Technical University. The conference theme is "Interaction between Theory and Practice in Civil Engineering and Construction".

Encouraged by the success of the ongoing ISEC Conference series, the Conference aims to bring together current academic research and development studies and initiatives, and the industrial practices in all aspects of Structural Engineering and Construction. It will include one day of pre-conference reception, two days of concurrent technical sessions and a banquet, one day for technical tours, and two days of cultural tours.

Istanbul, located at the intersection of two continents, Europe and Asia, has been home to three empires, namely Roman, Byzantine and Ottoman. The almost 3,000 year old city, with its ancient name of Constantinople, is a cultural magnet of east and west civilizations and offers many tourism attractions that include the gorgeous Bosphorus, museums, historical monuments and buildings, grandeur mosques, as well as its famous bazar for shopping.

If you have any queries, please feel free to contact the ISEC Secretariat on euro.med.sec@gmail.com or the local organizing committee. We look forward to welcoming you to EURO-MED-SEC-1, and hope to assist you to make your experience with our conference an enriching and memorable one for delegates and accompanying persons.

We look forward to welcoming you to EURO-MED-SEC-1 in Istanbul.

Kind regards,

Ruveyda Komurlu, PhD
Conference Chair
ruveydakomurlu@gmail.com

Asli Pelin Gurgun, PhD
Conference Co-Chair
pgurgun@gmail.com

Committees

Advisory Committee

- Amarjit Singh, President - ISEC Society, University of Hawaii at Manoa, USA
- Frank Yazdani, Secretary - ISEC Society, North Dakota State University, USA
- Dejan Dinevski, University of Maribor, Slovenia
- Takashi Hara, Tokuyama College of Technology, Japan
- Indubhushan Patnaikuni, RMIT University, Australia
- Swapan Saha, University of Western Sydney, Australia
- Vanissorn Vimonsatit, Curtin University, Australia

Local Advisory Committee

- Ismail Yuksek, President, Yildiz Technical University, Turkey
- Sadettin Hulagu, President, Kocaeli University, Turkey
- Ahmet Demir, Dean of Civil Engineering Faculty, Yildiz Technical University, Turkey
- Mehmet Berilgen, Head of Civil Engineering, Yildiz Technical University, Turkey

Local Organizing Committee

- Ruveyda Komurlu, Chair, Kocaeli University, Turkey
- Asli Pelin Gurgun, Co-Chair, Yildiz Technical University, Turkey
- Mehmet Sukru Ozcoban, Yildiz Technical University, Turkey
- Ali Cosar, Yildiz Technical University, Turkey
- Ozgur Ekincioglu, Istanbul Technical University, Turkey

Technical Committee

- Hakim S. Abdelgader, Tripoli University, Libya
- Syed Faiz Ahmed, Saudi Oger Ltd., Saudi Arabia
- Bulent Akbas, Gebze Technical University, Turkey
- Gulay Altay, Bogazici University, Turkey
- David Arditi, Illinois Institute of Technology, USA
- Gorun Arun, Yildiz Technical University, Turkey
- Senay Atabay, Yildiz Technical University, Turkey
- Neslihan Turkmenoglu Bayraktar, Kocaeli University, Turkey
- Saadet Arzu Berilgen, Yildiz Technical University, Turkey
- Talat Birgonul, Middle East Technical University, Turkey
- Andrea Boeri, University of Bologna, Italy
- Ilknur Bozbey, Istanbul University, Turkey
- Cenk Budayan, Yildiz Technical University, Turkey
- Necati Catbas, University of Central Florida, USA
- Haluk Cecen, Yildiz Technical University, Turkey

- Sai On Cheung, City University of Hong Kong, China
- Mehmet Sait Culfik, International Project Management Institute, Turkey
- Seval Pinarbasi Cuhadaroglu, Kocaeli University, Turkey
- Hans De Backer, Ghent University, Belgium
- Ahmet Dogan, Yildiz Technical University, Turkey
- Tuncer Edil, University of Wisconsin-Madison, USA
- Ilgin Gokasar, Bogazici University, Turkey
- Arie Gottfried, Politecnico di Milano, Italy
- Jack Steven Goulding, University of Central Lancashire, United Kingdom
- Murat Gunaydin, Istanbul Technical University, Turkey
- Erhan Guneyisi, Gaziantep University, Turkey
- Esra Mete Guneyisi, Gaziantep University, Turkey
- Suat Gunhan, University of Texas at San Antonio, USA
- Mustafa Gursoy, Yildiz Technical University, Turkey
- Makarand Hastak - Purdue University, USA
- Klaus Holschemacher, University of Applied Sciences, Germany
- Dean Kashiwagi Arizona State University, USA
- Jamal Khatib, University of Wolverhampton, UK
- Emre Kishali, Kocaeli University, Turkey,
- Zeynep Isik, Yildiz Technical University, Turkey
- Mohamed Marzouk, Cairo University, Egypt
- R. Edward Minchin Jr., University of Florida, USA
- Ruta Miniotaite, Kaunas University of Technology, Lithuania
- Cagri Mollamahmutoglu, Yildiz Technical University, Turkey
- Fuad Okay, Kocaeli University, Turkey
- Emre Otay, Bogazici University, Turkey
- Kutay Ozaydin, Yildiz Technical University, Turkey
- Pelin Tohumcu Ozener, Yildiz Technical University, Turkey
- Didem Ozevin, University of Illinois at Chicago, USA
- Hasan Ozkaynak, Beykent University, Turkey
- Hulusi Ozkul, Istanbul Technical University, Turkey
- Turan Ozturan, Bogazici University, Turkey
- Dieter Schlagbauer, Graz University of Technology, Austria
- Issam Srour, American University of Beirut, Lebanon
- Ali Touran - Northeastern University, USA
- Oral Yagci, Istanbul Technical University, Turkey
- Y. X. (Sarah) Zhang, University of New South Wales at the Australian Defense Force Academy, Australia

EURO-MED-SEC-1
The First European and Mediterranean Structural Engineering and Construction Conference
Istanbul, Turkey, May 24-29, 2016
Interaction between Theory and Practice in Civil Engineering and Construction

About Keynote Speaker - Dr. David Arditi

DAVID ARDITI, Ph.D

Professor and Director Construction Engineering and Management Program
Department of Civil, Architectural, and Environmental Engineering
Illinois Institute of Technology
Chicago, IL 60616

Dr. David Arditi is currently Professor of Civil and Architectural Engineering at Illinois Institute of Technology. He has been with the faculty of IIT since 1981. He is the founder and Director of the Construction Engineering and Management Program. He also served as department chair for six years. He has over 40 years of international development and construction project management experience.

Professor Arditi holds a PhD degree from Loughborough University, UK. His area of expertise covers all aspects of construction project management, engineering and support. In his 40-year academic career, he provided consulting services to several US firms, and several local and international agencies. He conducted several funded research projects sponsored by federal and state agencies. He supervised the research work of a large number of PhD and MS students from diverse countries. Many of his former PhD students serve as professors in their respective countries, while many of his former MS students occupy respectable positions in industry all over the world. Dr. Arditi and his research associates have published over 300 research papers in refereed national and international journals and in peer reviewed conference proceedings. He is currently serving on the editorial board of nine journals in the field of project management, and regularly reviews research proposals, papers and books for several research agencies and publications.

Dr. Arditi is an active member of several professional societies including CMAA, ASCE, AACE, and PMI. He has served over the years in many international, national and regional committees and was on several organizing committees of national and international conferences. He delivered several invited lectures and keynote lectures at different venues. He is also the recipient of multiple awards, inclu=ding being elected to the College of Fellows of CMAA in 2013.

Contact -

TEL# (312) 567-5751
FAX# (312) 567-3519
E-MAIL arditi@iit.edu
URL www.iit.edu/~arditi

Sponsors -

SOYAK
REAL ESTATE | ENERGY | CEMENT | CASTING

emay
www.emay.com

About Keynote Speaker - Dr. Tuncer B. Edil

TUNCER B. EDIL, Ph.D, P.E., D.GE

Department of Civil and Environmental Engineering
University of Wisconsin–Madison
34220 Istanbul, Turkey

Dr. Edil is a Distinguished Member of the American Society of Civil Engineers and a Diplomate of Geotechnical Engineering. He has been an active researcher and educator over 40 years at the University of Wisconsin-Madison. Professor Emeritus Edil is currently serving as Director of the new Recycled Materials Resource Center (RMRC-3G). His expertise includes use of recycled materials and sustainable construction, landfill liners and chemical transport, coastal erosion and landslides, soft soils and peat engineering, ground improvement, and use of geosynthetics.

Dr. Edil has published over 400 papers and book chapters in the field of geotechnical, geoenvironmental and pavement engineering. He has supervised approximately 125 MS and PhD students. Dr. Edil is a former Editor-in-Chief of the ASCE's Journal of Geotechnical Engineering and current Editor-in-Chief of Geotechnical and Geological Engineering journal. He served as President of the U.S. Universities Council on Geotechnical Engineering & Research and President of the ASCE Wisconsin Section as well as chairing important ASTM and ASCE committees. He is recipient of numerous personal and team/project awards from ASCE, ASTM, TRB and other organizations. Most notably, he is holder of 2007 Special Science Award from the Scientific and Technological Research Council of Turkey and received 2013 Thomas A. Middlebrooks and Terzaghi Awards and the rank of Distinguished Member from American Society of Civil Engineers. Additionally, he has received six team and project awards and numerous paper awards. He has served as a consultant to governmental agencies and engineering firms internationally on variety of issues.

Co-sponsors

CIOB
THE CHARTERED INSTITUTE OF BUILDING

The Chartered Institute of Building

TÜRKİYE HAZIR BETON BİRLİĞİ

The Turkish Ready Mixed Concrete Association

UPYE ULUSLARARASI PROJE YÖNETİMİ ENSTİTÜSÜ
INTERNATIONAL PROJECT MANAGEMENT INSTITUTE

The International Project Management Institute (UPYE)

SOYAK
REAL ESTATE | ENERGY | CEMENT | CASTING

SOYAK

emay
www.emay.com

EMAY

Interaction between Theory and Practice in Civil Engineering and Construction
Edited by Komurlu, R., Gurgun A. P., Singh, A., and Yazdani, S.

PREVENTING UNETHICAL AND ILLEGAL PRACTICES IN CONSTRUCTION

DAVID ARDITI

Dept of Civil, Architectural, and Environmental Engineering, Construction Engineering and Management Program, Illinois Institute of Technology, Chicago, IL, USA

Despite the fact that all professional associations, including ASCE, CMAA, PMI, AACE, and NSPE have codes of ethical conduct, the majority of the stakeholders in the construction process have encountered unethical behavior at one time or another. This is a contentious issue because there is always disagreement about what is and is not ethical. This paper looks into two practices that are most common in construction contracting, namely unbalanced bidding, and pre-award bid shopping. Both practices are considered to be totally acceptable by some, and totally unethical by others. Both practices can be stopped if preventive measures are put in place. The nature of these behaviors is examined, their impact is assessed, and preventive measures are formulated. In addition, in rare instances, illegal practices come to light. Collusive behavior is one such practice that is difficult but not impossible to detect. The nature of bid rigging is examined, its impact is assessed, and recommendations are made to detect parties that are involved in collusive practices. In addition to professional associations' efforts to publish standards for ethical conduct, the construction industry needs to be proactive in creating awareness about ethics issues, clear the confusion about what is ethical and what is not, and in promote ethical values throughout the entire spectrum of professions involved in construction.

Keywords: Ethics, Pre-award bid shopping, Unbalanced bidding, Bid rigging.

1 INTRODUCTION

Ethics is a set of accepted moral principles and values governing the appropriate conduct of an individual or group. Moral principles and values involve standards of what is right or wrong conduct. The importance of ethical conduct is recognized by all organizations. Administrators believe that good ethics is good business. Professional ethics, be it in science, medicine, law, or construction, is considered essential to the functioning of the profession (Loo 2002).

In the construction industry, the National Society of Professional Engineers (NSPE), the American Society of Civil Engineers (ASCE), the Construction Management Association of America (CMAA), the Project Management Institute (PMI), the Association for the Advancement of Cost Engineering (AACE), and many other associations have issued their respective standards of ethical conduct and consistently promote these principles to their members. The Accreditation Board for Engineering and Technology (ABET) has consistently expected an ethics dimension across the civil engineering curriculum in all accredited programs in the nation.

Despite these efforts to promote ethical conduct in the professions involved in the construction activity, there is a widespread perception of unethical conduct in the construction industry. A survey of owners, contractors, subcontractors, designers, and construction managers found that (1) nearly two thirds of the respondents believed that the industry is tainted by unethical actions; (2) more than 80% had personally encountered within the previous year acts or transactions they considered unethical; (3) fewer than one third of respondents' firms had an ethics policy that was formal, well understood by everyone in the firm, and enforced by top management; and (4) more than one third believed that unethical and illegal activities generate additional costs that amount to 2% or more of the total job budget (Doran 2004).

It has to be noted that even though most ethical decisions are rational, some are more emotional than rational. The trolley problem is a case in point that is cited by psychologists. Suppose a trolley is rolling down the track toward five people who will die unless you pull a lever that diverts it into another track where unfortunately lies one person who will die instead. Most people find this to be an easy call as it minimizes the loss of life. Now suppose the only way to save the five people is to push a bystander whose body will bring the trolley to a halt before it hits the others. It is still a one-for-five swap, but most people think that this is wrong. Similarly, in construction, one person's "unethical" may be another person's "tough competition". This paper takes up two controversial practices in construction, namely unbalanced bidding, and post-award bid shopping, practices that are justified as legitimate business practices by some and unethical behaviors that need to be stopped by others.

One should not confuse "unethical" with "unlawful". Ethics are governed by societal norms, whereas laws and regulations spell out what is legal or illegal. There are statutes in every country that spells out what constitutes a crime. Bribery and fraud are examples of unlawful action. After reviewing the evidence and hearing witnesses, a court of law decides whether the action is illegal or not. A conviction will result in sentencing that can include fines and/or incarceration. It is difficult to detect and prosecute unlawful conduct. This paper discusses one such practice, namely bid rigging, and proves that it is possible to detect and prevent collusive behavior.

2 UNBALANCED BIDDING

Unbalanced bidding is a serious problem in the construction industry. The owner may end up paying more money if the bid is unbalanced by the contractor. The owner has the right to reject an unbalanced bid. A bidder unbalances a bid by inflating the unit price of some line items and reducing the unit price of other line items. Frontloading is the most common practice where a bidder can mathematically unbalance a bid by overstating the unit price of line items scheduled to be performed early in the project and understating the unit price of line items performed later. A bidder can also overstate the unit price of a line item whose quantity was somehow underrated by the engineer. If the owner proves that a mathematically unbalanced bid costs more to perform, the bid is said to be materially unbalanced, in which case the owner can reject the bid.

While early linear programming models of unbalancing bids is relatively easy to detect by the owner, Nassar's (2004) research aims to unbalancing a bid and not be caught in the process. Cattell *et al.* (2007) summarize methods of unbalancing bids and argue that a client is given full information of a contractor's item pricing and that the client is given the choice to select among the contractor's competitors, implying that there are no ethical implications of unbalancing a bid. In other words, if an owner suffers the high cost of an unbalanced bid, it is rather the owner's fault to have selected the contractor who unbalanced the bid and not the contractor's fault who unbalanced the bid. However, according to a survey of 270 owners, architects, engineers, construction managers, general contractors, and subcontractors about ethical practices in the construction industry, unbalancing a bid was considered unethical by 84% of the respondents (Doran 2004). Also, New York City's Procurement Ethics Guide specifically asks contractors not to engage in unbalanced bidding.

A model was developed by Arditi and Chotibhongs (2009) that formalizes and automates the process of detecting mathematically and materially unbalanced bids by comparing line item prices with the engineer's estimates or the average prices offered by the bidders. This model allows owners to detect and reject unbalanced bids, and deters bidders from unbalancing their bid.

Because the extra cost of an unbalanced bid cannot be justified by the owner, and because owners have the right to legally reject unbalanced bids, owners should be able to stop and prevent unbalanced bids. A bidder's line item prices can be compared to the engineer's estimates to see if there are significant differences, an indication that the bid is potentially unbalanced. One can also compare a bidder's line item prices to the average line item prices of all bidders. Two separate models are therefore proposed but only one is described below. Since unbalancing can take the form of frontloading or adjusting the unit price of a line item whose quantity was understated by the engineer, both models are designed to deal with these situations. Both models are completely automated using MS Excel.

The proposed methodology is presented in Figure 1. Once all the bids are received from the bidders, the bottom line offers are compared with each other. Assuming that the bidders are qualified, the lowest offer is a candidate for contract award.

Starting from the first line item, the prices of each line item in the lowest offer are compared with the engineer's estimates (Figure 1) or the average prices of the respective items in all bids. If the differences are within acceptable limits, this bid is not unbalanced. The limits can be set by the owner depending on the type of project and the historical precedents. For example, Texas DOT has defined different limits for major and minor line items, while Wisconsin DOT's limits are based on project size. If the price difference for an item is beyond the acceptable limit, the analyst needs to check whether the quantity of the item was understated by the engineer.

- If the price of a line item is inflated by the bidder compared to the engineer's estimate (or the average prices of all bids) and the quantity involved in the line item was somehow understated by the engineer, this bid is mathematically unbalanced. In such cases, it is justified for the owner to recalculate the bids using the new (bigger) quantities but the same unit prices originally proposed by the bidders in order to establish if the lowest bid is materially unbalanced. If after recalculating the bids using the new quantities but the original unit prices proposed by the bidders, the original lowest bidder remains the lowest bidder, the contract is awarded to this bidder. But if the original lowest bidder is not the lowest bidder any more, then the original lowest bid is rejected on the basis that it was mathematically and materially unbalanced. The next lowest bidder becomes the new candidate for award of contract. The process described in this bullet point is repeated until an award is made or all bidders are exhausted, in which case the bid is re-advertised.

- If the price of a line item is inflated by the bidder compared to the engineer's estimate (or the average prices of all bids) and there are no doubts about the quantity involved in the line item, then the analyst needs to check if this bid was frontloaded. If the line item is scheduled early in the project, then the bid is frontloaded, i.e., it is mathematically unbalanced. The analyst then needs to check if this frontloading affects the present worth of the payments to the contractor by using an appropriate work schedule, discount rate, and economic analysis principles. If after calculating the present worth of the payments, the original lowest bidder remains the lowest bidder, the contract is awarded to this bidder. But if the original lowest bidder is no longer the lowest bidder, then the original lowest bid is rejected on the basis that it was mathematically and materially unbalanced. The next lowest bidder becomes the new candidate for award of contract. The process described in this bullet point is repeated until an award is made or all bidders are exhausted, in which case the bid is re-advertised. Back-end loading is not considered in this study because it is not likely to take place in the current low inflation environment in the U.S.

Unbalanced bidding is a serious problem for the construction owner. Unbalancing a bid is considered by many as unethical. Unbalanced bids can be rejected if caught by the owner. If awarded, the cost of these contracts to the owner is unjustifiably increased. Despite these facts, most researchers appear to be interested in the optimization of a contractor's cash flow by unbalancing a bid and not be caught in the process (e.g., Nassar 2004, Cattell *et al.* 2007). Detecting an unbalanced bid is normally difficult and has become even more difficult thanks to the efforts of these researchers. Given the current literature that aggressively tries to teach contractors the various methods of unbalancing a bid without getting caught, it is time to provide a sensible tool that allows owners to detect and reject unbalanced bids.

FHWA and a number of state DOTs in the U.S. have spelled out general principles to protect their interests with respect to unbalanced bids. Many DOTs routinely but informally check bids for unbalancing using different processes. Some DOTs have a formal process in place but each use different approaches and different assumptions. The proposed model represents a marked improvement on existing practice because it is an attempt to develop a thorough methodology that systematically covers all aspects of unbalancing a bid. The proposed model is fully automated. It institutionalizes the process of detecting unbalanced bids and is expected to deter bidders from unbalancing their bids.

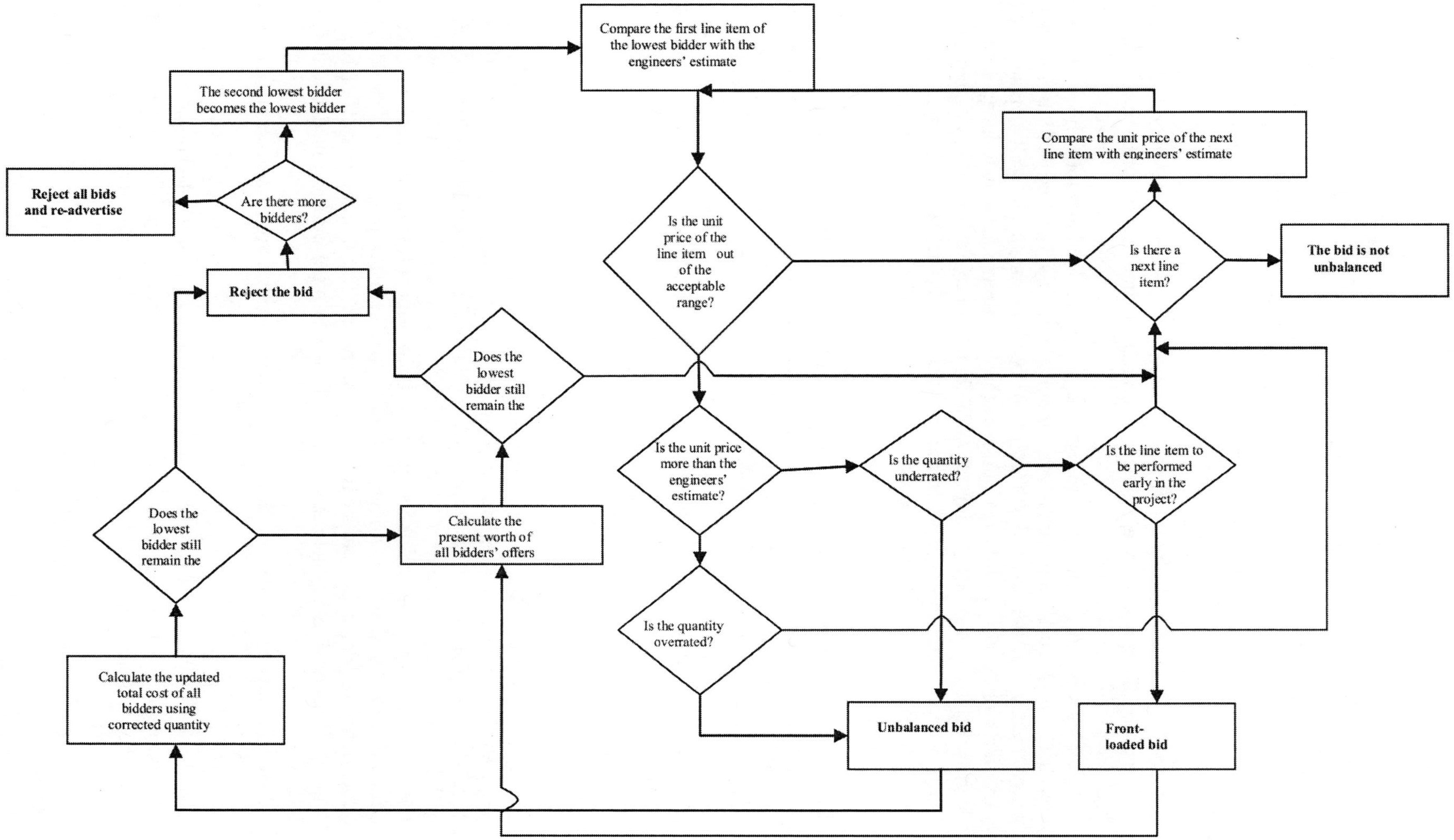

Figure 1. Detection and analysis of unbalanced bids using the engineer's estimates.

3 POST-AWARD BID SHOPPING

Pre-bid bid shopping is a way of life in the construction industry. General contractors typically shop bids before submitting a bid to the owner. If bid shopping occurs before a contract is awarded, the general contractor may or may not benefit of the practice, depending on the number of the other competing bidders who receive the same reduced quotations. The owner receives the benefit of the practice unless quality standards are reduced to accommodate the lower subcontractor quotations. The subcontractor who is selected for the project as a result of submitting the lowest price is the apparent beneficiary of bid shopping.

Post-award bid shopping is initiated by a general contractor who is awarded a contract. A general contractor who indulges in this practice will try to get the subcontractor to decrease the submitted bidding price or will engage another subcontractor who is willing to perform the specified job at a reduced price. The doctrine of promissory estoppel binds a subcontractor to perform in accordance with its bid if accepted by the general contractor. Conversely however, the contractor is not bound to use the subcontractor to perform the work, but is free to shop for a lower bid in an attempt to get a better price.

Any reduction in the subcontractor's price made after the contract is awarded, directly increases the general contractor's profit but may harm owners as well as subcontractors. It is likely that a bid that has been shopped is often awarded to an unqualified subcontractor who entered into the process late in the bid phase and has not adequately planned for the job. Consequently, post-award bid shopping may promote lower standards of work performance, reduce overall project quality, create an adversarial relationship between the parties, provoke legal disputes, foster unfair competition, be conducive to subcontractor insolvencies, and reduce overall jobsite safety (Clough *et al.* 2005). Post-award bid shopping and is considered unethical by most parties (Doran 2004, Arditi and Chotibhongs 2005).

How often do general contractors shop bids to select their subcontractors after the contract is awarded? On a 0-3 scoring system, the average frequency is 1.42 for subcontractors, 0.18 for general contractors and 1.36 for owners. The difference between the frequencies reported by subcontractors and owners is not statistically significant but the differences between the frequencies reported by general contractors and the other two groups are statistically significant at $\alpha = 0.05$. It is not surprising that the average frequencies of the responses from subcontractors and owners are high because post-award bid shopping is a major problem for them. On the other hand, general contractors appear to deny that post-award bid shopping is at all happening possibly because this practice is considered to be a legal but unethical practice in the industry.

What can be done to improve general contractors' selection practice of subcontractors? The distribution of the responses in Figure 2 shows that according to the large majority of owners (91%) and subcontractors (78%), a bid listing required by owners appears to be the best way to eliminate post-award bid shopping (Arditi and Chotibhongs 2005). Some states have enacted statutes that require general contractors to list their subcontractors in their bid to owners. In these states, the general contractor cannot change subcontractors if its bid is accepted. On the other hand, it is not surprising that only 14% of general contractors agree with this recommendation.

Indeed post-award bid shopping is clearly in the interest of general contractors since it helps them achieve their primary objective of profit maximization, albeit by means that are considered by most parties involved in the construction activity to be unethical.

About half of the general contractors (58%) and owners (49%) believe that subcontractors should submit fair quotations to stop bid shopping whereas only 17% of subcontractors agree with this assessment. It appears that some general contractors and owners attribute the necessity for post-award bid shopping to higher-than-expected quotations on the part of subcontractors, whereas subcontractors justify higher-than-expected quotations in order to better position themselves in the inevitable bid shopping that will occur after the award.

Some respondents (52% of subcontractors, 24% of general contractors, and 33% of owners) suggested that trade associations (e.g., AGC, ABC, ASA, etc.) should seriously discourage bid shopping in order to prevent adversarial relationships between subcontractors and general contractors. Post-award bid shopping is clearly disturbing to all subcontractors whenever general contractors shop bids for any reason other than scope changes in the subcontracted work. The fact that owners do not see the benefit of post-award bid shopping while only general contractors benefit from it makes it unacceptable for both subcontractors and owners.

Many subcontractors (66%) and general contractors (55%) indicate that subcontractors should consider the history of bid shopping of the general contractor before sending in a quotation.

4 COLLUSIVE PRACTICES

Collusion is a serious unethical and illegal practice in competitive bidding. In this scheme, a cartel bidder is allowed by other cartel bidders to win a particular bid and obtain the greatest possible benefit from it. Collusion among cartel bidders creates a non-competitive bidding environment (Zarkada-Fraser and Skitmore 1998). Collusive bidding behavior also increases standard market costs and causes economic damage to non-cartel bidders because of unfair competition (Zarkada-Fraser and Skitmore 1998). According to the U.S. Department of Justice, some potential signals of collusion include the same bidder winning many bids, the same group of bidders submitting bids, the bidders in the group taking turns being the lowest bidder, bids being much higher than the engineer's estimate, fewer bidders bidding compared to normal bidding instances, bidder bidding significantly higher on some bids, and bid prices drop whenever a new or infrequent bidder submits a bid.

The factors that affect collusion, the reasons why contractors are tempted to enter into collusive agreements with other contractors, the legal implications of collusion for contractors and construction owners have long been investigated by a small number of researchers. Zarkada-Fraser and Skitmore (1998) conducted a survey of 72 Australian estimators to explore the factors that affect collusive intent and found that the welfare of a respondent's company overrides abstract philosophical principles, and that the law, industry norms and direct orders rank higher than moral concerns, or personal values. Porter (2005) discusses the factors that facilitate or inhibit collusive schemes. According to Porter and Zona (2008), cartel members may bid high or refrain from bidding in anticipation of the opportunity to obtain higher profits, contemporaneously or in the future. Doree (2004) asserts that a tougher public sector procurement policy and continued reliance on lowest bid prices may not contribute to a collusion-free environment. According to Doree (2004), an alternative approach allowing for a balance of competition and collaboration with a wider number of selection criteria would create a more dynamic and iterative competitive process.

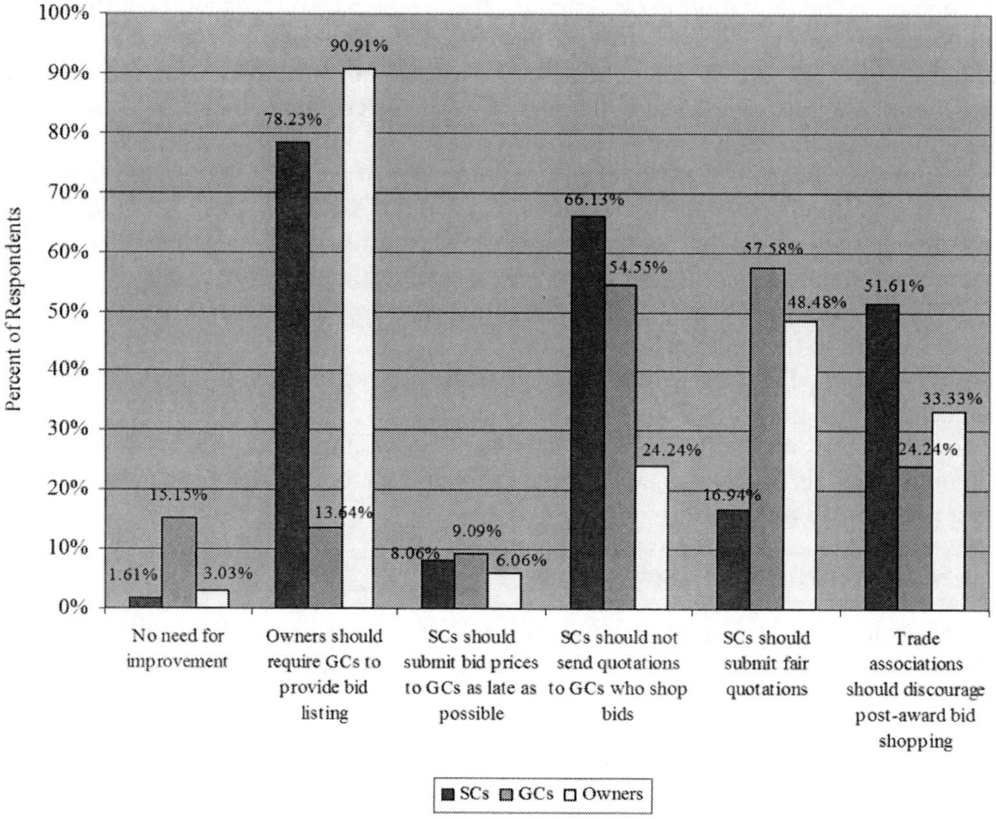

Figure 2. Ways to improve subcontractor selection practice.

It is difficult to tell how common collusion is in construction contracting even though Zarkada-Fraser and Skitmore (1998) believe that collusion is an "endemic malaise" in construction bidding. Because the marketing function of construction companies revolves around competitive bidding in public contracts, it is likely that there will be occasional cases of collusion even though it is considered illegal in most countries. For example, many writers exposed collusive practices in Japan, the U.S., the Netherlands, South Africa, and Australia.

Because collusion is a secret conspiracy between cartel bidders, its detection requires extensive police investigation in the form of collection of legal evidence such as recordings of meetings between cartel bidders and witness testimonies. Often collusion is detected by means of cooperation agreements with colluders in exchange for leniency. Another way to detect collusion is to analyze the abnormalities in the bids. For example, Porter and Zona (1993) investigated the differences in the cost structures between known cartel bidder and non-cartel bidder groups in federal highway construction projects. Pesendorfer (2000) investigated the differences in cost structures in sealed bid first-price auctions. Bajari and Ye (2003) proposed statistical methods to detect collusion in highway contracts. Despite these efforts to detect collusion, using these methods, it is difficult to detect cartel bidders with certainty, and the applicability of these methods is limited.

The differences in the bidding behavior of potential cartel bidders and non-cartel bidders were analyzed by Chotibhongs and Arditi (2012a, 2012b). In this study, the collusive behavior of construction companies bidding in response to the solicitations of a public agency during a 10-year study period 2001-2010 was investigated. After a list of potential cartel bidders was determined, three additional analyses were performed to confirm/refute the suspicion by adding a measure of certainty to the decision.

By using the residual and the cost structure stability tests in the first step (Figure 3), six bidders were categorized as potential cartel bidders since they failed both tests. The residual test assesses the correlations between the residuals of the regression equations. The cost structure stability test investigates whether bidders behave the same way when they are confronted with the same cost structure. These six bidders possess almost 50% (47.75%) of the public agency's market during the study period 2001 to 2010. They bid against each other 12 times on average during this time.

The first analysis in the second step (Figure 3) involved the analysis of bid distributions. This analysis provided an important outcome that the bids predicted by the potential cartel bidder model are higher than the bids predicted by the non-cartel bidder model. This result agrees with the assumption that cartel bidders increase their bids over the normal competitive level. The second analysis in the second step (Figure 3) involved the analysis of model dispersion. This analysis showed that the bid clustering of potential cartel bidders is denser than non-cartel bidders' because cartel bidders try to regulate and control the competitive bidding environment by forcing other bidders to bid higher but not much higher than theirs offers. The third analysis in the second step (Figure 3) involved the comparison of the cost structures. It showed that the independent variables impacted the bid prices of the potential cartel and non-cartel bidders differently. All these analyses, individually and jointly, reinforce the suspicion that the six potential cartel bidders identified in Step 1 have indeed been involved in potentially collusive activities.

Of the six potential cartel bidders who failed both the residual and the cost structure stability tests, two were audited by the public agency relative to bid fraud, while another was found guilty by a court of law for bid-rigging and banned from doing business with the public agency. None of the remaining bidders' names were mentioned in the news as having been involved in bidding irregularities. This information was found after the model was set up and run, quite independently of the study reported in this paper. It was welcome news as it partially supported the findings of the study.

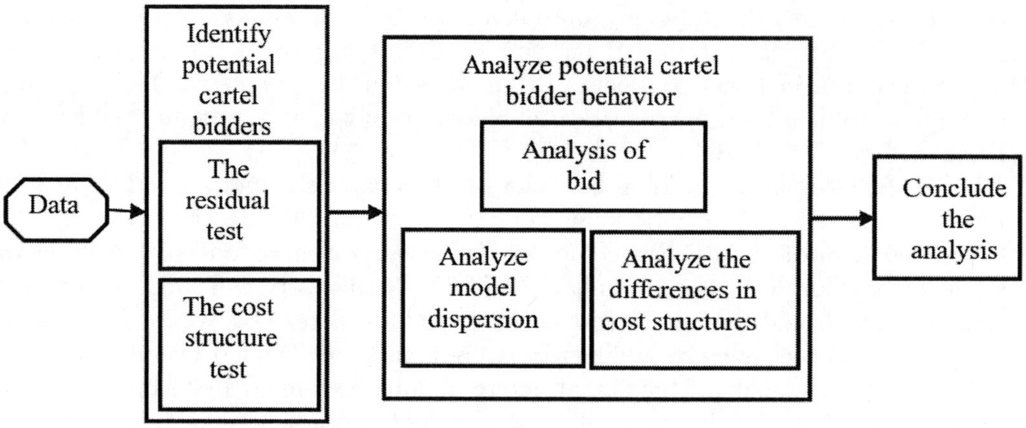

Figure 3. Collusion analysis model.

References

Arditi, D. and Chotibhongs, R. (2009). Detection and Prevention of Unbalanced Bids, *Construction Management and Economics*, Vol. 27(8), 721-732.

Arditi, D. and Chotibhongs, R. (2005). Issues in Subcontracting Practice, *Journal of Construction Engineering and Management,* ASCE, Vol. 131(8), 866-876.

Bajari, P. and Ye, L. (2003). Deciding Between Competition and Collusion. *Review of Economics and Statistics*, 85(4), 971-989.

Cattell, D.W., Bowen, P.A., and Kaka, A.P. (2007). Review of Unbalanced Bidding Models in Construction. *Journal of Construction Engineering and Management*, ASCE, 133(8), 562-573.

Chotibhongs, R. and Arditi, D. (2012a). Detection of Collusive Behavior, *Journal of Construction Engineering and Management*, ASCE, Vol. 138(11), 866-876.

Chotibhongs, R and Arditi, D. (2012b). Analysis of Collusive Bidding Behavior, *Construction Management and Economics*, Vol. 30(3), 221-231.

Clough, R.H., Sears, G.A. and Sears, S.K. (2005). *Construction Contracting*. Wiley.

Dorée, A. G. (2004). Collusion in the Dutch construction industry: An industrial organization perspective. *Building Research and Information*, 32 (2), 146-156.

Doran, D. (2004). *FMI/CMAA Survey of Construction Industry Ethical Practices*. FMI Corporation, Raleigh, NC.

Nassar, K. (2004). "Using Spreadsheets to Optimally Unbalance a Construction Bid." *Cost Engineering*, AACEI, 46(12), 28-32.

Pesendorfer, M. (2000). A Study of Collusion in First Price Auctions. *Review of Economic Studies,* 2000(67), 381-411.

Porter, R. H. (2005). Detecting collusion. *Review of Industrial Organization,* 26(2), 147-167.

Porter, R.H. and Zona, J.D. (1993). Detecting Bid Rigging in Procurement Auctions, *Journal of Political Economy,* 101(3), 518-538.

Zarkada-Fraser, A.Z. and Skitmore, M. (2000). Decisions with Moral Content: Collusion. *Construction Management and Economics*, 2000(18), 101-111.

Interaction between Theory and Practice in Civil Engineering and Construction
Edited by Komurlu, R., Gurgun, A. P., Singh, A., and Yazdani, S.

INNOVATIVE SOFT SOIL IMPROVEMENT BY VACUUM DE-WATERING AND DYNAMIC COMPACTION

TUNCER B. EDIL

Dept of Civil & Environmental Engineering, University of Wisconsin-Madison, Madison, Wisconsin, USA

Recently, an innovative soft soil improvement method was advanced in China by integrating and modifying vacuum consolidation and dynamic compaction ground improvement techniques in an intelligent and controlled manner. This innovative soft soil improvement method is referred to as "High Vacuum Densification Method (HVDM)" to reflect its combined use of vacuum de-watering and dynamic compaction techniques in cycles. Over the past ten years, this innovative soft soil improvement technique has been successfully used in China and Asia for numerous large-scale soft soil improvement projects, from which enormous time and cost savings have been achieved. In this presentation, the working principles of the HVDM will be described. A discussion of the range of fine-grained, cohesive soil properties that would make them ideal for applying HVDM as an efficient ground improvement method will be discussed. The economic benefits and environmental benefits of HVDM are elucidated.

Keywords: Ground improvement, Soft clays, In-situ improvement.

1 INTRODUCTION

In-situ improvement of soft cohesive soils is one of the main challenges facing geotechnical engineers and contractors alike. In countries such as China, India, and other emerging countries in Asia where the population is large and infrastructure development is proceeding at a heightened pace, the need for a fast, economical in-situ improvement for soft cohesive soils in a large-scale is clearly evident. The traditional methods of soft cohesive soil treatment include the use of the following techniques: (a) prefabricated vertical drains (PVDs) and fill preloading, (b) vacuum consolidation together with PVDs, (c) stone columns, (d) rammed aggregate piers, (e) thermal treatment, (f) chemical mixing, (g) electro-osmosis, and (h) deep dynamic compaction. Despite the availability of various methods of in-situ improvements listed above, the method of incorporating PVDs with fill preloading appears to be the most widely used technique throughout the world, even though the vacuum consolidation method has gained some interest recently. In large-scale applications — such as land reclamation using dredged materials, port facility construction, economic zone development along coastal areas, petro-chemical plants near shorelines, steel mills, power plants, airport runways, and highways — the areas to be treated could be excessively large and the availability of usable earth for fill preloading could be scare. Therefore, there is a great interest in developing a more effective way of treating soft cohesive soils in a large area where preloading fill cannot be economically obtained.

The application of vacuum to facilitate consolidation in saturated fine-grained soils has been used either alone by means of PVDs or in combination with the static surcharge load using fill materials (Kjellman 1952, Holtz 1975). The effectiveness of vacuum consolidation with or without surcharge loading is highly dependent upon soil permeability and the efficiency of the vacuum system. The desired degree of soil improvement and the allowable time duration for completion can also play an important role in determining if vacuum consolidation can be a viable soil improvement method for a project. The use of deep dynamic compaction technique in saturated fine-grained soils has not been widely accepted, even though the validity of its working mechanisms has been demonstrated (Menard 1975).

Due to rapid infrastructure development in China, an innovative soft cohesive soil treatment technique was developed in 2000 and had since been rapidly applied in China and other countries in Asia. The core of this innovative, in-situ, cohesive soil treatment method was termed as "High Vacuum Densification Method" (HVDM), and it was granted a series of international patents and registered in more than 25 countries. The success of HVDM was quite remarkable in a sense that the technique blends two well-known soil improvement methods, vacuum consolidation and deep dynamic compaction, into an intelligent yet efficient soft soil treatment method that can treat a large area within a relatively short time period.

In this paper, the working principles of this ground improvement method are described. The distinguishing features of this method and its advantages and limitations are elucidated. It should also be noted that this method is limited to treating soft soils to depths up to 8 to 10 meters. For treating the soft soils with depth greater than 8 meters or for achieving much higher improved strength, the HVDM technology for shallow soft soils has been extended to two additional innovative techniques or the conventional surcharge loading with prefabricated vertical drains (PVD) is required.

2 HVDM CONSTRUCTION SEQUENCES

HVDM can be described as a fast ground improvement technology utilizing drainage, consolidation, and densification principles. HVDM is generally executed in a controlled manner based on feedback of on-site monitoring data collected for quality assurance and quality control (QA/QC) purposes. Figure 1 provides a schematic drawing of HVDM using vacuum consolidation and deep dynamic compaction. The HVDM consists of the following steps:

Step 1: Conduct detailed geotechnical investigation at the project site. Evaluate and determine the soil profile at the site with detailed knowledge of the depth and thickness and distribution pattern of soft soils requiring treatment. Obtain important basic soil properties, including gradation curves, Atterberg limits, water content, hydraulic conductivity, compressibility, and coefficient of consolidation. Conduct in-situ tests, such as cone penetration testing (CPT) or standard penetration test (SPT) to establish baseline values prior to commencing the HVDM in the field. Understand and establish performance criteria of ground treatment. Perform preliminary design to provide plans for optimum spacing and depth of vacuum pipes, energy level of deep dynamic compaction (DC) and number of drops and grid spacing of tamper, time needed for vacuum consolidation between cycles of dynamic compaction, etc. However, it should be emphasized that the initial plans will generally need to be modified based on on-site monitoring data and the expected final performance criteria.

Figure 1. Schematic showing the HVDM method.

Step 2: Install vertical vacuum pipes and horizontal drainage pipes. The vertical vacuum pipes can be installed using several different methods, such as using a vibratory hammer and a mandrel or employing a hydraulic system to directly push vacuum pipes into ground. It is noted that vacuum pipes are steel pipes, typically 2.5 to 3-cm outside diameter, and 0.3-cm in thickness. The vacuum pipes contain perforated holes and are wrapped with a geotextile fabric for filtration purposes. The horizontal drainage pipes are typically polyvinyl chloride (PVC) pipes, which are connected to steel vacuum pipes through an elbow connector. Figure 2 shows an array of horizontal drainage pipes connected to vertical vacuum pipes at a project site.

Figure 2. Array of vacuum pipes and horizontal drainage pipes.

Step 3: Apply the first cycle of vacuum to reduce water content in the influence zone. In this phase, vacuum-induced dewatering of cohesive soils takes place. Generally, the net effect of this phase of vacuum dewatering is an increase of effective stress up to about 50 to 80 kPa, depending upon the efficiency of vacuum consolidation. It is noted that the highest vacuum pressure that can be exerted on the pore water in the soil is 1 atmosphere pressure (100 kPa). The undrained strength gain of normally consolidated soft clays corresponding to 50 to 80-kPa effective stress increase is roughly 15 to 25 kPa. Therefore, this phase of vacuum dewatering is primarily used for making the site accessible for equipment to carry out the next phase of work (i.e., deep dynamic compaction). The time required for completing this cycle of vacuum consolidation is dictated by the spacing of vertical vacuum pipes and by the horizontal hydraulic conductivity of the soils. In addition, smearing effects (soil disturbances due to installation of vertical vacuum pipes) need to be taken into account. This phase of work is typically completed within seven days before proceeding to the next phase of work.

Step 4: Apply deep dynamic compaction to create a crater and to generate positive pore water pressure. The direct impact from the heavy tamping creates a crater, resulting in displacement of soils and a corresponding reduction in void ratio (direct densification), while producing positive pore pressure in the influence zone. Previous studies indicated that deep dynamic compaction in cohesive soils can cause a rapid increase in both pore water pressure and gas pressure, whether the soil is fully saturated or not, due to the presence of micro air bubbles. The important controlling parameters of dynamic compaction are the weight, dimension, drop height, grid spacing, and number of tamper drops per spots. Decisions regarding these parameters need to be made based on site monitoring results to ensure that the soils underneath the bottom of the crater do not suffer from undrained shear failure or the so-called "rubber soil" phenomenon. A typical dimension for a tamper is about 1 to 1.5 meters in diameter, and the weight can vary from 20 to 70 tons. The tamper drop height varies from 10 meters to approximately 20 meters. Useful correlations between crater depth, soil properties, influence zones, and tamper energy are available and could be used in the preliminary selection of the controlling parameters. The duration of this phase of work can be accomplished within 7 days for a typical 10,000-m^2 coverage area.

Step 5: Apply the second cycle of vacuum to facilitate a rapid dissipation of pore pressure and to further reduce the water content and void ratio of the soils in the influence zone. The combined efforts of vacuum-generated negative pore water pressure and the deep dynamic compaction generated positive pore water pressure will create a very high pore pressure gradient, which in turn helps to facilitate accelerated dissipation of the pore water pressure resulting in reduced water content. The duration of this phase is generally seven days or less.

Step 6: Evaluate the soil properties after completing Step 5. In particular, the water content, pore pressures, ground water elevation, ground subsidence, and in-situ test results (such as cone resistance of CPT or *N* values of STP), need to be determined to assess the results of the first cycle (Steps 4 and 5) of the HVDM process. Evaluation of the outcome of ground improvement at this stage would allow for adjusting the operational parameters (spacing and depth of vacuum pipes, dynamic compaction energy level and grid spacing of tamping points, etc.) in the next cycle of the HVDM process.

Step 7: Repeat Steps 4 through 6 until the performance criteria are satisfied. It should be noted that, in general, two cycles of HVDM process are typically sufficient to achieve the required performance criteria, such as the strength (as determined by CPT or SPT) and the post-treatment settlement.

HVDM utilizes the combination of active drainage, consolidation, and densification principles in. In order to be successful, this ground improvement method needs to be executed in a controlled manner based on the feedback obtained using on-site monitoring data.

In summary, the aforementioned method is a repeated process that uses a combination of vacuum consolidation/dewatering and dynamic compaction, applying increased tamper impact energy with each successive cycle to achieve the desired density and the required depth of treatment. In addition, on-site monitoring plays a key role to ensure that not only the soil properties before and after ground improvement are monitored but also that the operational parameters (e.g., the spacing of vacuum pipes, the energy of the tamper, and the duration between each stage of vacuum consolidation) are optimized. Past case histories indicate about 30% cost savings compared to conventional methods and about 50% reduction in construction period again compared to conventional methods. HVDM is a mechanical process, which adds no new materials into the soil thus it is environmental safe. Further details can be found in Liang *et al.* (2015).

3 TECHNOLOGICAL INNOVATION

The HVDM method embodies at least four technological innovations that are worthy of mention. First, the HVDM method successfully utilizes an intelligent combination of cycles of well-designed vacuum dewatering and dynamic compaction to create not only a very high pore water pressure gradient to expedite pore pressure dissipation, but also to provide active drainage conduits through an innovative, airtight vacuum pipe system. HVDM technology essentially extends the applicable range of a vacuum well drainage method into highly impermeable soils with permeability in the order of 1×10^{-6} to 1×10^{-7} cm/sec.

The second distinctive innovation of the HVDM method is that it breaks the barrier limiting the use of dynamic compaction in soft, saturated cohesive soils.

The third distinctive feature of the HVDM is the actual densification achieved due to dynamic compaction, which in turn creates a very hard and over-consolidated top layer with thickness in the order of 3 to 4 m. Thus, the stresses transmitted to the underlying soil layer are reduced, which would place less stringent requirements on the soil improvement for this underlying soil layer.

The fourth breakthrough hinges on the ability to retrieve the vacuum pipes during and after the ground improvement at the site. With the production of hard, over-consolidated clay, which is essentially impervious, in conjunction with the retrieval of vacuum pipes (in contrast to leaving the PVDs in place) the post treatment water drainage path is restricted to the horizontal direction. As a result, even with additional pore pressure generation due to surface structure loads, the rate of pore pressure dissipation under this restrictive drainage condition would be very slow. Thus, the

HVDM can improve a soft clay site with a system illustrated that clearly minimizes the post treatment total and differential settlements.

4 ADVANTAGES AND LIMITATIONS

Some of the interesting advantages of using a combination of vacuum consolidation and dynamic compaction include (a) enhancing the vacuum well drainage techniques in fine-grained soils with relatively low permeability due to the enhanced pore pressure gradient created by combining the dynamic compaction-induced positive pore pressure and vacuum-induced negative pore pressure, (b) overcoming the common obstacle that dynamic compaction could not be applied to saturated cohesive soils due to the ability to lower the groundwater table and the creation of unsaturated soil zones through vacuum dewatering/consolidation, and (c) expediting pore pressure dissipation due to the creation of a high pore pressure gradient as a result of the combination of negative vacuum pore pressure and positive pore pressure generated by the dynamic compaction. The anticipated results include the following: (a) the creation of a highly over-consolidated clay layer near the ground surface by the deep dynamic compaction effort and the influence zone and (b) eliminating the post-treatment vertical drainage path as a result of withdrawing the vacuum pipes from the ground after completion.
There are some limitations of the vacuum consolidation/dynamic compaction method. The treatment depth is generally limited to 10 m due to the limit of the influence zone of deep dynamic compaction and the loss of vacuum efficiency when exceeding that depth. Also, fine-grained soils that contain a large portion of organic materials would not be suitable for this method due to the pronounced secondary compression (creep) that could not be treated. The range of permeability of the fine-grained soils that can be suitable for this treatment method is limited to a minimum of about 10^{-6} cm/sec.

HVDM technology for shallow soft soils has been extended to two additional innovative techniques: 1) Preloading plus HVDM for deep soft soils and 2) HVDM plus Composite Foundation for job sites requiring high bearing capacity.

5 SUMMARY AND CONCLUSIONS

In this paper, recent advances in ground improvement techniques for fine-grained soils involving the combined use of vacuum consolidation/dewatering and deep dynamic compaction (HVDM) were described. The mechanisms of HVDM technology for ground improvement are as follows: Dynamic compaction in nearly saturated contractive soils generate positive pore water pressure, which when coupled with vacuum induced negative pore pressure, can expedite drainage of water from the soil mass, thus increasing density and undrained shear strength of the treated soil.

Acknowledgments

This technology was developed and patented by Dr. Mr. Xu Shi-Long of Geoharbour Group in Shanghai, China. This paper was prepared with the input of Dr. Shi-Long and Prof. Robert Liang of the University of Akron.

References

Holtz, R. D., Preloading by Vacuum Consolidation: Current Practice, *Transportation Research Record*, No. 548, 26-29, Transportation Research Board, National Research Council, Washington, D. C., 1975.

Kjellman, W., Consolidation of Clay Soil by Means of Atmosphere Pressure." *Proc. Conf. on Soil Stabilization*, MIT, Cambridge, 258-263, 1952.

Liang, R., Xu, C., and Edil, T. B., Innovative Soft Soil Improvement Method through Intelligent Use of Vacuum De-Watering and Dynamic Compaction Techniques, *Geotechnical Engineering Journal of the SEAGS & AGSSEA,* 46 (4), 2015.

Menard, L., Theoretical and Practical Aspect of Dynamic Consolidation." Geotechnique, 25(1), 3-17, 1975.

Interaction between Theory and Practice in Civil Engineering and Construction
Edited by Komurlu, R., Gurgun, A. P., Singh, A., and Yazdani, S.

THE USE OF POLYPROPYLENE FIBERS AGAINST PLASTIC SHRINKAGE CRACKING

HASAN NURI TURKMENOGLU, HAKAN NURI ATAHAN, and CENGIZ SENGUL

Civil Engineering Dept, Istanbul Technical University, Istanbul, Turkey

Plastic shrinkage cracking (PShC) occurs within a few hours after fresh concrete formed into the molds and it takes part on the surface of the concrete. When concrete formed into the molds, the aggregates settle because of the gravity and in contrast, water bleeds. If the rate of evaporation is higher than the rate of bleeding, surface of concrete starts to shrink. However, under the surface, the fresh concrete cannot shrink as much as the surface. Because of this condition, cracking occurs on the surface of concrete. In this respect, the quantity of PShC majorly depends on the temperature of concrete, temperature of air, rate of relative humidity and the wind velocity. Use of short fibers in concrete is one of the most effective ways to prevent concrete from PShC. The aim of this research was to evaluate the effect of polypropylene fibers having different geometries and content on PShC. For this purpose, the principles of ASTM C 1579 standard was considered. Results have shown that the use of polypropylene fibers in limiting PShC is obvious. Moreover, fibrillated fibers has shown better performance than monofilament fibers.

Keywords: Crack reduction ratio, Crack width, Fibrillated fiber, Fiber dosage, Monofilament fiber, Image processing.

1 INTRODUCTION

Plastic shrinkage cracking (PShC) is one of the problems encountered in application. It typically occurs on the surface of floor slab or bridge deck, which are exposed to hot, and windy environmental conditions, Boshoff and Combrinck (2013). If the tensile stress on surface of concrete reaches its tensile strength, several cracks occur. These cracks become a canal for external deteriorating agents and accelerate the ingression of harmful gasses and liquid solutions into the concrete. Consequently, the durability and service life of concrete is affected adversely, (Sivakumar *et al.* 2007, Boshoff and Combrinck 2013, Leemann *et al.* 2013). Regarding to PShC, many researches have been reported for decades. To minimize the risk of PShC, several parameters, such as mixture design, curing methods, chemical and mineral admixtures, surface finishing methods and secondary reinforcement have been investigated both on conventional and high performance concretes.

Fiber reinforcement is one of the most effective ways to reduce the risk of PShC, Rahmani *et al.* (2012). The use of short fibers in concrete and mortar is increasing in recent years. In literature, performance of steel, glass, synthetic and natural short fiber types against PShC has been broadly investigated with considering different combinations between dosage, geometry and type of fibers.

Steel, glass and polymeric fibers that have different shape and size are added to increase the durability of concrete, Pelisser *et al.* (2010). Apart from these, nature origin fibers such as flax and cellulosic ones are also being used to avoid PShC (Soroushian and Ravanbakhsh 1998, Boghossian and Wegner 2008). Several researches examining the synthetic fiber, which have same geometry, have shown that thinner and longer fibers have better performance on holding the cracks. The finer the fibers, the smaller the crack values for same fiber dosage (Ma *et al.* 2002, Banthia and Gupta 2006). However, the effectiveness of length of fiber reduces after a definite fiber length, Soroushian and Ravanbakhsh (1998).

In this research, the effect of polypropylene fibers, having different geometries and content, on limiting the PShC was investigated. The reference concrete mixture used in this research is C35. Average 28-day cube strength was found as 48.6MPa.

2 MATERIALS USED

CEM I 42.5 R class Portland cement with specific gravity of 3.17 gr/cm^3 was used in concrete mixtures. A commercially available superplasticizer with a specific gravity of 1.19 gr/cm^3 was used to obtain desired workability. 0-2 mm natural sand with specific gravity of 2.64 gr/cm^3 and 0-4 mm crushed sand with specific gravity of 2.76 gr/cm^3 were used as fine aggregates. A crushed stone with a maximum size of 16 mm and specific gravity of 2.76 gr/cm^3 was used as coarse aggregate. Monofilament polypropylene (MPP) and fibrillated polypropylene (FPP) fibers with 12 mm length were used as fiber reinforcement. Denier of MPP fiber is between 2.5 and 6 (diameter = 20–30 μm), and denier of FPP fiber is 1200.

3 MIXTURE PROPORTIONING

The design of concrete mixtures on which PShC tests has been performed is given in Table 1. In these mixtures, water/cement ratio was kept constant at 0.55. Three different dosages of polypropylene fibers were selected as 600, 900 and 1100 gr/m^3. Concrete mixtures with and without fiber reinforcement were denoted as REF, MPP600, MPP900, MPP1100, FPP600, FPP900 and FPP1100. In order to minimize the effect of superplasticizer on setting times of concretes, its amount for all mixtures was kept constant at 0.9% by the weight of cement.

Table 1. Mixture proportions (kg/m^3) of concretes.

(kg/m^3)	REF	MPP600	MPP900	MPP1100	FPP600	FPP900	FPP1100
Cement				350			
Natural Sand				303			
Crushed Sand				614			
Crushed Stone				931			
MPP	-	0.6	0.9	1.1	-	-	-
FPP	-	-	-	-	0.6	0.9	1.1
Water				192,5			
Superplasticizer				3,15			

4 MIXING AND PLACING

Concrete mixtures were mixed using a rotary pan mixer of 50 liters capacity. Initially, coarse aggregate, fine aggregate and cement were put into the pan and mixed dry for 30 seconds. By reserving some of it, water added to mixer slowly. After that superplasticizer was mixed into the reserved water and added to the mixer. Finally, the fibers were added in the mixture. Workability of concrete was determined by using slump test. The concrete was placed into the molds and vibrated on a vibration table. A smooth steel screed was used to finish the concrete surface. Screeding procedure was done by making sawing motion, on the long edge direction of the mold.

5 SPECIMEN DIMENSIONS AND ENVIRONMENTAL CONDITIONS

Laminated board molds with dimensions of 355x560x95 mm were used. According to ASTM C 1579, molds have three stress risers. But to achieve greater crack areas and provide the criteria of minimum average crack limit of 0.5 mm two additional restraints added to system. Photo and cross-section of mold and environmental chamber are given in Figure 1.

Figure 1. Photo and cross-section of mold used (left), environmental chamber (right).

The concrete specimens were placed into the environmental chamber after casting. To obtain sufficient evaporation rate, a fan with air velocity of 16~17 km/h and an infrared heater were used. At the end of the experiment, temperature and relative humidity on the surface of samples were measured as 43±3°C and 17±3%, respectively. Evaporation rate was measured as 1.09 $kg/m^2.h$ which achieved the criteria of 1.0 $kg/m^2.h$.

The specimens were kept in the environmental chamber for 4 hours. After first crack became visible, photos of specimen were captured in 30 minutes intervals and used to measure total crack area, total crack length and maximum crack width.

6 RESULTS AND DISCUSSION

6.1 Slump Test, Setting Time Test and Crack Initiation Times

Slump values and crack initiation times are given in Table 2. Addition of fibers reduced slump values significantly. However, the consistency of the fiber reinforced mixtures were good enough to obtain a well compaction. All the mixtures could easily be formed into the molds by the vibration of four seconds.

Table 2. Slump values and crack initiation times.

Mixture ID	Slump (cm)	Crack Initiation (min.)
REF	16	90
MPP600	4	85
MPP900	1.5	102
MPP1100	3	98
FPP600	4	96
FPP900	2	94
FPP1100	3	100

Final setting time was determined only for reference concrete mixture and measured as 135 minutes. As seen in Table 2, crack initiation times extended with increasing fiber dosage.

6.2 Plastic Shrinkage Test

For all of the mixtures two specimens were tested and the crack characteristics were determined by means of image processing. The crack maps for plain, MPP and FPP fiber reinforced mixtures for different dosages are shown in Figure 2. Average of total crack area, total crack length, average crack width, maximum crack width measurements and calculated crack reduction ratio (CRR) values are given in Table 3. Crack reduction ratio of total crack area and average crack width calculated with formula given below.

$$CRR = \left(1 - \frac{\text{Average crack width or total crack area of fiber reinforced mixture}}{\text{Average crack width or total crack area of reference mixture}}\right) x100 \qquad (1)$$

Table 3. Average of total crack area, total crack length, average crack width, maximum crack width measurements and calculated CRR values.

Specim en ID	Fiber Dosage (gr/m^3)	Total Crack Area (mm^2)	Total Crack Length (mm)	Average Crack Width (mm)	Max. Crack Width (mm)	CRR of Total Crack Area (%)	CRR of Average Crack Width (%)
REF	-	188	450	0.42	0.98	-	-
MPP600	600	134	425	0.31	0.79	28.9	24.7
FPP600	600	113	424	0.27	0.74	39.9	36.2
MPP900	900	67	410	0.16	0.44	64.1	60.6
FPP900	900	90	468	0.19	0.48	52.1	53.9
MPP110	1100	43	342	0.13	0.47	76.9	69.5
	1100	33	347	0.09	0.34	82.5	77.3

The effect of fiber type and dosage on the average crack width, maximum crack width and total crack area can be seen in Figure 3. From these figures, it is obvious that, for both fiber types, as the fiber dosage increases average and maximum crack widths decreases significantly. CRR of "average crack width" for MPP fiber mixtures are 24.7, 60.6, and 69.5% for the fiber dosages of 600, 900 and 1100 g/m^3, respectively. "Maximum crack width" values of MPP fiber mixtures, on the other hand, were reduced from 0.98mm to 0.79, 0.44 and 0.47 mm for the fiber dosages of 600, 900 and 1100 g/m^3, respectively. Similar trend can also be observed for the FPP fiber mixtures. CRR of "average crack width" for FPP fiber mixtures are 36.2, 53.9, and 77.3% for the fiber dosages of 600, 900 and 1100 g/m^3, respectively.

"Maximum crack width" values of FPP fiber mixtures, on the other hand, were reduced from 0.98 mm to 0.74, 0.48 and 0.34 mm for the fiber dosages of 600, 900 and 1100 g/m^3, respectively. Based on the results obtained, it can be concluded that for a given dosage, the performance of FPP is likely better than MPP.

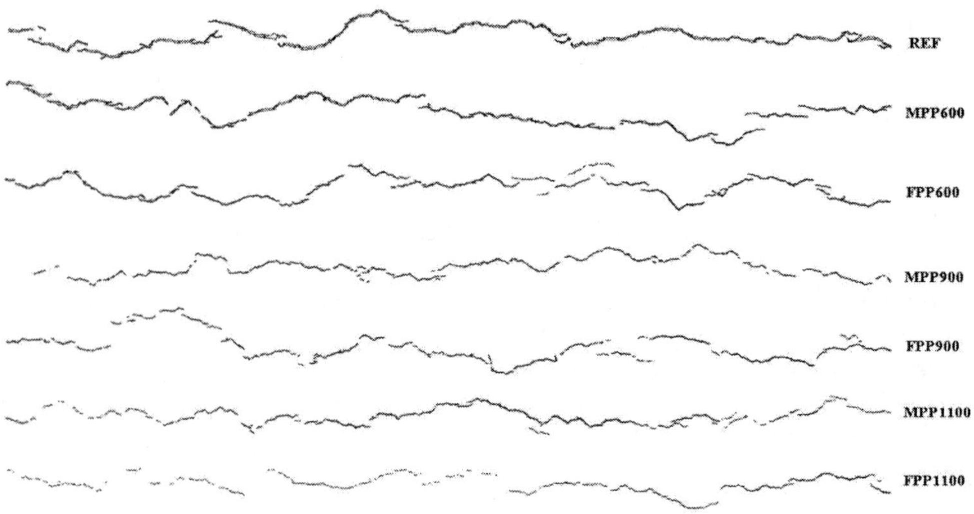

Figure 2. Crack maps of mixtures.

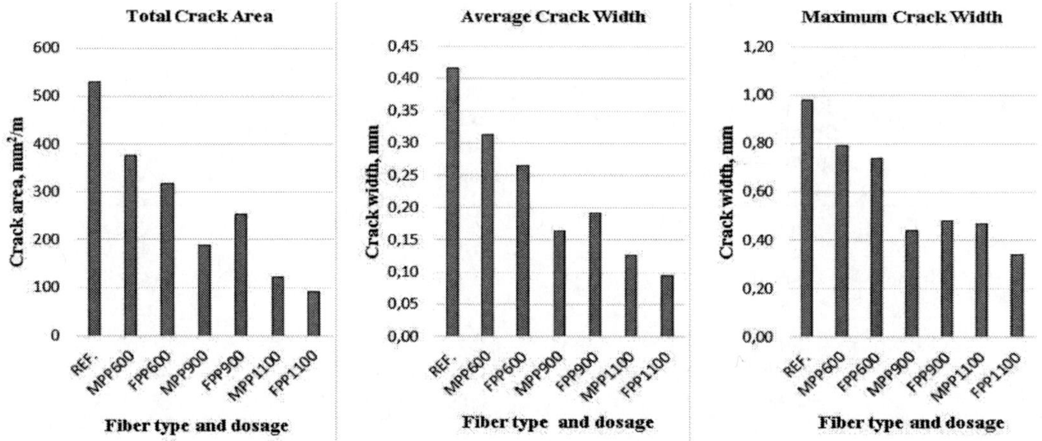

Figure 3. The effect of fiber type and dosage.

7 CONCLUSIONS

Within the limits of experimental work, the following conclusions can be drawn:

- Addition of fibers significantly reduced slump values of fresh concretes.
- For both fiber types, as the fiber dosage increases average and maximum crack widths decreases significantly.
- It is found that, for a given dosage, the performance of FPP is likely better than MPP.
- It is recommended that it will be better to isolate experimental setup from air conditions with converting the setup into the closed system. It is important for repeatability of testing.

References

ASTM C 1579, Standart test method for evaluating plastic shrinkage cracking of restrained fiber reinforced concrete (Using a steel form insert), 2013.

Banthia, N. and Gupta, R., Influence of polypropylene fiber geometry on plastic shrinkage cracking in concrete, *Cement and Concrete Research*, 2006.

Boghossian, E. and Wegner, L. D., Use of flax fibres to reduce plastic shrinkage cracking in concrete, *Cement and Concrete Composites*, 2008.

Boshoff, W. P. and Combrinck, R., Modelling the severity of plastic shrinkage cracking in concrete, *Cement and Concrete Research*, 2013.

Leemann, A., Nygaard, P., Lura, P., Impact of admixtures on the plastic shrinkage cracking of self-compacting concrete, *Cement and Concrete Composites*, 2013.

Ma, Y., Tan, M., Wu, K., Effect of different geometric polypropylene fibers on plastic shrinkage cracking of cement mortars, *Materials and Structures*, vol. 35, 2002.

Pelisser, F., Neto, A. B. S. S., Rovere, H. L., Pinto, R. C. A., Effect of the addition of synthetic fibers to concrete thin slabs on plastic shrinkage cracking, *Construction and Building Materials*, 2010.

Rahmani, T., Kiani, B., Bakhshi, M., Shekarchizadeh, M., Application of different fibers to reduce plastic shrinkage cracking of concrete, *7th RILEM International Conference on Cracking in Pavements*, 635-42, 2012.

Sivakumar, A. and Santhanam, M., A quantitative study on the plastic shrinkage cracking in high strength hybrid fibre reinforced concrete, *Cement and Concrete Research*, 2007.

Soroushian, P. and Ravanbakhsh, S., Control of plastic shrinkage cracking with specialty cellulose fibers, *ACI Materials Journal Technical Paper*, Title no. 95-M40, 1998.

Interaction between Theory and Practice in Civil Engineering and Construction
Edited by Komurlu, R., Gurgun, A. P., Singh, A., and Yazdani, S.

BEHAVIOR OF HYBRID STEEL FIBER REINFORCED HIGH STRENGTH CONCRETE

EMDAD K. Z. BALANJI, M. NEAZ SHEIKH, and MUHAMMAD N. S. HADI

School of Civil, Mining and Environmental Engineering, University of Wollongong, NSW, Australia

High strength concrete has higher strength but lower ductility. Inclusion of single type of fibers into concrete has been proven to improve the behavior of concrete to a limited extent. However, recently it was found that the behavior of concrete can be improved more with the addition of hybrid fiber *i.e.*, a combination of different types of fiber. This paper presents the results of an experimental investigation on the behavior of Hybrid Steel Fibre Reinforced High Strength Concrete (HSFR-HSC). A total of eight cylinder specimens with 150 mm in diameter and 300 mm in height were cast and tested under uniaxial compression. Three different combinations of HSFR-HSC specimens and reference specimens without steel fibers were prepared. The first combination of HSFR-HSC included 1.5% Micro Steel (MS) fibers and 1% Deformed Steel (DS) fibers. The second combination included 1.5% MS fibers and 1.5% Hooked-end Steel (HS) fibers. The third combination included 1% DS fibers and 1.5% HS fibers. The experimental results showed that the addition of hybrid steel fibers improved the strength and ductility of high strength concrete compared to the reference specimens. The results also showed that the specimens reinforced with different hybrid steel fibers failed in a ductile manner, while the reference specimens failed in a brittle manner.

Keywords: High strength concrete, Micro steel fibers, Deformed steel fibers, Hooked-end steel fiber, Hybrid steel fibers.

1 INTRODUCTION

High Strength Concrete (HSC) possesses improved mechanical properties including higher compressive strength and higher durability compared to Normal Strength Concrete (NSC). However, HSC tends to be brittle and lacks plastic deformation capacity under high axial loadings (Sheikh *et al.* 2010). The inclusion of steel fiber into concrete was found to improve the mechanical properties of concrete including flexural toughness, tensile strength and resistance to crack propagations (Song and Hwang 2004, Thomas and Ramaswamy 2007, Yazici *et al.* 2007). The properties of concrete reinforced with a single type of fiber can be improved to a limited extent. However, recently it was found that combination of different types of fiber (Hybrid Fiber) can enhance the mechanical properties of concrete (Prisco *et al.* 2009).

Several studies investigated the behavior of concrete with the inclusion of a single type of steel fiber (Fanella and Naaman 1985, Ezeldin and Balaguru 1992, Nataraja *et al.* 1999, Mansur *et al.* 1999, Hadi 2007, Hadi 2009, Oliveira *et al.* 2010, Balanji *et al.*

2015). However, limited information is available about the behavior of the HFRC in the literature. Qian and Stroeven (2000) investigated the influence of fiber type (Polypropylene and Steel) and volume contents on the compressive strength and tensile strength of HFRC. Yao *et al.* (2003) observed the mechanical properties of HFRC in terms of compressive strength and splitting tensile strength. Chi *et al.* (2014) presented stress-strain behavior of Hybrid Fibre Reinforced Normal Strength Concrete (HFR-NSC). Most of these studies focused on the improvement of strength of HFRC. Only Chi *et al.* (2014) studied the stress-strain behavior of HFR-NSC. The complete stress-strain behavior of HSFR-HSC has not yet been adequately investigated.

As part of a research program aimed at investigating the strength and ductility capacity of steel fiber reinforced HSC columns, the objective of this study is to investigate the stress-strain behavior of HSFR-HSC under uniaxial compression. The parameters investigated in this study are the fiber type (Micro Steel, MS; Deformed Steel, DS; and Hooked-end Steel, HS), volume content and aspect ratio of the steel fibers.

2 EXPERIMENTAL PROGRAM

2.1 Material Details

A local supplier provided HSC with a compressive strength of 65 MPa. Eight cylinder specimens of 150 mm x 300 mm were cast and tested. Three different types of steel fibers *i.e.,* MS (6 mm in length, 0.2 mm in diameter and nominal tensile strength of 2600 MPa), DS (18 mm in length, 0.55 mm in diameter and nominal tensile strength of 800 MPa) and HS (35 mm in length, 0.75 mm in diameter and nominal tensile strength of 1225 MPa) were used. The specimens were divided into four groups. Two specimens were used in each group. Group R (Reference group) included plain HSC without steel fibers. Group MD included 2.5% hybrid steel fibers (1.5% of MS and 1% of DS). Group MH included 3% hybrid steel fibers (1.5% of MS and 1.5% of HS). Finally, Group DH included 2.5% hybrid steel fibers (1% DS and 1.5% of HS).

2.2 Casting, Curing and Testing

The HSC was placed directly from the concrete chute into the concrete mixer in order to add the steel fibers into the concrete. As per the design mix proportions, different steel fibers were mixed in a tray and then sprinkled into the concrete. The mix drum was filled to the correct height so the total volume for each group of the HSFR-HSC was 0.01 m^3. The mixer was switched on for 1.5 to 2 minutes until the fibers were uniformly mixed in the concrete. The slump of HSFR-HSC was measured by using a slump cone apparatus. The slump values ranged between 65 mm to 105 mm depending on the hybrid steel fibers type and volume content. Eight cylinder specimens were cast to study the behavior of HSFR-HSC. After casting, the specimens were covered with plastic sheet and cured for 24 hours at room temperature. Afterwards, the specimens were demolded and cured in a water tank until the day of testing.

After 28 days, all the cylindrical specimens were tested until failure in the 5000 kN Denison Universal Testing Machine at the University of Wollongong, Australia. Prior to testing, the specimens were capped with a hard plaster to ensure uniform loading faces. Both ends of the specimens were wrapped using Fibre Reinforced Polymer (FRP) sheet for a width of 35 mm in order to prevent premature failure. To measure the axial deformation at the middle half of the cylinder, Linear Variable Differential Transducer (LVDT) was used. Displacement controlled loads were applied at 0.5 mm/min. An electronic data acquisition system was used to record the load and the corresponding axial deformation. Figure 1 shows a specimen in the testing machine.

Figure 1. Test setup.

3 EXPERIMENTAL RESULTS AND DISCUSSION

3.1 Effect of Hybrid Steel Fibers

Eight cylindrical specimens were tested to study the stress-strain behavior of HSFR-HSC under uniaxial compression. The major parameters defining a stress-strain curve are peak stress (f'_{cf}) and strain corresponding to the peak stress (ε'_{cf}). The average values of the results of the two specimens in each group are shown in Table 1. The average peak stress (f'_{cf}) for Groups MD and DH were slightly lower than the average peak stress of Group R. However, the average peak stress for Group MH was higher by 8% than the average peak stress of Group R. It was also observed that the strain corresponding to the peak stress (ε'_{cf}) was increased with the inclusion of hybrid steel fibers into the HSC. The average strain corresponding to the peak stress was increased by 19%, 21%, and 48% for Groups MD, DH, and MH, respectively.

The complete stress-strain curves of the eight specimens are presented in Figure 2. Figure 2 clearly shows that the ascending branches of HSFR-HSC are slightly affected by the addition of different types of hybrid steel fibers. However, the descending branches of the HSFR-HSC are significantly influenced by the inclusion of different types of hybrid steel fibers. The slope of the descending branches of the stress-strain curve of Group MH was less steep than the slopes of Groups MD and DH. This is due to the high volume content of fiber in Group MH (3%) compared to the lower volume contents of fiber in both of Groups MD and DH (2.5%). In addition, MS fiber enhances the pre-peak stress by controlling the small cracks at an early stage. After the cracks became wider HS fiber reduced crack propagation by carrying loads and transferring them to other fibers. Thus, the ductility of Group MH increased. The slope of the descending branches for Group DH was less steep than the slope of Group MD. This is due to the high proportion of the longer steel fiber (1% of DS and 1.5% of HS) in Group DH. Consequently improvement was achieved in post-peak behavior for Group DH compared with Group MD.

The average of ductility of the HSFR-HSC was calculated and shown in Table 1. The ductility of the specimen was calculated as a ratio of the ultimate strain to the yield strain. It was observed that the ductility of the HSFR-HSC is increased with the inclusion of different types of hybrid steel fibers. The ductility was increased by 61%, 109% and 180% for Group MD, Group DH and Group MH respectively compared to the ductility of Group R.

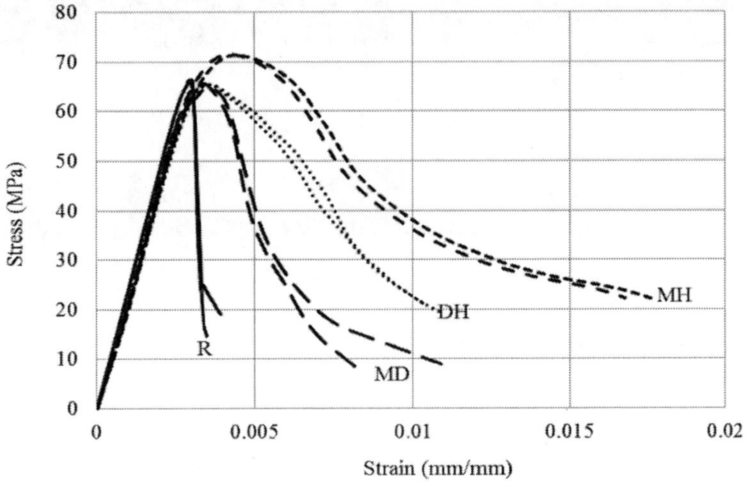

Figure 2. Stress-strain behavior of HSC with and without hybrid steel fibers.

Table 1. Average experimental results.

Group designation	Hybrid steel fibers volume content (V_f%)	Experimental results			
		f'_{cf} (MPa)	ε'_{cf} (mm/mm)	Ductility	Ductility relative to the Group R
R	0	66.3	0.0029	1.43	1
MD	3	65.2	0.00345	2.31	1.61
MH	2.5	71.4	0.0043	3.00	2.09
DH	2.5	65.4	0.0035	4.01	2.8

3.2 Failure Modes

The typical failure modes of the specimens are shown in Figure 3. The failure mode for Group R specimens was brittle. However, with the inclusion of different hybrid steel fibers into high strength concrete, the failure mode became a combination of shear failure and bugling of the specimens in the lateral direction. The failure mode of Group

MD specimens was due to the propagation of cracks parallel to the loading direction. This means that Group MD was effective in reducing early cracks propagation and contributes little to post peak behavior. The failure mode of Group MH specimens was due to shear failure and bugling of the specimens in the lateral direction. This means that Group MH was effective in reducing early cracks propagation and in improving the post peak behavior of the specimens. The failure mode of Group DH specimens was due to bugling of the specimens in the lateral direction. This behavior shows that Group DH was more effective in the post peak behavior and has less contribution to arresting early cracks propagation.

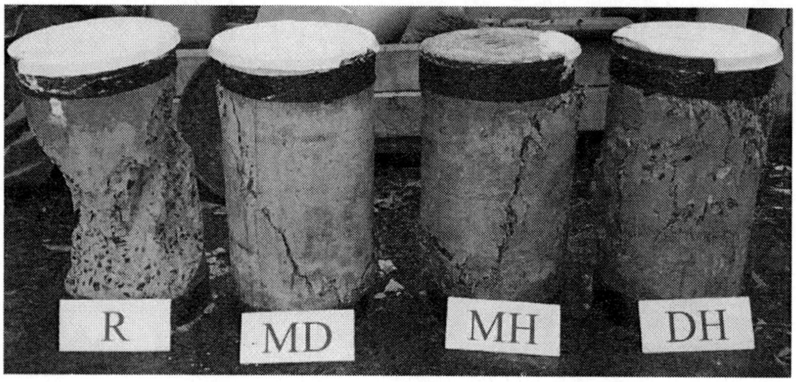

Figure 3. Typical failure modes.

4 CONCLUSION

In this study, a total of eight cylinder specimens of 150 mm x 300 mm were tested under uniaxial compression to investigate the behavior of different HSFR-HSC. The following conclusions can be drawn from the experimental results:

- The addition of hybrid steel fibers into HSC matrix changes the stress-strain behavior of HSC. The ascending branch of the stress-strain curve is slightly influenced, but the descending branch of the stress-strain curve is significantly influenced by the inclusion of hybrid steel fibers.
- The addition of hybrid steel fibers into HSC increases the average strain corresponding to the peak stress by 19%, 21% and 48% for Groups MD, DH, and MH, respectively. However, the average peak stresses of Groups MD and DH were slightly deceased, while the average peak stress was increased by 8% for Group MH compared to Group R.
- The inclusion of hybrid steel fibers into high strength concrete leads to significantly increase in the ductility of high strength concrete. The ductility was increased by 61%, 109% and 180% for Group MD, Group DH and Group MH, respectively compared to the ductility of Group R.
- The combination of the MS and HS fibers provides the highest enhancement in terms of strength and ductility compared to other combinations of the hybrid steel fibers and reference specimens without of fibers.

References

Balanji, E., Sheikh, M. N. & Hadi M. N.S. Models for Predicting the Axial Compressive Sterngth of Steel Fibre Reinforced Normal and High Strength Concrete Square Columns, *in the 12th International Symposium on Fiber Reinforced Polymers for Reinforced Concrete Structures & The 5th Asia-Pacific Conference on Fiber Reinforced Polymers in Structures Joint Conference*, Nanjing, China, 14-16 December 2015.

Chi, Y., Lihua X. & Yuanyuan Z., Experimental Study on Hybrid Fibre–Reinforced Concrete Subjected to Uniaxial Compression, *Journal of Materials in Civil Engineering*, 26(2), 211-218, 2014.

Ezeldin, A. S. & Balaguru, P. N., Normal- and High-Strength Fiber-Reinforced Concrete Under Compression, *Journal of Materials in Civil Engineering,* 4(4), 415-429, 1992.

Fanella, D. A. & Naaman, A. E., Stress-Strain Properties of Fibre Reinforced Mortar in Compression, *ACI Journal*, 82 (4), 475-483, 1985.

Hadi, M.N.S., Using Fibers to Enhance the Properties of Concrete Columns, *Construction and Building Materials*, 21(1), 118-125, 2007

Hadi, M.N.S., Reinforcing Concrete Columns with Steel Fibers, *Asian Journal of Civil Engineering*, 10(1), 79-95, 2009.

Mansur, M. A., Chin, M. S. & Wee, T. H., Stress-Strain Relationship of High-Strength Fiber Concrete in Compression, *Journal of Materials in Civil Engineering*, 11(1), 21-29, 1999.

Nataraja, M. C., Dhang, N. & Gupta, A. P., Stress-Strain Curves for Steel-Fiber Reinforced Concrete Under Compression, *Cement and Concrete Composites*, 21(5-6), 383-390, 1999.

Oliveira Junior, L. Á. D., Borges, V. E. D. S., Danin, A. R., Machado, D. V. R., Araújo, D. D. L., El Debs, M. K. & Rodrigues, P. F., Stress-Strain Curves For Steel Fiber-Reinforced Concrete in Compression, *Matéria (Rio de Janeiro)*, 15(2), 260-266, 2010.

Prisco, M. D., Plizzari, G., & Vandewalle, L., Fibre Reinforced Concrete: New Design Perspectives. *Materials and Structures,* 42 (9), 1261–1281, 2009.

Qian, C. X. & Stroeven, P., Development of Hybrid Polypropylene-Steel Fibre-Reinforced Concrete, *Cement and Concrete Research*, 30(1), 63–69, 2000.

Sheikh M. N., Tsang H. & Lam N. T. K. Yield curvature for seismic design of circular reinforced concrete columns, *Magazine of Concrete Research*, 62 (10), 741–748, 2010.

Song, P. S. & Hwang, S., Mechanical Properties of High-Strength Steel Fiber-Reinforced Concrete, *Construction and Building Materials*, 18(9), 669-673, 2004.

Thomas, J. & Ramaswamy, A., Mechanical Properties of Steel Fiber-Reinforced Concrete, *Journal of Materials in Civil Engineering*, 19(5), 385-392, 2007.

Yao, W., Li, J. & Keru, W., Mechanical Properties of Hybrid Fibre-Reinforced Concrete at Low Fibre Volume Fraction, *Cement and Concrete Research*, 33(1), 27–30, 2003.

Yazici, S., Inan, G. & Tabak, V., Effect of Aspect Ratio and Volume Fraction of Steel Fiber on the Mechanical Properties of SFRC, *Construction and Building Materials,* 21(6), 1250-1253, 2007.

Interaction between Theory and Practice in Civil Engineering and Construction
Edited by Komurlu, R., Gurgun, A. P., Singh, A., and Yazdani, S.

PERFORMANCE OF HIGH STRENGTH CONCRETE COLUMNS REINFORCED WITH HYBRID STEEL FIBER UNDER DIFFERENT LOADING CONDITIONS

EMDAD K. Z. BALANJI, M. NEAZ SHEIKH, and MUHAMMAD N. S. HADI

School of Civil, Mining and Environmental Engineering, University of Wollongong, NSW, Australia

The strength and ductility of high strength concrete columns improve with the addition of steel fiber. This paper reports the behavior of circular High Strength Concrete (HSC) columns reinforced with Hybrid Steel Fibers (HSF) under different loading conditions. In this study, HSF consisted of a combination of macro steel fibers and micro steel fibers. A total of eight circular specimens of 205 mm diameter and 800 mm height were cast and tested. All specimens were reinforced with same amount of steel reinforcements. The specimens were divided into two groups of four specimens. Group RC (reference group) contained no steel fibers. Group HSF (hybrid steel fibers) contained 2.5% by volume of HSF. From each group one specimen was tested under concentric loading, one under 25 mm eccentric loading, one under 50 mm eccentric loading, and one under four-point loading. The results showed that the specimens reinforced with HSF achieved higher strength and ductility compared to RC specimens under different loading conditions. It was also observed that the presence of HSF delayed the spalling of the concrete cover.

Keywords: Macro steel fibers, Micro steel fibers, Columns, Ductility, Eccentric loading, Concentric loading, Hybrid steel fibers.

1 INTRODUCTION

In structural applications, the use of High Strength Concrete (HSC) is on the rise due to its high compressive strength and high durability. One of the most common structural applications of HSC is in the columns of high rise buildings, where Normal Strength Concrete (NSC) results in larger cross sectional area and higher costs. However, lower ductility and brittle failure are obstacles for the application of HSC. The inclusion of fibers into HSC improves its structural performance, such as flexural strength, impact resistance, tensile strength, flexural toughness and ductility (Mohammadi *et al.* 2008). Adding a single type of fiber can improve the properties of Fiber Reinforced Concrete (FRC) to a limited level. However, the optimum improvement in mechanical properties of the FRC can be achieved by combinations of two or more different types of fiber namely as (hybrid fibers) in concrete (Chi *et al.* 2014). A number of researchers have studied the effect of a single type of fiber on concrete columns under concentric loads (Mangat and Azari 1985, Ganesan and Murthy 1990, Paultre *et al.* 2010, Palanivel and

Sekar 2013, Balanji *et al.* 2015), under cyclic load (Lee 2007, Tokgoz 2009) and a few studies were conducted under eccentric load (Hadi 2008, Trapko 2014). Based on the existing research studies, it was found that the strength and ductility of the columns vary based on different parameters such as distribution and tensile strength of the longitudinal steel bars, configuration, spacing and tensile strength of the transverse reinforcements. For FRC columns, fiber type, volume content, aspect ratio and orientation of the fibers play an important role in enhancing the strength and ductility of the concrete columns. Using hybrid steel fibers with a combination of micro and macro fibers in HSC columns may further improve in the strength and ductility. Thus, the main objective of this study is to investigate the behavior of circular HSC columns reinforced with hybrid steel fibers under different loading conditions.

2 EXPERIMENTAL PROGRAM

2.1 Test Specimens

A total eight FRC specimens were cast and tested. The dimensions of the specimens were chosen to be 205 mm in diameter and 800 mm in height. According to the Australian Standards AS-3600 (2009), all specimens were reinforced with the same amount of longitudinal and spiral reinforcements. For longitudinal reinforcement, 6 deformed steel bars of 12 mm diameter (6N12) with 500 MPa nominal tensile strength were used. For spiral reinforcement, plain steel bars of 10 mm diameter and 60 mm pitch (R10 @ 60 mm) with 250 MPa nominal tensile strength were used. The specimens cover was maintained on 20 mm. The specimens were divided into two groups of four specimens. Group RC contained no steel fibers. Group HSF contained 2.5% by volume of HSF *i.e.,* a combination of 1% by volume of macro steel fibers and 1.5 % by volume of micro steel fibers.

In each group, one specimen was tested under an axial concentric load and designated as 0. The second specimen was tested under an axial load with 25 mm eccentric load and designated as 25. The third specimen was tested under an axial load with 50 mm eccentric load and designated as 50. The fourth specimen was tested as a beam under four-point loading and designated as B. All specimens were tested at the laboratories of the School of Civil, Mining and Environmental Engineering, University of Wollongong, Australia.

2.2 Material Properties

High Strength Concrete (HSC) provided by a local supplier was used to form the specimens. The average compressive strength was 60 MPa at 28-day. Two different steel fibers of deformed macro steel fibers (18 mm length, 0.55 mm diameter and nominal tensile strength of 800 MPa) and straight micro steel fibers (6 mm length, 0.2 mm diameter and nominal tensile strength of 2600 MPa) were used to reinforce the specimens.

2.3 Casting and Curing Procedure

In order to form the specimens, PVC formworks with an inner diameter of 205 mm and a length of 800 mm were used. HSC was placed directly from the concrete chute into the concrete mixer in order to include HSF into the concrete. The mix drum was filled to the required height so the total volume for each group of specimens was 0.12 m^3. The mixer was switched on for 90 to 120 seconds until the fibers were uniformly mixed in the concrete. All the specimens were cured in the formwork for seven days by covering them with wet hessian and a plastic sheet to prevent moisture loss. After seven days the specimens were removed from the formwork and cured in the same moist condition for a further 21 days.

3 TEST SETUP

Prior to testing, the six specimens under axial compression were capped with a hard plaster to ensure uniform loading faces. Both ends of specimens were wrapped with 75 mm width Carbon Fiber Reinforced Polymer (CFRP) sheet in order to prevent premature failure. The eccentric load was applied through two special loading heads with the grooves at 25 mm and 50 mm off center. Two monitoring systems were used to measure the axial and lateral deformations of the specimens. For concentric loading, two Linear Variable Differential Transducers (LVDT) were fixed directly to the 5000 kN Denison Universal Testing Machine (UTM) to measure the axial deformation of the specimens during the test. For eccentric loading, in addition to the two LVDTs, a laser triangulation was installed at mid-height of the specimens to measure the lateral deformation of the specimens. Four-point loading was applied to the two beam specimens with clear span of 705 mm to investigate the bending capacity of the specimens with and without HSF. A laser triangulation was placed in the mid-span on the tension side of the specimen to determine the mid-span deflection. All specimens were tested under displacement control with a loading rate of 0.3-0.5 mm/min. Figure 1 shows a typical testing setup for the compression and four-point loading.

(a) (b)

Figure 1. Test setup: (a) axial compression and (b) four-point loading.

4 EXPERIMENTAL RESULTS AND DISCUSSION

Results of all the specimens tested under different eccentric loading are presented in Table 1. Results of the beam specimens are presented in Table 2. Ductility is a parameter that used to investigate the performance of the specimens under different loading conditions (Hadi and Schmidt 2002, Sheikh and Legeron 2014). In this study, the ductility of the specimens was measured by using the deformation at 80% of the maximum load (Δ_{80}) in the descending part of the axial load and axial deformation curve divided by the deformation at yield load (Δ_1). The yield load was assumed to be the load corresponding to an approximation of the limit of the elastic behavior.

Table 1. Results of specimens under different eccentric loading.

Specimens	Load (kN)		Deformation (mm)				Ductility
	Yield	Maximum	At yield load		At maximum load		
			Axial	Lateral	Axial	Lateral	
RC-0	1681	1827	2.16	-	2.57	-	3.15
HSF-0	1990	2196	2.70	-	3.36	-	3.51
RC-25	1140	1140	2.37	1.48	2.37	1.48	1.65
HSF-25	1168	1270	2.16	1.01	2.65	1.62	2.52
RC-50	661	697	2.11	1.83	2.39	2.25	1.53
HSF-50	803	857	2.13	1.97	2.44	2.56	2.52

Table 2. Results of specimens under four-point loading.

Specimens	Load (kN)		Mid-span deflection (mm)		Ductility
	Yield	Maximum	At yield	At maximum	
RC-B	228	294	4.38	16.66	5.49
HSF-B	302	398	5.48	30.84	6.47

4.1 Specimens under Axial Loading and Four-Point Loading

Figure 2 shows the load-deformation curves of the specimens under different loading conditions. Figure 2 (a) presents the axial load versus axial deformation of the specimens tested under concentric axial loading. The load of Specimen RC-0 suddenly dropped after spalling of the concrete cover, followed by buckling of the longitudinal steel bars and rupture of the spiral reinforcement. For Specimen HSF-0, the dilation in the lateral direction was observed without spalling of concrete cover due to the contribution of HSF. The maximum load of Specimen HSF-0 was 20% higher than Specimen RC-0. The ductility of Specimen HSF-0 was increased by 11% compared with Specimen RC-0. This indicates that the addition of HSF into HSC makes the specimens more ductile due to the beneficial effects of macro and micro steel fibers. While micro steel fibers arrested the micro crack, macro steel fibers arrested wider cracks.

Figure 2 (b) and (c) present the axial load and deformation of the specimens in axial and lateral directions for specimens tested under 25 mm and 50 mm eccentric loading, respectively. Specimens RC-25 and RC-50 failed by sudden loss of the concrete cover at the maximum load and buckling of the longitudinal steel bars in the compression zone. No cover spalling was observed for Specimens HSF-25 and HSF-50, and the test was stopped after a loud noise. For all specimens under different eccentricities the formation and development of horizontal cracks in the tension face and vertical cracks in the compression face were observed. The maximum load for Specimen HSF-25 increased by 11% compared with Specimen RC-25. The inclusion of HSF led to improvement in ductility of the specimens. The ductility increased by 50% for Specimen HSF-25 compared with Specimens RC-25. For the specimens under 50 mm eccentricities, Specimen HSF-50 showed the largest increase in the maximum load compared with Specimen RC-50. Maximum load was increased by 23% for Specimen HSF-50. The enhancement in ductility was about 60% for Specimen HSF-50 compared to Specimen RC-50.

Figure 2 (d) presents four-point loading and corresponding mid-span deflection for the specimens. It can be seen that the maximum load was significantly increased for Specimen HSF-B compared to Specimen RC-B. The maximum load of Specimen HSF-B increased by 35% compared to Specimen RC-B. It was found that the ductility of the specimens improved by the inclusion of HSF. The ductility was increased approximately by 18% for Specimen HSF-B, compared to Specimen RC-B.

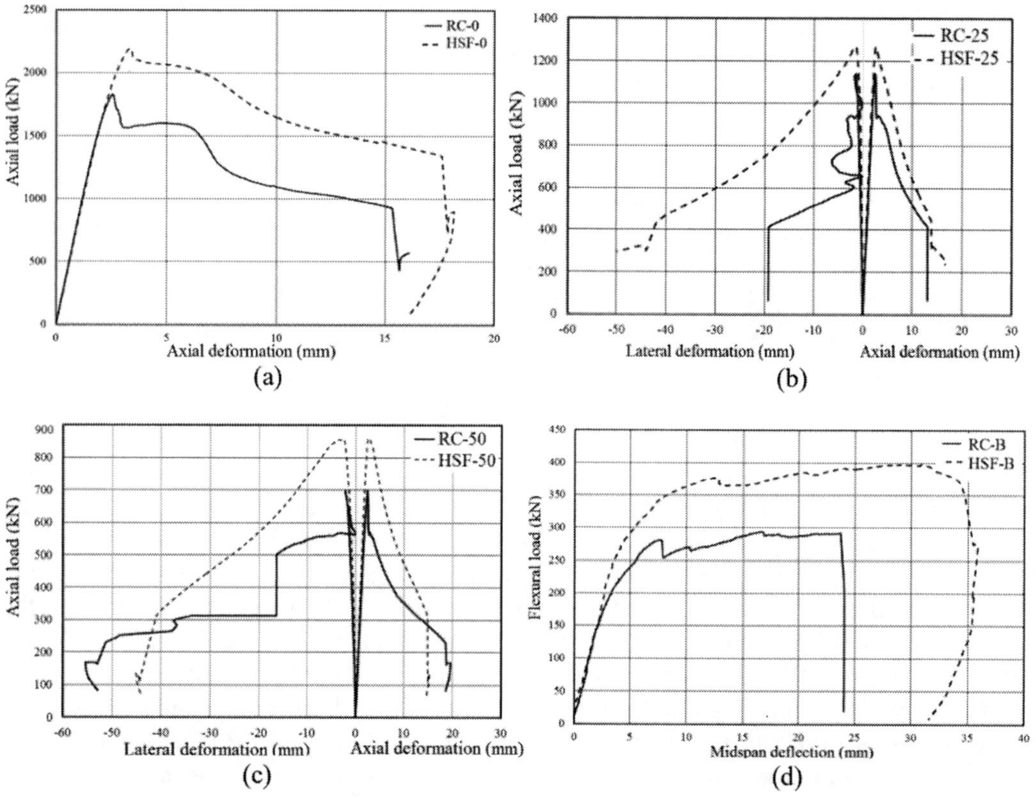

Figure 2. Test results for the specimens under (a) concentric compression (b) 25 mm eccentric compression (c) 50 mm eccentric compression and (d) four-point loading.

5 CONCLUSIONS

The behavior of circular HSC specimens with and without Hybrid Steel Fibers (HSF) was experimentally investigated. Six specimens were tested as columns under concentric, 25 mm and 50 mm eccentric loadings and two specimens were tested as beams under four-point loading. The experimental results showed that the maximum load of Specimens HSF were considerably higher than Specimens RC. Specimens HSF exhibited higher ductility than Specimens RC. The enhancements in the strength and ductility of the Specimens reinforced with HSF are due to the functions of macro and micro steel fibers.

References

Balanji, E., Sheikh, M. N. & Hadi M. N.S., Models For Predicting The Axial Compressive Strength of Steel Fiber Reinforced Normal and High Strength Concrete Square Columns, *in the 12th International Symposium on Fiber Reinforced Polymers for Reinforced Concrete Structures & The 5th Asia-Pacific Conference on Fiber Reinforced Polymers in Structures Joint Conference*, Nanjing, China, 14-16 December 2015.

Chi, Y., Xu, L. & Zhang, Y., Experimental Study on Hybrid Fiber-Reinforced Concrete Subjected to Uniaxial Compression, *Journal of Materials in Civil Engineering,* 26(2), 211-218, 2014.

Ganesan, N. & Murthy, J. V., Strength and Behavior of Confined Steel Fiber Reinforced Concrete Columns, *ACI Materials Journal,* 87(3), 221-227, 1990.

Hadi, M. N. & Schmidt, L. C. X, Use of Helixes in Reinforced Concrete Beams", *ACI Structural Journal,* 99(2), 2002.

Hadi, M. N. S, Behavior of Fiber RC Columns Wrapped with FRP Under Eccentric Loads, *Australasian Structural Engineering Conference (ASEC),* Melbourne, Australia, 26 – 27 June 2008.

Lee, H. H., Shear Strength and Behavior of Steel Fiber Reinforced Concrete Columns Under Seismic Loading, *Engineering Structures,* 29(7), 1253-1262, 2007.

Mangat, P. S. & Azari, M., Influence of Steel Fiber and Stirrup Reinforcement on the Properties of Concrete in Compression Members, *International Journal of Cement Composites and Lightweight Concrete,* 7(3), 183-192, 1985.

Mohammadi Y., Singh P. & Asushik S.K., Properties of Steel Fibrinous Concrete Containing Mixed Fibers in Fresh and Hardened State, Construct and Build Material, 22, 956–65, 2008.

Palanivel, S. & Sekar, M., Constitutive Behavior of Lateral Ties Confined Polyolefin Fiber Reinforced Concrete Under Monotonic Axial Compression, *Applied Mechanics and Materials*, 253-255, 2013.

Paultre, P., Eid, R., Langlois, Y. & Ĺvesque, Y., Behavior of Steel Fiber-Reinforced High-Strength Concrete Columns Under Uniaxial Compression, *Journal of Structural Engineering,* 136(10), 1225-1235, 2010.

Sheikh, M. N. & Légeron, F., Performance Based Seismic Assessment of Bridges Designed According to Canadian Highway Bridge Design Code, *Canadian Journal of Civil Engineering,* 41(9), 777-787, 2014.

Tokgoz, S., Effects of Steel Fiber Addition on the Behavior of Biaxial Loaded High Strength Concrete Columns, *Materials and Structures,* 42(8), 1125-1138, 2009.

Trapko, T., Behavior of Fiber Reinforced Cementitious Matrix Strengthened Concrete Columns Under Eccentric Compression Loading, *Materials & Design,* 54, 947-954, 2014.

Interaction between Theory and Practice in Civil Engineering and Construction
Edited by Komurlu, R., Gurgun, A. P., Singh, A., and Yazdani, S.

REPAIR OF CORRODED RC BEAMS

NABIL AL-AKHRAS[1], WALID AL-KUTTI[2], and ABDULAZIZ ALJABER[2]

[1]*Civil Engineering Dept, Jordan University of Science and Technology, Irbid, Jordan*
[2]*Construction Engineering Dept, University of Dammam, Dammam, KSA*

Twelve prototype rectangular RC beams (120 x 150 x 1000 mm) were constructed and evaluated using flexural test to investigate the employment of advanced composites in repairing corroded reinforced concrete (RC) beams. Three concrete mixtures with effective w/c ratio of 0.4 and cement content of 370 kg/m^3 were utilized in the study: control ordinary Portland cement concrete, silica fume (SF) and ground granulated blast furnace slag (GGBFS) concrete mixes. The RC beams (reinforced with two steel bars having diameter of 12 mm) were immersed (after 28 days of curing) in 2.5% NaCl solution and exposed to accelerated corrosion process using impressed electrical current. The corroded RC beams were repaired using advanced composite of carbon fiber reinforced polymer (CFRP) sheets. The CFRP sheets measuring 300-mm width by 0.131-mm thickness were used to rehabilitate and restore the mechanical behavior of the corroded and damaged RC beams. The investigation results confirmed that the corrosion of steel reinforcement caused significant deterioration and reduction of flexural capacity. The corroded SF and GGBFS beams showed higher flexural capacity compared to the corroded OPC beams. The repaired SF and GGBFS beams showed higher ductility and performance gain in the flexural capacity compared to the repaired OPC beams.

Keywords: Silica fume, Slag, Corrosion, Chloride, Concrete.

1 INTRODUCTION

Reinforcing steel is normally well protected against corrosion by embedment in high-quality concrete. The concrete high alkaline environment induces passive film around the steel reinforcement that protects the steel from corrosion. However, the penetration of chloride ions in sufficient amount (above threshold level) to the depth of the reinforcing steel breaks the passive film and initiates corrosion (Alonso *et al.* 2000). Different sources of chloride may come from deicing salts, seawater and some chemical admixtures (such as set accelerator). Marine structures such as bridges, piers, off-shore structures, etc. are exposed to high chloride environment that come from sea water that can penetrate with time into the reinforced concrete (RC) structures and causes severe corrosion (Scott and Alexander 2007). The development of corrosion products cause volume increase that consequently cause expansion, cracking and spalling of the concrete cover. Additionally, the corrosion cause reduction in the cross sectional area of the steel bars that jeopardizes the structural safety of the RC structures exposed to corrosion (Vu *et al.* 2005).

The penetration of chloride into the concrete depends on many factors including: permeability of concrete, w/c ratio, presence of mineral admixtures, and presence of cracks in the RC structure. Three factors should be present for the corrosion to occur including: moisture, oxygen and chloride. Variations in the amount of chloride, moisture, or oxygen at different points along the reinforcing steel bar, or between different steel bars, develop voltage differentials that greatly increase the rate of corrosion (Mechers and Li 2006)

The rehabilitation of damaged RC members such as beams, slabs and columns are well recognized to be practical and feasible solutions to restore the structural capacity of the deficient structural members. Many factors may induce damage and deterioration in the RC members including over-load, sulfate attack, alkali-silica reaction and corrosion of steel reinforcements. Several techniques may be applied to repair and retrofit the deteriorated structural members including: bonding of steel plate, enlarge the member's section, use of ferrocement and using advanced composites of fiber-reinforced polymers (FRP). The FRP is considered an innovative technique to repair, retrofit and restore the serviceability of the deteriorated structural elements (Chen and Teng 2003, Bousselham and Chaallal 2004, Adhikary and Mutsuyoshi 2006, Aslam *et al.* 2015).

The use of mineral admixtures such as silica fume (SF), pulverized fly ash (PFA) and ground-granulated blast furnace slag (GGBFS) in concrete mixtures is well established to have a very beneficial influence that improve significantly the mechanical properties and the durability of concrete. This study investigates the effect of SF and GGBFS on the performance and repair of corroded and deteriorated RC beams using advanced composite materials.

2 DESCRIPTION OF TEST PROGRAM

2.1 Materials

Type I Ordinary Portland cement (OPC) was used in the study. Micro-silica having specific gravity of 2.22 and Blaine fineness of 19,200 m^2/kg was employed in this study. The GGBFS used has specific gravity of 2.91 and specific surface area (using Blaine method) of 419 m^2/kg. Crushed limestone coarse aggregate with maximum aggregate size of 20 mm, bulk specific gravity of 2.62 and absorption of 2.3% was utilized in the study. Dune sand fine aggregate with bulk specific gravity of 2.6 and absorption of 2.2% was utilized in preparing the concrete mixture. High range water reducer (Superplasticizer) was used in all concrete mixtures (at 1.3% by weight of cement) to obtain workable fresh concrete with slump ranged from 22 to 26 cm. Steel reinforcement having diameter of 12 mm of Grade 60 steel was employed for the main reinforcement of the RC beams with yield stress of 420 MPa. An advanced composite material consists of carbon fiber reinforced polymers (CFRP) sheets measuring 300 mm width by 0.131 mm thickness were used to rehabilitate and restore the mechanical behavior of the corroded and damaged RC beams.

2.2 Concrete Mixtures

Three concrete mixtures with effective water to binder (w/b) ratio of 0.40 were prepared and utilized in the study. The cement content was 370 kg/m^3, the coarse aggregate content was 1100 kg/m^3 and the fine aggregate content was 70% of the coarse aggregate content. Table 1 shows the concrete mixture components per cubic meter.

Table 1. Concrete mix design, slump, and strength.

Mix	Cement (kg)	SF (kg)	GGBFS (kg)	Water (kg)	CA (kg)	FA (kg)	Slump (cm)	f_c' (MPa)
OPC	370	0	0	147	1100	765	22	30
SF	333	37	0	147	1100	765	23	42
GGBFS	148	0	222	147	1100	765	26	24

CA = Coarse aggregate, FA = Fine aggregate, f_c' = compressive strength at 28 days

2.3 Corrosion Exposure

An accelerated corrosion procedure using impressed electrical direct current (DC) of 2 mA/cm^2 of the surface area of the reinforcing steel was utilized. The beams were immersed in a water tank containing 2.5% NaCl solution. The steel bars were connected to the positive DC current and a steel plate immersed in the tank was connected to the negative DC current. The corrosion process was monitored periodically every 12 hours using visual observations of the beam.

2.4 Repairing Procedure

The rehabilitation process of the corroded beams was pursed as following: The concrete surface was cleaned thoroughly using a brush. The CFRP fabric sheets were cut into pieces of one meter in length. The epoxy resin was prepared by mixing two parts using laboratory mixer for two minutes. The epoxy was spread over the bottom surface and the sides of the beam. The CFRP fabric sheets were placed onto the resin coating and squeezed using a plastic roller until the epoxy was distributed evenly over the whole CFRP fabric surface. A thin layer of the epoxy was spread over the surface of the fabric sheets. The epoxy was cured at laboratory temperature of about 21 C for 3 days.

3 RESULTS AND DISCUSSION

3.1 Concrete Strength Properties

As shown in Table 1, the SF concrete compressive strength increased significantly from 30 MPa for the OPC to 42 MPa with percent increase of 40%. However, the GGBFS concrete compressive strength reduced from 30 MPa to 24 MPa with percent reduction of 20%. The increase in the strength of SF concrete is due to the high reactivity of the SF. Nevertheless, the decrease in the strength of the GGBFS is due to the low reactivity of GGBFS as pozzolanic material and the high content of GGBFS (60%) as a replacement of cement used in the concrete mix.

3.2 Intact Beams

Figure 1 shows the load-deflection behavior of the intact RC beams before exposed to corrosion. As shown in the figure, both the SF and GGBFS reinforced concrete beams showed higher flexural capacity performance compared to the OPC beams. The stiffness of the SF and GGBFS beams increased significantly compared to that of OPC beams. Also, the SF beams reached higher ultimate load of 46 kN compared to the GGBFS that reached 39 kN and the OPC that reached 40 kN which is about 1.15 times the results of OPC and GGBFS. This behavior may be due that the SF concrete has higher 28 days compressive strength (42 MPa) compared to that of the OPC concrete (30 MPa) and GGBFS concrete (24 MPa). Both the OPC and GGBFS reached almost the same ultimate maximum load. The SF beams were also more ductile and reached higher ultimate deflection of 4.3 mm compared to 4 mm for the OPC and 3.6 mm for the GGBFS beams.

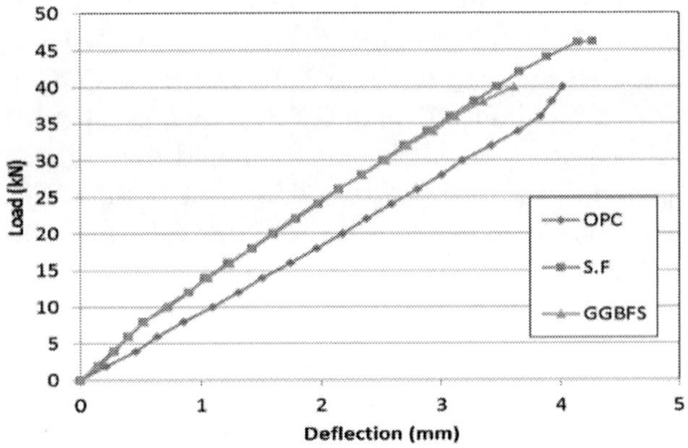

Figure 1. Load-deflection behavior for the RC beams before corrosion.

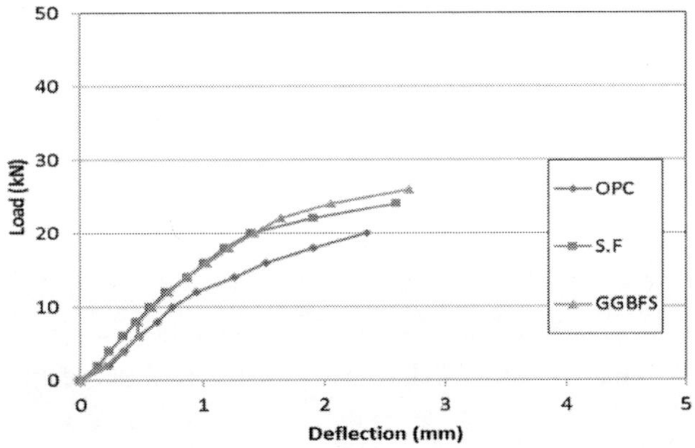

Figure 2. Load-deflection behavior for the corroded RC beams.

3.3 Corroded Beams

Figure 2 shows the load-deflection behavior of the corroded RC beams under flexural loading test. It is obvious that the RC beams containing SF and GGBFS showed higher performance to corrosion compared to the OPC beams. The ultimate load of the GGBFS and SF beams reached 26 and 24 kN, respectively, compared to 20 kN for the OPC beams. The percent reduction in the ultimate flexural load capacity for the OPC, SF and GGBFS were 50%, 48% and 33%, respectively. The use of 60% GGBFS replacement by cement concrete improved the performance of RC beams against corrosion compared to the OPC and SF concrete. The superiority performance of GGBFS in resisting the corrosion of reinforcing steel may be explained that the GGBFS particles effectively bind the chloride ions that initiate the corrosion of the steel bars.

3.4 Repaired Beams

Figure 3 presents the load-deflection behavior of the repaired beams. The RC beams containing mineral admixtures shows higher ultimate flexural load capacity compared to the OPC beams. The ultimate load capacity of the SF and GGBFS beams were 22 and 18 kN compared to 14 kN for the OPC beams. The GGBFS beams showed the highest deflection value of 3 mm compared to 1.9 mm for the OPC beams and 2.5 mm for the SF beams. The repaired SF and GGBFS beams showed higher ductility compared to the OPC beams. The SF beams exhibited the highest ultimate flexural load of 22 kN with percent gain of 48% from the intact original flexural load (before corrosion and deterioration) compared to 18 kN ultimate flexural capacity for the GGBFS with percent gain of 45% and 14 kN for the OPC with percent gain of 35%. The repaired SF and GGBFS beams showed higher performance gain in the flexural capacity compared to the repaired OPC beams.

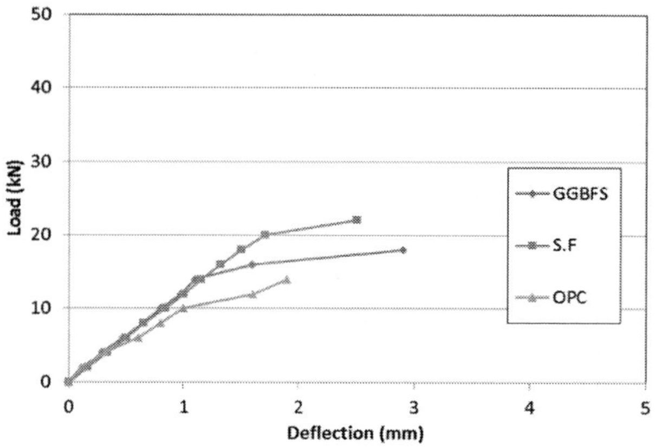

Figure 3. Load-deflection behavior for the repaired RC beams.

4 CONCLUSIONS

The followings can be concluded from the results of the study:

1. Both the SF and GGBFS reinforced concrete beams showed higher flexural capacity performance compared to the control beams. The stiffness of the SF and GGBFS beams increased significantly compared to those of the control beams.
2. The GGBFS beams showed the highest performance to corrosion with 33% reduction in the ultimate load, then the SF beams with 48% reduction. The control RC beams had 50% reduction in the ultimate flexural load capacity.

3. The rehabilitated SF and GGBFS beams exhibited higher ductility and flexural capacity gain of 48% and 45%, respectively, compared to the rehabilitated OPC beams that gained 35% of the ultimate flexural load capacity.

References

Adhikary, B. and Mutsuyoshi, H. Shear strengthening of reinforced concrete beams using various techniques, *Construction and Building Materials*, 20, 366–373, 2006

Alonso, C., Andrade, C., Castellote, M., and Castro, P., Chloride Threshold Values to Depassivate Reinforcing Bars Embedded in a Standardized OPC Mortar, *Cement and Concrete Research*, 30, 1047-1055, 2000.

Aslam M, Shafigh P, Jumaat M, Shah S, Strengthening of RC beams using prestressed fiber reinforced polymers – A review, *Construction and Building Materials*, 82(1) 235-256, 2015

Bousselham A, Chaallal O. Shear strengthening reinforced concrete beams with fiber-reinforced polymer: assessment of influencing parameters and required research, *ACI Structural Journal*, 101(2):219–27, 2004.

Chen J F and Teng J G. Shear capacity of fiber reinforced polymer-strengthened reinforced concrete beams: fiber reinforced polymer rupture, *Journal of Structural Engineering*, ASCE, 129(5):615–25, 2003

Mechers, R. E. and Li, C. Q., Phenomenological Modeling of Reinforcement Corrosion in Marine Environment, *ACI Materials Journal*, 103(1), 25-32, 2006.

Scott, A. N., and Alexander, M. G., The Influence of Binder Type, Cracking and Cover on Corrosion Rates of Steel in Chloride-Contaminated Concrete, *Magazine of Concrete Research*, 59(7), 495-505, 2007.

Vu, K., Stewart, M. G., and Mullard, J., Corrosion-Induced Cracking Experimental Data and Predictive Models, *ACI Structural Journal*, 102(5), 719-726, 2005.

Interaction between Theory and Practice in Civil Engineering and Construction
Edited by Komurlu, R., Gurgun, A. P., Singh, A., and Yazdani, S.

EFFECTS OF ALTERNATIVE AGGREGATES ON CONCRETE STRENGTH

DANIELA IONESCU, JOE PETROLITO, JOHN FORREST, SAMUEL GERMAN, and SAM PINNUCK

Dept of Engineering, La Trobe University, Australia

Worldwide, large quantities of solid waste are generated by construction and demolition (C&D) activities and the car industry. At the same time, the availability of landfill space is diminishing, and most industrialized nations are actively promoting recycling procedures to reduce the amount of waste going to landfill. Governments have also set up regulatory bodies to provide standards and protocols for these procedures. Environmental benefits and reduced concrete production costs are some of the advantages of using recycled materials. Given these worldwide trends, it is essential to consider recycling some of the solid waste from the construction industry and used tires from the car industry to supplement natural aggregates in the production of concrete for the construction industry. This papers discusses the on-going research at La Trobe University on the use of recycled aggregates such as crushed concrete, crushed brick and rubber chips in the production of concrete. The effects of partial or total replacement of normal-weight aggregates with recycled aggregates on the strength of concrete were investigated. Different types and proportions of aggregate replacements were studied to establish which protocols provided the best options. It was found that concrete mixes produced with recycled aggregates produced acceptable concrete with reduced production costs.

Keywords: Construction and demolition waste, Crushed bricks, Crushed masonry, Recycled masonry, Recycled tires, Rubber chips.

1 INTRODUCTION

Concrete is the most preferred construction material for civil infrastructure due to its versatility, strength and durability. However, concrete is not an environmentally friendly material as it requires significant quantities of natural aggregates in its production. Hence, there is an increasing demand and interest in aggregates from non-traditional sources, such as from recycled industrial (scrapped tires) and construction and demolition (C&D) wastes.

In Australia, scrapped tires and C&D waste represent a significant proportion of the solid waste disposed at landfills. In 2010-2011, 32% of the total of 63 million tons of solid waste was C&D waste (Randell *et al.* 2013). The recovery rate of the masonry materials was 66%. During the same period, the Australian car industry produced over 500,000 tons of waste rubber tires, with only 16% of this being recycled (Hyder Consulting 2012). Furthermore, only 3% of the recycled tires are currently used in civil engineering applications, which is well below the rate for other developed countries (Mohammadi *et al.* 2014).

As sustainability is a vital part of the construction industry, the use of recycled aggregates in concrete is one possible solution to reduce the environmental impacts of concrete production. Currently, recycled masonry materials are mainly used in pavement applications (Hyder Consulting 2011). Ionescu (2010) and Cavalline and Weggel (2013) reported that recycled masonry can successfully be used in concrete production as replacements for natural aggregates. Recycled rubber from scrapped tires is mainly used as a partial replacement for sand (Li *et al.* 2014, Mohammadi *et al.* 2014), while rubber chips may be used as a partial replacement of coarse aggregates in concrete production (Somerville *et al.* 2015). The results of an ongoing study into the use of waste products (masonry materials and scrap tires) as either partial or total aggregate replacements in the production of light-weight concrete are reported in this paper.

2 MATERIAL REQUIREMENTS

Neville (2005) reviewed the major properties of aggregates that affect the behavior of fresh and hard concrete, namely strength, hardness, toughness, durability, porosity, volume change, grains shape and texture, chemical reactivity and relative density. Past research into the possible uses of crushed masonry has suggested that crushed concrete and bricks have most of the required properties, and hence they may be used as either partial or total replacements for naturally occurring aggregates in concrete production (Oikonomou 2005, Ionescu 2010, Ionescu *et al.* 2015).

A good concrete mix shows evidence of equal amounts of bond failure and fracturing of the coarse aggregates (Neville 2005). Hence, aggregates that have angular grains with a rough surface texture and a broad gradation produce the minimum void space in the concrete matrix. Proportions of flaky and elongated particles and aggregates with high water absorption should also be avoided (Neville 2005).

3 EXPERIMENTAL PROGRAM

The physical characteristics of the materials used in the study were determined in accordance with relevant Australian Standards (primarily AS 1012, AS 1289, AS 2758). The governing factors were the compliance with current specifications and the feasibility of obtaining an acceptable material.

3.1 Concrete Aggregate

Considering the economic aspects, it was decided that the recycled masonry aggregates would be used as supplied with no additional crushing, sieving or washing. Materials were supplied by All Stone Quarries (ASQ) in Bendigo, Hymix Quarry in Axedale (HQA) and Tyrecycles in Melbourne. Independent of the material type used as aggregates, coarse and fine fractions combine best for a target Gradation 2 (C&CAA 1976, AS 2758.1 2014). The following aggregate combinations were used:

- HQA coarse basalt and blended washed river sand (B-CA + BWRS-FA)
- ASQ coarse recycled brick and HQA blended washed river sand (Br-CA + BWRS-FA)

- ASQ coarse and fine recycled brick (Br-CA + Br-FA)
- HSQ coarse basalt and ASQ fine recycled brick (B-CA + Br-FA)
- ASQ coarse recycled concrete and HQA blended washed river sand (CC-CA + BWRS-FA)
- ASQ coarse and fine recycled brick (CC-CA + CC-FA)
- HQA coarse basalt with 5%, 10% and 20% replacement by volume with Tyrecycles rubber chips and blended washed river sand (B&XRC + BWRS-FA, where X is 5, 10 and 20, respectively).

Table 1 gives the gradation characteristics of the nine aggregate combinations. Fines (grains < 0.425 mm) were tested for any clay content. A summary of the consistency tests (AS 1289 2005) and the clay and silt contents (AS 1141.33 1997) is given in Table 2. The finer fractions were classified as non-plastic/low plasticity silts. The physical properties of the aggregates were determined in accordance with the relevant specifications (AS 1141.6.1 2000), and they are summarized in Table 3. As expected, recycled brick and concrete (CA and FA) showed significantly higher water absorption rates compared with the currently-used aggregates, and this was accounted for in the mix design. The recycled CA concrete, brick and rubber chips were 18%, 20% and 60%, respectively, lighter than the basalt, whereas the recycled FA concrete and brick were only 4% to 6% lighter than the river sand.

Table 1. Grading characteristics of combined aggregates (after Ionescu *et al.* 2015).

Material	D_{10} (mm)	D_{30} (mm)	D_{60} (mm)	C_u	C_c
B-CA + BWRS-FA	0.51	3.75	8.20	16.1	3.4
Br-CA + BWRS-FA	0.41	2.90	11.05	27.0	1.9
Br-CA + Br-FA	0.85	5.71	10.34	12.2	3.7
B-CA + Br-FA	0.53	3.60	7.80	14.7	3.1
CC-CA + BWRS-FA	0.26	3.70	11.30	43.5	4.7
CC-CA + CC-FA	0.38	2.65	9.98	26.3	1.9
B&5RC + BWRS-FA	0.32	2.85	8.15	25.5	3.1
B&10RC + BWRS-FA	0.27	2.57	8.05	29.8	3.0
B&20RC + BWRS-FA	0.10	2.02	8.00	80.0	5.1

Table 2. Fine aggregates consistency and silt content characteristics (after Ionescu *et al.* 2015).

Material	Liquid limit (%)	Plastic limit (%)	Plasticity index (%)	Linear shrinkage (%)	Silt content (%
BWRS-FA	18	14	4	2	6
Br-FA	19	16	3	1	7
CC-FA	19	17	2	2	8

Table 3. Physical properties of aggregates used (after Ionescu *et al.* 2015).

Material type	B-CA	Br-CA	CC-CA	RC-CA	BWRS-FA	Br-FA	CC-FA
Water absorption (%)	2.23	7.00	6.02	0.40	0.60	4.60	3.80
Particle density (SSD) (t/m^3)	2.72	2.20	2.24	1.08	2.61	2.44	2.50

3.2 Concrete Properties

The concrete mixes were based on a characteristic compressive strength of 32 MPa. A 50-mm slump was used (C&CAA 2002), and the aggregates were used in the air-dried condition. The quantities of materials used per 1 m^3 are summarized in Table 4.

Table 4. Average properties for 28 days old concrete (after Ionescu *et al.* 2015).

Mix type \ Parameter	Mix quantities c/w/FA/CA/RC (kg)	Compressive strength (MPa)	Flexural strength (MPa)	Indirect tensile strength (MPa)	Density (kg/m^3)
B-CA + BWRS-FA	370/200/620/1090/0	53.0	5.9	3.5	2400
Br-CA + BWRS-FA	370/216/620/950/0	43.0	5.6	3.3	2240
Br-CA + Br-FA	370/258/530/950/0	41.0	5.9	3.2	2180
B-CA + Br-FA	370/208/530/1090/0	49.5	6.0	3.8	2320
CC-CA + BWRS-FA	370/206/700/1130/0	41.2	4.6	3.2	2280
CC-CA + CC-FA	370/232/610/1190/0	28.9	3.7	2.5	2200
B&5RC + BWRS-FA	370/200/620/1040/19	42.3	5.7	2.8	2380
B&10RC + BWRS-FA	370/190/620/980/38	35.9	5.3	2.4	2330
B&20RC + BWRS-FA	370/180/620/870/76	25.5	4.7	1.9	2250

Note: c = cement, w = water, FA = fine aggregates, CA = coarse aggregates and RC = rubber chips.

3.2.1 *Density*

The concrete specimens were water cured prior to testing. The average density of concrete at 28 days is given in Table 4. The alternative aggregates produced a slightly lower density concrete. The Br-CA + Br-FA combination caused the highest reduction in density (9%), followed by the CC-CA + CC-FA, Br-CA + BWRS-FA, CC-CA + BWRS-FA and B-CA + Br-FA combinations, with density reductions of 8%, 7%, 5% and 3%, respectively. The rubberized concrete had a density reduction of 1%, 3% and 6%, for a replacement by volume of 5%, 10% and 20%, respectively.

3.2.2 *Compressive strength*

Concrete produced with alternate aggregates resulted in a consistently lower compressive strength, with reductions between 6% and 52%. Concrete produced with recycled brick resulted in slightly better concrete than that produced with recycled concrete. The replacement of basalt and river sand with recycled brick (CA + FA) gave a 23% reduction in strength, whereas the replacement by recycled concrete (CA + FA) gave a 45% reduction in strength. This is believed to be due to the large percentage of dust (grains < 0.075 mm) present in CC-FA. The partial replacement of coarse aggregates with recycled masonry caused a reduction in strength of 19% and 22% for crushed brick and crushed concrete, respectively. The partial replacement of fine aggregates with crushed brick caused only a 6% reduction in strength.

The reduction in the strength of rubberized concrete increased with the percentage of aggregate replacement. The strength reduced by 20%, 32% and 52% for replacements of 5%, 10% and 20%, respectively. These results are attributed to a lesser bond between the cement paste and rubber chips.

3.2.3 *Flexural strength*

The 28-day flexural strengths are summarized in Table 4. The trends observed for the compressive strength were also observed for the flexural strength. Concrete produced with alternate aggregates resulted in strengths between 3% and 37% lower than that of the control batch. The effects of total and partial replacements of CA and FA were more critical for concrete produced with crushed concrete, which caused a 22% to 37% strength reduction, as opposed to only a 5% strength reduction for concrete produced with recycled bricks. These results are attributed to the presence of more dust (grains < 0.075 mm) in CC-FA. The partial replacement of aggregates (e.g., only FA) with crushed brick caused a slight improvement in the strength, which is perhaps due to the rough texture of crushed brick compared to river sand.

The rubberized concrete showed reductions in strength of 3%, 10% and 20%, respectively, when the percent of basalt replacement varied from 5% to 20%. The presence of rubber chips slowed the fracture propagation, and the two fragments of the failed test beams remained attached by the rubber chips.

3.2.4 *Indirect tensile strength*

As shown in Table 4, the use of alternate aggregates resulted in lower indirect tensile strengths with reductions between 6% and 46%. Concrete produced with recycled brick performed better (only 6% to 9% strength reduction) than that produced with crushed concrete (9% to 29% reduction in strength). Partial replacement of aggregates (e.g., only FA) with crushed brick caused an 8% improvement in the strength, which is believed to be due to the rougher grain surfaces of crushed brick.

The rubberized concrete showed reductions in strength of 20%, 31% and 46%, respectively, when the percent of basalt replacement varied from 5% to 20%. These results do not follow the trends observed for the flexural strength, although they are consistent with the trends observed for the compressive strength.

4 CONCLUSIONS

The effects of alternate aggregates for concrete production were discussed in this paper. The use of crushed brick CA and FA resulted in the lowest density concrete, and the strength of the concrete was reduced as well. Similar outcomes were obtained when the concrete was produced with coarse and fine recycled concrete aggregates. Partial replacement of natural aggregates also caused reductions in the concrete strength, although the resulting strengths were acceptable. Rubberized concrete resulted in significant strength reductions, although the flexural strength was less affected by the rubber chips.

Further investigation is required to evaluate the effects of alternate aggregates on the elastic properties and durability of concrete. In addition, research into methods of improving the overall properties of concrete with alternate aggregates without greatly affecting the production costs.

Acknowledgments

The authors gratefully acknowledge the help provided by All Stone Quarry, Bendigo, Hymix Quarry, Axedale and Tyrecycles, Melbourne with the supply of materials. The assistance with preparation and testing provided by Mr. David Osborne is also recognized.

References

Cavalline, T. L. and Weggel, D. C., Recycled Brick Masonry Aggregate Concrete, *Structural Survey*, Emerald Group, 31(3), 160-180, March, 2013.

Cement & Concrete Association of Australia (C&CAA), Guide to Concrete Construction, C&CAA T41, Standards Australia, Sydney, 2002.

Cement and Concrete Association of Australia (C&CAA), *Design, Control and Characteristic of Concrete*, UDC 620.1, C&CAA Press, Sydney, 1976.

Hyder Consulting, Encycle Consulting and Sustainable Resource Solutions, Construction and Demolition Waste Status Report – Management of Construction and Demolition Waste in Australia, October, 2011.

Hyder Consulting, COAG Standing Council on Environment and Water, Study into Domestic and International Fate of End-of-life Tyres, May 2012. Retrieved from: http://www.scew.gov.au/sites/www.scew.gov.au/files/resources/023a5607-0964-47e0-92cd-78fc0e6bb14e/files/hyder-end-life-tyres.pdf on October 3, 2015.

Ionescu, D., Mainali, B., Petrolito, J., Al Abadi, H., Davies, S. and Dickens, B., Construction Potential of Recycled Masonry for Concrete, *Implementing innovative ideas in in Structural Engineering and Project Management*, Saha, S., Zhang, Y., Yazdani, S., and Singh, A. (eds.), 619-624, ISEC Press, Fargo, USA, 2015.

Ionescu, D., Sustainable Civil Engineering Materials, *Incorporating Sustainable Practice in Mechanics of Structures and Materials*, Fragomeni, S., Venkatesan, S., Lam, N. T. K. and Setunge, S. (eds.), 833-838, CRC Press/Balkema, London, UK, 2010.

Li, L., Ruan, S. and Zeng, L., Mechanical Properties and Constitutive Equations of Concrete Containing a Low Volume of Tire Rubber Particles, *Construction and Building Materials,* Elsevier, 70, 291-308, 2014.

Mohammadi, I., Khabbaz, H. and Vessalas, K., In-depth Assessment of Crumb Rubber Concrete (CRC) Prepared by Water-soaking Treatment Method for Rigid Pavements, *Construction and Building Materials,* Elsevier, 71, 456-471, 2014.

Neville, A. M., *Properties of Concrete*, Pearson Prentice Hall, Harlow, England, 2005.

Oikonomou, N. D., Recycled Concrete Aggregates, *Cement & Concrete Composites*, Elsevier, 27, 315-318, 2005.

Randell, P., Pickin, J. and Grant, B., Waste Generation and Resource Recovery in Australia, Reporting Period 2010/11, 4 February 2014.

Somerville, S., Dhanasekar, R., Frame, T. and Suntaravadivel, T. G., An Investigation into the Techniques for Improving the Properties of Crumb Rubber Concrete, *Implementing innovative ideas in in Structural Engineering and Project Management*, Saha, S., Zhang, Y., Yazdani, S., and Singh, A. (eds.), 613-618, ISEC Press, Fargo, USA, 2015.

Standards Australia, Aggregates and Rock for Engineering Purposes - Concrete Aggregates, AS 2758.1, Standards Australia, Sydney, 2014.

Standards Australia, Methods of Sampling and Testing Aggregates, AS 1141.6.1, AS 1141.33, Standards Australia, Sydney, 2000.

Standards Australia, Methods of Testing Concrete, AS 1012, Standards Australia, Sydney, 2000.

Standards Australia, Methods of Testing Soils for Engineering Purposes, AS 1289, Standards Australia, Sydney, 2005.

Interaction between Theory and Practice in Civil Engineering and Construction
Edited by Komurlu, R., Gurgun, A. P., Singh, A., and Yazdani, S.

INVESTIGATING THE EFFECT OF NANO SILICA ON THE MECHANICAL PROPERTIES OF HARDENED CEMENT PASTE

SAMEER HAMOUSH, MIGUEL PICORNELL, and TAHER ABU-LEBDEH
Civil, Architectural, and Environmental Engineering Dept, North Carolina A&T State University, Greensboro, North Carolina, USA

This study investigated the compressive strength of hardened cement paste and the formation of calcium silicate hydrate (C-S-H) with the addition of nano silica (SiO_2). Through this search, the development of the concretes strength was determined to better understand the process of cement hydration. Compressive strength testing was performed using MTS and Forney testing machines to determine stress-strain curves and elastic modulus of materials. The hydration process and formation of C-S-H and calcium hydroxide (CH) was examined using Scanning Electron Microscopy (SEM) and Fourier Transform Infrared Spectroscopy (FTIR). This study also incorporates the use of vacuum curing, in comparison to that of the traditional water curing method. Results indicate an increase in compressive strength using 1%, 3% and 5% of nano silica to cement replacement by volume in comparison to the control mix (without nano silica). The optimum cement replacement to yield maximum strength was of the 1% nano silica content. The formation of C-S-H increases significantly during the early testing days which correspond with the drastic increase in compressive strength. The hydration process continues to increase throughout the 56 day trails at a moderate rate. The traditional water curing method proves to be more efficient and beneficial than of the vacuum curing method. However, vacuum cured results showed only about a 5% reduction in strength after 56 day tests in comparison to the water curing method.

Keywords: Nano-silica, SiO_2, Mechanical properties, FTIR monitoring, SEM monitoring, CSH growth, Curing methods.

1 INTRODUCTION

In recent years, environmental aspects of material conversion have induced research to modify Portland cement to support the increasing demands for sustainability in the construction division. Such change can be obtained by using additives to change the chemical composition of the cement (Ylmén *et al.* 2009). Nano silica is an emerging field of interest in civil engineering application. Nano silica has been proven to be more efficient in enhancing concrete strength than that of silica fume (Bi *et al.* 2012). Experimental results of Givi and his co-researchers showed that nano-particles blended with concrete increased the compressive, flexural, and tensile strength at all curing ages (Givi *et al.* 2010). Due to pozzolanic reaction and a reduction of calcium hydroxide, nano silica has been proven to improve performance of cement-based materials through the increase production of C- S-H gel. It has the capability to react with the free lime during the cement hydration process and forms additional C-S-H gel, giving strength, impermeability, and durability to concrete. It has also been found that the rate of pozzolanic reaction caused by nano silica is proportional to the amount of surface area that is available for reaction (Jo *et al.* 2007). The main objective of this study is to investigate the effect of nano silica in cement on the hydration process and on the mechanical properties of cement paste over the course of 56 days. The study investigated the hydration process of Portland cement with increments of nano silica addition by cement replacement, monitored by Fourier Transform Infrared Spectroscopy (FTIR) and by Scanning Electron Microscope (SEM). The results of FTIR spectroscopic signatures were compared to the represented samples examined under SEM. Correlation of the spectroscopic features to the development of concrete strength was also determined. Strength testing was carried out by using Forney Universal Testing and MTS machines for compressive strength and modulus of elasticity. The investigation also studied the effects of water curing in comparison to vacuum curing. Concrete cured in airless environments or vacuum curing, is a new area of interest and was examined in comparison to the universally used water curing approach.

2 EXPERIMENTAL PROGRAM

Approximately 211 specimens were prepared to conduct several tests using 4 in. by 8 in. cylinders and 2 in. cubes. The cementitious materials used for this study were Portland cement Type 1 and nano silica. Polycarboxilate polymer based superplasticizer was added in conjunction with the nano silica. Nano silica is silicon oxide nanoparticles (SiO2) synthetic product of porous and nearly spherical particles. With the combination of superplasticizer, nano silica has been proven to perform as a viscosity modifying agent to produce high performing concrete. The particle size of nano silica ranges from 15-20 nm. For the experimental setup, four (4) mix designs were created. The first mix, M1, was the control mix with cement and water. The next three mixes, M2, M3, and M4, consisted of a 1, 3, and 5 percent of cement replacement by volume of nano silica respectively. The water to cement (w/c) ratio was 0.27 using superplasticizer. For the control mix, M1, eight (8) cylinders and six (6) cubes specimens were prepared for each curing condition. Five (5) cylinders and six (6) cubes were manufactured for M2 and M3. For mixes M2 and M3, four of the cylinders were placed under water curing and one cylinder was placed under vacuum curing. For M4, two (2) cylinders were manufactured, each under separate curing, and six cubes. All of the cubes manufactured were separated in half for each curing method, three cubes under water and three under vacuum. The cylinders were comprised of plastic molds, 4 inches in diameter and 8 inches in depth, in compliance with ASTM C-470. The cube modes were comprised of brass, exactly 2 inches in size, and able to form three cubic specimens within each mold.

2.1 Mechanical Properties

The Forney machine was used for cylinder compression and the MTS used for cube compression. MTS 810 Landmark Servohydraulic Testing system was used. These systems include MTS software, FlexTest controls, MTS servohydraulic technology, and a complete selection of accessories. The load rate for the MTS machine was 0.01 in/s. The Forney Universal Testing Machine was used to carry out all cylinders testing, including compression and modulus of elasticity. The Forney machine has a 400 kip load capacity and a load rate of 12,000 lbs/min.

2.2 Scanning Electron Microscope (SEM)

SEM EVO LS made by ZEISS, was used for microstructure characterization and analysis of the hardened cement paste specimens. EVO LS has full environmental capabilities to capture nano scale interactions of samples under various pressures, temperatures, and humidities. During experimentation, the SEM was set into the environmental mode (EP) which allows atmosphere in the chamber since the sample is based on hydration. The scanning speed was set to four, which automatically changed the scanning cycle to 731 ms. The EP target set to 50 Pa, the KV value was set to 10 KV and the beam current to 46.0 μA.

2.3 Fourier Transfer Infrared Spectroscopy (FTIR)

FTIR examines the chemical properties of the material and monitors the changes in mechanical properties as the reactions take place. In this study, Thermo Scientific Nicolet iS10 FT-IR Spectrometer and OMNIC Specta software were used to perform FTIR monitoring. Several steps were followed to carry out FTIR testing. This includes: sample preparation, background check to subtract traces of dissolved gases and solvent molecules, placing the specimen onto the FTIR plate and collect at least 5 spectrums for each sample, the average of all the spectrums collected were recorded and used for results and discussion. After each sample spectrum has been obtained, data analysis was carried out and observed using the OMNIC software. For each sample, the absorption frequency bands assigned to each sample spectrum was observed for normal modes of vibrations within the molecules. These spectrums were later analyzed and compared within the discussion.

3 EXPERIMENTAL RESULTS

In this study, the hydration process of Portland cement with additives of nano silica has been monitored. With the use of SEM and FTIR, signatures of C-S-H which produce most of the concretes' strength, has been determined and examined from 3 to 56 days. Also, using Forney and MTS testing equipment, the overall strength of the harden cement pastes at all testing ages were determined and recorded for analysis.

3.1 Compressive Strength

Stress-strain curves were developed from each cube tested with the MTS Landmark Servohydraulic testing system. As stated in the experimental procedure, each material was casted into three cubes; control 0%, 1%, 3% and 5% of nano silica to cement replacement for water and vacuum curing. The stress-strain curves created from the three cubes were averaged together to obtain a single averaged stress-strain curve for each material tested. Although more than twenty figures were produced, Figure 1 displays (a) the stress-strain curves for the control samples for all days and water cured, and (b) the 3% of nano silica to cement replacement for the water cured samples. The Figure shows an increase in strength from the 3 day tests, ascending to 56 day tests. Figure 1(b) shows that with the incorporation of nano silica, the stresses at all testing ages are much greater than that of the control samples. The stresses for the 3% nano silica samples range between 12,300 psi for the 3 day results and 14,346 psi for the 56 day results. The Forney Universal Testing Machine was use to conduct all cylinder compressive strength tests for each sample. Again, several graphs were produced but Figure 2(a) shows the average compressive strengths for all water cured testing cylinders at ages 3, 7, 14, 28 and 56 day tests. As shown in the figure, the 1% and 3% nano silica samples perform significantly higher than that of the controlled. Figure 2(b) displays the average cylinder strength of the 0% controlled samples that were water and vacuum cured at all testing ages.

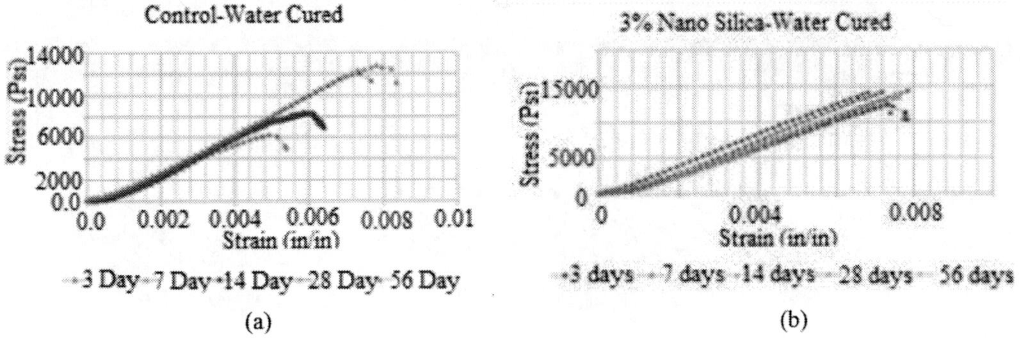

Figure 1. Stress-strain curves: (a) controlled-water cured and (b) 3% nano silica-water cured.

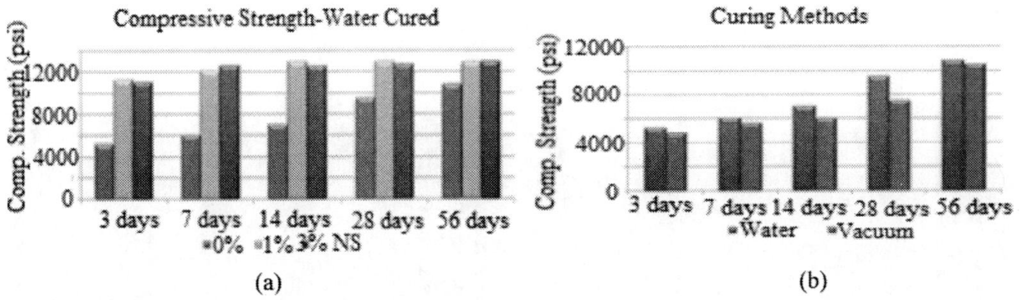

Figure 2. Compressive strength results for: (a) all water cured samples and (b) curing methods.

3.2 Scanning Electron Microscope (SEM)

The SEM was used to obtain a microscopic image of samples at different ages. From the images obtained, a further analysis using MatLab was conducted to filter the images using a Gaussian filter and to determine the percentages of chemical products for which that image portrays. From grey level imaging, four main phases were identified: porosity (P), calcium hydroxide (CH), calcium silicate hydrate (C-S-H), and unhydrated products (UP). Other products such as limestone and other hydrated products were not distinguished. Figure 3(a) shows the original SEM image of harden cement paste sample containing 1% nano silica after water curing and tested at 7 day. Figure 3(b) shows the same sample after filtered using the Gaussian filter. Features from this image are lighter and more distinguished. From the results gathered from the MatLab analysis of all testing samples, a table was generated to portray the results obtained. Results show an increase in C-S-H formation from 3 day to 56 day for the control and all nano silica mixtures. Results show a maximum amount of C-S-H formation in the 5% nano silica samples than that of the 1% and 3% nano silica samples. Also, at all testing ages, samples with nano silica out performed or contains more C-S-H content than the control. Also, results from vacuum cured specimens show similar results to the water cured ones. The amount of C-S-H found in the vacuum cured control samples and the 3% and 5% nano silica samples recorded nearly exactly the same percentage during the 56 day results, in comparison to the water cured samples at 56 days.

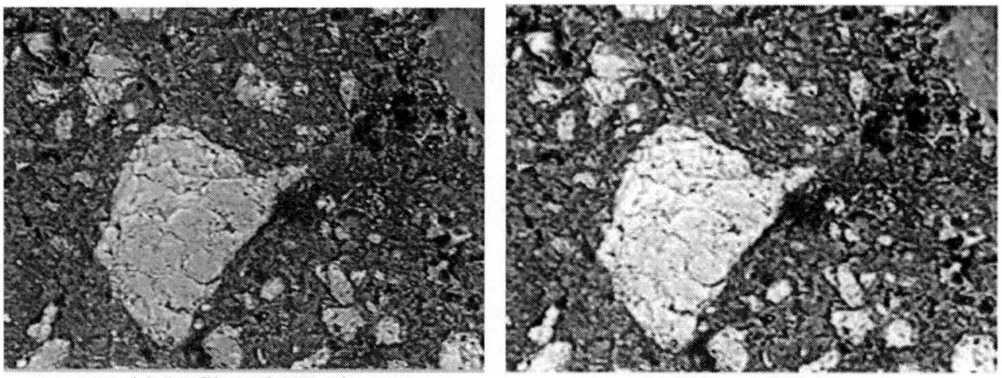

(a) Unfiltered SEM image (b) Filtered SEM image

Figure 3. SEM image of a7 days water cured 1% nano silica section.

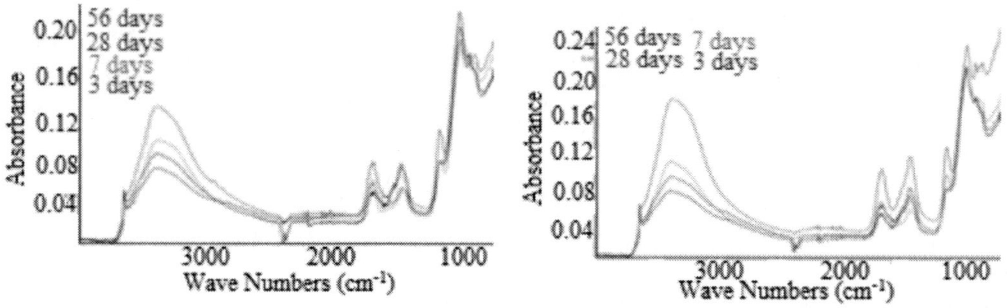

(a) (b)

Figure 4. Absorbance of control: (a) water cured and (b) vacuum cured samples, 3 to 56 day tests.

3.3 Fourier Transfer Infrared Spectroscopy (FTIR)

The hydration process was monitored for 3, 7, 28, and 56 days, by acquiring an FTIR spectrum for each sample under water and vacuum curing. The recorded spectra for the water and vacuum cured samples are displayed in Figures 4(a) and 4(b) respectively. The curing days increase from 3 to 56 days, from bottom to top of the spectra. From previous literature, signatures of hydration can be seen within the 900 to 1200 cm^{-1} region. Also notice the large peak in the area of 3400 cm^{-1}, which can be associated with hydrogen bond (O-H) or capillary water within that region. The remaining spectra for the 1%, 3% and 5% nano silica spectrums for both water and vacuum cured are collected but not shown. Once obtaining the possible C-S-H formation from each sample, further analysis was conducted. The amount of C-S-H found within each sample various amongst the different percentages of nano silica. Results show a drastic increase in C-S-H found from the 28 day to 56 day control samples. Figure 5(a) illustrates the change in area of C-S-H of 1%, 3% and 5% of nano silica replacement water cured control, while Figure 5(b) shows the change of C-S-H area in samples cured in vacuum. With the incorporation of nano silica, the hydration process tends to follow a similar pattern to that of the control mix. The data shows a significant decrease in hydration after 3 day and from there the hydration tends to increase at a slow pace leading up to the 56 day. The same occurrence takes places during the vacuum cured samples as well.

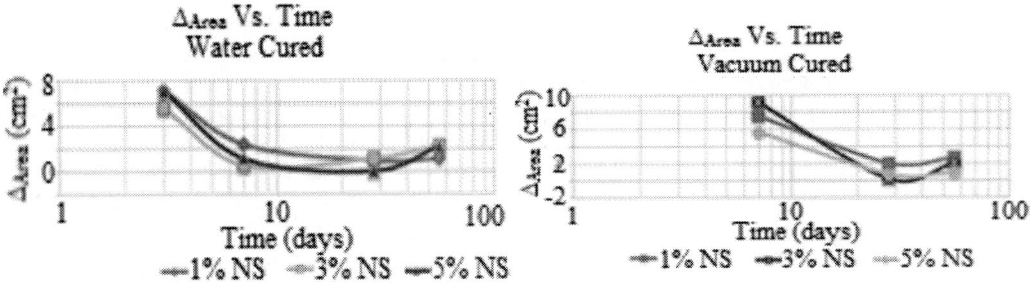

(a) (b)

Figure 5. Change in area of C-S-H for 1%, 3% and 5% of nano silica (a) water and (b) vacuum.

References

Bi, J., I. Pane, B. Hariandja and I. Imran, The Use of Nanosilica for Improving of Concrete Compressive Strength and Durability, App. Mechanics and Materials, pp 4059-4062, 2012.

Givi, N., S. Rashid, F. Aziz and M. Salleh, Experimental Investigation of the Size Effects of SiO2 Nano-particles on the Mechanical Properties of Binary Blended Concrete, Composites Part B: Engineering 41(8): 673-677, 2010.

Jo, B.-W., C.-H. Kim, G.-h. Tae and J.-B. Park, Characteristics of Cement Mortar with Nano-SiO2 Particles, Construction and Building Materials, 21(6): 1351-1355, 2007

Ylmén, R., U. Jäglid, B.-M. Steenari and I. Panas, Early Hydration and Setting of Portland Cement Monitored by IR, SEM and Vicat Techniques, Cement and Concrete Research 39(5): 433-439, 2009.

HYDRATION KINETICS OF CEMENT PASTE WITH SILICA FUME AND FLY ASH

MIGUEL PICORNELL, SAMEER HAMOUSH, and TAHER ABU-LEBDEH

Civil, Architectural, and Environmental Engineering Dept, North Carolina A&T State University, Greensboro, North Carolina, USA

This research study investigates the effect of fly ash and silica fume on the cement paste hydration. A total of 350 samples of different percentages of each additive were tested and compared with the controlled cement paste without additives. Testing method includes water curing and vacuum curing conditions and involves the use of Forney Universal Testing Machine and MTS Landmark Servohydraulic Testing System (MTS) for compressive strength; Fourier Transfer Infrared Spectroscopy (FTIR) monitored the hydration with spectra; and Scanning Electron Microscope (SEM) generated images for regional analysis. Compressive strength testing demonstrated that silica fume replacement had the highest overall strength under water curing. Replacement of fly ash exhibited the highest overall strength under vacuum curing. The hydration process was monitored with the use of FTIR and SEM. Signatures of CSH which produce most of the concretes' strength, has been determined and examined from 3 to 56 days. FTIR and SEM testing showed an increase in the change of CSH area with age. SEM testing revealed the formation of pores, CSH, and CH in images at all ages. The area of CSH grows most in early ages and diminishes over time. It is clear that the method of curing makes a difference in hydration. Results indicated that the area at which the possible formation of CSH was determined from each sample, has increased with respect to time; signifying the increase in strength over the course of testing days.

Keywords: Mechanical properties, MTS testing, FTIR monitoring, SEM monitoring, CSH growth, Curing methods.

1 INTRODUCTION

From the hydration process of cement and water, several products are formed. The main four products are: calcium silicate hydrate, calcium hydroxide, calcium aluminates and calcium sulfoaluminates. Calcium Silicate Hydrate (CSH) is the main and most important constituent of cement paste. Its hydration forms most of the new solid phases that give hardened cement paste its strength. It is known that the addition of fly ash materials have an adverse effects on the compressive strength of the cement. The replacement of cement with fly ash by percentage affects the properties, those being both the filler and dilution effects. Fly ash contains pozzolans, a siliceous material possessing cementitious properties, and consumes CH in the pozzolanic reaction (Narmluk and Nawa 2011). Zhang (2000) found that less CH will decelerate the hydration reaction. On the other hand, silica fume additive contains pozzolans which are siliceous and aluminous material which have no cementitious properties. However it will react with calcium hydroxide once hydrated. A pozzolanic reaction between silica fume and lime formed during cement hydration does not occur during the early stages of hydration. A study by Zelic *et al.* (2000) shows that the pozzolanic reaction between silica fume and lime formed during cement hydration does not occur until three days of hydration (Zelic *et al.* 2000). The presence of silica fume slightly increases the rate of cement hydration during the first few hours of hydration. Likewise, this indicates that silica fume affects hydration by the nucleating effect of its surface when it still exists as chemically inert filler. After silica fume hydrates increase in strength and hydration heat, pozzolanic activity takes over (Kadri and Duval 2009).

2 EXPERIMENTAL PROGRAM

In an effort to better comprehend the mechanisms underlying cement hydration, the use of new technology is needed to enhance the prediction of properties. This present work investigates the mechanical properties of cement paste as it goes under several stages of hydration, monitored by SEM and FTIR. The objective is to investigate the effect on cement hydration of supplemental fly ash or silica fume on the mechanical properties of cement paste. The integration of these two additives were studied to analyze their effect in accelerating C3S. Percentages of each additive will replace the cement by volume at 5% and 10% to be studied at 3, 7, 14, 28, and 56 day ages. Testing were carried out on 350 samples using a Forney Universal Testing Machine, MTS Landmark Servohydraulic Testing System (MTS), Fourier Transfer Infrared Spectroscopy (FTIR), and Scanning Electron Microscope (SEM). The testing methods should portray the compressive strength, modulus of elasticity, spectra, and images of the samples. This study will provide an understanding and insight towards the roles of fly ash or silica fume in the hydration properties of cement paste at various stages of hydration. The cement used in this study was of Portland cement Type 1. There are two commonly used classifications of fly ash, Classification F and C. Class F fly ash has good susceptibility to reducing the alkali-silica reaction (ASR). This classification will mitigate this issue better than class C fly ash. Classification F was used in this research. Silica fume is very fine noncrystalline silica produced in electric arc furnaces as a byproduct of the production of elemental silicon or alloys containing silicon. Silicon dioxide makes up a large percentage of silica fume.

For the experimental setup, five (5) mix designs were created. The first mix, M1, was the control mix with cement and water. The next four mixes, M2, M3, M4, and M5, consisted of a 5 and 10 percent of cement replacement by volume of fly ash and silica fume respectively. The water to cement (w/c) ratio was 0.40 for all mixes. For all mixes, eight (8) cylinders and six cubes were created for each age (hydration stage). The cylinders were 4 inches in diameter and 8 inches in height, while the cubes were exactly 2 inches cubed. All cylinders and cubes were unmolded after 24 hours and then half were cured in water and the other half vacuum cured for 3, 7, 14, 28, and 56 days until age of testing. Water curing involved the placement of the cement paste specimen into a water tank, completely submerged in water. For vacuum curing, a FoodSaver appliance series 2200 was used. The cement paste specimens were incased in vacuum seal bags and remained completely sealed until age of testing. After curing, the cylinders and cubes were dried in air for approximately one hour before testing.

2.1 Mechanical Properties

The Forney Universal Testing Machine was used to carry out all cylinders testing for compression and modulus of elasticity. It has 400 kip load capacity and a loading rate of 12000 lbs./in. Cylinders were capped with either rubber pads or a capping compound in accordance to ASTM C-617. Forney captured the maximum peak load applied to each cylinder under testing. The MTS Landmark Servohydraulic Testing machine was used to carry-out compressive strength testing for 2 inch cement paste cubes and produce stress-strain curves for each test. The loading rate for the MTS machine was 0.01 in/s. The system includes MTS software, FlexTest controls, MTS servohydraulic technology, and a complete selection of accessories. Once each cube was tested till breakage, the data results were stored. Test results include displacement in inches, loading in pounds, and the time in seconds. The data was later transferred to Excel to calculate the stress and strain, and further portrayed in the results.

2.2 Scanning Electron Microscope (SEM)

The high definition SEM EVO LS 10 made by ZEISS, was used for microstructure characterization and analysis of the hardened cement paste specimens. EVO LS 10 has full environmental capabilities to capture nano scale interactions of samples under various pressures, temperatures, and humidities. It includes EVO HD, which further increases an image resolution and contrast using a high definition beam source technology with low acceleration voltages. During experimentation, the SEM was set into the environmental mode (EP) which allows atmosphere in the chamber since the sample is based on hydration. The scanning speed was set to four, which automatically changed the scanning cycle to 731 ms. The EP target set to 50 Pa, the KV value was set to 10 KV and the beam current to 46.0 μA. SEM testing revealed the formation of pores, CSH, and CH in images at all ages.

2.3 Fourier Transfer Infrared Spectroscopy (FTIR)

The Thermo Scientific Nicolet iS10 FT-IR Spectrometer was used herein. Its many features includes creating standard operating procedures (SOPs) and suitable tests, verifies the quality of materials, identifies unknowns or mixtures, and can quantify mixture ingredients. With the use of OMNIC Specta software, the Nicolet iS10 can manage results and provide accurate and valid answers. The Nicolet iS10 consists of System Performance Verification (SPV) which tests the instrument against ASTM E-1421 method to insure quality performance. SPV is a tool used to ensure the spectrometer is performing as expected, at all times. For this experimental testing the Nicolet iS10 FTIR Spectrometer was used to identify and compare the unknowns within the cement paste mixtures. The FTIR spectrometer was used to identify the chemical properties during the hydration process.

3 EXPERIMENTAL RESULTS

As aforementioned, the hydration process of Portland cement with additives of fly ash or silica fume silica has been monitored. With the use of FTIR and SEM, signatures of

CSH which produce most of the concretes' strength, has been determined and examined from 3 to 56 days. Also, using Forney and MTS testing equipment, the compressive strength of the hardened cement pastes at all testing ages were determined and recorded for analysis.

3.1 Compressive Strength

Figure 1(a) shows the average compressive strengths for specimens that were cured in water. This figure represents the 0%, 5% and 10% of fly ash and silica fume to cement replacements, against the 3, 7, 14, 28 and 56 day tests. As shown in the figure, the 10% of silica fume consistently outperforms the 5% silica fume at all ages except three days. Figure 1(b) show the average compressive strengths for all vacuum cured testing specimens. In both the water and vacuum curing, the compressive strength increased with age. As stated in the experimental procedure, each material was casted into three cubes; 0%, 5% and 10% of fly ash and silica fume to cement replacement for water and vacuum curing. The stress-strain curves created from the three cubes were averaged together to obtain a single averaged stress-strain curve for each material tested at each age. Sample of the stress-strain curves is shown in Figure 2.

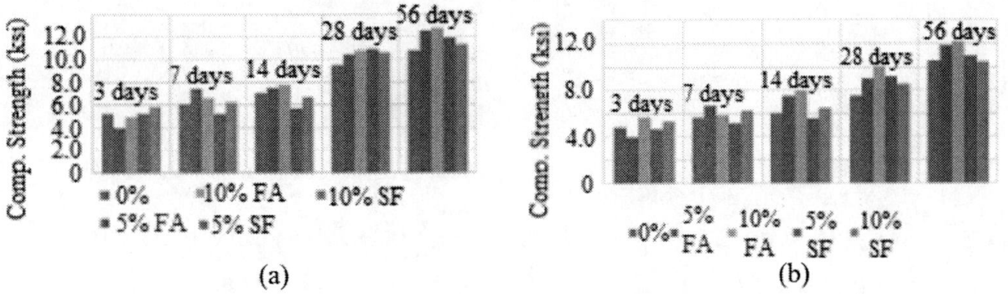

Figure 1. Average compressive strength: (a) water cured and (b) vacuum cured.

3.2 Scanning Electron Microscope (SEM)

The images in Figure 3 illustrate the cement hydration of 28 day cement paste at 1300x magnification without and with a filter, respectively. The filtered image has gone through both the convolution and median filters.

Analyzing SEM images show that the 10% fly ash replacement has a higher percentage of porosity, 7% more CSH, but less CH than 5% replacement. Silica fume at 10% replacement has a higher percentage of CSH by two percentage and lower porosity and CH than 5% replacement. Comparing fly ash and silica fume averages at 10%, silica fume had lower porosity, CH, and higher CSH. The control specimen had lower porosity than fly ash and higher than silica fume at 14%. Control specimen had 65% CSH average which was lower than fly ash and silica fume. In case of vacuum cured specimens, fly ash of 10% replacement has lower percentage of porosity, higher CSH by 4% than 5% replacement but same CH. Silica fume 10% replacement has a higher percentage of CSH by 8% and lower porosity and CH than 5% replacement. Comparing fly ash and silica fume averages at 10%, silica fume had lower porosity, CH, and higher CSH. The control had higher porosity than fly ash and silica fume but silica fume at 5% was the same at 16%. The CH of the control was higher than silica fume but lower than fly ash averages. Overall areas for water and vacuum curing have CSH percentages that continually grow larger with age. Figure 4 shows percentage of CSH area for specimens cured in (a) water and (b) in vacuum.

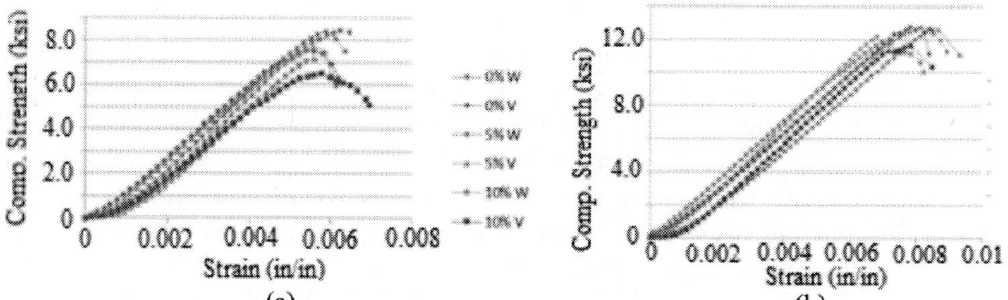

(a) (b)

Figure 2. Stress-strain curves: (a) 7 days fly ash and (b) 56 days silica fume.

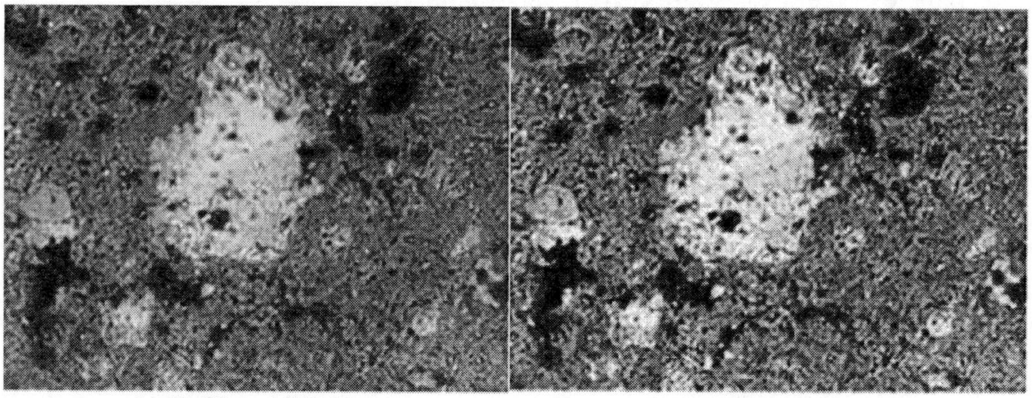

(a) Unfiltered SEM image (b) Filtered SEM image

Figure 3. SEM image of a 28 day water cured control section.

3.3 Fourier Transfer Infrared Spectroscopy (FTIR)

Fourier Transform Infrared Spectroscopy (FTIR) was used to monitor the hydration process and to approximate the amount of CSH formation within all samples. In this study, the hydration process was monitored for 3, 7, 14, 28, and 56 days, by acquiring an FTIR spectrum for each sample cured under water and vacuum. Figure 5 shows the absorbance spectrum of the 5% silica fume of cement replacement hydrated for 3, 7, 28 and 56 days after water curing.

As seen in Figure 5, the two major peaks occur around the wavelength numbers of 877 cm^{-1} and 1100 cm^{-1}. From the chemical composition of Portland cement and from past literature for possible assignment for peaks observed on cement spectrum; the peaks can possibly be assigned to the CaO and the SiO_2 for the two peaks respectively. Also notice the large peak in the area of 3400 cm^{-1}, which can be associated with hydrogen bond (O-H) or capillary water within that region. It should be noted that

Calcium silicate hydrate is a weak bond structure which appears around 1100 cm^{-1} and can be associated with the peak in that region.

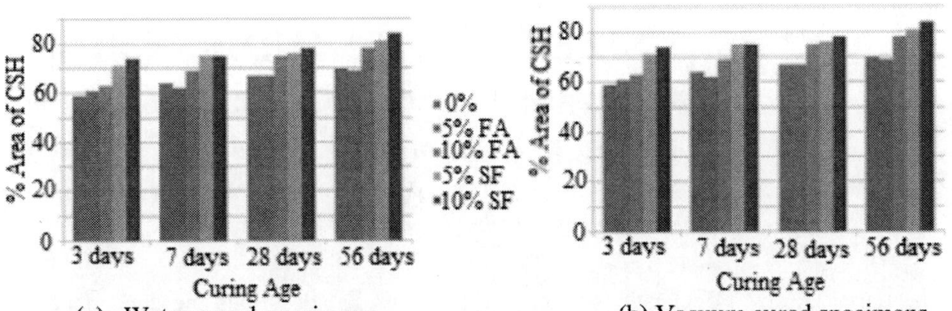

(a) Water cured specimens (b) Vacuum cured specimens

Figure 4. Growth of CSH with age.

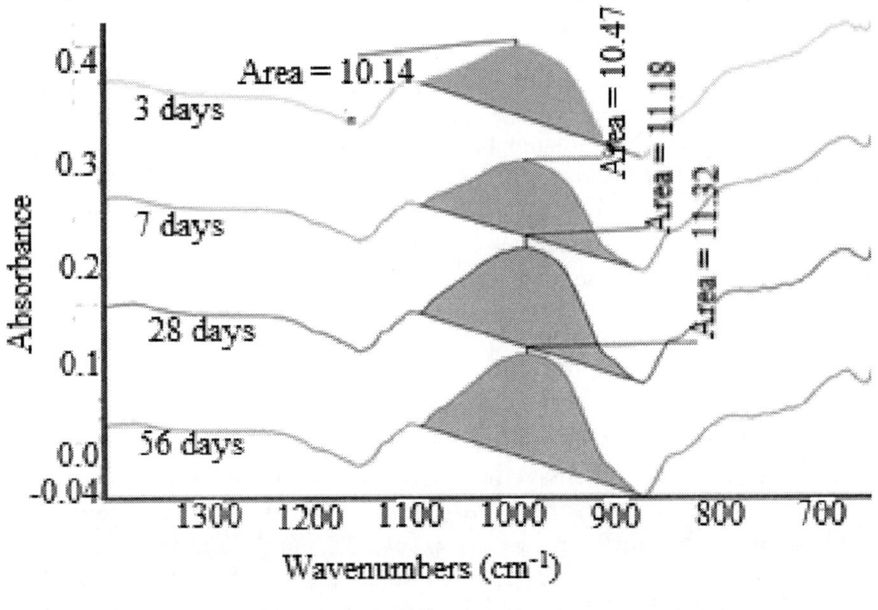

Figure 5. Absorbance of 5% silica fume spectra of water cured samples.

References

Kadri, E. and Duval, R., Hydration heat kinetics of cement paste with silica fume, Cement and Cement paste Research, pp: 3388-3392, 2009.

Narmluk, M. and Nawa, T., Effect of Fly Ash on the Kinetics of Portland Cement Hydration at Different Curing Temperature, Cement and Cement paste Research, pp: 579-589, 2011.

Zelic, J. *et al.*, The role of silica fume in the kinetics and mechanisms during the early stage of cement hydration, Cement and Cement paste Research, pp: 1655-1662, 2000.

Zhang, Y. M. *et al.*, Hydration of High-Volume Fly Ash Cement Paste. Cement and Cement Paste Composites, pp: 445-452, 2000.

Interaction between Theory and Practice in Civil Engineering and Construction
Edited by Komurlu, R., Gurgun, A. P., Singh, A., and Yazdani, S.

UNIVERSAL SIZE EFFECT STUDIES USING THREE POINT BEAM TESTS

SIDDIK ŞENER[1] and KADİR CAN ŞENER[2]

[1]*Dept of Civil Engineering, Bilgi University/Santral, Istanbul, Turkey*
[2]*Dept of Civil Engineering, Purdue University, W. Lafayette, IN, USA*

The universal size effect law for concrete is a law that describes the dependence of nominal strength of specimen or structure on both its size and the crack (or notch) length, over the entire of interest, and exhibits the correct small and large size asymptotic properties as required. The main difficulty has been the transition of crack length from 0, in which case the size effect mode is Type 1, to deep cracks (or notches), in which case the size effect mode is Type 2 and fundamentally different from Type 1. The current study is based on recently obtained comprehensive fracture test data from three-point bending beams tested under identical conditions. This paper presents a studying to improve the existing universal size effect law using the experimentally obtained beam strengths for various different specimen sizes and all notch depths. The updated universal size effect law is shown to fit the comprehensive data quite well.

Keywords: Concrete fracture, Scaling of strength, Fracture testing, Statistical testing, Notch depth effect, Flexural strength, Quasi-brittle fracture.

1 INTRODUCTION

The preceding conference article (Şener *et al.* 2014a) presented an introduction to the problem and reported comprehensive test data for fracture of concrete specimens. The experimental program, also described in (Şener *et al.* 2014b), consisted of 80 three-point bend beams with 4 different depths 40, 93, 215 and 500mm, corresponding to a size range of 1:12.5. Five different relative notch lengths, a/D = 0, 0.02, 0.075, 0.15, 0.30 were cut into the beams. A total of 20 different geometries (family of beams) were tested. The present paper will use these data to analyze the effects of size, crack length.

The Scientific and Technological Research Council of Turkey (TUBITAK) provided funding to carry out comprehensive fracture tests of beam specimens made from the almost the same age and same concrete mix to investigate the influence of size and notch length on specimen strength.

2 REVIEW OF SIZE EFFECT AND CRACK LENGTH EFFECT

The nominal strength of geometrically similar structures, defined with Eq. (1) is

$$\sigma_N = c_N \frac{P_u}{bD} \tag{1}$$

independent of structure size D (P = maximum load; b = structure width; and c_N = dimensionless constant chosen for convenience). Size effect – defined as any dependence of σ_N on D – is a phenomenon typical in fracture or damage mechanics.

According to linear elastic fracture mechanics (LEFM) theory, which applies to homogeneous perfectly brittle materials, and for geometrically similar structures with similar cracks, $\sigma_N \propto D^{-1/2}$, which is the strongest possible size effect. For quasi-brittle materials such as concrete, two simple types of size effects can be seen in Eq. (2).

$$\sigma_N = \frac{Bf_t}{\sqrt{1 + D/D_0}} \tag{2}$$

Here B and the transitional structure size D_0 are empirical parameters to be identified by data fitting and f_t = tensile strength of concrete introduced for convenience. Eq. (2) was derived (Bazant 1984) by simple energy release analysis and later by several different approaches such as by asymptotic matching based of the asymptotic power scaling laws for very large and very small D (Bazant and Planas 1998). In the standard size effect plot of log σ_N versus log D, Eq. (2) gives a smooth transition from a horizontal asymptote to an inclined asymptote of slope -1/2 (Figure 1).

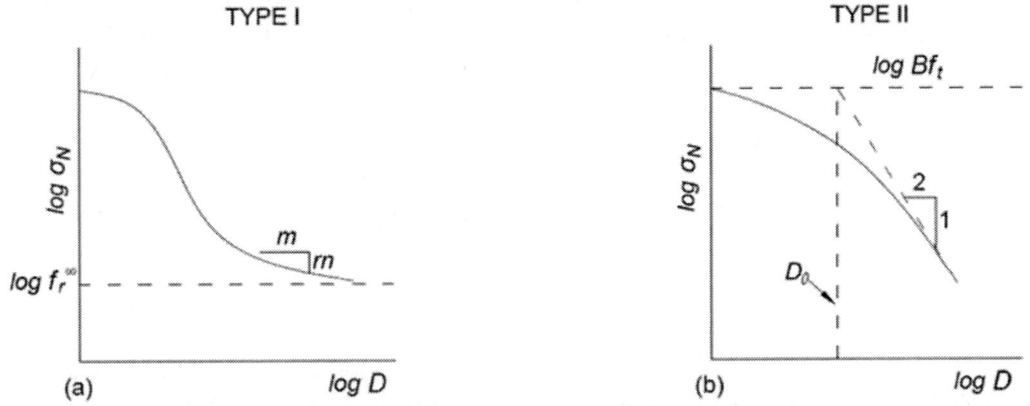

Figure 1. Dependence of □$_N$ on structure size D of beams with (a) no notched and (b) deep notch.

In Eq. (2),

$$Bf_t = \sqrt{\frac{E'G_f}{g_0{}'c_f}}, D_0 = \frac{c_f g_0{}'}{g_0} \tag{3}$$

where $g_0 = g(\alpha_0)$; $g_0' = g'(\alpha_0)$; $\alpha = a/D$ = relative crack length; $\alpha_0 = a_0/D$ = initial value of α; $g(\alpha) = k^2(\alpha)$ = dimensionless energy release rate function $g(\alpha)$ of LEFM; $k(\alpha) = b\sqrt{D}K_I/P$ where K_I = stress intensity factor, P = load; $g'(\alpha) = dg(\alpha)/d\alpha$; $E' = E$ = Young's modulus for plane stress and $E' = E/(1-\nu^2)$ for plane strain (where ν= Poisson ratio); G_f = initial fracture energy = area under the initial tangent of the cohesive softening stress-separation curve; c_f = characteristic length, which represents about a half of the FPZ length. Eq. (2) may be rewritten as shown in Eq. (4).

$$\sigma_N = \sqrt{\frac{E'G_f}{g_0 D + g_0' c_f}} \tag{4}$$

Because function $g(\alpha)$ or $k(\alpha)$ embodies information on the effects of crack length and structure geometry, Eq. (4) is actually a size effect law for Type 2 failures.

The Type 1 size effect, σ_N approaches, for large D, a constant value (a horizontal asymptote in the size effect plot), since the Weibull statistical size effect (Weibull 1939) is unimportant. For three point bend beams, it is indeed unimportant. Because the zone of high stresses is rather concentrated, even do not exist along a notch. This prevents the critical crack from forming at widely different locations of different random local strength (for the same reason, the statistical size effect is negligible in Type 2 failures also).

The large size asymptote for Type 1 size effect is, in the log-log plot, a downward inclined straight line of a slope $-n/m$, which is much milder than the slope of -1/2 for LEFM (Weibull 1939) (Figure 1); here m = Weibull modulus and n = number of spatial dimensions of fracture scaling (n = 2 for the present tests). The small size asymptote is also a horizontal line and, for medium sizes, the size effect is a transition between these two asymptotes. In absence of the statistical size effect, Eq. (5) was used by Hoover and Bazant (2014).

$$\sigma_N = f_r^{\infty}\left(1 + \frac{rD_b}{D + l_p}\right)^{1/r} \tag{5}$$

Here f_r^{∞}, D_b, l_p, and r are empirical constants to be determined from tests; f_r^{∞} = nominal strength for very large structures, assuming no statistical size effect (in the special case of very large beams, f_r^{∞} represents the flexural strength, also called the modulus of rupture); and D_b = depth of the boundary layer of cracking (roughly equal to the FPZ size). In all previous works, D = same characteristic structure size as used for the Type 2 size effect [Eq.(4)]. Furthermore, l_p = material characteristic length, which is related to the maximum aggregate size d_a. If the structure is larger than $10l_p$, one can set $l_p \approx 0$, which corresponds to the original formulation of the Type I law.

It was further shown that the Type 1 and 2 SELs satisfy the large-size and small size asymptotic properties of the cohesive crack model applied to Type 1 and 2 failures. Furthermore, it was experimentally confirmed that, within the range of inevitable experimental scatter, the SEL of Type 2 gives about the same values of fracture energy G_f when applied to notched fracture specimens [e.g., compact compression test (Barr *et al.* 1998)].

3 APPLICATION OF USEL BY FRACTURE TESTS

To calibrate the deterministic USEL, the mean of the data was computed separately for each family of identical specimens from comprehensive fracture tests (Şener *et al.* 2014a, 2014b, Çağlar and Şener 2015). The surface of the optimized Universal Size

Effect Law (USEL) is plotted in Figure 2. In this Figure 2 size effect curves were given for only α= 0, 0.15 and 0.3. Transition from these curves for calibrating USEL is just used with smooth curves. The studies on these transition curves are still on going.

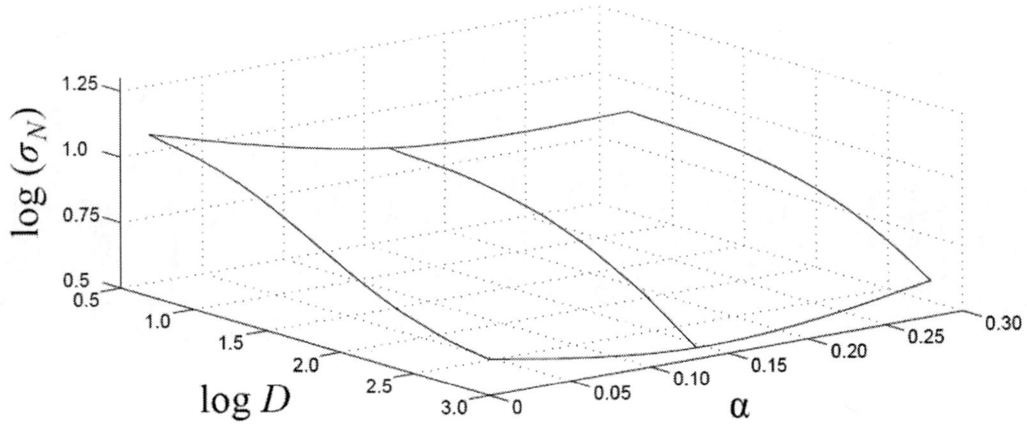

Figure 2. Entire Universal Size Effect law surface.

In particular, the fracture parameters G_f and c_f should not be influenced by the data for beams with no notches (Type I data) or shallow notches and f_r^{∞}, D_b, l_p, and r should not be influenced by the data for deep notches. Therefore, these parameters were determined first by separate fitting of specimens with deep notches (α = 0.30 or 0.15) and specimens with shallow or no notches (α=0). Only the nonstatistical USEL in Eq.(6) was considered. Nonlinear fitting of the Type I SEL [Eq.(5)] to the notchless (α = 0) beams gave (Şener *et al.* 2014a, b) values in Eq.(6) with coefficient of variation of fit 9.4%.

$$D_b = 90 \text{ mm}, l_p = 50 \text{ mm}, f_r^{\infty} = 4 \text{ MPa}, r = 1/2 \qquad (6)$$

These values are different from than the studies by Hoover and Bazant's (2014) D_b=73.2 mm, l_p=126.6 mm, $\square_{\square}^{\infty}$ =5.27 MPa. The differences between some of the parameters were in the order of two for especially l_p value. The size range 1:12.5 was large enough to identify all the fracture parameters in Eq.(5). The USEL can be drawn for a fixed α, which gives a size effect plot of log(σ_N) versus log D (Figure 2).

In Figure 3, this plot is created and compared with the data from Şener *et al.* (2014a, b). Also in Figure 3, for α=0.3 and 0.15, Type II (at top) size effect was used, for unnotched specimen α=0, Type I (at bottom) size effect was used. For the α=0.02 and 0.075 USEL (at center) should proposed.

4 CONCLUSIONS

(1) The Type 2 size effect in specimens with deep notches or cracks does not give a correct transition to of Type 1 in specimens with no notch or crack.

(2) The size effect data from deeply notched specimens (α = 0.3 and 0.15), and parameters f_r^{∞}, D_b, l_p, and r were determined separately by fitting only the size effect data for un-notched specimens (α = 0).

(3) USEL fits the measured nominal strength quite well.

(4) Both Type I and II sizes were effect observed in this study and confirmed the need to be account for size effect in design codes.

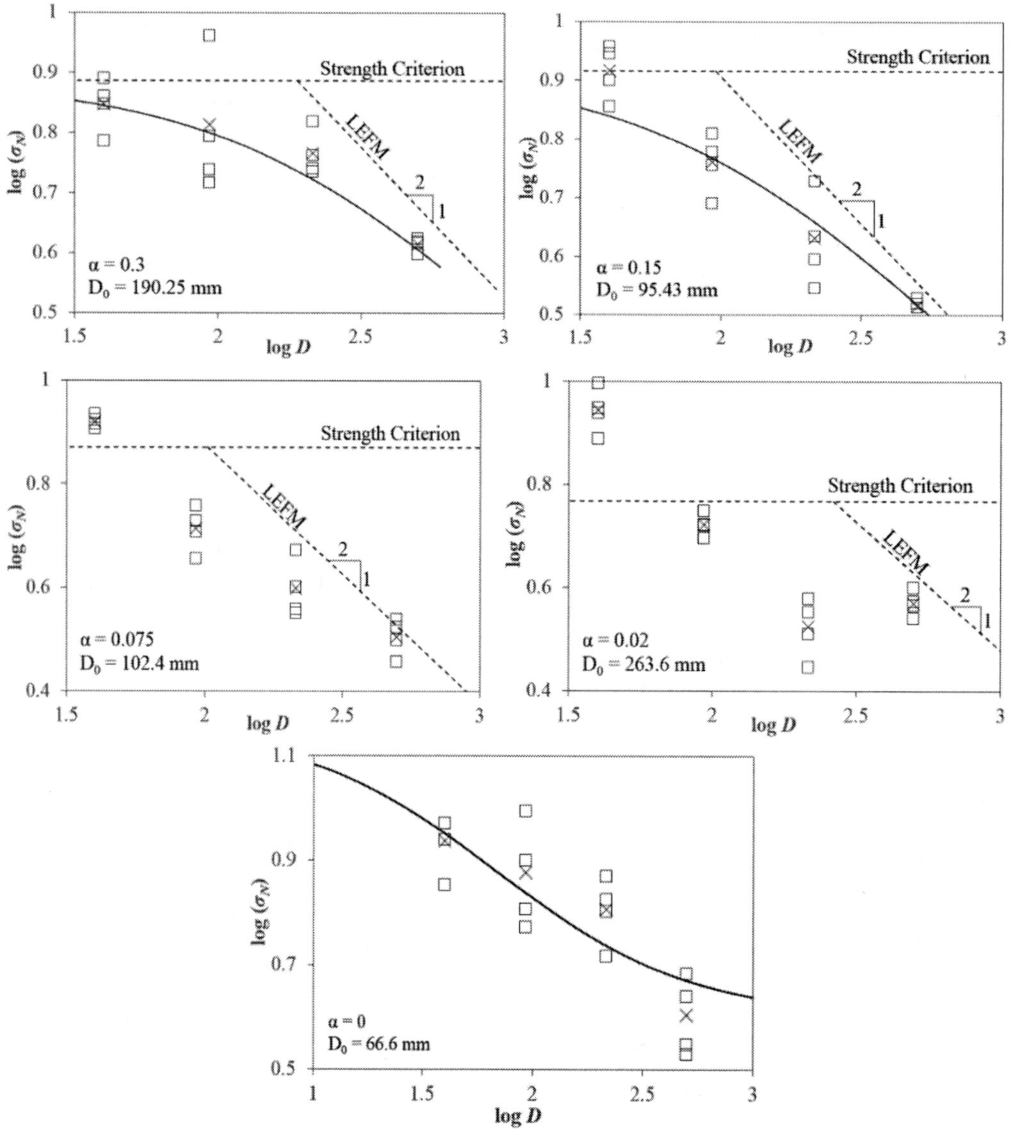

Figure 3. Effect of structure size on the nominal strength of the data from Şener *et al.* (2014a).

Acknowledgments

Financial support from The Scientific and Technological Research Council of Turkey (TUBITAK) was provided through Grant No.111M374 from the Gazi University and Istanbul Bilgi University, is gratefully appreciated.

References

Barr, B. I. G., Abusiaf, H. F., Şener, S., Size effect and fracture energy studies using compact compression specimens, *RILEM,Materias and Construction,* 31(1), 36-41, Jan.-Feb., 1998.

Bazant, Z. P., Size effect in blunt fracture: Concrete, rock, metal, *Journal of Engrg. Mech., ASCE*, 110, 518-535, 1984.

Bazant, Z. P., Planas, J., *Fracture and size effect in concrete and other quasibrittle materials,*CRC Press, Boca Raton, FL, 1998.

Çağlar, Y. and Şener, S., Size effect for notched, unnotched concrete beams, *XIX. National Mechanics Conf.,* 1-8, Trabzon, Turkey 2015.

Hoover, C. G. and Bazant, Z. P., Universal size-shape effect law based on comprehensive concrete fracture tests, *J. Engrg. Mechanics, ASCE,* 140, 473-479, March 2014.

Şener, S., Çağlar, Y., and Belgin, Ç. M., Size effect tests for Type I and Type II, *11th Inter. Conf. on Advance . in Civil Engrg.,* 51, Istanbul, Turkey 2014a.

Şener, S., Belgin, Ç. M., Çağlar, Y., Hasanpour, R., Topgül, S., Boduroğlu, V. B., Negin, M., and Çağlar, N. M., Size effect tests, A., Fracture mechanics analysis of concrete with carbon nano tubes. *Technical report for TUBITAK,* 111M374, Ankara, Turkey, 2014b.

Weibull, W., The phenomenon of rupture in solids, *proc. R. Swedish Inst. Eng. Res.,* 153, 1-55, 1939.

Interaction between Theory and Practice in Civil Engineering and Construction
Edited by Komurlu, R., Gurgun, A. P., Singh, A., and Yazdani, S.

REEVALUATION OF THE EFFECT OF COVERING SHEETS FOR REDUCING BUGHOLES ON TUNNEL LINING CONCRETE

TOMOYUKI MAEDA[1], SARI HARADA[2], MAI MORIUCHI[2], and ISAMU YOSHITAKE[2]

[1] *Inst. of Tech., Penta-Ocean Const., Tokyo, Japan*
[2] *Dept of Civil and Environmental Engineering, Yamaguchi University, Ube, Yamaguchi, Japan*

Most tunnel lining concretes have sidewall of negative angle, so bugholes on the concrete surface frequently appear. The bugholes are caused from entrapped air in concrete placing; the part of bugholes remains on concrete surface even after adequate vibrating for consolidation. In the previous investigation, the authors have studied on the generation mechanism of bugholes in a visible test and examined the reducing effects by covering sheet for concrete forms. According to the investigation, it was found that the water-permeable sheet causes color irregularity of concrete surface while the sheet is effective to decrease the bugholes. Hence, the present study also focuses on the effect of sheets reducing bugholes. The foci of this study are to develop an effective sheet for aesthetic of concrete surface and to reevaluate the effectiveness for reduction of bugholes. Fundamental test using a model form and various covering sheets was performed by referring to the previous investigation. The test confirms that the developed sheet adequately contributes on decrease of bugholes without the color irregularity of concrete. The observations may be useful for most concrete construction as well as tunnel lining concrete.

Keywords: Tunnel lining, Bughole, Covering sheet, Gas permeability, Water permeability, Breathable-waterproof material.

1 INTRODUCTION

It is well known that the durability of concrete structures is affected by surface quality. In addition, the surface quality is strongly related to aesthetic of concrete structures. Figure 1 demonstrates typical degradations for aesthetic; (a) concrete spalling, (b) water stains, (c) bugholes, and so on. This study focuses on improvement of surface quality of tunnel lining concrete. Yoshitake *et al.* (2012) reported the influencing factors for aesthetic of tunnel lining concrete. In particular, bugholes often occur on concrete surface of sidewall of tunnel lining even in an appropriate construction. However, it has been difficult to decrease bugholes effectively. The previous studies investigated the generation mechanism of bugholes using a visible testing form, and examined the reducing effects by covering sheet for concrete form (Maeda *et al.* 2014; Harada *et al.* 2015). The study reported that the color irregularity of concrete was observed while the water permeable sheet is an effective material to reduce bugholes (Harada *et al.* 2015). The foci of this experimental investigation are to confirm the reducing effect by covering sheets and to develop a suitable sheet for reduction of color irregularity as well as decrease of bugholes. Based on a laboratory test, the paper discussed and reevaluates the surface-quality improvement effect of the covering sheet.

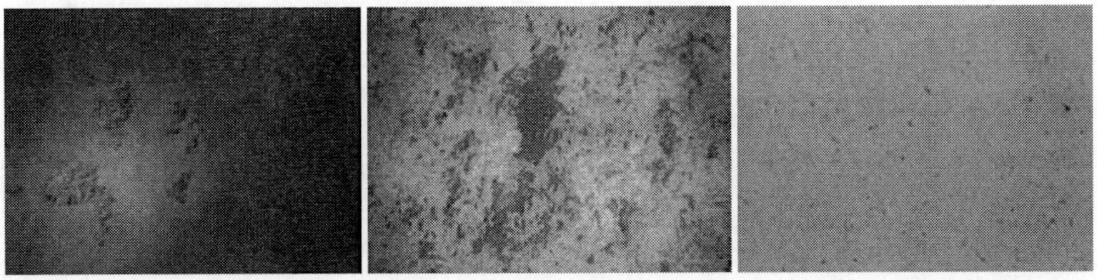

(a) concrete spalling (b) water stains (c) bugholes

Figure 1. Typical degradations in tunnel lining concrete.

2 METHODOLOGY

2.1 Concrete

Table 1 gives a mixture proportion of concrete. The concrete mixture is referred to the mixture proportion used in the previous study (Harada *et al.* 2015). The specified compressive strength of concrete was 21 MPa, and slump value was 150 mm. Further information of the concrete were described in the previous study.

Table 1. Mixture proportion.

w/cm	Unit weight (kg/m^3)							
	Water	BBC[a]	Aggregate				Water reducing agent	AE agent
			Fine	40-20 mm	20-15 mm	15-5 mm		
0.59	165	280	726	455	342	342	0.42-2.94	0-0.021

a: blast furnace slag cement (type B) defined in JIS R 5211

2.2 Covering Sheets

As mentioned earlier, color irregularity was observed in concrete made with the water permeable sheet. It was found that the color gradation was caused from discharged water via the permeable sheet. The previous study confirmed that breathable-waterproof materials sheets can decrease bugholes while the function was not better than the performance of water permeable sheet. Hence, four kinds of breathable-waterproof materials sheet were additionally employed to decrease the color irregularity.

Table 2 summarizes test parameters. The tested sheets having variable Gurley number and water bearing pressure was prepared.

Table 2. Test parameters.

No.	Water bearing pressure (kPa)	Gurley number (sec/100cc)	Water-repellent	Remover
1	200	10	No	Use
2	200	10	No	N/A
3	100	12	Yes	Use
4	100	12	Yes	N/A
5	30	4	No	Use
6	7	1	No	Use

2.3 Test Specimens and Procedure

Concrete specimens, dimensions of 750 x 300 x 300 mm, were prepared to simulate sidewall of tunnel lining. Figure 2 shows a special concrete form as well as the previous investigations (Maeda *et al.* 2014; Harada *et al.* 2015). Concrete is cast into the form with an angle of 30 degree. All concrete specimens were consolidated by using a vibrator of 220-280 Hz at the central-section of the specimen for 60 seconds.

(a) schematic of lining concrete (b) front view (c) side view

Figure 2. Steel form simulated tunnel lining.

2.4 Evaluation Method for Bugholes

An image analysis directly using colored photographic image was used to detect bugholes in addition to the previous study (Harada *et al.* 2015). Figure 3 presents the image analysis system. In this system, RGB values in each pixel of the photographic image are compared to the reference number for each color; and a pixel is estimated as an element of photographic image of bughole when all RGB values of the pixel are lower than the reference criterion. It should be noted that bugholes of 0.7 mm or less, negligible size for aesthetic, were omitted in the evaluation.

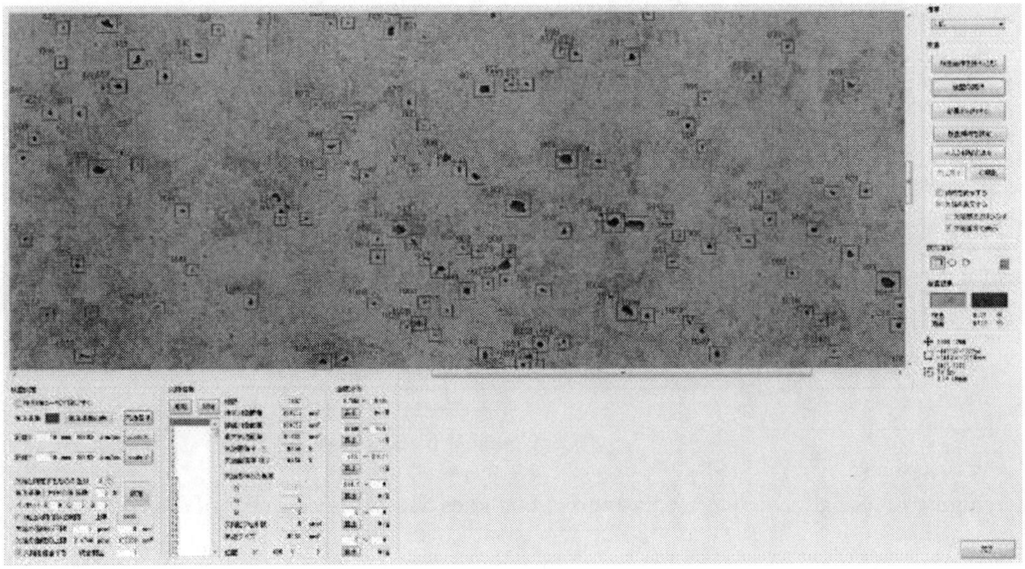

Figure 3. Image analysis using colored photographic image.

3 TEST RESULTS AND DISCUSSION

Figure 4 presents a relation between area-ratio of bugholes and Gurley number in this investigation. A regression line is also presented in the graph. The result shows that area-ratio of bug holes increases in accordance with Gurley number. The increasing ratio, that is the coefficient of the regression line, was almost 0.07 %/(sec./100cc). It is noteworthy that the area-ratio of bughole of concrete using the sheet without remover (unfilled-markers) were higher than the ratio of concrete using the sheet with remover (solid-markers). The observation may imply that the air/water drainage-effects of the sheet surface were promoted by the remover.

Figure 5 presents the area-ratio of bugholes of concrete made with the various sheets having different water bearing pressure. The test results of No.3 and 4 were significantly greater than other results because of water-repellent function in the sheet. Except for the test results of No.3 and 4, a regression line is obtained from the results. The increase of the ratio in the regression line was not remarkable compared with the result shown in Fig.4, while the area-ratio of bughole was also increased in accordance with the water bearing pressure.

Figure 6 demonstrates the concrete surface in the test of No.6. The result confirm that the covering sheet adequately contributes on decrease of bugholes without the color irregularity of concrete. Further investigation, such as actual applications, should be conducted to determine the most suitable covering sheet for improvement of aesthetics of tunnel lining concrete.

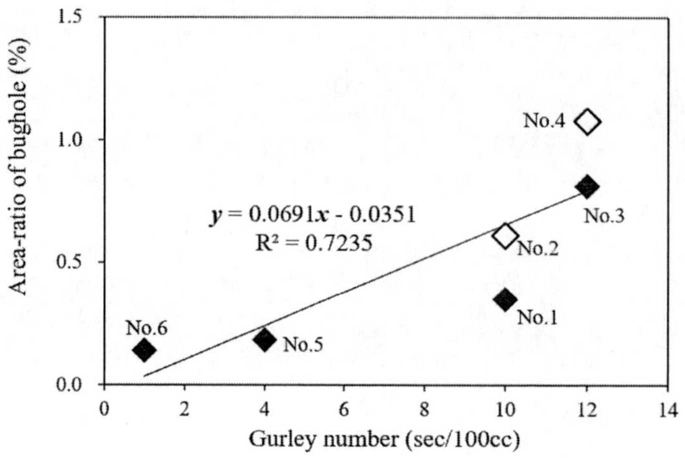

Figure 4. Relation between area-ratio of bugholes and Gurley number of covering sheet.

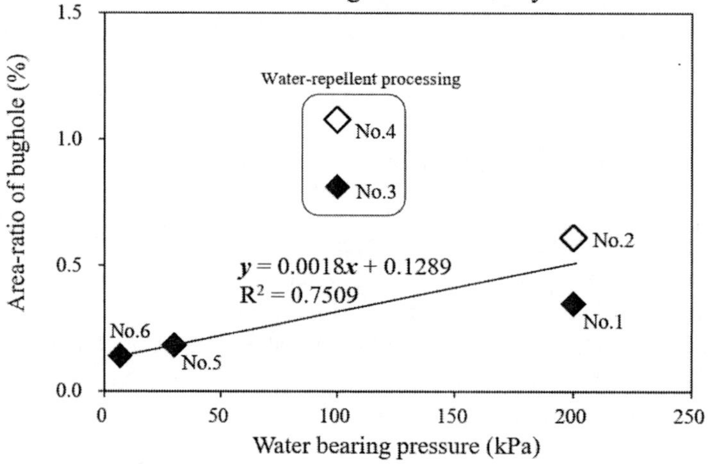

Figure 5. Relation between area-ratio of bugholes and water bearing pressure of covering sheet.

Figure 6. Concrete surface condition without color irregularity (No.6).

4 CONCLUSIONS

The foci of the study were to reevaluate the effect of breathable-waterproof materials sheet and to develop a suitable sheet for reduction of color irregularity as well as decrease of bugholes. The laboratory test confirmed the relation between bugholes and properties of the covering sheet. The conclusions of this fundamental investigation are as follows:

- The number of bugholes is almost proportional to the increase of Gurley number of the covering sheet.
- The area-ratio of bughole of concrete using the sheet without remover were higher than the ratio of concrete using the sheet with remover.
- The area-ratio of bughole was increased in accordance with the water bearing pressure of the covering sheet, however, the increase was not remarkable compared with the relation between bugholes and Gurley number.
- The laboratory test confirmed that the breathable-waterproof materials sheet contributes reduction of bugholes and color irregularity.

Acknowledgments

The authors thank to Mr. Ito and Mr. Muraki (Nittodenko Co.), and Mr. Hirano (Chuden Technology Consultants Co.) for their assistance.

References

Harada, S., Maeda, T., Hirano, M. and Yoshitake, I., Reducing Bug-Holes on Tunnel Lining Concrete by Using Covering Sheet, *Proceeding of ISEC-8*, Saha, S., Lioyd, N., Yazdani, S., and Singh, A. (Eds.), November 2015.

Maeda, T., Honma, H., Hirano, M. and Yoshitake, I. Permeability of Tunnel Lining with Air/Water Bubbles on Concrete Surface, *Proceeding of ASEA-SEC2*, Yazdani, S., Vimonsatit, V., and Singh, A. (Eds.),321-325, November, 2014

Yoshitake, I., Komure, H., Kakehi, M. and Inagawa, Y Questionnaire Investigation if Surface Deterioration of Lining Concrete in NATM Tunnel, *Proceedings of ASEA-SEC-1*, Vimonsatit, V., Singh, A. and Yazdani, S. (Eds.), 221-225, November 2012.

Interaction between Theory and Practice in Civil Engineering and Construction
Edited by Komurlu, R., Gurgun, A. P., Singh, A., and Yazdani, S.

THE EFFECTS OF VISUAL STABILITY INDEX ON PROPERTIES OF SELF CONSOLIDATING CONCRETE UNDER ACCELERATED CURING

MOHAMED KHALAFALLA and JOSEPH OWINO

Civil and Environmental Engineering, University of Tennessee at Chattanooga, USA

Self-consolidating concrete, also known as self-compacting concrete (SCC), is a highly flowable concrete that spreads into place and fills formwork without the need for mechanical vibration. SCC reduces the time and labor cost needed for concrete placement. This study is part of the proposed project by Tennessee Department of Transportation (TDOT) carried out by University of Tennessee at Chattanooga (UTC) to develop four new SCC mixtures (two Class P-SCC (precast) and two Class A-SCC (general use), and insure they meet the minimum strength and durability requirements for TDOT Class P and Class A mixtures. The objectives of the study are to analyze effects of visual stability index (VSI) on both fresh and hardened properties of Class P-SCC concrete under the accelerated curing condition. In addition, the relationship between VSI and fresh segregation of SCC is investigated. A total of 24 concrete mixtures varying in VSI values were produced for the study. Different sizes of coarse aggregates materials were used during the mixing process, as well as different kinds of sands. A number of fresh and hardened properties tests were performed on the concrete mixtures to assess the performance of the mixes. The SURE CURE system is used to accelerate the curing process of the concrete. Finally, the results of this study are analyzed according to the coarse aggregate sizes and evaluated to recommend performance specifications for Class P-SCC for TDOT adoption of SCC standard operating procedures of the precast elements.

Keywords: Fly ash, Slump flow, J-ring, Fresh segregation.

1 INTRODUCTION

The concept of SCC was proposed by Professor Hajime Okamura of Kochi University of Technology, Japan, in 1986 as a solution to the growing concrete durability concerns of the Japanese government. During his research, Okamura found that the main cause of the poor durability performances of Japanese concrete in structures was the inadequate consolidation of the concrete in the casting operations. By developing concrete that self-consolidates, he eliminated the main cause for the poor durability performance of their concrete. By 1988, the concept was developed and ready for the first real scale tests (Vachon 2002). Generally, SCC is made with conventional concrete materials with the addition of chemical admixture such as viscosity-modifying admixture (VMAs) to enhance cohesion and control the tendency of segregation resulting from the highly flowable SCC (ACI 2007, Elhassan 2014). Also, the fine aggregate content in SCC is higher than that for conventional concrete in order to provide better lubrication for course aggregates to enhance workability of the mixture (Adekunle 2012). The use of SCC can result in increased productivity in construction, improved jobsite safety, and improved hardened properties. Although SCC can potentially decrease total project costs through minimization of labor, SCC can result in a higher per unit cost compared to conventional concrete due to increased Cementious materials, chemical admixtures, more strict specifications and the need for the advanced technical expertise. The proper selection of materials and mixture proportions is an important factor to ensure that the advantageous properties of SCC can be achieved economically. The effects of individual constituents and changes in mixture proportions are often greater in SCC than in conventional concrete (Koehler 2007).

2 MATERIALS

Materials used in SCC are the same materials used in conventional concrete. Although SCC can be made with a wide range of materials, the proper combination of the materials is crucial for optimizing SCC. Compared to conventional concrete, SCC is more sensitive to changes in material properties (Koehler 2007). Cementious materials, Aggregates and Chemical admixtures were the main components used to produce SCC, in addition of water. SCC mixtures that produce satisfactory results typically have between usually have 650 lb/yd^3 to 840 lb/yd^3 (385 kg/m^3 to 500 kg/m^3) cementious materials content (Mata 2004). Portland cement (ASTM C150) Type I and Class F (ASTM C618) fly ash were the only cementitious materials used in this study. SCC typically contains both types of aggregate fine and coarse aggregate. For Fine aggregate or sand, all normal concreting sands are suitable for SCC. Both manufactured (crushed) and natural (rounded) sands were used in the study. For Coarse Aggregates, all types of aggregates are suitable. The normal maximum size is generally 16-20 mm; however particle sizes up to 40 mm or more have been used in SCC. The aggregate sizes used in this study were ASTM C 33 #67 Stone, #7 Stone, and #89 Stone. All the coarse aggregates had bulk specific gravity of 2.74 and absorption of 0.62 % (EFNARC 2002). Two types of chemical Admixtures used in the study to meet the requirements of ASTM C494, Mid-Range water reducer MasterPolyheed 900 which was used for the conventional concrete and High-Range water reducer (HRWR) ADVA® Cast 575 was used for SCC to maintain a relatively low water cement ratio (w/cm) while increasing fluidity. The deformability of the paste is increased by reducing the viscosity. HRWRA can provide a highly flowable concrete without a significant reduction in cohesiveness and improve the resistance to segregation (ACI 2007, Khayat *et al.* 2004).

3 MIX DESIGN

Two sets of mixtures were developed to assess the effect of VSI on both hardened and fresh properties of Class-P SCC. Each set consist of four Trial mixtures groups. Portland cement was used only as cementitious materials on the first batch, whereas portland cement plus Class-F fly ash is used in the second set. Fly ash was designed to be 20% of the cementitious materials in the second batch. Mixture proportions are provided in Tables 1 and 2.

Table 1. TDOT Class A mixtures with Portland cement.

Mixture No	1	2	3	4	5	6	7	8	9	10	11	12
VSI	1	2	Conv.	1	2	Conv.	1	2	Conv.	1	2	Conv.
Cement	678	678	678	678	678	678	678	678	678	678	678	678
Class F-Ash	0	0	0	0	0	0	0	0	0	0	0	0
# 67 stone	1551	1551	1735	1550	1550	1735	0	0	0	0	0	0
# 7 stone	0	0	0	0	0	0	1551	1551	1735	0	0	0
# 89 stone	0	0	0	0	0	0	0	0	0	1551	1551	1735
Natural sand	1470	1470	1295	0	0	0	1470	1470	1295	1470	1470	1295
Man. sand	0	0	0	1550	1550	1364	0	0	0	0	0	0
Water	304	304	304	304	304	304	304	304	304	304	304	304
H/MRWR (oz./cwt)	7	9	4	7	9	4	7	9	4	7	9	4
w/cm ratio	0.45	0.45	0.45	0.45	0.45	0.45	0.45	0.45	0.45	0.45	0.45	0.45
Sand ratio by volume	0.5	0.5	0.44	0.5	0.5	0.44	0.5	0.5	0.44	0.5	0.5	0.44

Table 2. TDOT class A mixtures with 20% cement replacement of class F fly ash.

Mixture No	13	14	15	16	17	18	19	20	21	22	23	24
VSI	1	2	Conv.	1	2	Conv.	1	2	Conv.	1	2	Conv.
Cement	543	543	543	547	547	547	543	543	543	543	543	543
Class F-Ash	135	135	135	131	131	131	135	135	135	135	135	135
# 67 stone	1536	1536	1720	1536	1536	1720	0	0	0	0	0	0
# 7 stone	0	0	0	0	0	0	1536	1536	1720	0	0	0
# 89 stone	0	0	0	0	0	0	0	0	0	1536	1536	1720
Natural sand	1455	1455	1280	0	0	0	1455	1455	1280	1455	1455	1280
Man. sand	0	0	0	1535	1535	1350	0	0	0	0	0	0
Water	304	304	304	304	304	304	304	304	304	304	304	304
H/MRWR (oz./cwt)	7	9	4	7	9	4	7	9	4	7	9	4
w/cm ratio	0.45	0.45	0.45	0.45	0.45	0.45	0.45	0.45	0.45	0.45	0.45	0.45
Sand ratio by volume	0.5	0.5	0.44	0.5	0.5	0.44	0.5	0.5	0.44	0.5	0.5	0.44

4 ACCELERATED CURING

Accelerated curing is a way to achieve high early age strength. This practice is most common in the precast industry, where there is a need for concrete elements to be casted and transported as quickly as possible to the site. The idea behind the accelerated curing is by increasing the concrete temperature, the rate of hydration increases and a larger portion of the later-age properties of the concrete can be attained during the short curing period compared with standard temperature curing. The concrete specimens were cured under the accelerated curing for 18 hours. The rate of temperature increments was set to be less than 50 °F per hour to avoid cracking, starting from 80°F up to 155°F in a time of 18 hours.

Table 3. Testing protocol.

Fresh Concrete Testing	
Slump Flow and Visual Stability Index (ASTM C1611)	1 per batch
Consolidating ability by J-Ring (ASTM C1621)	1 per batch
Static Segregation by Column Test (ASTM C1610)	1 per batch
Unit Weight and Gravimetric Air Content (ASTM C138)	1 per batch
Air Content by Pressure Method (ASTM C231)	1 per batch
Hardened Concrete Testing	
Compressive Strength 1 (ASTM C 39)	2-6x12 inch cylinders per test time
Static Modulus of Elasticity1 (ASTM C 469)	The 2-6x12 compressive strength cylinders will also be used for modulus per test time
Splitting Tensile Strength 1(ASTM C 496)	2-6x12 inch cylinders per test time

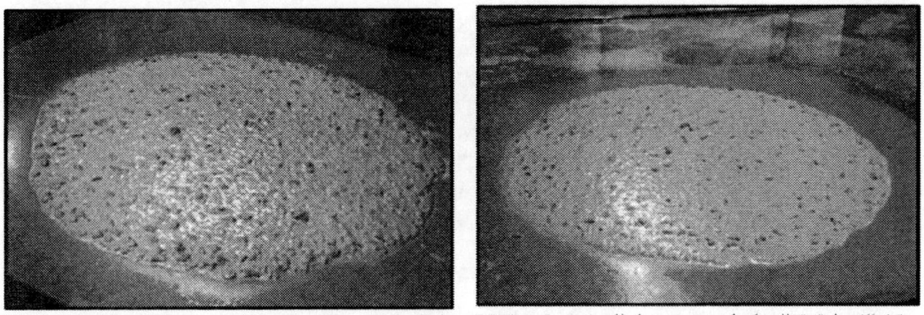

VSI = 1 – No evidence of segregation and slight bleeding observed as a sheen on the concrete mass

VSI = 2 – A slight mortar halo # 0.5 in.(# 10 mm) and/or aggregatepile in the of the concrete mass

Figure 1. Visual stability index criteria.

5 FRESH AND HARDENED PROPERTIES TESTS

Twelve Batches of each mixture was undergone standard fresh and hardened property testing as outlined in Table 3. The variety of fresh consistencies coupled with segregation and strength results provided data to determine the acceptance/rejection threshold that TDOT will ultimately adopt.

6 VISUAL STABILITY INDEX

The Visual Stability Index (VSI) is a method for determining the segregation stability of the mixture, and to evaluate the relative stability of batches of the same SCC mixture. The VSI is determined through visually rating apparent stability of the slump flow patty based on specific visual properties of the spread. The SCC mixture is considered stable and suitable for the intended use when the VSI rating is 0 or 1, and a VSI rating of 2 or 3 gives an indication of segregation potential (ACI 2007). The VSI required for the mixtures in this study were VSI of 1 and 2. The desirable VSI values were achieved by varying the amount of HRWR used during mixing. Values of VSI of 1 and 2 were recorded according to (ASTM C1611/C1611M) as shown in Figure 1.

Table 4. The results of fresh properties tests.

Mix	VSI	Slump (in)	J-ring (in)	HRWR (oz./cwt)	Air (%)	Unit weight (lb./ft3)	Col. Seg. (%)
1	1	19	17	2.99	3.60%	142.9727	9.2%
2	2	25	21	3.89	4.70%	141.3549	23.3%
3	Conv.	5.25		0.00	3.40%	0	
4	1	22.5	17.5	3.44	5.30%	144.186	15.0%
5	2	24.75	19.75	6.58	1.50%	148.0283	7.7%
6	Conv.	3.25		6.21	3.90%	0	
7	1	23	18.75	4.49	3.10%	143.3771	12.3%
8	2	25.5	24	6.88	1.80%	145.3994	18.2%
9	Conv.	3		10.36	3.60%	0	
10	1	20.5	16	4.19	6.20%	139.1304	10.1%
11	2	22.5	18.25	5.09	5.60%	139.9393	8.3%
12	Conv.	3.5		4.14	3.00%	0	
13	1	23	18.75	7.18	5.60%	143.1749	1.3%
14	2	26.5	25	8.68	4.30%	145.3994	20.9%
15	Conv.	5		0.00	2.20%	0	
16	1	23	18	9.28	3.20%	143.3771	20.3%
17	2	25	21.75	11.37	3.20%	144.9949	9.7%
18	Conv.	4.25		6.21	3.10%	0	
19	1	20	18.25	3.39	4.40%	147.0172	8.0%
20	2	28.25	28	3.59	2.40%	143.5794	16.4%
21	Conv.	4		0.00	3.20%	0	
22	1	19	17	1.80	6.10%	140.546	10.1%
23	2	25	24.5	4.49	7.10%	139.1304	14.8%
24	Conv.	2		4.97	3.60%	0	

7 RESULTS

The results of the fresh properties test are presented in Table 4. The results of the hardened properties tests are shown in Table 5. The representations of the results are shown as a comparison between VSI 1 and VSI 2 for both fresh and hardened properties.

8 CONCLUSION

- In general, the #7 and #89 stone mixtures have better fresh properties than #67 stone mixtures. And have good hardened properties when combined with OPC only.
- The test results indicated that Class F fly ash mixtures have poor hardened properties (compressive strength less than 4000 psi), this could be attributed to the slow reaction process of the fly ash.
- The coarse aggregate #7and #89 mixtures are more convenient for making Class-P SCC mixtures when cement was used as the sole cementious materials.

Table 5. The results of hardened properties tests.

		18-hrs			28 days		
Mix	**VSI**	**Comp. (psi)**	**Tensile. (psi)**	**E. (ksi)**	**Comp. (psi)**	**Tensile. (psi)**	**E. (ksi)**
1	1	3540	360	4850	5580	460	5222
2	2	3830	255	5740	6310	351	4940
3	Conv.	3250	276	4972	5700	310	6063
4	1	4380	272	5093	6310	452	4827
5	2	5140	390	5618	7080	477	4757
6	Conv.	4405	338	4981	7015	457	8046
7	1	5005	254	4163	7280	386	6467
8	2	5185	289	4850	7750	398	6548
9	Conv.	4455	391	5174	7185	406	6467
10	1	4615	359	6871	6610	428	4975
11	2	4710	356	4293	7105	370	6007
12	Conv.	4250	341	4238	6230	298	5337
13	1	4825	304	4320	7360	536	5206
14	2	4640	322	7996	6225	457	5340
15	Conv.	2835	232	3537	6860	441	6013
16	1	3270	270	3442	5505	414	5128
17	2	3875	358	3125	6780	447	5908
18	Conv.	3335	261	4005	5860	435	5293
19	1	2520	190	3395	5205	385	4357
20	2	1845	205	3601	5010	341	4258
21	Conv.	3145	214	4378	5950	464	4980
22	1	2460	294	3631	4950	340	3858
23	2	2710	186	3229	5165	313	4341
24	Conv.	3565	259	4514	6490	345	5047

References

ACI, Self-Consolidating Concrete. "*ACI Committee 237R-07.*" American Concrete Institute: Farmington Hills, MI, USA, 30, 2007.

Adekunle, Saheed Kolawole. *A study on developing Self-Consolidating Concrete (SCC) utilizing indigenous natural and industrial waste materials*. Diss. KING FAHD UNIVERSITY OF PETROLEUM AND MINERALS (SAUDI ARABIA), 2012.

ASTM, "*C1611/C1611M-05 standard test method for slump flow of SCC.*", 2005.

ASTM, "*C1621 (2009). Standard Test Method for Passing Ability of Self-Consolidating Concrete by J-Ring. West Conshohocken, PA, 2003, DOI: 10.1520/C1621.*", 2009.

EFNARC, Specification. "*Guidelines for self-compacting concrete.*" London, UK: Association House, 32-34, 2002.

Elhassan, Ammar Elfatih Abdelssamd. "*The effects of Visual Stability Index (VSI) on fresh segregation of self consolidating concrete (SCC) using fly ash class C and F.*", 2014.

Khayat, K. H., Assaad, J., & Daczko, J. Comparison of field-oriented test methods to assess dynamic stability of self-consolidating concrete. *ACI Materials Journal, 101*(2), 2004.

Koehler, Eric Patrick. *Aggregates in self-consolidating concrete*. ProQuest, 2007.

Mata, Luis Alexander. "Implementation of Self-Consolidating Concrete (SCC) for Prestressed Concrete Girders.", 2004.

Vachon, Martin. "ASTM Puts Self-Consolidating Concrete to the Test*." ASTM Standardization News(USA) 30.7*, 34-37, 2002.

Interaction between Theory and Practice in Civil Engineering and Construction
Edited by Komurlu, R., Gurgun, A. P., Singh, A., and Yazdani, S.

EVALUATION OF NORMAL AND MODIFIED ASPHALT ACCORDING TO CONVENTIONAL AND MODERN GRADING SYSTEMS

MARAM SAUDY and SAFWAN KHEDR

Construction Engineering Dept, American University in Cairo, Cairo, Egypt

Asphalt plays a significant role in pavement quality. The need for high-performance pavements with long service life and low maintenance requirements is the motive behind thorough research and studies of asphalt characteristics. This research focuses on studying all sources of Egyptian asphalt over a span of time using both conventional and Superpave grading techniques in order to characterize asphalt performance and also to answer the question whether the Egyptian asphalts need modification. The results of this research indicate that all Egyptian normal (virgin) 60/70 asphalt samples from different sources failed to meet penetration grading requirements, viscosity grading standardsAC-20 (high quality); with minor exceptions, viscosity grading system AC-20 (low quality), and both AR-8000 and AR-1000 Aged Residue grading systems. When Superpave grading system was employed, results indicate that all normal asphalt samples failed to meet the basic requirements (without traffic adjustment) according to the Egyptian climatic requirements for high reliability projects (PG70-10 and PG76-10). The testing results accommodate Superpave requirements for lower levels of reliability and/or lower level of conservativeness. This emphasizes the flexibility and reliability of Superpave grading system as compared to conventional grading systems. On the other hand all modified asphalt samples, using an SBS modifier, passed according to the base high reliability projects and/or high level of conservativeness requirements of the Superpave grading system. Finally it is concluded that Egyptian asphalt should be modified in order to provide satisfactory performance especially for high reliability projects in hot regions with high and/or slow traffic.

Keywords: Superpave, Penetration, Viscosity, AR viscosity, SBS modification, Performance grade.

1 INTRODUCTION

The pavement construction industry has been having problems with the asphalt and asphalt mixtures used over the years in Egypt. Similar problems with asphalts are not uncommon in various part of the world. The need to modify and improve the asphalts has been recognized by roadway agencies around the world.

Traditional methods of evaluation of bituminous materials were developed to assess the quality of normal asphalts. They are not suitable for assessing modified asphalts. This was one of the major motivations to develop the recent SUPERPAVE technology. Research efforts have been conducted to evaluate locally produced bituminous materials; e.g., Al-abdulWahhab *et al.* (1997) who studied the chemical and physical characterization of Arab bitumen and evaluated the suitability of the used regional normal bitumen on the basis of the SHRP performance-based specifications. The study concluded that locally produced bitumen needed modification to suit the different Gulf temperature zones. Aflaki and Nader (2009) investigated the properties of locally produced asphalt in Iran with suitable performance grade through modification techniques. Eleven normal bitumen samples, including 40–50, 60– 70, and 85–100 penetration grades, were obtained from seven asphalt production plants. The study indicated that grades of four out of five PGs were met through modifications. Al-Abdul Wahhab *et al.* (1998) determined Arab Gulf asphalt performance based characteristics and assessed their needs for modification. He concluded that the Gulf asphalts satisfy the low temperature performance requirement but need to be modified to be able to perform adequately in the Gulf countries' hot environmental conditions.

The main objective of this study is to assess the Egyptian locally produced asphalt quality using different grading systems. In order to achieve this objective, various asphalt samples obtained from the three Egyptian sources; 60/70 produced by Alexandria, El Nasr, and Suez Oil Companies, were the subject of investigation. Environmental considerations were pertinent to the Egyptian conditions, and Sealoflex (SBS modifier) was the asphalt modifier considered for investigation. These binders, as original (normal without modification) collected over one year span of time, were subjected to full characterization using two different approaches; conventional (including penetration, viscosity, and AR viscosity) and modern (superpave) grading systems. The SBS modifier used in this research is called Sealoflex. Its contents and morphology are granted by the producer (Ooms Co). A sample of the binder from each source was modified by the SBS modifier, Alex asphalt had two modified samples. The conventional grading systems were inadequate to characterize the modified asphalt. Only the superpave grading system was used to characterize the modified asphalt. Twenty different asphalt samples were obtained to accomplish the asphalt characterization task, sixteen samples for the normal asphalt and four modified samples.

2 EVALUATION OF EGYPTIAN ASPHALT USING CONVENTIONAL GRADING SYSTEMS

Sixteen samples of normal asphalt (5 samples produced by Alex Oil Company, 7 samples produced by El Nasr Oil Company, and 4 samples produced by Suez Oil Company) were tested against the requirements of 60/70 penetration, AC-20 viscosity (with both high and low qualities), and AR-1000 and AR-8000 aged residue viscosity grading systems. A conclusion of the final status (either failed or passed) to all tested asphalt samples according to the conventional grading systems is illustrated in Table 1. It was found that although some asphalt samples passed for individual test (requirement), when the whole grading system is considered asphalt samples could not meet the whole requirements and then considered to be "failed". Detailed asphalt test results (each test result for each sample for all grading systems considered) are available in Khedr *et al.* 2010).

3 EVALUATION OF EGYPTIAN ASPHALT USING SUPERPAVE GRADING

In order to use the modern SUPERPAVE grading system, the prevailing Egyptian climatic conditions have to be determined in order to specify the required performance grade (PG) for asphalt to be used in Egypt. Then, the performance grade (PG) of the locally produced asphalt has to be determined. Determining the required asphalt PG pertinent to the Egyptian climatic conditions was the subject of a research by Khedr *et al.* (2014). The results of this research indicated that for high scale projects (with reliability more than 98%) the required base (without traffic adjustments) PG for Egyptian conditions ranged from 70-10 to 76-10; whereas for low scale projects (with reliability more than 50%) base PGs requirements ranged from 52-10 to 64-10. In this research, base PG for high scale projects are considered as the specification requirement to characterize asphalt for use in Egypt employing the superpave technique. This consideration may also account for higher traffic volumes, Khedr *et al.* (2014).

Table 1. Characterization of all Egyptian asphalt sources and types using conventional grading systems.

Asphalt Source	Asphalt Type	Sample ID	Grading system		
			Penetration (60/70)	Viscosity AC-20 (low and high qualities)	AR-1000 AR8000
Alex	Normal	AN1	Failed	Passed, low quality	Failed
		AN2	Failed	Failed	Failed
		AN3	Failed	Failed	Failed
		AN4	Failed	Failed	Failed
		AN5	Failed	Failed	Failed
	Modified	AM1	N.A	N.A	N.A
		AM2	N.A	N.A	N.A
El Nasr	Normal	NN1	Failed	Passed, low quality	Failed
		NN2	Failed	Failed	Failed
		NN3	Failed	Failed	Failed
		NN4	Failed	Failed	Failed
		NN5	Failed	Failed	Failed
		NN6	Failed	Failed	Failed
		NN7	Failed	Failed	Failed
	Modified	NM1	N.A	N.A	N.A
Suez	Normal	SN1	Failed	Passed, low quality	Failed
		SN2	Failed	Failed	Failed
		SN3	Failed	Passed	Failed
		SN4	Failed	Failed	Failed
	Modified	SM1	N.A	N.A	N.A

3.1 Egyptian Asphalt Characterization According to Superpave Requirements

The results of the tested samples show the status of the asphalt sample either passed or failed as per the criteria explained above to meet the Egyptian climatic requirements is summarized in Table 2. Detailed results for each individual test are available in Khedr *et al.* (2010).

3.2 Determining Performance Grade (PG) for Egyptian Asphalt

In order to fully apply the superpave grading system in characterizing the locally produced asphalt and after finding that it fails according to the Egyptian climatic conditions (for high reliabilities). It is suitable to determine the Egyptian asphalt performance grade (PG) to find out whether it could be fitted for use in lower reliability requirements. The high-end and low-end performance grades were determined for all asphalt samples using the dynamic shear test as recommended in the literature, Freddy *et al.* (1991). The final performance grades (PGs) for all samples are summarized in Figure 1. It is observed from Figure 1 that the performance grades of the normal Alex and the normal Suez asphalt samples ranged from PG58-40 to PG64-40; whereas the normal El Nasr asphalt samples gave higher PGs ranging from PG64-40 to PG70-40. When the modified asphalts for the three sources were considered, it was found that the performance grade of both modified Alex and modified Suez asphalt were PG82-40, which was noticeably higher than the performance grades provided by the normal asphalt of these two sources. As for the modified El Nasr asphalt, it resulted in the highest performance grade of all Egyptian asphalt samples at PG94-40. It can be stated that Nasr asphalt as normal or modified gave the higher performance grades compared to the other two asphalt sources.

Table 2. Final characterization of all Egyptian asphalt sources and types using superpave grading system.

Asphalt Source	Asphalt Type	Sample ID	Superpave @ (PG70-10 and PG 76-10)
Alex	Normal	AN1	Failed
		AN2	Failed
		AN3	Failed
		AN4	Failed
		AN5	Failed
	Modified	AM1	Passed
		AM2	Passed
El Nasr	Normal	NN1	Failed
		NN2	Failed
		NN3	Failed
		NN4	Failed
		NN5	Failed
		NN6	Failed
		NN7	Failed
	Modified	NM1	Passed
Suez	Normal	SN1	Failed
		SN2	Failed
		SN3	Failed
		SN4	Failed
	Modified	SM1	Passed

4 EGYPTIAN ASPHALT PG "SUPPLY" VERSUS EGYPTIAN ASPHALT CLIMATIC REQUIREMENTS "DEMAND"

The Egyptian asphalt PG "supply" was determined in the previous section, and the required asphalt performance grades according to the Egyptian climatic requirements "Demand" were determined by Khedr *et al.* (2014). The question to be raised is "Does the supply satisfy this demand?". In other words, "Do the performance grades of the normal asphalts locally produced in Egypt can satisfy the wide range of Egyptian asphalt performance grade requirements?", and if not "Do modified ones can meet these required PGs?" The following paragraphs should answer these questions.

Figure 1. Determining performance grades (PGs) of all used Egyptian asphalt samples.

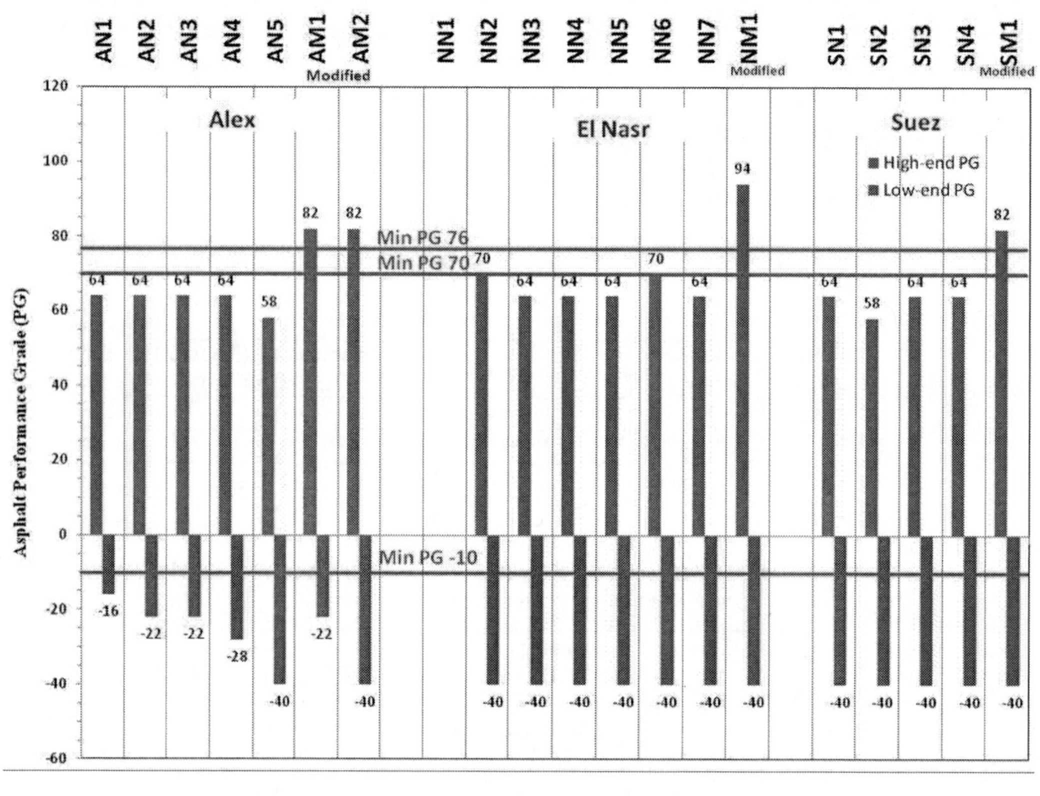

For high scale projects (with reliability more than 98%), when high conservative level of analysis is applied, Egypt is divided into two zones, Khedr *et al.* (2014), zone I with requirements of PG70-10 which cannot be met by any normal asphalt, except some samples of El Nasr, and zone II with PG76-10 which cannot be met by any normal asphalt and the move to the modified asphalt is a must.

For lower reliability projects (50% reliability), when moderate level of conservation in analysis is considered, Egypt is divided into 3 zones, Khedr *et al.* (2010). According to this zoning, zone I requires PG 52-10, zone II requires PG58-10, and zone III requires PG64-10. Accordingly, in this case, there is no need for the modified asphalt since many different normal asphalt samples can work efficiently all over Egypt.

It can be stated that normal asphalt can be used in moderate climatic conditions at lower reliability projects with low to moderate traffic levels. In hot regions with higher traffic levels especially at higher reliability projects modified asphalt becomes an obligatory need.

5 CONCLUSIONS

Based on the work carried out in this research, the following can be concluded:

(1) All normal Egyptian asphalt samples fail according to conventional grading systems with minor exceptions.

(2) All normal Egyptian asphalts fail according to high conservative level of analysis and high reliability projects of Superpave grading system requirements.

(3) The modified asphalt does satisfy all Egyptian PG requirements

(4) The actual PG of the normal Egyptian asphalt could not satisfy the Egyptian asphalt requirements for high reliability projects; whereas it may fit for lower reliability projects and/or less conservative analysis.

(5) Superpave grading system is a more fundamental more reliable system and is more flexible in its requirements. It even accommodates Superpave requirements for lower levels of reliability and/or conservativeness; while conventional grading requirements reject most normal asphalt samples.

(6) When using modified asphalt, conventional grading systems are not applicable and superpave grading system must be used.

References

Al-Abdul Wahhab, A, Al-Dubabe, I., and Asi, M., *Performance-based Characterization of Arab Asphalt*, Building and Environment, Volume 33, Issue 6, November 1998, Pages 375-383.

Al-Abdul Wahhab, I., Ibrahim M., Ibrahim A., and Al-Dubabe, M., *Development of performance-based bitumen specifications for the Gulf Countries*, Construction and Building Materials, Volume 11, Issue 1, February 1997, Pages 15-22.

Aflaki S., and Nader T., *Proposals for modification of Iranian bitumen to meet the climatic requirements of Iran*, Construction and Building Materials, Volume 23, Issue 6, June 2009, Pages 2141-2150.

Freddy, R., Kandhal, S., and Brown, E., *Hot Mix Asphalt Materials, Mixture Design and Construction*, second edition, NAPA Education Foundation Lanham, Maryland, 1999.

Khedr, S., Saudy, M., and Khafagy , M., , *Development of Asphalt Binder Performance Grades for Egypt*, The Second Australasia and South East Asia Structural Engineering and Construction Conference (ASEA-SEC-2), Sustainable Solutions in Structural Engineering and Construction, pp393-398, Bangkok, Thailand, November 2014.

Khedr, S., Saudy, M., Khafagy, M., AbdElAziz, R., Magdy, M., and Boraie, Y., *Modified Asphalt and Superpave Application in Egypt*, Research Project by AUC and GARBLT, GARBLT library, Ministry of Transportation, Cairo, Egypt, 2010.

MACRO DEFECT FREE CEMENTS: A REVIEW

OZGUR EKINCIOGLU[1], M. HULUSI OZKUL[1], and SILVIA PATACHIA[2]

[1]*Dept of Civil Engineering, Istanbul Technical University, Istanbul, Turkey*
[2]*Dept of Product Design, Mechatronics and Environment, Transilvania University of Brasov, Brasov, Romania*

The usage of polymers in different sectors has been increasing in recent decades, and even our current age may have been defined as polymer age. When concrete is considered polymers are also widely used to modify the properties of both mortar and concrete and the usage of polymers in concrete dates back to 1920's. On the other hand, Macro-defect free (MDF) cements are one particular type of cement-polymer composites and developed and patented by scientists at Imperial College at the beginnings of 1980's. MDF cements are produced by mixing cement (commonly calcium aluminate cement) with small amounts of polymer (usually polyvinyl alcohol acetate) and water. High shear, relatively low pressure (about 5 MPa) and moderate temperature (about 80-100 °C) are applied during the production of this material. MDF cements, although consist of more than 80% by weight of cement, show 20-30 times higher flexural strength comparing to ordinary Portland cements. However, MDF cements show a considerable reduction in strength when they are exposed to water even for a short time. Many studies have been conducted to solve the water sensitivity of MDF cements for over 30 years. In this study, production, basic properties and the current state of the art of MDF cements are explained, and the future research works are suggested.

Keywords: MDF composites, cement, polymer, calcium aluminate cement, PVA, Water sensitivity.

1 INTRODUCTION

Concrete, mainly made with aggregates, cement and water, is the most consumed man-made material in the world. In addition to main ingredients, many additives are used in the production of concrete for different purposes. Notably, various types of polymers are used in concrete and mortar production to enhance the fresh and hardened state properties. The first usage of polymers in concrete dates back to 1924 (Ohama 1995). Polymer modification of cementitious materials has been witnessed to the innovation of polymer-modified materials with the improvement in polymer technology, and many different cement-polymer composites were produced since then.

Macro-Defect Free (MDF) cements are one example of a very broad field of cement-polymer composites, and had been developed and patented by Birchall *et al.* (1981, 1983) at Imperial College. The MDF cement can be prepared by using cement, polymer and water. 80-90% of this composite consists of cement by weight, while 10-15% is water and 1-7% is polymer (Ekincioglu 2009). Many researches had been conducted about MDF cements in last 30 years. In this study, a brief review of MDF cements is given. The main materials used, production processes, properties and limitations are summarized in detail. Also, some promising attempts to improve the durability properties are mentioned.

2 MDF CEMENT PRODUCTION PROCESSES

2.1 Materials

Different cement and polymer combinations have been used for the production of MDF cements. Calcium aluminate cement (CAC) (Russell 1991, Ekincioglu 2009) and poly(vinyl alcohol) (PVA) (Birchall 1983, Santos 1999, Ekincioglu *et al.* 2012a) are the most widely utilized materials for the production of MDF cements. Besides CAC, Portland cements and slag cements were also used (Sinclair and Groves 1984, Poon and Groves 1988, Titchell 1991, Poon *et al.* 1998, Santos *et al.* 1999) in productions. Polyacrylamide (Sinclair and Groves 1985, Poon and Groves 1988, Ohama 1993, Poon 1998), cellulosic products (Eden and Bailey 1984, Drabik *et al.* 1994, Drabik *et al.* 1999, Drabik *et al.* 2001, Mojumdar 2001) and alcohol soluble polymers such as phenol resin precursors (Hasegawa *et al.* 1995, Pushpalal *et al.* 1997, Walberer and McHugh 1998, Pushpalal *et al.* 1999) were proposed as polymers for the productions of MDF cement. The approximate flexural strength of most typical MDF cement combinations can be seen in Figure 1.

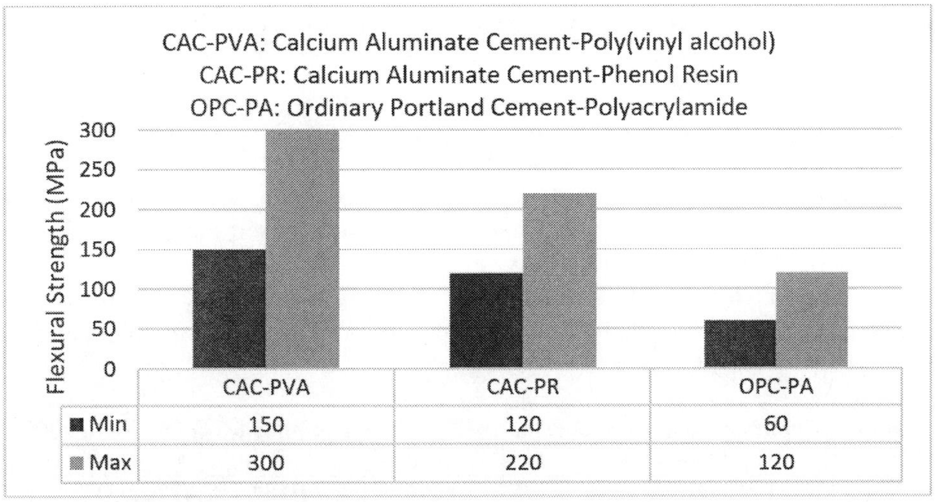

Figure 1. Approximate flexural strength of different MDF cement combinations.

2.2 Production Methods

Birchall *et al.*'s (1981, 1983) production method by applying high shear was generally accepted for as the standard MDF cement production procedure. This process was optimized by Russell (1991) and later Ekincioglu (2009). Another production method, Banbury type mixer for high shearing, was introduced by Tan (1992). Basic steps followed in the production and testing of MDF cements by Ekincioglu (2009) can be seen in Figure 2.

Figure 2. Basic steps to produce and test MDF cements (Ekincioglu 2009).

3 PROPERTIES OF MDF CEMENTS

3.1 Mechanical Properties

Cement-based materials are known with their high compressive strength, but low tensile strength. On the other hand, MDF cements have superior mechanical properties both in compression and tension. However, they can be produced with various cement and polymer types, and their properties largely depend on the used materials. Comparison of some general properties of OPC, CAC-PVA and CAPR MDFs are shown in Table 1. As can be seen in Table 1, flexural strengths of MDF cements are 20-30 times higher than those of conventional (plain) cement pastes and nearly equal to that of an ordinary steel.

Table 1. Typical properties of OPC and MDF cement composites (Ekincioglu 2009).

Property	OPC	CAC-PVA MDF*	CAPR MDF**
Flexural Strength, MPa	5-10	150-300	120-200
Compressive Strength, MPa	40-60	380	300
Young's Modulus, GPa	20-30	40-50	32-45
Density, g/cm^3	2,3	2,5	2.2-2.3

*Polymer was PVA and the cement was alumina cement (Russell 1991).
**Polymer was phenol resin and the cement was alumina cement (Pushpalal *et al.* 1999).

The high strengths of MDF composites are obtained by both the eliminating of macropores in the composite during processing and also the crosslinking formed between cement and polymer. Studies (Rodger *et al.* 1984, 1985, Popoola *et al.* 1991) showed that crosslinking reactions between polymer chains and metal ions generated during the hydration reaction of cement are responsible for achieving such high strengths. On the other hand, MDF cements have serious durability problems. High strength losses are observed when they are exposed to water.

3.2 Durability Properties

The water sensitivity of MDF cements limits the usage of them in commercial applications. Therefore, many researchers have focused on this issue. Several different additives are used in MDF cement productions to prevent this problem. Some of the used additives are; alkali metal silicates (Lynn and Durey 1992), cross-linking chemicals (Desai 1992, Liutkus and Kovac 1988, Lewis and Boyer 1995), gypsum (Brown 1996), $CaCl_2$ and $ZnCl_2$ (Poon 1998), silica fume (Santos *et al.* 1999), activated carbon (Chowdhury 2004a, Chowdhury 2004b), epoxy resin (Ekincioglu *et al.* 2011) and TiO_2 (Ekincioglu *et al.* 2012b). All these studies provided a better understanding of MDF cements, although none of them ultimately reduced the moisture sensitivity.

MDF cements are mostly produced with alumina cements and poly(vinyl alcohols), and the effects of the type of these most common ingredients were also investigated recently by Ekincioglu (2009). In a first study, MDF cements were produced with seven different types of poly(vinyl alcohol-co-vinyl acetate) (PVA) copolymers with various hydrolysis degrees (between 79.6% and 99.1%) and their effects on the mechanical properties were investigated. All specimens prepared were adversely affected by moisture. However, the lowest strength losses were obtained with the least hydrolysis degree PVA (79,6 mole%). Increasing the degree of hydrolysis of PVAs increased the strength loss of MDF specimens which were stored in water. MDF cements could not be produced with fully hydrolyzed PVA or carboxylated PVA. In the second part of the study, the effect of Al_2O_3 content of alumina cements on the moisture sensitivity was investigated. Decreases in strength after 28 days water storage for w/c ratios between 0.09 and 0.19 shown in Figure 3. It seems that there is a minimum strength loss between 49% and 70% alumina contents of cement. Optimum Al_2O_3 content for the lowest strength loss was found approximately 58%.

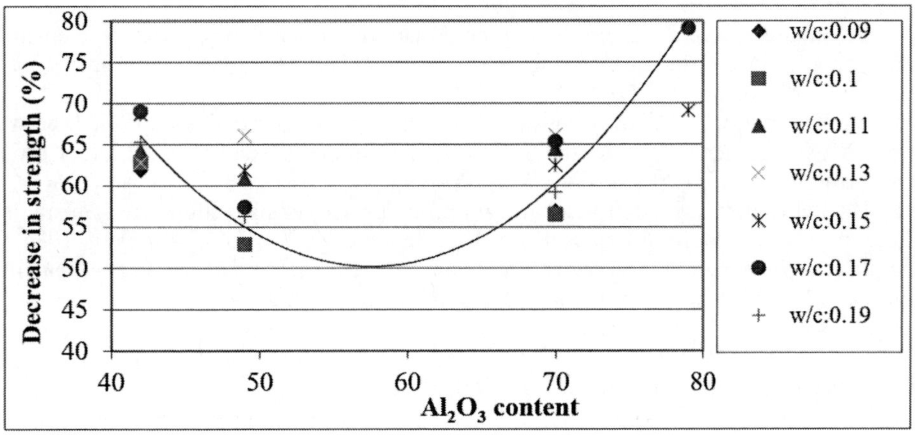

Figure 3. Strength loss versus Al_2O_3 content in cement for 28 days water storage.

4 CONCLUSIONS

MDF cements are cement-polymer composites and show superior mechanical properties. Special production techniques by using calendering are necessary to obtain high mechanical strengths. However, it is a brittle material, and especially the durability problems in water have limited its usage in any commercial application. Research activities about MDF cements have also declined in recent years due to durability problems. Many different additives, such as the addition of epoxy resin or TiO_2, were proposed to overcome the water sensitivity of MDF cements. Although some of them are promising, none of the proposed solutions completely solved this problem. A better understanding of crosslinking between the polymer and cement may allow us to produce water-resistant MDF cements. One area, which has not been sufficiently researched, is that coating MDF cements with a protective layer. MDF cements have superior properties and deserve further studies.

References

Birchall, J. D., Howard, A. J., and Kendall, K., Flexural strength and porosity of cements, *Nature,* 289(29), 388-390, 1981.

Birchall, J. D., Howard, A. J., Kendall, K., and Raistrick, J. H., Cement composition and product, *United States Patent,* No: 4410366, 1983.

Brown, J. L., Cement products and a method of manufacture thereof, *United States Patent,* No: 5514744, 1996.

Chowdhury, B., Investigations into the role of activated carbon in a moisture-blocking cement formulation, *Journal of Thermal Analysis and Calorimetry,* 78(1), 215–226, Oct, 2004a.

Chowdhury, B., Method for using activated carbon for producing moisture-blocking durable cement, composition for the same and method of characterizing the same, *United States Patent,* No: 6797052, 2004b.

Desai, P. G., *Cement Polymer Interactions in Macro-Defect-Free Composites,* M.S. Thesis, University of Illinois at Urbana-Champaign, Il, USA, 1992.

Drabik, M., Galikova, L., Kubranova, M., and Slade, R. C. T., Studies of Model Macroscopic-Defect-Free Materials Part 1-Investigations of the System $4CaO.AI_2O_3.Fe_2O_3$-$4CaO.3AI_2O_3.SO_3$-Hpmc-H_2O by X-Ray, Thermoanalytical and NMR Techniques, *Journal of Materials Chemistry,* 4(2), 265-269, Feb, 1994.

Drabik, M., Galikova, L., and Zimmermann, P., Attack by Moisture on Advanced Cement-Based Macroscopic Defect-Free Materials: A Thermoanalytical Study, *Journal of Thermal Analysis and Calorimetry,* 56(1), 117-124, Jul, 1999.

Drabik, M., Mojumdar, S. C., and Galikova, L., Changes of Thermal Events of Macro Defect Free (MDF) Cements Due to the Deterioration in the Moist Atmosphere, *Cement and Concrete Research,* 31(5), 743-747, May, 2001.

Eden, N. B., and Bailey, J. E., The mechanical properties and tensile failure mechanism of a high strength polymer modified Portland cement, *J Mater Sci,* 19(8), 2677–90, Aug, 1984.

Ekincioglu, O., *Investigations of Moisture Sensitivity in Macro Defect Free Cements,* PhD Thesis, Istanbul Technical University, Istanbul, Turkey, 2009.

Ekincioglu, O., Ozkul, M. H., Ohama, Y., Patachia, S., and Moise, G., Effect of Epoxy Resin Addition on the Moisture Sensitivity of Macro Defect Free Polymer-Cement Composites, *Key Engineering Materials,* 466, 65-72, Jan, 2011.

Ekincioglu, O., Ozkul, M. H., Struble, L. J., and Patachia, S., Optimization of Material Characteristics of Macro-Defect Free Cement, *Cement and Concrete Composites,* 34(4), 556-565, Apr, 2012a.

Ekincioglu, O., Ozkul, M. H., Patachia, S., and Moise, G., Effect of TiO_2 Addition on the Properties of Macro Defect Free Cement, *in 7th Asian Symposium on Polymers in Concrete, ASPIC2012,* 297-308, Ozkul, M. H., *et al.* (eds.), Istanbul, Turkey, 2012b.

Hasegawa, M., Kobayashi, T., and Pushpalal, G. K. D., A New Class of High Strength, Water and Heat Resistant Polymer-Cement Composite Solidified by an Essentially Anhydrous Phenol Resin Precursor, *Cement and Concrete Research,* 25(6), 1191-1198, Aug, 1995.

Lewis, J. A., and Boyer, M. A., Effects of an organotitanate cross-linking additive on the processing and properties of macro-defect-free cement, *Journal of Advanced Cement Based Materials,* 2(1), 2-7, Jan, 1995.

Liutkus, J. J., and Kovac, C. A., Polysiloxane modified cement, *United States Patent,* No: 4780754, 1988.

Lynn, M. E., and Durey C. A., Cementitious products, *United States Patent,* No: 5147459, 1992.

Mojumdar, S. C., Processing-Moisture Resistance and Thermal Analysis of Macro-Defect-Free Materials, *Journal of Thermal Analysis and Calorimetry,* 64(3), 1133-1139, Jun, 2001.

Ohama, Y., Improvement in water resistance of macro defect free cements using ordinary portland cement *in Proceedings of the 3rd Japan Congress on Materials Research,* The Society of Materials Science, 192-195, Kyoto, Japan, 1993.

Ohama, Y., *Handbook of Polymer-Modified Concrete and Mortars-Properties and Process Technology,* Noyes Publications, New Jersey, 1995.

Poon, C. S., and Groves G. W., The microstructure of macrodefect free cement with different polymer contents and the effect on water stability, *Journal of Materials Science,* 23(2), 657-660, Feb, 1988.

Poon, C. S., The influence of Admixtures on the Microstructure and Water Stability of Macro-Defect Free Cement, *Journal of Materials Science Letters,* 17(18), 1593-1595, Sep, 1998.

Popoola, O. O., Kriven, W. M., and Young, J. F., Microstructural and Microchemical Characterization of a Calcium Aluminate-Polymer Composite (MDF Cement), *Journal of American Ceramic Society,* 74(8), 1928-1933, Aug, 1991.

Pushpalal, G. K. D., Kobayashi, T., and Hasegawa, M., High Alumina Cement-Phenol Resin Composite: Water Resistivity and Effect of Post Hydration of Unreacted Cement on Durability, *Cement and Concrete Research,* 27(9), 1393-1405, Sep, 1997.

Pushpalal, G. K. D., Kobayashi, T., Kawano, T., and Maeda, N., The Processing, Properties, and Applications of Calcium Aluminate–Phenol Resin Composite, *Cement and Concrete Research,* 29(1), 121–132, Jan, 1999.

Russell, P. P., *Processing Studies of Macro-Defect-Free Cement and Investigation of Chemical Modifiers to Improve the Water Resistance of the Composite,* M.S. Thesis, University of Illinois at Urbana-Champaign, Il, USA, 1991.

Santos, R. S., Rodrigues, F. A., Segre, N., and Joekes, I., Macro-defect free cements influence of poly(vinyl alcohol), cement type, and silica fume, *Cem Concr Res,* 29(5), 747–751, May, 1999.

Sinclair, W., and Groves G. W., The microstructure of high strength cement pastes in *Proceedings of the Materials Research Society Symposia,* Young, J.F. (ed.), 42, 31-37, Boston, Massachusetts, USA, 1984.

Sinclair, W., and Groves, G. W., High Strength Cement Pastes - Part 1: Microstructures, *Journal of Materials Science,* 20(8), 2846-2852, Aug,1985.

Tan, L. S., *Processing-property interactions in macro-defect-free cement,* MSc Thesis, University of Illinois at Urbana-Champaign, Il, USA, 1992.

Titchell, I., Environmental degradation of macrodefect free cements, Part I: Mechanical properties investigation, *Journal of Materials Science,* 26(5), 1199-1204, Jan, 1991.

Walberer, J. A., and McHugh, A. J., Processing/Property/Structure Interactions in a Calcium Aluminate-Phenol Resin Composite, *Advanced Cement Based Materials,* 8(3-4), 91-100, Oct, 1998.

Interaction between Theory and Practice in Civil Engineering and Construction
Edited by Komurlu, R., Gurgun, A. P., Singh, A., and Yazdani, S.

SUPERPAVE VERSUS CONVENTIONAL TECHNIQUES FOR DESIGNING NORMAL AND MODIFIED ASPHALT MIXES

SAFWAN KHEDR and MARAM SAUDY

Construction Engineering Dept, American University in Cairo, Cairo, Egypt

Due to the empirical nature and drawbacks of the conventional procedures, the Strategic Highway Research Program (SHARP) has developed a Superior Performance Asphalt Pavements (SUPERPAVE) mix design procedure. The main objective of this research is to study the applicability of the Superpave in Egypt. This is done by studying aggregate characteristics using both the Superpave and the conventional techniques, investigating the normal (virgin) and SBS modified asphalt characteristics using Superpave, and designing asphalt mixtures comprised of the characterized materials using both the Superpave and the conventional Marshall design methods. Results indicate that Superpave is applicable to Egyptian aggregate with a more restrictive supervision of crushing aggregates and gradations (some gradations may need modifications). Mix design results indicated two main findings; first, most optimum asphalt contents (OAC) determined by the Superpave mix design method are consistently less than OAC determined by the Marshall Mix design method. Second, modified asphalt mixes result in less OAC than normal asphalt mixes according to both Marshall and Superpave mix design methods for both binder and surface layers.

Keywords: Aggregate properties, Aggregate gradation, OAC, SBS modified asphalt, Marshall, and Superpave.

1 INTRODUCTION

Current specifications used for road construction in Egypt do not properly account for pavement performance. Applying more mechanistic design method is essential to accommodate the rapid growth of road network and traffic volumes, and to improve network performance and safety. Superpave practices have been widely used by both developed and many developing countries. The Superpave system is now spreading over many parts of the world. The main objective of this research is to study the applicability of SUPERPAVE techniques in Egypt. In order to achieve this objective, various materials, asphalt and aggregates, obtained from Egyptian sources are used to design mixes using both Marshall and Superpave mix procedures. Environmental considerations are pertinent to Egyptian conditions and Sealoflex (SBS modifier) was the asphalt modifier considered for investigation.

Khedr and Breakah (2010) investigated the application of Superpave in Egypt. They examined aggregate graduation requirements in the Egyptian code in view of the Superpave aggregate structure. They also designed mixes using Egyptian materials with gradations that satisfy both the Egyptian code of practice and Superpave aggregate structure. They compared the results of Superpave mix design to Marshall mix design. The investigation established a relationship between resilient and residual properties of asphalt concrete made of Egyptian materials. They concluded that most of the Egyptian gradation bands do not fit the Superpave gradation limits. Then, in order to start applying Superpave in Egypt, these bands should be reviewed in view of the Superpave 0.45-power charts. Superpave shows lower optimum asphalt content as compared to Marshall Mix design for the studied mixes. The general trend found in this study indicates that the samples prepared using the Superpave gyratory compactor have higher values in stability and flow than those prepared using the Marshall method.

Asi (2007) conducted a comprehensive evaluation of the locally available aggregate in Jordan commonly used in the asphalt concrete mixtures, to ensure that these materials conform to the new mix design procedures developed by SUPERPAVE. Swami *et al.* (2004) compared the design of asphalt mixtures by the Superpave and Marshall methods of mix design in India. A detailed laboratory study was carried out using local aggregate and local bitumen. From the analysis of design of mixtures, Superpave mixes fulfilled all the criteria for easy and good construction at lesser binder content than the Marshall mixes. The study recommended that Marshall mix design be replaced by Superpave mix design for the Indian national highways.

2 MIX DESIGN USING CONVENTIONAL AND SUPERPAVE METHODS

Mix designs are performed according to conventional Marshall and SUPERPAVE methods using the Egyptian materials. The variables involved in this task were; types of asphalts, asphalt modification, types of aggregates materials, aggregates structures, traffic conditions, and climatic regions. The mix design program is limited to three sources of asphalts; Alexandria, Suez, and El Nasr Oil companies, two types of asphalts; normal and SBS modified asphalts, one type of coarse aggregate; limestone from Ataqua quarry, one type of fine aggregate; natural sand from Al-haram quarry, one type of mineral filler; limestone filler, two aggregate structures: 3D (binder course) and 4B (surface layer) as per the Egyptian specifications, and heavy traffic. This resulted in 18 mix designs.

2.1 Aggregate Characterization

The used Egyptian coarse and fine aggregates have been characterized using both the conventional and Superpave approaches. The full characterization results are illustrated in Table 1. It is clear from Table 1 that, the used coarse and fine aggregates satisfy all requirements except the fine aggregate angularity, and the coarse aggregate angularity only for surface layer requirements and marginally satisfy the clay lumps and friable materials requirements.

The shortage in the angularity percentage of fine aggregates was expected since the used fine aggregate is natural sand. The natural sand was used since it is the fine aggregates used in the majority of projects implemented in Egypt. Although these deficiencies were encountered, the aggregates were used in designing asphalt mixtures.

Table 1. Aggregate characterization using conventional and Superpave procedures.

Aggregate Type	Test	Result	Specification	Status
Coarse Aggregate	Nominal Max. Size (NMS)	1inch (25mm)	.....	......
	Los Angeles Abrasion at 100 cycles, %	5.1	< 10%	Passed
	Los Angeles Abrasion at 500 cycles, %	25.5	< 40%	Passed
	Bulk Specific Gravity	2.475	....	....
	Apparent Specific Gravity	2.643	....	....
	Absorption, %	2.49	<2.5%	Passed
	Clay Lumps and Friable Materials, %, Supe.	1.06	<1%	**Marginal**
	Flakiness Index	20.25	<25%	Passed
	Elongation Index	8.85	<25%	Passed
	% of Flat and/or elongated, Superpave	2	<10%	Passed
	Angularity, %, (Binder course), Superpave	85.5% have one fractured face and 93.74% have two	Min 80% have one fractured face and Min 75% have two*	Passed
	Angularity, %, (Surface layer), Superpave	85.5% have one fractured face and 93.74% have two	Min 95% have one fractured face and Min 90% have two**	**Failed**
	Chemical Soundness (%materials lost)	0.25	<12%	passed
Fine Aggregate	Bulk Specific Gravity	2.659	....	......
	Apparent specific Gravity	2.68	....	......
	Absorption, %	0.3	<2.5%	Passed
	Clay Lumps and Friable Materials, %, sup.	1%	<1%	**Marginal**
	Chemical Soundness (%materials lost)	0.5%	<10%	Passed
	Angularity, %, Superpave	30.3	Min 40%***	**Failed**
	Sand Equivalent, Superpave	76%	Min 45%****	Passed
Filler	Apparent Specific Gravity	2.7	....	....

2.2 Aggregate Gradation

Two different layers gradations, binder course and surface layer, were designed using both Marshall and Superpave procedures with the normal and the modified asphalts. The first step was to select the aggregate gradation. Binder gradation 3D according to the Egyptian pacifications was selected as shown in Figures 1. It is clear from figure 1 that 3D gradation fits easily into the Superpave control points. Many difficulties were encountered in selecting a surface gradation which can best fit with the Superpave requirements. A modified 4B surface gradation was selected for the surface layer as shown in Figure 2. It was modified to satisfy the requirements of the Superpave gradation and voids ratios.

2.3 Binder Characterization

Table 2 illustrates the final results of both the normal and the modified Egyptian asphalt obtained from the three available sources according to the Superpave grading system.

3 MIX DESIGN

Once materials used (aggregates and binder) have been fully characterized, the next step is designing the mixture using the normal and the modified asphalts for two different layers (gradations), binder course and surface layer. The Marshall and Superpave methods were used to design the mix.

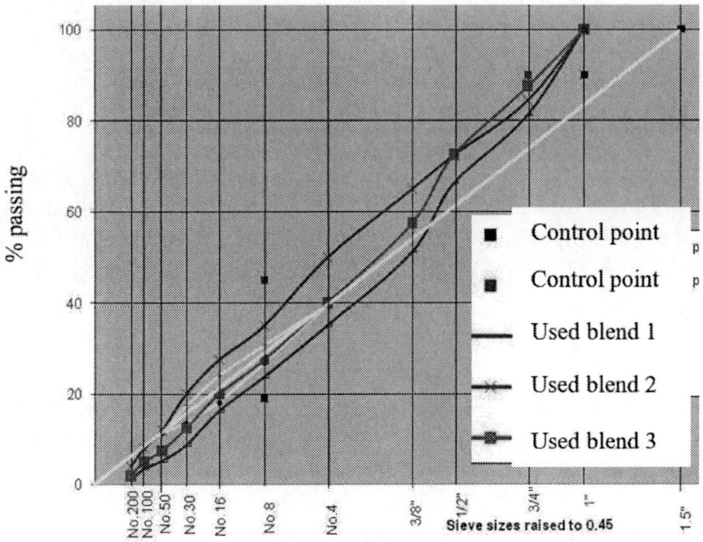

Sieve size raised to 0.45

Figure 1. 3D binder gradation used for Superpave mix design method.

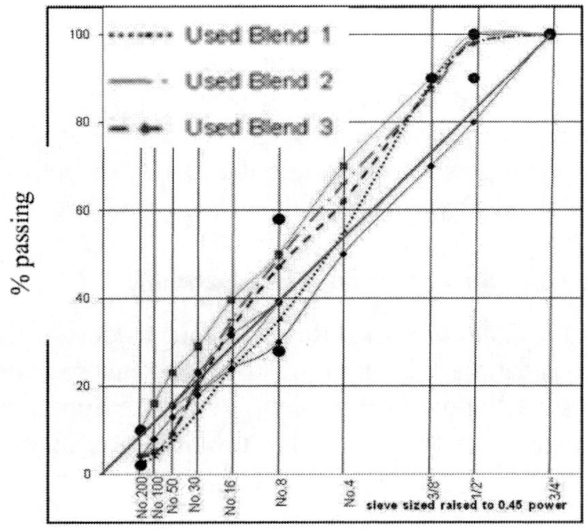

Sieve size raised to 0.45

Figure 2. Modified 4B surface gradation used for Superpave mix design method.

Table 2. Binder characterization for different Egyptian asphalt sources and types using Superpave grading system.

Asphalt Source	Asphalt Type	PG
Alex Oil Company	Normal	PG64-22
	Modified	PG82-40
El Nasr Oil Company	Normal	PG70-40
	Modified	PG94-40
Suez Oil Company	Normal	PG58-40
	Modified	PG82-40

3.1 Mixture Design using Marshall Method

The Marshall mix design method was applied only to the binder course to compare Mix OAC to the Superpave mix OAC, see Figure 3. It can be noted from Figure 3 that modified asphalt results in less OACcompared to the normal asphalt for both Alex and Suez binders which may indicate that using the Sealoflex modifier provides better lubrication and coating to the aggregates results in reducing the amount of the optimum asphalt required for the mix when using the Marshall method. This was not observed in the case for asphalt El Nasr and this may be explained by the unexpected high viscosity (RV = 3.8 Pa.s), according to Khedr *et al.* (2010), encountered by El Nasr asphalt which may require a higher percentage of the optimum asphalt content to sufficiently coat the aggregates.

3.2 Mixture Design using Superpave Method

3.2.1 *Binder course with normal and modified asphalts*

The normal asphalt was first used to design the mixtures with the Superpave to select the optimum blends (aggregate structures) and to select the optimum asphalt content. The results indicate that the optimum blend was blend # 1 within the 3D gradation (see Figure 1). Then, the modified asphalt was used to design the mixture of the binder course using the Superpave method. The selected blend was blend # 2 within the 3D gradation (see Figure 1). Figure 4 illustrates the OAC for both normal and modified asphalt selected for the three Egyptian asphalt sources used.

3.2.2 *Surface layer with normal and modified asphalts*

The typical Superpave procedure was followed again to design the surface layer with gradation 4B surface gradations according to the Egyptian specifications. Difficulties were encountered to fit the standard 4B gradation with the Superpave requirements then 4B surface gradation was used but with some modification in sieves numbers 30, 50, 100 and 200 as illustrated in Figure 2. Mixture was designed with both normal and modified asphalt using the Superpave procedure. The selected blend was blend # 1and Blend # 3 for normal and modified asphalt, respectively as illustrated in Figure 2. Also Figure 4 shows the final selected OAC for surface layer for the three sources used.

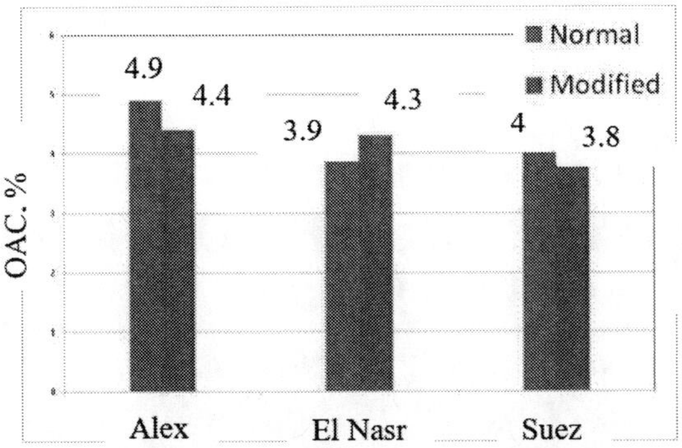

Figure 3. Marshall optimum asphalt contents for different asphalt mixtures.

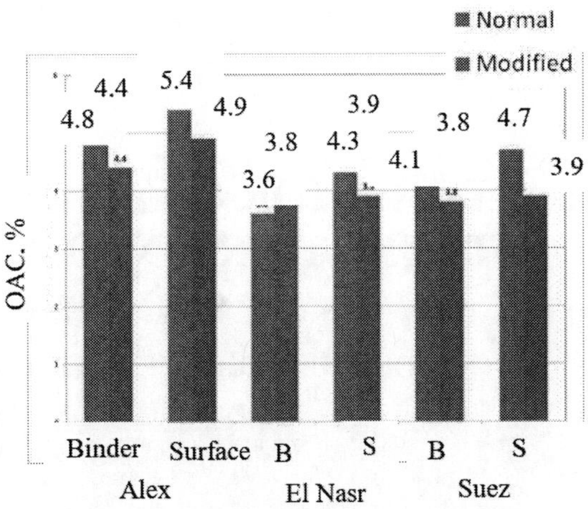

Figure 4. Superpave optimum asphalt content for different asphalt mixtures.

4 CONCLUSIONS AND RECOMMENDATIONS

Based on the work carried out in this research, the following can be concluded:

- The used Egyptian fine and coarse aggregates marginally satisfy the requirements for clay lumps and friable materials. The used Egyptian fine aggregates fail in satisfying the requirement of the fine aggregates angularity. Also, the used coarse aggregates fail in satisfying the Superpave angularity requirements only for the surface layer.
- For binder course, gradation within standard 3D binder according to the Egyptian specifications satisfies Superpave requirements for mix design. Also, for Surface course, standard 4B surface gradation according to the Egyptian specifications need some modifications to satisfy Superpave requirements for surface layer mix design.
- Most OAC determined with the Superpave mix design method are consistently less than OAC determined by the Marshall Mix design method for both binder and surface Courses. Superpave mix design method is applicable with Egyptian materials, gradations and environment.

References

Asi, M., *Performance evaluation of SUPERPAVE and Marshall asphalt mix designs to suite Jordan climatic and traffic condition*, Construction and Building Materials, Volume 21, Issue 8, pp. 1732-1740, August 2007.

Khedr, S. and Breakah, T., *Evaluation of the Readiness of Egypt for Superpave Implementation*, Middle East Society for Asphalt Technologist (MESAT) First Conference, Beirut, Lebanon, July, 2010.

Khedr, S., Saudy, M., Khafagy, M., AbdElAziz, R., Magdy, M., and Boraie, Y., *Modified Asphalt and Superpave Application in Egypt*, Research Project by AUC and GARBLT, GARBLT library, Ministry of Transportation, Cairo, Egypt, 2010.

Swami, L., Mehta, A., and Bose, S., *A Comparison of the Marshall and Superpave Design Procedure for Materials Sourced in India*, International Journal of Pavement Engineering, Volume 5, Number 3, pp. 163-173, September 2004.

Interaction between Theory and Practice in Civil Engineering and Construction
Edited by Komurlu, R., Gurgun, A. P., Singh, A., and Yazdani, S.

BEHAVIOR OF PLAIN AND STEEL FIBER HIGH STRENGTH REINFORCED CONCRETE COLUMNS CONFINED WITH CFRP

DUYGU ERTURKMEN[1], CENGIZ DUNDAR[1], and SERKAN TOKGOZ[2]

[1]*Civil Engineering Dept, Cukurova University, Adana, Turkey*
[2]*Civil Engineering Dept, Mersin University, Mersin, Turkey*

High strength reinforced concrete columns have been increasingly used in structural applications. It is known that high strength reinforced concrete columns exhibit brittle behavior under applied loads. Therefore, steel fibers can be added into the core concrete to improve the ductility and deformability of high strength reinforced concrete columns. In addition, reinforced concrete columns can be wrapped with carbon fiber reinforced polymer (CFRP) materials for strengthening the columns. In this work, experimental and analytical studies on plain and steel fiber reinforced concrete columns and CFRP confined columns have been presented. The experimental strength capacities, load-deflection relations and failure modes of the column specimens have been observed. The effects of the parameters of concrete compressive strength, load eccentricity, steel fibers, and carbon fiber polymer material on the structural behavior of high strength reinforced concrete columns have been examined. It is concluded that the inclusion of steel fibers has considerable effect on column ductility and deformability, but has insignificant effect on column strength capacity. The experimental study exposed that wrapping CFRP material on plain and steel fiber high strength reinforced concrete columns has provided significant improvement on the column strength capacity and confinement. The tested columns have been analyzed with a theoretical method considering the experimental stress-strain relations of CFRP confined concrete; and the analysis shows good agreement with the experimental results.

Keywords: Carbon fiber polymer, Confinement, Deflection, Ductility, Strength capacity, Deformability.

1 INTRODUCTION

Carbon fiber polymer (CFRP) sheet wrapping to the reinforced concrete columns presents significant advantages in the field of civil engineering applications and constructions. Hence it is important to understand the behavior of plain and steel fiber reinforced concrete columns confined with CFRP.

A number of research studies were conducted to determine the behavior of CFRP confined eccentrically loaded slender reinforced concrete columns and mechanical behavior of fiber reinforced polymer confined concrete (Mirmiran and Shahawy 1997, Samaan *et al.* 1998, Sheikh 2002, Hadi 2009, Punurai *et al.* 2013). But there is lack information about steel fiber reinforced concrete columns confined with CFRP sheets.

So in this research a total of 16 both plain and steel fiber reinforced concrete column specimens were wrapped with CFRP material and experimentally tested to determine the column structural behavior. In addition, the tested column specimens have been analyzed to predict the ultimate strength capacity and load–deformation behavior of CFRP confined slender reinforced concrete columns. The experimental stress–strain relation of the CFRP confined concrete is used in the analysis procedure. To obtain the concrete stress-strain relations of the each CFRP wrapped column specimens, one layer and two layers of CFRP wrapped standard cylinder specimens tested under uniaxial compressive loading. In conclusion, the analysis and experimental results of tested columns have been achieved in good accuracy in this study.

2 EXPERIMENTAL STUDY

The experimental study was focused on the determination of the behavior of CFRP confined plain and steel fiber reinforced concrete columns. For this purpose, a total of 16 column specimens were prepared and experimentally tested under biaxial bending and axial load.

The yield strength of the longitudinal bars and lateral reinforcement was 550 MPa and 630 MPa, respectively. Steel fiber concrete was prepared with using RC 65/35 BN-type hooked steel fibers. The steel fiber column specimens were constructed at dosages of 50 and 60 kg/m^3 steel fiber composites. For strengthening of the columns, bi-directional CFRP material namely SikaWrap Hex 300C 0/90 was used. This material applied on the columns surface with Sikadur-330 adhesive. All the details of the specimens are shown in Figure 1.

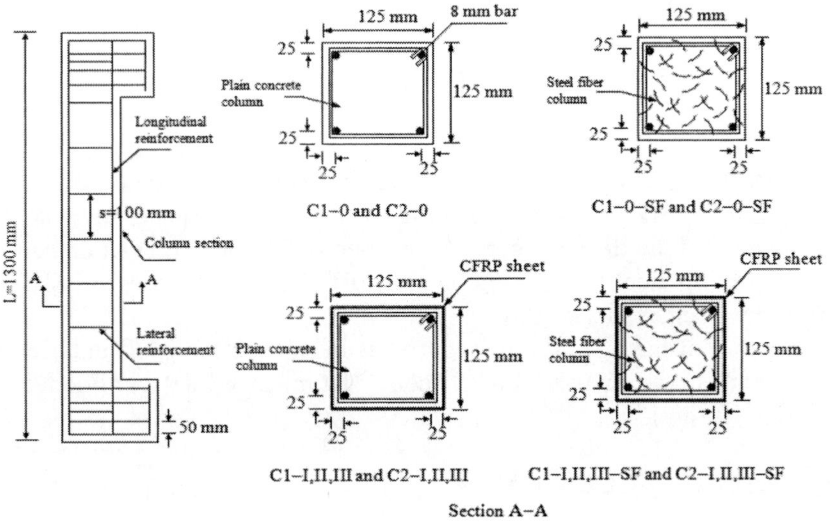

Figure 1. Details of test specimens and reinforcement configuration.

Eight specimens were plain high strength reinforced concrete columns (C1-0,I,II,III and C2-0,I,II,III) and other specimens were steel fiber high strength columns (C1-0,I,II,III-SF and C2-0,I,II,III-SF). The control plain and steel fiber column specimens of C1-0, C2-0, C1-0-SF and C2-0-SF were not wrapped with CFRP material. But the other columns were strengthened by one layer and two layers of sheets, respectively. The column specimens of C1-III, C1-III-SF, C2-III and C2-III-SF were strengthened by two layers of CFRP sheets similarly.

In the tests, a data acquisition system was established to record the digital measurements during the tests. Four linear variable displacement transducers were applied horizontally on each face of mid-height of the column specimens. A load cell was used to measure the load values. The column specimens were vertically tested using a universal testing machine with pinned conditions at both ends in the Structural Laboratory, at Cukurova University in Turkey. More details for the experimental work can be obtained from the study reported by Dundar *et al.* (2015).

3 ANALYTICAL RESEARCH

A theoretical method to analyze the reinforced concrete columns under combined axial compression and biaxial bending have been previously presented in Dundar *et al.* (2007) and Tokgoz (2009). The method has been modified by Dundar *et al.* (2015) to determine the behavior of CFRP confined plain and steel fiber slender reinforced concrete columns. In this study, the experimental stress-strain relations have been used to analyze the plain and steel fiber high strength CFRP confined slender columns.

The fundamental equilibrium equations for the axial load N, and the bending moments M_x and M_y of CFRP confined reinforced concrete columns are described as follows:

$$N=\sum A_c \sigma_c + \sum A_s \sigma_s + \sum A_{fp} \sigma_{fp} \tag{1}$$

$$M_x=\sum A_c \sigma_c y_c + \sum A_s \sigma_s y_s + \sum A_{fp} \sigma_{fp} y_{fp} \tag{2}$$

$$M_y=\sum A_c \sigma_c x_c + \sum A_s \sigma_s x_s + \sum A_{fp} \sigma_{fp} x_{fp} \tag{3}$$

where A_c, A_s and A_{fp} are the elemental area of concrete segment, area of reinforcing steel bar and elemental area of CFRP sheet, respectively; σ_c, σ_s and σ_{fp} are the concrete, reinforcing steel and CFRP material stresses, respectively; (x_c, y_c), (x_s, y_s) and (x_{fp}, y_{fp}) indicate the distance between, respectively, the center of elemental area of concrete, reinforcing steel bar and the center of the elemental area of CFRP element, and the geometric center in x–y plane.

The algorithm of the analysis procedure has been reported by Dundar et al. (2015). The stress-strain curves of the CFRP, steel and typical CFRP confined concrete are shown in Figure 2 (a-c).

4 RESULTS AND DISCUSSION

The structural behavior of plain and steel fiber reinforced concrete columns strengthened by CFRP sheets were investigated in this study. It was observed that the unconfined plain high strength reinforced concrete columns (C1-0, C2-0) behaved more brittle under applied load. These specimens failed suddenly and in an explosively manner. On the other hand, the steel fiber column specimens (C1-0-SF, C2-0-SF) exhibited ductile behavior particularly after the peak load. It was concluded that, the inclusion of steel fibers considerably developed column ductility and deformability.

The failure was observed near the mid-height of the tested column specimens. For the CFRP confined columns, some of the longitudinal sheets were fractured in tension side at the time of failure. Lateral stiffness has been provided for the plain and steel fiber reinforced concrete columns strengthened with CFRP material owing to the transverse sheets (Dundar *et al.* 2015). The typical failure mechanism of the tested CFRP confined slender reinforced concrete column specimens is shown in Figure 3 and the typical comparative experimental axial load–lateral deflection diagrams of column specimens (C2-0-SF, C2-I-SF and C2-II-SF) are illustrated in Figure 4.

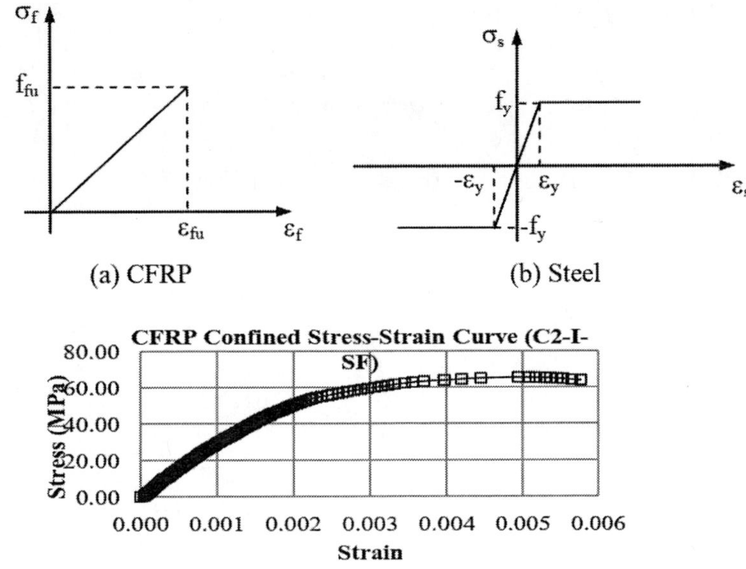

(a) CFRP (b) Steel

(c) CFRP confined concrete

Figure 2. (a-c) Typical stress-strain curves of CFRP, steel and confined concrete specimens.

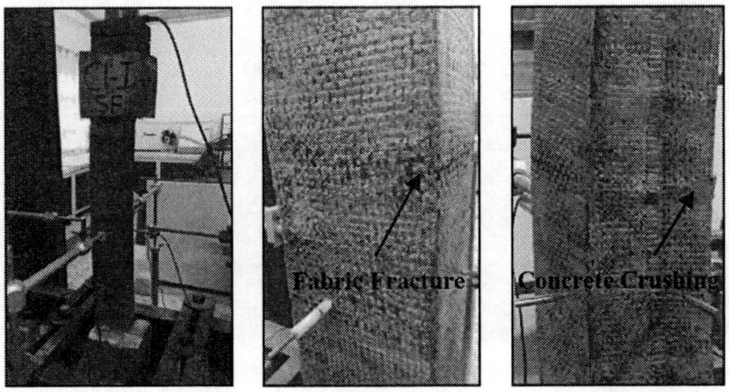

Figure 3. Experimental setup and typical failure mode of column specimens.

In this study, the plain and the CFRP confined slender reinforced concrete columns were analyzed by using the experimental stress–strain curves determined from plain and CFRP wrapped concrete standard cylinder specimens. The column specimen properties of the concrete compressive strengths and the load eccentricities are given in Table1. In addition, the experimental strength capacities (N_{test}), the analysis loads (N_u), and the comparative strength results of the column specimens are shown in Table 1.

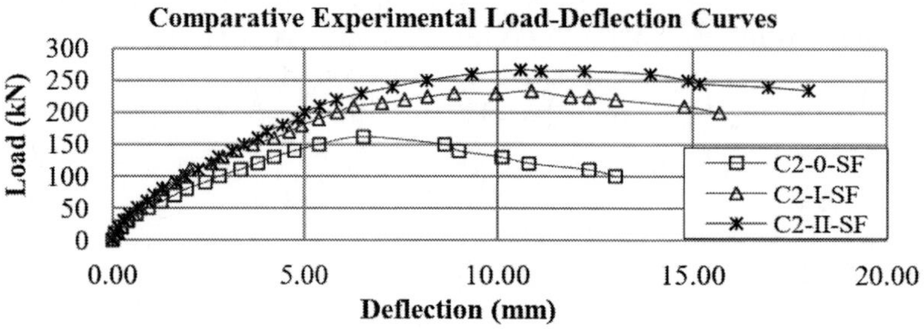

Figure 4. Comparative experimental axial load-lateral deflection diagrams.

Table 1. Test and ultimate strength results of column specimens.

Specimen	f_c (MPa)	$e_x = e_y$ (cm)	N_{test} (kN)	N_u (kN)	M_{ux}&M_{uy} (kN cm)	N_u / N_{test}
C1-0	66.77	4	203	194.08	881.12	0.956
C1-I	59.86	4	258	251.13	1145.19	0.973
C1-II	61.62	4	350	311.87	1402.93	0.891
C1-III	69.00	5	254	249.74	1365.47	0.983
C1-0-SF	65.20	4	220	198.52	916.32	0.902
C1-I-SF	63.72	4	299	280.66	1288.97	0.939
C1-II-SF	67.03	4	365	322.56	1512.97	0.884
C1-III-SF	78.66	5	273	314.68	1812.20	1.153
C2-0	76.76	5	155	155.17	926.96	1.001
C2-I	77.88	5	255	237.80	1342.13	0.933
C2-II	73.65	5	283	282.82	1578.28	0.999
C2-III	78.27	6	222	235.33	1535.10	1.060
C2-0-SF	67.00	5	162	151.88	870.86	0.938
C2-I-SF	65.55	5	234	232.98	1350.11	0.996
C2-II-SF	65.30	5	267	300.59	1838.46	1.126
C2-III-SF	69.59	6	225	236.92	1581.28	1.053
Mean Ratio						**0.987**

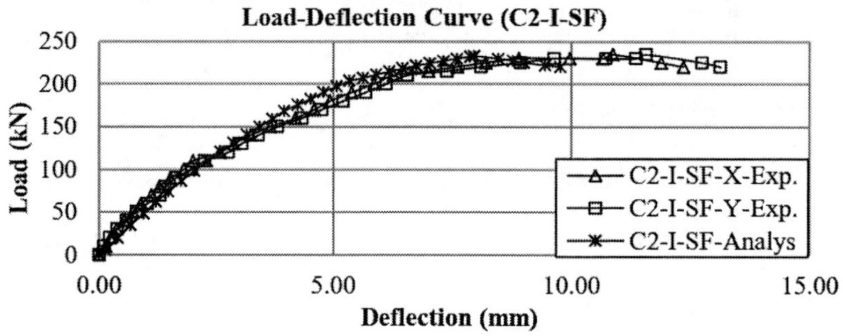

Figure 5. Typical experimental and analytical axial load-lateral deflection diagram.

5 CONCLUSIONS

The use of CFRP sheets have improved confinement, ductility, deformability and the ultimate strength capacity of plain and steel fiber reinforced concrete columns. In addition, the ultimate load and the deformation capacity of plain and steel fiber reinforced concrete columns have increased with increasing CFRP layers. The results show that the ultimate load capacities of reinforced concrete column specimens are significantly influenced by load eccentricity, concrete compressive strength, and slenderness effect. Finally, the comparisons between the test and the analysis results have indicated good accuracy in this study.

References

Dundar, C., Tokgoz, S., Tanrikulu, A. K., Baran, T, Behaviour of Reinforced and Concrete-Encased Composite Columns Subjected to Biaxial Bending and Axial Load, *Building and Environment*, 43(6):1109-20, 2007.

Dundar, C., Erturkmen, D., Tokgoz, S., Studies on Carbon Fiber Polymer Confined Slender Plain and Steel Fiber Reinforced Concrete Columns, *Engineering Structures*, 102:31-39, 2015.

Hadi, M. N. S., Behaviour of Eccentric Loading of FRP Confined Fibre Steel Reinforced Concrete Columns, *Construction and Building Materials*, 23(2):1102-8, 2009.

Mirmiran, A., Shahawy, M., Behavior of Concrete Columns Confined by Fiber Composite, *Journal of Structural Engineering*, ASCE, 123(5):583-590, 1997.

Punurai, W., Hsu, C. T. T., Punurai, S., Chen, J., Biaxially Loaded RC Slender Columns Strengthened by CFRP composite Fabrics, *Engineering Structures*, 46:311-321, 2013.

Samaan, M., Mirmiran, A., Shahawy, M., Model of Concrete Confined by Fiber Composites, *Journal of Structural Engineering*, ASCE, 124(9):1025-1031, 1998.

Sheikh, S. A., Performance of Concrete Structures Retrofitted with Fibre Reinforced Polymers, *Engineering Structures*, 24:869-79, 2002.

Tokgoz, S., Effects of Steel Fiber Addition on the Behaviour of Biaxially Loaded High Strength Concrete Columns, *Materials and Structures*, 42(8):1125-1138, 2009.

Interaction between Theory and Practice in Civil Engineering and Construction
Edited by Komurlu, R., Gurgun, A. P., Singh, A., and Yazdani, S.

LONG TERM BEHAVIOR OF CONCRETE MEMBER STRENGTHENED WITH INTERNAL PRESTRESSING SYSTEM

ISAMU YOSHITAKE[1], TATSUHIKO MIMOTO[2], TAKUYA SAKAKI[1], and JUN MIZUSHIMA[1]

[1] *Dept of Civil and Environmental Engineering, Yamaguchi University, Yamaguchi, Japan*
[2] *Kyokuto Kowa Corporation, Hiroshima, Japan*

To maintain and retrofit appropriately concrete structures, various strengthening materials and methods have been developed and applied. One of the effective and reliable strengthening method for concrete is an application of the prestressing system. The present study focuses on a strengthening system using an internal anchorage and a prestressing tendon. The strengthening system is acceptable even in relatively narrow workspaces, and also applicable for joints between existing and additional concrete members. In our previous investigations, a static push-out and pull-out tests were performed to examine the load-bearing capacity of the prestressing tendon embedded in the wedge anchor. The test confirmed that the prestressing tendon can be anchored firmly in an internal wedge hole filled with high-strength mortar. Long-term behavior of the strengthened member should be examined to confirm the applicability of the system. The load and deformation of a concrete member subjected to sustained force by the prestressing bar were measured for 1 year. This paper reports the long-term loading test, and discusses the time-dependent properties of the strengthened concrete member. The test result confirms that the loss of prestressing force is negligible for actual applications.

Keywords: Strengthening, Loss of prestress, Internal anchor, Prestressing bar, Time-dependent behavior.

1 INTRODUCTION

Numerous concrete structures for civil infrastructure, such as bridges and so on, have been constructed for decades in Japan. Civil engineers have to maintain and retrofit appropriately these structures to sustain the modern infrastructures. Various strengthening materials and methods have been developed and applied to concrete members. The most effective and reliable method for concrete structures is an application of a prestressing system. The strengthening system has been investigated in previous studies. Kim *et al.* (2010) examined the strengthening effect of prestress by using a carbon fiber-reinforced polymer (CFRP) laminate. Michels *et al.* (2013) developed an anchorage system for prestressing CFRP strips to strengthen concrete members.

The anchorage system is a key to ensure the strengthening effect using a prestressing bar. The anchor system for reinforcing materials has been discussed in various previous studies. Kim *et al.* (2013) conducted a pull-out test of post-installed anchorage systems drilled in concrete members. Marti-Vargas *et al.* (2013) focused on and examined the anchor length of pre-tensioned prestressed concrete members. Yilmaz *et al.* (2013) investigated tensile behavior of post-installed chemical anchors embedded in low-strength concrete blocks.

The study focuses on a strengthening system using an internal wedge-shaped anchorage for a prestressing bar (Fig.1). The prestressing bar can be firmly anchored in the wedge hole filled with high-strength mortar. The strengthening system is acceptable even in relatively narrow workspaces, and also applicable for joints between existing and additional concrete members. Mimoto *et al.* (2015) conducted a push-out test to examine the load-bearing capacity of the wedge hole, and Sakaki *et al.* (2015) performed a pull-out test to confirm the yielding and ultimate loads of the prestressing bar embedded in the wedge anchor of full-size concrete member. These tests indicated the adequate strength of the anchor system for strengthening concrete members.

Figure 1. Internal anchorage system.

It should be noted that the tensile load in the pull-out test was statically increased up to the ultimate strength of the prestressing bar. Long-term behavior of the concrete member under a sustained load should be examined to confirm the applicability of the system. An objective of the experimental investigation is to examine the time-dependent properties of the concrete member strengthened with this system. The loss of prestressing force and deformation of the concrete member subjected to the tensile load (design load for strengthening) for 1 year are reported in this paper.

2 STRENGTHENING SYSTEM USING AN INTERNAL ANCHORAGE

This section outlines the strengthening system developed in the previous studies (Mimoto *et al.* 2015; Sakaki *et al.* 2015). Figure 2 shows the strengthening process of the post-installed anchorage system. To create a wedge-shaped hole (wedge anchor), a tip of hole drilled in concrete member is enlarged by using a special drilling device (Fig.3). The wedge anchor is filled with high-strength mortar after installing a prestressing bar as shown in Fig.1. The strengthening work is implemented as follows: Step 1: Core-drill into existing concrete and enlarge the hole tip; Step 2: Install a prestressing tendon (bar) and grout a filler (high-strength mortar) into the enlarged hole tip; Step 3: Place new concrete to increase size if necessary. Pull the tendon for prestressing; Step 4: Grout mortar into the hole.

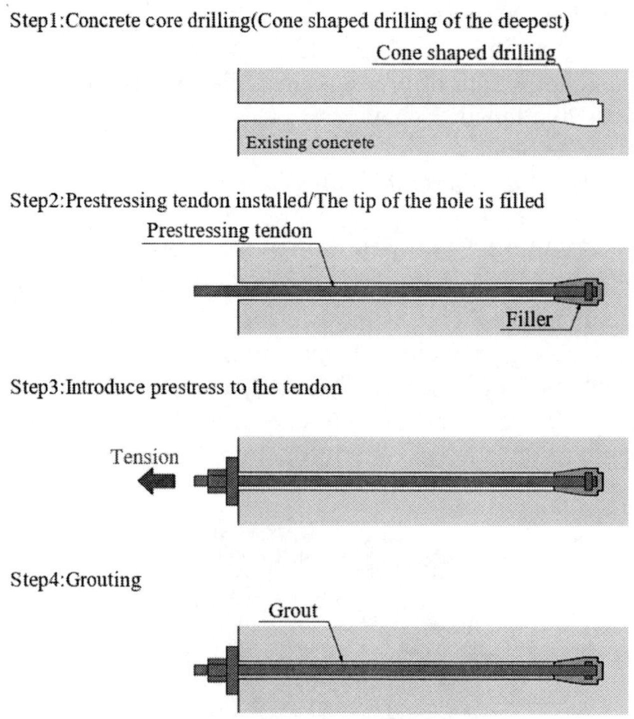

Figure 2. Process of the strengthening system.

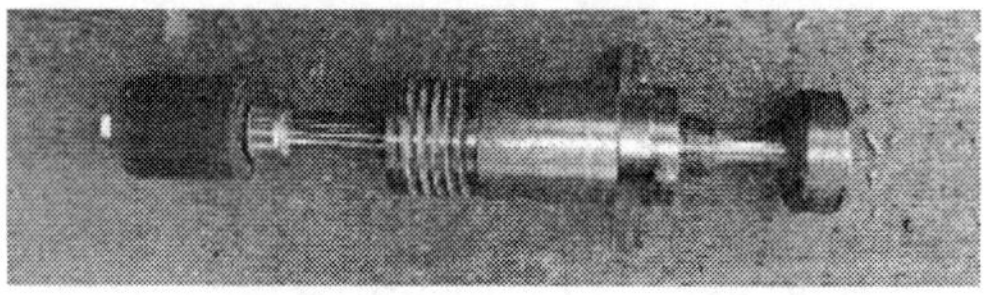

Figure 3. Special drilling device.

Table 1. Mixture proportion of concrete and properties of reinforcing materials.

w/cm	Water	Cement	Fine agg.	Coarse agg.	Admixture	Air
0.53	175 kg/m^3	330 kg/m^3	887 kg/m^3	978 kg/m^3	1.16 kg/m^3	4.5 %
Reinforcing bar		Bar diameter: 13 mm; Yield strength: 345 MPa; Young's mod.: 206 GPa				
Prestressing tendon		Bar diameter: 23 mm; Yield strength: 1080 MPa; Young's mod.: 202 GPa				

3 EXPERIMENTAL PROGRAM

3.1 Materials

Table 1 summarizes a mixture proportion of concrete and properties of strengthening materials. The specified strengths of concrete were 35.9 MPa (Comp.), 2.6 MPa (Split. tens.) and 34.5 GPa (Young's mod.). In addition, the mechanical properties of filling mortar were 90.2 MPa (Comp.), 4.4 MPa (Split. tens.) and 39.8 GPa (Young's mod.). The detail of used materials is described in the previous report (Sakaki *et al.* 2015).

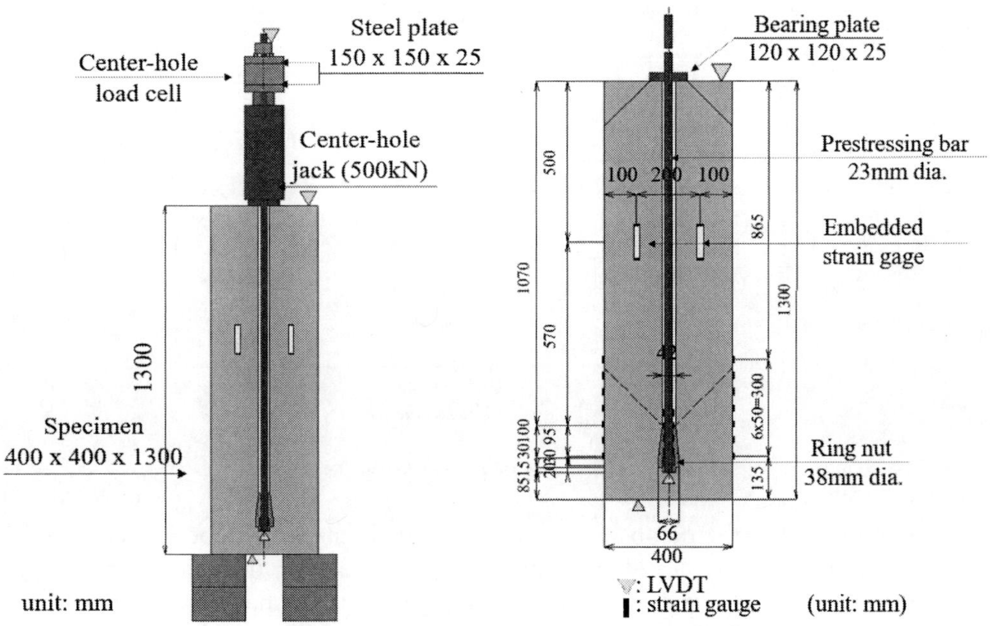

Figure 4. Schematic of test specimen.

Figure 5. Test specimen covered with drying prevention and thermal insulating sheets.

3.2 Test Specimens

Figure 4 presents the schematic of specimen for the sustained loading test. The concrete specimen was full-size concrete for actual strengthening work, and was same to concrete specimens for static loading test (Sakaki *et al.* 2015). The initial load of 314.1 kN was applied to the test specimen at the concrete age of 92 days. The day after tensioning, the hole in concrete was filled with cement grout to ensure a solid connection between the prestressing bar and the concrete. The applied load was determined as 70 % of the ultimate tensile force (448.7 kN) of prestressing tendon. It should be noted that the maximum loads (489.6 kN, 490.9 kN, 490.5 kN) observed in the static test achieved the ultimate force (Sakaki *et al.* 2015).

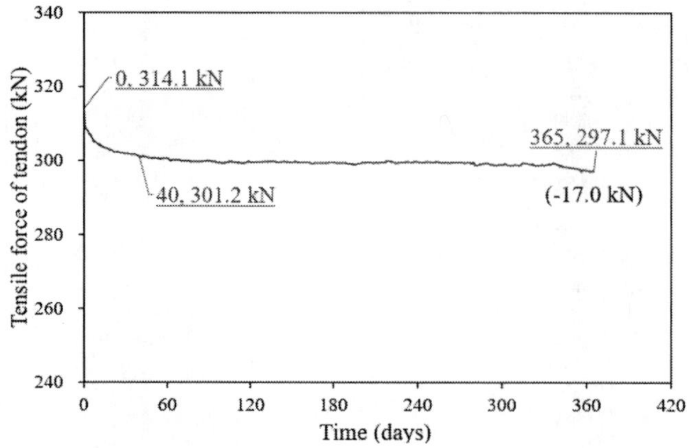

Figure 6. Loss of prestressing force.

This study uses the superposition principle to estimate time-dependent strain due to the sustained load. Hence, a full-size concrete block made at the same time was placed in the laboratory to determine concrete shrinkage and thermal strain, as well as the loaded specimen. Both specimens were covered with drying prevention and thermal insulating sheets to prevent volume change of concrete. Figure 5 demonstrates test specimens placed in the laboratory. The load of prestressing bar and deformation were measured by using a data acquisition system up to 1 year-loading.

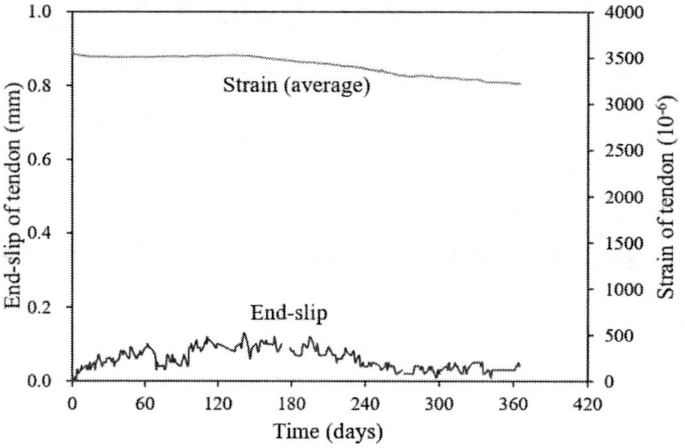

Figure 7. End-slip and strain of prestressing tendon.

4 TEST RESULTS AND DISCUSSION

Figure 6 illustrates the tensile force of prestressing tendon in the sustained loading test. The tensile force gradually decreased with time, but stabilized in around 300 kN after 40 days approximately. The loss of prestressing force was -17.0 kN (5.4 %) during 365 days. The observation implies that the loss of prestressing force is almost equal to or lower than the loss in conventional prestressed concrete structures. End-slip and averaged strain of prestressing tendon were presented in Fig.7. The end slip hardly increased while the strain gradually decreased due to shrinkage and creep of concrete. The test results confirm the prestressing bar was anchored firmly in the internal wedge hole filled with mortar even under the sustained loading. Further research should be conducted to examine durability of the strengthened member, such as fatigue.

5 CONCLUSIONS

This investigation focused on the long-term behavior of the prestressing tendon anchored into the wedge hole embedded in full-size concrete specimen. A 1-year loading test was conducted to measure loss of prestress and deformation of concrete. The conclusions of this experimental investigation are as follows:

- The prestressing force decreased from 314.1 kN to 297.1 kN in the 1-year loading test. The loss of prestressing force (5.4 %) was equal to or lower than the loss in conventional prestressed concrete structures.
- The test confirms that the end slip hardly increased with time and the maximum slip was negligible for practical application.

Acknowledgments

The authors thank to Dr. Kawakane, Mr. Mihara (Kyokuto Kowa, Co.), and Mr. Harada (Chuden Engineering Consultants Co.) for their assistance.

References

Kim, Y. J., Green, M. F. and Wight, R. G., Effect of Prestress Levels in Prestressed CFRP Laminates for Strengthening Prestressed Concrete Beams: A Numerical Parametric Study, *PCI Journal*, Prestressed Concrete Institute (PCI), 55(2), 96-108, 2010.

Marti-Vargas, J. R., Serna, P., Hale, W. M., Strand Bond Performance in Prestressed Concrete Accounting for Bond Slip, *Engineering Structures*, Elsevier, 51, 236-244, June 2013.

Michels, J., Sena-Cruz, J., Czaderski, C., and Motavalli, M., Structural Strengthening with Prestressed CFRP Strips with Gradient Anchorage, *Journal of Composite for Construction*, ASCE, 17(5), 651-661, 2013.

Mimoto, T., Yoshitake, I., Harada, M., and Sakaki, T., An Internal Prestressing System to Strengthen Existing Concrete Members, *III Convegno di Ingegneria Forense, VI Convegno su Crolli, Affidabilità Strutturale, Consolidamento (IF CRASC'15)*, Bontempi, F. (Ed.), 515-523, Rome, Italy, 2015.

Sakaki, T., Mimoto, T., and Yoshitake, I., Experimental Investigation of an Internal Prestressing System to Strengthen Concrete, *The 2015 International Conference on Innovative Structural Engineering and Mechanics (ISEM15)*, Choi, CK. and Lee, P-S. (Eds.), Incheon, Korea, 2015.

Yilmaz, S., Ozen, M., Yadim, Y., Tensile Behavior of Post-Installed Chemical Anchors Embedded to Low Strength Concrete, *Construction and Building Materials*, Elsevier, 47, 861-866, October 2013.

Interaction between Theory and Practice in Civil Engineering and Construction
Edited by Komurlu, R., Gurgun, A. P., Singh, A., and Yazdani, S.

THE EFFECT OF REINFORCING ARRANGEMENT ON THE BEHAVIOR OF REINFORCED CONCRETE FLANGED DEEP BEAMS WITH WEB OPENINGS

ABBAS HILO ALI

Iraqi Ministry of Water Resource, Baghdad, Iraq

The objective of the work is to investigate the structural *behavior* of reinforced concrete flanged deep beams with web openings under repeated loading through experimental tests which were carried out on simply supported T- beams , subject to two point loads. This work deals with the results of experimental work with three test specimens with square opening 25% of web depth at mid high through critical shear path. The effect of steel reinforcement arrangement for web opening is investigated. The experimental tests of sections under repeated loading showed that reinforcement of web opening increased the cracking and ultimate loads. It is observed that reinforcement for opening reduced the deflection and the crack width. The inclined reinforcement results are better than orthogonal.

Keywords: Reinforced concrete, Web reinforcement, Repeated loads, Flange.

1 INTRODUCTION

Reinforced concrete deep beams are widely used in many structural engineering applications, such as, transfer girders piles cap, offshore structures (caisson), shear wall, wall footings, floor diaphragms, and complex foundation system. When the span to depth ratio of simply supported beam is less than 2, or less than 2.5 for any span of a continuous beam, then these beams are considered as deep beams (Park and Paulay 1975).The ACI Building Code 318-08 Deep beams are members loaded on one face and supported on the opposite face so that compression struts can develop between the loads and the supports, and have either:

- Clear spans, l_n, equal to or less than four times the overall member depth; or
- Regions with concentrated loads within twice the member depth from the face of the support.

Deep Beams are structural element loaded as beams but having a large depth/thickness ratio and a shear span/depth ration not exceeding 2 for concentrated load and 4 for distributed load, where the shear span is the clear span of the beam for distributed load. In many instances reinforced concrete slab together with the beams behave as a T section, stronger compression zone provided by flanges in T section generally precludes shear compression failure in T beams and as a result has shear strength higher than the shear strength of rectangular sections (Hawy 2005).

1.1 Behavior of Deep Beams

Deep beams differ from the common flexural members in that they behave as two-dimensional rather than one-dimensional members and are subjected to a two-dimensional state of stress. This difference in *behavior* is mainly attributed to the significant effects of vertical normal stresses and shear deformations in these members (Kani 1969). In deep beams the stress distribution with depth is nonlinear even at the elastic stage (Subedi 1986). The strength of deep beams is usually controlled by shear, rather than flexure, provided that normal amount of longitudinal reinforcement are adequately used. The main parameters governing the deep beam *behavior* are the span/depth ratio, shear span/depth ratio, slenderness (depth/thickness) ratio, main and web reinforcement, and the loading and supporting condition (ACI 2008).

1.2 Effects of Web Opening

Maximum crack width at failure will be greater when the opening center is located at the center of the shear zone than at any other position. So location of the opening center at this point is undoubtedly the maximum damaging situation in the web region. The strength of the beam increases when the opening is located away from what can be called the loaded quadrant to the unloaded quadrant and vice-versa. The location of the web opening is therefore a major factor influencing the strength of the beam. It should be remembered that the deflection in deep beams are substantially influenced by shear and, as such, location of the opening in the region of high shear and intercepting the critical path is understandable. The openings should invariably be provided with some loop reinforcement in their periphery to avoid possible stress concentration. The web reinforcement controls crack widths and deflection. Of all types of web reinforcement, the inclined type placed perpendicular to the plane of rupture (critical diagonal crack) has been found to be the most effective arrangement. The next practical and effective type is the horizontal web steel which with nominal vertical web steel may further increase the effectiveness of the beam and so its strength. It was observed that in beams with web openings, horizontal web reinforcement distributed equally on either side of the opening location showed better results (Kong 2002).

2 EXPERIMENTAL TEST PROGRAM

2.1 Materials

All materials used in experimental investigations are available in local market, which include cement, natural gravel, natural silica sand, water, and deformed reinforcement bars. All materials are tested according to Iraqi specifications.

2.2 Concrete Mix

The quantities of cement, fine aggregate and course aggregate based on dry weights were as follows: (Cement 1:Sand 1.5:Gravel 3) Water/ cement ration (W/C ratio= 0.55) The mix was designed to give a compressive strength of about 30 N/mm2 at age of 28 days and slump of about 80 mm. The same concrete mix was used throughout the whole investigation. Tilting drum mixer was used; mixing was carried out according to BS 1881.

2.3 Testing the Properties of Concrete

During the casting of each group of beams, three (150 x150x150) mm cubes and three (100 x 200) mm cylinders were made. The cube compressive and splitting tensile strength were obtained by testing three cubes and three cylinders according to BS 1881.

2.4 Description of Tested Deep Beams

TB#1 is designed according to CIRIA Guide 2 (1977), which is summarized in Table 1, gives 4-ø10 mm for main reinforcement, ø6 at 150 mm for vertical web reinforcement and ø6 at 200 mm for horizontal web reinforcement. All beams have length of 1300 mm , depth of 400 mm, flange width of 300 mm, flange thickness of 60 mm , web thickness of 120 mm and the distance between the supports center / center are 1000 mm, with web square opening (100 x 100) mm (Figure 1). In this study, Two type of reinforcing arrangement for web opening used in this study, orthogonal for TB#2 and inclined for TB#3, by using same bar diameter all changes above were with same main reinforcement at TB#1.

Table 1. Details of reinforcement for specimens.

I	specimen Symbol	Main reinforcing	Web reinforcing Vertical	Web reinforcing Horizontal	Flange reinforcing	Opening reinforcing	Loading type
1	TB#1	4 ø 10 mm	ø 6@150 mm	ø 6@200 mm	ø 6@200 mm	None	repeated
2	TB#2	4 ø 10 mm	ø 6@150 mm	ø 6@200 mm	ø 6@200 mm	ø 6 Orthogonally	repeated
3	TB#3	4 ø 10 mm	ø 6@150 mm	ø 6@200 mm	ø 6@200 mm	ø 6 Inclined	repeated

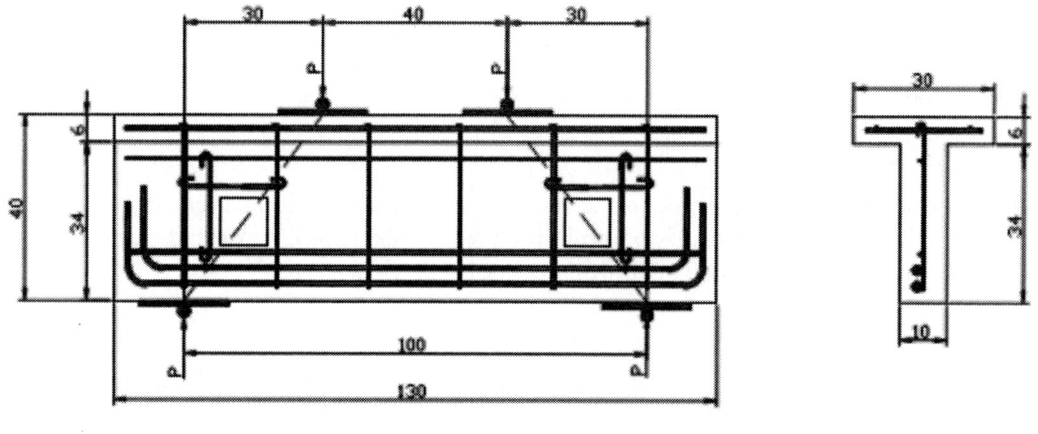

Figure 1. Description of tested specimens.

2.5 Fabrication and Casting of Beams

Timber forms with plywood face were used in casing beams, the interior face of forms were coated with oil prior casting and before the reinforcement cage was placed in position, plastic spacers were used to maintain the concrete covers, the concrete mix was then placed in layers and each layer compacted by means of Poker vibrator, the top surface of the beam was finished level by a hand trowel. After the mold is stripped, the beams were cured for seven day after casting by using damp canvas.

2.6 Testing Equipment and Instrumentations

Below is a list of equipment and instrumentations used during experimental work:

- Machine of load applying: Torsee Universal Testing Machine with a capacity of 2000 kN was used to apply the load.
- Electrical strain gauges: TML strain gauges.
- Data Acquisition System: The data acquisition system used in this study includes personal laptop computer, strain indicator which has 16 channels named Data Taker (data logger) DT85 – Australian made 2010.
- Measurement of deflection: The mid-span deflection of each beam was measured by using dial gauge with magnetic base. The accuracy of the dial gauge was 0.01 mm.
- Micro-Crack reader: Width of concrete cracks was monitored using hand microscope of accuracy 0.02 mm per division.

2.7 Test Procedure

- **Loading for first crack**: Loading initiated from zero with slow rate about 1 kN and recorded the data readings for strains and deflection until detected the first crack,. Then loading was stopped to measure the values of crack width by crack detector device and the load value. If that crack was diagonal crack then item 2 below performing, if not the loading continue until detecting the first diagonal crack.
- **Loading repeating**: When detecting the first diagonal crack, some cycles of loading and unloading performed during this stage for minimum five times, the loading and unloading performed in gradual increment and decrement in a suitable rate. During these individual steps, deflections, strains, crack width readings were recorded. If specimen did not, the loading increased by (25 kN), some cycles were performed about five times. If specimen did not, the loading increased by (25 kN), some cycles were performed for about five times. If specimen did not failed, the loading increased by (25 kN), some cycles were performed of about five times. If specimen did not failed, the loading increased by (25 kN), some cycles performed about five times, If specimen did not failed in this step the test continued to failure.

- **Loading to failure**: When the loading was increased to next loading step some specimens were fail and a test terminated when the total load applied on the specimens was started to drop off.

3 TEST RESULTS AND DISCUSSION

The test results can be seen on Figure 2 and Table 2.

3.1 Effect on Cracking and Ultimate Load

It is observed that cracking load for beam with web opening reinforcement arranged orthogonally TB#2 and arranged inclined TB#3 was increased by about 50% than without open reinforcement TB#1, but it was equal for TB#2 and TB#3.Ultimate load was increased by 14% for TB#2 and 48% for TB#3 than ultimate load for TB#1, and the ultimate load for TB#3 was more than for TB#2 by 26%.

Table 2. Experimental cracking load and ultimate loads for the tested specimens.

Specimen symbol	Cracking Load V_{cr} (kN)	Ultimate Load V_u (kN)	V_{cr} / V_u
TB#1	20	74	0.27
TB#2	30	87.5	0.34
TB#3	30	110	0.27

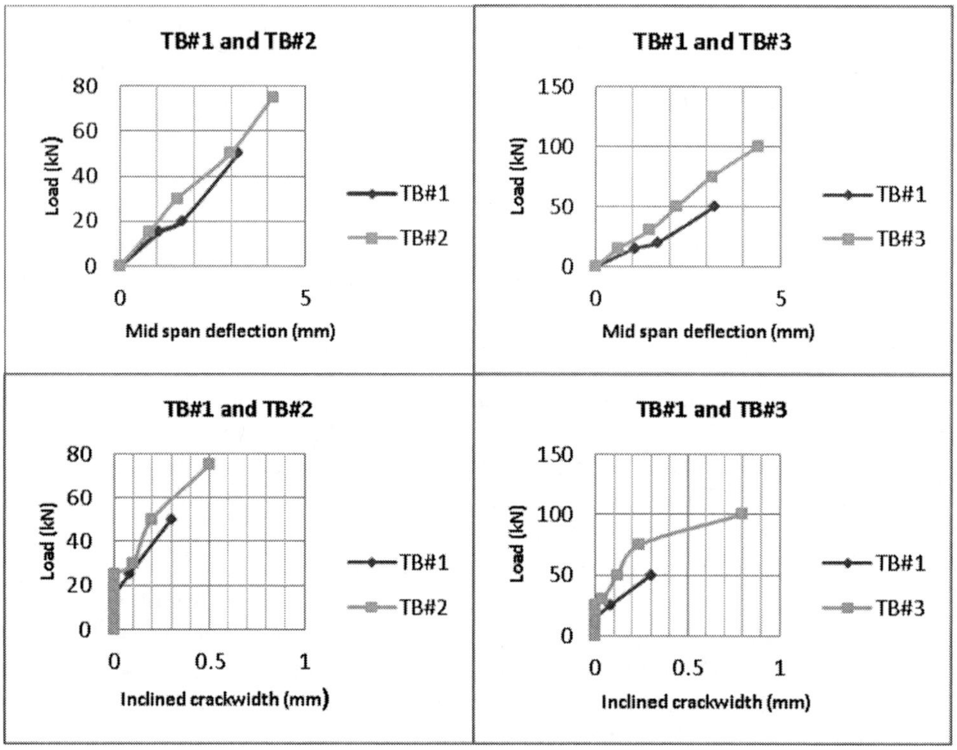

Figure 2. Charts of test results.

3.2 Effect on deflection

It is observed that reinforcing for opening was reduced the deflection, and the inclined reinforcing was gave better result than orthogonally web opening reinforcing under repeated loading.

3.3 Effect on Crack Width

It is observed that reinforcing for opening was reduced the crack width, and the inclined reinforcing gave better result than orthogonally reinforcing under repeated load.

3.4 Web opening Reinforcement Strain

Two electrical strain gauges were installed on web opening reinforcement bars for specimen TB#2, one on horizontal web opening reinforcement under web opening point number 5 and one on vertical web opening reinforcement near web opening point number 6, it is observed that the stain at these point was very close until inclined cracking, and after inclined cracking the values were varied, in stage of loading repeated on cracking load stage, the strain in vertical web reinforcement was higher, but when loading reached about 70% of cracking load stage the strain in horizontal web reinforcement was increased more than vertical until load reached to ultimate loading the strain in vertical web reinforcement return to be higher than in horizontal bars. One electrical strain gauge was installed on inclined web opening reinforcement point number 4 in specimen TB#3 top of opening it is observed that the strain at point 4 was increased more than strain in points (5 and 6) in TB#2 after inclined cracking.

4 CONCLUSIONS

The experimental tests of sections under repeated loading showed that reinforcement of web opening increased the cracking and ultimate loads and reduced the deflection and the crack width. The inclined reinforcement results are better than orthogonal.

References

ACI 318-08, *Building Code Requirements for Structural Concrete (318-08) and Commentary (318-08)*, American Concrete Institute, Detroit, 2008.

Construction Industry Research and Information Association (CIRIA), CIRIA Guide 2, London, 1977.

Park, R. and Paulay, T., *Reinforced Concrete Structures*, John Wily and sons, 1975.

Kong, F. K., *Reinforced Concrete Deep Beams*, Taylor and Francis Inc., 2002.

Kani, G. N. J., *A Rational Theory for the Function of Web Reinforcement*, ACI Journal, Vol.66, No.3, Mar., 1969.

Subedi, N. K., et al., *Reinforced Concrete Deep Beams _ Some Test Results*, Magazine of Concrete Research, Vol. 38, No. 137, Dec. 1986.

Interaction between Theory and Practice in Civil Engineering and Construction
Edited by Komurlu, R., Gurgun, A. P., Singh, A., and Yazdani, S.

ISBN: 978-0-9960437-2-4

SHEAR BEHAVIOR OF REINFORCED CONCRETE SLENDER BEAMS

MUHAMMAD K. KAYANI, WASIM KHALIQ, and MUHAMMAD K. SHEHZAD

National University of Sciences and Technology, Islamabad, Pakistan

Major factors contributing to the shear behavior in reinforced concrete (RC) beams have been identified as compressive strength of concrete, shear span to effective depth ratio, and longitudinal reinforcement. Though significant, few of these factors are not fully incorporated in ACI code provisions for design of minimum shear reinforcement. To investigate the effect of these parameters, an analytical and experimental study was undertaken on the shear behavior of ordinary strength RC slender beams with moderate longitudinal reinforcement. The experimental program consisted of testing of eight simply supported RC slender beams subjected to two concentrated loads at a shear span to depth ratio (*a/d*) of 2.5 and equipped with varying shear reinforcement according to four different criteria. Ultimate shear strengths obtained in this experimental program are compared to the analytical shear strengths calculated according to existing as well as proposed equations. Test results exhibit that, the modified equation proposed in this work gives more accurate prediction of shear capacity of RC beams.

Keywords: Minimum shear reinforcement, Shear strength prediction, Shear span, Analytical equation.

1 INTRODUCTION

Reinforced concrete (RC) flexural members are, almost in all cases, primarily designed for flexure to decide on the cross section and longitudinal reinforcement. However, in the absence of adequate shear reinforcement, members may fail in shear before attaining their flexural capacity. Theories and relationships for predicting shear behavior have been extensively studied; however, consensus over a single theory to predict the response of members under shear does not exist. Although ACI code provisions on shear have been developed on the basis of years of research, yet all significant factors have not been fully incorporated. Major parameters influencing the shear strength of beams have been identified by many researchers as; shear span to depth ratio (a/d), depth of members and shape of cross section, presence of axial force, type and ratio of longitudinal reinforcement and concrete compressive strength. However, ACI (ACI Committee 318-11 2011) equations for shear capacity and minimum shear reinforcement do not take into account all of these factors.

1.1 Zararis Theory of Critical Diagonal Cracks

Zararis (2003) and Zararis and Papadakis (2001) carried out analytical research on shear behavior of RC slender beams with and without shear reinforcement and evolved a theory which describes the diagonal shear failure in slender beams. In this theory, reason for shear failure has been identified as the loss of sheer force of main tension reinforcement, which occurs due to a horizontal splitting of concrete cover along the reinforcement (Zararis 2003). Failure in slender beams without shear reinforcement, loaded under two or single point loading, occurs due to critical diagonal crack which is composed of two distinct branches (Zararis 2003). First branch is an inclined shear crack having height almost similar to flexural cracks. The second branch initiates from the tip of first branch and propagates towards the load point crossing the compression zone, with its line meeting the support point. Figure 1 illustrates the forces acting in the cracked portion of the beam:

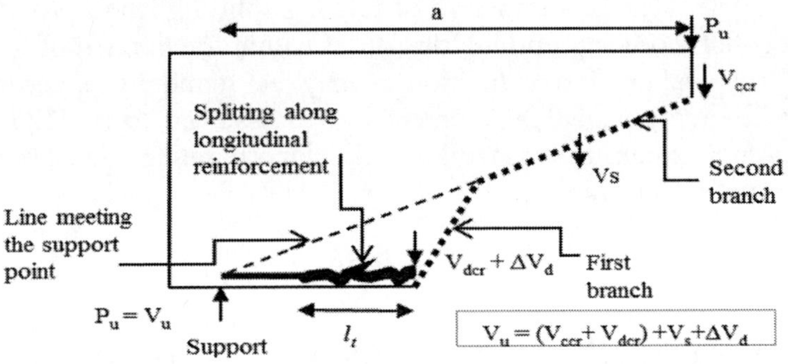

Figure 1. Distribution of forces in beam (based on Zararis theory).

By considering all these forces, ultimate shear force at the failure of a beam with stirrups can be given as under:

$$V_u = V_{cr} + V_s + \Delta V_d \tag{1}$$

The force V_d (in case of beams without stirrups) and ΔV_d (in case of beams with stirrups) are responsible for horizontal splitting between concrete and steel along the main reinforcement. Preventing this splitting can be helpful in avoiding the shear failure. It has been assumed that splitting is caused when the tensile stresses developed along the reinforcement in a distance l_t from the point of initiation of critical crack exceeds the tensile strength, f_t of concrete. It is believed (Zararis 2003, Zararis and Papadakis 2001) that splitting length, l_t has a constant value in any case which is about 0.5d. The equation proposed by Zararis (2003) for ultimate shear capacity (V_u) of beams with stirrups is as under:

$$V_u = [(1.2 - 0.2(\frac{a}{d})d)(\frac{c}{d})f_{ct} + (0.5 + 0.25(\frac{a}{d}))\rho_v f_{yv}]bd \tag{2}$$

Minimum shear reinforcement should be calculated in such a manner that it satisfies the following relationship:

$$\frac{\rho}{\rho_v} \leq 1.75(\frac{a}{d}) \tag{3}$$

When the condition mentioned in Equation 3 is not met, the shear failure is likely to be accompanied by quick and extensive splitting along longitudinal reinforcement and significant widening of critical diagonal crack (Zararis 2003).

1.2 Proposed Modification to Zararis Theory

In this experimental study, theory developed by Zararis (2003) has been considered being concise and incorporating all major factors influencing the shear strength of RC beams. However, it is believed that splitting length, l_t is linked with the development length, l_d of the bars. This concept is based on the fact that the factors influencing the development length are similar to those linked with the splitting along main reinforcement. According to ACI Committee 318-11 (2011), the development length is influenced by size, location and number of bars, concrete cover, coating on the reinforcement, confining reinforcement, yield strength of steel and compressive strength of concrete. Therefore, instead of relating splitting length, l_t with the depth of beam only, it is felt more appropriate to relate it with a fraction (α) of the development length, l_d. Exact value of this fraction α may be found by experimental studies, however, for the purpose of this research, it was assumed to be 0.25 and was later confirmed during experimental work. By incorporating the above mentioned assumption, Equation 3 becomes:

$$\frac{\rho}{\rho_v} \leq 0.89(\frac{l_d}{d})(\frac{a}{d}) \tag{4}$$

Similarly, using the value of splitting length $l_t = 0.25 l_d$, the Zararis equation for predicting ultimate shear strength of reinforced concrete slender beams can be modified as under:

$$V_u = [(1.2 - 0.2(\frac{a}{d})d)(\frac{c}{d})f_{ct} + 0.25((\frac{l_d}{d}) + (\frac{a}{d}))\rho_v f_{yv}]bd \tag{5}$$

2 EXPERIMENTAL INVESTIGATION

Eight full scale beams were cast and tested to study the shear behavior of RC beams. Concrete strength (f'_c) was selected as 28 N/mm^2 and cross section was 254 mm x 457 mm which was selected so that minimum shear reinforcement requirement is governed. Details of all the beam specimens are given in Table 1.

Table 1. Details of all the beam specimens.

Beams	f'_c, (N/mm^2)	Longitudinal reinforcement		Shear reinforcement		a/d	d, (mm)
		Rebar No	f_y, (kN/mm^2)	Rebar No	f_{yv}, (kN/mm^2)		
N Series	28	3 # 8	414	-	-	2.5	406
A Series	28	3 # 8	414	# 2 @ 190 mm c/c	276	2.5	406
Z Series	28	3 # 8	414	# 3 @ 152 mm c/c	276	2.5	406
M Series	28	3 # 8	414	# 3 @ 203 mm c/c	276	2.5	406

2.1 Materials

Type I cement conforming to ASTM C 150 – 04 and locally available sand with fineness modulus of sand was calculated to be 2.51. Grade 60 and grade 40 steel was used for longitudinal and transverse reinforcement respectively. Table 2 gives the mix design details used for the beam specimens.

2.2 Test Specimens

Specimens were cast as per ASTM C 31 (ASTM Standard C31 2012). Details of reinforcement for each type of beams are mentioned in Table 1. The specimens were cast in 25 mm thick plywood formwork which was removed from beams after 48 hours.

2.3 Test Setup and Loading

The testing facility of Civil Infrastructure Laboratory at NUST, Islamabad was used. The load was applied through a hydraulic jack having 1200 kN capacity. The supports comprised of 102 mm diameter solid steel bars, making the beam simply supported. A steel girder was used to apply two point loading at predefined shear span of 1016 mm. Three linear variable displacement transducers (LVDTs) were placed under the beams at mid span and at quarter points to measure the deflections. Deflections were also measured and recorded through the structural load analysis and data logging system. The load was applied in increments of 25 kN, deflections recorded and cracks marked at each load increment.

Table 2. Details of concrete mix design

Description	Details
Cement, Type I, ASTM C150	450 kg/m^3
Fine Aggregate	760 kg/m^3
Coarse Aggregate	1040 kg/m^3
W/C ratio	0.41
Superplasticizer (Ultra Superplast 470), ASTM C494	0.9 % by weight of cement
Average compressive strength at 28 days	27.57 N/mm^2
Average compressive strength at test day (90 days)	28.71 N/mm^2

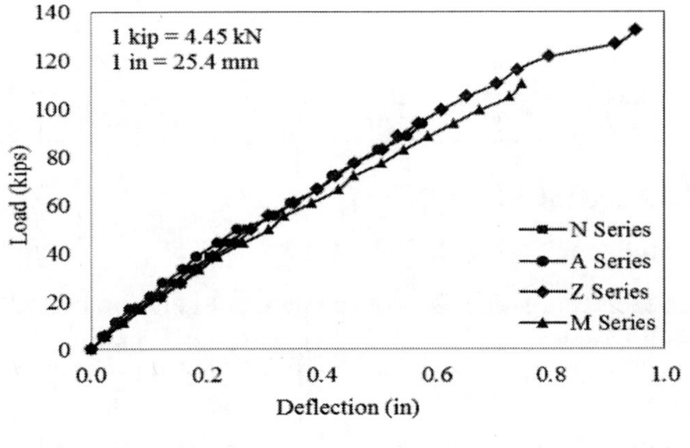

Figure 2. Average load – deflection plot for all beams.

3 EXPERIMENTAL RESULTS

Average load – deflection response of all series of beam specimens is shown in Figure 2. Initial flexural cracks in almost all cases occurred between 75 – 100 kN load. The location of significant diagonal crack was within the shear span. Both beams in N series exhibited sudden failure induced by diagonal cracking at 250 kN. In A-series beams A-1 and A-2, failure occurred at 414 kN and 424 kN respectively. The inclined cracks which lead to failure in Z-series beams appeared at relatively higher loads as compared to the other beams. Failure load was 589 kN and 586 kN for Z-1 and Z-2, respectively. Two distinct branches of the critical diagonal cracks were not witnessed, as was postulated by (Zararis 2003). Beams in M series also behaved much like those in Z-series; however, they failed at lesser loads as their failure occurred at 493 kN and 482 kN for M-1 and M-2, respectively. Since the longitudinal reinforcement and the a/d were kept constant, an increase in failure load and deflections was seen with the increase in amount of shear reinforcement.

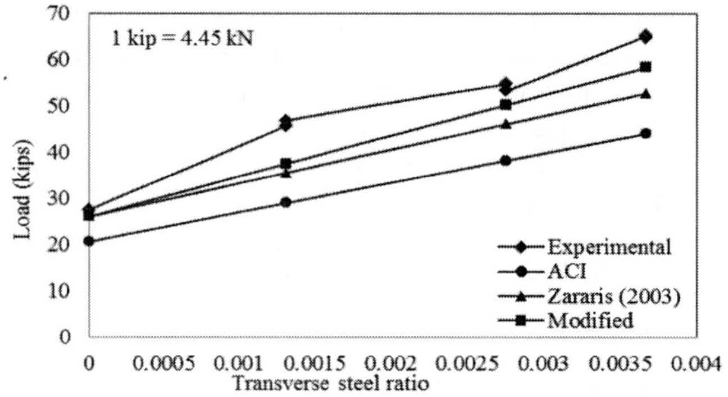

Figure 3. Comparison of shear strength predictions.

Table 3. Variation between experimental and ACI predicted shear strengths.

Beams	Experimental Strength, kN			ACI Predicted Strength, kN			$\gamma = V_u / f'_c$
	V_c	V_s	V_u	V_c	V_s	V_u	
N Series	123	0	123	92	0	92	2.67
A Series	169	37	206	92	37	129	3.67
Z Series	184	105	289	92	105	197	4.00
M Series	162	78	240	92	78	170	3.52

4 ANALYSIS AND DISCUSSION

4.1 Discussion on Shear Strength Calculation

Shear strength for each series of beams was calculated and predicted according to ACI Equation 11.2 (ACI Committee 318-11 2011), Zararis Equation 2 (Zararis 2003) and the modified form of Zararis Equation 5. Experimentally obtained ultimate shear strengths were compared to the theoretical predictions as illustrated in Figure 3. This shows that ACI Equation gives quite conservative results. Shear strength calculation by Zararis equation was claimed to be more accurate (Zararis 2003). However, present experimental study indicates that the modified form of Zararis equation is more appropriate in predicting the ultimate shear capacity of beams. Variation between experimental values and those predicted by ACI code Equation 11.2 is given in Table 3.

4.2 Proposed Modification to Zararis Equations

Equations developed by Zararis were modified to take into account the development length and value of α was assumed to be 0.25. This provided a value of 0.00275 for ρ_v and accordingly the M-Series beams were equipped with this amount of transverse steel. Experimental results revealed a value of α to be 0.21. By incorporating this value, Equations 6 and 7 represent modified Zararis equations as under:

$$\frac{\rho}{\rho_v} \leq 0.75(\frac{l_d}{d})(\frac{a}{d}) \tag{6}$$

and,

$$V_u = [(1.2 - 0.2(\frac{a}{d})d)(\frac{c}{d})f_{ct} + 0.21((\frac{l_d}{d}) + 0.25(\frac{a}{d}))\rho_v f_{yv}]bd \tag{7}$$

5 CONCLUSIONS

Based on the results of this experimental study, following conclusions are drawn:

1. ACI equation for prediction of shear strength in reinforced concrete beams is more conservative, while Zararis equation appears to be better in predicting the ultimate shear capacity. The proposed modified Zararis equation gives a better shear strength prediction in RC slender beams.
2. Initiation and propagation of shear cracks in general and critical diagonal cracks in particular do not follow the pattern of two distinct branches as was claimed by Zararis. These cracks follow approximately the same orientation on which they initially appeared.
3. Failure in all test beams provided with the shear reinforcement occurred due to excessive diagonal crack widths. Failure due to splitting along longitudinal reinforcement, as assumed by Zararis, was not witnessed.

References

ACI Committee 318-11 (2011). Building code requirements for structural concrete and commentary, American Concrete Institute, Farmington Hills, MI.

ASTM Standard C31 (2012). "Standard Practice for Making and Curing Concrete Test Specimens in the Field." ASTM International, West Conshohocken, PA.

Zararis, P. D. (2003). "Shear strength and minimum shear reinforcement of reinforced concrete slender beams." ACI Structural Journal, 100(2), 203-214.

Zararis, P. D., and Papadakis, G. C. (2001). "Diagonal shear failure and size effect in RC beams without web reinforcement." Journal of Structural Engineering, 127(7), 733-742.

Interaction between Theory and Practice in Civil Engineering and Construction
Edited by Komurlu, R., Gurgun, A. P., Singh, A., and Yazdani, S.

TESTS AND ANALYSES OF RC SHEAR WALLS WITH VERTICALLY ALIGNED OPENINGS

YASUSHI SANADA[1], HONG LIU[1], SUGURU SUZUKI[1], MOMOKO IIZUKA[2], SUSUMU TAKAHASHI[3], and TOSHIKATSU ICHINOSE[3]

[1]*Graduate School of Engineering, Osaka University, Osaka, Japan*
[2]*Graduate School of Engineering, Toyohashi University of Technology, Aichi, Japan*
[3]*Graduate School of Engineering, Nagoya Institute of Technology, Aichi, Japan*

This study investigates fundamental behavior of reinforced concrete (RC) shear walls with vertically aligned openings through laboratory tests and numerical analyses. A series of structural tests was conducted using scaled models representing typical multi-story RC shear walls in Japan. Three 1/6 scale shear wall specimens were designed and manufactured with different opening configurations: W0 with no opening, W1 with single-aligned openings, and W2 with double-aligned openings. The specimens were subjected to static cyclic lateral loads. The W1 and W2 specimens with openings failed with concrete crushing at the wall bottom and shear failure of the beams above openings, while the W0 failed in shear. The maximum strengths were 318 kN, 145 kN, and 93 kN for the W0, W1, and W2 specimens, respectively; hence, the experimental strength reduction factors due to the openings were 0.46 and 0.29 for the W1 and W2. These results were simulated by FEM analyses mainly to investigate detailed behavior of the components of the W1 and W2 specimens. The analytical results, in particular the maximum strengths and failure patterns, agreed well with the experimental results. It was numerically verified that high compression was applied to the wall bottom with the tensile boundary column and the beams above openings, as observed during the tests, which should be considered in practical design for this type of wall.

Keywords: Cyclic loading test, FEM analysis, Reinforced concrete, Seismic performance, Strength reduction factor.

1 INTRODUCTION

Reinforced concrete shear walls are effective building components to provide high seismic resistance; therefore, the structural performance has been investigated by a number of researchers in the world. Openings are occasionally planned in the shear walls for lighting, ventilating, and doorway. Although the openings significantly affect the structural performance, limited investigations have been reported in the previous studies. This may be attributed to structural design concepts recommended by major design codes/standards, in which this type of wall is often modeled by each component.

On the other hand, in Japan, the structural performance of RC shear walls with openings has been directly evaluated considering several performance reduction factors. Recently, a new strength reduction factor has been implemented in AIJ (Architectural Institute of Japan) design standards, which was presented to evaluate strength reductions due to vertically aligned openings (AIJ 2010). However, this factor has not been verified experimentally, because of a lack of test data from reliable experimental simulations on fundamental behavior of this type of wall.

Therefore, the objective of this study is to obtain fundamental data on structural behavior/performance of RC shear walls with vertically aligned openings.

2 EXPERIMENTAL PROGRAM AND RESULTS

2.1 Specimens

Three 1/6 scale shear wall specimens were designed and manufactured with different opening configurations: W0 with no opening, W1 with single-aligned openings, and W2 with double-aligned openings. Shear walls with aligned openings along height behave in a similar manner to coupled shear walls. Therefore, separated partial stubs were applied to the specimens instead of a typical massive stub which fully covers the top of specimens. Figure 1 shows the structural details of the specimens. Tables 1 and 2 give the material properties of concrete and reinforcements used for the specimens.

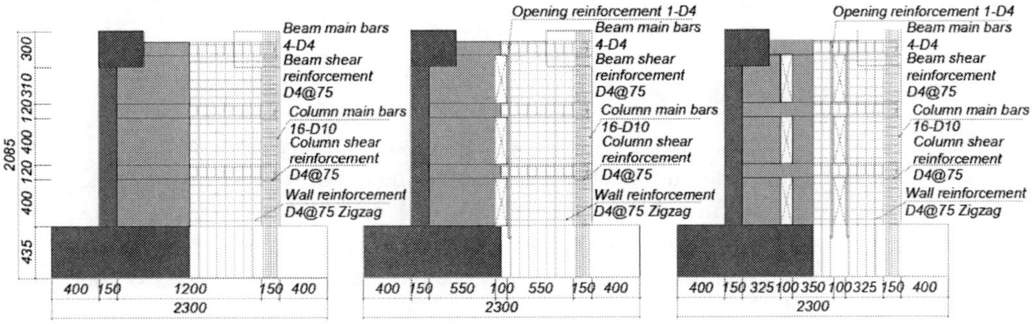

Figure 1. Structural drawing of specimens (W0, W1, and W2 from the left).

Table 1. Material properties of concrete.

Specimen	Compressive strength (N/mm^2)	Elastic modulus (kN/mm^2)	Strain at compressive strength (μ)
W0	25.8	25.6	1838
W1	26.5	26.3	1993
W2	26.6	25.7	1920

Table 2. Material properties of reinforcements.

Bar number	Yield stress (N/mm^2)	Elastic modulus (kN/mm^2)	Tensile strength (N/mm^2)
D10 (SD345)	379	174	479
D4 (SD295A)	323	153	458
ϕ17	1176	201	1273

2.2 Experimental Methods

The specimens were subjected to static cyclic lateral loads by two horizontal jacks connected to the separated stubs, as shown in Figure 2. Tension rods were provided between the stubs to prevent tensile failure at the stub-beam/wall boundaries. Axial loads were not applied to the specimens, while they were designed with additional longitudinal rebars in the boundary columns so that realistic compressive forces would act on the cross-sections under lateral loads.

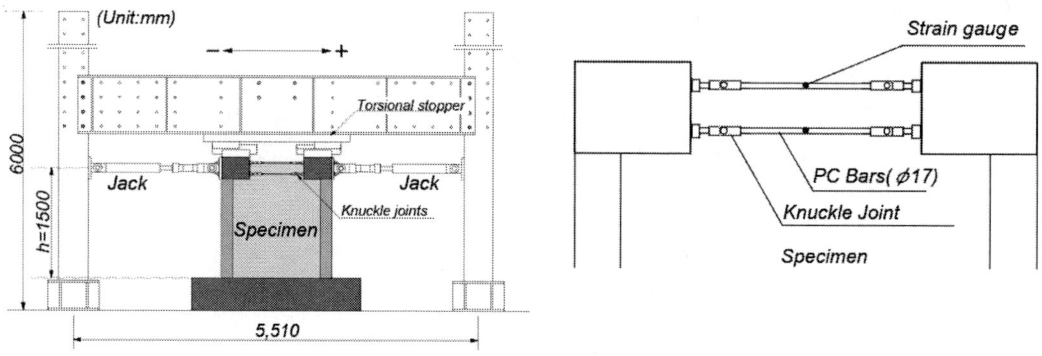

Figure 2. Experimental setup.

2.3 Experimental Results

The W1 and W2 specimens with openings failed with concrete crushing at the wall bottom and shear failure of the beams above openings, while W0 failed in shear, as shown in Figure 3. The failure mechanisms of W1 and W2 were similar to that of coupled shear walls. As a result, the seismic performance, in particular the maximum strengths of the specimens decreased due to the existence/higher number of openings: 318 kN, 145 kN, and 93 kN for the W0, W1, and W2, respectively, as shown in Figure 4; hence, the experimental strength reduction factors due to the openings were 0.46 and 0.29 for the W1 and W2.

ACI code describes that the seismic performance of shear walls with openings should be evaluated based on modeling of components assembling the walls (ACI Committee 318). FEM analyses were performed in the following to clarify key modeling of this type of wall for design analyses.

3 ANALYTICAL SIMULATIONS

3.1 Modeling of Specimens

Two-dimensional modeling was adopted for FEM analyses. Figure 5 illustrates the analytical modeling for W2. Concrete was replaced by 4-node quadratic isoparametric elements with a stress-strain relationship, as shown in Figure 6. Reinforcements were embedded in the concrete elements. A bilinear model was applied to the stress-strain relationship for the reinforcements. The material properties provided in Tables 1 and 2 were used for the stress-strain relationships.

As for the stress-strain relationship of concrete, Modified Ahmad model (Naganuma 1995) was adopted for the compressive stress-strain curve, as shown in Figure 6(a), while the Nakamura and Higai (1999) model was adopted in its stress-softening regions. Kupfer and Gerstle (1973) criterion was applied to evaluate failure under biaxial compression. Stiffness reduction due to cyclic stress was considered, as shown in Figure 6(b) (Naganuma and Ohkubo 2000). Furthermore, Al-Mahaidi (1979) model was adopted for evaluating shear transfer after cracking, as shown in Figure 6(c).

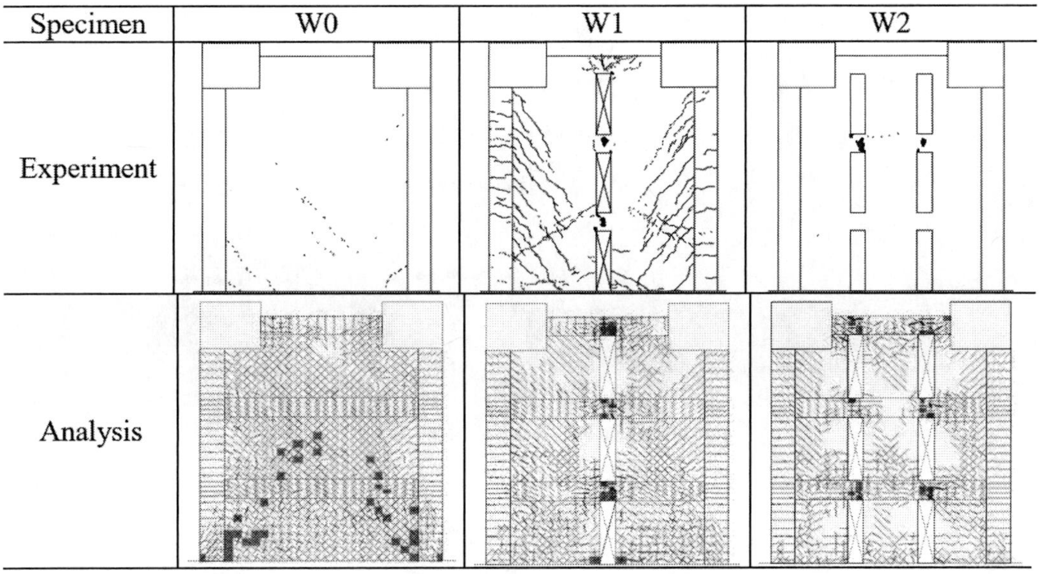

Figure 3. Experimental and analytical damage at 0.5% rad.

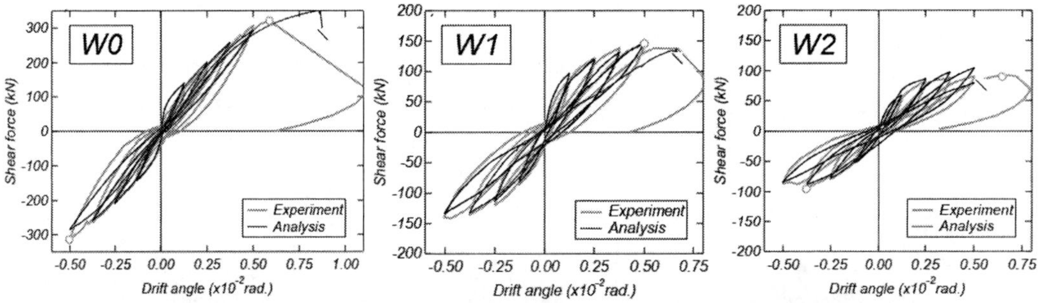

Figure 4. Experimental and analytical shear force-drift angle relationships.

3.2 Analytical Results

Figures 4 compares the shear force vs. drift angle relationships for all specimens with the experimental results, which show good agreements up to a drift angle of 0.5% rad. Figure 3 also compares the analytical damage at 0.5% rad. with the experiments, where gray color indicates elements with concrete beyond strain at the compressive strength. Focusing on the W1 and W2, the damage was significant at the boundary beams above openings for both specimens and the wall bottom with the boundary column on the tensile side for the W1, which means that these are key elements for macro modeling to evaluate inelastic behavior of this type of wall. Figure 7 additionally shows the minimum principle stress contours for all specimens at 0.5% rad. It was found for the specimens with openings that high compression was applied to the wall bottom with the tensile column and the boundary beams. Such higher stresses were consistent with the damage patterns provided in Figure 3.

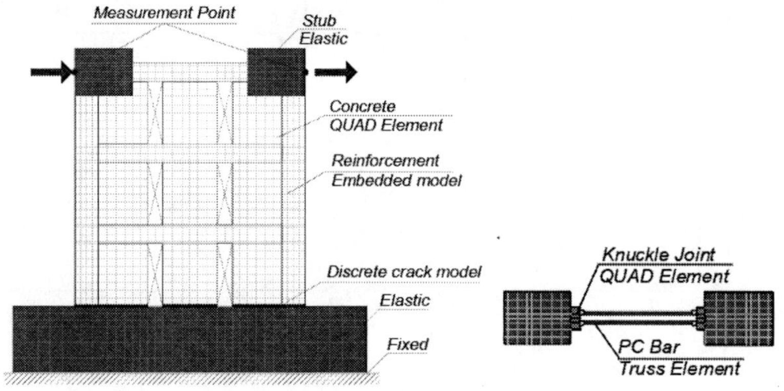

Figure 5. Modeling of W2 for FEM analysis.

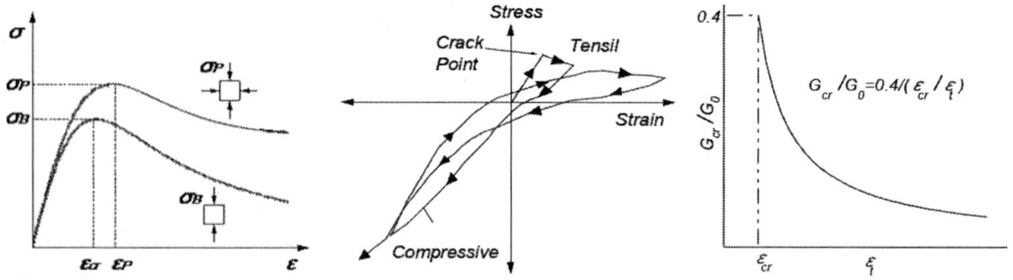

(a) Stress-strain relationship (b) Hysteresis (c) Shear Transfer

Figure 6. Constitutive laws for concrete.

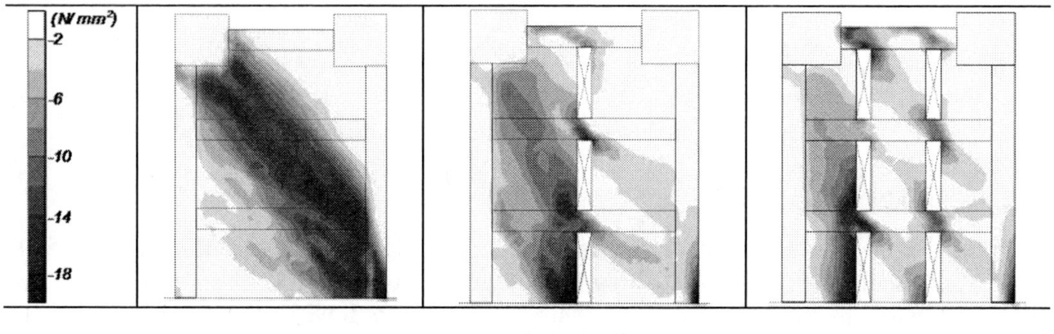

Figure 7. Minimum principle stress contours at 0.5% rad.

Figure 8 presents the transitions of the total axial and shear forces applied to the boundary beams. Higher axial compression was applied to the beams with increases of the wall drift angle; therefore, the beams were subjected to high shear forces which resulted in the shear failure of the beams, as shown in Figure 3. It means that the high axial compression applied to the boundary beam above openings must be considered in the design analyses for this type of wall to prevent its premature brittle failure.

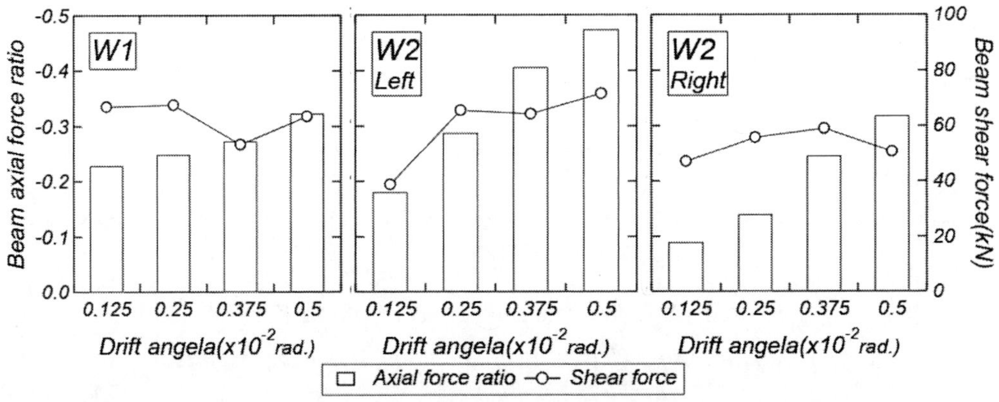

Figure 8. Transitions of total axial and shear forces applied to the boundary beams.

4 CONCLUDING REMARKS

This paper reported a series of laboratory tests conducted with three scaled shear wall specimens with/without vertically aligned openings. Consequently, the failure mechanisms of the specimens with openings (W1 and W2) were similar to that of coupled shear walls. The maximum strengths of the specimens decreased due to the existence/higher number of openings.

The observed behavior of the specimens was simulated by FEM analyses, to investigate key modeling of the walls with openings for design analyses. In conclusion, the damage patterns and shear force-drift angle relationships obtained from the analyses agreed well with the experimental results. The analytical results on the specimens with openings (W1 and W2) indicated that high compression was applied to the wall bottom with the boundary column on the tensile side and the boundary beams above openings. In particular, the high axial compression on the beams must be considered in the design analyses to prevent the premature brittle failure.

References

ACI Committee 318, Building Code Requirements for Structural Concrete (ACI 318-14) and Commentary (ACI 318R-14), American Concrete Institute, September, 2014.

Al-Mahaidi, R.S.H., Nonlinear Finite Element Analysis of Reinforced Concrete Deep Members, Report 79-1, Dept. of Structural Engineering, Cornell University, January, 1979.

Architectural Institute of Japan (AIJ), AIJ Standard for Structural Calculation of Reinforced Concrete Structures revised 2010 (in Japanese), February, 2010.

Kupfer, B. and Gerstle, H., Behavior of Concrete under Biaxial Stress, Journal of the Engineering Mechanics Division, ASCE, Vol. 99, No. EM4, pp. 853-866, August, 1973.

Naganuma, K., Stress-strain relationships for concrete under triaxial compression (in Japanese), Journal of structural and construction engineering, No. 474, pp. 163-170, August, 1995.

Naganuma, K. and Ohkubo, M., An Analytical Model for Reinforced Concrete Panels under Cyclic stresses (in Japanese), Journal of structural and construction engineering, No. 536, pp. 135-142, October, 2000.

Nakamura, H. and Higai, T., Compressive Fracture Energy and Fracture Zone Length of Concrete, Seminar on Post-peak Behavior of RC Structures Subjected of Seismic Load, JCI-C51E, Vol. 2, pp. 259-272, October, 1999.

DEFLECTION PREDICTION OF TWO-SPAN CONTINUOUS CONCRETE BEAMS REINFORCED WITH GFRP BARS

KARAM MAHMOUD and EHAB EL-SALAKAWY

Dept of Civil Engineering, University of Manitoba, Winnipeg, Manitoba, Canada

Continuous beams are main structural members in many reinforced concrete (RC) applications such as parking garages and bridges, which are vulnerable to harsh environments. The use of glass fiber-reinforced polymers (GFRP) bars in such structures has proven to be a good solution to overcome the steel corrosion problem. In GFRP-RC beams, deflection is a concern and may govern the design; thus, attempts have been made to propose models to predict the effective moment of inertia and consequently the deflection. The current prediction models were verified against deflection of simply-supported beams. In this paper, available predictions models are compared to the experimental deflections of two-span continuous beams. Eight rectangular beams (200×300 mm) continuous over two equal spans of 2,800 mm each were constructed and tested to failure. All beams were reinforced with GFRP bars and stirrups. The test variables included concrete strength, longitudinal reinforcement ratio and transverse reinforcement ratio. The comparison revealed that the investigated models reasonably predicted the deflection at service load level while they underestimated the deflection at higher load levels. These models require to be modified in order to yield better predictions of the deflection at lower and higher load levels than service load.

Keywords: Effective moment of inertia, High strength concrete, Fiber reinforced polymer, Modulus of elasticity, Cracking moment, Service moment.

1 INTRODUCTION

Fiber reinforced polymer (FRP) bars are a viable alternative to conventional steel reinforcement in structures, which are exposed to aggressive environment because of their superior corrosion resistance. However, the low modulus of elasticity of the FRP bars, compared to that of steel bars, results in much lower post-cracking flexural stiffness and consequently large deflections. This raised concerns about the performance of FRP-reinforced concrete (RC) members especially at service load. Extensive studies investigated the applicability of deflection prediction models used for steel-RC beams to FRP-RC beams and proposed modified versions of these models to reasonably predict the deflection behavior of such beams (Mota *et al.* 2006, Bischoff 2005, Bischoff and Gross 2011). However, all these studies focused on FRP-RC simply-supported beams and one-way slabs and no attention was given to continuous beams, which are main structural elements commonly used in structural applications. In this paper, the load-deflection behavior of eight two-span continuous concrete beams reinforced with GFRP bars is compared to the predictions of models available in the literature.

2 REVIEW OF MODELS FOR EFFECTIVE MOMENT OF INERTIA

In this paper, four models available in the literature are reviewed in this section (Table 1). These models were verified previously against results of simple beams with rectangular sections. The first model is adopted by ISIS design manual 3 (2007). This model was previously evaluated by Mota *et al.* (2006) and it was found that it gives consistently conservative results for simply-supported beams.

The second model is a modified version of Branson's expression recommended by the ACI 440.1R-06. The modification factor ($\beta_d = 0.2(\rho/\rho_b)$) accounts for the reduced tension stiffening in FRP-RC members. In the ACI 440.1R (2015), another expression for the effective moment of inertia is introduced which is based on a model proposed by Bischoff (2005). Also, a modification factor ($\gamma = 1.72\text{-}0.72\ (M_{cr}/M_a)$) is added to the equation to account for the variation of the stiffness along the member length where M_{cr} is the cracking moment and M_a is the moment at which deflection is to be calculated.

Bischoff and Gross (2011) introduced an equivalent moment of inertia based on integration of curvature to account for changes in member stiffness along the span. Closed-form solutions of integrated expressions for deflection are expressed in terms of an equivalent moment of inertia.

Table 1. Review of effective moment of inertia equations.

Reference	Effective moment of inertia	Related formulas
ISIS Design M3 (2007)	$I_e = I_t I_{cr} / \left(I_{cr} + \left(1 - 0.5 (M_{cr} / M_a)^2 \right) (I_t - I_{cr}) \right)$	$I_{cr} = \frac{bd^3}{3} k^3 + n_f A_f d^2 (1 - k^2)$
ACI 440.1R-06	$I_e = (M_{cr} / M_a) \beta_d I_g + \left[1 - (M_{cr} / M_a)^3 \right] I_{cr} \leq I_g$	$k = \sqrt{2 \rho_f n_f + (\rho_f n_f)^2} - \rho_f n_f$ $n_f = \frac{E_f}{E_c}$
ACI 440.1R-15	$I_e = I_{cr} / \left(1 - \gamma (M_{cr} / M_a)^2 \left[1 - I_{cr} / I_g \right] \right) \leq I_g$ where $M_a \geq M_{cr}$	$M_{cr} = \frac{0.62 \lambda \sqrt{f_c'} I_g}{y_t}$
Bischoff and Gross (2011)	$I_e' = I_{cr} / \left[1 - \gamma \eta (M_{cr} / M_a)^2 \right] \leq I_g$ where $\eta = 1 - (I_{cr} / I_g), \gamma = 1.7 - 0.7 \left(\frac{M_{cr}}{M_a} \right)$	$I_g = \frac{bh^3}{12}$

M_{cr}=cracking moment; M_a=applied moment; I_g=I_t=gross moment of inertia; I_{cr} =moment of inertia of transformed cracked section; k=ratio of the neutral axis depth to reinforcement depth; E_c=modulus of elasticity of concrete; E_f= modulus of elasticity of reinforcement; b= width of the beam; h=overall height of the beam; d= effective depth of the beam; ρ_f = reinforcement ratio; f_c'= concrete strength; A_f = reinforcement area; y_t = distance from cantorial axis of gross section to tension face; λ= factor accounts for concrete density.

3 EXPERIMENTAL INVESTIGATION

Eight large-scale RC continuous beams were constructed and tested to failure. All test beams had a rectangular cross (200×300 mm) and were continuously supported over two equal spans of 2,800 mm each. All beams were reinforced with GFRP bars and stirrups. The longitudinal reinforcement ratio was 0.8% in two beams, while it was 1.2% in the other six beams. The shear reinforcement ranged from half to twice the minimum shear reinforcement as specified by the Canadian standards CSA/S806-12 (CSA 2012). Dimensions and details of the test specimens are shown in Figure 1.

The beams were tested under a two-point loading system in each span. The strains in the concrete, longitudinal reinforcement, and stirrups were measured at critical locations using electrical strain gauges. The deflection was measured using linear variable differential transducers (LVDTs) at mid-span, and at each loading point. Moreover, two load cells were used to measure the reactions at the exterior supports.

The specimen designation can be explained as follows: The first letter indicates the concrete strength ("N" for normal strength and "H" for high strength). The second number indicates the percentage of the longitudinal reinforcement ratio (0.8% and 1.2%). The third number indicates the amount of transverse reinforcement ("1.0" for minimum and "0.5" and "2.0" for half and twice the minimum, respectively).

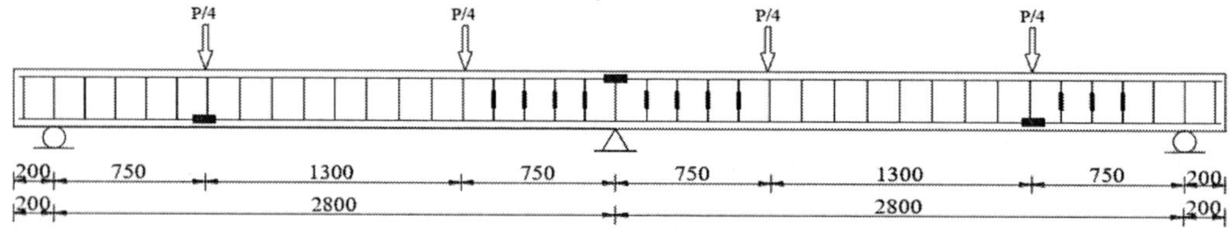

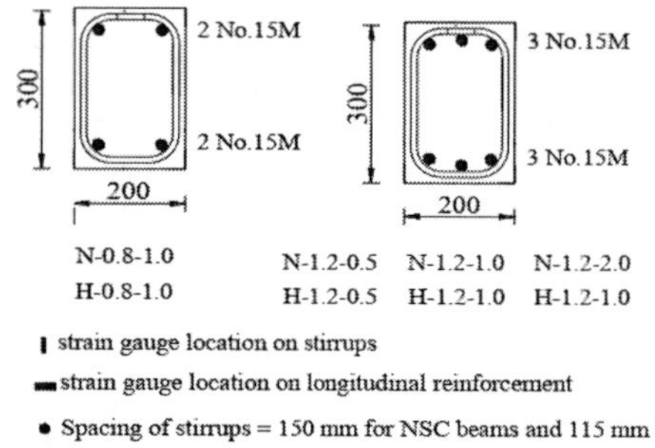

Beam	Concrete strength (MPa)	Flexural reinforcement ratio (%)	Transverse reinforcement	
			Ratio (%)	Stirrup dia. (mm)
N-0.8-1.0	43	0.8	0.48	9.5
N-1.2-0.5	43	1.2	0.21	6.3
N-1.2-1.0	42	1.2	0.48	9.5
N-1.2-2.0	43	1.2	0.83	12.7
H-0.8-1.0	81	0.8	0.63	9.5
H-1.2-0.5	80	1.2	0.28	6.3
H-1.2-1.0	81	1.2	0.63	9.5
H-1.2-2.0	80	1.2	1.10	12.7

• All dimensions are in mm

Figure 1. Details of test beams.

The properties of the used sand-coated GFRP bars are reported in Table 2. All beams were constructed using normal weight concrete. The average concrete strength obtained on the test day (f_c') is reported in Figure 1.

Table 2. Properties of the reinforcing bars and stirrups.

Bar type	Diameter (mm)	Area (mm^2)	f_{fu} (MPa)	E_f (GPa)	ε_{fu} (%)
Straight bar	15.9	198	1442	67	2.1
Stirrups	6.3	32	1383	53	2.6
	9.5	72	1195	45	2.7
	12.7	127	1328	53	2.5

f_{fu} = tensile strength of the bar; ε_{fu} = rupture strain of the bar.

4 TEST RESULTS

Test results and detailed discussion of cracking behavior, deflection, strains in longitudinal reinforcement and concrete, moment redistribution and shear capacity of the test beams can be found elsewhere (Mahmoud and El-Salakawy 2014, 2016). In this paper, the discussion focuses on the deflection behavior and the comparison between experimental and predicted deflections.

4.1 Load-deflection Response

The load-deflection relationship at midspan for all test beams is shown in Figure 2. The typical load-deflection graph can be defined by two distinct stages; pre-cracking, characterized by small deflection, and post-cracking, in which significant deflection was observed. In the post-cracking stage, the flexural stiffness of the beam was mainly dependent on the axial stiffness of the reinforcing bars. Therefore, at the same load level, the measured deflection in beams with 0.8% longitudinal reinforcement ratio was higher than that in beams with 1.2%. Also, beams made of high strength concrete (HSC) had higher flexural stiffness; therefore, the deflection graphs for HSC beams were steeper than those of the normal strength concrete (NSC) ones.

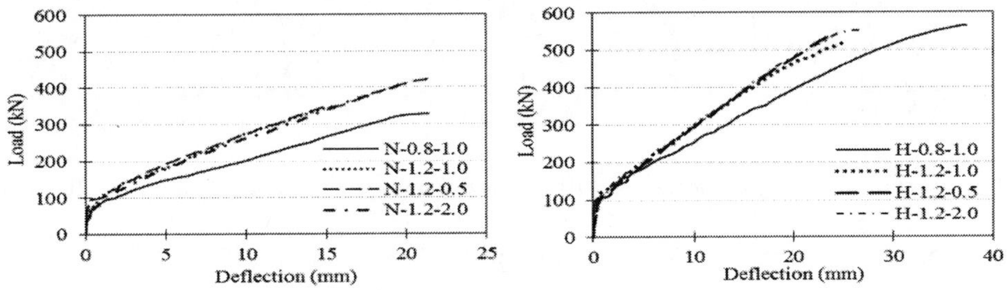

Figure 2. Load versus mid-span deflection of test beams.

5 COMPARISON OF EXPERIMENTAL AND PREDICTED DEFLECTION

Figure 3 shows the experimental and predicted load-deflection relationships for representative beams with different longitudinal and transverse reinforcement ratios and concrete strength. It can be seen that all investigated models reasonably predicted the mid-span deflection up to approximately 50% of the ultimate load; however, they highly underestimated the deflection at higher load levels. The difference between the experimental and predicted deflections increased in beams with lower reinforcement ratio (0.8%) as well as for HSC beams compared to those with 1.2% reinforcement ratio and NSC, respectively. For all test beams, the predicted deflections were close to the experimental ones near the service load level; however, the slope of the post-cracking stage is shallower than that predicted by all the evaluated models of effective moment of inertia. Also, these models yielded poor predictions of the deflection at load levels less than service load. Most structures encounter higher loads than service load before reaching failure. In case of steel-RC members the deflections do not increase significantly because of the high modulus of elasticity of the steel bars; however, FRP-RC members exhibit large deflections with a slight increase in the load beyond their service load (Figure 3). This large deflection may cause damage to other elements attached to such members. Thus, it is required to modify these models to reasonably predict the deflection behavior for the full range of loading not only at service load.

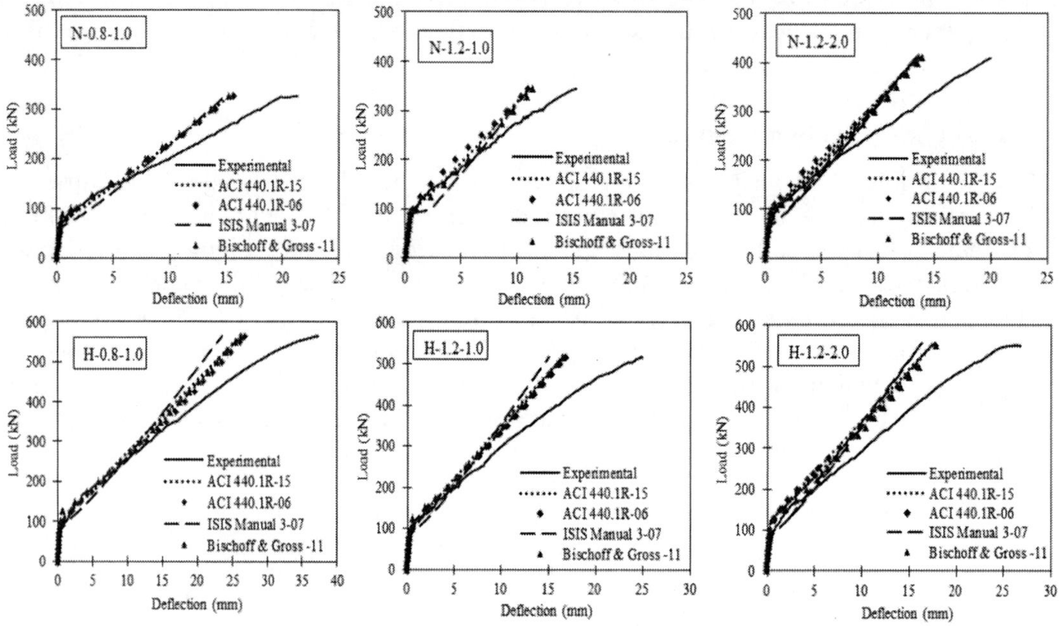

Figure 3. Load-deflection relationship for representative beams.

Table 3. Statistical comparison of experimental-to-predicted deflection.

Reference	@Cracking load		@1.2 P_{cr}		@Service load		@0.8 P_u	
	Mean	COV	Mean	COV	Mean	COV	Mean	COV
ACI 440.1R-15	2.59	0.33	1.42	0.24	1.17	0.13	1.32	0.04
ACI 440.1R-06	1.51	0.31	1.61	0.12	1.36	0.08	1.31	0.04
ISIS Manual 3 (2007)	0.51	0.29	0.67	0.24	0.94	0.17	1.33	0.06
Bischoff & Gross (2011)	1.65	0.28	1.12	0.23	1.05	0.12	1.26	0.04

Also, a statistical comparison of the experimental and predicted deflection considering the results of all tested beams at different load levels is presented in Table 3. The load levels considered herein are the cracking load (P_{cr}), 1.2 P_{cr}, service load taken as 43% of the ultimate load (Mousavi and Esfahani 2012), and at 80% of the ultimate load. It can be seen that the models proposed by ACI 440.1R (2006 and 2015) and Bischoff and Gross (2011) underestimated the deflection at P_{cr} and 1.2 P_{cr} while better predictions were obtained at the service load level. Also, the newly adopted model by ACI 440.1R -15 yielded better predictions compared to that of the ACI 440.1R-06, especially at the service load level. On the other hand, the model adopted by ISIS design manual 3 (2007) overestimated the measured deflection; highly at P_{cr} and 1.2 P_{cr} and slightly at the service load level. With further increase in the load beyond the service load level, all models underestimated the deflection where the average experimental-to-predicted deflection ratio ranged from 1.26 to 1.33.

6 CONCLUSIONS

Based on the comparison between the experimental and predicted deflection by different models available in literature, the following conclusions can be drawn.

1) All evaluated models fairly predicted the mid-span deflection of GFRP-RC continuous beams up to the service load level. Also, these models underestimated the deflection of beams with low reinforcement ratios and HSC beams at high levels of loading.
2) At service load level, both models proposed by Bischoff and Gross (2011) and ISIS design manual 3 (2007) yielded the best predictions with a mean experimental-to-predicted deflection ratio of 1.05 and 0.94, respectively. The models proposed by ACI 440.1R (2006 and 2015) underestimated the deflection where the mean experimental-to-predicted deflection ratio was 1.36 and 1.17, respectively.

Acknowledgments

The financial support received from the Natural Science and Engineering Research Council of Canada, through the Canada Research Chairs program is sincerely appreciated as well as the help received from the technical staff of Structural Laboratory at the University of Manitoba.

References

ACI (American Concrete Institute), Guide for the design and construction of structural concrete reinforced with FRP bars, *ACI 440.1R-06*, Farmington Hills, MI, 2006.

ACI (American Concrete Institute), Guide for the design and construction of structural concrete reinforced with FRP bars, *ACI 440.1R-06*, Farmington Hills, MI, 215.

Bischoff, P., Reevaluation of deflection prediction for concrete beams reinforced with steel and fiber reinforced polymer bars, *Journal of Structural Engineering*, 131(5), 752-767, 2005.

Bischoff, P. and Gross, S, Equivalent moment of inertia based on integration of curvature, *Journal of Composites for Construction*, 15(3), 263-273, June, 2011.

CSA (Canadian Standard Association), Design and construction of building structures with fibre-reinforced polymers, *S806-12*, Rexdale, ON, Canada, 2012.

ISIS Canada Corporation, Reinforcing concrete structures with fiber reinforced polymers, *ISIS Canada: Design Manual No.3*, The Canadian Network of Centers of Excellence on Intelligent Sensing for Innovative Structures, Winnipeg, Manitoba, Canada, 2007.

Mahmoud, K., and El-Salakawy, E., Shear strength of GFRP-reinforced concrete continuous beams with minimum transverse reinforcement, *Journal of Composites for Construction*, ASCE, 18(1), 04013018, Feb., 2014.

Mahmoud, K., and El-Salakawy, E., Effect of transverse reinforcement ratio on the shear strength of GFRP-RC continuous beams, *Journal of Composites for Construction*, 20(1), 04015023, Feb., 2016.

Mota, C., Alminar, S., and Svecova, D., Critical review of deflection formulas for FRP-RC members, *Journal of Composites for Construction*, 10(3), 183-194, June, 2006.

Mousavi, S. and Esfahani, M. Effective moment of inertia prediction of FRP-reinforced concrete beams based on experimental results, *Journal of Composites for Construction*, 16(5), 490-498, October, 2012.

Interaction between Theory and Practice in Civil Engineering and Construction
Edited by Komurlu, R., Gurgun, A. P., Singh, A., and Yazdani, S.

STRENGTHENING OF PRESTRESSED HOLLOW CORE SLABS WITH OPENINGS USING NSM-CFRP STRIPS

KARAM MAHMOUD[1], STEVEN FOUBERT[2], and EHAB EL-SALAKAWY[1]

[1]*Dept of Civil Engineering, University of Manitoba, Winnipeg, Manitoba, Canada*
[2]*Stantec Consulting Ltd., Winnipeg, MB, Canada*

Precast, prestressed, hollow-core (HC) slabs are extensively used in many structures such as parking garages, commercial and industrial buildings. It is common for such structures to undergo changes while still in service, which may result in a reduction of their strength. One of these changes is to add in-situ openings at different locations along the slab to accommodate services such as intake/exhaust ducts or utility conduits. Near surface mounted (NSM) strengthening technique has proved to be an adequate technique that can be employed to restore or enhance the performance of the original structure. In this paper, the effect of openings and the efficiency of using NSM carbon fiber reinforced polymer (CFRP) strips as strengthening technique are investigated. A total of three full-scale prestressed HC slabs were tested; one without opening, one with an opening in the pure flexural zone and one with strengthened opening. Test results showed the presence of opening along the flexure span had a significant adverse effect on the post-cracking flexural stiffness, cracking load and the ultimate capacity of the slab. However, strengthening the opening with two strips of NSM-CFRP effectively enhanced the post-cracking stiffness, increased the ductility of the member and restored the flexural strength deficit incurred as a result of cutting the opening, and provided a net increase in flexural capacity.

Keywords: Near surface mounted, High strength concrete, Carbon fiber reinforced polymer (CFRP), Restoration, Post-cracking stiffness, Ductility, Flexural capacity.

1 INTRODUCTION

Precast prestressed hollow core (HC) slabs are extensively used in many structures such as parking garages and commercial and industrial buildings. It is common for such structures to undergo changes while still in service, which may result in a reduction of their strength or require them to resist an increased applied load. One of these changes is to add in-situ openings in several locations within the slab span to accommodate intake/exhaust ducts along industrial building roofs, or utility conduits along parking garage floors. Strengthening techniques are commonly employed to restore or enhance the performance of the original structure. Near surface mounted (NSM) strengthening technique, using FRP composites, has proved to be an adequate technique to strengthen reinforced concrete (RC) structures (De Lorenzis and Nanni 2002, El-Hacha and Rizkalla 2004, Teng *et al.* 2006). The non-corrodible and durable nature of fiber reinforced polymer (FRP) composites can successfully mitigate the effects of inclement conditions, and effectively improve the structures performance.

The behavior of one-way and two-way RC slabs with openings strengthened with FRP materials has been previously studied (Tan and Zhao, 2004, Enochsson *et al.* 2007, Smith and Kim 2009, Seliem *et al.* 2011). Both externally bonded (EB) laminates or NSM strips were used to strengthen these slabs. It was found that providing CFRP strengthening for slabs with openings effectively enhanced the stiffness and the ultimate load capacity of the slab where comparable behavior to the control slab was observed. On the other hand, the carbon (C) FRP strengthened slabs exhibited a brittle failure in contrast to a more ductile behavior for the unstrengthened slabs (Tan and Zhao 2004). Also, it was reported that all slabs with CFRP strengthened openings exhibited an increased load-carrying capacity compared to their counterparts without strengthening (Enochsson *et al.* 2007). Moreover, slab with opening strengthened with NSM-CFRP strips reached 92% of the slab capacity (compared to 82% in slab with unstrengthened opening); however, the NSM strips were ineffective in restoring the stiffness of the slab (Seliem *et al.* 2011). This study represents, up to the authors' knowledge, the first attempt to investigate the behavior of prestressed HC slabs with openings unstrengthened and strengthened with NSM-CFRP strips.

2 EXPERIMENTAL PROGRAM

2.1 Test Specimens

Three full-scale prestressed HC slabs were tested to failure. All slabs had a thickness of 203 mm with a concrete gross sectional area of 140,194 mm^2. The slabs had originally an internal prestressing steel reinforcement ratio of 0.00274 (7-9 mm strands). The test specimens include one slab without opening to serve as reference, one slab with opening located at mid-span along the constant moment region and one slab with identical opening strengthened with NSM CFRP strips. The opening had a rectangular shape measuring 308 × 600 mm. It is worth mentioning that the middle strand was cut in slabs with openings. Details of test specimens are shown in Fig. 1.

2.2 Materials

Normal weight, high-strength concrete with a target compressive strength of 28 MPa at 18 hours and 45 MPa at 28 days were used to cast the specimens. The compressive strength, at the day of testing for each slab, is listed in Table 1. The internal prestressing reinforcement was 7-wire low-relaxation high-strength steel strands. The strands were of grade 1860 and were initially stressed to $0.75f_{pu}$. CFRP strips (2×16 mm rectangular section) were used to strengthen the specimens. The laminates had a modulus of elasticity of 131 GPa and an ultimate tensile strength of 2068 MPa. The structural adhesive used to bond the NSM-CFRP laminates to the concrete substrate was Kemko 038 with a tensile strength of 62 MPa, and a flexural modulus of 38 GPa.

2.3 Cutting Openings and Strengthening Procedure

A hand-held saw was used to cut the openings and the longitudinal grooves along the inverted soffit of the slabs. The longitudinal grooves were terminated 200 mm from each support. The target depth of the grooves was 22 mm while the thickness was 6 mm. After cutting the grooves, the structural adhesive was injected along the groove, using specialized mixing and dispensing equipment until the groove was ¾ full, leaving an adequate space for the CFRP laminate. Strips then were pushed into grooves, in a vertical orientation with the larger dimension perpendicular to the concrete surface.

Table 1. Details of test specimens.

Specimen	Concrete strength (MPa)	Internal Reinforcement		Opening Location	Number of CFRP Laminates
		Configuration	Ratio		
S1	64.0			None	None
S2	64.0	7 No. 9	0.00274	Flexure span	None
S3	56.5			Flexure span	2

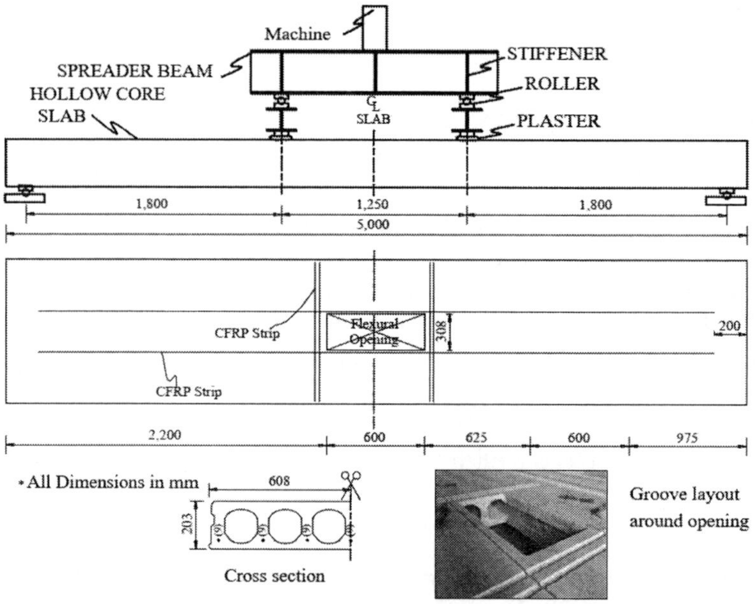

Figure 1. Details of test setup, opening location, cross section and groove layout.

2.4 Test Set-up and Instrumentation

The slab specimens were subjected to four-point bending up to failure. Figure 1 illustrates a schematic drawing of the test setup components along with key dimensions.

A 5000-kN MTS hydraulic machine was used to apply a concentrated load on a rigid system of steel spreader beams that was used to deliver a distributed line load along the full width of the slab. The slabs had a clear span of 4,850 mm, divided into a constant bending moment zone and two shear spans. During testing, the load was applied at a stroke-controlled rate of 1.0 mm/min and was increased to 2.0 mm/min after cracking.

Linear variable displacement transducers (LVDTs) were used to monitor deflection along the span length of each specimen. Also, electrical strain gauges were installed on the internal steel strands and the NSM-CFRP reinforcement to measure the induced tensile strains. The strain gauges were attached to the steel strand in one side of the opening and to the CFRP strip in the other side of the opening. A data acquisition system monitored by a computer was used to collect readings of the instrumentation.

3 TEST RESULTS

3.1 Mode of Failure

Figure 2 shows photos of the mode of failure of all test slabs. The control slab (S1) and the unstrengthened slab with opening (S2) exhibited pure flexural behavior where they experienced steel yielding, followed by concrete crushing, and ultimately failed as individual steel strands began to rupture. In strengthened slab (S3), near failure, a pair of transverse strips delaminated from the concrete substrate (Failure occurred in the surrounding concrete substrate as well as the surrounding adhesive matrix). Following the delamination of the transverse strips, the longitudinal strips experienced a partial delamination failure at several locations but did not fully de-bond from the substrate. This occurred within the maximum moment zone. Further increase in the load led to the outermost tensile fibers of the CFRP strips to fray. Finally, concrete crushing was observed at failure.

Figure 2. Mode of failure of test specimens.

3.2 Deflection

The load-deflection relationship at mid-span of the tested slabs is depicted in Fig. 3. All the tested slabs demonstrated similar behavior where the load-deflection curve can be divided into pre-cracking and post-cracking stages. The pre-cracking stage is characterized by linear relationship and small deflections in all slabs. In the post-cracking stage, excessive deformations can be seen reflecting the reduced flexural stiffness after cracking. Slab without strengthening showed large deformations without any significant increase in the load. The addition of an opening within the slab reduced the post-cracking flexural stiffness. Strengthening the opening with NSM CFRP strips improved the postcracking stiffness, as indicated by the increasing slope of the curve until failure.

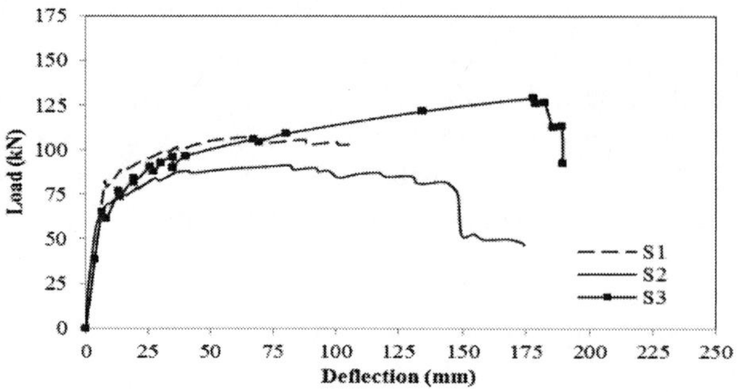

Figure 3. Load-deflection relationship of test slabs.

3.3 Strains

In general, the strains in the strands were small up to the yielding point; then, it increased rapidly until failure. In addition to the strain induced in the strands due to prestressing force, the measured strain at failure was approximately 9,920 με in slab S1 while the strains in slab S2, failed at lower load capacity, reached approximately 3,300 με. In S3, the measured strain at mid-span at failure was 10,130 με. Regarding the measured strain in the CFRP strips, it was insignificant before cracking of the slab. However, considerable increase was observed with increasing the load until failure. At failure, the measured strain at mid-span was approximately 9,370 με.

3.4 Capacities of Test Specimens

Cracking, yielding and ultimate moments for each slab are presented in Table 2. The cracking moment of the control specimen was approximately 73 kN.m. In slabs with openings (S2 and S3), the cracking moment decreased significantly, by approximately15%. The moment corresponding to yielding of strands was approximately 80 kN.m for slab S1; however, it decreased by 13% in the unstrengthened slab with opening (S2). Strengthening the opening enhanced the yielding moment slightly by 8% (form 71 kN.m in slab S2 to 77 kN.m in slab S3).

The presence of an opening in the pure flexural region decreased the ultimate capacity 17% when compared to the control specimen. On the other hand, strengthening openings with two strips of NSM-CFRP effectively restored the flexural strength deficit incurred as a result of cutting the openings. The addition of strengthening reinforcement resulted in a significant capacity enhancement of 40% compared to slab with unstrengthened opening while this increase is approximately 23% higher the capacity of the reference slab without opening.

Table 2. Summary of test results.

Specimen	Cracking moment, (kN.m)	Yielding Moment, (kN.m)	Ultimate Moment, (kN.m)
S1	73.0	80.0	97.0
S2	62.0	71.0	83.0
S3	63.0	77.0	116

4 CONCLUSIONS

Based on the limited test results presented above, the following conclusions can be drawn:

1) The presence of an opening in the flexural zone of prestressed hollow-core slab decreased the cracking load, the post-cracking stiffness and the load at failure.
2) Slab with an opening exhibited similar failure mode to that observed in the reference slab without openings. This could be attributed to the presence of the opening in the constant bending moment region.
3) Strengthening the opening with two strips of NSM-CFRP not only restored the flexural strength deficit incurred as a result of cutting the openings but provided additional flexural capacity.

It is worth mentioning that more experimental work is required to verify these findings.

Acknowledgments

The authors would like to thank Canada Research Chair Program (CRC), Tetra Tech Inc., Lafarge Canada Inc. and Vector Construction Ltd. for their financial support and industry perspectives. Also, the assistance received form the technical staff at McQuade Structures laboratory is acknowledged.

References

De Lorenzis, L. and Nanni, A., Bond between Near-Surface Mounted Fiber Reinforced Polymer Rods and Concrete in Structural Strengthening, ACI Structural Journal, 99(2), 123-132, March, 2002.

El-Hacha, R. and Rizkalla, S., Near-Surface-Mounted Fiber-Reinforced Polymer Reinforcements for Flexural Strengthening of Concrete Structures, *ACI Structural Journal*, 101(5), 717-726, October, 2004.

Teng, J. G., De Lorenzis, L., Wang, B., Li, R., Wong, T. N. and Lam, L., Debonding Failures of RC Beams Strengthened with Near Surface Mounted CFRP Strips, *Journal of Composites for Construction*, ASCE, 10(2), 92-105, April, 2006.

Tan, K. H. and Zhao, H., Strengthening of Openings in One-Way Reinforced-Concrete Slabs using Carbon Fiber-Reinforced Polymer Systems, *Journal of Composites for Construction*, ASCE, 8(5), 393-402, October, 2004.

Enochsson, O., Lundqvist, J., Taljsten, B., Rusinowski, P., and Olofsson, T., CFRP Strengthened Openings in Two-Way Concrete Slabs – An Experimental and Numerical Study, *Construction and Building Materials*, Elsevier, 21(4), 810-826, April, 2007.

Smith, S. T. and Kim, S. J., Strengthening of One-Way Spanning RC Slabs with Cutouts using FRP Composites, *Construction and Building Materials*, , 23(4), 1578-1590, April, 2009.

Seliem, H., Seracino, R., Sumner, E., and Smith, S., Case Study on the Restoration of Flexural Capacity of Continuous One-Way RC Slabs with Cutouts, *Journal of Composites for Construction*, ASCE, 15(6), 992–998, December, 2011.

Interaction between Theory and Practice in Civil Engineering and Construction
Edited by Komurlu, R., Gurgun, A. P., Singh, A., and Yazdani, S.

CONSIDERATION ON TRUSS AND SPLICE JOINT DESIGN MODELS IN EXTREMAL LOADING

LILITA OZOLA

Dept of Structural Engineering, Latvia University of Agriculture, Jelgava, Latvia

In this study the behavior of elements and joints in hard loaded Warren trusses is analyzed theoretically with the purpose of disclosing the sections and elements which possibly can become extremely stressed under the design load proposed being real before the ultimate limit state. The significance of truss topology selected has been brought forth, since unfavorable loading of a tensioned chord connection is discussed in this paper. Three types of tensioned splice joints have been examined considering expected elongations of bolts in order to determine additional stresses induced due to the redistribution of tensile forces. An assumption of a continuous beam model on elastically deformed discrete supports has been adopted as a design model for the examination of stresses possibly caused in the sections of bolts. It has been proved that a joint with extended end-plates over both flanges of chord I profile may be accepted as the one which is safe and robust enough for use in tensile chords of bearing structures. The joint type with end-plates extended over a more tensioned flange only and the one with connecting bolts all hidden between flanges has been subjected to sharp criticism due to the extremely nonuniform behavior and the overloading of bolts. The results of this numerical case study promote a deeper understanding and help assessing the end-plate joint behavior since they lack the uniqueness of the solution recommended by the building codes, which is particularly significant when hard-loaded structures have been designed for covering the spans of public building areas.

Keywords: Steel structures, Truss topology, Tensile joints, End-plate connections, Bolt force, Prying action.

1 INTRODUCTION

During the recent few decades the end-plate connections in hard loaded tension chords made of I-profiles have not been chosen as a priority of research themes, since hollow section profiles prevail in construction. Nevertheless Kim and Madugula (2010) studied flanged connections and proved the T-stub model design equations for leg members of guyed lattice towers. Mohamadi-shooreh and Mofid (2008) performed FEM-based parametric analysis of bolted end-plate splices of beams. Some research works have been devoted to the analysis of high strength steel end-plate connections (Coelhoa and Bijlaard 2007, Yamaguchia *et al.* (2004) focusing on the ductility effects.

Failure of the roof trusses in Riga (2013) raised some disputable topics in the truss design area dealt with the conformity of the design model. The methodology proposed for the modeling of end-plate connection behavior has been detailed in previous work by Ozola (2015). In this study three types of tensioned splice joints are compared, considering possible additional force (the so called "prying effect") due to the deformation of a bolt and the action of bending moment.

2 CONSIDERATION ON TRUSS DESIGN MODEL

Unjustified assumptions involved in the calculation model usually are caused by the simplifications representing the joint behavior. A detailed analysis of different static models proposed for the simulation of the behavior of the bearing structure becomes more significant for hard loaded trusses, and particularly when the junction has been put into a heavily loaded tensioned chord element. It is a routine in design to define the design model of the truss representing the web members as a pin connected to the continuous chords. In most cases no more static trials have been carried out and generally not needed. Nevertheless, such simplification leads to overlooking some important portion of stresses in some special cases.

The opinion suggested in this current paper is illustrated by a practical example of a hard loaded Warren truss with an effective span of 18 m, height 2 m, and vertical design load applied directly to the upper chord through concrete panels q= 65,7 kN/m.

Welded connections between chord members and diagonals provide rigidity of a joint capable of preventing any rotation of the ends of members when the structure endures the load. The model shown in Figure 1 is applicable for the determination of a full scale internal force assembly: bending moments, shear and axial forces which must be all taken into consideration. Using the conventional model some portion of internal forces is ignored. Particularly significant are the differences of axial force values for tensioned diagonals at the supports depending on the calculation model used, see joint loading schemes A' (from simplified model) and A (from correct model) in Figure 1. Besides, it is important to note that bending moment produces an additional portion of normal stress perpendicular to the throat of the filled weld at the joint.

Another phenomenon, usually not considered in practice, is the option to insert or remove the middle vertical bar called zero one in static sense, see joint C in Figure 1. A middle bar promotes an effect of obstruction of the deformation of truss contour, and thereby internal forces result in a close relationship with the section stiffness's of the elements involved in both the middle bar and bottom chord elements. The loading situations around the middle point of the bottom chord are of particular interest (see joint C Figure 1) when one solves the connection problem.

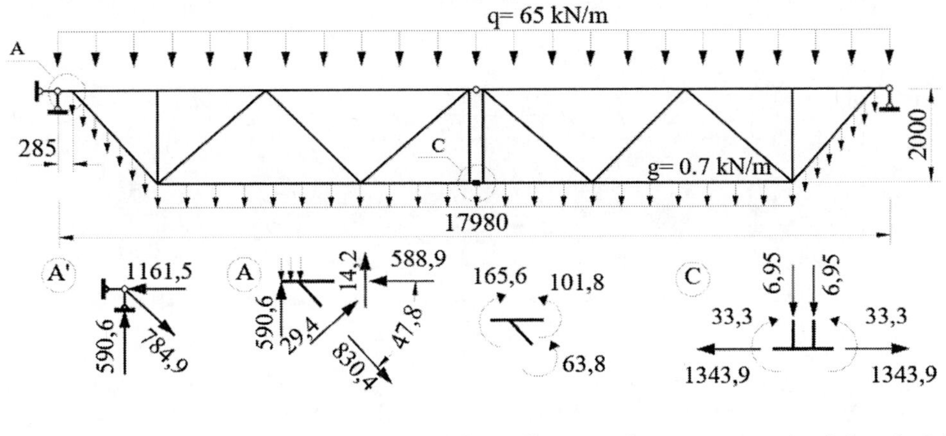

Figure 1. Rigid model of Warren truss and force diagrams for more attractive joints (axial forces in kN, bending moments in kNm).

3 TYPES AND MODELS OF END PLATE CONNECTIONS

Different variants may be chosen for joining together the tensioned bottom chord elements made of single I sections, see Figure 2(a). The variant with extended end plates over both chord flanges (ExTC) was used many decades ago but nowadays it is subjected to criticism due to the unpleasant impression from an architectural point of view. The splice joint variant with end plates extended over a bottom flange only (ExT) in order to compensate an additional tensile force due to bending moment may be another option, but a variant with joining bolts hidden between flanges (CL) is becoming more acceptable by architects.

In this study the end-plate is modeled as a continuous beam on discrete elastically deformed supports which represent joining bolts, see Figure 2(b), and loaded by concentrated forces transferred by the flanges and the web of the I-section.

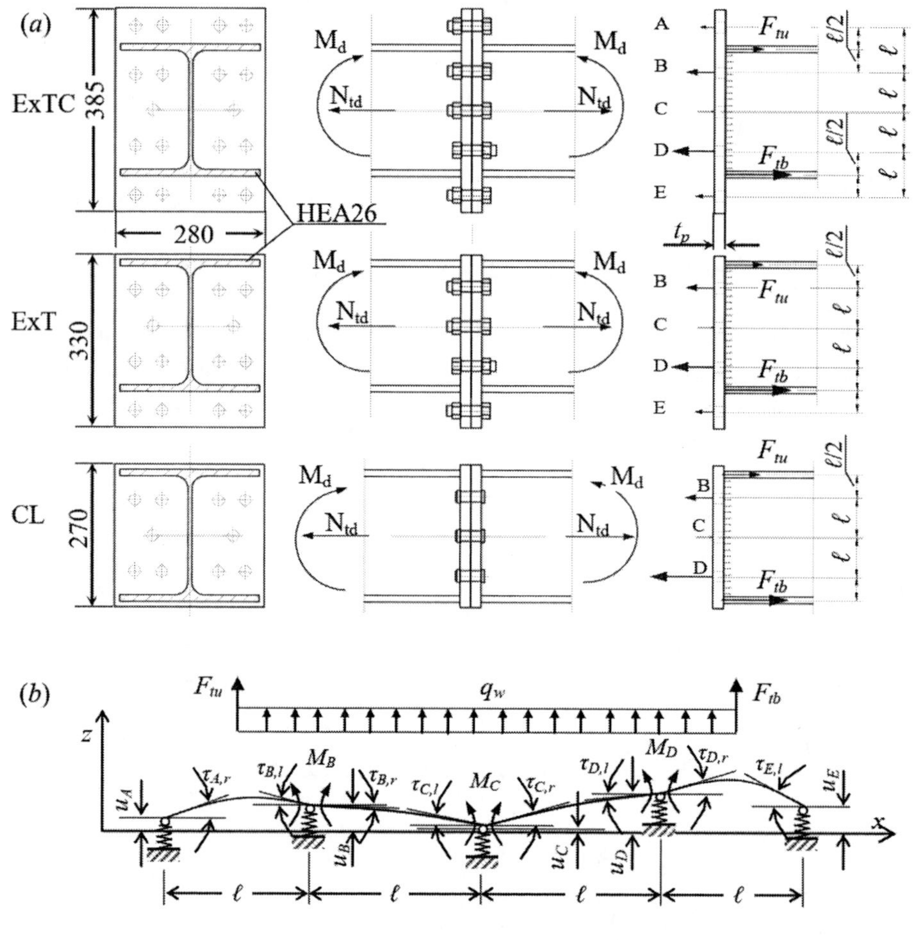

Figure 2. Tensioned splice joints: (a) structural variants ExTC, ExT, CL of splice joints (Ftu= 401 kN, Ftb=682 kN, qw=11.6 kN/m); (b) design model of behavior of joint ExTC.

4 NUMERICAL TESTS

First-order structural analysis and dimensioning of truss, have been performed according to conditions stated by the Eurocodes, see design model in Figure1.

Numerical tests of end-plate connections under extremal loading have been performed basing on a continuous beam theory (Marti 2013). Continuous beam on the rigid supports, represented by bolts, has been defined as an initial design model of the end-plate loaded by forces transferred by flanges mainly as well as by web of I-profile. Forces in bolts determined in a rigid model using the three-moment equation method have been taken as input data for the further calculation cycles assuming bolt elongations as elastic deformation of supports. Support settlement values of the end-plate have been determined as proposed elongations of an individual bolt in the row due to axial force action according to Hooke's law. Then a new equation system for a corresponding type of connections (Eq.(1) for ExTC, Eq.(2) for ExT, and Eq.(3) for type CL) has been formulated and solved taking into consideration the additional rotations of the beam axis due to different settlements of the supports, see Figure 2(b). Calculation cycles have been repeated until the bending moment values of the two last cycles differ insignificantly.

$$\begin{cases} \frac{2\ell}{3EI}M_B + \frac{\ell}{6EI}M_C + \frac{u_C - u_B}{\ell} - \frac{u_B - u_A}{\ell} + \frac{1}{EI}\left(\frac{F_{tu}\ell^2}{16} + \frac{q_w\ell^3}{8}\left(\frac{3}{16} + \frac{1}{3}\right)\right) = 0 \\ \frac{\ell}{6EI}M_B + \frac{2\ell}{3EI}M_C + \frac{\ell}{6EI}M_D + \frac{u_D - u_C}{\ell} - \frac{u_C - u_B}{\ell} + \frac{q_w\ell^3}{12\,EI} = 0 \\ \frac{\ell}{6EI}M_C + \frac{2\ell}{3EI}M_D + \frac{u_E - u_D}{\ell} - \frac{u_D - u_C}{\ell} + \frac{1}{EI}\left(\frac{F_{tb}\ell^2}{16} + \frac{q_w\ell^3}{8}\left(\frac{3}{16} + \frac{1}{3}\right)\right) = 0 \end{cases} \tag{1}$$

$$\begin{cases} \frac{2\ell}{3EI}M_C + \frac{\ell}{6EI}M_D + \frac{u_D - u_C}{\ell} - \frac{u_C - u_B}{\ell} + \frac{q_w\ell^3}{12\,EI} - \frac{\ell}{6EI}\left(\frac{F_{tu}\ell}{2} + \frac{q_w\ell^2}{8}\right) = 0 \\ \frac{\ell}{6EI}M_C + \frac{2\ell}{3EI}M_D + \frac{u_E - u_D}{\ell} - \frac{u_D - u_C}{\ell} + \frac{1}{EI}\left(\frac{F_{tb}\ell^2}{16} + \frac{q_w\ell^3}{8}\left(\frac{3}{16} + \frac{1}{3}\right)\right) = 0 \end{cases} \tag{2}$$

$$\frac{2\ell}{3EI}M_C + \frac{u_D - u_C}{\ell} - \frac{u_C - u_B}{\ell} + \frac{q_w\ell^3}{12\,EI} - \frac{\ell}{6EI}\left(\left(F_{tu} + F_{tb}\right)\frac{\ell}{2} + \frac{q_w\ell^2}{4}\right) = 0 \tag{3}$$

where ℓ is distance between bolt rows, see Figure 2, E is the modulus of elasticity of steel, I is the moment of inertia of the end-plate section, u_A, u_B, u_C, u_D, u_E are the displacements of end-plate supports (bolt elongations), M_B, M_C, M_D are the bending moments supposed to be active at the corresponding plate sections, F_{tu} and F_{tb} are the tensile forces transferred to the end plate by upper and bottom flanges correspondingly, as well as q_w is the load transferred by the web assumed uniformly distributed.

5 RESULTS AND DISCUSSION

It has been proved by the results of the current study that for any of the joint types bending moments generated in the tensile bottom chord sections around the middle bar even of reasonably moderate values itself affects significantly the force distribution between the bolts depending on the location.

Geometric sizes, maximal expected stress values and steel consumption for each joint type are presented in Table 1. It has been found that in the case of the variant ExT solution the increase of tensile force in bolts near the flanges may be expected 1.7 times more due to the non-uniform elongations of bolts between rows B, C, D, E, see Figure 2, in comparison with the force distribution assuming the initial rigid stage, see Table 2.

The joint type with extended end-plates (ExTC) is highly recommended as in the case of failure of extremely stressed bolts at level D the tension force will be transferred by bolts at E and C levels, and due to the assumption for robustness it would be likely to use the same number of bolts at the middle level C as it is near the flanges.

Architectural assumptions often result in the use of end-plate connections of type CL when all the bolts must be placed between flanges. It is clear that this type is detrimental because a large increase of tension forces can occur due to bending moment, moreover in the limit state it is a highly unsafe connection without any contingency elements for short term transferring of forces released by failed bolts.

It has been revealed by numerical tests that in the range of stresses up to 100 MPa the variation of tensile force in the joint type ExT has been affected significantly by the ratio of moments of inertia of plate and bolt's sections (I_plate/I_bolts), see Fig.3.

A phenomenological approach may be preferred for use by the designer when the definition of an adequate structural model has become the subject for decision making but experimental tests are quite expensive. Indeed, it is not possible to achieve a complete similarity between the assumed design model and a real structure. However, a characteristic tendency will be disclosed.

Table 1. Characteristic indicators for tested joints.

Specification	**Values for variants**		
	ExTC	**ExT**	**CL**
Bolt diameter, mm	18	22	24
Maximal tensile stress expected in a bolt section, MPa	68	73	93
Thickness of end-plate, mm	20	28	36
Maximal bending stress expected in an end-plate section, MPa	59	71	74
Steel consumption for bolts and end-plates, kg	38	46	48

Table 2. Reactive forces in bolts assuming rigid and elastically deformed bolts.

Bolt row	**Reactive forces, kN, per bolt row assuming rigid/ elastic supports**								
	Variant ExTC			**Variant ExT**			**Variant CL**		
	rigid	**elastic**	**ratio**	**rigid**	**elastic**	**Ratio**	**rigid**	**elastic**	**ratio**
A	160.2	170.9	1.07	-	-	-	-	-	-
B	406.5	368.4	0.91	769.7	854.7	1.11	835.6	926.3	1.11
C	-79.1	-21.5	0.27	-335.4	-577.3	1.72	-731.0	-912.4	1.25
D	599.7	556.1	0.93	663.8	892.4	1.34	1257.1	1347.8	1.07
E	274.4	287.8	1.05	263.7	191.9	0.72	-	-	-

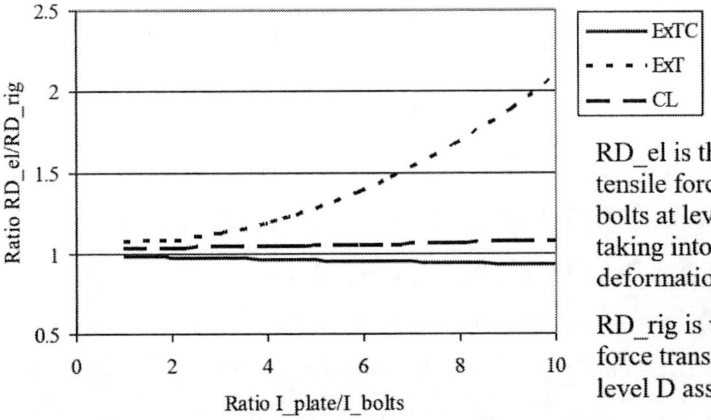

RD_el is the final value of tensile force transferred to the bolts at level D (see Figure 2) taking into account elastic deformation of bolts

RD_rig is value of tensile force transferred to the bolts at level D assuming rigid model.

Figure 3. Relationship of force variation in bolts versus ratio of moments of inertia.

6 CONCLUSIONS

- It has been proved by this numerical study that serious consequences may be expected due to the disability of traditional design models to reveal more unfavorable internal forces in specific elements of a structural system.
- It is necessary to include recommendations for a detail design of end-plate connections between tensioned I profile chord elements in structural codes.
- Splice joint variant with end-plates extended over flanges (ExTC) may be acknowledged as proper solution regarding both normal behavior under service loads and providing some robustness in the case of overloading.

References

Coelhoa, A. M. G. and Bijlaard, F. S. K., Experimental behavior of high strength steel end-plate connections, *Journal of Constructional Steel Research*, Elsevier, 63, 1228–1240, 2007

Kim, Y.J. and Madugula, M. K. S., Behavior of bolted circular flange connections subject to tensile loading, *International Journal of Steel Structures*, Springer, Vol.10(1), 65-71, 2010.

Marti, P. *Theory of structures: fundamentals, framed structures, plates and shells*/ Peter Marti; transl. by Philip Thrift, Ernst&Sohn, Berln, 2013

Mohamadi-shooreh, M.R. and Mofid, M., Parametric analyses on the initial stiffness of flush end-plate splice connections using FEM, *Journal of Constructional Steel Research*, Elsevier, 64, Issue 10, 1129–1141, 2008

Ozola, L., The analysis of splice joint behavior in heavily loaded tensile chords of warren trusses, *Proceedings of the 1st Pan-American Congress on Computational Mechanics and XI Argentine Congress on Computational Mechanics*, Idelsohn, S.R., Sonzogni, V., Coutinho, A., Cruchaga, M., Lew, A., Cerrolaza M. (Eds.), 507-517, Buenos Aires, 2015.

Yamaguchi, T., Suzuki Y., Kitada T. and Sugiura K., Effect of tensile and shear force on mechanical behavior of high strength bolted tensile joints, *Journal of Constructional Steel Research*, Elsevier, 60, 1545–1560, 2004

Interaction between Theory and Practice in Civil Engineering and Construction
Edited by Komurlu, R., Gurgun, A. P., Singh, A., and Yazdani, S.

EMBODIED CARBON EFFICIENT INDUSTRIAL BUILDINGS WITH OPTIMIZED LONG SPAN SANDWICH PANELS

MARTIN D. HEYWOOD, DIMITRIOS MOUTAFTSIS, and RAYMOND G. OGDEN

Faculty of Technology, Design and Environment, Oxford Brookes University, Oxford, UK

A series of studies have been undertaken to quantify potential steelwork and embodied carbon savings in single-storey industrial buildings with long span sandwich panel roof cladding. It was found that trussed-roof frames with north lights offer the greatest potential steelwork saving (38%-60%) against traditional portal frame construction when used with long span panels. However, in order to achieve the necessary span between adjacent frames, the structural performance of the sandwich panels needed to be improved. A subsequent study was undertaken to develop revised specifications for long span roof sandwich panels to increase their spanning capability, while minimizing the increase in embodied carbon. The mechanical resistance of the panels was evaluated using a combination of theoretical analysis and structural testing and a Pareto-optimal set of solutions was found. Finally, the optimized structure-envelope assembly was reviewed holistically in terms of its embodied carbon. The results showed that savings of up to 7.3% may be achieved for the frame when compared against traditional portal frame construction.

Keywords: Pareto-optimization, Structural testing, Cladding, Steel structures, North lights, Steelwork reduction.

1 INTRODUCTION

In Europe, buildings are responsible for 40% of energy consumption and are therefore at the core of efforts to meet climate change targets set by the EU and its member states. In an attempt to reduce the energy and CO_2 associated with operating buildings, regulations have demanded ever increasing insulation thicknesses within the building envelope. This increase in the amount of insulation material has been resulting in higher levels of embodied carbon (i.e., the lifecycle greenhouse gas emissions occurring during manufacture and transport of construction materials and components, construction processes and building end-of-life aspects). In the case of composite insulated (sandwich) panels, where the strength of the panel relies on the composite action between the insulation core and bonded metal faces, the increased insulation thickness has also improved the panel's stiffness and bending resistance, theoretically allowing the panels to span further. This presents an opportunity to make greater structural use of the envelope, permitting the removal of some structural elements and reducing the overall level of embodied carbon within the building.

Moutaftsis *et al.* (2015a) have determined the structural forms that are best able to utilize the structural capability of sandwich panels and examined the corresponding potential steelwork savings. Their study concluded that the greatest potential benefit arises from the use of long span envelope systems. Moutaftsis *et al.* (2015b) extended this analysis by reviewing the structure-envelope options in terms of their embodied carbon. A reduction in the embodied carbon associated with the structural frame was demonstrated in all schemes considered, but in order to maximize these savings, the sandwich panels were required to span over 8.0m (for the medium and large buildings), beyond their current spanning capability.

The present study was undertaken to extend the previous research and identify improved specifications for long span roof sandwich panels with steel faces and a polyisocyanurate (PIR) core capable of spanning the optimum spacing between frames. Crucially, while it was necessary to improve the bending resistance and stiffness of the panels, it was also important to limit the increase in the panels' embodied carbon to ensure that savings in the embodied carbon of the frame were not offset by increased embodied carbon in the envelope. The focus of the present study was on single-story industrial buildings.

2 METHODOLOGY

Three generic building sizes were chosen as representative of modern single-storey industrial buildings and these are presented in Table 1. For each building size, a base scheme was defined for the structural frame to reflect current construction practice in the UK, together with a series of re-engineered structural frame schemes designed to exploit the structural capability of the cladding. The frames were designed by elastic analysis using the structural Eurocodes for an assumed location of Oxford, UK. This initial study took no account of the spanning capability of the sandwich panels.

The mechanical resistance of typical roof sandwich panels was then evaluated using a combination of theoretical analysis and a structural testing program. This led to the optimization study during which adjustments were made to the mechanical and geometrical properties of the panels to improve their bending resistance in order to achieve the desired spacing between structural frames, while minimizing any increase in embodied carbon. A Pareto-optimal set of solutions was found and single solutions were selected for the required load capacities. This element of the study followed a similar methodology to research by Kurpiela and Lange (2013). Finally, the structure-envelope assemblies were reviewed holistically in terms of their embodied carbon, using a 'Cradle-to-gate' approach and data by Jones and Hammond (2008).

3 BUILDING STRUCTURE

Four structural frame options were studied as shown in Table 2. The first scheme represents the current practice in the UK for single-storey industrial buildings, while the others are re-engineered to favour long-span cladding systems. A duo-pitch portal frame without purlin, a re-oriented portal frame and a north light truss are shown in Figure 1. For each scheme, the optimum frame spacing and the percentage reduction of steelwork weight (referred to as "steelwork reduction" and including the primary frame, bracing and secondary members) against the base case for the same frame spacing distance are presented in Table 2.

Table 1. Generic industrial building dimensions.

Building size	Width	Length	Height to eaves	Area	Roof pitch
	m	m	m	m^2	degrees
Small	25	40	4	1,000	6
Medium	50	80	6	4,000	6
Large	80	125	6	10,000	6

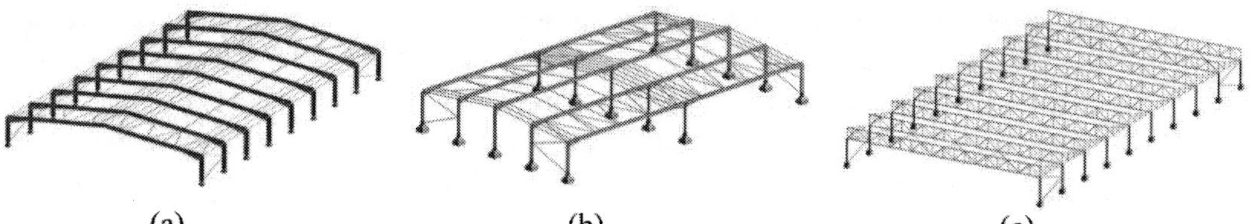

Figure 1. (a) Duo-pitch portal frames without purlins (b) Re-oriented portal frames (c) Truss roof with north lights.

4 OPTIMIZATION OF SANDWICH PANELS FOR LONG SPAN ROOFS

The study described above suggested that trussed roof frames with north lights provide the greatest potential for steelwork reduction for an optimum frame spacing of 6.67m for small buildings and 8.00m for medium and large buildings. As typical roof sandwich panels available today can already span 6.67m, the focus of the subsequent study was the re-engineering of the panels to achieve the 8.0m span required for the medium and large buildings.

4.1 Structural Testing

The structural testing program was devised to determine (a) the effect of PIR core density on its mechanical properties and (b) the compressive resistance of fully and lightly profiled steel sheets with varying geometries. All tests were undertaken according to EN 14509:2013 (CEN, 2013). The properties determined included: Density (ρ_C); Shear modulus (G_C) and strength (f_{Cv}); Compression modulus (E_{Cc}) and strength (f_{Cc}); Tension modulus (E_{Ct}) and strength (f_{Ct}); Poisson ratio (ν). Linear relationships between core density and the mechanical properties were determined based on statistical analysis. The test results to determine the compressive resistance of the steel sheets were compared against those obtained from existing analytical methods.

The study showed that, for the given slenderness ratios, the fully profiled sheets may be regarded as fully effective with the compressive resistance of the sheets equal to the yield strength. For the lightly profiled sheets, the model of Pokharel and Mahendran (2003) was found to give the best results when the depth was in excess of 3mm, while the formulas of Stamm and Witte (1974) were ideal for rib depth below 3mm. These methods were therefore chosen for the optimization model.

Table 2. Steelwork reduction against base case and optimum frame spacing.

Scheme	Steelwork reduction	Optimum frame spacing
Duo-pitch portal frame with purlins	N/A (base case)	Small: 6.67m Medium: 8.00m-13.3m Large: 10.00m-13.3m
Duo-pitch portal frame without purlins	Small: 6% Medium: 1%-13% Large: 4%-19%	Small: 6.67m Medium: 8.00m-13.34m Large: 10.00m-13.33m
Re-oriented portal frames	Small: 13% Medium: 51%-53% Large: 34%-42%	Small: 6.25m Medium: 6.25m-8.3m Large: 10.00m-13.3m
Truss roof with north lights	Small: 20%-38% Medium: 46%-53% Large: 49%-55%	Small: 6.67m Medium: 6.67m-10.00m Large: 8.00m

4.2 Optimization of Roof Sandwich Panels with Fully Profiled Faces

The problem was defined as maximizing the resistance of the panel while minimizing the embodied carbon. When the required resistance is known for the given span, a single optimal solution can be found. The problem may be described by Eq. (1):

$$minimize\, f(x) = [f_1(x), f_2(x)^T] \tag{1}$$

Where $f_1(x)=1/q$, q being the maximum resistance and $f_2(x)=k_{PIR}W_{PIR}+k_{Steel}W_{Steel}$ is the embodied carbon function using the carbon coefficients of PIR (k_{PIR}) and galvanized steel (k_{Steel}) and W_{PIR} and W_{Steel} refer to the weight of PIR and steel respectively. The results are expressed in $kgCO_2e/m^2$.

The variable vector x includes all the geometrical and mechanical properties which influence the strength and stiffness behavior of a sandwich panel with one face fully profiled and one face lightly profiled. Since relationships between the core density and its mechanical properties were earlier established, the variable vector can be reduced to:

$$x = [D, d_1, h_1, h_2, b_1, b_2, \rho_C, f_{y1}, f_{y2}, t_{F1}, t_{F2}] \tag{2}$$

Where D is the depth of the panel excluding the height of the full profile d_1 at the outer face; h_i is the height and b_i the width of the light profiles at each face; $t_{f,i}$ is the thickness and $f_{y,i}$ the yield strength of each steel face; subscripts *1* and *2* refer to the outer and inner faces of the panel respectively.

The constraints were associated with the design checks in terms of strength and stiffness according to EN 14509:2013 (CEN 2013). Temperature effects were also taken into account, considering medium-color for the outer face of the panel. The following ranges of properties were adopted for the study:

$$100mm \leq D \leq 150mm;\ 31.3mm \leq d_1 \leq 61.3mm;\ h_1=h_2=3mm,\ 23.8mm \leq b_2 \leq 33.8mm; \tag{3}$$

$$0.45mm \leq t_{F1} \leq 0.75mm;\ 0.35mm \leq t_{F2} \leq 0.65mm; f_{y1} = f_{y2} = 350N/mm^2$$

The solutions are shown in Figure 2 and include optimal and non-optimal solutions. The lower bound series is the Pareto-optimal solution. For desired characteristic loads of +0.60kN/m^2 in pressure and -0.90kN/m^2 in suction for the assumed load conditions in the study, a single optimal solution can be determined on the Pareto-front. This gives a panel specification with the variables and embodied carbon shown in Eq. (1). The re-engineered panel has 13.9% more embodied carbon compared to the currently available panels with the same thermal performance (U-value = $0.15W/m^2K$).

$$x = [D=150mm,\ d_1=61.3mm,\ h_1=,\ h_2=3mm,\ b_2=23.8mm,\ \rho_C=38kg/m^3 \tag{4}$$

$$f_{y1}=f_{y2}=350N/mm^2,\ t_{F1}=0.44mm,\ t_{F2}=0.34mm];\ f_2(x) = 38.21CO_2e/m^2$$

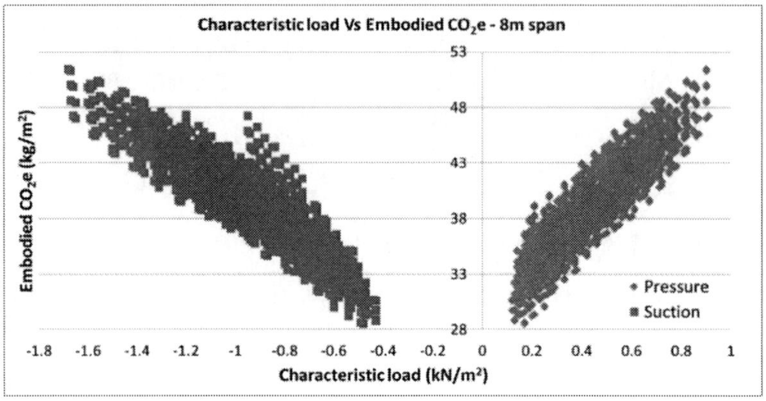

Figure 2. Solutions for 8m single-span sandwich panels, medium-color group.

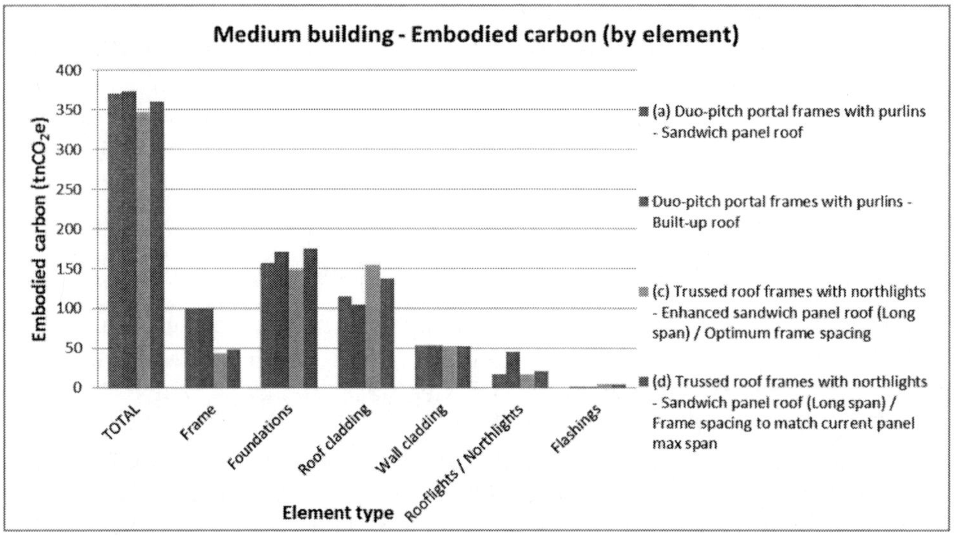

Figure 3. Embodied carbon for various schemes (medium-sized building).

5 EMBODIED CARBON OF STRUCTURE-ENVELOPE ASSEMBLIES

An embodied carbon review was undertaken for the combined structure and envelope using a ‘Cradle-to-gate’ approach and data by Jones and Hammond (2008). The schemes selected for comparison were traditional portal frames with purlins using (a) sandwich panels, (b) built-up twin skin metal cladding and trussed-roof frames with north lights using long span roof sandwich panels at (c) optimum frame spacing comprising the enhanced panels and (d) with frame spacing at the maximum span possible using current technology. Sandwich panel wall cladding systems were selected for all schemes, together with appropriate specifications for rooflights / north lights glazing and flashings. Figure 3 shows the results for the medium-sized building.

It is apparent that scheme (c) gives a reduction of 6.5% against scheme (a) and 7.3% against scheme (b) in terms of total embodied carbon. Schemes (c) and (d) have a much lesser impact in terms of emission associated with the frame due to the reduced steelwork weight, but their impact is higher for the roof cladding due to the enhanced sandwich panel specifications and increased envelope area. There is an overall higher superstructure weight for schemes (b) (due to the heavier built-up cladding) and (d) (due to non-optimum frame spacing). Finally, the study showed that the selection of rooflight / north light glazing system has a considerable impact on the total embodied carbon of the structure. Similar results were also found for the large building size, while the differences were only very small for the small building.

6 CONCLUSIONS

This study has demonstrated that a significant saving in embodied carbon is possible by exploiting the spanning capability of modern sandwich panel roof cladding to reduce the weight of the structural frame. However, in order to take full advantage of this potential saving, it is necessary to make minor changes to the panels to optimize their performance. An enhanced sandwich panel specification has been presented that meets this requirement.

Acknowledgments

The authors would like to thank the Engineering and Physical Sciences Research Council (EPSRC) and TATA Steel Colors for the financially sponsoring the research project.

References

European Committee for Standardization (CEN), *EN 14509:2013: Self-supporting double skin metal faced insulating panels – Factory made products – Specifications*, Brussels, 2013.

Jones C and Hammond G (2008) Inventory of Carbon & Energy (ICE) (Version 1.6a). Bath: Sustainable Energy Research Team (SERT), University of Bath.

Kurpiela A., Lange J., Optimierung von Sandwichwandbauteilen (Optimisation of wall sandwich panels), *Stahlbau*, 82(11), p. 843-850, 2013.

Moutaftsis, D., Heywood, M. D., Ogden, R. G. Improved Structural Efficiency in Steel Single Storey Buildings using Sandwich Panels, Proceedings of the 8th International Conference on Advances in Steel Structure, Lisbon, Portugal, 21-24 July, 2015a.

Moutaftsis, D., Resalati, S., Heywood, M. D., Ogden, R. G., Optimised Low and Operational carbon solutions for single storey industrial buildings, Proceedings of the Eighth International Structural Engineering and Construction Conference: Sustainable Solutions in Structural Engineering and Construction, Sydney, Australia, 23-28 November, 2015b.

Pokharel N. and Mahendran M., Experimental investigation and design of sandwich panels subject to local buckling effects, *Journal of Constructional Steel Research*, Volume 59, Issue 12, p. 1533-1552, 2003.

Stamm, K. and Witte, H., *Sandwichkonstructionen ± Berechnung, Fertigung, Ausfuhrung (Sandwich Construction - Calculation, Manufacture and Use)*, Springer-Verlag, Wien and New York, 1974.

Interaction between Theory and Practice in Civil Engineering and Construction
Edited by Komurlu, R., Gurgun, A. P., Singh, A., and Yazdani, S.

INFLUENCE OF SPAN LENGTH AND CROSSBEAM ON LOAD DISTRIBUTION FACTOR FOR GIRDER BRIDGES

HYO-GYOUNG KWAK and JOUNGRAE KIM

Dept of Civil Engineering, KAIST, Daejeon, Korea

Load distribution factor at concrete girder bridges and steel girder bridges are analyzed with finite element method to see effect of span length and cross beam to load distribution factor. Span lengths of analyzed bridge models are 30m, 40m, 50m and 60m. The number of intermediate cross beam is increased from one to until distance between cross beams becomes 5m. The finite element analysis results show that concrete girder and steel girder can use same load distribution factor and span length doesn't affect to load distribution factor. Even though load distribution factor in interior girders is not influenced by cross beam, in exterior girders it is influenced by cross beam. Effect of cross beam in exterior girder is influenced by the number of lanes and distance from exterior girder to curb. Since design code introduces conservative load distribution factor, economically improved load distribution factor is proposed. The proposed load distribution factor includes cross beam effect with the number of lanes and distance from exterior girder to curb. The proposed equation is compared with AASHTO code and grillage method which is well-known method to calculate load distribution. The comparison results showed that the proposed equation is more efficient and useful than AASHTO and safer than the grillage method.

Keywords: Effect of cross beam, AASHTO code, Grillage method.

1 INTRODUCTION

Since bridge is a complex structure, calculation of load distribution is very complicated. Therefore design codes suggest load distribution factor for efficient bridge design. Regardless of girder material, AASHTO LRFD (2012) suggests different shear load distribution factor to interior girder and exterior girder (Zokaie 2000). For girder bridge with crossbeam or bracing, AASHTO introduces different load distribution factor based on rigid body analysis. Since it ignores many factors such as span length, crossbeam spacing or number of lanes which affect the load distribution, load distribution factor gives conservative result. It causes economical waste with less efficiency (Bishara *et al.* 1993, Barr and Amin 2006, Yousif and Hindi 2007).

2 AASHTO LRFD LOAD DISTRIBUTION FACTOR

Load distribution factor without crossbeam or bracing is calculated based on girder spacing(S). Load distribution factor to interior I-shaped girder is shown in (1)

$$LDF = 0.2 + \frac{S}{3600} - (\frac{S}{10700})^{2.0} \quad (1)$$

where S is distance between adjacent girders(mm).

Load distribution factor to exterior I-shaped girder is calculated based on load distribution factor of interior girder. It is shown in (2).

$$LDF = (0.6 + \frac{d_e}{3000}) \times LDF_{interior} \quad (2)$$

where d_e is distance from exterior girder to curb(mm).

Load distribution factor of exterior girder with crossbeam or bracing is derived based on rigid body analysis. It is shown in (3)

$$LDF = (\frac{N_L}{N_b}) + \frac{X_{ext} \sum^{N_L} e}{\sum^{N_b} x^2} \quad (3)$$

where N_L is the number of loaded lanes under consideration, e is eccentricity of a design truck or a design lane load from the center of gravity of the pattern of girders(ft), x is horizontal distance from the center of gravity of the pattern of girders to each girder(ft), X_{ext} is horizontal distance from the center of gravity of the pattern of girders to the exterior girder(ft), N_b is the number of beams or girders.

Keating *et al.* (1997) found that since the design code doesn't include effect of span length or crossbeam properties such as spacing and position, it gives very conservative result. Therefore for economical design, effect of span length and crossbeam properties should be considered to improve load distribution factor.

3 GEOMETRIC AND STRUCTURAL PROPERTY

I-shape girder brides with fixed girder spacing and various span lengths are chosen to obtain same load distribution factors from AASHTO Code. Girder spacing is 2.5m and span length is increased by 5m from 30m to 60m. Six different bridge cross sections are selected based on the number of girder and distance of exterior girder to curb. The properties of bridge model are shown in Table I.

The number of intermediate crossbeam is varied from one to six. So crossbeam spacing varies from 5m to 17.5m. Since AASHO LRFD uses same load distribution factor for concrete girder and steel girder, concrete girder and steel girder are selected to compare material property effect. Boundary condition of bridge is simply supported

Table 1. Properties of bridge model.

Case	1	2	3	4	5	6
Number of girder	3	3	4	5	6	6
Number of lane	2	2	3	3	4	4
Distance of exterior girder to curb(m)	0.5	0.8	0.8	0.3	0.3	0.5

4 FINITE ELEMENT METHOD

Finite element method (FEM) is used to analysis load distribution with commercial finite element software, ABAQUS. Concrete girder FEM model is from Barker and Puckett (2013) and steel girder FEM model is from Mabsout *et al.* (1997) and Chung and Sotelino (2006). Girders, crossbeam and slab are modeled by solid elements (ABAQUS C3D20).

5 RESULT

Figure 1~4 are results of FEM about span length effect to interior and exterior girders. Figure 1 and Figure 3 are results of concrete girder and Figure 2 and Figure 4 are results of steel girder. Variations of load distribution factor of all cases are less than 2%. Since load distribution factor doesn't change as span length increase, it can conclude that span length doesn't affect to load distribution factor.

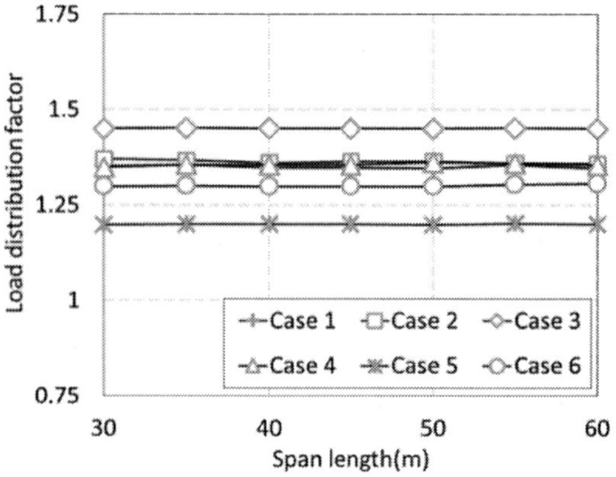

Figure 1. Load distribution factor of interior concrete girder about span length.

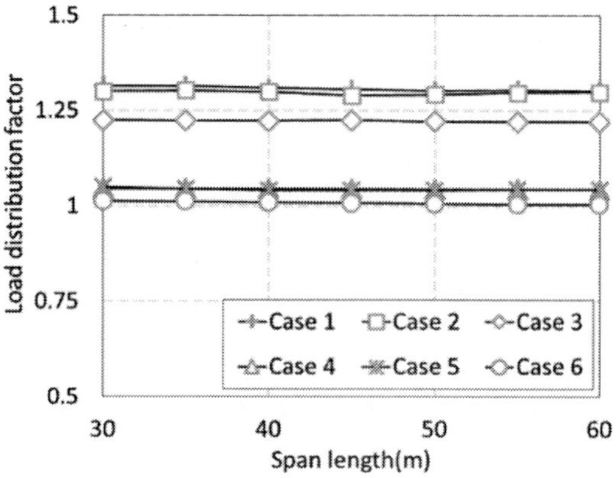

Figure 2. Load distribution factor of interior steel girder about span length.

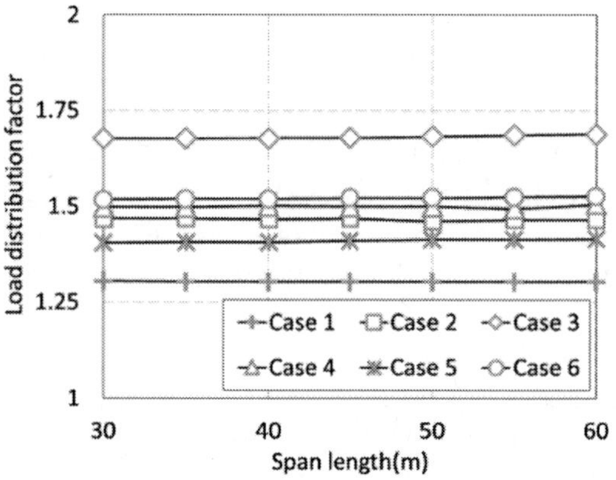

Figure 3. Load distribution factor of exterior concrete girder about span length.

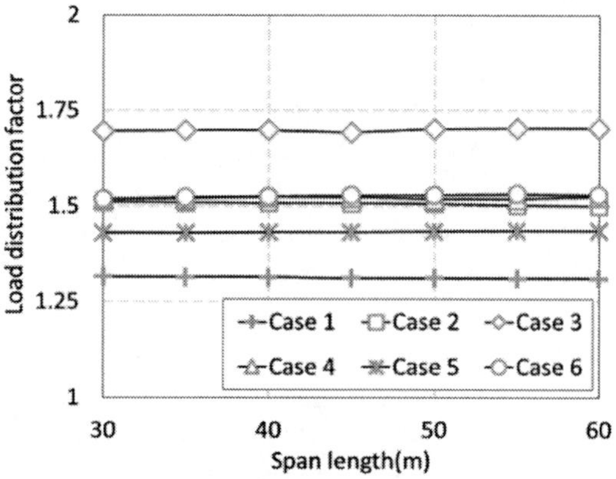

Figure 4. Load distribution factor of exterior steel girder about span length.

Figure 5~8 are results of FEM about crossbeam spacing effect to interior and exterior at 35m span length. Figure 5 and Figure 7 are results of concrete girder and Figure 6 and Figure 8 are results of steel girder.

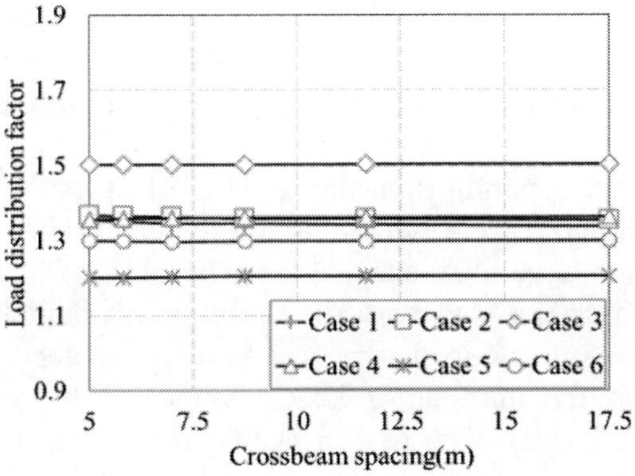

Figure 5. Load distribution factor of interior concrete girder about crossbeam spacing.

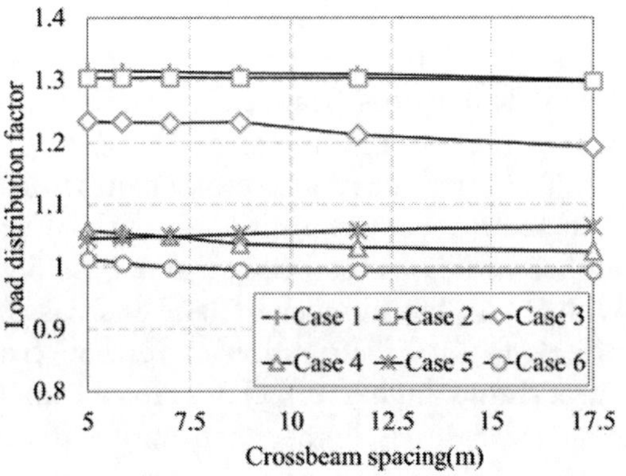

Figure 6. Load distribution factor of interior steel girder about crossbeam spacing.

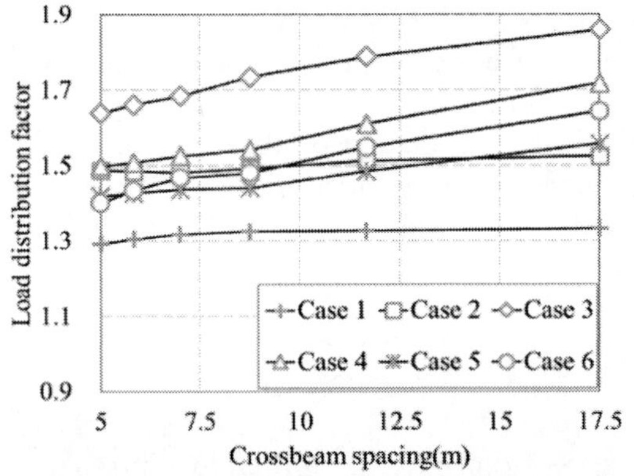

Figure 7. Load distribution factor of exterior concrete girder about crossbeam spacing.

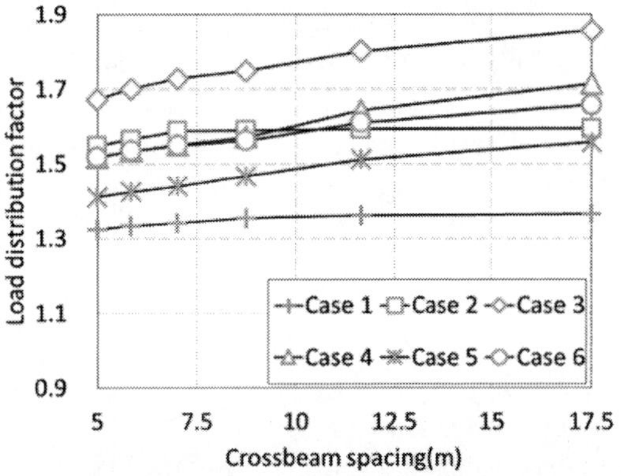

Figure 8. Load distribution factor of exterior steel girder about crossbeam spacing.

Figure 5 and Figure 6 show load distribution factors of interior girder with different crossbeam spacing. Variations of load distribution factors in interior girder are less than 2%. Therefore for interior girder, effect of crossbeam can be ignored as suggested by AASHTO.

In AASHTO, load distribution factor of interior girder is calculated only with girder spacing. So six cases should have same load distribution factor. However in Figure 5 and Figure 6, load distribution factor is rather affected by the number of girders, distance from exterior girder to curb and girder material. Even though load distribution factors of interior girder are different from case to case, all of them are below AASHTO code which is 1.66

Figure 7 and Figure 8 show load distribution factors of exterior girder with different crossbeam spacing. Case 1 and Case 2 show little variations which are less than 3%. Case 3~6 show higher variations of load distribution factor which is at least 10%. Case 1 and Case 2 are bridges supported by three girders. Therefore it can conclude that crossbeam spacing affects load distribution factor of exterior girder in the case of the four or more girders bridge.

The effectiveness of crossbeam is different depending on the number of lanes and distance from exterior girder to curb. Case 3 and Case 4 which have three lanes, show higher variations than Case 5 and Case 6 which have four lanes. This is because of slab width. As the number of lane increases, so as the slab width. Therefore wide slab affects to load distribution more than crossbeam. Also distance of exterior girder to curb shows higher effect to crossbeam than the number of lane. Slope of shorter distance cases is higher than longer distance cases. The reason is that if distance of exterior girder to curb is longer, the more vehicle load can be applied at outside of exterior girder, so effect of cross beam between exterior and interior girder is reduced.

6 PROPOSED LOAD DISTRIBUTION FACTOR

The number of crossbeam is effective to load distribution factor at exterior girder of four or more girder bridges. As spacing between crossbeams increases, load distribution factor increases. Also distance from exterior girder to curb and the number of lanes affect to slope of variation of load distribution factor. As distance from exterior girder to curb and the number of lane decrease, variation of load distribution factor due to cross beam is increased. Proposed load distribution factor includes all these factors. Therefore proposed load distribution factor is shown in (4).

$$LDF = \left(\frac{0.75}{N_l^{1.5}} + \frac{18}{d_e}\right) \times s^{0.355} + 2 \times \left(0.6 + \frac{d_e}{3000}\right) \times \left(0.2 + \frac{S}{3600} - \left(\frac{S}{10700}\right)^{2.0}\right). \quad (4)$$

where N_l is the number of lane, de is distance from exterior girder to curb, s is distance between cross beam (m) and S is distance between girder(mm). Last term in the right equation is from AASHTO LRFD which is for non-crossbeam.

Figure 9 and Figure 10 are comparisons between FEM, Grillage method, AASHTO LRFD and proposed equation for concrete and steel girders. AASHTO LRFD gives the highest load distribution factors for both cases. Grillage method is traditional method of analyzing load distribution. Even though grillage method shows cross beam effect, it gives the lowest load distribution factors. Therefore, proposed equation is more efficient and useful than AASHTO, and safer than the grillage method.

7 CONCLUSION

Even though span length is important design factor, it doesn't affect to load distribution. Therefore span length effect can be neglected at load distribution factor. From this research, it can figure out that crossbeam is effective at exterior girder with more than four girders. As the number of crossbeam increases, vehicle load is more equally distributed to girders which is purpose of crossbeam so that the load distribution factor is decreased. AASHTO LRFD introduces very conservative load distribution factor compare to FEM results and uses different load distribution factor depending on presence of crossbeam. The proposed equation includes load distribution factor of non-crossbeam case for convenient use so it can be used without considering crossbeam presence. Therefore proposed equation gives more efficient load distribution factor than AASHTO, and safer load distribution factor than the grillage method.

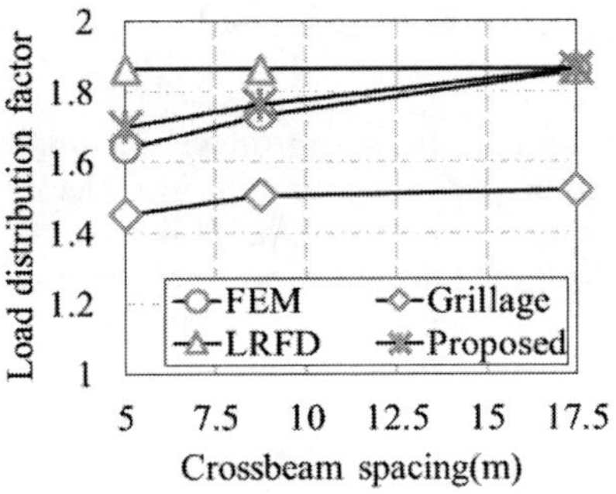

Figure 9. Comparison with methods at concrete girder.

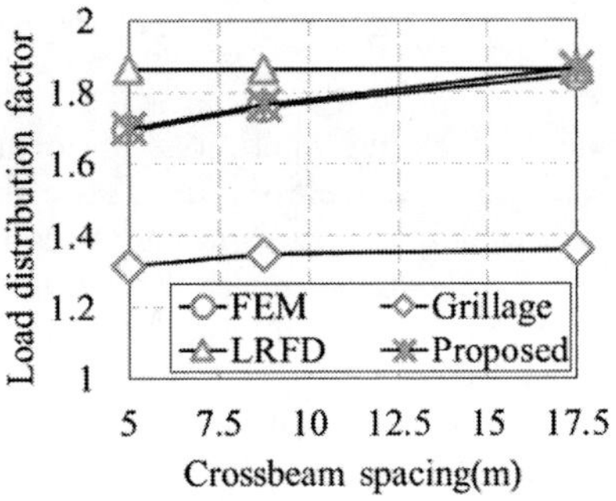

Figure 10. Comparison with methods at steel girder.

Acknowledgments

This research was supported by a grant (13SCIPA02) from Smart Civil Infrastructure Research Program funded by Ministry of Land, Infrastructure and Transport (MOLIT) of Korea government and Korea Agency for Infrastructure Technology Advancement (KAIA) and a grant (16RDRP-B076268-03) from R&D Program funded by Ministry of Land, Infrastructure and Transport of Korean government.

References

AASHTO, *LRFD Specifications for Highway Bridges, 6th ed*, Washington, DC ,2012.Barker, R. M., and Puckett, J. A., *Design of Highway Bridges : An LRFD Approach, 2nd ed*., Wiley, New York, 2013.

Barr, P. J., and Amin, MD. N., Shear live-load distribution factors for I-girder bridges, *Journal of Bridge Eng*., 11(2), 197-204, March, 2006.

Bishara, A. G., Maria, C. L., and El-Ali, N. D., Wheel load distribution on simply supported skew I-beam composite bridges, *Journal of Structural Eng*., 119(2), 399-419, 1993.

Chung, W., and Sotelino, E. D., Three-dimensional finite element modeling of composite girder bridges., *Engineering Structures*., 28(1), 63-71, Jan, 2006.

Keating, P. B., Saindon, K. C., and Wilson, S. D., *Cross Frame Diaphragm Fatigue and Load Distribution Behavior in Steel Highway Bridges*., Texas Transportation Institute, Texas, 1997.

Mabsout, M. E., Tarhini, K. M., Frederick, G. R., and Tayar, C., Finite-element analysis of steel girder highway bridges., *Journal of Bridge Eng*. 2(3), 83-87, Aug, 1997.

Yousif, Z., and Hindi, R., AASHTO-LRFD live load distribution for beam-and-slab bridges: Limitations and applicability, *Journal of Bridge Eng*., 12(6), 765-773, Nov, 2007.

Zokaie, T., AASHTO-LRFD live load distribution specifications, *Journal of Bridge Eng*., 5(2), 131-138, May, 2000.

Interaction between Theory and Practice in Civil Engineering and Construction
Edited by Komurlu, R., Gurgun, A., Yazdani, S., and Singh, A.

USE OF THE HOLE DRILLING METHOD TO DETERMINE RESIDUAL WELD STRESSES IN BRIDGE CONSTRUCTIONS

EVY VAN PUYMBROECK, WIM NAGY, and HANS DE BACKER

Dept of Civil Engineering, Ghent University, Ghent, Belgium

Complex welding operations in orthotropic steel bridge decks introduce residual stresses near the weld region. To estimate fatigue failure of this type of bridge deck, tensile residual yield stresses are usually assumed around the weld region. However, to estimate the residual stress distribution near a weld connection more precisely, a test setup is developed. The weld connection of a closed longitudinal trapezoidal stiffener with the deck plate of an orthotropic bridge deck is investigated. The incremental hole-drilling technique is used to measure the residual stresses with strain gauge rosettes. Strain gauge rosettes are positioned on the deck plate and on a longitudinal stiffener of the orthotropic steel deck. A small hole is drilled through the center of the strain gauge rosettes and strains are measured at incremental depths. The residual stresses are calculated and based on these experimental measurements a distribution of the residual stresses is obtained. Compressive residual stresses exist near the longitudinal stiffener-to-deck plate weld. On the deck plate, the compressive residual stresses are equal to 60% of the yield strength while the compressive residual stresses on the stiffener are 42% of the yield strength. There are tensile residual stresses on both sides of the weld region. However, more research is necessary to confirm this distribution since it is contradictory to expected stresses in literature.

Keywords: Orthotropic steel deck, Strain gauge rosette, Residual stress measurement, Stress analysis, Fatigue failure.

1 INTRODUCTION

Orthotropic steel decks are stiffened in two mutually perpendicular directions with longitudinal and transversal stiffeners welded to the deck plate. Complex welding operations are required that introduce residual stresses in the orthotropic steel deck (OSD). Residual stresses are often ignored in the design of OSDs. This is acceptable when only focusing on the stress variations, thus eliminating the initial stress state. However, residual stress distributions can have large stress gradients due to their non-uniform behavior. These stress gradients make it necessary to perform residual stress measurements in order to correctly estimate premature fatigue failure and stress corrosion cracking (Schajer 2013).

Around the weld region, tensile residual stresses comparable to the yield stress are usually assumed to be present in order to estimate fatigue failure of an OSD (Xiao 2008, Xiao 2006). However, accurate fatigue design of an OSD requires an evaluation of the residual stress distribution. Depending on the sign and location of the residual stresses, the contribution of the residual stresses to the load-induced stresses can be either harmful or beneficial. Tensile residual stresses caused by manufacturing processes such as welding operations tend to be harmful due to their contribution to fatigue failure and stress-corrosion cracking. On the other hand, compressive residual stresses are usually beneficial because they increase wear and corrosion resistance and prevent the initiation of fatigue cracks (Rossini *et al.* 2004).

2 INCREMENTAL HOLE-DRILLING TECHNIQUE

The residual stresses around the weld connection of the deck plate with a longitudinal stiffener of an OSD are evaluated making use of the incremental hole-drilling technique. This is a semi-destructive measuring method where a small hole is drilled into the test material through the center of a strain gauge rosette. These strain gauge rosettes are used to measure the relieved surface strains caused by the introduction of a hole that is formed by drilling in a series of small steps. The measured strains are used to calculate the residual stresses according to the calculation principles specified in ASTM E837-13a 2015. The test method only applies when material behavior is linear-elastic. Therefore, reliable measurements are achieved by limiting the residual stresses to 80% of the yield stress of the material (ASTM E837-13a 2015).

To evaluate the residual stresses near the weld connection of the deck plate with a longitudinal stiffener, the RS-200 Milling Guide, Figure 1, from Vishay Measurements Group was used. To start the strain measurements, zero depth has to be established first. Therefore, the test surface material has to be exposed by drilling only through the material of the strain gauge rosette. The initial uncertainty of the separation of the cutter from the outer surface by the strain gauge rosette and coating can be disregarded by establishing this zero depth (Vishay Measurements Group 2011).

Figure 1. RS-200 Milling Guide (left) and hole-drilled strain gauge rosette (right).

3 TEST SETUP

Residual stress measurements were performed on an OSD with steel quality S235. The deck plate has a total length of 8.2 m and is simply supported at 100 mm from its edges. It spans over a width of 4.1 m and is equipped with six longitudinal and three transversal stiffeners. The deck plate has a thickness of 15mm and the longitudinal stiffeners are 6mm thick. The closed trapezoidal longitudinal stiffeners are 300 mm high, 300 mm wide on top and 125 mm at the lower soffit. The weld toes have a width of 7 mm. For the residual stress measurements, the connection of the deck plate with a longitudinal stiffener is considered, Figure 2.

Strain gauge rosettes are placed on the top and bottom of the deck plate and at the exterior of the longitudinal stiffener in the vicinity of the weld region, Figure 2. There will be no strain gauge rosettes located at the inside of the stiffener because the hole drilling rig has to be perpendicular to the strain gauge rosette and this is not possible at the inside of a stiffener. To measure strains adjacent to the weld, strain gauge rosettes type CEA-13-062UM-120 are used. This is a type B rosette with all three strain gauges located at the same side of the hole, allowing for measuring strains adjacent to the weld. On all other locations, strain gauge rosettes type CEA-13-062UL-120 are used because these type A rosettes are more accurate than type B rosettes.

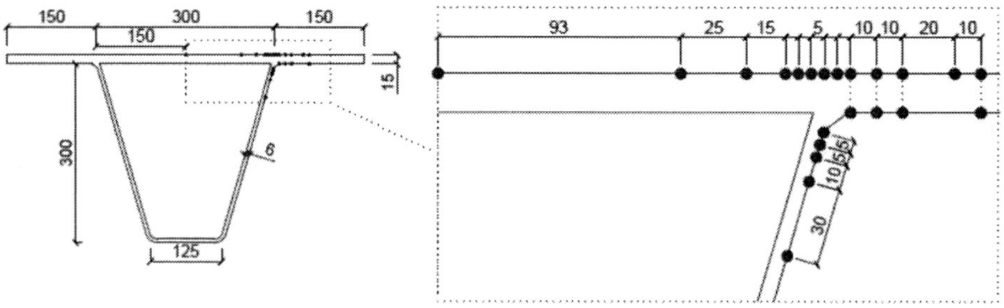

Figure 2. Weld connection of the deck plate with a closed trapezoidal longitudinal stiffener of the test setup with the position of the strain gauge rosettes (right) (dimensions in mm).

A total of 22 strain gauge rosettes are positioned on both the deck plate and longitudinal stiffener and their positions are indicated in Figure 2. A minimum distance between the strain gauge rosettes needs to be respected because the drilling of a hole into the surface results in relaxation effects which extend beyond the boundaries of the strain gauge rosette (ASTM E837-13a 2015, Schajer 2013). This minimum distance is equal to minimum six times the hole diameter. A number of strain gauge rosettes are necessary in the close proximity of the weld connection of the deck plate with the longitudinal stiffener to evaluate the residual stresses near this connection detail. Therefore, the residual stress measurements are performed in different cross sections every 50 mm, starting from the middle of the first span of the OSD. The diameter of the borehole for all strain gauge rosettes is approximately to 2 mm, corresponding to the diameter of the drill. In order to comply with the requirements for linear-elastic material behavior, the maximum drilling depth is to 1 mm (ASTM E837-13a 2015).

4 RESULTS

For each strain gauge rosette, strains are measured at incremental depths of 0.05 mm. The calculation software H-DRILL provided by the manufacturer is used to convert the measured strains into residual stresses according to the principles of ASTM E837-13a (2015). The residual stress distribution into the depth of the material is obtained for every location where a strain gauge rosette is positioned. However, to evaluate the initial stress state caused by welding the longitudinal stiffener to the deck plate, a single value for the residual stress that is representative for the considered strain gauge rosette is chosen. Near-surface stresses are unreliable due to surface preparations of the OSD because the results show large stress peaks higher than yield stress near the surface. It is not clear what the progress of the residual stresses is going to be when drilling deeper into the material and therefore, only the residual stresses at the final hole depth of 1 mm are considered. These residual stresses are used to make a representative distribution of the residual stresses in relation to their distance to the weld.

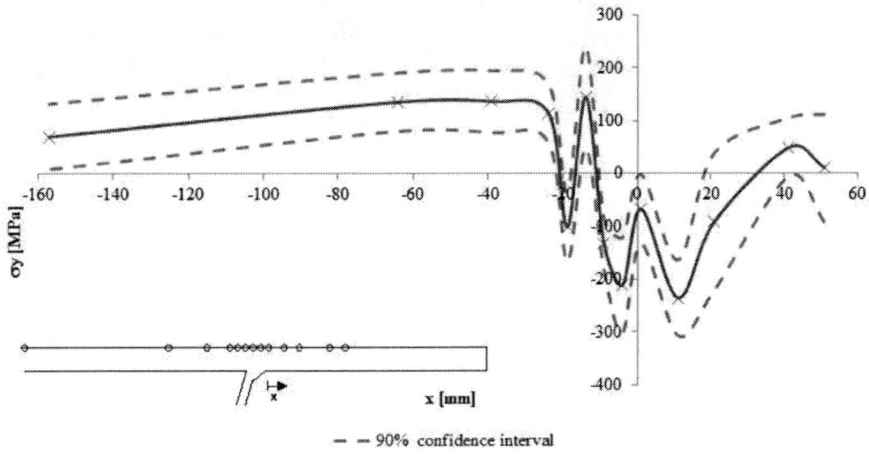

Figure 3. Transversal residual stress distribution surface of deck plate.

A residual stress distribution for the longitudinal and transversal direction is established in relation to the position towards the weld. Only the transversal direction is shown here because the residual stresses in the longitudinal direction are similar in sign and magnitude near the stiffener-to-deck plate weld. The 90% confidence intervals are also plotted and they indicate the expected stress range that has a 90% probability of containing the actual residual stresses. The position where the residual stress was measured is indicated by the distance *x* to the weld toe. The transversal residual stress distribution on top of the deck plate is shown in Figure 3, the distribution for the bottom of the deck plate in Figure 4 and for the longitudinal stiffener in Figure 5.

When the residual stresses on the top, Figure 3, and bottom of the deck plate, Figure 4, are compared, it is clear that the results have a relatively good correspondence at the right-hand side of the weld. Therefore, a uniform residual stress distribution throughout the thickness of the deck plate is assumed.

To evaluate the residual stresses near the longitudinal stiffener-to-deck plate weld, a distinction is made between compressive and tensile zones. The residual stress values assigned to these zones are calculated by averaging the residual stresses of the strain gauge rosettes located in the considered zone. These zones are evaluated in relation to their position towards the weld region.

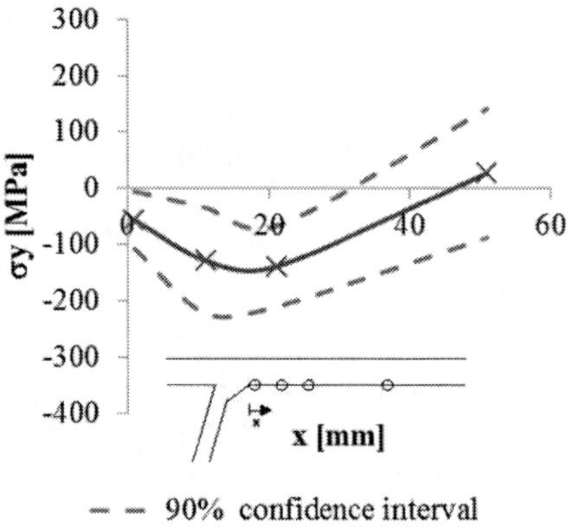

Figure 4. Transversal residual stress distribution bottom side deck plate.

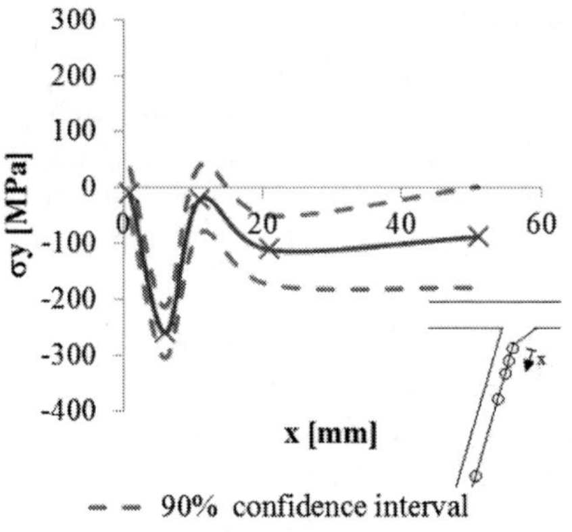

Figure 5. Transversal residual stress distribution longitudinal stiffener.

On the deck plate, two tensile zones and a compressive zone are present. An average tensile residual stress of 121 MPa is present in the deck plate between the two welded webs of the stiffener. This corresponds to 51% of the yield strength. In this zone, one strain gauge rosette that indicated a compressive residual stress is left out. Near the location of the weld on the deck plate, a compressive zone with an average residual stress of 141 MPa or 60% of the yield strength exists. On the right hand side of the weld region, again a tensile residual stress zone with an average value of 34 MPa or 14% of the yield strength is found. For the stiffener, compressive residual stresses near the weld with an average value of 99 MPa or 42% of the yield strength exist while at the weld toe the residual stress is close to zero. This general transverse residual stress distribution is shown in Figure 6; positive values indicate tensile stress.

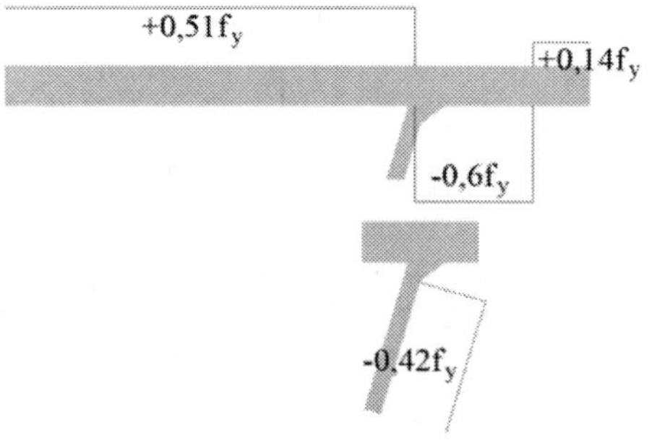

Figure 6. General transverse residual stress distribution (f_y= yield strength).

Residual stresses remain within the material even after the initial cause of the stresses has been removed. Consequently, they are self-equilibrating, i.e. the sum of all stresses in a local area creates zero resultant force and moment. However, the general transverse residual stress distribution, Figure 6, shows that the residual stresses obtained with the experimental measurements are not yet in equilibrium for the considered zone.

5 COMPARISON WITH A THEORETICAL RESIDUAL STRESS DISTRIBUTION

The established residual stress distribution is compared with a theoretical residual stress distribution according to FHWA-IF-12-027 (2012). This theoretical distribution predicts tensile yield stresses to be present near the location of the weld on the deck plate. In between these tensile yield stresses, compressive residual stresses up to 25% of the yield strength are assumed. This distribution for the deck plate is totally different from the measured residual stresses. For the measured residual stress distribution, compressive residual stresses equal to 60% of the yield strength exist instead of tensile yield stresses according to FHWA-IF-12-027 (2012). On both sides of the weld region, the measured residual stresses in the deck plate are tensile while the theoretical residual stresses are compressive. For the stiffener, the sign of the measured residual stresses is equal to the theoretically expected residual stresses, however the value of the measured residual stresses is twice as high as the theoretically expected values.

6 CONCLUSIONS

The residual stresses near the weld connection of the longitudinal stiffener with the deck plate of an OSD are evaluated according to the incremental hole-drilling technique. A residual stress distribution in the transverse direction is obtained. However, when comparing the measurement results with a theoretical residual stress distribution (FHWA-IF-12-027 2012), a large difference in sign and magnitude of the residual stresses was noted. It is possible that the measured strains are still suffering from strain gauge rosette preparation. Therefore, more research will be necessary to verify the distribution of the residual stresses near the longitudinal stiffener-to-deck plate weld obtained with the hole-drilling technique. Bigger strain gauge rosettes can be used to obtain residual stresses deeper into the depth of the surface. In addition, drilling to greater depths can provide a better indication of the residual stresses because the uncertain stresses near the surface can be eliminated.

References

ASTM E837-13a, Standard Test Method for Determining Residual Stresses by the Hole-Drilling Strain-Gage Method, ASTM International, 2015

FHWA-IF-12-027, Manual for Design, Construction, and Maintenance of Orthotropic Steel Deck Bridges, US Department of Transportation, Federal Highway Administration, 2012.

Rossini, N., Dassisiti, M., Benyounis, K., and Olabi, A., Methods of measuring residual stresses in components, *Materials & Design*, Elsevier, 35, 572-588, 2012.

Schajer, G.S., *Practical Residual Stress Measurement Methods*, John Wiley & Sons Ltd, United Kingdom, 2013.

Vishay Measurements Group, Model RS-200 Milling Guide Instruction Manual Version 2.0, Micro-Measurements, USA, 2011.

Xiao, Z.G., Yamada, K., Inoue, J., and Yamaguchi, K., Fatigue cracks in longitudinal ribs of steel orthotropic deck, *International Journal of Fatigue*, Elsevier, 28, 409-416, 2006.

Xiao, Z.G., Yamada, K., Samol, Y., and Zhao, X.L., Stress analyses and fatigue evaluation of rib-to-deck joints in steel orthotropic decks, *International Journal of Fatigue*, Elsevier, 30, 1387-1397, 2008.

Interaction between Theory and Practice in Civil Engineering and Construction
Edited by Komurlu, R., Gurgun A. P., and Singh, A.

SKEW PLACEMENT OF ARCHES FOR ROAD BRIDGES

HANS DE BACKER, AMELIE OUTTIER, KEN SCHOTTE, WIM NAGY, and MARCO DIVERSI

Dept of Civil Engineering, Ghent University, Ghent, Belgium

This article describes a parametric study using a detailed finite element model of an arch bridge consisting out of a single arch with varying arch and hanger arrangement. The objective is to investigate the influence of overall arch dimensions, these different hanger arrangements and a variation of the skewness angle between the arch axis and the longitudinal bridge deck axis. The influence on the stresses, normal forces and bending moments in the main girders, hangers and arch cross-section is examined. Finally, a buckling analysis is performed for all variations of the base model and the change in critical buckling load is discussed. The displacements in the bridge deck are found to be the same for small angles of skewness. Starting from an angle of ± 13°, the location of the maximum deflection in the bridge deck shifts towards the ends of the bridge deck instead of being in the middle. This can be explained by the influence of the wind load acting on the arch structure combined with the direction of the horizontal resultants of the hanger forces. This can cause partial relaxation, thus ensuring a small uplift at the longitudinal girders of the bridge deck. This is also the cause for the higher displacements in the arch. The torsion effect in the arch results in a counter clockwise rotation of the arch. When performing a buckling analysis on the changing angle of skewness, it is found that this initially has only a small impact on the critical load factor.

Keywords: Tied-arch, Buckling, steel, Eurocode, Stability, Design.

1 INTRODUCTION

Nowadays, bridges are generally constructed out of steel, concrete or a combination of both. The components which have to bear mostly compression stresses are preferably made with concrete, avoiding tensile stresses, while the structural elements in tension are constructed out of steel, avoiding buckling problems. This fundamental choice of materials, combined with the type of bridge has a major influence on the bridge design and the actual behavior. An arch bridge is generally used for medium span bridges. Most arch bridges consist out of two parallel arches connected with each other by wind bracings in order to safely bear the transversal loads. A single arch does not have this characteristic and has to rely on a stiff connection of the arch with the bridge deck in the transversal direction. A skew placement could help to bring a solution for this problem.

A skew placement of an arch bridge is a rather modern bridge design so not many bridges of this type have been constructed. Since this is a new design concept, the amount of research on this topic is rather small. Most of the research on arch bridges focus on an arch bridge consisting out of two parallel arches.

To have the largest buckling load, parabolic arches are chosen above catenary arches or circular arches, which has been investigated by Austin and Ross (1976). But not only the arch form has a large influence, also the rise-to-span ratio is an important parameter in arch bridge design. Most arch bridges have a ratio between 0.1 and 0.5. For which a small ratio corresponds to a large horizontal trust at the arch springs and a large ratio to a heavier construction. For a special-shape arch bridge the influence of this parameter is investigated by Wen-Liang *et al.* (2010) and found that the optimal value lies in the neighborhood of 0.35 for which this value corresponds to a maximum value of the stability coefficient. The rise-to-span ratio is also an important factor for the buckling capacity of an arch. According to Chao Dou *et al.* (2014), a bigger rise-to-span ratio (in the range of 0.025 to 0.25) results in a smaller axial force in the arch so the tendency of buckling is reduced. The buckling behavior of an arch is strongly depending on the arch length. When the arch has a bigger rise, a longer arch is associated and a smaller critical buckling load is obtained.

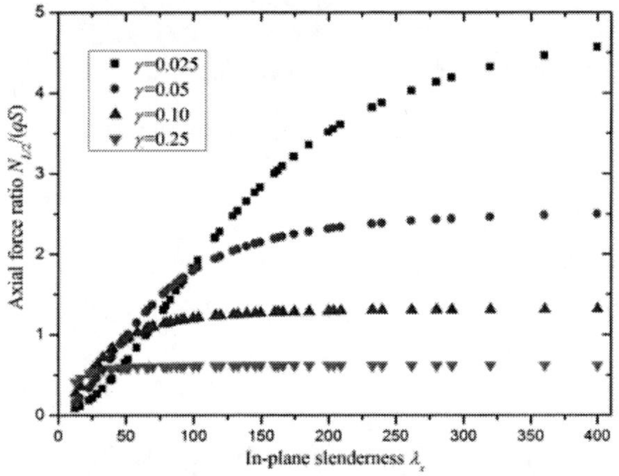

Figure 1. The influence of the slenderness of the arch cross-section on the axial force in the arch.

Another aspect investigated by Chao Dou *et al.* (2014) is the influence of the slenderness of the arch cross-section on the axial force in the arch. A cross-section associated with a larger slenderness, gives a larger axial force in the arch and thus buckling can happen easier. Both of the conclusions made by Chao Dou *et al.* (2014) are shown in Figure 1.

2 REFERENCE BRIDGE DESIGN

In the parametric study, the influence of every parameter is investigated in accordance with the reference design. This means that only this parameter has an impact on the results. The study starts with an arch bridge with a span of 90 m and a width equal to 9 m, shown in Figure 2. This circular arch has a height of 25 m, which results in a rise-to-span ratio of 0.278. The cross-section of the arch is a rectangle with a web height and flange width of 1500 mm and a flange- and web thickness of 45 mm. The hangers are able to sustain a design force of 2385 kN.

The two arch springs are modelled as clamped constraints and do not sustain any movements and rotations. The bridge deck is supported in the vertical direction at both ends. The longitudinal movements of the deck are restraints at one side and tolerated at the other. The movements in the transverse direction are restraint at only one side of the end beams.

To have relevant results, it is important to start with a realistic design. That is why some elements of this model are verified and compared to realistic arch behavior.

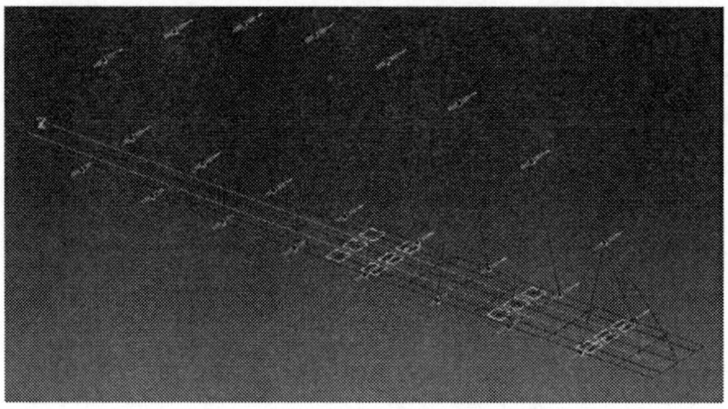

Figure 2. Arch bridge design.

3 PARAMETRIC STUDY

3.1 Width of the Bridge Deck

The elongation of the cross-girders from 9 m to 19 m in steps of 1 m, result in a bigger dead load of the bridge deck and in a larger surface on which all loads can be applied. The longer length of the beams gives larger deflections and bending moments so the yield strength and elastic moment of the girders are easily reached. The rise in tension in the cables is linear with the widening of the bridge deck. Of course, the extra forces acting in the hangers are brought to the arch structure, which leads the loads to its springs. The stresses in the arch undergo a raise, especially in the arch. The normal forces in the arch have to be smaller than the buckling force. This only results in collapse by buckling for bridges wider than 19 m. The buckling behavior of the arch gives a first buckling mode, which is associated with a half-sine function formed out of the arch plane. The critical load factor for a 9 m wide bridge deck has a value of 10.04, what means that all loads can be multiplied by this factor before buckling occurs. This factor decreases to 7.33 when the deck is widened to 19 m.

3.2 Rise-to-Span Ratio

As already stated in the literature study, the rise-to-span ratio has a large influence on the behavior of the arch bridge. The displacements in the arch and bridge deck have a parabolic distribution, so an optimum value can be found which has a magnitude of 0.2. The cable stresses decrease when a larger ratio is applied. The explanation in this development can be found in the angle between the bridge deck and the cable. Not only the displacements have a parabolic behavior, also the stresses in the arch tend to behave in that way. Again an optimum value can be found. The minimum tensile stresses in the arch occur at a rise-to-span ratio of 0.15 while the minimum pressure stresses happen at a ratio of 0.25. A changing rise-to-span ratio is associated with a changing arch length, which has a large influence on the critical buckling behavior of the arch. When comparing the normal forces in the arch, it can be seen that the out-of-plane buckling behavior is more critical than the in-plane behavior. Which corresponds to what is stated in previous sections. Another important consequence of the changing height of the arch is the direction of the resulting force vector at the arch springs. When constructing a lower arch, the impact of the horizontal trust will be much more important than the vertical resultant. An important remark here is that the second buckling mode is not an out-of-plane full sine, but an in-plane buckling mode for very low arches. This is because a lower arch will tend to behave like a horizontal beam, which has a critical buckling mode in its plane.

Figure 3. First buckling mode.

3.3 Arch Cross-Section

The deflections in the bridge deck as well as in the arch top decrease when installing a thicker arch cross-section. Because of the increased moment of inertia in the arch, the displacements and bending moments in the arch decrease. The larger surface of the cross-section corresponds to a decrease in the stresses in the arch, but with an increasing force as result because of the larger dead load of the arch. The increasing normal force is not a threat to the buckling behavior of the arch because with the increasing moment of inertia, the critical buckling load has a steeper increase. The critical load factor of the arch also increases when applying a thicker flange- and web thickness of the arch. The influence on the stresses in the bridge deck is minor when compared to the stress differences in the arch. The thickness of the arch doesn't have a direct influence on the bridge deck.

3.4 Skewness Angle and Hanger Arrangement

The displacements in the bridge deck stay the same for small angles of skewness. It is starting from an angle of ± 13°, that the position of maximum deflection in the bridge deck shifts towards the end of the bridge deck instead of staying in the middle. This is explained by the direction of the wind load acting upon the arch structure and the direction of the horizontal resultant of the hanger force vector. When both of these act in the same direction, the hanger will be more relaxed than when the horizontal resultant of the hangers is opposite to the wind loads. This results in a smaller uplift at the longitudinal girders of the bridge deck. The torsion effect in the arch results in a counter clockwise rotation of the arch. When performing a buckling analysis on the changing angle of skewness, it is found that this actually only has a small impact on the critical load factor. So placing the arch in a skewed position over the bridge deck does not mean that the arch will be submitted to an extra buckling load.

Different cable configurations are investigated to look for an optimal situation. These configurations are combined with 5 load models. Four different cable arrangements are investigated to look for the optimum amount of cables and to have sufficient available height for traffic. The first configuration uses cables at each intersection between the cross-girders and the longitudinal girders connected to the arch at equal horizontal distances. This configuration results in strong restrictions to the traffic height. The second configuration connects all cables to the inner 50% of the arch length. Not all cross-girder ends are being connected. The cables, which would cross the complete bridge deck in a rotated situation, are removed. This is also the case for configuration 3 and 4. In configuration 3, the cables are kept perpendicular in the horizontal plane to the longitudinal girder whereas in configuration 4, the cables are always connected perpendicular to the arch.

The normal forces in the arch increase when approaching the arch springs, as shown in Figure 4. The different jumps in the graph are located at the positions where hangers are attached. The bigger the hanger force, the bigger the jump. An important aspect of the skew placed arch are the direction of the hangers. These hangers pull the arch out of its plane, so large vertical bending moment around the vertical axis can be observed. The higher the angle of skewness, the bigger this effect and the higher the clamping moments at the arch springs.

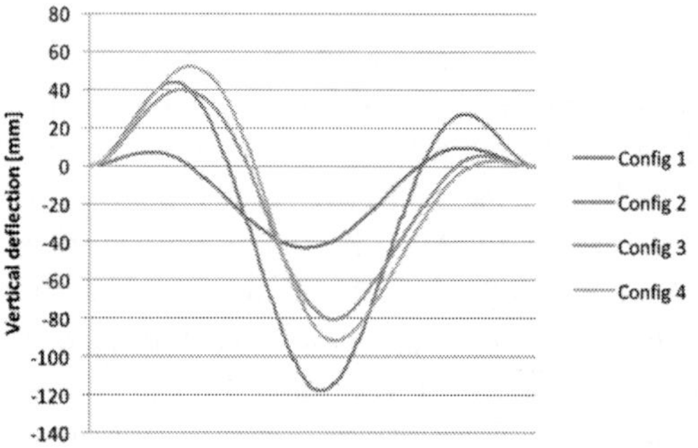

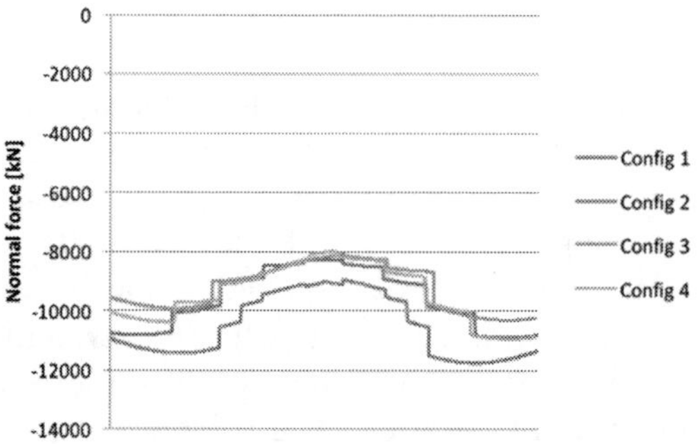

Figure 4. Vertical deflections (left) and normal forces (right) in the arch for different hanger configurations.

When looking at the critical buckling load of the different configurations, only small differences can be observed when using the same load model and angle of skewness. When changing the angle of skewness, some changes can be noticed. When increasing the angle from 0° to 20°, only minor differences occur. But when increasing the angle up to 40°, a strong decrease up to 30% can be noticed. An important difference between the difference configurations is the amount of hangers and location of the hangers. In configuration 1, 16 hangers are applied whereas for configuration 3 only 8. It is obvious that the stress in the hangers in configuration 3 will have to bear higher stresses since the same bridge deck has to be supported. In configuration 1 and 2, no difficulties with regard to pressure in the hangers can be found. This is not the case for configuration 3 and 4. Once a hanger is under pressure, it does not contribute to the system.

4 CONCLUSIONS

As already mentioned in the literature study, the rise-to-span ratio has an important impact on the bridge design. An optimal value of the ratio can be found in the range of 0.20-0.25. The stresses, deflections and forces in the deck and arch display a minimum value at this ratio. A widening bridge deck mostly influences the stresses in the transverse girders in the bridge deck. Reinforcement of these beams will be necessary. The extra dead load and applied loads on the deck result in extra hanger stresses and larger stresses in the arch structure as well. Changing the wall thickness of the arch results in smaller stresses and bending moments in the arch, but goes hand in hand with an increase of the dead load of the arch, larger normal forces will appear. The problem of the extra load and dead load also has a major influence on a longer bridge span. The longer the span, the larger the stresses in the arch. A stronger arch will be needed. Changing the skewness of the arch only has an influence on the end zones of the bridge deck. Extra deflections and stresses are observed in these sections. The buckling mode of an arch bridge stays the same for all models and show a half-sine function out of its plane. The critical load factor does not keep a constant value. The changing span decreases the load factor largely. The load factor for a changing rise-to-span ratio again shows an optimum value for a ratio of approximately 0.25. The angle of skewness does not have a major influence, so no buckling problems should be expected.

References

Austin, W.-J. and Ross, T.-J., Elastic Buckling of Arches under Symmetrical Loading, *Journal of Structural Division*, ASCE, 102, 1085-1095, 1976.

Chao, D., Yan-Lin, G., Yong-Lin, P. and Si-Yuan, Z., *Flexural-Torsional Buckling and Ultimate Resistance of Parabolic Steel Arches Subjected to Uniformly Distributed Vertical Load,* American Society of Civil Engineers, 2014.

Wen-Liang, Q., Chin-Sheng, K. *et al.*, Stability Analysis of Special-Shape Arch Bridge, *Tamkang Journal of Science and Engineering*, 13(4), 365-373, 2010.

Interaction between Theory and Practice in Civil Engineering and Construction
Edited by Komurlu, R., Gurgun, A. P., Singh, A., and Yazdani, S.

FINITE ELEMENT MODELING OF THERMAL LOADING ON STEEL BOX GIRDERS

MARCO DIVERSI and HANS DE BACKER

Civil Engineering Dept, Ghent University, Ghent, Belgium

Thermal loads are not often considered during the design of steel box girders, but their influence can be quite important. When the thermal gradient within the steel box girders reaches considerable values, a number of other effects are influenced. High thermal gradients will reduce the cohesion of waterproofing layers and surface layers. Therefore, the quantification of the thermal loads acting on the system becomes essential. This is primarily done using detailed finite element models of steel box girders. Modelling all the thermal fluxes within this system, including solar radiation, radiation with the environment, mutual radiation, convective airflow, etc. it becomes possible to study the temperature variations on the bridge structure. The thermal loading is based on the actual revolution of the sun, while the bounds and the orientation angles are the result of a detailed calculation of the position of the sun. The same is true for the respective fluxes. All of these variations are studied along a one year period and this on hourly basis.

Keywords: Bridge codes, Health monitoring, Temperature effects, Eurocode, Temperature load.

1 INTRODUCTION

The Vilvoorde Bridge is a landmark bridge located in Belgium in the suburbs of the capital Brussels. In this paper we will study the temperature effects on comparable steel box girder bridges. These bridges are constantly exposed to a non-uniform, constantly varying, solar radiation and the resulting temperature domain variations, during the seasons as well as daily, can be quite impressive. This can cause problems and stresses that are not included in the initial design, especially when combined with the other loads. The inclination of the sun with respect to the different surfaces of the box girder becomes more complicated when considering yearly cycles of daily variations transmitted to the supporting steel structure. The natural ventilation because of the manholes in these types of girders can never be influential enough to create a cooling airflow substantial enough to balance this effect. This paper describes an accurate model of these temperature distributions, related variations and a quantification of the thermal effect. The objective of the proposed research is to reach new and more fundamental perspectives for the given problem and for the behavior of closed steel box girders with orthotropic steel decks, subjected to a variable thermal load (De Backer *et al.* 2008; 2009).

2 DESIGN LOADS FOR TEMPERATURES

2.1 Differential Temperature

A finite element model has been created (Figure 1), based on the cross-section of the Vilvoorde Bridge in Belgium. It has geometry for the box girder that is 8 m wide and 5 m high, with side flanges with a width of 5 m each. This geometry was introduced into the SAMCEF finite element software. The thickness of the steel plates varies between 18 mm for the deck plate to 48 mm for the bottom plate.

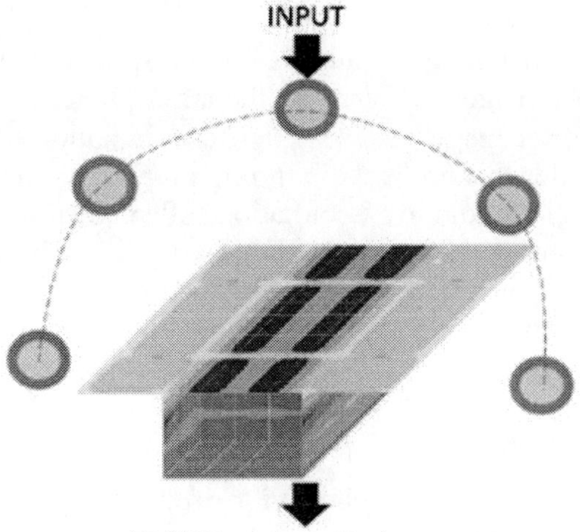

Figure 1. Input and outputs.

An epoxy asphalt layer with a thickness of 15 mm is assumed to be installed on the entire upper surface of the box girder for this calculation (De Backer *et al.* 2008). The thermal effects that need to be studied are:

- Solar radiation, resulting in a positive heat flux on the steel box girder;
- Heat exchange between the internal or external surfaces by convection;
- Heat loss because of the radiation between the structure and the environment;
- Mutual radiation between the external and internal surfaces of the box girder.

2.2 Solar Flux Calculation Method

The sun is actually a continuous fusion reactor. The solar constant is estimated to be a value that averages around 1367 W/m^2 (Kim 2009), although this is just a theoretical value. Ultimately, the estimated transmittance for beam radiation under the standard clear atmosphere on a horizontal surface of the box girder bridge can be calculated as following:

$$G_{cnb} = 1339 \, X \, \tau_b \tag{1}$$

Where:

$$\tau_b = a_0 + a_1 exp(-k/\cos\Theta_z) \tag{2}$$

A is the altitude of the site in kilometres, the constants a_0, a_1 and k are respectively,

$$a_0 = 0.4237 - 0.00821(6 - A)^2 \tag{3}$$

$$a_1 = 0.5055 + 0.00595(6.5 - A)^2 \tag{4}$$

$$k = 0.2711 + 0.01858(2.5 - A)^2 \tag{5}$$

Beam radiation, the major component of solar radiation, is the solar radiation received from the sun that is not scattered by the atmosphere. It is often referred to as direct solar radiation (Recknagel 1974). Diffuse radiation is the solar radiation received from the sun after the atmosphere through a process of scattering changed its direction. Ground-reflected radiation is radiation reflected from the ground cover and bodies of water on the surface of the earth. Ultimately the values that are obtained for the 1-day and 1-year calculations are extremely accurate and assigned as functions to the finite element model giving a realistic overview of the real thermal situation of the bridge (Duffie *et al.* 1974).

2.3 Results for Daily Approach

In this paragraph we will focus on the behavior of the bridge on a daily base, several nodes of the structure are studied but just the most interesting locations are reported in this paper. The day considered is June 21st since it is the day when the solar noon occurs at 13.44 and since it is, theoretically, the hottest day of the year for Brussels.

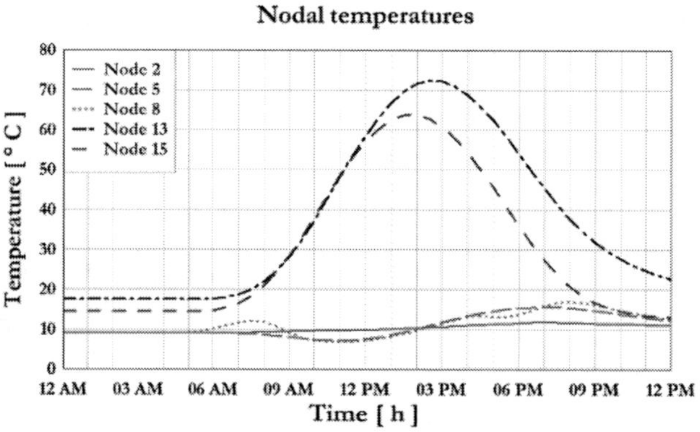

Figure 2. Temperature variations on the side of the box girder for 1-day calculation.

During this day (Figure 2), there are 16.31 hours of daylight and the component of the filtered solar flux on a horizontal plane averages a value of 710 W/m^2. The finite element model is composed of a structure of 2501 nodes, in this paper only 9 are selected nodes from the central cross-section of the steel box girder shown in Figure 3. The reason why we chose these nodes is because the central section of the bridge is the section where the highest values are reached, all the values are along the same line and more easy and visible to compare. There are 6 nodes laying on the central sides which are interesting because they are showing different values since the sun rises east and sets west, and there is a node laying in the bottom and two on the top in between the asphalt layer and the structure. The variation of the temperatures on the sides of the bridge between the sides is mainly due to the heating process that starts earlier on the eastern side where the sun rises. It is quite clear that the two nodes on the asphalt on the top are delivering the maximal values for the temperatures. However, the maximum value does not occur at the same instance as the minimum value because different factors such as position, mutual radiation and heat exchange between the constituting parts are varying the temperature field in the box girder.

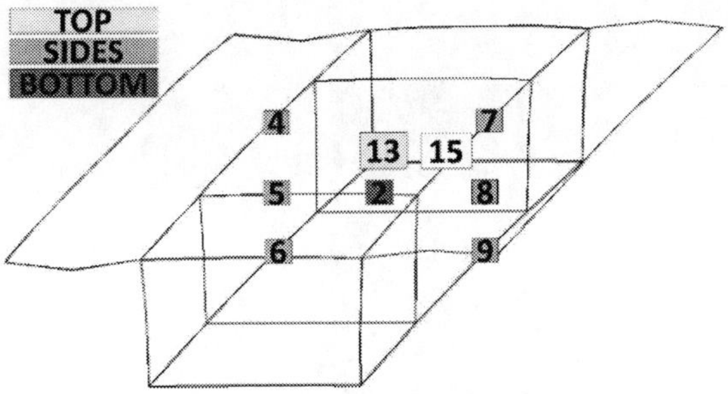

Figure 3. Overview of the position of the more significant nodes of the finite element model of the box girder of the bridge.

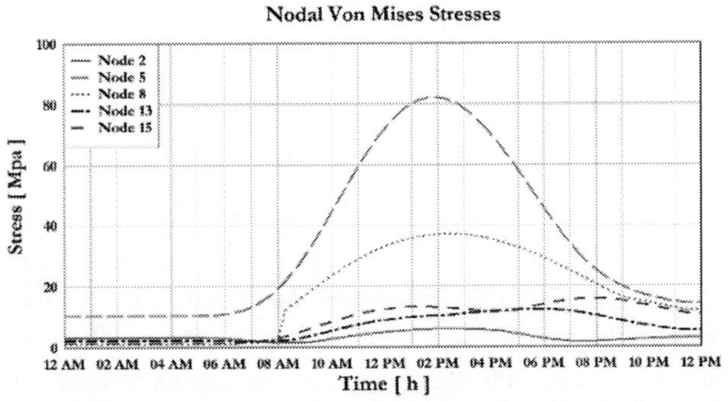

Figure 4. Stress variation considered nodes for a daily calculation.

Also interesting is the fact that the temperatures at the end of the cycle are not going back to the value registered in the morning (Figure 4). This gives a good indicator that the model works in a proper way because the bridge will heat during the day and it will cool down during the night to a temperature that in practice cannot be the same as the one registered in the morning. The model thus correctly represents the influence of successive hot days as an overall heating of the average temperature of the structure. The influence of the diaphragms in the box in conducting the heat flow away from the deck plate is also quite obvious, by the substantially lower temperature of the deck plate. The deck plate (Figure 5) also remains a bit cooler halfway between the sides of the box, since the influence of the secondary radiant heat sources, being the inner surface of the sides of the box girder, is smaller (De Backer *et al.* 2009).

The resulting additional stresses seem to be smallest in the deck plate zone above the diaphragms and above the sides of the box girder, since these parts of the structure help to disperse the solar heating to the cooler parts of the structure, thus resulting in a smaller thermal loading. The study of the temperatures and stresses are processed firstly through a thermal computation and then a mechanical computation that uses the previous thermal values as constraints. We can note that Von Mises stress is at maximum towards the fixed end of the beam, and the value is around 80 MPa. This is less than the yield point value of mild steel. So the design is safe. In short, an engineer task will be to keep the maximum value of Von Mises stress induced in the material less than its strength.

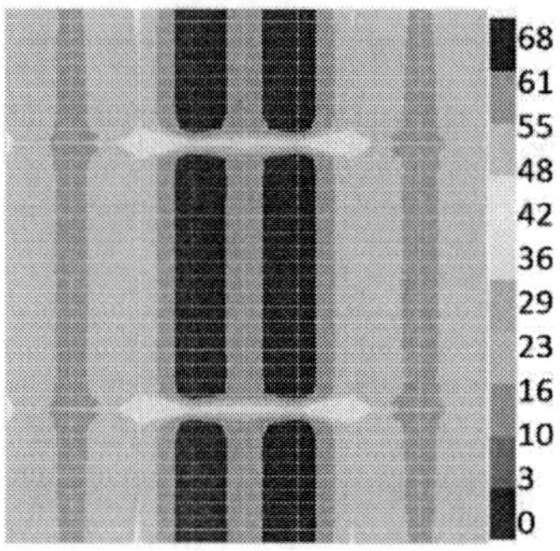

Figure 5. Top view of the temperature distribution in the deck plate during the hottest hour of the day.

2.4 Results for the Yearly Approach

The issue becomes more interesting and complicated for the yearly study, in this paragraph we will focus on the behavior of the bridge on a yearly base. The left part of the structure takes more time to cool down since it is on the eastern side and the sun starts to heat up from early in the morning, both patterns in the winter and in the summer show a consistent variation even though in Belgium the seasonal variations are not as big as in other locations. However, given the large temperature variations found here, it will be even more interesting to re-calculate the solar fluxes for another city with an another climate, be it much hotter or colder.

During the seasonal changes the trends are visible in Figure 6 where the node 13 on the central area of the top deck plate of the box girder bridge is the one that gets more radiations due to the blackened color of the asphalt and its elevated position. During both periods, yearly as well as daily period, the strains increase while the temperatures are rising. In July values of 37 °C, are clearly visible in Figure 6, however the maximal values are not in the same nodes as in the daily calculation. In this calculation other nodes get hotter because the heat will be a sum of all the previous days showing rising temperatures.

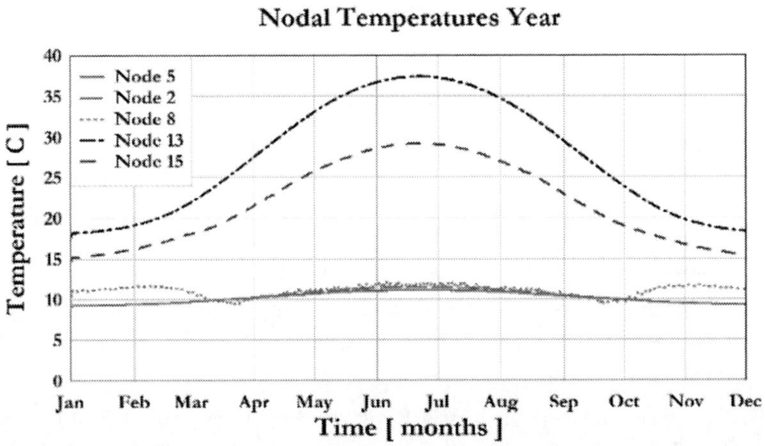

Figure 6. Distribution of the average temperature during a one year period.

3 CONCLUSION

This paper mainly focuses on the importance of temperature effects on structures. The variations of these effects are related to the location of the structure. They are well represented for the Vilvoorde case. The differences between the outer face and inner face are giving a clear map of the situation on the bridge and it is possible to retrieve data and conclusions that confirm the importance of the temperature effects. These effects will have a considerable importance on the cohesion of the different layers of bridge such as the asphalt and the wearing courses, also the stresses combined to the actual loads can be significant. Further research can be done to find a new design code, new materials and connections between the parts of the bridge that allow the displacements and rotations due to thermal effect.

References

De Backer, H., and Outtier, A., Quantification of thermal loads in steel box girders, *11th East Asia-Pacific Conference on Structural Engineering and Construction (EASEC-11): Building a sustainable environment*, Yang, Y., Leu, L., Chen, C., and Su, P. (eds.), 178-179, Taipei, Taiwan, 2008.

De Backer, H., Outtier, A, & Van Bogaert, P. (2009). Numerical and experimental assessment of thermal stresses in steel box girders. Nscc2009, 65.

Duffie, J & Beckman, W. (1974) Solar Engineering of thermal processes. Second edition.

Kim, S.-H., Cho, K.-I., Won, J.-H., & Kim, J.-H. (2009). A study on thermal behaviour of curved steel box girder bridges considering solar radiation. Archives of Civil and Mechanical Engineering, 9(3), 59–76.

Recknagel, Sprenger, Hönman & Schramek (1974). Le Recknagel, Manuel pratique du génie climatique, tome 1.

Interaction between Theory and Practice in Civil Engineering and Construction
Edited by Komurlu, R., Gurgun, A. P., Singh, A., and Yazdani, S.

REHABILITATION AND LOAD TESTING OF A BRIDGE

KAWAN G. ALANI[1] and ABDULLAH S. AHMED [2]

[1]*Alani & AlShamma Consultancy Bureau , Baghdad, Iraq*
[2]*Civil Engineering Dept , Al-Mansour University College , Baghdad, Iraq*

Al-Fatha Bridge is constructed on Tigris River at Baiji town, about 220 km north of Baghdad. Beside its main purpose of connecting the main highway west of Tigris River with the Kirkuk city, its deck structure is purposely designed to carry pipes conveying crude oil from Kirkuk oil fields to Baiji refinery and to the oil exporting network of the country. During the war on Iraq in spring 2003, the bridge was subjected to an air strike. In addition to the damages caused by the impact and explosion of the munitions, the great fire of the crude oil erupted and lasted several days caused much heavier damages in different parts of the structure. The parts of the bridge damaged beyond repair, were replaced, while the parts which were found to have sufficient structural integrity were repaired. This paper addresses the repair and testing of the most affected span exposed to fire. The efficiency of the repair work had been evaluated by carrying out a load test on the above mentioned span. The deflection at different stages of loading were recorded at specified points and compared with theoretical results. Good agreement was obtained between the theoretical and measured deflections, which prove the adequacy of the repair work. This paper introduces briefly an assessment of damages of the bridge, the rehabilitation work, analysis and load test results.

Keywords: BS5400, Grillage method, Prestressed girder, Fire, Deflection, Repair.

1 INTRODUCTION

Al-Fatha Highway Bridge consists of ten equal simply supported spans of 42m long each. Each span consists of two separate, side by side, super structures supported by common wall type piers and cellular type abutments.

The two super structures are separated by about 50mm wide gap. Each super structure contains 7.1m wide carriageway, 2.3m side walk and 0.4m wide raised median curb. The super structure is made of three 3.5m deep bulb-T precast prestressed concrete girders and 30cm thick cast in situ R.C. deck slab on top of the girders. The spacing between each two adjacent girders is 3.2m. In these spaces a number of pipes carrying crude oil pass. The bulb of girders is 0.8m wide and 0.6m high and two 0.3m high inclined sides which end with a 0.22m wide web. The top flange is 1.22m wide with 0.2m end thickness and 0.25m thickness at the junction with the web. The bulb houses four pre stressing straight tendons. Another two parabolic tendons start at 2.73m and 3.37m at ends of girders and drop to 0.59m and 0.78m above the bottom of girders at mid span. Each tendon is made of 19-ϕ10.5mm grade

1670MPa strands complying with BS5896-80. The concrete cover for the straight tendons is about 150mm. While the concrete cover for the parabolic tendons within the web is about 60mm. All the information about the deck structure mentioned above were gathered from site measurement of damaged girders, since no document was available about the bridge after most of the important government offices were vandalized by arson during the invasion. The air attacks on the bridge, besides causing heavy damages due to the impact and explosion of weapons, they ignited a great fire of the crude oil contained in the pipes passing through the deck structure.

2 ASSESSMENT OF DAMAGES

Many missions of inspection for the bridge took place. The first one was made by the General Corporation of Roads & Bridges (GCRB) engineers. Then after the rehabilitation contract was awarded, many inspection missions were made for the bridge. The reason for these multiple inspection was the difficulty of accessing the deck structure from below because the space between the girders was full of distorted steel pipes and their steel supports. See Figure 1. The obvious damages recorded immediately after the attack, were the following, noting that span numbering starts from west (Baiji) end of the bridge:

- A complete collapse of span nine.
- Large opening in the middle of the upstream carriageway of span six with heavy damage of its middle girder.
- Excessive deflection of the downstream carriageway of span eight.
- The other spans were affected to a varying degree, where the fire had caused peeling of concrete cover and the exposure of reinforcing bars. Cracking appeared in some of the main girders.
- The two abutments show only minor damages.
- Piers P1, P2 and P3 were in good conditions except for their crossheads which suffered moderate to minor damages. Piers P6, P7, P8 and P9 suffered heavy damages throughout their exposed parts above water line. While piers P4, P5 suffered moderate damages at parts above the mid height of piers.
- Most of the expansion joints were damaged.
- After thorough study of the results of inspections, it was decided to replace spans six to ten and demolishing the exposed parts of piers P6, P7, P8, and P9, then rebuild them to the original design.
- As for the rest of the bridge parts, which mainly suffered from exposure to fire, a decision was taken to restore them to their original capacity by carrying out engineering repair work. Among these, span five being closed to span six, which was subjected to the direct attack, and hence suffered more from the fire caused by the attack. See Figure 2. Thorough inspection of span five was conducted and the state of each girder was documented. In general, the elements of the deck structure did not exhibit signs of distress, like excessive deflection and/or heavy cracking.

3 REPAIR WORK

The repair work was mainly concentrated on the replacement of the affected surface concrete by good quality repair material. In addition, the cracks should be sealed with suitable filling material. Therefore, the repair scheme used for the deck structure included the following:

- The removal of all the loose materials until a layer of sound concrete was reached.
- Then a water jet was applied to the exposed areas of concrete in order to remove the dust and the remaining loose materials, thereby leaving clean and sound surface. See Figure 3.
- The next step was to fill all visible cracks with structural quality injection material, i.e., low viscosity high strength epoxy sealing system conforming to ASTM C881 Specification. Injection material was selected and applied according to the instructions of the manufacturer. This type of material can be used to seal cracks up to 50mm width.
- After completion cracks injection work, the following task was to apply the replacement for the damaged surface layers. For this task, high quality shrinkage compensated cementitious material with polymer fiber reinforcement in order to reduce cracking, and with good bonding properties with underline concrete was chosen. A suitable material was chosen and applied by pressure sprayer. See Figure 4. Because the areas which were covered were large, they were divided to small panels by narrow guiding strips made of the same material to the required thickness. Then these panels were sprayed and troweled to the accurate level of the guiding strips. See Figure 5.
- At one location, a small area of the surface concrete had to be removed deeper than thickness of concrete cover. In this area, normal concrete with polymer fiber was casted against vertical formwork to bring it to the level of adjacent clean surfaces. Then finally it was covered by spraying repair mortar with the rest of the web.

4 EVALUATION OF SPAN 5 STRENGTH

As a conservative assumption, only the four straight prestressing cables are assumed to be effective in resisting the applied moments in the girders. Calculations of the resisting moment were carried out according BS5400-P4. The calculation of maximum bending moments and shear forces due to factored loads are less than the calculated resisting capacity of the girder. From these calculations, it was concluded that additional external reinforcement will not be necessary, Alani & AlShamma Consultancy Bureau Report (2010).

5 LOAD TEST

After completing the repair work of span five, a load test was carried out to verify the adequacy of the repair work conducted on the girders and deck slab of the upstream carriageway.

5.1 Loads and Loads Arrangement

A carriageway 7.1m wide, according to BS5400-P2, consists of two notional lanes. For design purposes, either the two notional lanes are loaded with full HA load, or one lane is loaded with full HA load while the other is loaded with HB load. For an effective span for girders of 41.44m, the total span load which gives maximum deflection at mid-span was determined to be about 2000kN. To simulate the effect of this load, eight trucks were selected each weighs 250kN. See Figure 6. This load was applied in four increments; each consists of two trucks, i.e., 500kN per increment.

Figure 1. Damages of bridge.

Figure 2. Damages of span five.

Figures 3, 4, & 5. Different stages of repair work.

Figure 6. Load test (full load stage).

The deck structure was modeled as grillage using composite section properties for the members in the main span direction and rectangular elements having a thickness equal to that of the deck slab. The analysis was carried out for the same load increments and loading positions as per the load test. The theoretical deflections at three points along each of the three girders (mid-span and quarter points) were calculated.

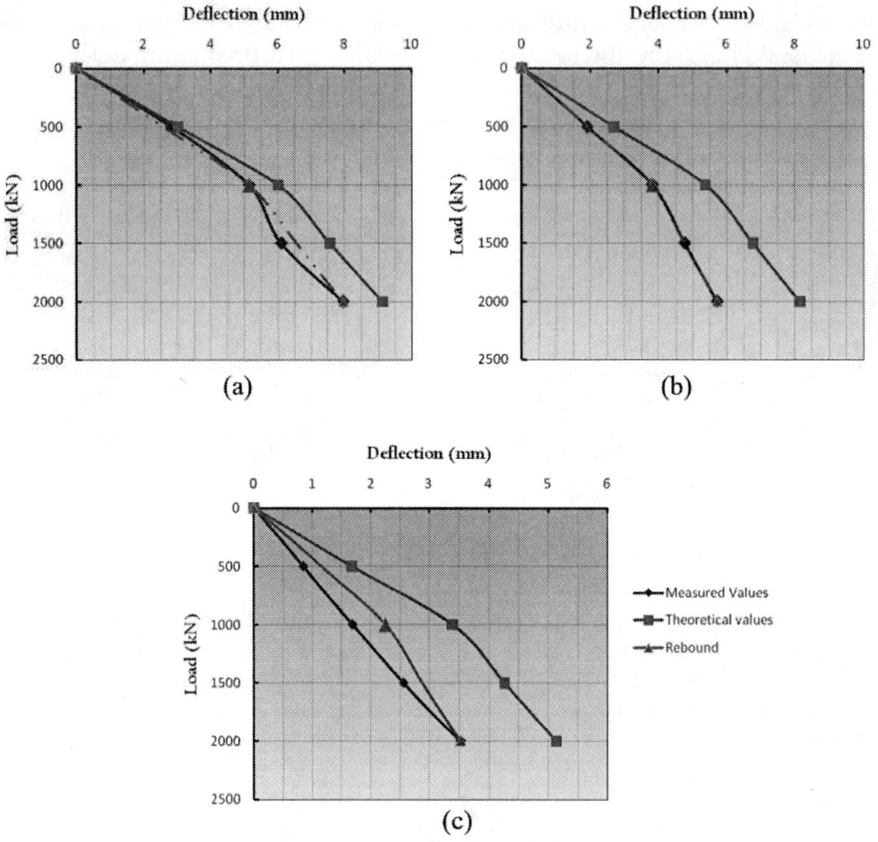

Figure 7. Load-deflection at midspan (a) girder 1. (b) girder 2. (c) girder 3.

5.2 Deflection Measurements and Evaluation

A load test is carried out on October 10th, 2010. Special platforms were designed, constructed and erected on the piers at the two ends of the span at a level of about 2m below the bottom of girders. These platforms were used by the surveying crews.
Six dial gauges were placed under the ends of the girder for measuring end settlement (elastomeric vertical deformation). Three measuring sticks were fixed to the soffit of each of the three girders (mid-span and quarter points).

The locations of the left rear tires of the eight trucks were marked on the road surface for accurate positioning of the trucks during the load test. Then, the trucks were moved to their designated locations at four stages. Readings were recorded after the deck structure was stabilized under each load increment.

After reading the deflections under the full test load, the four trucks closer to the ends of the span were ordered to leave the span, then the deflections were recorded after the structure was stabilized. Finally the last four trucks were moved out of the span, and the deflections were recorded.

6 RESULTS AND CONCLUSIONS

Graphs of measured and calculated deflections at mid-span of the three girders are plotted against loading. See Figure 7. In this Figure, the theoretically calculated deflections are higher than the measured values. The graphs of measured deflections show complete recovery of deflection after removal of loading.
The above results reflect the following:

- The complete recovery of deflection after removal of the applied load means that the deck structure was acting in the elastic range of the material (elastic behavior).
- The measured maximum deflections of girders are smaller than the calculated values for the same loading conditions may be attributed to that the approximation used in the modeling of the structure is on the conservative side and that the repair material properties are appreciably higher than the concrete used in the original structure, i.e., the repair material has little bit higher modulus of elasticity (E_c).

References

Alani & AlShamma Consultancy Bureau, *Evaluation of Span Five Girders of Al-Fatha Bridge*, An evaluation report submitted to Iraqi General Corporation of Roads & Bridges, September 2010.

American Society for Testing and Materials, *C881: Standard Specification for Epoxy-Resin Based Bonding Systems for Concrete,* 2010.

British Standards for steel, Concrete and Composite Bridges BS5400-Part 2*: Specifications for Loads,* 1978.

British Standards 5896-80, *Specifications for High Tensile Steel Wire and Strands for Prestressing of Concrete,* 1980.

British Standards for steel, Concrete and Composite Bridges BS5400-Part 4*: Code of Practice for Design of Concrete Bridges,* 1990.

Interaction between Theory and Practice in Civil Engineering and Construction
Edited by Komurlu, R., Gurgun, A. P., Singh, A., and Yazdani, S.

ANALYSIS OF HAGIA SOPHIA USING NONDESTRUCTIVE MESUREMENT

TAKASHI HARA[1] and KENICHIRO HIDAKA[2]

[1]*Dept of Civil Engineering and Architecture, National Institute of Technology, Tokuyama College, Shunan, Japan*

[2]*Tokyo University of the Arts, Tokyo, Japan*

In this paper, the structural behavior of Hagia Sophia was analyzed by use of vibration records of the structure. The vibration records were measured by the micro tremor meters. The recorded data was utilized not only to define the mode shape but also to evaluate the structural problems. Firstly, the vibration characteristics of the main building were analyzed by combining the vibration records. Secondary, the deformation characteristics and the structural functions of the second cornice were investigated in the same manner as the main building. In case of main building, from the frequency domain analyses, the natural frequency of the main building was analyzed and the obtained natural frequencies were compared with the results of the previous researches. The vibration modes were investigated as well. From these results, the characteristics of the total structure were clarified and the structural problems were detected. In case of the second cornice, the base at the corner where was the junction of the main pier and the sub-dome connecting to the semi-dome showed the movement of the embedded stone. To investigate such phenomena, the same non destructive measurements were applied. Then, applying the finite element method to solve such structures, the structural behaviors were clarified and the health monitoring was performed.

Keywords: Vibration, FDD, FEM, Measurement, Structural behavior.

1 INTRODUCTION

Hagia Sophia (Figure 1), Istanbul, was constructed in 537 as the church and was changed the usage as the mosque since 1453. Since 1935, Hagia Sophia has been opened as the museum. Their structural characteristics and the function of each structural element have been studied by many researchers (Mainstone 1988, Çakmak, *et al.*, 1995, Hara *et al.*, 2015, etc.). The building shows particular characteristics. Huge dome was supported by the main arches, the tympanums and the piers. To resist the horizontal thrust from the main dome, the semi domes, the buttress and the pendentives were arranged. These were also supported by the complex structure.

In spite of many earthquake attacks, Hagia Sophia has been survived and has shown the ancient status as the world heritage. However, the deformation of the structure has grown after completion of the building and several structural problems have been found.

To maintain such structure in the future, the appropriate renovations or the restorations are the important matter. The structural assessments are also required by use of nondestructive methods.

Figure 1. Hagia Sophia.

In this paper, the structural behavior of Hagia Sophia was analyzed by use of vibration record of the structure. The vibration record was measured by the micro tremor meters. Firstly, the vibration characteristics of the main building were analyzed by use of FDD. Secondary, the deformation characteristics and the structural functions of the second cornice were investigated in the same manner as the main building. In case of main building, the natural frequency of it was analyzed and the obtained natural frequencies were compared with the results by other researches (Çakmak, *et al*, 1995 and Hidaka 2004). Then, the vibration modes were investigated. From these results, the characteristics of the total structure were clarified and the structural problems were obtained as well. In case of the investigation of the second cornice, the base at the corner where was the junction of the main pier and sub-dome connecting to the semi-dome shows the movement of the embedded stone. To investigate such phenomena, the same nondestructive measurements were carried out. Combining the finite element analyses, the structural behaviors were clarified.

2 DYNAMIC ANALYSIS OF MAIN BUILDING

From the drawing by Van Nice (1965), the height of the roof top is about 56m and the diameter of the main dome in N-S and EW directions are 31.805m and 30.855m, respectively. The main dome is supported by four main arches and the arches are placed on four main piers. The thrust induced into main arches and the piers is reacted by the buttresses in N-S direction. The thrust induced into arches and the piers in E-W direction is also reacted by the semi-domes on the second cornice. The structure of Hagia Sophia consists of many structural elements and provides huge spaces in the dome (see Figure 2). The structure is constructed using the bricks and stones up to the gallery layer and the masonries are used for building upper structures.

The building has been attacked by the strong earthquake in several times and by the penetration of the water and moisture. Therefore, several problems, such as cracking of the walls and the domes, inclination of the columns and water penetration, are detected.

In addition, due to the different depth of the bed rocks and the large deformation of the materials, the building structure shows a lot of damages on the surfaces and represents the incredible structural deformation. To restore and to conserve such a huge heritage structure, the detail structural inspections are required.

Figure 2. Structure of Hagia Sophia (Mainstone 1988).

Figure 3. Micro tremor meter.

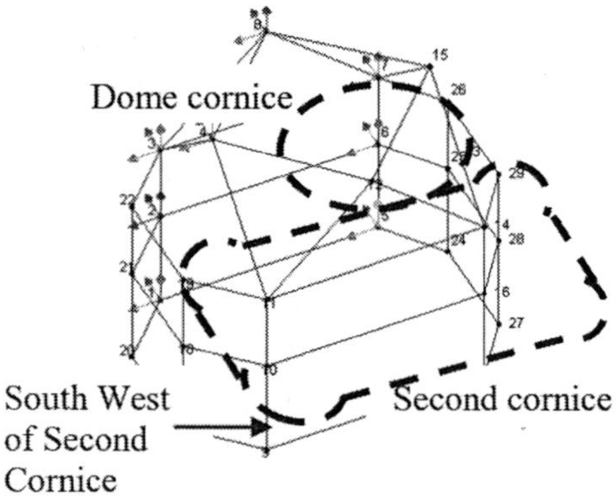

Figure 4. Measuring model.

2.1 Measurement Procedure

The purpose of the dynamic measurement is to acquire the structural data of existing building and to access the structural performance or structural deficiencies. To obtain the structural data, the acceleration tremors were measured. Then the natural frequencies and the vibration modes of the main building were determined. Eight micro-tremor meters (Figure 3) were applied in this analysis. Each tremor meter consists of three acceleration pickups and measures the accelerations in orthogonal direction simultaneously.

Figure 4 shows the measuring model and nodal points in this investigation. The main structure shown in Figure 2 is transformed into skeleton as shown in Figure 4. Measuring point numbers are denoted at each node. Each tremor was measured 40Hz sampling frequency and 3600 second time duration. One micro-tremor meter is fixed on base floor and the other 7 micro-tremor meters are moved on point by point.

Firstly, the tremors at the base floor, the gallery level, the second cornice and the dome cornice on the northern side of main building were measured. Secondary, tremors on the southern side of main building were measured as the same manor. Finally, the piers at the entrance gate and the piers under apse dome were measured. The tremors are totally measured at 29 nodes. All the tremor data obtained in this measurement were collected and were analyzed by FDD (Frequency Domain Decomposition) and the natural frequency and vibration mode were obtained.

2.2 Dynamic Characteristics of Main Building

Using the FDD analysis, the frequency domain expressions of four measuring record sets were combined and represented. From the measurement, the first two frequencies 1.85Hz and 2.09Hz were obtained. These were the same results obtained by Hidaka (2004) and Çakmak (1995). 1.85Hz and 2.09Hz are natural frequency for the predominant vibration in N-S direction, respectively. However, in addition, in this analysis, the peak frequencies 2.93Hz and 3.06Hz, were obtained respectively. The vibration modes of 2.93Hz is shown in Figures 5. In case, the building at south west pier moves, even it is a column base. Çukmak (1995) reported that the southern portion of the building has deeper foundation than the northern portion. It may cause the column movement mode.

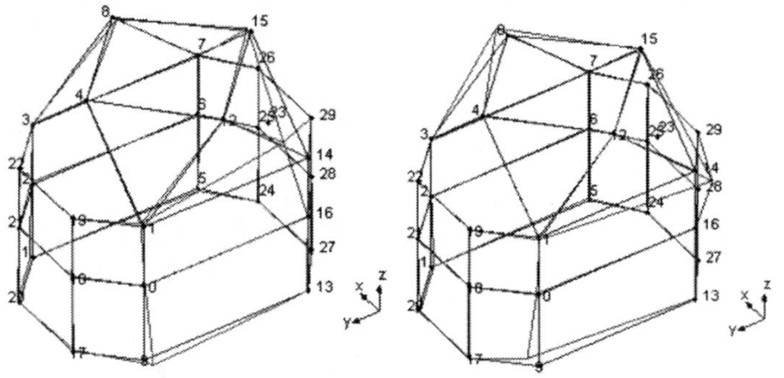

Figure 5. Vibration mode shape (2.93Hz).

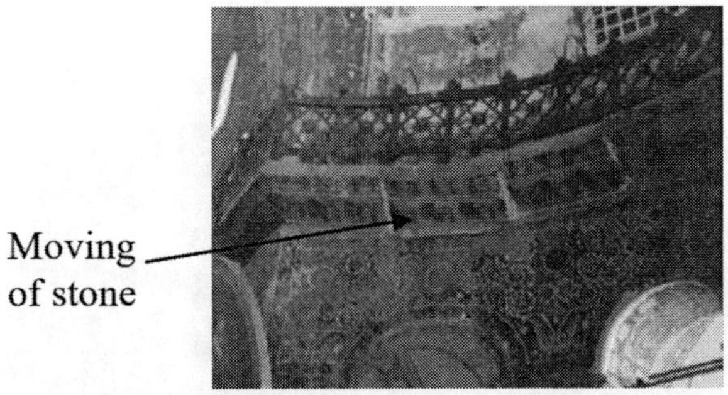

Figure 6. Problems at the south west second cornice.

3 DYNAMIC ANALYSIS OF THE SECOND CORNICE

In Hagia Sophia, there is the main dome at the center and the main dome is supported on four main arches. Two arches are supported by the buttress in northern southern direction. Also, other arches are supported by the semi-domes in eastern western direction. Each semi-dome has two sub-domes. The second cornice is attached to the sub-dome placed on the exedra. At the eastern corner on the south west cornice (see

Figure 6), the deformation of the second cornice and the movement of the composing stones were detected. From the observation, the same phenomena were detected at each corner of the second cornice on the main piers. There are particular deformations. Therefore, the movement of the cornice at south west corner was investigated as an example. Figure 6 shows the picture of the problem of the south western cornice.

3.1 Measurement of Acceleration Tremor

In this analysis, firstly, six tremor meters were placed along the cornice with equidistance. Each tremor was measured 40Hz sampling frequency and 3600 second time duration. Then, the movement of each stone on the cornice was measured.

The total movement of the cornice shows large amplitude at the center with the natural frequency 9.70Hz (Figure 7(a)). Also, at the east end of the south west cornice, three stones vibrate independently with natural frequency 2.91Hz (Figure 7(b)). The movement of the stones was detected at the east end of the south west cornice.

The local deformations shown on the south west cornice incorporate with the global deformation bring us information to understand and to maintain the building structure.

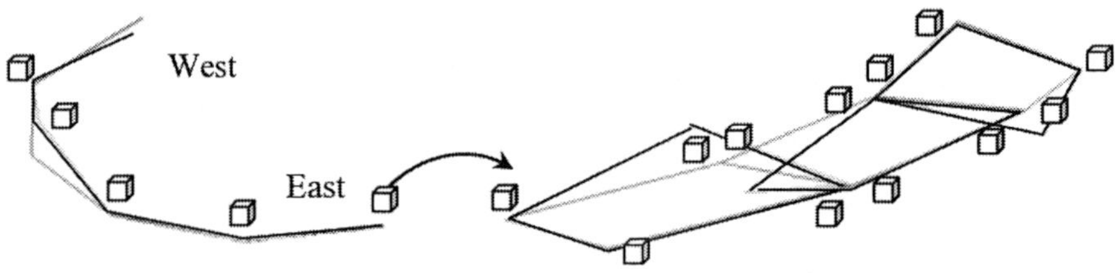

(a) Total vibration (b) Brick vibration at east

Figure 7. Vibration of second cornice (▢ tremor meter).

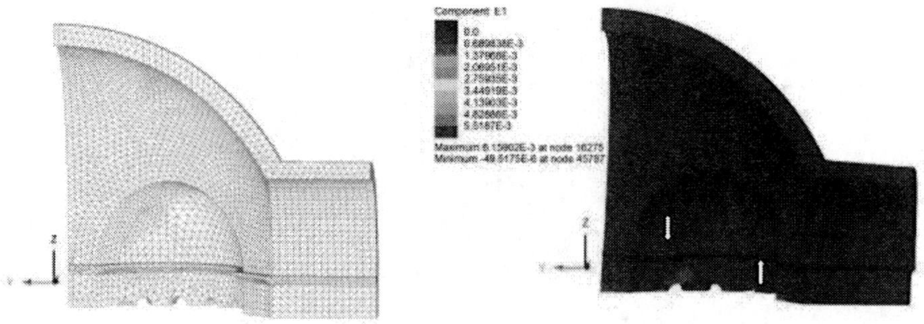

(a) FE Mesh (b) Equivalent stress by FE Analysis

Figure 8. FEM analysis.

3.2 Numerical Deformation Analysis of the Cornice

To represent the deformation and stress states of the cornice around the south west sub-dome under complex stress states, the finite element analyses were performed (Hara *et al*, 1913, 1914, 2015). Figure 8(a) shows the FEM model adopted in this analysis.

Considering the symmetry of N-S and EW directions, a quarter of the model was adopted. Applied load was the self weight and the lateral load on the x-z plane that was applied as the dome thrust. The boundary condition at the bottom was fixed in the vertical direction and the symmetrical boundary conditions were applied to other cutting edges. Figure 8(b) shows numerical results as well. The cornice did not deform at west end and moved upward on the western part of the cornice and sank downward on the eastern cornice. The cornice did not move at the east end by the main pier. Therefore, the cornice deforms vertically as like S shape. Also, the deformation shape was similar to the deformation obtained from the vibration mode shown in Figure 6(a). The cornice at the left end on the exedra shows higher equivalent stress (Figure 8(b)).

4 CONCLUSIONS

In this study, the vibration characteristics of the main building and the second cornice were analyzed experimentally. From the analysis, following conclusions are obtained.

1. FDD analyses of recorded data show the vibration characteristics of the structure precisely.
2. Hagia Sophia shows the global type of deformation of the structure. Also, several particular deformations and the local failures have been detected on the cornice.
3. The local deformations appear on the south west cornice incorporate with the global deformation.

The recorded data will be utilized to obtain the response of structure precisely by using FE Analysis with combining the earthquake record. The local deformations shown on the south west cornice bring us the information to understand and to maintain the building structure.

References

Çakmak A., Moropoulou A., and Mullen C. L., Interdisciplinary study of dynamic behavior and earthquake response of Hagia Sophia, *Soil Dynamics and Earthquake Engineering*, 14,125-133, 1995.

Mainstone, R. J., *Hagia Sophia, Architecture, Structure and Liturgy of Justinian's Great Church.*, Thames & Hudson, London, 1988.

Hara, T., Hidaka, K., and Seta R., Structural characteristics of ancient dome in Jerash *Proceedings of Applied Structural Engineering and Mechanics, Jeju, Korea,* 1302-1307, 2013.

Hara, T. and Hidaka, K., Numerical evaluation of the deformation of West Bath dome in Jerash. *Proceedings of the IASS-SLTE 2014 Symposium, Brasilia, Brazil,* 1-8, 2014.

Hara, T. and Hidaka K., Dynamic behaviour of masonry structure. *Proceedings of the IASS Symposium 2015, Amsterdam, The Netherland* 1-10, 2015.

Hidaka, K. and Sato T., *Architectural-Structural Survay of Hagia Sophia,* Chuokouronbijyutsu Press 2004.

Van Nice, R. J., *Saint Sophia in Istanbul, an architectural survey.*, Dumbarton Oaks, Washington, 1965.

Interaction between Theory and Practice in Civil Engineering and Construction
Edited by Komurlu, R., Gurgun, A. P., Singh, A., and Yazdani, S.

PROBABILISTIC BEHAVIOR OF JOINT FORCES IN MECHANISMS

BOŠTJAN HARL[1], MARKO KEGL[1], NENAD GUBELJAK[1], and DEJAN DINEVSKI[2]

[1]*Faculty of Mechanical Engineering, University of Maribor, Maribor, Slovenia*
[2] *Faculty of Medicine, University of Maribor, Maribor, Slovenia*

This paper discusses the influence of the clearance in joints on the joint reaction forces in mechanisms. By using mathematical programing, the optimal parameters of kinematic chains can be efficiently obtained by using the deterministic approach. However, the situation becomes more sophisticated if random effects of tolerances of the arm lengths and the random pin positions have to be considered. In this work the influence of clearances on joint forces is calculated by using the Taylor approximation and the Monte Carlo method. Using the two methods was necessary, because the Taylor approximation usually yields satisfying results only for small values of clearances and for this reason it makes sense to compute the required quantities also by another independent method. The implementation of the model is illustrated with two examples. The first example considers a closed loop chain, representing a four-bar mechanism being an actual part of a hydraulic support, employed in mining industry. The hydraulic support must fulfill some requirements so it's very important to have influence of clearances on joint forces in control. The second example considers joint reaction forces of car wiper mechanism. It will be shown, that the clearance in joints have some influence on the joint reaction forces.

Keywords: Kinematic chain, Stochastic model, Mathematical programming.

1 INTRODUCTION

Kinematic chains are widely used in industry, which is also reflected in extensive literature and studies available, see e.g.,,, Yoshikawa (1985), Conkur and Buckingham (1997). For the analysis and synthesis of kinematic chains, optimization methods have proved to be a valuable tool. Consequently, since the 1960s, various formulations for optimization of problems related to kinematic chains have been developed, see e.g.,, Arora and Wang (2005). Some good examples are the optimal track problem of a closed kinematic chain Oblak *et al.* (2000) and the problem related to the impact of tolerances of arm lengths and random pin positions, discussed in Dhande *et al.* (1973), Lee and Gilmore (1989).

In this paper, the study is conducted on two examples. The first one is the kinematic chain with the optimal length of the links, obtained in Oblak *et al.* (2000) by employing optimization methods, and the second example is a car wiper mechanism. For both examples the influence of the clearances in the joints on the joint reaction forces is studied. This is achieved by an adequate mathematical model that reflects the effects of tolerances of arm lengths and random pin positions. The influence of clearances on joint forces is calculated by the Taylor approximation and the Monte Carlo method.

The first example is a hydraulic support, Figure 1a, that is a part of the mining industry equipment in the mine Velenje - Slovenia, used for protection of working environment in the gallery. It consists of two four-bar mechanisms *FEDG* and *AEDB* as shown on Figure 1b. The mechanism *AEDB* defines the path of the coupler point C and the mechanism *FEDG* is used to drive the support by a hydraulic actuator. The original support design exhibited large transversal displacements of the point C and relatively very high joint forces, which reduced its employability. This was the reason to determine the best values for the most problematic parameters of leading four-bar mechanism *AEDB* by employing the methods of mathematical programming Kegl *et al.* (1991). In this work, the influence of the clearance in joints on the joint reaction forces will be studied for the four-bar mechanism *AEDB* of the support, Figure 1b.

(a)

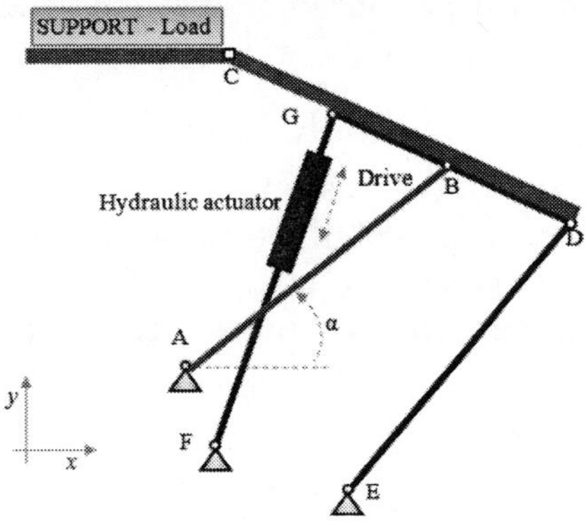

(b)

Figure 1. (a) Hydraulic support engagement in a mine;
(b)The schematic model of the hydraulic support.

The car wiper mechanism consists of two four-bar mechanisms *ADCB* and *DGFE* as shown in Figure 2. The clearance in joints on the joint reaction forces will be studied for the *ADCB* mechanism.

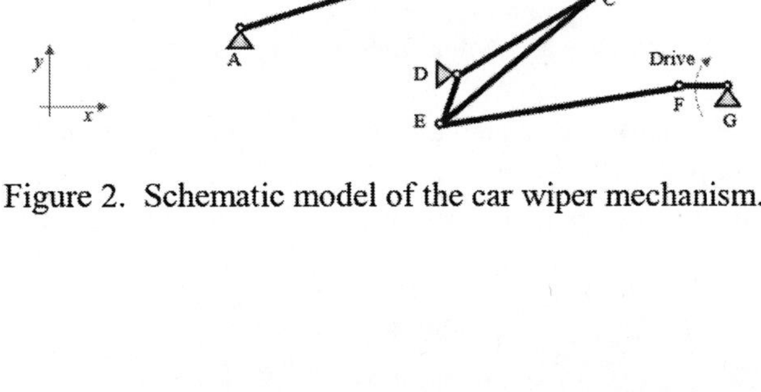

Figure 2. Schematic model of the car wiper mechanism.

2 THE STOCHASTIC MODEL OF THE MECHANISM

The foundation of our considerations is an appropriate mechanical model of the mechanical system – kinematic chain, which will enable the computation of the kinematic and dynamic quantities Harl *et al.* (2004). We assume here that this model is given by a system of m response equations, written as

$$h_i(\mathbf{b},\mathbf{u})=0, \quad i=1,2,\cdots,m \tag{1}$$

The response equations allow us to compute the vector of response variables (u) in dependence on the vector of some design variables (b). This implies that $\mathbf{u}=\tilde{\mathbf{h}}(\mathbf{b})$, where $\tilde{\mathbf{h}}$ denotes a vector function establishing the relationship between the vector of design variables and the response of our mechanical system.

In a stochastic model, the vector b is treated as a random vector $\mathbf{B}=[B_1,B_2,...,B_n]^T$, meaning that the vector u of response variables is also a random vector $\mathbf{U}=[U_1,U_2,...,U_m]^T$, where is $\mathbf{U}=\tilde{\mathbf{h}}(\mathbf{B})$. It is supposed that the design variables $B_1,B_2,...,B_n$ are mutually independent, at least from the probability point of view.

The probability distribution function of the random vector U, that is searched for, depends on the probability distribution function of the random vector B and it is mostly practical impossible to compute. Therefore, the random vector U will be described either by the help of "number characteristics", that can be estimated by a Taylor approximation of the function $\tilde{\mathbf{h}}$ at the point $\mathbf{b}$, or by using the Monte Carlo method described in Harl *et al.* (2015).

3 CLEARANCE IN JOINTS

In this work, the effective link length model is used Lee and Gilmore (1989). This model considers the uncertainties due to pin location, link length tolerance, and radial clearance tolerance as an effective length variation of the link. Therefore, the local configuration of each pin joint does not affect the model. The effective link length model is shown in Figure 3.

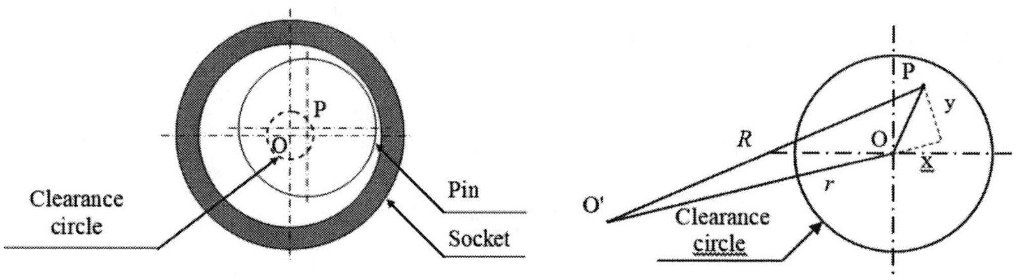

Figure 3. Mechanical model of the pin clearance.

As the pin moves inside the inner circle of the socket, the pin center moves on or inside the clearance circle. The radius of the clearance circle, i.e., the radial clearance, is determined by the tolerances of the pin and socket diameters. Since most tolerances follow a normal distribution at the worst case, the probability distribution of the pin position is assumed to follow such distributions in order to analyze the pin joint successfully. This model, the clearance model, readily accounts for uncertainties in a pin joint such as tolerance of the radial clearance and random location of the pin. Hence, the origin of such a frame coincides with the socket center O and the axis x is parallel to the nominal link direction. The line $\overline{O'P}$ represents the effective link and its length is given as

$$R = \sqrt{(r+x)^2 + y^2} \tag{2}$$

where x and y are the coordinates of the random pin location in the local coordinate frame and r is random link length.

4 NUMERICAL EXAMPLES

Two numerical examples will be considered. The influence of the clearance in joints A, E, D, B of the hydraulic support (Figure 1b) and for joints A, D, C, B of the car wiper (Figure 2) on the joint reaction forces will be calculated.

The symbols F_j, where $j = 1,\ldots,4$, represent the norms of the joint force vector. By using the Taylor approximation, the influence for all four joints can be estimated by the standard deviations

$$\sigma_j = \sqrt{\sum_{i=1}^{4} (\frac{\partial F_j}{\partial u_i})^2 (\frac{\Delta r_i}{6})^2 + (\frac{\partial F_j}{\partial x})^2 (\frac{\Delta x}{6})^2 + (\frac{\partial F_j}{\partial y})^2 (\frac{\Delta y}{6})^2} \tag{3}$$

In the above equations, Δr_i are the tolerances of the link lengths of $AEDB$ and $ADCB$ mechanisms, while Δx and Δy are the tolerances of the pin and the socket.

During numerical investigation, it turned out that the Taylor approximation usually yields satisfying results only for small values of Δr_i, Δx, and Δy. For this reason it makes sense to compute the required quantities also by another independent method. In our case the Monte Carlo method was engaged for this purpose. In this case the influence of clearances on joint forces can be computed as

$$\mu_j = \frac{\sum_{i=1}^{n} F_{ji}}{n} \text{ and } \sigma_j = \frac{\sum_{i=1}^{n} (\mu_j - F_{ji})^2}{n-1} \tag{4}$$

For the first numerical example - hydraulic support, the carrying capability is 1600kN. Both four-bar mechanisms $AEDB$ and $FEDG$ must fulfill the following requirements:

- They must exhibit minimal transversal displacements of the point C,
- They must provide sufficient side stability.

The starting parameters of both four-bar mechanisms (Figure 1b) were the values obtained in the earlier work by optimization. These values are (in millimeters)

$\left[\overline{AE},\overline{AB},\overline{BD},\overline{DE},\overline{EF},\overline{FG},\overline{GD}\right]^T=\left[676,1361,382,1310,400,(1325+d),1251\right]^T$. The parameter d is a displacement of the hydraulic actuator with a maximal value of 925 mm. During operation, the parameter d varies, so that the angle α between links $\overline{AB}$ and $\overline{AE}$ can is varied in the range from 76.8° to 94.8°. For the selected pin and socket, the maximal values of x and y are $x=y=2\,\text{mm}$. For calculating the clearances, the tolerances of the links of the four-bar mechanism *AEDB* are needed and the values, in millimeters, are $\left[\Delta r_1,\Delta r_2,\Delta r_3,\Delta r_4\right]^T=\left[0.01971,0.00868,0.001,0.00933\right]^T$ meanwhile, the tolerances of the clearance are $\left[\Delta x,\Delta y\right]^T=\left[0.001,0.001\right]^T$.

By using this data the deterministic joint forces in the hydraulic support, during the variation of *d*, can be computed. Their histories are shown in Figure 4a, where: F_1 (full line – blue), F_2 (dotted line – green), F_3 (dashed line – yellow) and F_4 (dash-dotted line - red). It is now possible to estimate the effect of clearances on joint reaction forces, obtained by the Taylor approximation. The variations of these forces are shown in Figure 4b, where: ΔF_1 (full line – blue), ΔF_2 (dotted line – green), ΔF_3 (dashed line – yellow) and ΔF_4 (dashed-dotted - red).

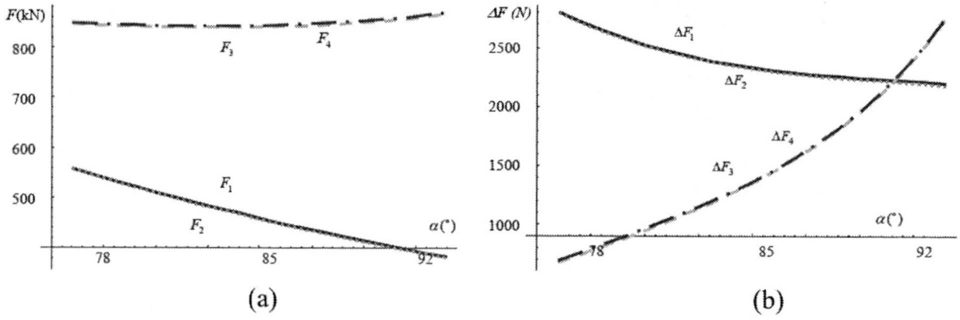

Figure 4. (a) Joint forces of the mechanism; (b) The effect of clearances on the joint forces of the mechanism *AEDB* obtained by the Taylor approximation.

The car wiper, Figure 2, is loaded by a force of 20N. The starting parameters, in mm, are $\left[\overline{AB},\overline{BC},\overline{CD},\overline{AD},\overline{DE},\overline{EF},\overline{FG},\overline{DG}\right]^T=\left[200,40,100,125,42.5,110,11.5,220\right]^T$. The drive of the car wiper mechanism is defined as $f(t)$ and can vary the angle α between the links $\overline{IH}$ and $\overline{IJ}$ from 10° to 90°. For the selected pin and socket, the maximal values of x and y are $x=y=0.2\,\text{mm}$. To compute the clearances, the tolerances of the links of the four-bar mechanism *ADCB* are needed. Their values, in millimeters, are $\left[\Delta r_1,\Delta r_2,\Delta r_3,\Delta r_4\right]^T=\left[0.01,0.01,0.01,0.01\right]^T$. The tolerances of the clearance are $\left[\Delta x,\Delta y\right]^T=\left[0.001,0.001\right]^T$.

The joint forces during the movement are shown in Figure 5a, where: $F_1\approx F_2\approx F_3$ (full line – blue) and F_4 (dashed-dotted - red). Figure 5b shows the effect of clearances on joint reaction forces, obtained by the Taylor approximation, where: $\Delta F_1 \approx \Delta F_2 \approx \Delta F_3$ (full line – blue) and ΔF_4 (dashed-dotted - red).

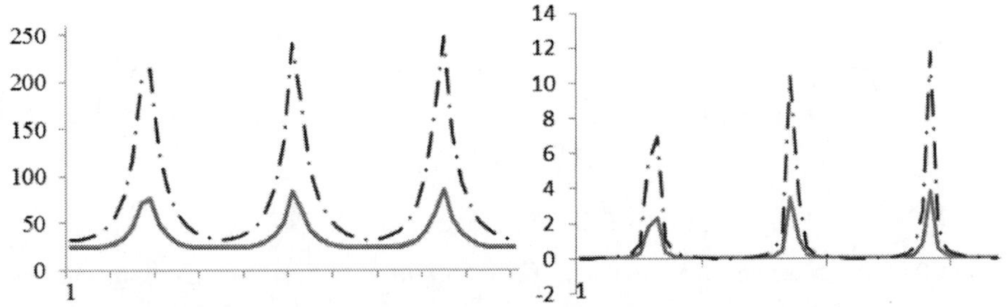

(a) (b)

Figure 5. (a) Joint forces of the mechanism *AEDB*; (b) The effect of clearances on the joint forces of the mechanism *AEDB* obtained by the Taylor approximation.

5 CONCLUSIONS

By utilizing an adequate mathematical model of the mechanical system the effects of random parameters on forces in a kinematic chain can be estimated. In this paper a hydraulic support and a car wiper were considered. It was shown that the clearance in joints have some influence on the joint reaction forces. T his influence varies as the kinematic chain operates. However, for the considered examples the magnitudes of these variations were rather minor except maybe at some position of the car wiper mechanism. So, one can say that for usual mechanisms with rather large security factors the influence of considered random parameters may probably be neglected. However, for highly loaded mechanisms the presented approach might provide valuable additional information on the actual loads emerging during operation.

References

Arora, J.S.; Wang, Q. Rewiew of formulations for structural and mechanical system optimization. *Structural and Multidisciplinary Optimization.* 30, 251 – 272, 2005.

Conkur E.S.; Buckingham R. Clarifying the definition of redundancy as used in robotics. *Robotica*, 15, 583 – 586, 1997.

Dhande S.G.; Chakraborty J. Analysis and Synthesis of Mechanical Error in Linkages – A Stohastic Approach. *ASME Journal of mechanical design*, 672 – 676, 1973.

Harl, B., Oblak, M., Butinar, B. Minimization of joint reaction forces of kinematic chains by a multi-objective approach. *Struct. multidiscipl. optim.*, 27, 4, 243-249, 2004.

Harl, B., Gubeljak, N., Kegl, M. Probabilistic behaviour of joints on joint forces in mechanisms. *Technical gazette*, 113-117, 2015

Kegl M.; Butinar B.; Oblak M. Optimization of mechanical systems: On strategy of non-linear first-order approximation. *Int. J. Numer. methods Eng.*, 33, 223 – 234, 1991.

Lee S.J.; Gilmore B.J. The determination of the probabilistic properties of velocities and accelerations in kinematic chains with uncertainty. *ASME Journal of mechanical design*, 225 – 232, 1989.

Oblak, M.; Harl, B.; Butinar, B. Optimal design of hydraulic support. *Structural optimization*, 20, 1, 76-82, 2000.

Yoshikawa T. Manipulability of Robotic Mechanisms. *The International Journal of Robotic Research.* 4, 3 – 9, 1985.

Interaction between Theory and Practice in Civil Engineering and Construction
Edited by Komurlu, R., Gurgun, A. P., Singh, A., and Yazdani, S.

VIBRATION CONTROL OF STRUCTURES USING MAGNETO-RHEOLOGICAL DAMPERS WITH H_∞ METHOD

HUSEYIN AGGUMUS and SABAN CETIN

Dept of Mechanical Engineering, Yildiz Technical University, Istanbul, Turkey

In this study, the vibration mitigation of the structures under the earthquake excitation using magneto-rheological (MR) damper is researched. For this purpose, one MR damper is installed to an eight story structure and the system performance is investigated. MR damper is installed between the second floor and the ground. The MR damper is a semi-active control device that can be controlled by only the voltage transmitted to the electromagnetic coils. Since the MR damper is a semi active device, damping force can be changed by applied voltage. Therefore, determination of the applied voltage is important for vibration reduction. A robust controller is designed for determination of the MR damper voltage. H_∞ method and model reduction approach are used for robust control design and application. The performance of the proposed method is tested by numerical simulation on the semi active structural system with MR damper. The performance of the control algorithm and the effect of the MR damper arrangement is evaluated using earthquake (El-centro earthquake) data. The results are compared with situations in which MR damper is not connected and when the MR damper is situation of semi-active controlled. The evaluations are based on the displacement and acceleration responses. The simulation results showed that the arrangement of the MR damper connected between the second floor and the ground is effective in reducing the vibration amplitudes.

Keywords: Structural control, MR damper layout, H_∞ robust control, Simulation study.

1 INTRODUCTION

The control of linear and nonlinear systems equipped with semi-active devices is an interesting subject that can be studied both experimentally and theoretically (Atabay and Ozkol (2013); Castao *et al.* (2011); Cetin *et al.* (2011); Cornejo and Alvarez-Icaza (2011); Paksoy *et al.* (2014); Cetin *et al.* (2011)). Active and semi-active systems can be controlled electronically. Passive systems are simple and classical systems that are currently used. Semi-active systems can act as both active and passive systems. Their performance is similar to active systems. Because they can act as a passive system in the case problems of control systems and they are more reliable than active systems.

A number of studies are conducted regarding the control of the MR damper and the arrangement of multiple MR dampers. Dyke *et al.* (1998) employed MR dampers on the first floor of a three-story structure. Dyke *et al.* (1996) applied a clipped optimal control algorithm, which was previously tested via simulations (Dyke *et al.* 1998), by connecting the MR damper to the test model between the ground floor and first floor.

Jansen and Dyke (2000) implemented MR dampers in a building model with six DOF; the MR damper was laid out on the first two floors in a parallel configuration. Aldemir (2009) developed a causal sub-optimal control, placing an MR damper on the first floor of the base-isolated structure with 2 DOF. The results were compared with instantaneous optimally controlled and uncontrolled situations, and the presented control algorithm was shown to be effective in reducing the effects of earthquakes on the structure. Bitaraf *et al.* (2010) used two MR dampers, which were placed on the first and second floors of the structure, and applied two control methods which are direct adaptive control based on a simple adaptation technique and a genetic-based fuzzy control. Cetin *et al.* (2011) used a six-story steel structural model, and an MR damper was implemented on the first floor. A nonlinear adaptive controller based on the Lyapunov technique, which can balance parametric uncertainties, was used to command the MR damper voltage. This study reports the control and arrangement of one magneto-rheological (MR) damper that is used to reduce structural vibrations. H_∞ method and model reduction approach are used for robust control design and application. The simulation results showed that the arrangement of the MR damper connected between the second floor and the ground is effective in reducing the vibration amplitudes.

2 PROBLEM FORMULATION

In this study, MR damper layout in an eight-story steel structure is investigated, as shown in Figure 1. The mathematical movement of the building model is shown in Figure 1 and can be given as

$$M_s\ddot{x}(t)+C_s\dot{x}(t)+K_sx(t)=-Hf(t)-M_sL\ddot{x}_g(t) \tag{1}$$

where *f(t)* is the damping force of the MR damper and M_s, C_s, and $K_s \in \Re^{8x8}$ are the mass, damping and stiffness matrices, respectively. $\ddot{x}(t)$, $\dot{x}(t)$ *and* $x(t) \in \Re^{8x1}$ are the acceleration, velocity and displacement vectors, respectively.

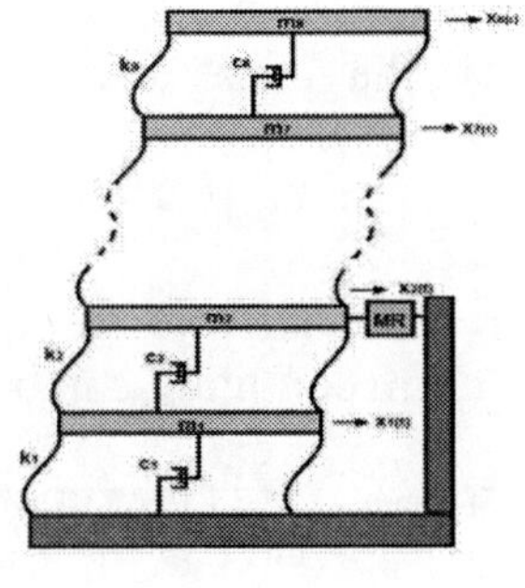

Figure 1. The MR damper is located between the second floor and the ground.

Unidirectional horizontal movement is considered in this model. The relative displacement vector for the model is $x=[x_1 \;\; x_2 \;\; x_3 \;\; x_4 \;\; x_5 \;\; x_6 \;\; x_7 \;\; x_8]^T$.The *H* vector indicates the placement of the control units. The MR damper is placed between the second floor and the ground, so $H_{MR2} = [0 \ 1 \ 0 \ 0 \ 0 \ 0 \ 0 \ 0]^T$. The seismic input vector is $L = [1 \ 1 \ 1 \ 1 \ 1 \ 1 \ 1 \ 1]^T$. $\ddot{x}_g(t)$ is the earthquake ground acceleration.

3 ROBUST CONTROLLER DESIGN

Considering the structural system modeled in the previous section, the full-order model of the system in physical coordinates can be written in the state space model as follows:

$$\dot{x}_f = A_f x_f + B_f u, \ y_r = C_f x_f \tag{2}$$

where A_f, B_f, C_f and D_f, are the linear system matrices and given respectively, by

$$A_f = \begin{bmatrix} 0 & I \\ -M_f^{-1}K_f & -M_f^{-1}C_f \end{bmatrix}, \ B_f = \begin{bmatrix} 0 \\ M_f^{-1}F_f \end{bmatrix}, \ C_f = [C_y \quad 0], P_f(s) = \begin{bmatrix} A_f & B_f \\ C_f & 0 \end{bmatrix} \tag{3}$$

For the model order reduction, the system must be transformed from the physical space to the modal space. The equation in modal coordinates can be written for full order and reduced order models as

$$\ddot{\eta} + C_f\dot{\eta} + K_f\eta = H_f f \ , \ \ddot{\eta}_r + C_r\dot{\eta}_r + K_r\eta_r = H_r f \ . \tag{4}$$

When structural control is considered, controlling the first two modes provide good results for the earthquake hazard mitigation of structural systems and reduce the amplitudes. Differences between real systems and dynamic models are modeled as $\Delta_t(s) = P_f(s) - P_r(s)$. The two primary transfer functions in this control system are *S(s)* is defined as the sensitivity transfer function, and *T(s)* is the complementary sensitivity transfer function. When *T(s)* and Δ_t(s) are considered stable, using a W_T filter and provided that the upper limit of Δ_t(s) satisfies

$$|\Delta_t(j\omega)| \le |W_T(j\omega)| \tag{5}$$

a norm W_T condition from ω to z_2, the feedback system can be stable

$$\|W_T T(s)\|_\infty < 1 \tag{6}$$

The second aim of the H_∞ controller is to improve the performance of the feedback control system. The H_∞ norm condition can be written as

$$\|S(s)\|_\infty = \sup \bar{\sigma}[S(s)], \ (\bar{\sigma} \text{ is maximum singular value of } S(s).) \tag{7}$$

where W_S is the filter for the system output. Thus, specifying the W_T(s) and W_S(s) filters in the control system satisfies both robust stability and response performance. This type of H_∞ controller is called a mixed sensitivity problem and is defined as

$$\left\| \begin{bmatrix} W_S S \\ W_T T \end{bmatrix} \right\|_\infty < \gamma \tag{8}$$

where γ is a design parameter that is positive. An important step in H_∞ control is to determine the frequency shape filters. Additive uncertainty is used to select W_T. A filter should cover the uncertainty to provide robust stability. In this manner, frequency shape filters take the following form:

$$W_T = k_w \left(\frac{s^2 + 2\xi_{nm}\omega_{nm}s + \omega_{nm}^2}{s^2 + 2\xi_{dm}\omega_{dm}s + \omega_{dm}^2} \right) \tag{9}$$

where ω_{nm} is the frequency of the last controlled mode and ω_{dm} is the frequency of the first uncontrolled mode. The open- and closed-loop responses of the full order model is shown in Figure 2.

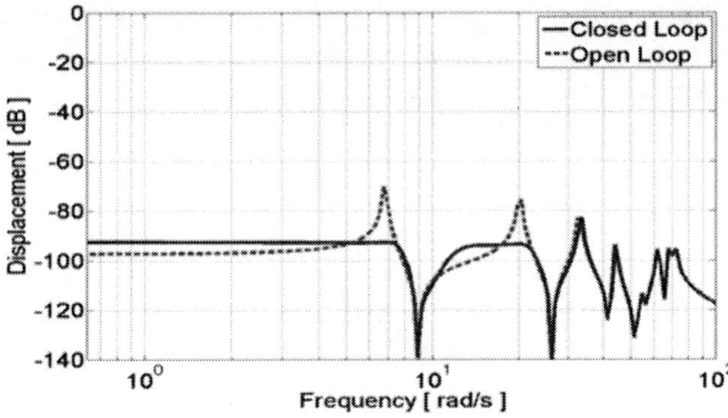

Figure 2. Open- and closed-loop responses of system.

3.1 Controller Application in Semi-Active System

The voltage of the MR damper is selected as follows (El-Kafafy and El-Demerdash, 2012; Lam and Lio, 2002):

$$\begin{aligned} &\text{If} \quad G(f_c - f_d)\,\text{sgn}(f_d) > V_{max},\ v = V_{max} \\ &\text{or} \ G(f_c - f_d)\,\text{sgn}(f_d) < V_{min},\ v = V_{min} \\ &\text{otherwise},\ v = G(f_c - f_d)\,\text{sgn}(f_d) \end{aligned} \tag{10}$$

where V_{max} is the maximum voltage in the MR damper, V_{min} is the minimum voltage in the MR damper, f_c is the force necessary for the system and is determined by the controller, and f_d is the force formed by the MR damper and is measured by the system. Finally, G is the MR damper control gain.

3.2 Simulation Results

The mass value of system for each floor is 107.5, and the mass matrix is $M_f = diag[107.5]_{8x8}$. Considering the connection with 8 bars made of spring steel, the stiffness of the system for each floor is evaluated as $k_{1-8} = 8*12EI/l^3 = 145152N/m$

(Cetin *et al.*, 2011). According to the Rayleigh damping principle, if $\alpha_0 = 0.0265$ and $\beta_0 = 0.00011431$, $[C_s] = \alpha_0[M_s] + \beta_0[K_s]$, the damping coefficient is $C_{1,2,3,4,5,6,7,8} = 16.59 N.s/m$, and the controller gain $G = 0.04$. The maximum voltage V_{max} is 2 V, and the minimum voltage V_{min} is 0 V.

The time responses for the situation of the MR damper's layout (MR_2) is shown in Figure 3. The acceleration response is illustrated in Figure 4. The passive (MR damper disconnected) and H_∞ controlled situation is compared to determine the performance of the controller. The case in which the MR damper is only on the second (MR_2) is examined. The displacement responses of all floors are shown in Figure 3. Comparisons of uncontrolled and controlled cases for the MR damper's layout is depicted in Figure 3, respectively. The connection of the MR damper reduces the vibration of each floor. The amplitudes of the displacement in situation corresponding to the MR with the controller is lower than the displacement amplitude in the situation corresponding to the passive system (without the MR damper). The acceleration responses of each floor for the MR damper layout is shown in Figure 4. The connection of the MR damper and the application of the robust controller improve the acceleration responses of each floor.

4 CONCLUSIONS

In this study, an H_∞ robust controller is designed to command MR damper voltage, by placing an MR damper on second floor to reduce building vibrations during earthquakes. The designed controller is tested in an eight-story structural model. By simulation studies the performances of the controller and MR damper are investigated. Comparison is made among case when the MR damper is not connected and controlled with H_∞. The evaluations are based on the displacement and acceleration responses. The arrangement of the MR damper connected between the second floor and the ground is effective in reducing the vibration amplitudes. Furthermore, the designed controller clearly improves the system performance.

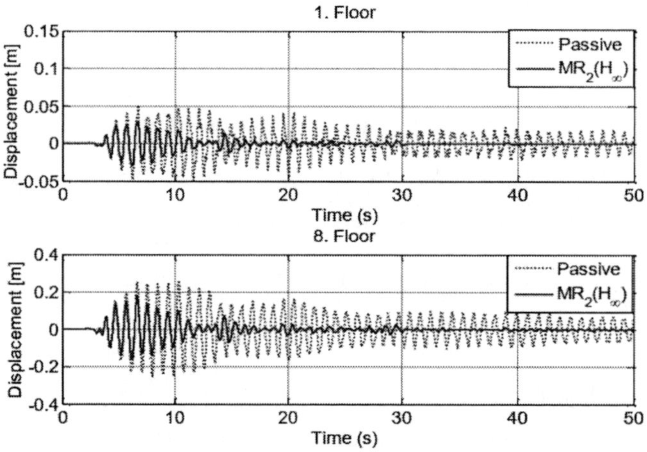

Figure 3. Displacements of (1-8) floors for the combinations of the MR damper layout: MR_2.

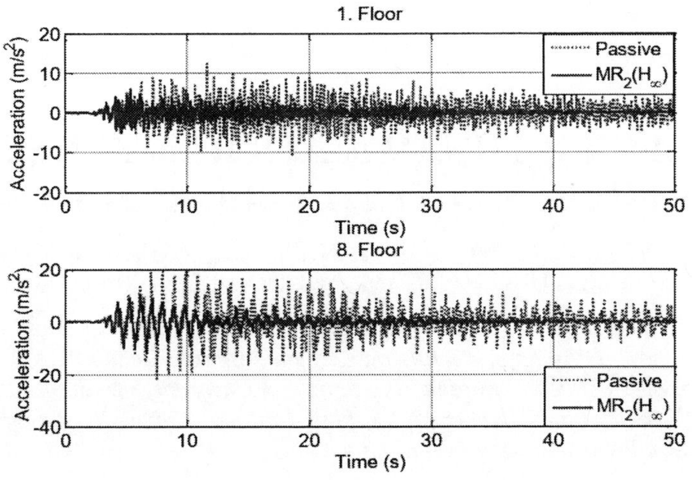

Figure 4. Accelerations of (1-8) floors for the combination of the MR damper layout: MR_2.

References

Aldemir U (2009) Causal semiactive control of seismic response. *Journal of Sound and Vibration* 322: 665-673.

Atabay E and Ozkol I (2014) Application of a magnetorheological damper modeled using the current–dependent Bouc–Wen model for shimmy suppression in a torsional nose landing gear with and without freeplay. *Journal of Vibration and Control* 20: 1622-1644.

Bitaraf M, Ozbulut OE, Hurlebaus S and Barruso L (2010) Application of semi-active control strategies for seismic protection of buildings with MR dampers. *Engineering Structures* 32: 304-3047.

Cornejo C and Alvarez-Icaza L (2012) Passivity based control of under-actuated mechanical systems with nonlinear dynamic friction. *Journal of Vibration and Control* 18: 1025-1042.

Cetin S, Zergeroglu E, Sivrioglu S and Yuksek I (2011) A new semiactive nonlinear adaptive controller for structures using MR Damper: Design and experimental validation. *Nonlinear Dyn* 66: 731–743.

Cetin S, Sivrioglu S, Zergeroglu E, Yuksek I (2011) Semi-active H_∞ robust control of six degree of freedom structural system using MR damper . Turk J Elec Eng & Comp Sci, 19 , 797-805.

Dyke SJ, Spencer Jr BF, Sain MK and Carlson JD (1996) Modeling and Control of Magneto-rheological Dampers for Seismic Response Reduction. *Smart Mater Struct* 5: 565- 575.

Dyke SJ, Spencer Jr BF, Sain MK and Carlson JD (1998) An Experimental Study of MR Dampers for Seismic Protection. *Journal of Smart Materials and Structures: Special Issue on Large Civil Structures* 7(5): 693-703.

El-Kafafy M, El-Demerdash SM and Rabeih AAM (2012) Automotive ride comfort control using MR fluid damper. *Engineering* 4(04): 179-187.

Jansen LM and Dyke SJ (2000) Semiactive control strategies for MR dampers: comparative study. *Journal of Engineering Mechanics* 126(8): 795–803.

Lam AHF and Lio WH (2002) Semi-Active Control of Automotive Suspension Systems with Magnetorheologi-cal Dampers. *International Journal of Vehicle Design* 33: 50-75.

Paksoy M, Guclu R, and Cetin S (2014) Semiactive Self-Tuning Fuzzy Logic Control of Full Vehicle Model with MR Damper. *Advances in Mechanical Engineering* Volume 2014.

Interaction between Theory and Practice in Civil Engineering and Construction
Edited by Komurlu, R., Gurgun, A. P., Singh, A., and Yazdani, S.

MULTIPLE TUNED MASS DAMPERS FOR SEISMIC-EXCITED TALL BUILDING

YOUNG-MOON KIM[1], KI-PYO YOU[1], JANG-YOUL YOU[2], SUN-YOUNG PAEK[3], BYUNG-HEE NAM[3], and DOO-KIE KIM[4]

[1]*Dept of Architecture Engineering Chonbuk National University, Jeonju, Korea, Long-Span steel Frame system Research Center*
[2] *Dept of Architecture Engineering Songwon University, Gwangju, Korea*
[3]*Dept of Architecture Engineering Chonbuk National University, Jeonju, Korea*
[4]*Dept of Civil and Environmental Engineering, Kunsan National University, Kunsan, Korea*

Multiple tuned mass dampers (MTMD) is used for suppressing seismic excitation of tall building. The performance of MTMD is compared with that of a single tuned mass system (TMD). Optimum parameters of TMD/MTMD for minimizing the variance response of the damped primary structure derived by Krenk, Igusa, and Jangid were used. The optimally designed MTMD system is more effective than that of a single TMD system for reducing seismic excitation of a tall building.

Keywords: MTMD, Tuned mass system, Tall structures.

1 INTRODUCTION

Tuned Mass Damper (TMD) is a classical vibration control device consisting of a mass, a spring and a damper supported at the primary vibrating structure (Patil 2011). The optimum tuning frequency and damping ratio for the undamped primary structure under harmonic and white noise random excitations was derived by Den Hartog, Warburton and Ayroinde (Den Hartog 1985, Ayroinde 1980). Krenk derived the optimum parameters of TMD for the damped primary structure (Krenk 2008). And MTMD having distributed natural frequencies around the natural frequency of system has been utilized (Xu 1992, Patil 2011). Xu & Igusa carried out parametric studies of closely spaced natural frequencies of MTMD under a wide-band random excitation (Xu 1992).

In this study, the performance of MTMD for suppressing seismic-excitation of a tall building under the El-Centro 1940 N-S component earthquake ground motion is investigated.

2 EQUATIONS OF MOTION

Dynamic analysis procedure can be simplified if the contribution of higher modes of the primary structure is ignored (Kareem 1995). Therefore, tall building-MTMD system can be modeled as the first- mode generalized SDOF/MTMD system as shown in Figure 1.

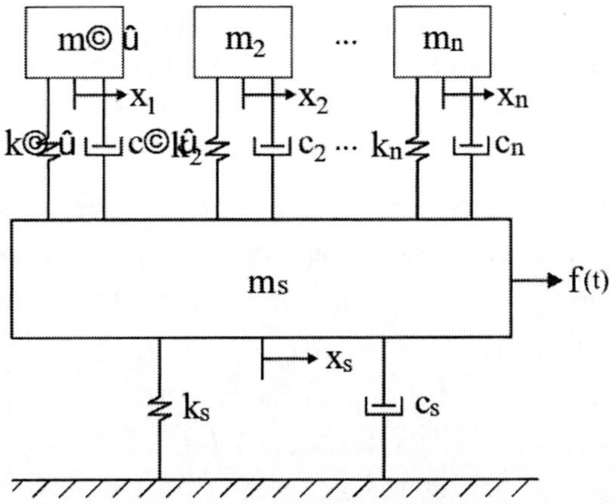

Figure 1. Tall building-MTMD model.

Dynamic equations of motion of the system can be written in matrix form of Eq. (1):

$$M\ddot{X} + C\dot{X} + KX = f(t) \tag{1}$$

where

$$X = [x_s \quad x_1 \quad x_2 \quad \cdot\cdot \quad x_n] \tag{2}$$

$$M = \begin{bmatrix} m_s & 0 & 0 & 0 & \cdot & 0 \\ 0 & m_1 & 0 & 0 & \cdot & 0 \\ 0 & 0 & m_2 & 0 & \cdot & 0 \\ 0 & 0 & 0 & m_3 & \cdot & 0 \\ \cdot & \cdot & \cdot & \cdot & \cdot & 0 \\ 0 & 0 & 0 & 0 & 0 & m_n \end{bmatrix} \tag{3}$$

$$C = \begin{bmatrix} C_s + \sum_{j=1}^{n} c_j & -c_1 & -c_2 & -c_3 & \cdot & -c_n \\ -c_1 & c_1 & 0 & 0 & \cdot & 0 \\ -c_2 & 0 & c_2 & 0 & \cdot & 0 \\ -c_3 & 0 & 0 & c_3 & \cdot & 0 \\ \cdot & \cdot & \cdot & \cdot & \cdot & 0 \\ -c_n & 0 & 0 & 0 & 0 & c_n \end{bmatrix} \tag{4}$$

$$K = \begin{bmatrix} K_s + \sum_{j=1}^{n} k_j & -k_1 & -k_2 & -k_3 & \cdot & -k_n \\ -k_1 & k_1 & 0 & 0 & \cdot & 0 \\ -k_2 & 0 & k_2 & 0 & \cdot & 0 \\ -k_3 & 0 & 0 & k_3 & \cdot & 0 \\ \cdot & \cdot & \cdot & \cdot & \cdot & 0 \\ -k_n & 0 & 0 & 0 & 0 & k_n \end{bmatrix} \tag{5}$$

Where M, C, K = the mass, damping and stiffness matrix of tall building with MTMD, *f(t)* = ground acceleration; m_s c_s, k_s,= the first-mode modal mass, damping and stiffness constant of tall building, m_i c_i, k_i,= the mass, damping and stiffness constant of each TMD of MTMD:

x_s , $\dot{x}_s$, $\ddot{x}_s$ = displacement, velocity and acceleration response of a tall building; x_i , $\dot{x}_i$, $\ddot{x}_i$ = displacement, velocity and acceleration response of TMD,

Dynamic equations of motion of the system can be rewritten as a state-space formulation as

$$\dot{Z} = AZ + Hf(t) \tag{6}$$

where

$$Z = [x \quad x_1 \quad x_2 \quad \cdots \quad x_n \quad , \quad \dot{x} \quad \dot{x}_1 \quad \dot{x}_2 \quad \cdots \quad \dot{x}_n] \tag{7}$$

$$A = \begin{bmatrix} 0 & I \\ -M^{-1}k & -M^{-1}C \end{bmatrix} \tag{8}$$

$$H = \begin{bmatrix} 0 & M^{-1} \end{bmatrix}^T \tag{9}$$

Z = state vector of primary structure with MTMD system; A = system matrix; H = the location vector of *f(t)*.

3 OPTIMUM PARAMETERS OF TMD

Krenk derived the optimum parameters of TMD for the damped primary structure (Krenk 2008).

$$f_{opt} = \frac{1}{1+\mu} \tag{10}$$

$$\xi_{opt} = \frac{\sqrt{\mu}}{2} \tag{11}$$

where f_{opt}= optimum tuning frequency; ξ_{opt}= optimum damping ratio; and μ=mass ratio.

4 OPTIMUM PARAMETERS OF MTMD

4.1 Optimal Number of MTMD

Patil & Jangid proposed the optimal number of MTMD is 5. Therefore 5 MTMD are used in this study and the sum of all MTMD's masses is the same as the mass of a single TMD (Patil 2011).

4.2 Optimal Frequency Band Width

The natural frequencies of MTMD are distributed uniformly around their average natural frequency which is the same value of the fundamental natural frequency of the primary structure (Xu 1992).

The natural frequency of the jth TMD ω_j is expressed as

$$\omega_j = \omega_T\left[1+\left(j-\frac{n+1}{2}\right)\frac{\omega_s}{n-1}\right] \tag{12}$$

where $\omega_T = \sum_{j=1}^{n}\omega_j/n$ is the average natural frequency of all MTMD; $\omega_s = \dfrac{\omega_n - \omega_1}{\omega_T}$ is the non-dimensional frequency spacing of the MTMD; ω_j=the natural frequency of jth TMD; and n = the total number of MTMD. The ratio of the total MTMD's mass to the primary structure's mass is defined as the mass ratio μ ;

$$\mu = \frac{\sum_{j=1}^{n} m_j}{m_s} = \frac{m_t}{m_s} \tag{13}$$

where m_t=total mass of MTMD; m_s=mass of the primary structure; and m_j = mass of the jth TMD. Patil & Jangid proposed the optimal frequency band width is 0.3 when the number of MTMD is 5 and the mass ratio is 1.0% (Patil 2011).

4.3 Optimal Tuning Frequency Ratio

Optimal tuning frequency ratio of MTMD is similar to a single TMD's, which is in Eq. (10) as $1/1+\mu$. Patil & Jangid proposed that optimal tuning frequency ratio is 1.0 when the number of MTMD is 5 and the mass ratio is 1.0% (Patil 2011).

4.4 Optimal Damping Ratio

If each TMD of MTMD has the same stiffness and damping constant for easy manufacturing MTMD, the stiffness constant k_T and damping constant c_T of each TMD can be evaluated as

$$k_T = \mu m_s \Big/ \left(\sum_{j=1}^{n}\frac{1}{\omega_j^{\,2}}\right) \tag{14}$$

$$c_T = 2k_T\zeta_T / \omega_T \tag{15}$$

where ξ_T = average damping ratio of the MTMD expressed as

$$\xi_T = \sum_{j=1}^{n}\frac{\xi_j}{n} = \frac{\omega_T c_T}{2k_T} \tag{16}$$

where ξ_j = damping ratio of the jth TMD. The mass of the jth TMD are expressed as

$$m_j = k_T / \omega_j^{\,2} \tag{17}$$

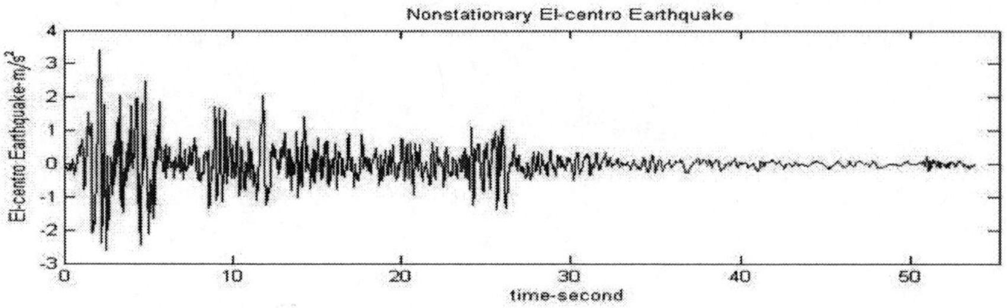

Figure 2. Accelerogram of El-Centro 1940 N-S component earthquake ground motion.

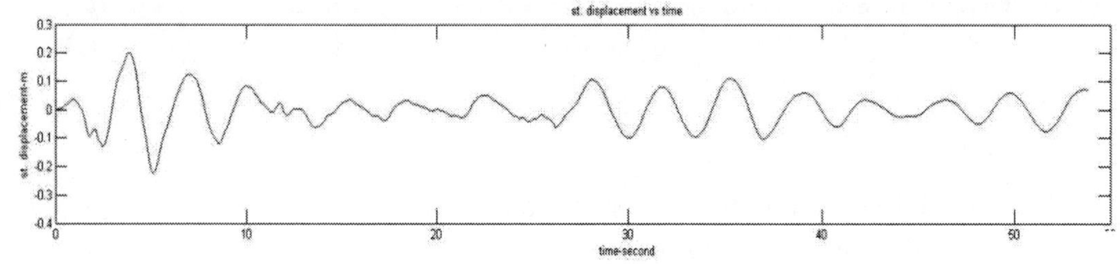

Figure 3. Dynamic Responses without TMD (rms=0.0595m).

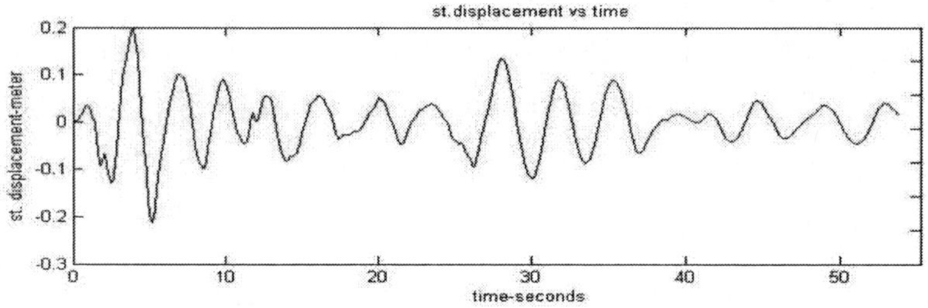

Figure 4. Dynamic Responses with TMD (rms=0.0586m).

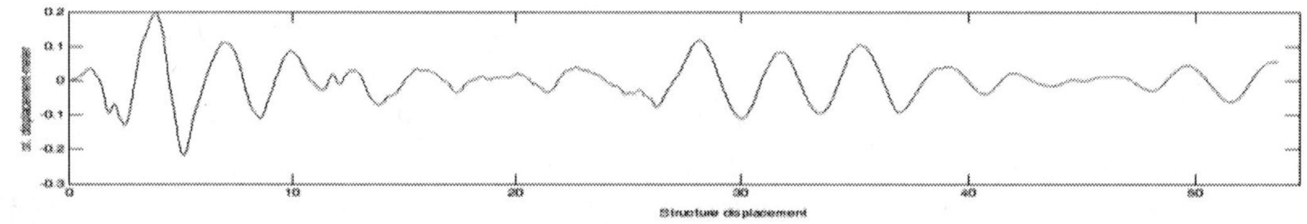

Figure 5. Dynamic Responses with MTMD (rms=0.0573m).

5 NUMERICAL EXAMPLE

This numerical example is from "Numerical Examples" in reference (Solari 1993). The tall building's height H = 180m, width B = 60m, depth D = 30m. The modal mass, damping and stiffness constants are $2.4*10^7 k_g$, $1.220832*10^6$ N-s/m, and $6.9001425*10^7$ N/m. Natural frequency f_1 = 0.27Hz, damping $r_a$tio ξ_1 = 0.015. The optimum parameters of MTMD were used. Figure 2 shows the actual accelerogram of El-Centro 1940 N-S component earthquake ground motion. Dynamic rms response of tall building without TMD is 0.0595m, shown in Figure 3. And dynamic rms response with a single TMD is 0.0586m shown in Figure 4, which shows the reduction effect is about 2%. Dynamic rms response of a tall building with MTMD is 0.0573m, shown in Figure 5, which shows that 4% reduction effect could be obtained.

6 CONCLUSIONS

The performance of MTMD and a single TMD for suppressing seismic excitation of tall building is investigated. Optimum parameters of MTMD for minimizing the variance response of the damped main structure were used. Suppressed rms response with MTMD is smaller than that of a single TMD. However, that effectiveness is not significant as known before.

Acknowledgment

This work was supported by the National Research Foundation of Korea Grant funded by the Korean Government (NRF-2014R1A2A1A10049538).

References

Ayorinde E.O. & Warburton G.B., Minimizing structural vibrations with absorbers, *Earthquake Engineering and Structural Dynamics* 8: 219-236, 1980.

Den Hartog, J.P, Mechanical Vibration, 4th edn. McGraw-Hill, New York, 1956. (Reprinted by Dover, NewYork, 1985)

Kareem, A & Kline, S, Performance of multiple mass dampers under random loading, *Journal of Structural Engineering ASCE* 121(2):348-361, 1995.

Krenk S & Hogsberg J., Tuned mass damped structures under random load, *Probabilistic Engineering Mechanics*, 23:408-415, 2008.

Patil,V B & Jangid R S., Optimum multiple tuned mass damper for the wind excited benchmark building, Journal of Civil Engineering and Management 17(4): 540-557, 2011.

Solari, G., Gust buffeting, Journal of Structural Engineering ASCE 119(2): 383-398, 1993.

Xu, K, Igusa,T., Dynamic characteristics of multiple substructures with closely spaced frequencies, *Earthquake Engineering and Structural Dynamics* 21(12): 1059-1070, 1992.

Interaction between Theory and Practice in Civil Engineering and Construction
Edited by Komurlu, R., Gurgun, A. P., Singh, A., and Yazdani, S.

EFFECTS OF A CENTRAL HOLE ON COMPRESSIVE BEHAVIOR OF CIRCULAR FIBER-REINFORCED RUBBER BEARINGS

SEVAL PINARBASI and FUAD OKAY

Dept of Civil Engineering, Kocaeli University, Kocaeli, Turkey

The aim of this study is to investigate the effects of the existence of a central hole on the main design parameters of a circularfiber-reinforced rubber bearing; namely compression modulus and shear strain due to compression. Since the compressive behavior of a multi-layered rubber bearing is mainly governed by the behavior of a single interior rubber layer in the bearing bonded to the reinforcing sheets at its top and bottom faces, the study is concentrated on uniformly compressed "bonded" rubber layers. The related compression problem is formulated using the "pressure" method. After deriving the closed-form expressions for compression modulus and shear strain, the effects of the hole on compressive behavior are investigated for fiber-reinforced bearings with different initial shape factors (a kind of aspect ratio for individual rubber layers) and rubber compressibility. It is shown that the compression modulus of a fiber-reinforced bearing may decrease considerably as the size of the central hole or the flexibility of the reinforcement increases, especially if the shape factor of the bearing is high and the compressibility of the rubber is not negligible. The study also shows that the effect of the hole on maximum shear strain reaches its most striking value when the hole size is smaller than 10% of the outer radius of the bearing.

Keywords: Compression modulus, Elastomer, Hollow circular bearing, Isolator, Pressure method, Seismic isolation.

1 INTRODUCTION

Multi-layered steel-reinforced rubber bearings have widely been used in various engineering applications, including isolation of machines from detrimental effects of vibrations, of bridges from adverse effects of thermal expansion and contraction, or of buildings from devastating effects of earthquakes. Rubber is, indeed, an ideal material for such isolation applications since it inherently has considerably low shear modulus, which ranges from 0.30 to 2.22 MPa (Kelly 1997). However, if a rubber bearing is subjected to high compression or bending loads, the excessive bulging occurred at the bulge-free surfaces of the bearing can lead to the failure of rubber under hydrostatic tension, which has to be prevented to fully utilize the favorable mechanical properties of rubber. The bulging of rubber, in turn, the hydrostatic tension in rubber, can be reduced if it is layered and reinforced using relatively rigid plates. In fact, the primary function of thin steel plates bonded to top and bottom faces of thin rubber layers in a multi-layered elastomeric bearing is to enhance the compressive and bending behavior of the bearing by controlling the bulging of interior rubber layers.

In the last three decades, many structures with sensitive equipment, historical value or post-earthquake importance have been isolated using multi-layered steel-reinforced rubber bearings. On the other hand, the researchers have realized that the use of rubber bearings for isolation of regular (residential, office, etc.) buildings or of structures in developing countries is rather limited. This phenomenon is mostly attributed to the fact that the conventional steel-reinforced bearings are usually considerably large, heavy and expensive due to the existence of steel plates as reinforcing elements. Kelly (1999, 2002) verified through analytical and experimental studies that both the cost and weight of the bearings can be reduced if fiber-reinforcement (in the form of two-directional sheets with large openings) is used instead of steel reinforcement. In the last fifteen years, many studies have been conducted on fiber-reinforced bearings (e.g., Kelly and Takhirov 2001, Mordini and Strauss 2008, Pinarbasi and Mengi 2008, Toopchi-Nezhad *et al.* 2008, Osgooei *et al.* 2014). However, most of these studies have been concentrated on either long rectangular strips or solid circular bearings. On the other hand, the use of hollow-circular isolators is also very common in practice. For this reason, it is essential to investigate the effects of the existence of a central hole on behavior of a fiber-reinforced bearing under compression/bending.

The main objective of this study is to investigate the effects of the existence of a central hole on the main design parameters of a circular fiber-reinforced rubber bearing; namely compression modulus and shear strain due to compression. Since the compressive behavior of a multi-layered rubber bearing is mainly governed by the behavior of a single interior rubber layer in the bearing bonded to the reinforcing sheets at its top and bottom faces, the study is concentrated on uniformly compressed "bonded" rubber layers. The related compression problem is handled using the well known "pressure" method. After deriving the closed-form expressions for compression modulus and shear strain, the effects of the hole on compressive behavior are investigated for fiber-reinforced bearings with different shape factors (a kind of aspect ratio for individual rubber layers) and rubber compressibility.

2 UNIFORM COMPRESSION OF A RUBBER LAYER BONDED TO FIBER-REINFORCED SHEETS AT ITS TOP AND BOTTOM FACES

The uniform compression of a typical interior rubber layer in a multi-layered circular fiber-reinforced rubber bearing with a central hole is shown in Figure 1. The disc has an inner radius of a, outer radius of R, rubber thickness of t and equivalent reinforcement thickness of t_f. Under a concentric compressive load of P, the top surface of the disc approaches to the bottom with a relative vertical displacement of Δ. This compression problem is formulated using the pressure method modified to include the extensibility of the flexible reinforcing sheets (Kelly 1999). It is convenient to define a cylindrical coordinate system (r, θ, z) with its origin located at the center of the disc (Figure 1).Since the deformed shape of the disc is axisymmetric, the displacement component in θ direction vanishes and those in r and z directions (i.e., u and w) are independent of θ. Using the pressure method, the nonzero displacement components can be expressed as $u(r,z) = u_0(r)(1-4z^2/t^2)+u_1(r)$ and $w(r,z) = w(z)$. The constitutive equation for the rubber disc can be written, in terms of the strain components (ε_{rr}, $\varepsilon_{\theta\theta}$ and ε_{zz}), mean pressure (p) and bulk modulus (K) of rubber as follows:

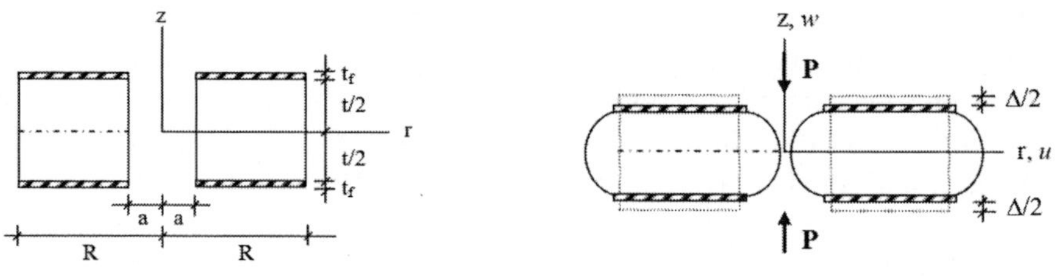

Figure 1. Uniform compression of a circular fiber-reinforced rubber layer with a central hole.

$$\varepsilon_{rr}+\varepsilon_{\theta\theta}+\varepsilon_{zz}=-p/K \tag{1}$$

ε_{rr}, $\varepsilon_{\theta\theta}$and ε_{zz} can be expressed in terms u_0, u_1 and w,using axisymmetric strain-displacement relations. Substituting these relations into Eq. (1), then integrating the resulting equation through the rubber thickness, Eq. (1) can be rewritten in terms of the three unknowns of the problem: u_0, u_1 and p. The second governing equation for the problem, the equilibrium equation in radial direction for the rubber disc, can be written, in terms of the stress components σ_{rr}, $\sigma_{\theta\theta}$,$\sigma_{zz}$and τ_{rz}, as follows:

$$\sigma_{rr,r}+\tau_{rz,z}+\frac{\sigma_{rr}-\sigma_{\theta\theta}}{r}=0 \tag{2}$$

Using the "pressure" assumption and stress-strain relation for shear stress and strain (i.e., $\tau_{rz}=\gamma_{rz}G$, where G is the shear modulus of rubber), Eq. (2) can be expressed in terms of the unknown. Finally, the third governing equation is obtained from the equilibrium of the reinforcing sheets, which are subjected to bonding shear stresses (τ^{-}_{rz} and τ^{+}_{rz}, see Figure 2), in radial direction. Assuming that the sheets are in the state of plane stress, the equilibrium equation can be written, in terms of the internal forces per unit lengthin the sheet(N_{rr} and $N_{\theta\theta}$) and bonding shear stresses, as follows:

$$N_{rr,r}+\frac{N_{rr}-N_{\theta\theta}}{r}+\tau^{-}_{rz}-\tau^{+}_{rz}=0 \tag{3}$$

Using the linear stress-strain relations, N_{rr} and $N_{\theta\theta}$can be expressed in terms of u_1, elasticity modulus$E_f$and Poisson's ratio ν_f of the reinforcement. Similarly, bonding shear stresses can be expressed in terms of u_1 and u_0.Thus, Eqs. (1) to (3) constitute three governing equations for the studied problem. Since there is no applied load/stress at the inner/outer lateral faces of the layer or the edges of the sheets, the boundary conditions for the problem can be expressed as $p=0$ and $N_{rr}=0$ at $r=a, R$. Using these conditions, the unknown displacement functions (u_0 and u_1) and the mean pressure (p) can be obtained. For details, Pinarbasi and Okay (2011) is to be referred.

The compression modulus for the studied layer (see Figure 1) can be computed from $E_c=(P/A)/(\Delta/t)$, where P is obtained by integrating the mean pressure (p) over the bonded area and $A=\pi(R^2\text{-}a^2)$. Thus, the compression modulus for a multi-layered fiber-reinforced elastomeric bearing (denoted as $E_{c,HC}$) has the following form:

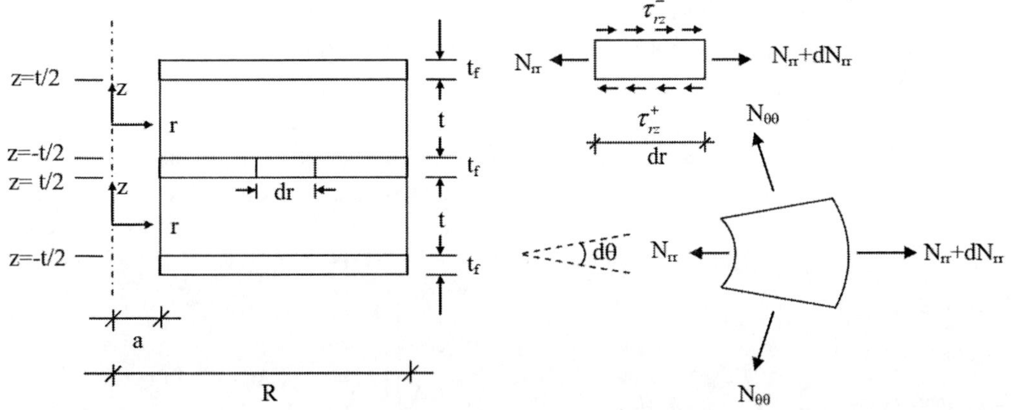

Figure 2. Forces on infinitesimal area of a reinforcing sheet (Pinarbasi and Okay 2011).

$$E_{c,HC}=\frac{k_f}{t}\frac{(1+\nu_f)}{2}\frac{\left\{\begin{aligned}&\left(1-\frac{a^2}{R^2}\right)\left[I_0(\beta^*R)-A_1K_0(\beta^*R)\right]\\&-2\left[\frac{I_1(\beta^*R)+A_1K_1(\beta^*R)}{(\beta^*R)}\right]+\frac{2a^2}{R^2}\left[\frac{I_1(\beta^*a)+A_1K_1(\beta^*a)}{(\beta^*a)}\right]\end{aligned}\right\}}{A_2-\frac{a^2}{R^2}A_3+\left(1-\frac{a^2}{R^2}\right)A_4} \quad (4)$$

$$(\beta^*)^2=\alpha^2+\lambda^2=\frac{12G}{k_f t}+\frac{12G}{Kt^2},\quad k_f=\frac{E_f t_f}{1-\nu_f^2} \quad (5)$$

In Eq. (4), I_i and K_i (i=1,2) are the modified Bessel functions of, respectively, first and second kinds of order i.The closed-form expressions for A_i (i=1 to 4) are presented in Pinarbasi and Okay (2011). The other design parameter for the disc, namely, the maximum absolute shear strain in the rubber layer, has the following closed-form:

$$\gamma_{\max}=\frac{6\beta^*}{\alpha^2 t}\frac{\Delta}{t}\frac{(1+\nu_f)}{2}\left(1-\frac{a^2}{R^2}\right)\frac{\left[I_1(\beta^*r)+A_1K_1(\beta^*r)\right]}{A_2-\frac{a^2}{R^2}A_3+\left(1-\frac{a^2}{R^2}\right)A_4} \quad (6)$$

3 DISCUSSIONS AND CONCLUSIONS

Figure 3 shows the variation of $E_{c,HC}/G$ as a function of initial shape factor (defined as $S_o=R/2t$) for two different values of reinforcement rigidity, k_f/Gt=30000 (which represents CFRP sheets used by Kelly (2002)) and 300,for three different values of radius ratio, $\beta=a/R$=0, 0.01, 0.1 and 0.5 andrubber compressibility, ν=0.5, 0.4995 (which represents natural rubber) and 0.49.As it is seen, $E_{c,HC}$ increases as$S_o$increases until it reaches an asymptotic value, which increases as $k_f/Gt\rightarrow\infty$ and $\nu\rightarrow0.5$. The presence of hole decreases$E_{c,HC}$ considerably if S_o, $k_f/Gt$and ν are large. For example, for

S_o=30 and k_f/Gt=30000, the ratio of $E_{c,HC}$ to $E_{c,C}$ (i.e., E_c for circular case) is 66% and 21% respectively when β=0.1 and 0.5 if ν=0.5, but 93% and 62% if ν=0.4995. Figure 4 shows the variation of $\gamma_{max}/\varepsilon_c$ (where $\varepsilon_c=\Delta/t$) as a function of β for various values of S_o, k_f and ν. As shown in the graphs, $\gamma_{max}/\varepsilon_c$ increases considerably as $\beta \rightarrow 0$. The increase is larger if S_o, k_f and ν are large. However, when β>0.1, $\gamma_{max}/6S_o\varepsilon_c$ (ratio to the incompressible circular steel-reinforced case) is less than 2.5 even when ν=0.5.

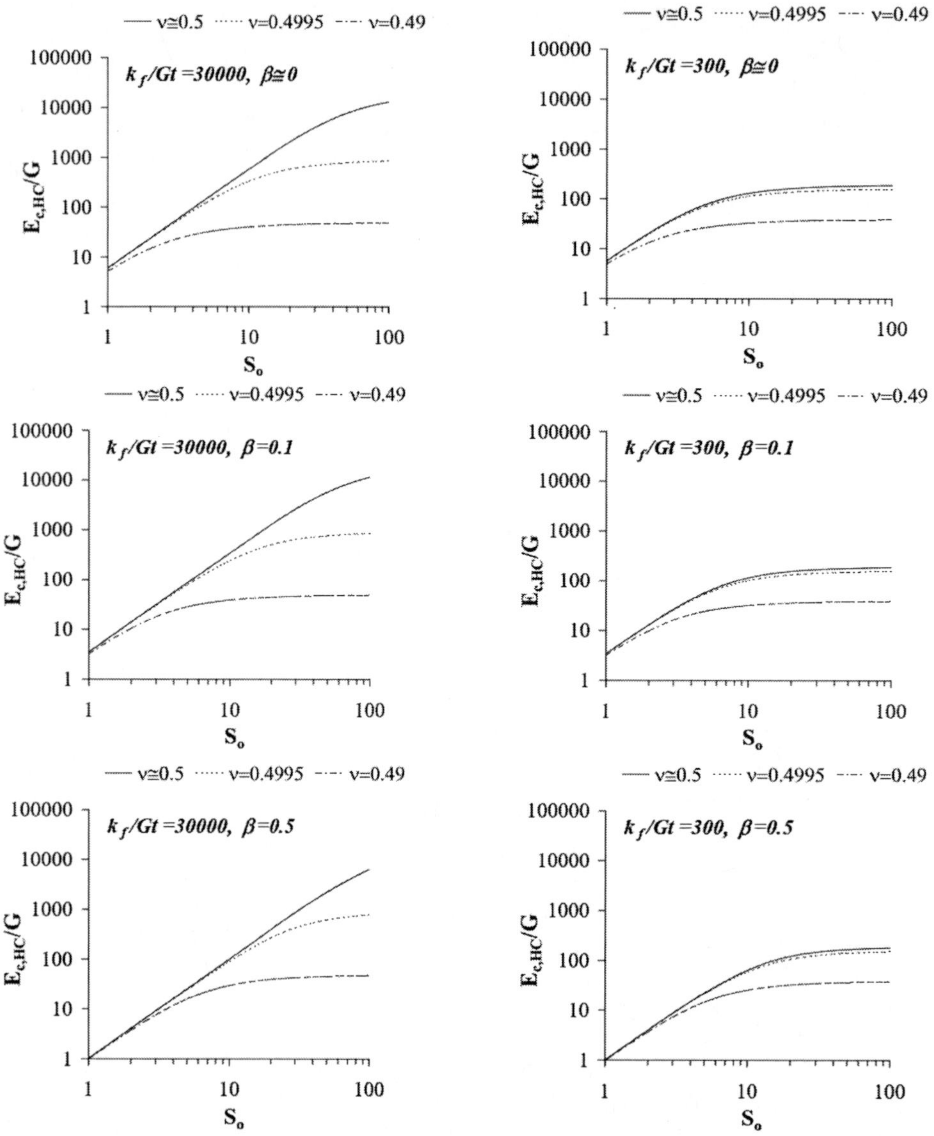

Figure 3. Effect of the existence of a central hole on compression modulus.

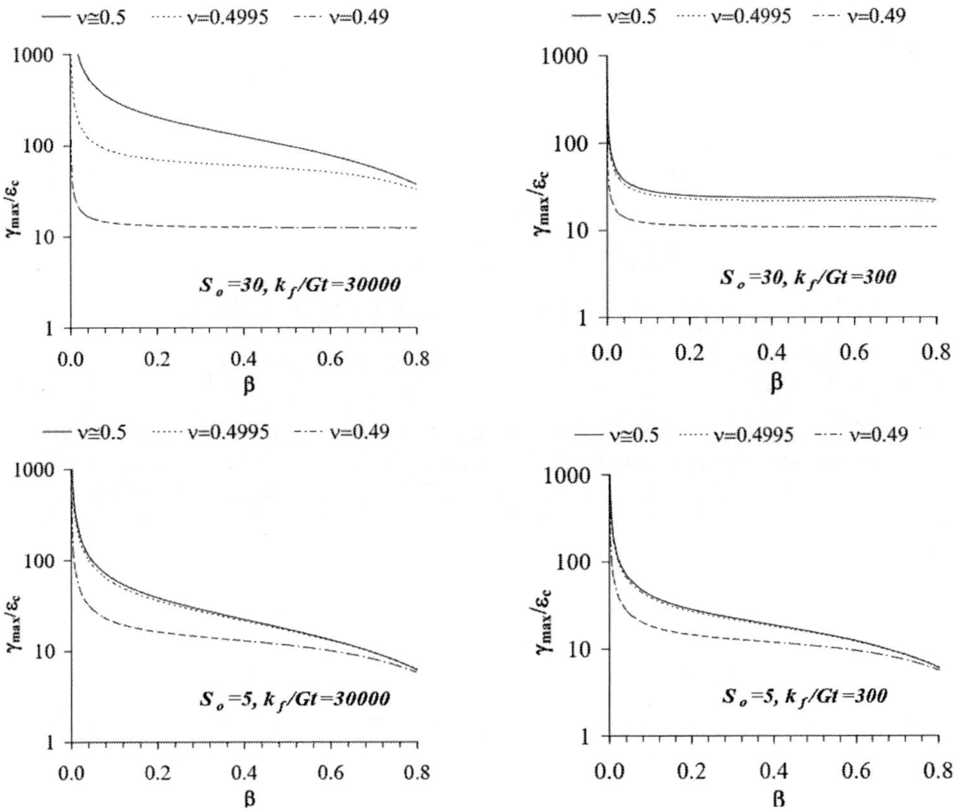

Figure 4. Effect of the existence of a central hole on maximum shear strain due to compression.

References

Kelly, J.M.,*Earthquake Resistant Design with Rubber*, Springer-Verlag London, 1997.

Kelly, J. M., Analysis of Fiber-Reinforced Elastomeric Isolators, *J. Seismolog. Earthquake Eng.*, 2 (1), 19-34, 1999.

Kelly, J. M., Seismic Isolation Systems for Developing Countries, *Earthquake Spectra*, 18(3), 385-406, 2002.

Kelly, J.M. and Takhirov, S.M.,*Analytical and Experimental Study of Fiber-Reinforced Elastomeric Isolators*, PEER Rep. No.2001/11, University of California, Berkeley, 2001.

Mordini, A. and Strauss, A., An Innovative Earthquake Isolation System using Fiber Reinforced Rubber Bearings", *Eng. Struct.*, 30 (10), 2739-2751, 2008.

Osgooei, P. M., Tait, M. J. and Konstantinidis, D., Three-Dimensional Finite Element Analysis of Circular Fiber-Reinforced Elastomeric Bearings under Compression, *Compos. Struct.*, 108, 191-204, 2014.

Pinarbasi, S. and Mengi, Y., Elastic Layers Bonded to Flexible Reinforcements, *Int. J. Solids Struct.*, 45(3-4), 794-820, 2008.

Pinarbasi, S. and Okay, F., Compression of Hollow-Circular Fiber-Reinforced Rubber Bearings, *Struct. Eng. Mech.*, 38(3), 361-384, 2011.

Toopchi-Nezhad, H., Drysdale, R. G. and Tait, M. J., Testing and Modeling of Square Carbon Fiber-Reinforced Elastomeric Seismic Isolators, *Struct. Control Health Monit.*, 15 (6), 876-900, 2008.

Interaction between Theory and Practice in Civil Engineering and Construction
Edited by Komurlu, R., Gurgun, A. P., Singh, A., and Yazdani, S.

ON THE VARIATION OF MAXIMUM ISOLATOR DISPLACEMENTS DUE TO GROUND MOTION DIRECTIONALITY

GÖKHAN ÖZDEMİR, BURAK ERŞEKER, and ÖZGÜR AVŞAR

Civil Engineering Dept, Anadolu University, Eskişehir, Turkey

In most of the cases, code specifications dictate the use of nonlinear response history analyses (NRHA) to estimate maximum isolator displacements (MIDs) of a seismically isolated structure (SIS). For this purpose, a set of ground motion records with similar characteristics needs to be selected. Then, the structure is analyzed bidirectionally by considering both orthogonal horizontal components of these records. However, there is not any provision regarding the ground motion directionality effect in the codes but simply use of as-recorded motions is encouraged. This study investigates the effect of ground motion directionality on variation of MIDs in case of bidirectional NRHA. Thus, a typical SIS, where the isolator units are composed of lead rubber bearings (LRBs), is subjected to ground motions rotated from their as-recorded original form by increments of 10° up to 360°. Here, LRBs are modelled by a deteriorating hysteretic representation in which the strength of the isolator reduces gradually due to the applied loading. In the analyses, first, the original as-recorded ground motion is applied to the SIS and the corresponding MID is noted. Then, the same structure is subjected to rotated versions of the same motion and again the MIDs are noted. To quantify the variation in the isolator displacement, analytically obtained MIDs are compared. Results showed that there is an amplification in MIDs due to change in ground motion direction.

Keywords: Incidence angle, Seismic isolation, Bidirectional analysis, Lead rubber bearing, Isolator displacement, Hysteretic deterioration.

1 INTRODUCTION

Maximum isolator displacement (MID) is one of the primary concerns during the design stage of a seismically isolated structure (SIS). Hence, estimation of MIDs is of crucial importance for design. Code specifications establish three methods to be used in estimating the MID of a SIS (ASCE 2010). These methods are Equivalent Lateral Force Procedure, Response Spectrum Procedure and Response History Procedure. Among these procedures, Response History Procedure is assumed to provide the so called "exact" solution and defined as the procedure that has to be followed in most of the cases. It assures the use of nonlinear force-deformation relation of isolators obtained from test results in analytical representation of isolators. In the analyses, the structural form composed of superstructure and isolation system shall be subjected to a set of ground motion records selected to represent the scenario earthquake.

Furthermore, in these analyses, the orthogonal horizontal components of the selected ground motions shall act simultaneously by means of bidirectional simulations.

According to current code specifications, bidirectional Nonlinear Response History Analyses (NRHA) of SISs are performed using either as-recorded or synthetic ground motion records. There is not any code provision regarding the ground motion directionality effect in the simulations. However, there are several studies that have discussed the effect of ground motion directionality on the structural response (Ghersi and Rossi 2001, Athanatopoulou 2005, Rigato and Medina 2007, Moschonas and Kappos 2013, Kalkan and Reyes 2015). Research outcomes of studies mentioned here revealed that variation in ground motion direction result in amplification in structural response quantities. Furthermore, it has been stated that the critical incidence angle differentiates highly depending on the response quantity. However, all of these studies employed fixed-base structures rather than SISs. The study presented herein aims to quantify the amount of variation in MID of a typical SIS when subjected to bidirectional ground motion excitations with different incidence angles. For this purpose, bidirectional NRHA were conducted with ground motion records rotated from their as-recorded original forms by increments of 10° up to 360°. The isolation system of the investigated structure is composed of lead rubber bearings (LRBs). The nonlinear hysteretic behavior of LRBs is represented by a deteriorating force-deformation relation proposed by Kalpakidis and Constantinou (2009). Thus, reduction in the strength of LRBs under cyclic motion is properly considered in the analyses.

2 STRUCTURAL MODEL

The analyzed structure is selected to be representative of a typical SIS. Accordingly, the hypothetical emergency operation center designed for National Earthquake Hazards Reduction Program (NEHRP 2006) is considered as the superstructure. The isolation system is redesigned such that all isolator units are composed of LRBs. It is a 3-story seismic isolated steel frame structure. The height of the building is 9m (each story height is 3m) with plan dimensions of 36mx54m. Span lengths in short and long directions are identical and equal to 9m. The structure is symmetric in both plan and elevation. Weight of the superstructure is 73000 kN. Figure 1 presents the geometrical features of the analyzed SIS.

3 MODELING OF LEAD RUBBER BEARINGS

For SISs where the only nonlinearity takes place at the isolation level, modeling of nonlinear hysteretic behavior of isolator units is crucial. Experimental studies revealed that any LRB subjected to cyclic motion undergoes a deterioration in its force-deformation relation. The main reason for such deterioration is found to be the temperature rise in the lead core of LRB. On the other hand, existing design approaches refer to non-deteriorating hysteretic representations rather than the actual deteriorating one. This is basically due to the lack of ability to model the actual behavior of LRBs. Such a modeling approach aims to define envelopes for response quantities of LRBs instead of simulating the actual response. In this study, the deteriorating hysteretic behavior of LRBs is employed in the analyses. For this purpose, a coupled material model that is capable of incorporating the effect of lead core heating on the hysteretic behavior of LRBs is used in NRHA. Coupled solution is also vital because, NRHA are performed bidirectionally.

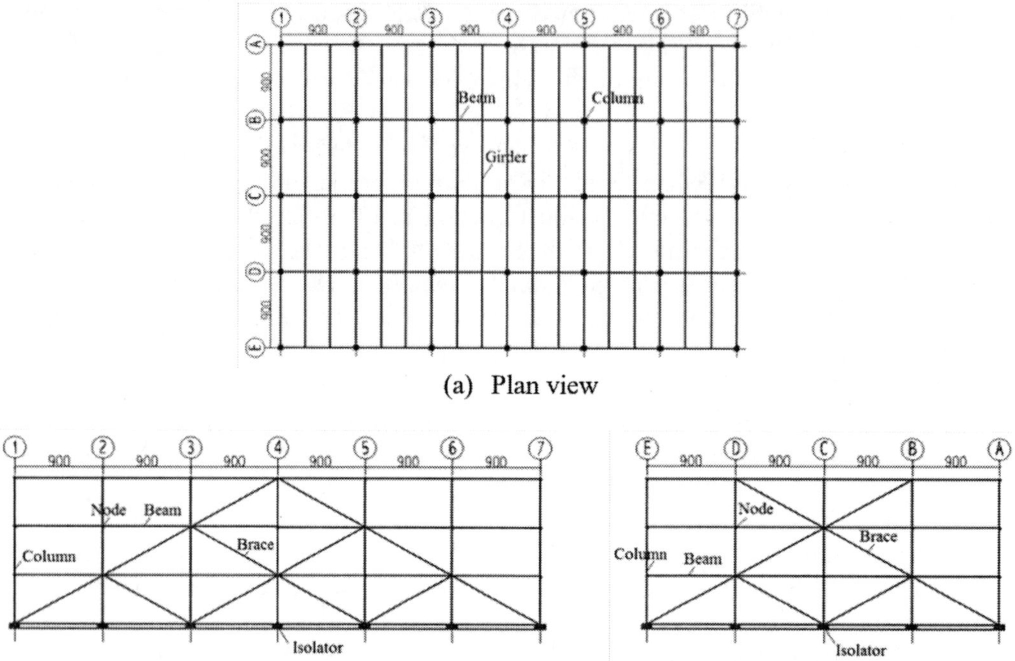

(a) Plan view

(b) Side view for B and D axes

(c) Side view for 2 and 6 axes

Figure 1. Properties of the analyzed seismically isolated structure (all units are in cm).

Effect of lead core heating on hysteretic behavior of LRBs is taken into account by means of Eq. (1)-(4) where, h_L, a, ρ_L, c_L and σ_{YL0} are the height, radius, density, specific heat and initial yield stress of the lead core, respectively; t_s is the total shim plate thickness, α_s is the thermal diffusivity of steel, k_s is the thermal conductivity of steel, t^+ is the dimensionless time, t is the time since the beginning of the motion, and E_2 is a constant that relates the temperature and yield stress. In Eq. (1), Z_x and Z_y are the hysteretic dimensionless quantities that varies between ±1, $\mathring{U}_x$ and $\mathring{U}_y$ are the relative velocities of the bearing. Once definition of hysteretic behavior of LRBs is established, the coupled solution of the isolator forces in both orthogonal horizontal directions is based on equations proposed by Park *et al.* (1986). Due to space limitations, related set of equations are not presented here.

Employed LRBs are selected so that they are representative of typical characteristics of SISs. Accordingly, characteristic strength to weight ratio (Q/W) of the analyzed LRB is 0.10 and the isolation period is 2.5s. Height of the isolator (h_L) is 290mm whereas the radius (a) and the total shim plate thickness (t_s) are 83.5mm and 84 mm, respectively.

$$\sigma_{YL}(T_L) = \sigma_{YL0} \cdot \exp(-E_2 T_L) \tag{1}$$

$$\dot{T}_L = \frac{\sigma_{YL}(T_L)\cdot\sqrt{Z_x^2+Z_y^2}\sqrt{\dot{U}_x^2+\dot{U}_y^2}}{\rho_L c_L h_L} - \frac{k_S\cdot T_L}{a\rho_L c_L h_L}\cdot\left(\frac{1}{F}+1.274\cdot\left(\frac{t_s}{a}\right)\cdot\left(t^+\right)^{-1/3}\right) \quad (2)$$

$$F = \begin{Bmatrix} 2\cdot\left(\frac{\tau}{\pi}\right)^{1/2} - \frac{\tau}{\pi}\cdot\left[2-\left(\frac{\tau}{4}\right)-\left(\frac{\tau}{4}\right)^2-\frac{15}{4}\left(\frac{\tau}{4}\right)^3\right], & \tau<0.6 \\ \frac{8}{3\pi}-\frac{1}{2(\pi\cdot\tau)^{1/2}}\cdot\left[1-\frac{1}{3\cdot(4\tau)}+\frac{1}{6\cdot(4\tau)^2}-\frac{1}{12\cdot(4\tau)^3}\right], & \tau\geq 0.6 \end{Bmatrix} \quad (3)$$

$$\tau = \frac{\alpha_S t}{a^2} \quad (4)$$

4 GROUND MOTION RECORDS

Ground motions used in this study have very similar characteristics in terms of Magnitude (M_w), closest distance to fault rupture (R), and soil classification. Average shear wave velocities of the selected ground motions at the upper most 30 m soil profile are ranging from 180 m/s to 360 m/s (soil class D per NEHRP). Peak Ground Velocity (PGV) values of selected ground motions are also very close to each other. Avsar and Ozdemir (2013) showed that there is a high correlation between PGV and MIDs. That correlation is also not sensitive to any change in isolator characteristics. Table 1 presents the characteristics of selected ground motions. In Table 1, PGA and PGD stand for peak ground acceleration and peak ground displacement.

Table 1. Ground motion characteristics.

#	Earthquake	Station	M_w	R (km)	Component	PGA (g)	PGV (cm/s)	PGD (cm)
1	Imperial Valley	Brawley Airport	6.5	10.4	315	0.22	38.9	13.6
					225	0.16	35.8	22.3
2	Imperial Valley	El Centro Array #10	6.5	6.2	050	0.17	47.5	31.1
					320	0.22	41.2	18.0
3	Northridge	Canyon Country W	6.7	12.4	270	0.48	44.9	12.6
					000	0.41	43.0	11.8

Figure 2 shows the response spectra of the ground motions listed in Table 1. In Figure 2, orthogonal horizontal components of ground motions are identified as "strong" and "weak" according to their PGVs. The component with higher PGV is named as "strong" component. In order to mimic different incidence angles, as-recorded forms of selected records are rotated through 360° with increments of 10°. This is achieved by Eq. (5) where $a_x(t)$ and $a_y(t)$ are the rotated forms of original motions $a_1(t)$ and $a_2(t)$.

$$\begin{Bmatrix} a_x(t) \\ a_y(t) \end{Bmatrix} = \begin{bmatrix} \cos\theta & -\sin\theta \\ \sin\theta & \cos\theta \end{bmatrix}\cdot\begin{Bmatrix} a_1(t) \\ a_2(t) \end{Bmatrix} \quad (5)$$

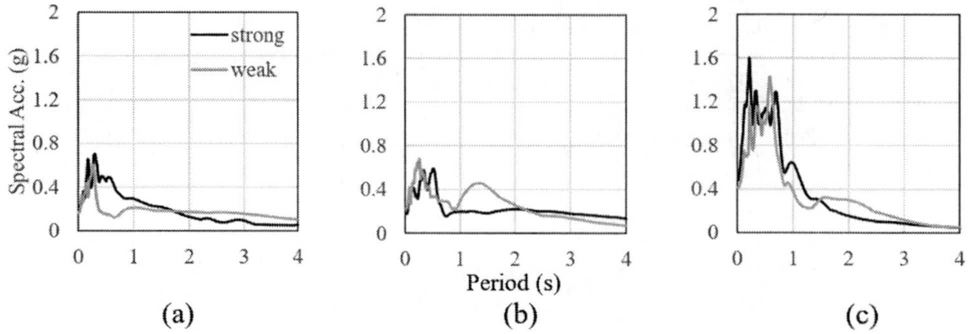

Figure 2. Response spectra for (a) EQ #1 (b) EQ #2 (c) EQ #3.

5 ANALYSES RESULTS

To identify the amount of amplification in MID of a SIS due to variation in incidence angle of ground motion records, as-recorded and the rotated versions of selected motions are subjected to model structure. Thus, a total of 108 NRHA have been performed and the corresponding MIDs were noted. MIDs obtained from analyses in which as-recorded motions were used, are called as MID_O. On the other hand, MIDs obtained from analyses conducted with rotated versions of original records are represented by MID_R. Figure 3 depicts the variation of MID_R's as a function of incidence angle. Corresponding MID_R/MID_O ratios are presented in Figure 4.

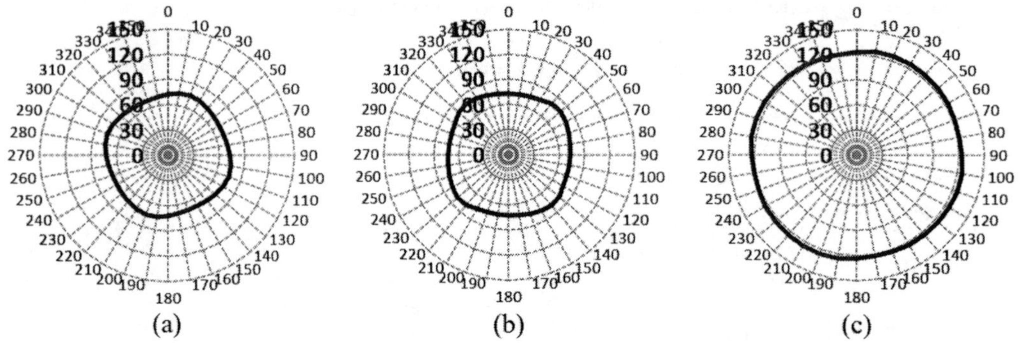

Figure 3. MID_R versus incidence angles for (a) EQ #1 (b) EQ #2 (c) EQ #3.

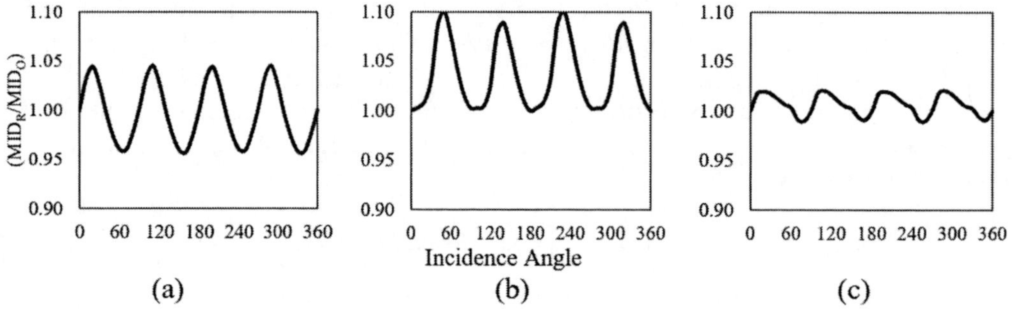

Figure 4. MID_R/MID_O ratios versus incidence angles for (a) EQ #1 (b) EQ #2 (c) EQ #3.

As can be seen from Figure 4, the amount of amplification in MIDs due to change in incidence angle is not constant and dependent of the ground motion characteristics. Specifically, for the selected ground motion records, calculated maximum amplifications are 5%, 10% and 3% for EQ #1, EQ #2 and EQ #3, respectively.

6 CONCLUSIONS

This study focused on the amplification in MIDs of a typical SIS due to variation in incidence angle of ground motions. For this purpose, bidirectional NRHA were conducted by considering both orthogonal horizontal components of ground motion records. Isolation units of the analyzed structure was composed of LRBs. Nonlinear hysteretic representation of LRBs were achieved by a deteriorating force-deformation relation. Hence, the gradual reduction in strength of LRBs under cyclic motion was taken into consideration. Analyses results showed that the amount of amplification in MIDs is not constant and depends on the ground motion characteristics.

Acknowledgments

This study was funded by the Turkish Scientific and Technical Research Council (TÜBİTAK) under grant number 114M259 which the authors gratefully acknowledge.

References

American Society of Civil Engineers, Structural Engineering Institute (ASCE/SEI), *Minimum Design Loads for Buildings and Other Structures*, Reston, VA., 2010.

Athanatopoulou, A. M., Critical Orientation of Three Correlated Seismic Components, *Engineering Structures*, 27(2), 301-312, January, 2005.

Avsar, O., Ozdemir, G., Resp onse of Seismic-Isolated Bridges in Relation with Intensity Measures of Ordinary and Pulse-Like Ground Motions, *Journal of Bridge Engineering* (ASCE), 18(3), 250-260, March, 2013.

Ghersi, A., Rossi, P. P., Influence of Bi-Directional Ground Motions on the Inelastic Response of One-Storey In-Plan Irregular Systems, *Engineering Structures*, 23(6), 579-591, June, 2001.

Kalkan, E., Reyes, J. C., Significance of Rotating Ground Motions on Behavior of Symmetric- and Asymmetric-Plan Structures: Part 2. Multi-story Structures, *Earthquake Spectra*, 31(3), 1613-1628, August, 2015.

Kalpakidis, I. V., Constantinou, M. C., Effects of Heating on the Behavior of Lead-Rubber Bearing. I:Theory, *Journal of Structural Engineering* (ASCE), 135(12), 1440-1449, December, 2009.

Moschonas, I. F., Kappos, A. J., Assessment of Concrete Bridges Subjected to Ground Motion with an Arbitrary Angle of Incidence: Static and Dynamic Approach, *Bulletin of Earthquake Engineering*, 11(2), 581-605, April, 2013.

National Earthquake Hazard Reduction Program (NEHRP), *Recommended Provisions: Design Examples*, Washington, 2006.

Rigato A. B, Medina R. A., Influence of Angle of Incidence on Seismic Demands for Inelastic Single-Storey Structures Subjected to Bi-Directional Ground Motions, *Engineering Structures*, 29(10), 2593-2601, October, 2007.

Park, Y. J., Wen, Y. K., Ang, A. H., Random Vibration of Hysteretic Systems Under Bi-Directional Ground Motions, *Earthquake Engineering and Structural Dynamics*, 14(4), 543-557, July/August, 1986.

Interaction between Theory and Practice in Civil Engineering and Construction
Edited by Komurlu, R., Gurgun, A. P., Singh, A., and Yazdani, S.

EFFECTS OF FOUNDATION SETTLEMENT ON IN-PLANE BUCKLING OF SHALLOW ARCHES

XIAOCHUN FAN[1], YU YU[1], YONG-LIN PI[2], and MARK ANDREW BRADFORD[2]

[1]*Schoool of Civil Engineering and Architecture, Wuhan University of Technology, Wuhan, China*

[2]*Center for Infrastructure Engineering and Safety, University of New South Wales, Sydney, Australia*

Foundation settlement of structures may take place and it may influence the structural behavior of structures. Shallow arches are used in engineering structures such as large span roofs and bridges. This paper is concerned with the effects of the foundation settlement on the in-plane buckling of shallow circular arches under a uniform radial load. Because the deformations of a shallow arch prior to buckling become nonlinear, their effects on the buckling also need to be considered. Differential equations of equilibrium for the nonlinear analysis are derived using the principle of virtual work. Analytical solutions for the nonlinear behavior and for the nonlinear buckling load are obtained. It is found that the foundation settlement plays an important role in the nonlinear structural response of the arch to the external load and that significantly influence the nonlinear buckling load of the arch.

Keywords: Analytical, Instability, Nonlinear, Structures.

1 INTRODUCTION

Foundation settlement of structures may take place from time to time, which influences the structural behavior of structures. Shallow arches are used in engineering structures such as large span roofs and bridges. It is known that shallow arches may buckle elastically in the plane of loading and fail in a limit point (snap-through) instability mode or in a bifurcation buckling mode (Pi *et al.* 2002). However, it is unknown if the foundation settlement influences the in-plane buckling behavior of shallow arches.

Because the deformations of a shallow arch prior to buckling become nonlinear, their effects on the buckling need to be considered in addition to the settlement. This paper investigates the influence of foundation settlements of shallow arches on their nonlinear equilibrium and buckling load. Differential equations of equilibrium for the nonlinear analysis are derived using the principle of virtual work. Analytical solutions for the nonlinear behavior and for the nonlinear buckling load are obtained and verified by comparisons with the finite element (FE) results.

2 NONLINEAR EQUILIBRIUM

The differential equations of equilibrium for non-liner analysis by using the principle of virtual work, which states that

$$\delta\Pi = \int_{-\Theta}^{\Theta} [-NR(\delta\tilde{w}' - \delta\tilde{v} + \tilde{v}'\delta\tilde{v}'\} - M\delta\tilde{v}'' - qR^2]\mathrm{d}\theta = 0 \tag{1}$$

where N and M are the axial compressive force and bending moment and they can be expressed as

$$N = AE(\tilde{w}' - \tilde{v} + \frac{1}{2}\tilde{v}'^2) \tag{2}$$

and

$$M = -EI\frac{\tilde{v}''}{R} \tag{3}$$

Integrating Eq. (1) by parts leads to the differential equations of equilibrium in the axial and radial directions as

$$N' = 0 \quad \text{and} \quad \frac{\tilde{v}^{iv}}{\mu^2} + \tilde{v}'' = Q \tag{4}$$

with

$$Q = \frac{qR - N}{N} \quad \text{and} \quad \mu^2 = \frac{NR^2}{EI} \tag{5}$$

and the boundary conditions can be written as

$$\tilde{v}'' = 0 \text{ at } \theta = \pm\Theta, \;\; \tilde{w} = 0, \tilde{v} = 0 \text{ at } \theta = -\Theta \tag{6}$$

and

$$\tilde{w} = \Delta\sin\Theta, \tilde{v} = \Delta\cos\Theta \quad \text{at} \quad \theta = \Theta \tag{7}$$

where Δ is the vertical settlement of the arch end at $\theta = \Theta$.

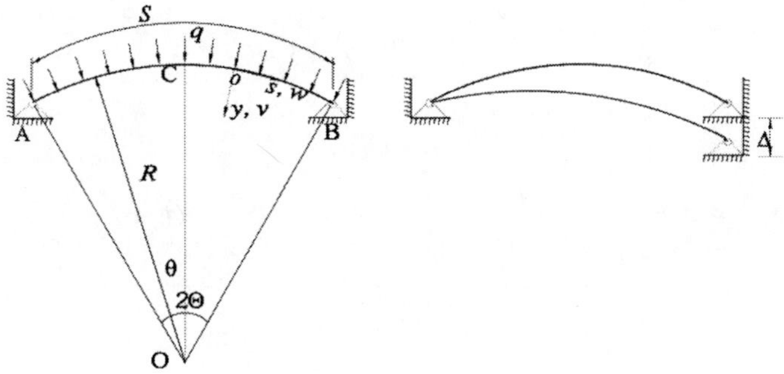

(a) Arch (b) Foundation settlement

Figure 1. Arch and its foundation settlement.

The first equation of Eq. (4) indicates the axial compressive force N is a constant along the arch length. Solving the second equation of Eq. (4) under boundary conditions given by Eqs. (6) and (7) leads to the analytical solutions for the radial displacement as

$$\tilde{v} = \frac{Q}{\mu^2}\left(\frac{\cos\mu\theta}{\cos\beta} - 1 + \frac{1}{2}\mu^2\theta^2 - \frac{1}{2}\beta^2\right) + \frac{\Delta\cos\Theta}{2}\left(1 + \frac{\theta}{\Theta}\right) \tag{8}$$

where the axial force parameter $\beta = \mu\Theta$ (Pi *et al.*2014).

It is noted the radial displacement is a function of the dimensionless external load Q and the axial force parameter β. Hence, an equation of equilibrium between Q and β is required. For this, substituting the solution for the radial displacement given by Eq. (8) into the axial force expression given by Eq. (2) and integrating it over the entire arch and considering the boundary condition for the axial displacement given by Eq. (7) leads to the equation of equilibrium between Q and β as

$$A_1 Q^2 + A_2 Q + A_3 = 0 \tag{9}$$

where

$$A_1 = \frac{1}{4\beta^2}\left(5 - 5\frac{\tan\beta}{\beta} + \tan^2\beta\right) + \frac{1}{6} \tag{10}$$

$$A_2 = \frac{1}{\beta^2}\left(1 - \frac{\tan\beta}{\beta}\right) + \frac{1}{3} \tag{11}$$

$$A_3 = \frac{\beta^2}{\lambda^2} + \frac{\Delta^2\cos^2\Theta}{8\Theta^4} - \frac{\Delta(2\sin\Theta + \cos\Theta)}{2\Theta^2} \tag{12}$$

where the modified slenderness λ is defined as (Bradford *et al.* 2002)

$$\lambda = \frac{S}{r_x}\frac{\Theta}{2} \tag{13}$$

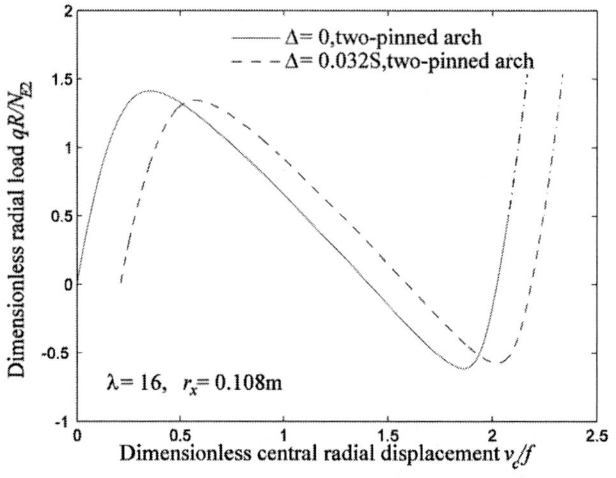

Figure 2. Variation of qR/N_{E2} with v_c/f.

The effects of the foundation settlement on the nonlinear equilibrium of shallow arches are shown in Figure 2 as variations of the dimensionless load qR/N_{E2} with the dimensionless central radial displacement v_c/f for an arch with $\lambda = 16$, where N_{E2} is the second mode flexural buckling load of a pin-ended column under uniform axial compression (Timoshenko and Gere 1961) and f is the rise of the arch..

It can be seen that the foundation settlement influences the nonlinear behavior of shallow arches. The influence is also seen in Figure 3, which shows variations of the dimensionless load qR/N_{E2} with the dimensionless axial force N/N_{E2} for the same arch.

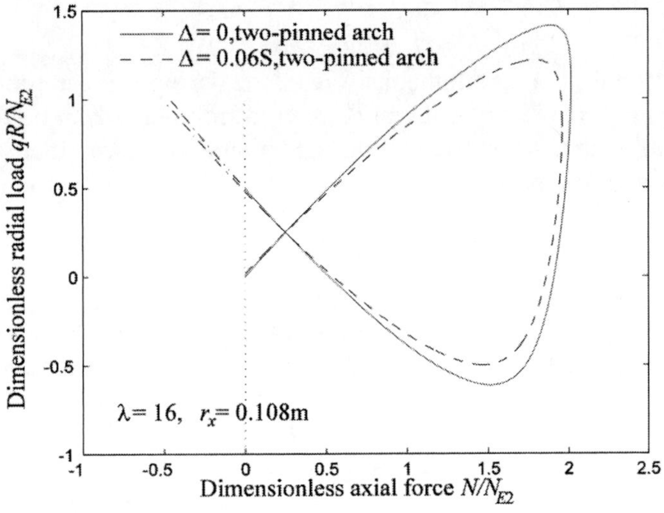

Figure 3. Variation of qR/N_{E2} with N/N_{E2}.

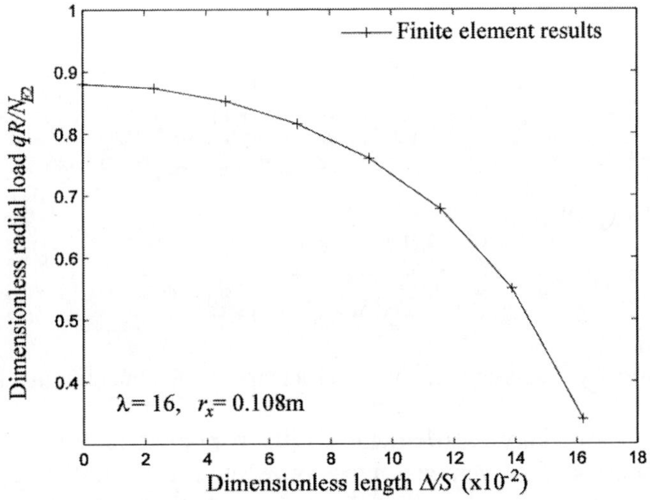

Figure 4. Effects of foundation settlements on the buckling load.

3 NONLINEAR BUCKLING

The non-linear buckling load is the local maximum and so differentiating the non-linear equilibrium equation given by Eq. (9) with respect to β leads to the equilibrium equation between Q and β at the limit point as being

$$B_1 Q^2 + B_2 Q + B_3 = 0 \tag{14}$$

where

$$B_1 = 2A_1 - \frac{\beta \partial A_1}{2\partial \beta}, \quad B_2 = 4A_1, \quad B_3 = A_2 - \frac{\beta_2}{\lambda^2} \tag{15}$$

Solving Eqs. (9) and (14) simultaneously leads to the non-linear limit point buckling load as shown in Figures 2 and 3. It can be seen from Figures 2 and 3 that the nonlinear buckling load of arches with foundation settlements is lower than that of the arch without foundation settlements.

4 INFLUENCE OF FOUNDATION SETTLEMENTS

The influence of foundation settlements on the in-plane buckling load is shown in Figure 4 as variations of the dimensionless buckling load with the settlement. It can be seen that the in-plane buckling load decreases as the foundation settlement increases.

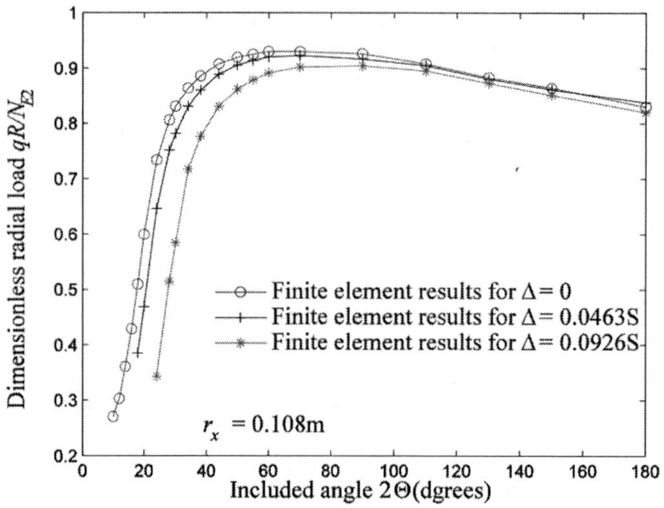

Figure 5. Effects of foundation settlements on the buckling load.

The influence of foundation settlements on the in-plane buckling load is also shown in Figure 5 as variations of the dimensionless buckling load with the included angle for different settlements. It can be seen that the influence of the foundation settlement on the in-plane buckling load is more significant for shallow arches but not significant for deep arches.

5 VERIFICATIONS

To verify the analytical solutions, the finite element results for an arch with a modified slenderness $\lambda = 16$ are compared with the corresponding analytical solutions in Figure 6 as the variations of the dimensionless load with the dimensionless central displacement. It can be seen that the agreements between them are extremely good.

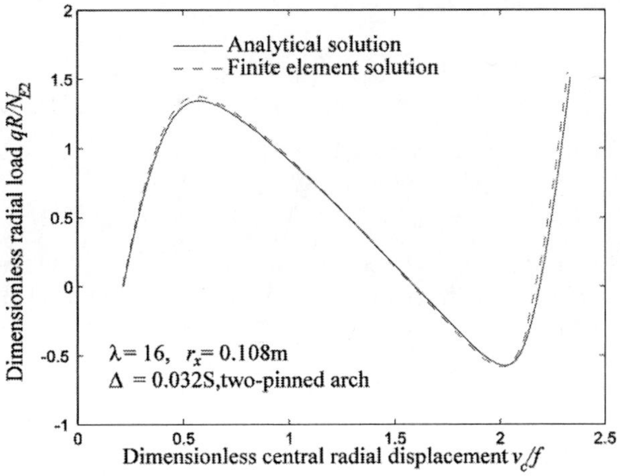

Figure 6. Comparison with FE results.

6 CONCLUSIONS

The influence of foundation settlement on the nonlinear in-plane equilibrium and buckling of shallow arches under a uniform radial load has been investigated. Analytical solutions for the nonlinear behavior and for the nonlinear buckling load were obtained. It was found that the foundation settlement plays an important role in the nonlinear structural response of the arch to the external load and that significantly influence the nonlinear buckling load of the arch.

Acknowledgments

This work has been supported by the Australian Research Council through Discovery Projects (DP140101887, DP150110446, DP160103919) awarded to third and fourth authors, and a Linkage Project (LP150101196) awarded to the fourth author.

References

Bradford, M. A., Uy, B., and Pi, Y.-L, In-plane elastic stability of arches under a central concentrated load, *Journal of Engineering Mechanics ASCE*, 128(7), 710-719, 2012.

Pi Y-L, Bradford MA, Luo K, Gao W, Non-linear analysis of three-pinned circular arches. *Sustainable solutions in structural engineering and construction*, Chantawarangul, K *et al.* (Eds), ISEC Press, Fargo, ND, US, 2014.

Pi, Y.-L., Bradford, M. A., and Uy, B., In-plane stability of arches, *International Journal of Solids and Structures*, 39, 105-125, 2002.

Timoshenko, S. and Gere, J. M., *Theory of elastic stability*, 2nd Edition, McGraw-Hill, New York, 1961.

Interaction between Theory and Practice in Civil Engineering and Construction
Edited by Komurlu, R., Gurgun, A. P., Singh, A., and Yazdani, S.

LINEAR AND NONLINEAR THERMOELASTIC ANALYSES OF BEAMS UNDER TEMPERATURE GRADIENT

YONG-LIN PI and MARK ANDREW BRADFORD

Centre for Infrastructure Engineering and Safety, The University of New South Wales, Sydney, NSW, Australia

In a structure, beams are often connected with other members such as columns, which provide considerable restraints against the rotation and extension of the beam ends. When a beam is subjected to an in-plane temperature gradient field, the temperature gradient tends to change the curvature of the beam in the transverse direction and expand the beam in the axial direction. The restrained actions will produce bending moments and compressive forces in the beam, which increase with an increase of the temperature differential and average temperature of the temperature gradient field. When these actions reach critical values, the elastically restrained beam may bifurcate from its primary in-plane equilibrium state to a lateral-torsional buckled equilibrium configuration. This paper carries out linear and nonlinear thermoelastic analyses of an elastically restrained beam of doubly symmetric open thin-walled cross-section that is subjected to a linear temperature gradient field over its cross-section. It is found that geometric nonlinearity influences the thermoelastic responses of the beam to the temperature changes significantly. The influence decreases with an increase of the stiffness of the elastic restraints.

Keywords: Thermal effects, Elastic, In-plane, Restrained, Effective centroid.

1 INTRODUCTION

In structures, the rotations of beam ends are often restrained by adjacent structural members. When a uniform straight beam with ends being restrained rotationally is subjected to an in-plane temperature gradient field that varies linearly from the bottom surface to the top surface of the cross-section of the beam, the expansion produced by the in-plane temperature gradient field is distributed linearly over the cross-section. The elastic end restraints with stiffness k_{ri} at both ends restrain the end thermal expansions and rotations and the restrained axial expansion will produce axial compressive action and the restrained rotation will produce bending action in the beam. When the combine bending and axial compressive actions produced reach certain values, the beam may suddenly deflect laterally and twist out of the plane of the temperature gradient and fail in a lateral-torsional buckling mode.

To perform corrects flexural-torsional buckling analysis, it is important to determine the distributions of internal actions accurately. The prebuckling analysis for homogeneous uniform beams under temperature gradient is quite different from that for the beams under transverse loads. Firstly, because the elastic modulus of a material is a function of the temperature (Pi and Bradford 2008, AS4100 2012) and the temperature at a material point of the beam in a linear temperature gradient field is a function of the coordinates of the point, the elastic modulus at the point is also a function of its coordinates. Hence, the effective centroid of the cross-section is different from its geometric one. The may cause an additional bending moment in the beams. Secondly, in classical theory, the prebuckling analysis for beams under transverse loads is linearized. However, it is not known if the nonlinear analysis is required for beams under the temperature gradient field.

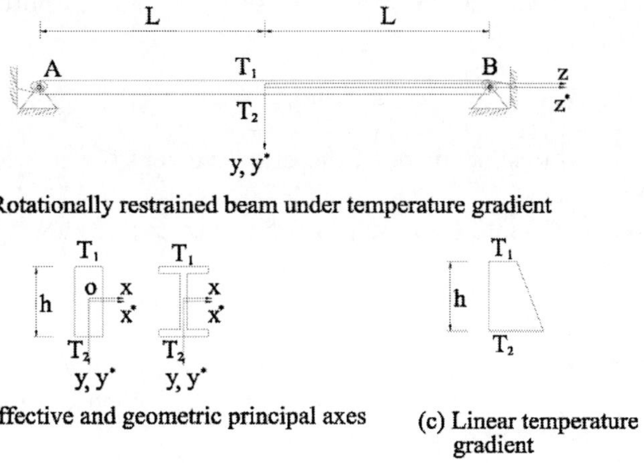

Figure 1. Elastically restrained beam.

The purpose of this paper is to carry out linear and nonlinear in-plane analyses for homogenous uniform beams with a doubly symmetric and in-plane having elastically restrained ends under a linear in-plane temperature gradient field.

2 EFFECTIVE PRINCIPAL AXES

The basic assumptions used in this investigation for thermoelastic analysis are:

1. Deformations are assumed to satisfy the Euler-Bernoulli hypothesis.
2. The temperature at an arbitrary point P can be expressed as (Figure 1)

$$T(y^*) = T_{ave} + \frac{y^* \Delta T}{h} \quad \text{with} \quad T_{ave} = \frac{T_1 + T_2}{2} \quad \text{and} \quad \Delta T = T_2 - T_1 \tag{1}$$

Because the modulus of elasticity is a function of temperature and the linear temperature gradient field $T(y^*)$ is a function of the coordinate y^*, the modulus of elasticity $E(y^*)$ of the beam is also a function of the coordinate y^*. In this case, when an axial force is applied in the direction of the geometric centroidal axis o^*z^*, the beam is subjected to combined axial compression and bending because the modulus of elasticity $E(y^*)$ varies over the entire cross-section. Hence, using the axis system $o^*x^*y^*z^*$ (Fig.1) consisting of the geometric principal axes o^*x^* and o^*y^* of the cross-section and the geometric centroidal axis o^*z^* of the beam is not convenient for the thermoelastic analysis. In order to simplify the analysis, an effective centroidal axis oz can be defined such that when an axial compressive force N is applied in the direction of the effective centroidal axis, the beam is subjected to uniform compression only, i.e. the following requirements are satisfied

$$\int_A \sigma_z \,\mathrm{d}A = N, \quad \int_A \sigma_z x \,\mathrm{d}A = 0, \quad \int_A \sigma_z y \,\mathrm{d}A = 0 \tag{2}$$

The axes *ox* and *oy* of the axes *oxyz* are parallel to the geometric principal axes o^*x^* and o^*y^* of the cross-section and is denoted as effective principal axes, and the axis *oz* is the locus of the effective centroids defined by Eq. (2) (Figure 1).

The longitudinal normal strain ε_{zz} and stress σ_{zz} due to pure uniform compression can be expressed as

$$\varepsilon_z = w' \text{ and } \sigma_z = E(y)w' \tag{3}$$

where *w* is the axial displacement of the effective centroid *o*, which is the origin of the axes *oxy*, ( )'=d( )/dz.

Substituting Eq (3) into the conditions given by Eq. (2) leads to

$$w'\int_A E(y)dA = w'\overline{EA} = -N, \quad w'\int_A E(y)xdA = 0, \quad w'\int_A E(y)ydA = 0 \tag{4}$$

From Eq. (4) and by replacing the coordinates *x*, *y* with $x = x^* + x_c$ and $y = y^* + y_c$, the coordinates x_c, y_c of the effective centroid *o* in the geometric principal axes $o^*x^*y^*$ of the cross-section can be obtained as (Figure 2)

$$x_c = \frac{\int_A E(y^*)x^*\mathrm{d}A}{\int_A E(y^*)\mathrm{d}A} \quad \text{and} \quad y_c = \frac{\int_A E(y^*)y^*\mathrm{d}A}{\int_A E(y^*)\mathrm{d}A} \tag{5}$$

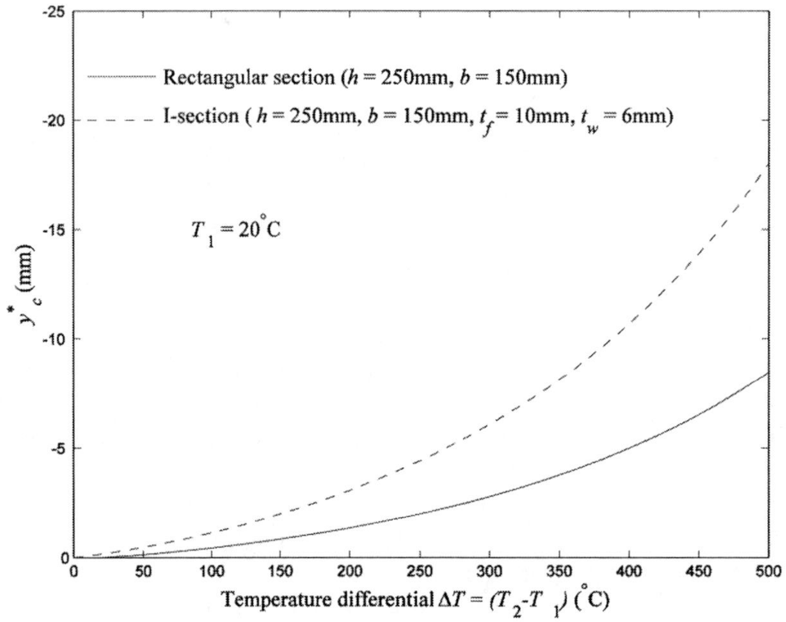

Figure 2. Effective centroids.

The typical differences between the effective centroid (x_c, y_c) and geometric centroid (x_c, y_c^*) are shown in Figure 2. It can be seen that the distance between the effective and geometric centroids increases with an increase of the temperature differential ΔT between the bottom and top surfaces.

The temperature T_o at the effective axis ox can be expressed as

$$T_o = T_{ave} + \frac{y_c^* \Delta T}{h} \tag{6}$$

3 LINEAR ANALYSIS

For the linear in-plane thermoelastic analysis of a beam having elastic axial and rotational restraints at its both ends, the differential equations of equilibrium can be derived by using the principle of virtual work based on the effective axes $oxyz$, which can be stated as

$$\Pi = \int_{-L}^{L} [-N\delta w' - M\delta v'']\mathrm{d}z + \sum_{i=\pm L} k_n v_i' \delta v_i' = 0 \tag{7}$$

where the axial force N and the bending moment M are given by

$$N = -\int_A \sigma \mathrm{d}A = -\overline{EA}(w' - \alpha T_o) \text{ and } M = \int_A \sigma y \mathrm{d}A = -\overline{EI}(v'' - \frac{\alpha \Delta T}{h}) \tag{8}$$

Integrating Eq. (7) by parts leads to the differential equations of equilibrium as

$$\overline{EA}w'' = 0 \text{ and } \overline{EI}v^{iv} = 0 \tag{9}$$

in the axial and transverse directions respectively; and to the static boundary conditions

$$EI(v'' + \frac{\alpha \Delta T}{h}) \pm k_r v' = 0 \text{ at } z = \pm L \tag{10}$$

The essential geometric or kinematic boundary conditions of end-supported beams are

$$w = 0 \text{ and } v = 0 \text{ at } z = \pm L \tag{11}$$

The solutions of Eq. (10) can then be obtained as ((Pi and Bradford 2015)

$$v = \frac{\alpha \Delta T (L^2 - z^2)}{2h(1+\eta_r)} \text{ and } w = \alpha T_o z \tag{12}$$

where the dimensionless stiffness of rotational restraints $\eta_r = k_r L / \overline{EI}$.

By substituting Eq. (12) into Eq. (8), the axial force N and bending moment M can be obtained as

$$N = \alpha T_o \overline{EA} \text{ and } M = -\frac{\overline{EI}\eta_r \alpha \Delta T}{h(1+\eta_r)} \tag{13}$$

which indicates that the linear thermal axial force and bending moment are uniform along the span of the beam.

4 NONLINEAR ANALYSIS

For nonlinear in-plane analysis, the strain can be expressed as

$$\varepsilon = w' + \frac{1}{2}v'^2 - yv'' \tag{14}$$

and the axial force becomes

$$N = -\int_A \sigma dA = -\overline{EA}(w' + \frac{1}{2}v'^2 - \alpha T_o) \tag{15}$$

Accordingly, the differential equations of equilibrium can be expressed as (Pi and Bradford 2015)

$$N' = 0 \quad \text{and} \quad M'' - (Nv')' = 0 \tag{16}$$

in the axial and transverse directions.

Hence, the axial force N = constant, and the differential equation of equilibrium in the transverse direction can be written as

$$v^{iv} + \mu^2 v'' = 0 \quad \text{with} \quad \mu^2 = \frac{N}{\overline{EI}} \tag{17}$$

The solution of Eq. (20) that satisfies the boundary conditions is then obtained as

$$v = \frac{L^2 \alpha \Delta T\left[\cos\theta - \cos(\theta z / L)\right]}{h\theta(\eta_r \sin\theta - \theta\cos\theta)} \tag{18}$$

where $\theta = \mu L$.

The nonlinear displacement is a function of the temperature gradient and the axial compressive force N (through θ). Hence, an equation of equilibrium of the temperature gradient and the axial force is needed. For this, substituting Eq. (18) into the axial force given by Eq. (15) and integrating over the entire beam leads to

$$\frac{\theta^2}{\lambda^2} = \alpha T_o + \frac{\alpha L^2 \Delta T^2 (\sin\theta\cos\theta - \theta)}{h^2\theta(\eta_r \sin\theta - \theta\cos\theta)} \tag{19}$$

where the slenderness λ of the beam is defined by

$$\lambda = \frac{L}{r} \quad \text{with} \quad r = \sqrt{\frac{\overline{EI}}{\overline{AE}}} \tag{20}$$

For a given beam under a given temperature gradient, the parameter θ can be obtained Eq. (19). The nonlinear transverse displacement v can then be calculated using Eq. (18).

Typical distributions of linear and nonlinear transverse displacement along the beam are compared in Figure 3. It can be seen that under the temperature gradient, the transverse displacements of pin-ended beam are larger than those of beams with end rotational restraints. As the dimensionless stiffness η_r of rotational restraints increases, the transverse displacements decrease. It can also be seen that the nonlinear transverse displacements are larger than the linear counterparts.

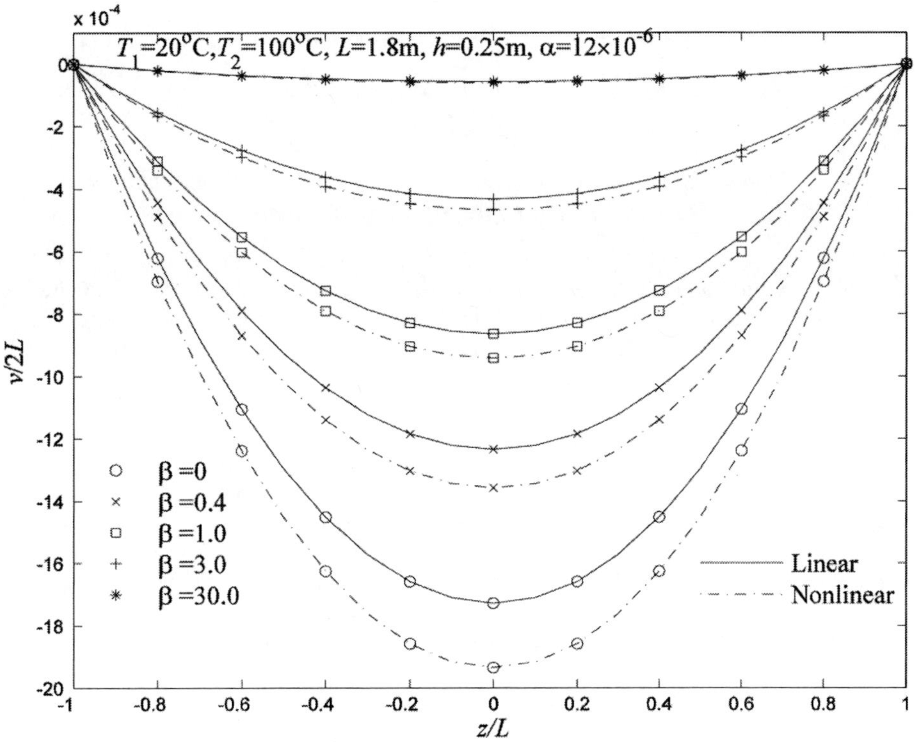

Figure 3. Distributions of transverse displacements.

5 CONCLUSIONS

This paper carried out linear and nonlinear thermoelastic analyses of an elastically restrained beam of doubly symmetric open thin-walled cross-section that is subjected to a linear temperature gradient field over its cross-section. It was found that geometric nonlinearity influences the thermoelastic responses of the beam to the temperature changes significantly. The influence decreases with an increase of the stiffness of the elastic restraints.

References

AS4100, *AS4100-1998, steel structures*, Standards Australia, Sydney, 2012.

Pi, Y.-L. and Bradford M. A., Thermoelastic lateral-torsional buckling of fixed slender beams under linear temperature gradient, *International Journal of Mechanical Science*, 50(7), 1183-1193, 2008.

Pi, Y.-L. and Bradford M. A., Out-of-plane buckling of beams having in-plane elastic restraints under gradient. *Journal of Engineering Mechanics*, 10.1061/(ASCE)EM.1943-7889.0000995, 04015080, 2015.

Interaction between Theory and Practice in Civil Engineering and Construction
Edited by Komurlu, R., Gurgun, A. P., Singh, A., and Yazdani, S.

YIELD SURFACE SHAPE ESTIMATED BY USING NATURAL STRAIN UNDER PRE-DEFORMATION OF TENSION AFTER SIMPLE SHEAR

YASUYUKI KATO

Dept of Mechanical Engineering, College of Science & Technology
Nihon University, Tokyo, Japan

Since the rigid body rotation can be clearly removed from the shearing strain component, and the additive law of strain can be satisfied, the Natural Strain is effective strain representation for expressing the stress-strain behavior under large elasto-plastic deformation. Based on the Natural Strain theory, the purpose of this study is to clarify the mechanism of development of anisotropy in the yield surface that occurs with the progression of deformation by examining the shape of yield surface obtained under large deformation. In our previous studies, using test pieces that the large pre-deformation of uniaxial tension, simple shear and the proportional loading of them are applied, proportional loading tests are carried out again by changing the ratio of tension and shear. In the present paper, another type of the pre-deformation that the deformation path is different but the final deformation state is the same as the pre-deformation of proportional loading of tension and shear, is chosen as the subject of this study. In particular, the deformation history of tension after simple shear is examined in this study. Then, the relationship between the development of anisotropy in yield surface and the deformation history is revealed in this paper.

Keywords: Finite strain, Anisotropy, Deformation history, Elasto-plastic analysis, Ductile materials, Yield criterion, Proof stress.

1 INTRODUCTION

The Natural Strain theory proposed in the present paper is the strain representation, which is obtained by integrating an infinitesimal strain rate on an identical line element along the whole process of deformation path. Since the associative law of strain on an identical line element can be satisfied, this strain can be represented from anywhere with reference to the intermediate deformed configuration. Moreover, since the shearing strain component of this strain is defined as a pure angular strain, the rigid body rotation can be clearly removed from the shearing strain component. Therefore, the Natural Strain theory proposed in this paper is effective strain representation.

On the basis of the Natural Strain theory, the purpose of this study is to reveal the mechanism of the development of anisotropy in yield surface by examining the shape of yield surface obtained after applying a large deformation. In a series of our previous study (Kato and Kazama 2013), using test pieces that the large pre-deformation of uniaxial tension, simple shear are applied, proportional loading tests have been carried out again by changing the ratio of tension and shear. Then, the shape of yield surface has been estimated by investigating the slope of the tangent in those deviatoric stress and deviatoric strain curves. Consequently, it was revealed that the shape of yield surface in a stress space becomes convex at the side of pre-deformation and becomes flat at the opposite side. Moreover, in the case of simple shear, it became clear that if the pre-deformation is applied in the opposite direction after it was applied in the forward direction, the anisotropy in the yield surface develops in the opposite direction. However, in these studies, the pre-deformation was limited to the uniaxial state of tension or shear. Hence, in subsequent work (Kato 2014), the shape of yield surface formed under the pre-deformations of combined loading of tension and shear with proportional loading state has been investigated.

However, as for the pre-deformation of combined loading of tension and shear, the experiments changing the loading path of pre-deformation had not performed yet in our previous studies. Hence, as the next step of this research, another type of the pre-deformation that the deformation path is different but the final deformation is the same as the pre-deformation of the proportional loading of tension and shear, is chosen as the subject of this research. In particular, the shape of the yield surface, which is obtained after applying the pre-deformation of tension after simple shear, is examined in this study. The effect of the deformation history on the development of anisotropy in the yield surface is revealed by comparing the results of this study with the previous study.

2 THE DEFORMATION HISTORY OF PRE-DEFORMATION

Figure 1 shows the shape of yield surface obtained by pre-deformation of the proportional loading of tension and shear that had been revealed in the previous study. In the present paper, the deformation, which the deformation path is different but the final deformation state is the same as the pre-deformation of proportional loading of tension and shear, is chosen. Especially, the pre-deformation of tension after simple shear is examined as the deformation history of pre-deformation. Figure 2 (a), (b) shows the differences in the deformation history between the pre-deformation of the present paper O→A→B and the pre-deformation of proportional loading O→B. As for the deformation path of this study, in the first stage, the simple shear is applied during O→A, and after that, in the second stage, tension is applied during A→B.

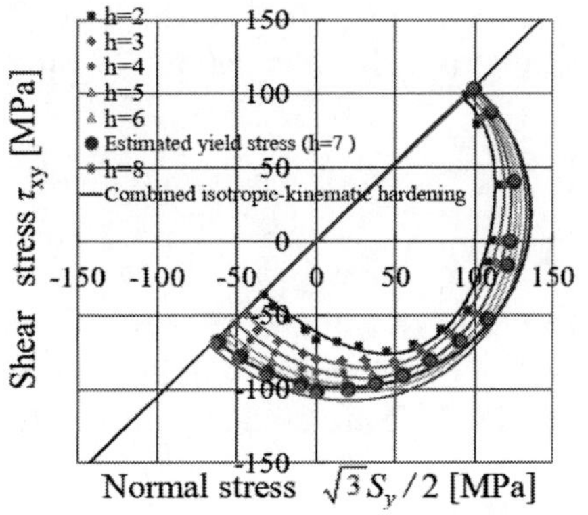

Figure 1. Distribution of h and yield surface in the case of proportional loading of tension and shear (the principal axis of pre-deformation; $\theta = 65$ [deg.] and the final stretch is $\lambda = 1.38$ [-]).

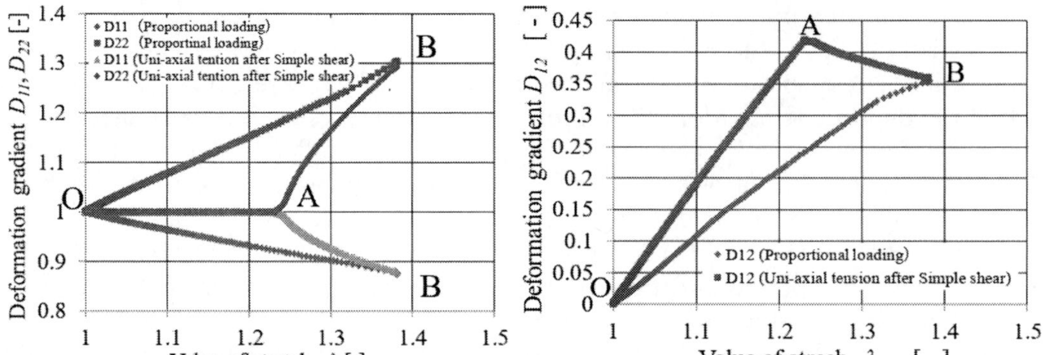

(a) Normal components D_{11}, D_{22} (b) Shearing component D_{12}

Figure 2. Deformation path during pre-deformation.

3 ESTIMATION METHOD OF YIELD STRESS

The method for determining the yield stress based on the slope of the tangent in the deviatoric stress and deviatoric strain curve is described in this chapter. Since the yield stress has been already found from the final stress condition of the pre-deformation, the slope of the tangent at the yielding point can preliminarily be determined. First, the experimental equation for the principal deviatoric stress, which becomes gentle curve around yielding point, is formulated as:

$$S_1 = a\left(1 - exp\left(b e_1\right)\right) + c e_1 + d \tag{1}$$

where, a, b, c, d are coefficients determined by using the *Levenberg-Marquardt Method*, which is one of the non-linear least-squares method. Moreover, in order to derive the slope of the tangent in the principal deviatoric stress-deviatoric strain curve, the following equation is obtained by differentiating Eq.(1), i.e.,

$$\frac{dS_1}{de_1} = -a\,b\,exp\left(b e_1\right) + c \qquad (2).$$

Hence, the slope of the tangent at yielding can be specified by using Eq.(2). On the other hand, the strain hardening modulus h is derived as follows.

$$h = \frac{2G}{\dfrac{dS_1}{de_1}} - 1 \tag{3}$$

As previously mentioned, since the slope of tangent at yielding is already specified, the value of the strain hardening modulus h at yielding can also be specified. Therefore, the shape of the yield surface can be estimated from the distributions of h.

4 EXPERIMENTAL METHOD

In the experiments, in order to apply a large pre-deformation to the specimens, the cylindrical specimens made of pure copper so called tough pitch copper, i.e., the gauge length 30 [mm] and purity 99.99%, are adopted. As for the experimental equipment, the multi-axial testing machine, which can be applied tension and torsion at the same time, is used. The displacement meters are used for measurements of longitudinal displacements, and the rotary encoders are used for measurements of twisting angles.

In the present study, the experiments are composed of two stages. Namely, the experiments, which apply the pre-deformation of tension after torsion to the specimens, are firstly conducted. Then, the experiments of the proportional loading for tension and torsion are conducted in arbitrary direction in the stress space. As for the experimental conditions of pre-deformation, the torsion, i.e., the principal stretch λ = 1.16 (shear deformation D_{12} = 0.3), is firstly applied to the specimens, then tension, i.e., the principal stretch in tension m = 1.28, is applied. However, the final deformation of each pre-deformation is the same value in all experiments, i.e., the principal stretch λ = 1.38 [-]. On the other hand, as for the experimental conditions for proportional loading tests, attaching the tri-axial strain gages to the specimens that the pre-deformation was already applied, the experiments are conducted with conditions that the principal axis of stress is fixed in arbitrary angle, i.e., 0, 10, 25, 35, 45, 55, 65, 70, 80, 90, 100, 110, 115, 120, 130, 135, 145, 150, 155, 170 [deg.]. Here, the principal strain in the proportional loading tests is measured from the extensional strain components of the tri-axial strain gauge in consideration of the Natural Strain theory, and the principal deviatoric stress-deviatoric strain curve in each tests are drawn. Then, according to the method explained in the previous chapter, the value of yield stress is determined. Moreover, the shape of the yield surface is estimated from the distribution of the yield stress.

5 EXPERIMENTAL RESULTS

Figure 3 shows the locus of stress points in each proportional loading tests, which are conducted after applying the pre-deformation to the specimens. Here, ① is the result of 0 [deg.], i.e., the case of uni-axial compression, ② is 10[deg.], ③ is 25[deg.], ④is 35[deg.], ⑤ is 45[deg.], i.e., the case of the simple shear, ⑥ is 55[deg.], ⑦ is 65[deg.], ⑧ is 70[deg.], ⑨ is 80[deg.], ⑩ is 85[deg.], ⑪ is 90[deg.], i.e., the case of uni-axial tension and it is the final direction of the pre-deformation, ⑫ is 100[deg.], ⑬ is 110[deg.], ⑭ is 120[deg.], ⑮ is 135[deg.], i.e., the case of the simple shear in the reversed direction, ⑯ is 145[deg.], ⑰ is 155[deg.] and ⑱ is 170[deg.]. As an example, the principal deviatoric stress- deviatoric strain curves in case of ①, ⑤, ⑧, ⑪, ⑭ and ⑮ are described in Figure 4 (a), (b), (c), (d), (e), (f), respectively. Here, the experimental equation, i.e., Eq. (1), is described by the solid curve in each figure. As is obvious from these figures, the stress and strain relation is a sharp curve in tension side of ⑪, and it becomes gradually gentle curve in order of ⑧, ⑭ ,⑤, ⑮, ①. Hence, in the case by proof stress, the yield stress is estimated smaller as the loading direction approaches compression side, i.e., the opposite direction to the final pre-deformation.

Figure 5 (a) shows the results of the estimated yield stress in all directions in a stress space. In this figure, the circular plots (●) indicate the estimated yield stress. Moreover, the distributions of the strain hardening modulus h near the yield surface are also described in this figure. As the loading direction approaches to the tension side that is the second stage of the pre-deformation, the distance of curves in h becomes narrow, however, it becomes larger as the loading direction approaches the compression side. Therefore, in contrast to the result in Figure 1, it can be found that the anisotropy in the yield surface is formed near the tension side.

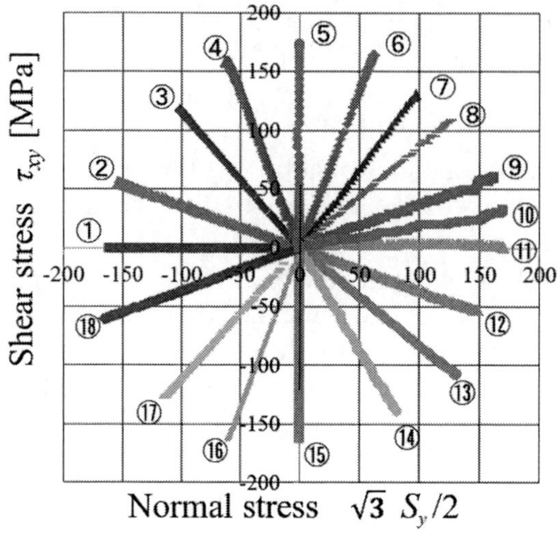

Figure 3. Stress points in each proportional loading tests.

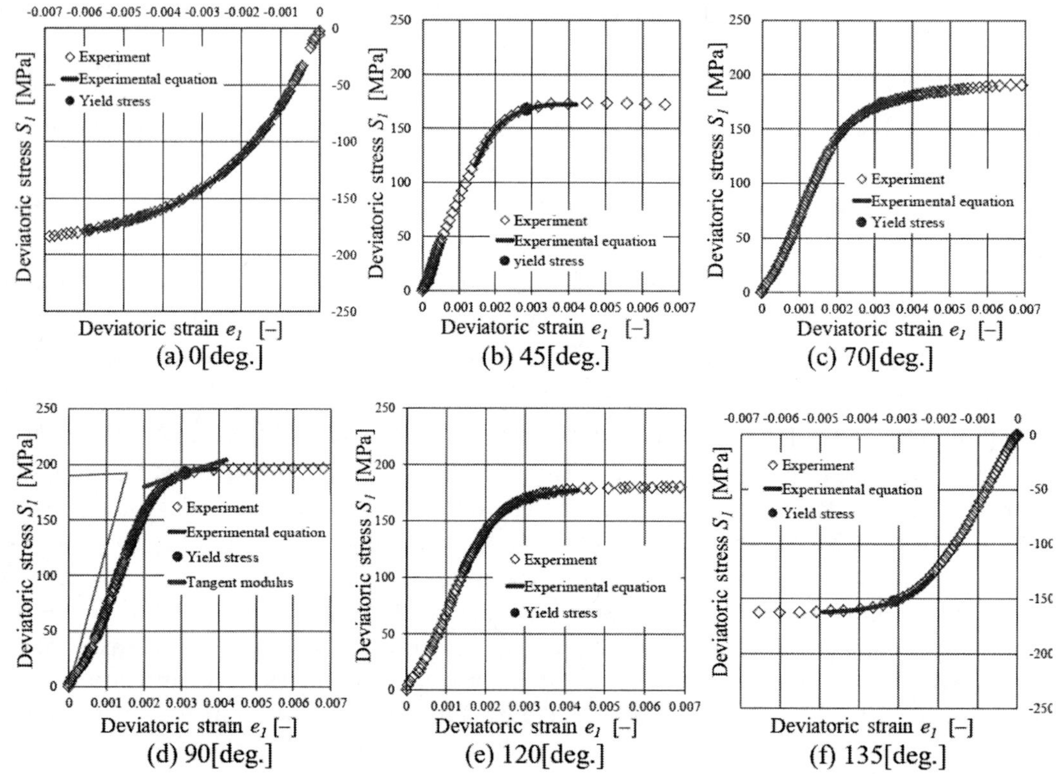

Figure 4. Principal deviatoric stress – deviatoric strain diagram.

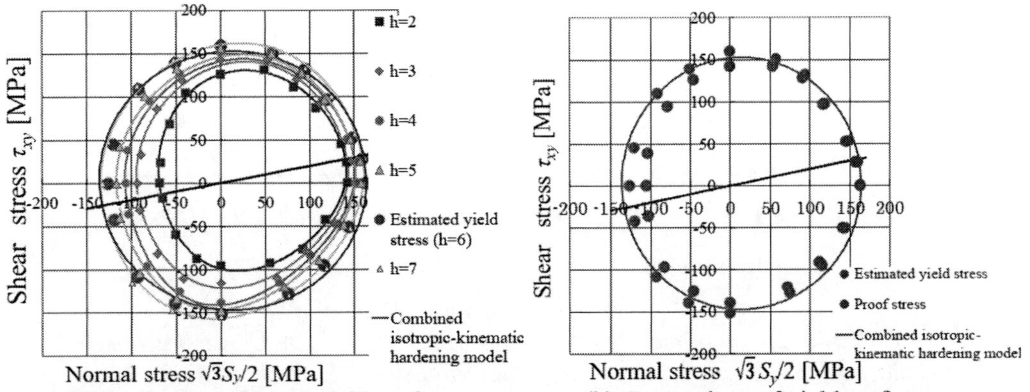

(a) Distribution of h and yield surface . (b) Comparison of yield surface.

Figure 5. Shape of yield surface and distributions of strain hardening modulus h.

Lastly, Figure 5 (b) shows the results of comparison between the estimated yield surface and conventional proof stress. The shapes of both yield surfaces almost coincide in the neighborhood of the tension side of 90 [deg.]. However, as approaches the compression side of 0 [deg.], the yield surface estimated by proof stress becomes smaller compared with the yield surface estimated in this study. Hence, in the case of conventional proof stress, the flattening tendency of yield surface is seen more compared to the result in this study.

6 CONCLUSIONS

In the present study, in order to examine the effect of deformation history on the anisotropy of yield surface, the yield stress was estimated by examining the slope of tangent for the principal deviatoric stress- deviatoric strain curve on the basis of the Natural Strain theory. As the result, the shape of the yield surface has a tendency to become convex near the tension side and become flat at the compression side. Hence, it was revealed that the anisotropy of the yield surface is formed closely related to the tension applied at the second stage of pre-deformation. As for the distributions of strain hardening modulus h, distance between adjacent curves of h is small at the tension side, however, at the compression side, the distance becomes large. Moreover, the shape of yield surface estimated in this study was compared with the shape estimated by conventional proof stress. Consequently, both yield surfaces almost coincide at the tension side. However, on the opposite side, the yield surface by proof stress becomes smaller compared with the yield surface estimated in this study.

References

Kato, Y., Kazama, A., Elasto-plastic Analysis using Natural Strain: Shape of Yield Surface generated after Pre-deformation of Large Simple Shear, Advanced Materials Research Vols. 875-877, pp 518-523, Feb, 2013.

Kato, Y., Natural Strained Yield Surface Shape Estimated under Pre-deformation of Tension and Torsion, Sustainable Solutions in Structural Engineering and Construction, Chantawarangul, K., Suanpaga, W., Yazdani, S., Vimonsatit,V., Singh, A. (ed.), 153-158, ISEC Press, 2014.

Interaction between Theory and Practice in Civil Engineering and Construction
Edited by Komurlu, R., Gurgun, A. P., Singh, A., and Yazdani, S.

NON-LINEAR FINITE ELEMENTS APPLICATION REMOVAL OF SOFT TISSUE STENOSIS

S. BORGERSEN[1], S. YAZDANI[2], and F. TAHMOORIAN[3]

[1]*SEBCorp, Eagan, USA*

[2]*Dept of Civil and Environmental Engineering, North Dakota State University, Fargo, USA*

[3]*Institute for Infrastructure Engineering, Western Sydney University, Sydney, Australia*

The non-linear finite element method has been used for the analysis of a variety of non-traditional materials involving large deformations such as rubber, plastics, and soft tissues. In these types of nonlinear analysis, strain energy density functions are usually postulated that may incorporate dissipative internal variables to indicate micro-tearing and/or damage. One particular challenging area is plaque accumulation in cardiovascular and other arterial lumens that may partially or completely occlude blood flow, leading to coronary events such as stroke, myocardial infarction, or death. This paper describes the efficacy of a proposed high speed rotational atherectomy device for the removal of arterial plaque. The prototype design is based on the use of Finite Element Methods to evaluate the efficacy of a proposed high speed rotational atherectomy device for the removal of arterial plaque. Correlation between the numerical FEA model results and real world test results was achieved using a prototype device, a simulated arterial lumen with and without a fully deployed stent in the vascular lumen. The test results obtained by using the prototype high speed rotational atherectomy device to remove plaque in an arterial lumen with fully expanded stent are also presented in this paper.

Keywords: Rotational, Atherectomy, Numerical, von Mises.

1 INTRODUCTION

Nonlinear finite elements analysis has been used to model large deformation of soft tissues. Such simulation tools provide engineers an ability to help design medical devices necessary to promote patients' health. The approach usually involves postulating the existence of a strain energy density function that would then serve as a potential function to obtain stresses using Lagrangian formulation (Fung 1993, Liu *et al.* 2004). To incorporate dissipative mechanisms in soft tissue behavior, internal variables are used within the general theory of continuum thermodynamics (Hokanson and Yazdani 1997).

Medical device companies face continuous pressure to accelerate innovation while maintaining quality and efficacy in the product design. Devices need to be smaller and smarter to support less invasive procedures. Evaluating new biomedical material and design behavior is complex, time consuming and expensive. These factors lead companies to seek methods for improving the way they select, develop, and test new materials and products. Virtual simulation by finite element method has become a powerful tool to reduce development cost and required prototyping time.

This study describes the proposed removal of arterial stenosis through the development and use of an atherectomy device, rotating at very high speeds about a central guide wire passing through the vascular channel. The atherectomy prototype is unusual in that it consists of a multi-strand, helically wound structure formed by mechanically deforming around an internal mandrel, the leading and trailing sections being reduced in diameter, and the central section having an expanded "waist" surrounded by a solid metal collar coated by abrasive material on its outer surface.

2 REMOVAL OF PLAQUE

A comparison of a normal artery with that of a partially occluded artery is provided in Figure 1. Plaque modification may be accomplished by any of several methods, depending on the type of plaque deposit, location of the accumulation and other factors. If the plaque deposit is sufficiently soft, the vascular blockage may be opened through the use of an angioplasty balloon and guide wire. The guide wire is passed through the blockage, and an angioplasty balloon is advanced along the guide wire to the preferred location along the blockage. The angioplasty balloon is then expanded, usually using a saline solution under pressure, until the desired lumen diameter is achieved.

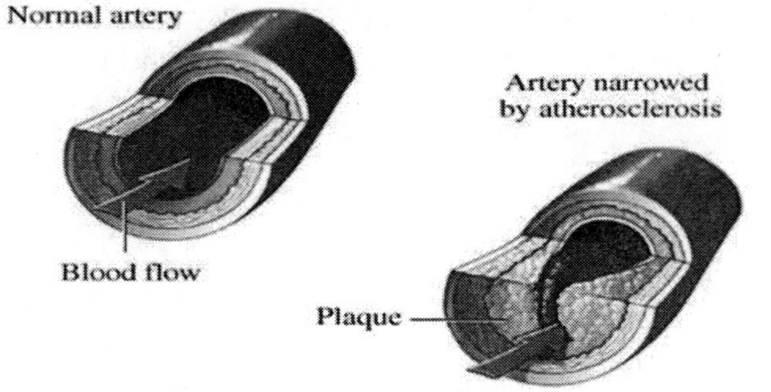

Figure 1. Comparison of clear and plagued artery.

In order to maintain the desired lumen, and prevent collapse of the vascular wall, a compressed stent is advanced along the guide wire and similarly expanded to maintain the arterial lumen. Angioplasty stents are available from a number of biomedical companies, have individual designs and are fabricated from a variety of materials, including stainless steel alloys, shape-memory alloys such as Nitinol, and non-metallic materials, including bio-absorbable materials.

A typical stent scaffold design is shown in Figure 2. The stent geometry shown is a typical cylindrical scaffold design, with member connections and material properties which allow the stent to be compressed onto an angioplasty balloon, and deployed along a central guide wire to the desired location. Once in position, the angioplasty balloon is then inflated, expanding the stent to a design diameter and using the expanded stent as a cylindrical scaffold to maintain the vascular lumen. Metallic stents are generally inserted as a permanent structure, but some bio-absorbable stent materials and designs allow the stent structure to be dissolved within the vascular lumen.

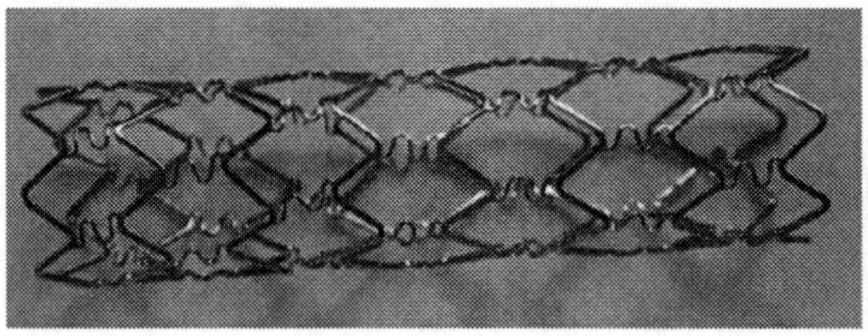

Figure 2. A typical stent scaffold design used in angioplasty.

Re-stenosis of the vascular lumen may occur after balloon angioplasty, with or without stent emplacement. Removal of the new plaque deposit may sometimes be accomplished by repeating the above procedure and installing additional stents.

A novel approach for the removal of arterial stenosis (plaque) is proposed in this paper by development and use of a rotational atherectomy device, rotating at very high speeds, about a central guide wire passing through the vascular channel. The proposed prototype is unusual in that it consists of a multi-strand, helically wound structure formed by mechanically deforming around a mandrel, the leading and trailing sections being reduced in diameter, and the central section having an expanded "waist", surrounded by a solid metal collar or spin-element, coated by abrasive material on its outer surface (Figure 3). The abrasive material on the outer surface of the cylindrical "waist" is intended to remove the plaque deposit through high speed abrasion of the plaque material.

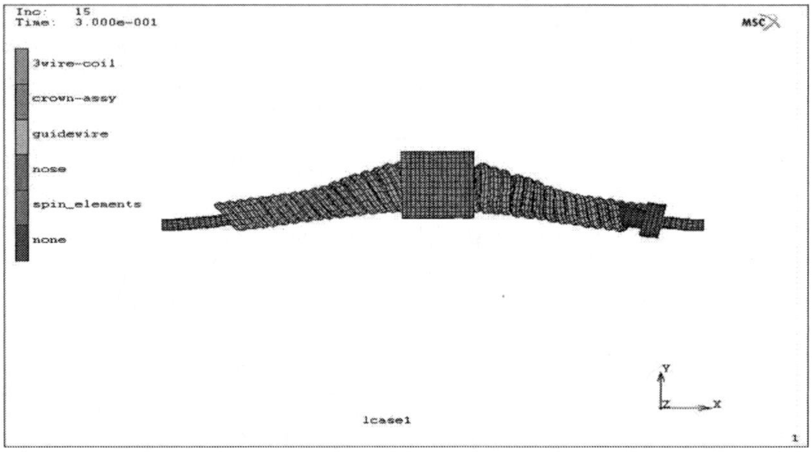

Figure 3. Finite element presentation of the rotational atherectomy device.

There are many advantages of the high speed rotational atherectomy device as designed. These include: a) removal of fatty and calcified deposits in arteries; b) with the rotation of 150,000 to 200,000 rpm very small particles are created that are smaller than red blood cells so that there would be no need to capture the particles; c) occluded vessels may be opened as long as guide-wire can pass through; and d) with very quick procedure in 1-2 minutes, greater patient comfort is provided.

The cutting path diameter depends on Rotational speed, stiffness of coil geometry and guide-wire, system center of rotational mass eccentricity, and fluid drag effects.

These all have been included in the finite element analysis presented in the next section. Fatigue life factors include parameters such as operational cycles, stress corrosion of guide-wire surface, fatigue of guide-wire, and safety factors.

3 FINITE ELEMENT ANALYSIS AND RESULTS

FEA Models of Three Filar Coil Shaft with Crown, Guide-wire and Tip were developed and contact normal forces between the coils and the guide-wire were evaluated (Figures 4, and 5).

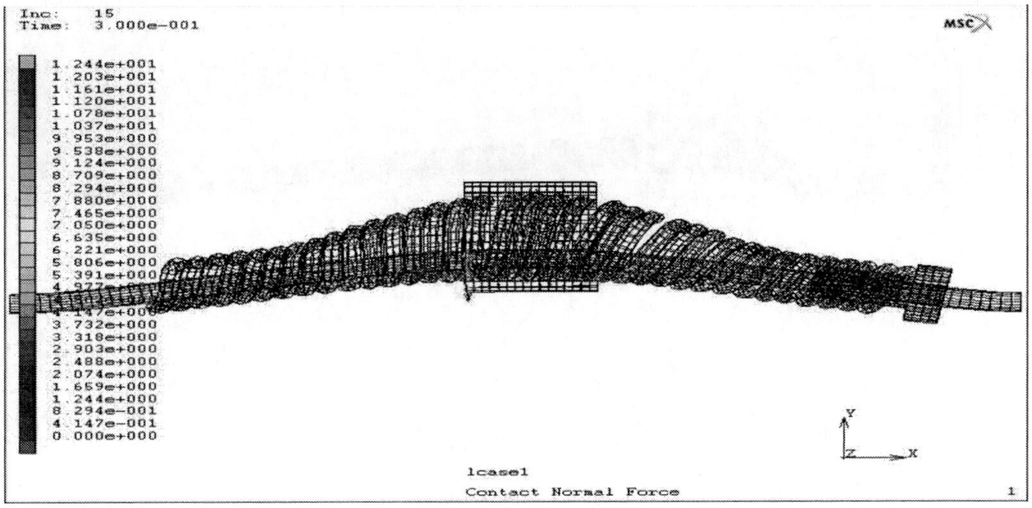

Figure 4. Contact normal force variation and intensity.

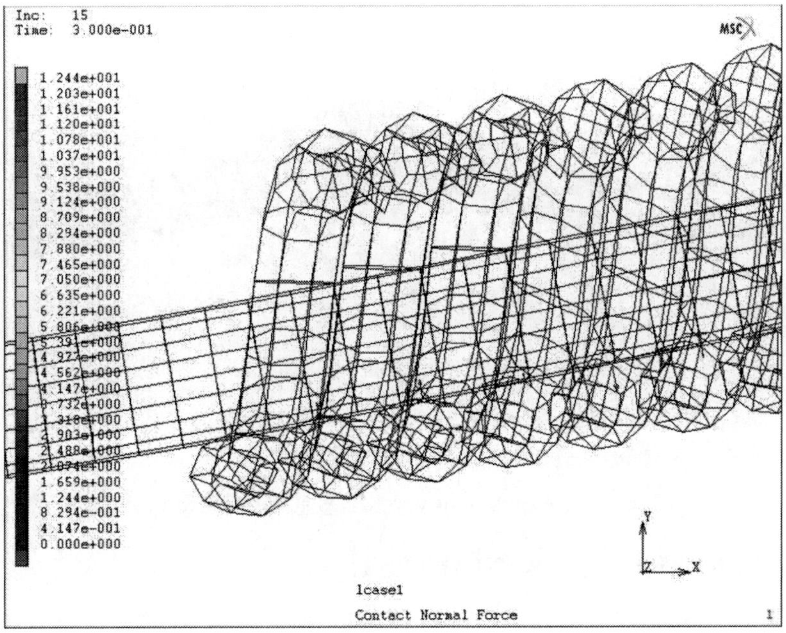

Figure 5. Guide wire and tail end.

Modified FEA models (Figure 6) were also used to evaluate the effect of variances in tip length and rotational speeds with respect to Crown diameter, cutting diameter (dynamic shift in center of gravity), contact stresses at the interface between the guide-wire/atherectomy device coil wires, and internal stresses in the design. These were used to predict fatigue life, factors of safety, and evaluate joint stress levels. Stress distribution within the proposed design was determined using the FEA model, equivalent von Mises stress distributions (Figure 7) was used as an evaluation factor.

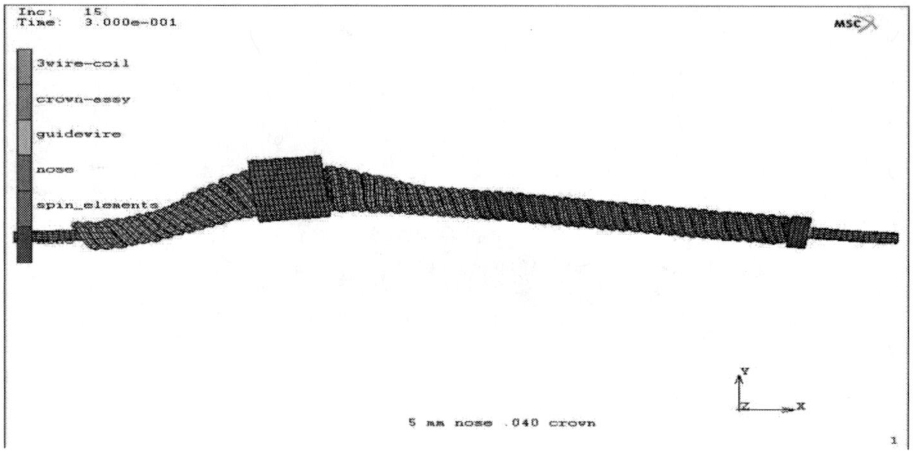

Figure 6. Extended nose length simulation in FEM model.

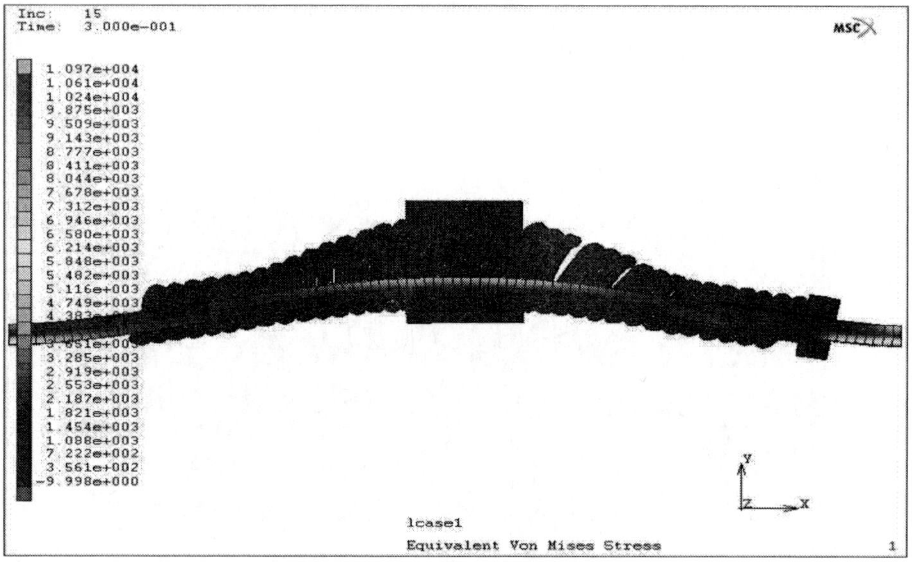

Figure 7. Equivalent von Mises stress distribution.

4 PLAQUE REMOVAL – BENCH TESTS

Preliminary rotational atherectomy design, analysis and testing results were highly successful. Therefore a determination was made to perform a series of final proof of concept tests using a Tygon tube to replicate a vascular artery/stent configuration. Initial testing was performed using a bare stent and a rotational speed which would simulate "cleaning" plaque from the interior surfaces of the stent. Examination of the interior stent surface after several such cleanings revealed that the stent member surfaces were gouged by the abrasive materials on the atherectomy crown's surface.

This type of gouging would possibly lead to weakening of the stent members, resulting in the possible collapse of the stent and the arterial lumen it was designed to support. A test guideline was then established to maintain an operational rotational speed which would not allow the atherectomy device crown to contact the stent members, but safely allow the maximum removal of plaque.

As a final bench test, the stent was expanded within the Tygon tubing (Figure 8) simulation an arterial lumen. The intent of the test was to insert the guide-wire along the arterial lumen, through the expanded stent and pass the rotational atherectomy device through the stent at various rotational speeds. During the test, it was found that at specific rotational speeds or frequencies, the guide-wire began to oscillate in conjunction with the rotating atherectomy device. This dynamic coupling unexpectedly causing the rotational diameter of the atherectomy/guide-wire assembly to contact the stent members. Stent members became entangled with the rotating atherectomy device, resulting in a dramatic failure.

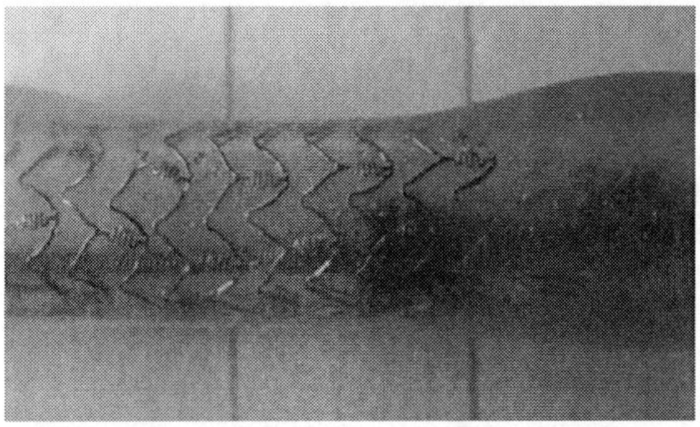

Figure 8. Stent/atherectomy device in the simulated arterial lumen.

5 CONCLUSION

Detailed FEA simulations are a very useful technique to evaluate a proposed prototype device design, there is no substitute for real life bench testing. A failure of this type in a cardiovascular artery during plaque removal from a stent using a rotational atherectomy device of this design, would most probably result in the death of a patient.

References

Fung, Y. C., *Biomechanics: Mechanical Properties of Living Tissues*, Second Edition, Springer-Verlag, New York, 1993.

Hokanson, J., and Yazdani, S., A Constitutive Model for Artery with Damage, *Journal of Mechanics Research Communications*, 24 (2), 151-159,1997.

Liu, Y., Kerdok, A. E., and Howel, R. D., A Nonlinear Finite Element Model of Soft Tissue Indentation, *Proceedings of Medical Simulations*, ISMS 2000, 67-76, Cambridge, Massachusetts, 2004.

Interaction between Theory and Practice in Civil Engineering and Construction
Edited by Komurlu, R., Gurgun, A. P., Singh, A., and Yazdani, S.

PROBLEMS IN RECONSTRUCTION OF PILE-RAFT FOUNDATION

MIROSLAV TODOROV

Structural Design Ltd., Sofia, Bulgaria

In practice, in many cases using combination of piles and large-slabs could be observed technological and structural defects before the start of their operation. These defects are dictated by the complex interaction between piles system and thin raft slabs. In order to establish the causes of defects can be imposed a wide range of studies covering research on the ground, the reinforced concrete structural elements and materials (piles and raft), the technology performance and others. In this study in order to identify the causes of defects are used a series of research methods for determining the properties of materials, environmental conditions and computational models to describe the interaction between ground and upper structure. With this set of methods are established causes and selected measures for appropriate reconstruction. The scope of these measures consists in the use of a combination of hardening by injection of the ground under the contact base of raft. With grouting the aims are to improve the skin friction and end bearing resistance of the piles. In the complex decision are sought the approaches to keep the completed structures - piles and raft. The focus of this work is to improve the interaction of the pile-raft system and ground base at the expense of reducing the load on piles. In this case, the geological formation, represented by Paleogene sediments with alternating layers of hard clay and clay sandstones largely supports the realization of the design measures.

Keywords: Renovation, Defects, Soil structure interaction, Grouting.

1 INTRODUCTION

Usually, in such cases the investigation of the problems starts with the analysis of the causes of the defects detected by visual and instrumentation devices. Due to the huge of engineering technologies such investigations are possible, but in most common cases the selection of the tools and procedure consequence depends on the financial and technical restrictions. In general, the examination of such complex situations, involved in the scope of soil structure interaction, requires research of the following:

- Ground, geological formation and its modifications in the building process;
- Climatic conditions during the implementation of pilot works and reinforced concrete structures;
- Quality of materials;
- Media conditions and measures for early stage of concrete curing.

The conventional approaches for increasing of the bearing capacity of the elements with good economic results have been prioritized, but they have to be examined in view of the general context of the issue.

The investigation of the complex of soil structure interactions could use the defects and damages reliably as a tool for verification of the analytical models. In the present task this feature is applied both for the formulation of suitable models that use the defects and as an indicator for development of some processes. These processes are described and examined by the means of different calculation procedures.

2 DESCRIPTION OF THE STRUCTURE SOLUTION

The structure made by the beginning of the present researches is a pile-raft system. The total area of the foundation is 2400 m^2 of round shape layout. The manufactured piles are of 0.9 m diameter of 10 m length, located in a regular grade of 4 m step. Reinforced slab of 0.6 m thickness is made on the pile. The slab has fixed thickness on the whole foundation area and it is placed on the compacted embankment of crushed rock of 0.3 m thickness.

For the execution of the different structure elements of the foundation, some conventional technologies are used. The piles are drilled and poured by the means of casing and drilling by Kelly bar without any protection against the bentonite suspension. The placement of the reinforcement is made before the concreting and the concrete is placed using the contraction technology.

The slab is manufactured according to the classical technology and the concrete is placed in a scheme with consequent zigzag movements of the concrete placer.

3 DESCRIPTION OF THE DETECTED DEFECTS IN THE STRUCTURE

One week after the concrete placement for the pile-raft some cracks in radial and ring directions have been detected in it. Based on the following researches made by drilling it is established that the cracks pass through the whole slab cross-section. Later these cracks expand up to 1-2 mm.

There is another main defect detected during the manufacturing of the piles. For the cases of piles of such relatively big diameter, the problem related to the provision of good back resistance is well known. By drilling of the piles throughout the cross-section of the reinforcement next to the bottom some caverns have been detected. These features assume relatively big initial settlements of the piles and sliding in the contact area of the pile and soil.

All these features mostly outline the directions of the needed researches, the comprehensive extent and the methods applicable for structure analysis.

3.1 Conducted Researches

The elaborated research program includes in situ CPTU and DPSH soil researches. All geological data currently collected from the drilling and laboratory teste are summarized. Some piles are drilled alongside their entire length and in the area under the bottom. The manner and type of the reinforcement of the pile-raft are determined by non-destructive methods and removal of the concrete covering. In order to determine the compressive strength of the concrete some samples from the concrete are provided.

Data based on the climatic monitoring of the region during concreting of slab until the moment of conducting of the researches are generalized.

High-accuracy geodetic measurements are made of the upper surface of the slab. These measurements were compared with the design assumptions and some displacements of the structure have been detected. Based on the collected number of data series of numerical analyses have been conducted, in order to track the processes that have been developed in the structure and caused the defects.

3.2 Geotechnical Researches

Geotechnical researches conducted within the framework of the whole research aim to verify the plausibility of the data from the previous geological investigations, which are performed during the design stage.

By the performed CPTU and DPSH, made from the level of the formed soil surface, a data set has been collected and used for the analyses. It has been compared with the laboratory data.

Table 1. Laboratory data.

		Specific density	Void ratio	Natural water content	Max water content	Dry density	Natural density	Plastic index	Compression parameters	Characteristics of the peak shear strength	
		ρ_s	e	W_n	W_r	ρ_d	ρ_n	I_p	C_c	c'	φ'
		g/cm^3		%	%	g/cm^3	g/cm^3			kPa	deg
Layer 3 - Powdered clay											
Average value	X_{avg}	2,69	0.66	18.76	25.09	1.63	1.92	13,26	0,074	23	23
LCV *		2.53	0.27	9.71	11.11	1.28	1.66		0.060	19	19
UCV**		2.85	1.05	27.81	39.06	1.97	2.18		0.089	28	28
Layer 4 - Soft clayey sandstone											
Average value	X_{avg}	2.66	0.87	29.11	32.66	1.43	1.84	25.79	0.104	60	19
LCV		2.52	0.59	18.27	22.97	1.24	1.71		0.052	31	11
UCV		2.79	1.15	39.95	42.35	1.62	1.96		0.155	88	27
Layer 5 - Powdered clay											
Average value	X_{avg}	2.67	0.77	22.86	28.79	1.52	1.85	13.87	0.078	30	25
LCV		2.58	0.50	14.86	19.08	1.30	1.61		0.029	24	20
UCV		2.75	1.04	30.86	38.50	1.73	2.09		0.127	35	30
Layer 6 - Clayey sandstone											
Average value	X_{avg}	2.71	0.75	25.23	27.76	1.55	1.90	22.90	0.067	52	21
LCV		2.47	0.56	18.59	20.45	1.14	1.40		0.049	39	15
UCV		2.94	0.95	31.87	35.06	1.95	2.40		0.085	66	26

**LCV - Low characteristic value, **UCV - Upper characteristic value*

The in situ investigations confirmed the stratigraphic features of the massif, but also some considerable differences between the physical and mechanical properties of the soil varieties have been detected. For comparison of the data in Tables 1 and 2 the results of the laboratory and field tests of the area are provided.

The processing of the single results is made pursuant to EN 1990. The use of this approach allows exclusion and adding of new tests to the number of results in order to achieve plausibility.

General information about the location of the stratigraphic units with a typical geological section is provided. It shows typical structure of the Paleogene sediments in the region.

Table 2. Correlated data from DPSH.

DPSH	Nspt	Terzaghi-Peck		Terzaghi-Peck (1967)	Mayne, 1998 Kulhawy and Mayne, 1990	Wolff, 1989	Bowles 1996	D'Appolonia *et al.* 1970	Michell & Gardner (1975)	Callanan & Kulhawy (1985)
		$I_{c,min}$	$I_{c,max}$	c_u	φ	φ	Young's molulus E_b			
N	pcs			kPa	deg	deg	kPa			
Layer 3 - Silty clay										
Average value	20-28	0.58	0.83	181	40	35	21566	39099	28132	42198
LCV		0.46	0.67	146	33	28	17420	31582	22723	34085
UCV		0.69	0.98	216	48	42	25712	46616	33540	50310
Layer 4 - Soft clayey sandstone										
Average value	22-36	0.60	0.85	172	41	36	20888	42150	26776	40165
LCV		0.36	0.61	139	30	26	16872	10039	21628	32443
UCV		0.84	1.09	205	53	46	24904	74260	31925	47887
Layer 5 - Silty clay										
Average value	35-50	0.75	1.00	327	48	41	32545	55568	50091	75136
LCV		0.55	0.74	241	35	30	23982	40947	36911	55367
UCV		0.95	1.26	414	60	51	41109	70189	63271	94906

3.3 Investigation of the Concrete Properties

Based on the research program a task for investigation of the mechanical concrete properties in the reinforced structures is established. The research procedure is conducted by the means of cylindrical sampling bodies of 150 mm (standard for such type of concrete). Lower indicators of the pressure strength have been established.

3.4 Calculation Analysis

The calculation analyses in this design situation are mostly used for identification of the causes of the defects occurred in the structure. The analyses have been made using two software products – Plaxis and SAP. Plaxis software is used for preliminary investigation and determination of the parameters of the soil structure interconnection. Thus, the Winckler parameters (coefficient of subgrade reaction) for all three spatial directions are determined. The used models are shown on Figure 1 & 2.

Based on the parameters of the soil structure interconnection a spatial model is established, which is investigated in elastic and elastic-plastic behavior of the materials.

The researches cover the loading conditions before the beginning of the operation of the facility and they include thermodynamic analysis of the concrete exothermity and registered loading conditions in the building process.

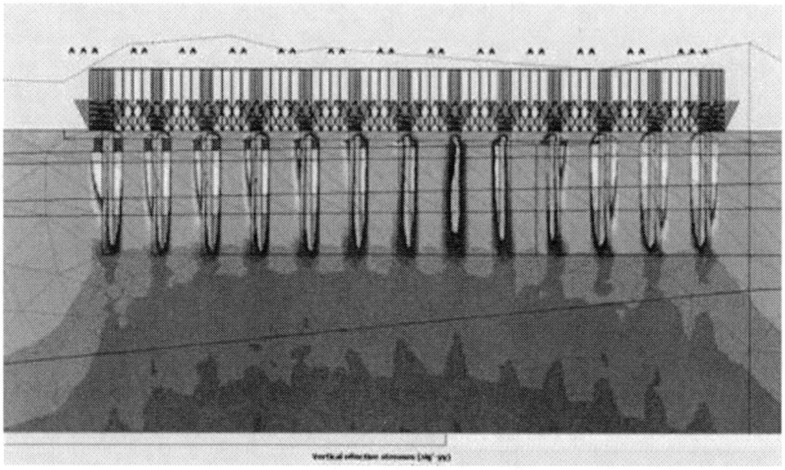

Figure 1. 2D model for research of the interconnection of the pile-slab system.

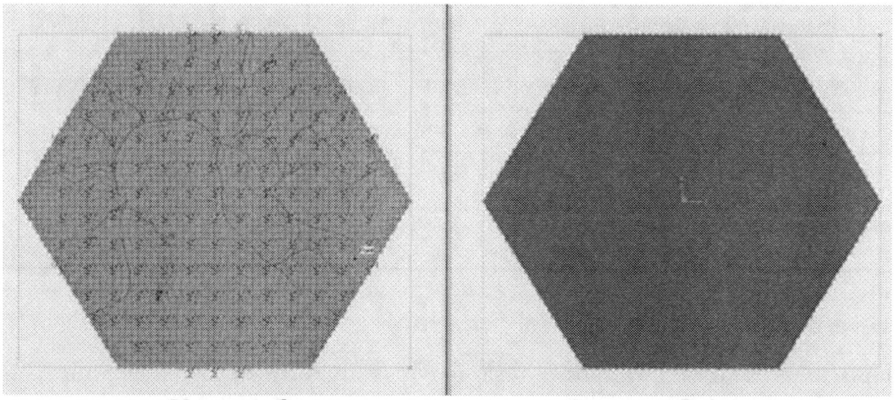

Upper surface Lower surface

Figure 2. 3D model of the upper pile slab with registered cracks.

The results of this model research are used for verification of the reinforced cross-sections for states of opening and development of cracks with material characteristics in early concrete age.

4 CONCLUSIONS

The main objective of this study is to select measures and to undertake solutions able to restore the design operation suitability and long life of the facility. The main conclusions made by the research, based on the implemented structure and model analyses, allow describing of the following main conclusions:

- For such kind of problems, it is important to formulate the calculation models, which describe reliably the limit states both during the construction process and during the operation. Due to the complicated nature of the structure operation, it would be difficult to implement this only by one calculation model. The complicated scheme of soil structure interaction imposes a risk of use of too complicated models that could not be verified.
- The use of large diameter piles and thin slab with big dimensions establish conditions for restriction of the temperature deformations. The soil properties and the piles rigidity should be considered. The selection of the design concept should pass series of preliminary analyses for selection of pile diameters.
- Special attention should be paid on the structure of the reinforcement in the slab. This reinforcement should be determined based on an analysis of the Serviceability Limit State (SLS) - opening and expanding of cracks in early stage of the concrete strength. The detected stress-strain conditions in the slab show development of big normal forces, which are the most probable cause for the cracking systems detected by the measurements.

Based on these conclusions in the concrete situation, some reconstruction measures are offered, including the following solutions:

Injection of clayey-silica content to an area under the pile-raft slab. The scope of the injection is determined by settlement analysis. Thus, the soil slab interaction as well as the allocation of the forces in the system is improved. The results of the model analyses with this intervention show equal allocation of the internal forces and settlements. The injected compound has been studied in a tested section and after the injection in this area an in situ test was made (DPSH). The results of the in situ tests show increasing of the mechanical parameters of the area after the intervention into the massif. The use of the injection compounds has been made with control of the pressures, maintained in order to prevent deteriorations of the soil natural structure.

Based on the solution a general conclusion might be drawn that the success of the work in such solutions could be achieved in conditions where the slab interaction forces are comparable with the forces determined for the piles. In such solutions the forces are optimally allocated and they do not give any hint about the development of the deformations causing disturbances of the operation conditions.

References

Bowles, J. E. Foundation Analysis and Design 5th Edition, *McGraw-Hill*, Singapore, 1996

D' Appolonia. D.L D' Appolonia. E & Brissette. RF, Closure. J. *Soil Mech. Fdns. Div. ASCE 96*. SM 2, 754-762, March, J970.

Kulhawy, F.H. & Mayne, P.W., Manual on estimating soil properties for foundation design. Report No. EL-6800. Palo Alto, CA: *Electric Power Research Institute*, 1990.

Terzaghi, K.T. & Peck R.B., Mersi Ch., Soil Mechanics in Engineering Practice, third ed. *A Wiley-Interscience Publication John Wiley & Sons, Inc,* 1996.

Wolff, T. F,. Geotechnical Judgment In Foundation Design. Foundation Engineering: Current Principles And Practice, Vol. 2, *American Society of Civil Engineers*, 903– 917, 1989.

Interaction between Theory and Practice in Civil Engineering and Construction
Edited by Komurlu, R., Gurgun, A. P., Singh, A., and Yazdani, S.

REASONING ON IMPLEMENTATION OF FROST PROTECTED SHALLOW FOUNDATIONS

LILITA OZOLA and GUNTIS ANDERSONS

Dept of Structural Engineering, Latvia University of Agriculture, Jelgava, Latvia

Implementation of frost protected shallow foundations (with insulation) in frost susceptible soils in the areas of the Baltic region is an item under discussion. A design method recommended by EN ISO 13793 has been applied. The design base is the EN 1997-1 which includes conditions and valid climatic data for the localities in the Republic of Latvia. The study contains the results of external air temperature data processing from the last 70 years, and consequently a determination of a freezing index value, and moreover pointing out the variation depending on the reference period taken. Results of temperature data processing for decades of winter seasons testify that quite distinctive frost protection levels may be defined for shallow foundation design parameters depending on the number of frost seasons sampled. The specific design results were obtained for eccentrically loaded columnar spread foundations of an unheated building insulated to reduce heat loss from the soil below the foundations keeping the subgrade soil unfrozen. As a result of the research the conclusion about the benefits expected has been presented, based on the comparison of concrete consumption and soil excavation volumes from different localities in Latvia. It has been concluded that the cost effectiveness of heated foundations correlates closely with the type of frost-heaving soil. Use of frost protected shallow foundations in clayey soils leads to an increase of ground volume to be excavated and filled back, and concrete consumption for foundations decreases. In silty sand soils, if the required foundation depth is less than some definite level, both reductions may be achieved by shifting the ground, and in concrete consumption as well.

Keywords: Data analysis, Freezing season, Gumbel distribution, Codes and standards.

1 INTRODUCTION

Material and labor consumption for building foundation construction relates closely with freezing depth of the frost-susceptible soil.

In this extended study by Andersons and Ozola (2013) the authors aim at the assessment of an effectiveness of implementation of thermally insulated foundations in construction practice of Baltic region regarding the local geological and climatic conditions and subgrade soil properties. In order to assess the parameters of freezing seasons the temperature data samples were processed for some geographically representative localities of Latvia: Riga, Liepaja, Aluksne and Daugavpils having not long distances between but significantly differences regarding climate conditions. The study provides background information for decision making when an innovative construction method has been proposed.

2 CHARACTERISTICS OF FROST SEASONS IN LATVIA

2.1 Analysis of Frost Season's Data

Data of external air temperature where collected for 70 winter seasons from 1945/1946 until 2014/2015 from records of Meteorology stations in Riga, Liepaja, Aluksne and Daugavpils (External Air Temperature Data, 2015), see Figure 1 and Table 1.

Normally the freezing season characterizes the winter period when the differences between the freezing point ($\theta_f = 0°C$) and the daily mean external air temperature ($\Theta_{d,j}$) remain positive: $(0-\Theta_{d,j}) \geq 0$. Temperature data of every winter season were displayed graphically in order to estimate the real freezing season period according to the specifications in Annex A of EN ISO 13793. There are great variations both in duration of the freezing season and in temperature range from year to year, as it is typical of coastal regions, see Table 1 and Figure 2(*a*) for illustration. The graph of freezing seasons' average temperatures displays some trend of decrease as it is found from the trend line equation a decrement is in a range from -0.01° up to -0.02° per year, and a decrement of duration of a freezing season varies between 0.4 and 0.7 days per year.

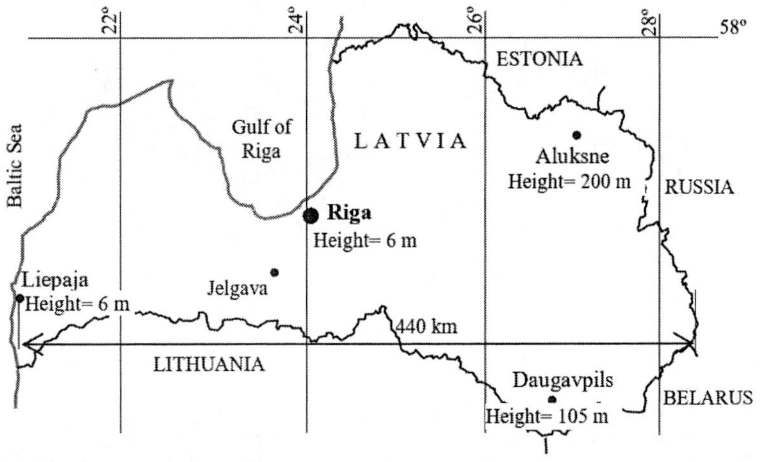

Figure 1. Displacement of meteorological stations monitored in the territory of Latvia.

Table 1. Characteristics of frost seasons for different periods.

Location	Maximal duration of frost season in last ... years, days			Minimal daily mean temperature (°) in frost season in last ... years			Minimal mean temperature (°) of frost season in last ... years		
	70	50	20	70	50	20	70	50	20
Riga	148	138	138	-9.14	-9.14	-8.79	-29.8	-29.8	-25.0
Liepaja	123	121	116	-9.85	-9.85	-9.85	-27.0	-25.4	-21.7
Aluksne	153	153	153	-11.5	-11.5	-11.5	-34.4	-32.6	--29.2
Daugavpils	153	153	153	-11.2	-11.2	-11.2	-36.0	-32.4	--27.9

The extreme value theory was used for estimation of confident parameters for design as the upper tails of distributions are of great significance. The Gumbel model is the most widely applied for extreme value distribution approximations (Kottegoda

2008). Data samples of temperature cumulates were verified using cumulative distribution function (*G(z)*) of the Gumbel model defined as follows:

$$G(z) = \exp\left(-e^{-\left(\Theta_{y,i} - Mod(\Theta)\right)/\beta}\right), \tag{1}$$

where $\Theta_{y,i}$ is the cumulate of differences between the freezing point and the daily mean external air temperature, $Mod(\theta)$ is mode of data sample, and β is a positive real number. As it shown in Figure 2(*b*) Gumbel distribution shapes produce sufficiently good compatibility with data observed, and consequently we can use the probabilistic parameters recommended by the standard EN ISO 13793 with good reason.

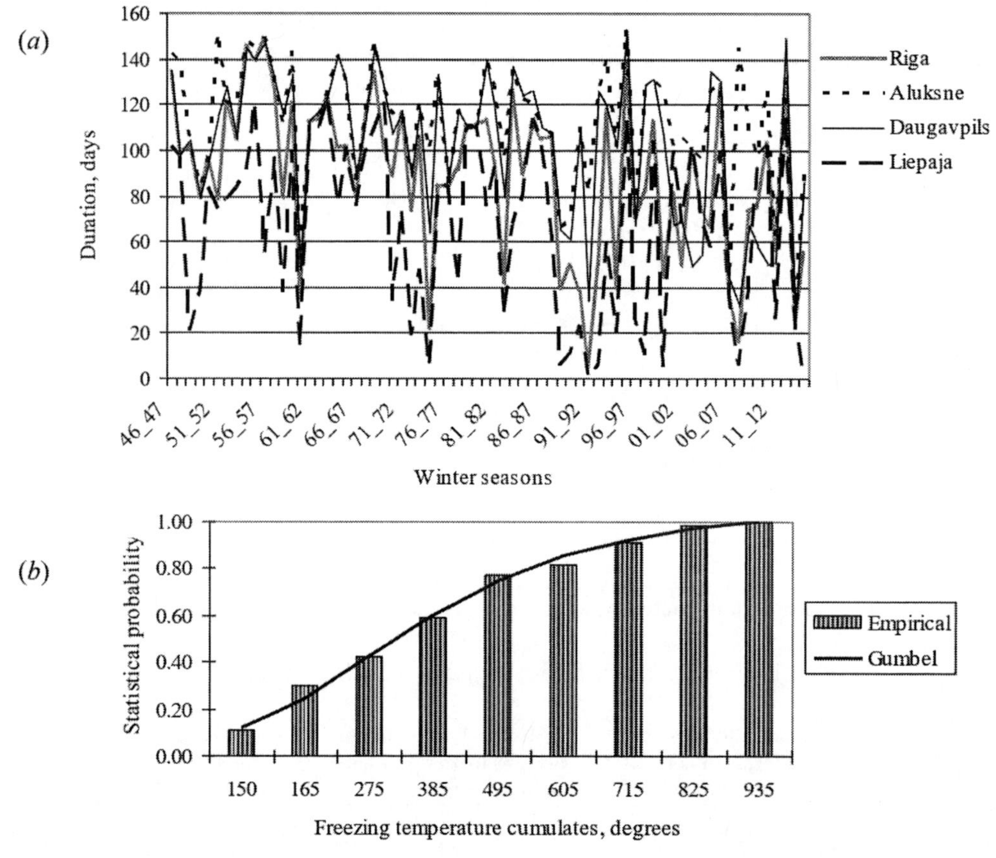

Figure 2. Frost seasons data displays: (*a*) duration of frost seasons; (*b*) cumulative distribution of freezing temperature cumulates (data sample of Riga station, Mode=330 °K, β=220).

2.2 Design Freezing Index for Reference Periods

For design purposes the severity of the winter is characterized by the freezing index which is expressed as an annual average of the sum of the differences between the freezing point and the daily mean external air temperature ($0-\theta_f$) in the freezing season multiplied by 24, and supplemented by a correction for the variation so that the value represented statistically may be exceeded once in n years for the locality concerned (EN ISO 13793). Thereby F_n has a 1 in n probability of being exceeded in a given winter. In this study the design freezing index F_n has been calculated using equation (1) using real external air temperature data from records in meteorological stations.

$$F_n = \frac{\sum_{i=1}^{m}\left(24\sum_{j=1}^{k}\left(0-\Theta_{d,i}\right)\right)}{m} + \frac{\sqrt{\sum_{i=1}^{m}\left(F_i-\overline{F}\right)^2/(m-1)}}{s_y}\left(y_n - y\right) \tag{2}$$

where $\theta_{d,j}$ is the daily mean external air temperature the average of several readings for day j, in °C; k is a number of days in the freezing season, m is a number of winter seasons included, s_y, y_n and y denote the reduced variable in the Gumbel distribution correspondingly to reference period n.

Table 2. Freezing index values for different temperature data reference period.

Location	Freezing index values in °K•h depending on data reference years						
	n= 70	n= 60	n= 50	n= 40	n= 30	n= 20	n= 10
Riga	26933	26581	25110	22782	20859	18983	16760
Liepaja	20327	20207	19184	17137	15130	13632	12300
Aluksne	36145	36079	34166	34244	30050	28495	26998
Daugavpils	33192	33552	31268	29015	27036	25570	23205

3 MODEL DESCRIPTION

A single storey unheated building of a size 19x31 m in plan has been taken as a model for numerical testing of a frost protected shallow foundation. The main load bearing structure of the building is a planar steel frame (span 18 m, space 6 m) performed by restrained steel columns and simply supported roof trusses. The sizes of the rectangular pad of the foundation depend on the subgrade soil resistance but no less than the required size at the section of column restraint- 600x600 mm.

The freezing depth of the soil depends on both climatic conditions in area and the properties of the soil. For clayey soils at the locations within Latvia the characteristic frost depth values range from 1.20 up to 1.35 m. For an unheated building tested consequently the design depth of foundation (D_f) for clayey soils is defined D_f = 1.3×1.1≈1.4 m, but for silty sand D_f= 1.3×1.1×1.2=1.7 m.

Characteristics of soils typical of locations in Latvia, such as clays, clay sands, sandy clays and silt sand were included. The design bearing capacity values of subgrade soils were determined according to Terzaghi's theory (Terzaghi 1996, Das 2006). The width of footing (B) has been found as optimal for the transmitting of forces from restrained columns. The capacity values of soil under footing to its width B= 1.2 m vary between 170 and 260 kPa depending on the soil type.

4 RESULTS

For this study extruded polystyrene Styrodur 4000CS (density 30 kg/m^3) has been chosen as the insulation material used by practitioners in European countries, Canada and the United States in recent decades. The verified acceptable long-term (50 years) compressive stress for this material is declared at 180 kPa, creep deformation less than 2% , compression strength at 10% deformation – 180 kPa (Technical Data… BASF).

The concrete consumption for shallow foundations to be built in freezing depth has been estimated and compared with the one for insulated foundations, and it is found that savings may be achieved even two times, see Figure 3.

Cost-effectiveness of building depends considerably on the bulk of ground excavation needed for the foundation. The comparison of soil volumes to be moved during construction for insulated foundations as the ratio to deep ones depending on the depth of foundation has been illustrated in graphs in Figure 4. It is clear that frost protected shallow foundations are effective at a depth less than approximately 0.8 m.

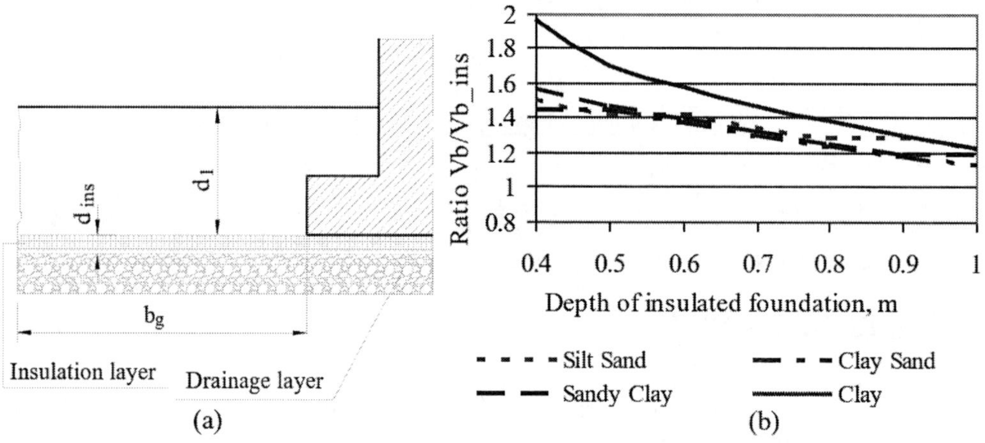

Figure 3. Frost protected foundation: (*a*) section; (*b*) relative concrete consumption.

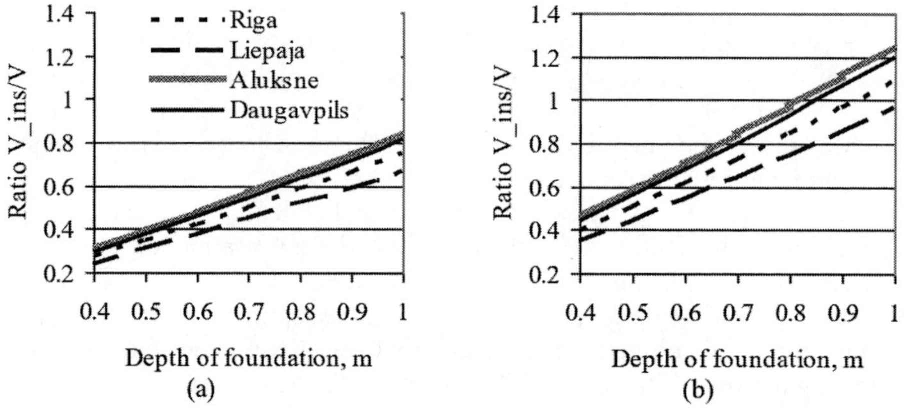

Figure 4. Ground excavation volumes versus depth: (*a*) for silty sands; (*b*) for clay soils.

The decrease of the depth of shallow foundations may be of an exceptional importance when the water table is high.

The use of decreased depth of foundations approved by building codes for heated buildings is an item under discussion when approaching the sustainability aspects of buildings. Normally there are expected periods in service life when the building remains unheated for some season(s). From this point of view frost protected foundations by insulation may be a good safety measure for the expected non-occupation periods of the building when heating may be interrupted due to a change of service conditions.

5 CONCLUSIONS

- Due to the changeable climatic conditions for locations in Latvia the freezing index values vary significantly depending on both temperature data variability and assumed duration of reference period consequently involving the uncertainty for insulation size (width) needed within the range of 20-25%.
- The use of heated foundations in clayey soils, including silty sand leads to an increase of soil volume to be excavated and filled back whereas but concrete consumption for foundations decreases.
- The decreased side area of foundations obtained using the insulation leads to a decrease of heave forces and consequently to more safety in construction regarding human errors in construction (if the backfill is non-quality or the material is off-grade, i.e., contains a clay fraction).
- Insulation of foundations may be recommended for any heated building, which due to consideration of thermal expansion has a reduced depth of foundations, to ensure the sustainability of the building for the possible periods of usage while the object may be left without owner.
- Thermal design may be recommended when groundwater level is high; the use of this method does not require lowering the water table during construction.

References

Andersons, G., and Ozola, L., Efficiency of thermal design of shallow foundations. *Civil Engineering'13. 4th International Scientific Conference Proceedings*, Vol.4., Part 1, 21-30, Jelgava, Latvia University of Agriculture, 2013.

Das, Braja M., *Principles of geotechnical engineering*/ Braja M. Das. - 6th ed. - Toronto, Ont., Canada, Thomson, 2006.

EN 1997-1. Eurocode 7 - Geotechnical design - Part 1: General rules

EN ISO 13793. Thermal Performance of Buildings - Thermal Design of Foundations to avoid Frost Heave.

External Air Temperature Data, retrieved from http://www.meteo.lv on June, 2015.

Kottegoda, Nathabandu T., *Applied statistics for civil and environmental engineers*/ Nathabandu T. Kottegoda, Renzo Rosso. Oxford, Malden, MA, Blackwell Publishing, 2008.

Recommended Applications. BASF, retrieved from http://www.styrodur.de on June, 2, 2015

Terzaghi, Karl, *Soil mechanics in engineering practice*/ Karl Terzaghi, Ralph B.Peck, Gholamreza Mesri. New York etc., Wiley, 1996.

Interaction between Theory and Practice in Civil Engineering and Construction
Edited by Komurlu, R., Gurgun, A. P., Singh, A., and Yazdani, S.

A CASE STUDY ON SOIL IMPROVEMENT WITH DEEP SOIL MIXING APPLICATION

M. INANC ONUR[1], H. ORKUN KILIC[2], MUSTAFA TUNCAN[1], and AHMET TUNCAN[1]

[1]*Civil Engineering Dept, Anadolu University, Eskisehir, Turkey*
[2]*Yerteknik Ltd.Şti, Eskisehir, Turkey*

Soil improvement is defined as the improvement of the some soil parameters by using different methods when the soil properties are insufficient to prevent static and dynamic loads. The Deep Soil Mixing (DSM) is an in situ soil treatment technology whereby the soil is blended with cementitious and/or other materials. The method has gained its popularity after 1990's in the world and last ten years in Turkey. In this study, a case study on soil improvement with deep soil mixing application in Eskisehir, Turkey is presented. The application procedure was created according to steps of usual geotechnical projects. Firstly, soil investigation was done and soil parameters were defined. And then, bearing capacity of deep soil mixing columns was calculated and modelled by "All Pile" software. After the site application, core samples were taken and the load test was applied to the core samples. Finally, deep soil mixing is becoming a popular as a soil improvement method and gives succeeding results.

Keywords: Cement mixing, Geotechnical project, Ground modification, Ground strengthening, Soilcrete, Soil investigation.

1 INTRODUCTION

Soil improvement is defined as modification of soil properties when they are insufficient for the construction of the superstructure (Das 1984). For example, increasing the soil bearing capacity, decreasing the consolidation settlement and prevention against liquefaction for the foundation design, increasing the stability of earth fill dams and slopes are the common reasons of the soil improvements. Main improvement methods are classified as mechanical modification, hydraulic modification, chemical additives and reinforcement usage (Nicholson 2015). Mechanical modification involves all compaction techniques and the aim is to densify the soils. Hydraulic modification consists forcing pre-consolidation and decreasing the permeability of soil by dewatering and drainage methods to control settlement and water flow. On the other hand, soil properties are modified by additives with mixing or grouting processes and geosynthetics, nails and other materials usage increases tensile strength and friction resistance.

Deep soil mixing (DSM) is creating soilcretes by cement injection and mixing it into the in-situ soil by blades. Soilcretes are soil-cement columns having high elasticity modulus. DSM method was firstly used in USA, 1954 and gain popularities after 1990's. Application process is cutting the soil by blades to the certain depth and then cement injection is started while blending the soil and cement. The aim of this method is to improve the soil properties, to support the excavations and hydraulic cut-off wall (Kilic 2013). The method is applied under the foundation to increase soil bearing capacity and decreasing liquefaction potential. DSM columns are formed as crossed columns for ground water flow and this application named as Hydraulic Cut-Off Wall. On the other hand, DSM is constructed as a wall for preventing horizontal deformations on the excavation surfaces. In-situ soil mixing process is given in Figure 1.

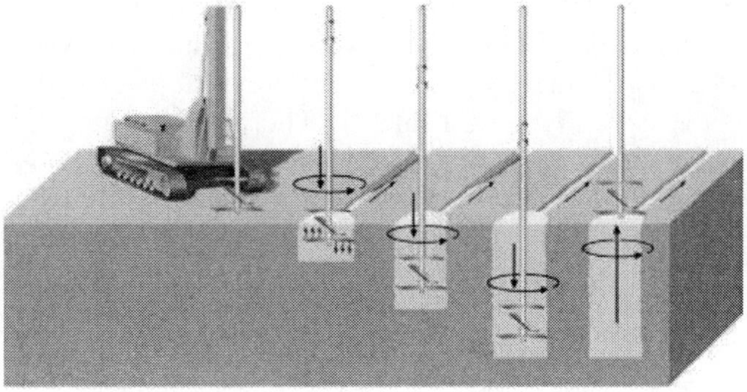

Figure 1. In-situ soil mixing process (Kilic 2013).

Onur (2007) investigated the soil deformations under earthquake loads and decreasing methods of these deformations by applying different improvement methods. He reported that the soil improvement methods show different effects due to the local soil conditions and earthquake characteristics. Andromalos *et al.* (2003) used the DSM columns for deep excavation supporting system and they observed that horizontal displacements are reduced and ground water level are controlled, easily. Madhyannapu *et al.* (2009) studied the DSM methods for the swelling soils and they obtained acceptable results. Holm (2005) gaves the history of the DSM method and past case studies. The research shows that the method can be applied to improve problematic soils, to prevent liquefaction potential and intolerable settlement all around the world. Fang (2006) modeled the improvement of the soil properties after DSM application by using numerical methods. Some differences between the field tests and numerical analysis results are found. Ye *et al.* (2006) compared the DSM method and drains with preloading for soft soils. The study showed that DSM method gives higher bearing capacity with cost and time consuming.

In this study, a case study on soil improvement with deep soil mixing application in Eskisehir, Turkey is presented. The application procedure was created according to steps of usual geotechnical projects. Firstly, soil investigation was done and soil parameters were defined. And then, bearing capacity of deep soil mixing columns are calculated and modelled by "All Pile" software. After the application, core sampling was done and the load tests were applied to the core samples. At the end, results were discussed and suggestions were given.

2 MODELING AND METHOD

DSM is an in-situ soil treatment technology whereby the soil is blended with cementitious and/or other materials. In this study, a deep soil mixing application was analyzed. 8 storey hotel building with 1228 m^2 of foundation area was constructed in the city of Eskisehir, Turkey. The total structural load was 6000 tons and foundation depth was -4.50 meters. The process was ordered according to Earthquake Resistant Design Code of Turkey.

2.1 Soil Investigation

Borings were done by Yerteknik Company. Total 5 boring hole were opened up to -15.00 meters. Standard penetration tests were performed at every 1.50 meters and undisturbed samples were taken by using Shelby tubes at every 3.00 meters. The SPT numbers with depths were given in Table 1. The ground water level was found at -4.50-meter depth.

Table 1. The variation of SPT numbers with depths.

Depth (m)	SPT Number	Soil Type
-3,00	17	Silty Clay
-4,50	7	Silty Clay
-6,00	8	Silty Sand
-7,50	17	Silty Sand
-9,00	23	Sandy Gravel
-10,50	43	Sandy Gravel
-13,50	R	Very Stiff Clay

Laboratory tests were conducted on the disturbed and undisturbed samples. Consolidation test, unconfined compression test, triaxial test were performed on the undisturbed samples. Specific gravity, sieve analysis, hydrometer test and Atterberg limit tests were performed on the disturbed samples. Therefore, all properties of the sub-soils were determined. Soils were classified by using Unified Soil Classification System. Soil strength parameters were given in Table 2.

Table 2. Soil strength parameters.

Soil Type	Unit Weight (t/m^3)	Int. Friction Angle (°)	Cohesion (t/m^2)
Silty Clay	1.8	20	10
Silty Sand	1.9	35	-
Sand Gravel	2.0	40	-

Soil exploration report was prepared and allowable soil bearing capacity was calculated as 9 t/m^2 and also liquefaction potential was observed at Silty Sand layer. Therefore, soil improvement is suggested in the exploration report.

2.2 Bearing Capacity Calculation of DSM Columns

The most preferred improvement methods are stone column, jet-grouting and deep soil mixing in the city of Eskisehir. Stone column method is inadequate for this soil and load conditions on the other hand, jet grouting method is not economical. Therefore, deep soil mixing method was chosen for this project. Designing the DSM application project was started with bearing capacity calculation of singular DSM column. For the calculation "All Pile" software was performed. "All pile" is a commercially available computer program using by many of geotechnical engineers.

The DSM columns were designed as 7 meters in length and 0.60 meters in diameter. And the location of the columns was between -4.60 and -11.60 meters in depth. Distance between the columns from center to center was 1.70 meters in x and y directions. Soil properties and the column dimensions are the inputs for the program and analysis was started with considering loads, stresses, settlement limits, axial and shear forces, moments and deformation limits. Allowable bearing capacity of one DSM column was found as 391.2 kN with 0.44 cm settlement with factor of safety 2. An example screen of the analysis was given in Figure 2.

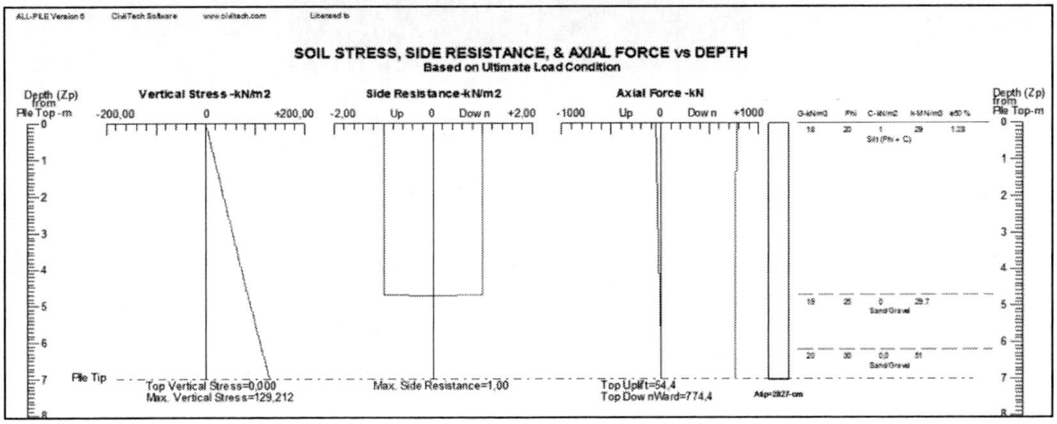

Figure 2. An example screen of the software.

2.3 Application of DSM

Firstly, a trial column was created at the construction site to control column diameter and to determine the application parameters. Some application parameters were given in Table 3. The trial column was excavated three days later and diameter was measured as 63 cm and it was given in Figure 3. The measured diameter was suitable so project was applied. Totally 476 DSM column was constructed in the whole area.

3 RESULTS

Quality control tests for the deep soil mixing applications are pile integrity testing or full length core sampling and compression tests on the cores. The integrity of the column is controlled by performing pile integrity testing or full length core sampling. The bearing capacity of the columns is controlled by performing compression tests on the cores. After the application, full length core sampling and compression tests on the cores were performed. The DSM cores can be seen from the Figure 4.

Full length core sampling showed that the DSM columns were formed between -4.60 and -11.60 meters in depth. And also, four core samples were taken from different columns to perform compression test. Average compression strength of the cores was found as 24.09 N/mm^2. So bearing capacity of a single column was calculated as much higher than the design requirements. The results showed that the soil improvement was successful and subsoil now can carry the upper structure loads safely.

Figure 3. Trial column diameter measurement.

Table 3. Some application parameters.

Injection Pressure	0-100 Bar
Penetration Speed	50-150 cm/min
Nozzle Number	10
Nozzle Diameters	2,5 mm
Rotation Speed	5-40 rpm
Dosage	450 kg/m^3

Figure 4. DSM cores.

4 CONCLUSION

In this study, a case study on soil improvement with deep soil mixing application in Eskisehir, Turkey is presented. The application procedure was created according to steps of usual geotechnical projects. Firstly, soil investigation was done and soil parameters were defined. And then, bearing capacity of deep soil mixing columns are calculated and modelled by All Pile software. After the application, core sampling was done and the load tests were applied to the core samples. At the end, results were agreeable and DSM method gave a good performance. Therefore, DSM method can be suggested to use as an improvement method against weak soils.

References

Andromalos, K.B, Bahner, E.W., The Application of Various Deep Mixing Methods for Excavation Support Systems, *Grouting and Ground Treatment, ASCE*, 515-526, 2003.

Das, B. M., *Principles of Foundation Engineering*, Brooks Cole Eng. Division, USA, 1984.

Fang, Z., *Physical and Numerical Modeling of the Soft Soil Ground Improved by Deep Cement Mixing Method*, The Hong Kong Politechnic University, Ph.D Thesis, 2006.

Holm, G., Deep mixing properties and applications, *Star Conference*, Cambridge, UK, 2005.

Kilic, H. O., Design and Application of Deep Soil Mixing, Anadolu University Msc Thesis, 2013.

Madhyannapu, R.S, Puppala, A.J, Bhadriraju, V., Nazarian, S., "Deep Soil Mixing (DSM) Treatment of Expansive Soils," *Work Shop on Ground İmprovement Technologies, ASCE*, 130-139, 2009.

Nicholson, P. G., *Soil Improvement and Ground Modification Methods*, Elsevier, USA, 2015.

Onur, M. I., Soil Deformations under Dynamic Effects, Anadolu University Msc Thesis, 2007.

Ye, G., Xu, C.A., Gao, Y., Improving Soft Soil Using Combined Cement Deep Mixing Column and Preloading with Prefabricated Vertical Drains, *Ground Modification and Seismic Mitigation, ASCE*, 152, 23-28, 2006.

Interaction between Theory and Practice in Civil Engineering and Construction
Edited by Komurlu, R., Gurgun, A. P., Singh, A., and Yazdani, S.

SUSTAINABLE BUILDING DELIVERY PRACTICES

SUAT GUNHAN

Dept of Construction Science, University of Texas at San Antonio, San Antonio, TX, USA

Today's clients both public and private have additional expectations from builders. Full cycle services capability that includes financing, planning, building, and facilities management started to be pronounced in larger projects. These expectations are mainly driven by sustainability requirements of clients. These projects necessitate a comprehensive planning approach. The comprehensive planning scope includes all project activities from design to post-construction services (facilities management) with collaboration of project stakeholders. There is no defined method of project delivery for sustainable projects yet. The majority of top green building contractors in the United States are experienced in Construction Management at Risk (CMR) and Design-Build (DB) delivery methods. This research study aims to explore the project delivery processes currently practiced in the United States' building construction industry. The survey questionnaire was conducted with the selected Top Green Building Contractors (the list of contractors is published in the Engineering News Record Magazine annually). The original list is composed of 100 national construction firms. The San Antonio and Austin firms were invited to participate to the survey questionnaire. The findings reveal the major green building providers' methods and procedures that are used during sustainable projects' delivery. The results indicate that the sustainability is seen as more of a common practice for all projects by most survey participants. There is no specific method for sustainable projects yet. A well-defined sustainable building delivery method should ideally originate from DB and/or CMR.

Keywords: Sustainability, Pre-construction, Construction, Post-construction, Design-build, CM at risk.

1 INTRODUCTION

Today's clients both public and private have additional expectations from builders. Full cycle services capability that includes financing, planning, building, and facilities management started to be pronounced in larger projects. These expectations are mainly driven by sustainability requirements of projects. These projects necessitate a comprehensive planning approach. The comprehensive planning scope includes all project activities from design to post-construction services, with collaboration of project stakeholders. There is no defined method of project delivery for sustainable projects. It is more complex and requires more communication among stakeholders (Klotz and Horman 2009). Architecture-engineering-construction (AEC) industry is highly fragmented and this situation creates silo effect (Robichaud and Anantula 2011). However, sustainable project delivery process requires a holistic approach where all stakeholders including subcontractors interact with each other in an intense manner starting from early design (Robichaud and Anantula 2011). The necessity and importance of intense communication have been highlighted by many scholars (Lapinski *et al.* 2006; Klotz and Horman 2010).The majority of top green building contractors in the United States are experienced in Construction Management at Risk (CMR) and Design-Build (DB) delivery methods (Tulacz 2014). These methods are well recognized with their processes which allow collaboration among stakeholders. The study by Korkmaz *et al.* (2010a) also revealed that design-bid-build is used in green office-building projects' delivery. The aim of this study is to explore major characteristics of sustainable building delivery process and understand how it differs from the delivery of conventional building projects. The study is also expected to provide a base for developing a detailed process map for sustainable building project delivery.

2 THE RESEARCH METHODOLOGY

The research study has explored the project delivery practices for sustainable projects by major construction companies' South Texas offices. Ten companies were identified in San Antonio and Austin to be included in this study. The survey questionnaire was conducted with the selected Top Green Building Contractors (the list of contractors is published in the Engineering News Record Magazine annually).The list is composed of 100 national construction firms. Their total revenue from green projects in 2012 was $44.67 billion (Tulacz 2012). The top 37 companies' revenue makes up to $35.72 billion which roughly corresponds to 80% of all the top 100 green contractors. Among the top 37 companies 10 companies (their total revenues constitute 44% of the top 37 companies), have been identified with local offices in San Antonio or Austin. The 10 companies were invited to participate in the survey questionnaire. Nine companies accepted to be part of the survey. The research study was reviewed and approved by the Internal Review Board. The questions were sent to the participants via e-mail. All participants were at executive levels with busy travel schedules therefore phone interviews took place instead of face to face interviews. The survey questionnaire has explored the preferred delivery methods used in sustainable building projects, and if they are modified for sustainable building delivery. In order to reveal the differentiating aspects, participants were specifically asked about pre-construction and planning phases' characteristics. Following pre-construction phase questions, the participants were also asked how construction and post-construction phases of sustainable building project delivery differed from the conventional buildings' delivery process.

3 THE SURVEY FINDINGS

The common project delivery practices for sustainable building projects take place with DB and CMR methods. Even though most participants indicated that they do not modify the delivery methods, timing and extension of major services change. Five out of nine companies indicated that their preferred method is DB. Three companies indicated that their preferred method is CMR. One company indicated that they equally prefer DB and CMR.

Six out of nine firms reported that they do not modify DB or CMR delivery methods due to sustainability requirements. Rather, most firms embrace sustainability requirements as a common practice.

The involvement intensity of construction firms in sustainable projects relative to conventional projects during the design process has been analyzed. The participants were asked if the timing and extent of constructability analysis, value engineering, and involvement of mechanical/electrical contractors' differ. In addition, they were also asked if the use of energy and building simulations are commonly used, the intensity of team communication, the differences in commissioning planning and agency involvement, and if they provide life cycle analysis services. The U.S. Green Building Council (USGBC) recognized the importance of commissioning and required its inclusion if the buildings provide required green performance (Elzarka 2009). Green building by definition is expected to enhance life cycle costs (Low and Wu 2010). Energy modeling contributes to life cycle cost savings and has a greatest impact on building energy use (Korkmaz *et al.* 2010b).

The answers to specific questions revealed the following findings: In general all firms reported that sustainable projects require more intense communication. The responses indicate that sustainable building projects require an additional level of early communication in order to collectively work to meet the sustainability goals and strategies for the project. The schedule is affected by additional documentation, meetings, submittals, and commissioning. Additional costs also occur due to staff to prepare the documentation, post documentation to the LEED online website (if the project is expected to be LEED certified).

There is mainly a minimal difference when it comes to extent and timing of constructability analysis and value engineering. Conducting constructability analysis and value engineering may take 10% longer in some projects. When it comes to timing and extent of mechanical and electrical contractors' involvement, the findings indicated that there is quite a bit difference. Their involvement is more extensive than conventional projects due to the selection of the type of energy saving equipment; coordination and selection of material and preparation of submittals; review by the commissioning agent, etc. All contractors reported that the timing is much earlier for commissioning planning and the level involvement is more intense compared to conventional projects. The process is especially more extensive for the mechanical engineer since coordination meetings are required between them as well as the commissioning agent and the Testing Adjustment & Balance (TAB) subcontractor. Meetings are also coordinated between the electrical engineer and the respective subcontractor regarding lighting; selection of equipment and so forth.

Technology applications diffuse into firms' practices when there are tangible benefits. Building Information Modeling (BIM) is a perfect example. In a majority of CMR and DB projects companies use BIM that is integrated into the project. It has been well accepted by the construction industry especially in sustainable projects. Its acceptance by both contractors and designers should be considered as an opportunity for collaboration in such a fragmented industry.

The findings indicate that contractors are not leading energy and lighting applications nor life cycle cost analysis. The contractors' responses indicated that use of energy and lighting simulations are usually handled by design consultants.

Several contractors reported that they provide life cycle cost analysis. Some reported that it is provided mainly by design teams. The responses were mix. Contractors could take more active role in leading life cycle cost analysis practices.

They initiate life cycle costs with first investment costs (building construction costs). There is an opportunity for them to develop cost expertise for post-construction phase of building projects.

The construction phase change dramatically, depending on the size of the project, and especially due to waste disposal/coordination, commissioning, and material tracking/invoicing. It also differs in the team interaction point of view. With hard bid work it is much regimented, and there is little interaction. It is much more interactive and team oriented with DB. The significant difference would be the time and effort that is put into place to properly document the construction activities that affect the project's ability to achieve prerequisites and credits to meet the sustainability goal. Segregation of construction waste is also a factor, co-mingled dumpsters allows basic construction processes to remain consistent whether or not the project has sustainable goals or requirements. The process differs since more items are required for submittal which increases that effort; more items have to be incorporated into the schedule such as commissioning which must take place after TAB. This requires more steps before achieving Substantial Completion or on federal projects in achieving Building Occupancy Date (BOD). Also, there is a tremendous increase in preparation and documentation of the sustainable process. There are greater quantities of materials that have to be tracked and stored under the sustainable process as well. If the project is LEED certified, submittal documentation collection adds more time.

The results indicate that more firms preferred using DB or modified version of DB vs. CMR or modified version of CMR while delivering sustainable building projects. When the Top 100 Green Contractors list has been analyzed, more firms are identified with leadership in CMR, even though it does not mean that they must be using CMR in sustainable project delivery. This is a pilot study and obviously more firms at the national level must be included in order to see if DB or CMR is more preferred.

According to the results, post-construction services do not dramatically differ from conventional projects. One firm reported that it provides facility management or maintenance for only very strategic clients. It is not yet a common practice for construction firms to provide facilities management services. However, it is highly important service area in the age of energy conservation.

4 CONCLUSIONS

Conducting more inclusive research at the national level is expected to reveal more detailed results. It is not possible to make universal recommendations based on the interviews conducted with nine firms. The concluding remarks can be summarized as follows:

- Sustainability is seen as more of a common practice for all projects by most survey participants.
- There is no specific method for sustainable projects yet.
- DB and CMR continue as the leading delivery methods for sustainable building projects.
- BIM is heavily used as the technology tool during sustainable buildings' delivery process.

- Design firms lead energy and lighting simulations.
- Contractors do not offer any specific services to sustainable building projects' post-construction phase. In fact there is a big potential in facilities management area.
- A well-defined sustainable building delivery method should ideally originate from DB and/or CMR.

This study comprises the first step of designing a process map for sustainable building projects delivery. Defining more precise sustainable building delivery principles will make a significant impact in the building construction industry.

Providing effective solutions to sustainable building construction processes is critical. The AEC industry is fragmented in nature but there is an immediate need for collaborative work solutions. This is especially true for sustainable project processes. This research study has explored the initial facts to design a roadmap for sustainable building delivery. The results provide a clearer image for the next phase which is conducting the research study at the national level.

Acknowledgments

This project was fully funded by the University of Texas at San Antonio, Office of the Vice President for Research.

References

Elzarka, H., Best Practices for Procuring Commissioning Services, *Journal of management in Engineering*, ASCE, 25(3), 155-164, 2009.

Klotz, L. and Horman, M., Counterfactual Analysis of Sustainable Project Delivery Processes, *Journal of Construction Engineering and Management*, ASCE, 136(5), 595-605, 2010.

Korkmaz, S., Riley, D., and Horman, M., Piloting Evaluation Metrics for Sustainable High-Performance Building Project Delivery, *Journal of Construction Engineering and Management*, ASCE, 136(8), 877-885, 2010.

Korkmaz, S., Swarup, L., Horman, M., Riley, D., Molenaar, K., Sobin, N., and Gransberg, D., *Influence of Project Delivery Methods om Achieving Sustainable High Performance Buildings,* Report on Case Studies, Charles Pankow Foundation, 2010.

Lapinski, A., Horman, M., and Riley, D., Lean Processes for Sustainable Project Delivery, *Journal of Construction Engineering and Management*, ASCE, 132(10), 1083-1091, 2006.

Robichaud, L. B., and Anantatmula, V. S., Greening Project Management Practices for Sustainable Construction, *Journal of Management in Engineering*, ASCE, 27(1), 48-57, 2011.

Tulacz, G. J., The Top Green Contractors, *Engineering News Record*, September 17, 1-6, 2012.

Tulacz, G. J., The Top Green Contractors, *Engineering News Record,* August 11/18, 85-90, 2014.

Wu, P., and Low, S., P., Project Management and Green Buildings: Lessons from the Rating Systems, *Journal of Professional Issues in Engineering Education and Practice,* ASCE, 136(2), 64-70, 2010.

EVALUATION OF LEED CERTIFIED GREEN BUILDINGS

HASAN GOKBERK BAYHAN and ASLI PELIN GURGUN

Dept of Civil Engineering, Yildiz Technical University, Istanbul, Turkey

Leadership in Environmental and Energy Design, LEED, is one of the most recognized green building certification systems in the world. Turkey is also one of the countries, where it is widely used. It offers different rating groups to address varying requirements of building and project types. LEED-v3 NC, the rating group for new construction, includes seven categories of credits: sustainable sites, water efficiency, energy and atmosphere, materials and resources, indoor air quality, innovation, and regional priority. The maximum achievable point is 110. Local conditions significantly affect the achievement of some of the credits, as LEED is originated in the USA. This study focuses on the assessment of 122 LEED-v3 NC certified buildings in Turkey, considering the points achieved in each category. The aim is to understand which categories are successfully fulfilled and more challenging for implementation. Analysis results show that buildings receive higher points from water efficiency, sustainable sites, innovation and regional priorities, when compared to indoor environmental quality, energy and atmosphere and material and resources categories in Turkey. The lowest category is identified as material and resources. This situation should lead the new projects taking precautions to this individual area especially at the beginning of "planning step" in construction process. Recognizing the successful implementations for earning points and the challenging issues in Turkish construction industry regarding green buildings can be of benefit for professionals in this area.

Keywords: Green building sector, v3-LEED NC, Success ratio, Material and resources.

1 INTRODUCTION

Use of Green Building certification systems is becoming more popular in construction industry in the world. These systems provide numerous environmental and economic benefits. As a developing country, there are efforts to develop a national certification system in Turkey like Çedbik, TSE and Seeb-TR (Aytaç 2013, Çetik 2014). Green buildings are designed to reduce the overall impact of the built environment on human health and the natural environment by efficiently using energy, water, and other resources, protecting occupant health, improving employee productivity, reducing waste, pollution and environmental degradation (Table 1).

Environmental-friendly projects are approved by several certification systems and the most commonly used certification system in Turkey is LEED. In this study, LEED-certified buildings in Turkey are analyzed to identify the achievement of LEED points. The aim is to make an assessment by implementations and highlight the challenges during certification.

Table 1. Potential benefits of green buildings.

Environmental Benefits	• Enhance and protect biodiversity and ecosystems • Improve air and water quality • Reduce waste streams • Conserve and restore natural resources
Economic Benefits	• Reduce operating costs • Create, expand, and shape markets for green product and services • Improve occupant productivity • Optimize life-cycle economic performance
Social Benefits	• Enhance occupant comfort and health • Heighten aesthetic qualities • Minimize strain on local infrastructure • Improve overall quality of life

2 ENERGY AND WATER CONSUMPTION IN TURKEY

As a growing Organization for Economic Co-operation and Development (OECD) member country, energy consumption in Turkey is increasing every day (Candemir *et al.* 2012, IEA 2010, IEA 2011). For example, Turkish electricity consumption was 242.4 billion kWh in 2012, in 2013, this number raised by 1.3% and became 245.5 billion kWh (Republic of Turkey Ministry of Energy and Natural Resources 2015). In 2010, 89.3% percentage of primary energy demand funded by fossil fuels in Turkey as; 31.9% natural gas, 26.7% petroleum, 16.6% coal, 14.1% lignite (Kıncay 2014). If this consumption trend tends to continue, statistics show that in 40 years fossil fuels could be insufficient to meet the increasing demand. The primary energy demand in Turkey will be raised by 160% in 2020, which is three times more than world average (ETKB 2009).

Given its growing population, rapid urbanization and industrialization, Turkey is expected to become a water stressed country by 2030 despite its water resources (DSİ 2010, WWF 2014). The amount of water per capita of Turkey is far below the average amount of water per capita in European countries (Doğan 2014). Water rich countries are those, which have 10,000 cubic meters of water per capita yearly. This is well above the 1,500 cubic meters per capita in Turkey (T. C. MINISTRY OF FOREIGN AFFAIRS 2015). Therefore, Turkey is obliged to use water resources sustainable, both at the national level and especially in the transboundary context because of its aggressive geopolitical position.

These statistics show the great need of saving from resources not only for today, but for future requirements as well. The first step of the saving policy should start with buildings, since buildings account for more than 40% of total energy consumption, and the construction sector as a whole is responsible for approximately 40% of all human-produced wastes (UNEP 2003). Green buildings can save around 30% of energy bills, around 35% of carbon emissions, 35% of water usage and nearly 70% of waste utilization (Capital E 2013). With all these facts, dissemination of green buildings is a must for Turkey like the rest of the world (Yeşil Bina 2013, Yeşil Bina 2014).

3 LEED IN TURKEY

Leadership in Energy & Environmental Design (LEED) is a green building certification program that recognizes best-in-class building strategies and practices, which has wide-spread usage around the globe (USGBC 2015). The major feature of this certification system is its applicability to various projects for different location and characteristics. In USGBC official website, there are nearly 83,000 projects around the globe and 441 projects in Turkey are certified or in a certification process. Today's certified project number is 143 inside the Turkish borders yet, 122 of them could be examined according to the availability of the project scorecards online from where it is the only basis of the data gathered (USGBC 2015).

Projects have been examined from 16 different cities in Turkey (İstanbul, Kocaeli, Ankara, Izmir, Mugla, Balikesir, Elazıg, Eskisehir, Cankırı, Konya, Hatay, Sakarya, Antalya, Kayseri, Balikesir, and Adana) and Istanbul is the city including most of the projects with 73 projects. The widely used version (v3-LEED2009) had 5 different subheadings in order to describe the building more accurately. LEED BD+C: New Construction (v2009) and LEED BD+C: Core and Shell (v2009) are the mostly used subheadings to describe Turkish LEED certified buildings. These subheadings apply to buildings that are being newly constructed or going through a major renovation. (New Construction, Core and Shell, Schools, Retail, Healthcare, Data Centers, Hospitality, Warehouses and Distribution Centers) In 122 different projects (11 platinum, 82 gold, 20 silver and 9 certified) that have been examined, 70 of them are certified according to LEED BD+C: New Construction and 28 of them are certified according to LEED BD+C: Core and Shell. This situation shows the majority of green certification usage is in new buildings and transformation of old buildings are not commonly used yet.

In this study, achievement of LEED categories of these buildings is analyzed. As most of the LEED certified buildings are certified with LEED BD+C: New Construction and LEED BD+C: Core and Shell, the study focuses on the characteristics of these two systems. The findings are shown in Table 2.

The analysis results showed that regional priority (88%) and water efficiency (77%) are the two categories that can be considered as the most successful ones. Regional priority had such high achievement related to its scoring which consists of other categories and saliency of the environmental issues. Water category includes three type of credit as water efficient landscaping, innovative wastewater technologies and water use reduction in LEED v2009. Among 70 projects certified with LEED BD+C: New Construction, water efficient landscaping (67.9%), innovative wastewater technologies (82.9%) and water use reduction (82.9%) had high achievements, among them both water use reduction and innovative wastewater technologies are successful headlines with the achievement of 82.9%. 28 projects were certified with LEED BD+C: Core and Shell, and the achieved percentages were determined for innovative wastewater technologies and water use reduction as 89.3% and (85.7%), respectively. The water efficient landscaping credit was calculated as 57.1%.

Water efficient building design has 5 components especially in residential projects; submetering, installation of pressure reducing valves, install water saving devices / use of water efficient labeled products, water substitution / recycling of used water and water efficient landscaping. As the main problem occurs in water efficient landscaping, using water efficient plants, reused water, water efficient irrigation systems and rainwater harvesting would help to have higher points in that field.

The two categories with lowest average were material & resources (39%) and energy & atmosphere (44%). Material and resources category includes eight type of credits as two of them about building reuse, construction waste management, materials reuse, recycled content, regional materials, rapidly renewable materials and certified wood. (USGBC 2015) Among 70 projects certified with LEED BD+C: New Construction, three categories were successful; regional materials (85%), recycled content (84.3%) and construction waste management (75%). Remaining 5 categories had relatively lower achieved percentages; certified wood (2.9%), materials reuse and building reuse - maintain interior nonstructural elements (4.3%), building reuse - maintain existing walls, floors and roof (9%) and rapidly renewable materials (10%). 28 projects were certified with LEED BD+C: Core and Shell. The credits, building reuse - maintain existing walls, floors and roof (0%), materials reuse (3.6%) and certified wood (10.7%) had very low percentage values. Construction waste management (73.2%), recycled content (94.6%) and regional materials (96.4%) had quite high achievement with respect to others.

Even though LEED system was developed in the U.S., the supply and use of sustainable materials are quite challenging for projects in countries both inside and outside of North America. For example; one of the highest rated projects in the world, Pixel from Australia got 105 points out of 110 (LEED BD+C: New Construction v3 - LEED 2009). In its scorecard, the missing 5 points are all from Material & Resources component (USGBC 2015, Greenbuildings.com Team 2012). Similar issues related sustainable materials are also seen in Turkey.

Several solutions can be addressed to receive higher point in Material & Resources such as, using new type – green concrete, utilizing from recycled materials and existing buildings. The selection of appropriate material can differ depending on project conditions such as; choosing local manufacturing products and try to find the material from local producers, evaluating the using eco-technology, evaluating the recyclability of materials and using durable materials, making sure that the material is not toxic over time, using certified materials and waste management and reduction should be managed wisely. The average of the achievement of the categories can be seen in Table 2.

Table 2. Results of achievement ratio according to 122 projects.

CATEGORY	CERTIFICATION LEVEL(%)				
	Platinium	Gold	Silver	Certified	Total Average
Sustainable Sites	88	77	73	59	76
Water Efficiency	89	80	66	52	77
Energy & Atmosphere	50	45	36	22	44
Material & Resources	50	39	33	35	39
Indoor Environmental Quality	76	51	30	33	49
Innovation	91	78	65	54	76
Regional Priority	98	91	76	67	88

4 CONCLUSION

The aim of this study was forming a database and evaluate the results about LEED-certified buildings in Turkey which would create a baseline for further and more comprehensive studies. According to analysis results, Turkish LEED Certified projects scored higher points in "Water Efficiency" with 77% percent compared to other credit categories. The results also present that, LEED Certified projects in Turkey received lower achievement percentage in "Material & Resources" with 39%. Especially the following subcriteria have developmental needs; building reuse - maintain existing walls, floors and roof, building reuse - maintain existing walls, floors and roof, building reuse - maintain interior nonstructural elements, materials reuse, rapidly renewable materials and certified wood. In addition to these, water efficient landscaping can also be improved to provide better solutions requiring less use of water.

Turkish green building sector has an important potential in the world related to growing Turkish construction sector. In 2013, the cost of electricity import was 56 billion dollars in Turkey (EPDK 2015). A new regulation for buildings to perform as a green building can save up to 30% of this expense. Green building construction dissemination would have greater benefits on a larger scale.

References

Capital E, *The Business Case for Green Building: A Review of the Costs and Benefits for Developers, Investors, and Occupants*, April 2013.

Candemir, B., Beyhan, B., and Karaata, S., *İnşaat Sektöründe Sürdürülebilirlik: Yeşil Binalar ve Nanoteknoloji Stratejileri*, November 2012.

Çetik, M., *The Governance of Standardisation in Turkish Green Building Certification Schemes*, Tilburg Sustainability Centre, Law and Economics Research Centre, University of Tilburg, Warandelaan 2, 5037 AB, Tilburg, the Netherlands, May 2014.

Doğan, A., *Water Efficiency in Sustainable Buildings Presentation*, Ceske Budejovice, Czech Republic, August 2014.

General Directorate of State Hydraulic Works (DSİ), *TURKEY WATER REPORT 2009*, January 2010.

ETKB (Ministry of Energy and Natural Resources), *Faaliyet Raporları* 2008, Feb. 2009, Retrieved from: www.enerji.gov.tr on June 11, 2015.

EPDK (Energy Market Regulatory Authority), *2013 Sektör Raporları*, 2015, Retrieved from: http://www.epdk.org.tr/ on June 13, 2015.

Greenbuildings.com Team, *Pixel in Australia: Does the Highest Leed Score Make it the Greenest Building in the World?*, February 2012. Retrieved from http://www.green-buildings.com/articles/pixel-in-australia-does-the-highest-leed-score-make-it-the-greenest-building-in-the-world/ on June 11, 2015.

International Energy Agency (IEA), International Monetary Fund, World Economic Outlook, *IEA Energy Statistics – Energy Balances for Turkey, 2010*, April 2011, Retreived from: http://www.iea.org/stats/index.asp on June 5, 2015.

International Energy Agency (IEA), *Turkey 2009 Review,* April 2010, Retreived from: http://www.iea.org/publications/freepublications/publication/turkey2009-1.pdf on June 5, 2015.

Kıncay, O., *Türkiye'nin ve Dünya'nın Enerji Profili*, 2014, Retrieved from: http://www.yildiz.edu.tr/~okincay/dersnotu/EnerjiProfili.pdf on May 15, 2015.

Sertaç, A., "*Zirve…*", Yeşil Bina Dergisi, 2013, Retrieved from: http://www.yesilbinadergisi.com/?pid=30162 on June 12, 2015.

T. C. Ministry of Energy and Natural Resources, *Renewable Energy General Directorate of 2014 Annual Report*, January 2015.
T. C. Ministry of Foreign Affairs, *Turkey's Policy on Water Issues*, January 2015 Retrieved from http://www.mfa.gov.tr/turkey_s-policy-on-water-issues.en.mfa on June 12, 2015.
United Nations Environment Programme (UNEP) Division of Technology, Industry and Economics, *Energy and Cities Sustainable Building and Construction*, June 2003.
USGBC, *Directory and resources*, June 2015, Retrieved from http://www.usgbc.org/ on June 15, 2015.
WWF Report, *Turkey's Water Footprint Report*, 2014. Retrieved from http://awsassets.wwftr.panda.org/downloads/su_ayak_izi_raporweb.pdf on June 5, 2015.
Yesil Bina (2014), *Sustainable Buildings Technology Magazine*, Jan.-Feb. 2014, Issue 23.
Yesil Bina (2013), *Summit on Quality in Construction*, Nov.-Dec. 2013, Issue 22.

Interaction between Theory and Practice in Civil Engineering and Construction
Edited by Komurlu, R., Gurgun, A. P., Singh, A., and Yazdani, S.

EXAMINATION OF USE OF WATER AND SOIL RESOURCES IN TERMS OF URBAN SUSTAINABILITY

NECATI GULBAHAR[1], IMAM BAKIR KANLI[2], and DIDEM ODABASI[1]

[1] *Dept of Civil Engineering, Istanbul Gelisim University, Istanbul, Turkey*
[2] *Dept of Public Administration, Marmara University, Istanbul, Turkey*

Cities are the organization of "collective consumption" and of welfare delivery and are the major consumers of renewable and non-renewable resources. The rapid urbanization process and its chaotic problems such as drastic urbanization and ecological challenges have been phenomenal. As the population rapidly rises in the world, the requirements for water resources also increase for various reasons at the same time. On the other hand, global warming, greenhouse effect and degradation of forests make worse the situation. However, these negative consequences show how we are going to face the tragic problems for the next two or three decades. While the term of sustainability is defined and connected only with the environment now it is explained with a comprehensive term undertaking not only the environment but also the social, economic, cultural and governance aspects of cities. Sustainability has been discussed at various meetings for years such as the UN meeting "the Earth Summit" which took place in Rio in 1992 and emphasized that cities should be liveable in terms of costs, social equity and the ecology, which are the three major principles of sustainable development, but concrete and applicable results have not been achieved yet. In this study use of basic resources in particular water and soil will be analysed and examined based on sustainability and its major components such as economic, sociologic, ecologic within the limits of the paper. The study will also be evaluated through the balance of protection and use of resources considering certain samples in the context of Marmara Watershed in Turkey and Istanbul in particular. Literature reviews, quantitative and qualitative information, reports prepared by the institutions, statistics, and observations have been chosen and used as the method. Solutions and proposals will also be suggested based on the findings.

Keywords: Urbanization, Ecosystem, Watershed planning, Water resources, Sustainable use, Tragic problems.

1 INTRODUCTION

Deterioration and pollution, as a result of pressure on the natural resources of the world's population increases in natural resources, has necessitated all countries, especially UN, to take precaution. Human and Environment Conference held in Stockholm in 1972 can be considered as the first study in this regard. In the following years on a global scale, many international meetings were held under the leadership of the UN and many important decisions were held. In 1992, important leading decisions were taken in this regard at Rio Conference and an action plan called Agenda 21 was adopted which aims progress and environmental protection for actualizing of the sustainable development concept and the expression of the highest level of the global reconciliation and analytic commitment. The EU member countries have also started necessary works and have taken binding decisions for member countries. Within this framework, necessary works were initiated in Turkey, which was aimed at environmental protection for sustainability.

In 1963, Eastern Marmara Region Planning Work, which is the first regional plan for Marmara Region, has not been approved by any institution. The plan remained as the planning work of Istanbul Metropolitan Municipality. The issues such as, Istanbul's growth is inevitable and this should be encouraged, are indicated in the plan target. The second regional plan called Marmara Regional Plan, has been initiated after the 1999 Marmara Earthquake. The main objective of this plan can be specified as; elimination of damage caused by the earthquake in the region, balancing the migration, land use planning; issues such as agriculture, industry, trade, tourism considering the seismic risk of settlements, control urban growth in order to protect the environment and place values. In the following years, İstanbul Development Agency (İSTKA) was established and prepared regional development plan 2010-2013, approved by the Ministry of Development, for İstanbul. As the basis for these plans, it is anticipated that Istanbul is having high value-added economy, finance and trade centre; innovation in the industry, a structure producing high technology and high added value, accessible education for all as a social goal, improving income distribution, supporting the children, elderly and disabled. In terms of the physical planning, sustainable urban development, urban identity compatible with the original design and quality of spatial planning were foreseen (İSTKA 2012). Despite the measures in the National Development Plan prepared by the State Planning Organisation (SPO), excessive migration, rapid population growth, unplanned industrialization and settlements, inadequate infrastructure are experienced in Istanbul as the largest city. Primarily, encouraging reverse migration and the prevention of migration, it is necessary to ensure equitable income distribution and to spread the prosperity to the base of community in the country. This means that the problems of Istanbul are solved in Agri, Mus, Sirnak, etc.

In this article, water and land resources in Marmara basin in terms of sustainability and urbanization will be discussed as the special case of Istanbul. Therefore, existing studies and reports of the basin were reviewed. Furthermore, some observations were made and the necessary data were collected. Finally, it was examined whether the problems in the basin and the existing solution methods are adequate in terms of sustainability and the balance of protection and use and some proposals were made.

2 BRIEF INFORMATION FOR MARMARA BASIN

Marmara Basin in Northwestern Turkey is like a strip that surrounds the Sea of Marmara. Its surface area is 2,338,525 hectares. It involves 8 provinces as Istanbul, Kocaeli, Bursa, Tekirdag, Kırklareli, Yalova, Balıkesir ve Canakkale (Figure 1).

The basin is an irregular and complex. Its population is 22,065,168 in 2014. Density is 514 people per km^2. It's higher than Turkey's average of 101 people per km^2. Regarding the climate it receives precipitation in all seasons and average is 600-700 mm. It shows climate transition feature. About half (51%) of the basin is forestland. The most forested land is Yalova with 59%. This rate is 45% in Istanbul. There are no large streams in the basin. The major streams are Biga and Gonen Streamlet and Karasu, Sazlidere, Alibeyler, Kagithane, Kirazdere, Kocadere, Kozanli. The major lakes are Buyukcekmece, Kucukcekmece and Lake Iznik (Figure 2).

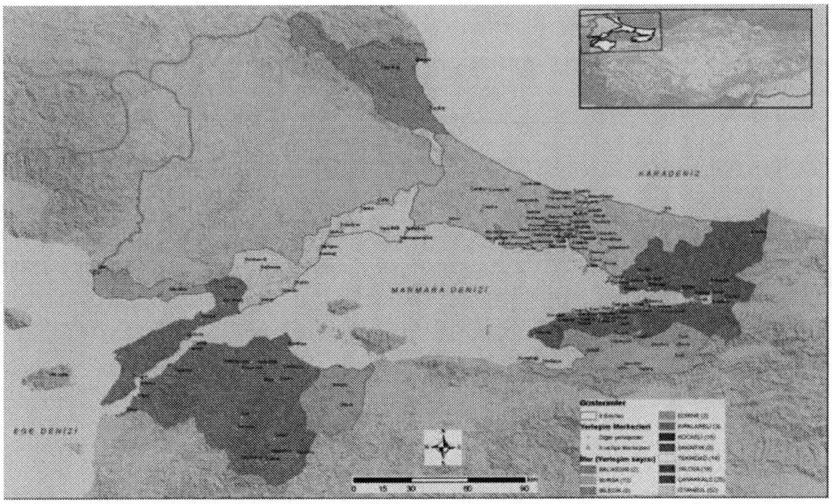

Figure 1. Marmara Basin (Tubitak-MAM 2010).

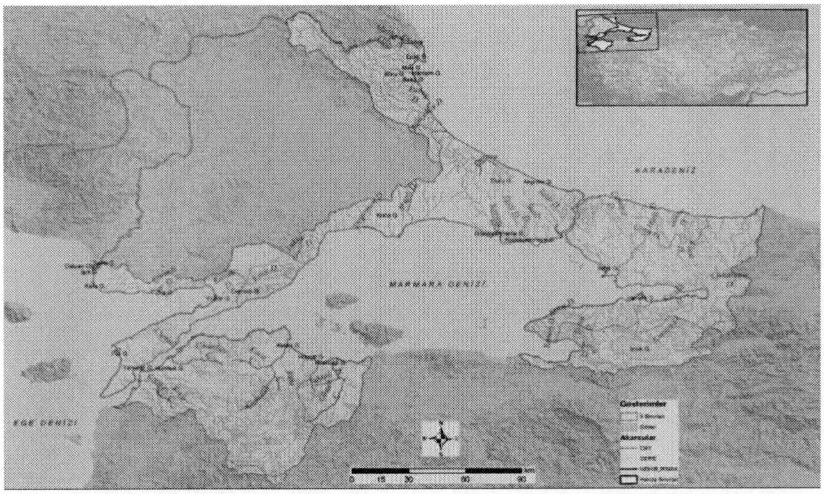

Figure 2. Surface water resources in the Basin (TUBITAK MAM 2010).

Regarding the land use, there has been major change in land use as a result of the intense industrialization in Kocaeli and Istanbul provinces in the basin. Hence, many arable land were allocated to settlement and industrial purposes. The basin is the vital area with the possibilities of industry, transportation, trade, history and tourism. However, industry comes first in all cities except Kirklareli. Agriculture and tourism come after the industry in the region. It is the region, which has natural resources under huge pressure due to industrial and commercial activities. According to TUIK 2015, the highest unemployment rates are in Istanbul, Kocaeli and Yalova with the sequential values of 11.2%, 11% and 10%. The lowest unemployment rate is in Balikesir with 6%. On the other hand, as the highest employment rate is in Tekirdag with 51.6%, the lowest rate is in Balikesir with 44%. When the population growth rate and density between 2007 and 2014 are examined by considering Turkey's average (13.3%), it's seen that the highest population growth rate is in Tekirdag with 36.2% then Yalova comes with 28.6%; the lowest population growth is in Kirklareli with 9.2% (Table 1). If density is examined, it's seen that the most thickly populated province is Istanbul with 2,767 people per km^2. The population density in other provinces except Balikesir, Canakkale and Kirklareli is higher than the average of Turkey, which is 101 people per km^2 (TUIK 2015).

When the migration rates between the years 2013-2014 are examined, it's seen that the highest migration rate is in the province of Tekirdag with 24.3% and Istanbul can't employ the migration increment with the rate of 1%. Kirklareli comes with the rate of 2.5% and net migration rate is lower in Tekirdag, Balikesir, and Yalova (Table2).

Table 1. Population growth between the years 2007-2014 in the Basin (TUIK 2015).

	Growth Rate of Population by Years, 2007-2014 (%)						
Provinces	2007-2008	2008-2009	2009-2010	2010-2011	2011-2012	2012-2013	2013-2014
Total	**13,1**	**14,5**	**15,9**	**13,5**	**12**	**13,7**	**13,3**
Balıkesir	10,6	8,6	10,7	1,7	5,5	1,7	22,4
Bursa	27,5	16,9	21,3	17,7	13,5	19,5	16,8
Canakkale	-2,8	6,2	26,2	-8,1	14,8	17,3	18,7
İstanbul	9,8	17	26	27,4	16,8	21,8	15,2
Kırklareli	11	-11,2	-1,2	22	3	-1,9	9,2
Kocaeli	35,8	21,3	24,5	26,3	20,4	25,1	27,4
Tekirdag	56,5	16,1	18,7	39	26,7	25,7	36,2
Yalova	82,6	25,6	6	13,6	25,2	38,5	28,6

Table 2. Movement of migration between the years 2012-2014 (TUIK 2015).

	Provincial in-migration, out-migration, net migration, rate of net migration, 1980-2014					
Period	Province	Total Population	In-migration	Out-migration	Net migration	Rate of net migration
2013-2014	Balıkesir	1 189 057	57 551	39 918	17.633	14,9
	Bursa	2 787 539	80 717	65 027	15.690	5,6
	Canakkale	511 790	27 429	20 540	6.889	13,6
	İstanbul	14 337 018	438 998	424 662	14.336	1
	Kirklareli	343 723	14 553	13 678	875	2,5
	Kocaeli	1 722 795	79 697	55 060	24.637	14,4
	Tekirdag	906 732	52 994	31 266	21.728	24,3
	Yalova	226 514	14 680	11 064	3.616	16,1
2012-2013	Balıkesir	1 162 761	38 710	39 688	-978	-0,8
	Bursa	2 740 970	75 518	61 744	13.774	5
	Canakkale	502 328	24 766	18 238	6.528	13,1
	İstanbul	14 160 467	437 922	371 601	66.321	4,7
	Kirklareli	340 559	14 120	13 297	823	2,4
	Kocaeli	1 676 202	71 159	54 742	16.417	9,8
	Tekirdag	874 475	45 313	31 681	13.632	15,7
	Yalova	220 122	14 670	9 808	4.862	22,3

3 ISTANBUL

Istanbul connects Europe and Asia continents. The city was the capital of Byzantine, Roman and the Ottoman Empire and in terms of history, architecture, culture trade, industry and economy. After 1950's, making investments here has accelerated the migration and has led to excessive population growth. Today, Istanbul shows unplanned, disorganized and irregular settlement and industrial growth. A provincial land asset is 520,042 Ha. There are 39 district municipalities and one metropolitan municipality. It is the largest city in Turkey with intensive industries and transportation networks and urban settlements as well as tourism. Of the industry 74% is in the Europe and the rest is in Asia. There are eight organized industrial zones. According to ISO 2010, 8,177 corporations are on the European side while there are 2,735 on the Asian side (TUBİTAK-MAM 2010).

3.1 Land Use

Arable lands along E5 highway which are located Kocaeli-Istanbul and Istanbul-Edirne were occupied by industrial, commercial and residential areas. Out of purpose land use, which are I, II, III, IV classes have reached 57,778 Ha. It is 52% as of total agricultural land (Haktanır *et al.* 2000) and the green areas were used for residential purpose due to improper development plans. While the ratio of green space to total urban areas is 1.5% in Istanbul, it is 41% in Hong Kong, 38.4% in London and 14% in Berlin (Kural 2015). Due to residential and industrial requirements, basins that provide water to Istanbul are under great pressure. The ratio of residential and industrial area to total basin area in Elmali is 26% while that is 12% in Buyukcekmece and in Omerli 10%. Moreover, the intensive residential area is seen in Kucukcekmece basin (Axelle and Menekşe 2013).

3.2 Water Resources

There are no large rivers supplying water to Istanbul. The groundwater level is also limited due to excessive exploitations together with salinization problems. The surface water resources of the province small streams and streamlets. Because these streams can't provide the water requirement of the province, a project has been prepared to deliver 1,047 m^3/year water with a pipeline system from Melen River to Istanbul and necessary works are under progress (Figure 3).

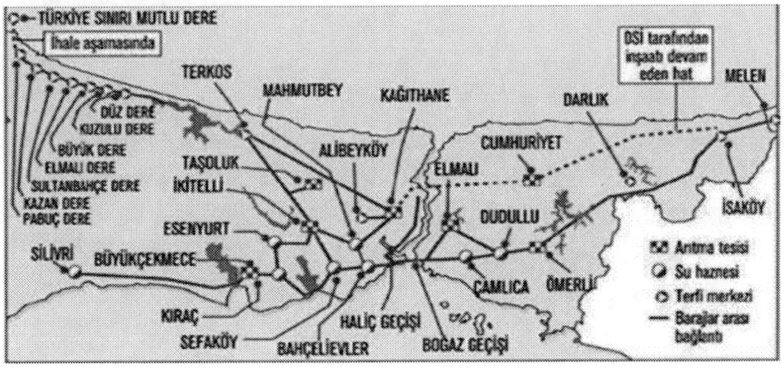

Figure 3. Water resources system of Istanbul (DSI 2014).

3.3 Wastewater Management

The management of clean water and wastewater system is carried out by General Management of Water and Sewage (İSKİ) in Istanbul. According to a survey conducted by TUBITAK-MAM sewage system in Istanbul is on a large scale connected to the wastewater treatment system. The entire industrial zones are connected to industrial wastewater treatment system in Istanbul. Individual industrial companies are under control of the Ministry of Environment and Urban Planning.

4 RESULTS & DISCUSSION

From the years experienced of foundation of the Republic and World, it reveals that the prosperity of the nation-wide distribution can be achieved by a comprehensive land reform especially in East and Southeast Anatolia where mostly the poor and landless farmers live. Naturally, this may help solve the other social problems arising in the region. Another important measure is to change the mission thus reduce population density of Istanbul. 5% of the Turkey population live in the city, which is an unbearable load. Therefore, within a plan, the industry companies in the city must be distributed to the appropriate geographic regions throughout the country. Naturally some technical and economic supporting measures must also be provided for them. However, it is important that the necessary transport links and infrastructure facilities be provided for these companies.

References

Axelle K. ve Kızıldere M., *İstanbuldaki İçme Suyu Havzalarında Arazi Kullanımı,* Çevre Mühendisleri Odası, TMMOB, 2013. www.cmo.org.tr Accessed date:18.08.15

Haktanır K, Cangir C. and Arcak S. *Toprak Kaynakları Kullanımı* ZMO. V. Teknik Kongresi, TMMOB 2000, Ankara, s. 203-229, www.zmo.org.tr, Accessed date: 29.08.15

İSTKA, *2014-2023İstanbul Bölge Planı*, İSTKA, www.istka.gov.tr, Accessed date: 30.08.2015

Kural O. *Kentleşme, Betonlaşma Yeşil Alanlar,* www.sideplayer.biz.tr, Accessed date: 23.08.2015

TUBİTAK-MAM, *Havza Koruma Eylem Planı Hazırlanması-Marmara Havzası Projesi,* İzmit, 2010, www.mam.tubitak.gov.tr, Accessed date:18.08.15

TUİK,Türkiye İstatistik Kurumu, *Results of General Population Censuses,* www.tuik.gov.tr, Accessed date: 15.08.15

Interaction between Theory and Practice in Civil Engineering and Construction
Edited by Komurlu, R., Gurgun, A. P., Singh, A., and Yazdani, S.

USING BAYESIAN NETWORKS MODELS TO IMPROVE BEHAVIOR TOWARDS WASTE MANAGEMENT IN CONSTRUCTION PROJECTS

AMAL BAKSHAN, ISSAM SROUR, and GHASSAN CHEHAB

Civil and Environmental Engineering Dept, American University of Beirut, Beirut, Lebanon

The effective implementation of waste management strategies in construction projects requires the identification of critical factors that affect the behavior regarding waste management as well as quantifying the relationships between causal factors. In this context, the Bayesian Networks (BN) analysis technique is an effective means that not only identifies the influencing factors but also establishes a probabilistic relational network among causal factors, thereby providing a methodology to identify potential strategies for improving behavior towards construction waste management. The comprehensive literature review, presented in this paper, is a vital source of information required to build a research model that includes key factors influencing behavior towards waste and to construct a structured questionnaire. As such, the research model, along with the data collected, is necessary to demonstrate the application of the proposed BN model in the real world.

Keywords: Construction waste management, Causal factors, Strategies.

1 INTRODUCTION

The construction industry has witnessed a significant growth over the past decade, leading to an excessive amount of construction waste (CW) generated worldwide. The huge amount of waste generated by construction activities ties down landfill space, leading to considerable loss in land resources for the expansion of existing landfills or the creation of new ones (Esin and Cosgun 2007, Yuan 2013). CW generation not only consumes land resources, but also has several harmful environmental impacts including air pollution, water pollution, depletion of natural resources, and risks to human health (Poon *et al.* 2003).

In addition to the importance of raising the awareness towards CWM, a study conducted by Srour *et al.* (2013) affirmed that the effective implementation of CWM also necessitates defining roles and responsibilities of the construction and governmental authorities. One of these fundamental responsibilities is to propose practical strategies that aim at improving the behavior of construction workers towards waste management (Merino *et al.* 2010). As such, the improvement of behavior towards waste management in construction projects not only requires the identification of factors that influence the behavior towards waste but also entails the quantification of the relationships between causal factors in order to propose effective strategies at the site level. For this reason, a comprehensive literature review is carried out in this paper to 1) shed light on the limitations of existing studies related to behavior towards waste management, 2) identify possible factors that might have influence on behavior towards waste minimization and/or recycling, and 3) suggest a new methodology that can be adopted to overcome the drawbacks of existing studies.

2 BEHAVIOR TOWARDS WASTE MANAGEMENT

Previous studies have used statistical analysis methodologies, particularly factor analysis, ranking analysis, regression analysis and structural equation modeling (SEM), to examine the possible factors that influence waste minimization and recycling for the purpose of improving the behavior regarding waste management. Factor analysis is a data reduction method that generates a set of factors (i.e., unobserved variables) that represent a set of correlated observed variables (Torres-reyna 2007a). For example, Wang and Yuan (2011) and Udawatta *et al.* (2015) explore the critical factors that have an influence on the behavior towards waste management by grouping the correlated observed variables. The ranking analysis, on the other hand, orders the variables under investigation based on their ranking values. The data are collected by asking respondents to rank alternatives based on their preferences (Coe 2002). Several studies (e.g., Kulatunga *et al.* 2006, Tam 2008) rank the factors based on their influence on the implementation of waste management strategies, where rank 1 indicates the most influential factor. The linear regression analysis method, applied by Begum *et al.* (2009) and Sidique *et al.* (2010), investigates the relationships between a dependent variable and its causal independent variables by testing the statistical significance of these linear relationships (Torres-reyna 2007b). At a more complex level, Ramayah *et al.* (2012) uses SEM to build a model that represents the relationships between the recycling behavior and its causal factors in addition to the relationships between the respective factors. SEM is an advanced prescriptive data-analytic technique that has become such an increasingly popular statistical analysis option due to various strengths.

In spite of its strengths, SEM, and generally regression analysis, has some limitations. One of these potential problems is multicollinearity that refers to high correlations between latent (i.e., unobserved) independent variables. When multicollinearity occurs, the variance of the coefficient estimates may be increased, making the estimates unstable and sensitive to minor changes in the model which consequently induce statistical non-significance of the estimates (Grewal *et al.* 2004). This situation, in turn, undermines the statistical power of the analysis making it more difficult to determine the correct model and to quantify the influencing factors affecting behavior towards waste management.

3 FACTORS AFFECTING BEHAVIOR

Factors influencing behavior towards waste minimization and/or recycling fall under two main categories, personal factors and corporate factors. Personal factors, also called individual factors, represent the psychological and cognitive factors that are involved in shaping the individual's decision-making. Corporate factors, also called organizational factors, act as part of the organizational context, which may either facilitate or constrain the success of individuals' initiatives to any behavioral change (Young *et al.* 2013).

3.1 Personal Factors

The *attitude* towards the behavior, as described by the theory of planned behavior (Ajzen 1993), refers to an individual's positive or negative evaluation of the behavior in question and is shaped based on personal beliefs and knowledge. Attitude is usually measured by asking individuals' to specify their feelings about performing the behavior, e.g., whether the behavior in question is good, useful, rewarding, etc. Begum *et al.* (2009) studies the relationship between attitude and behavior and concludes that contractor attitude towards waste management significantly affect the behavior regarding waste management. Contractors having positive attitude toward waste management also undergo satisfactory behaviors. In an effort to investigate the factors that positively affect attitude, Ramayah *et al.* (2012) found that employees' knowledge and awareness of environmental benefits are positively related to attitude, which in turn contributes to improving waste management behavior.

Consequences awareness refers to individuals' awareness towards benefits of waste management, including the environmental, economic and social benefits. Yuan (2013) highlights the critical role of enhancing workers' awareness about saving economic resources and environmental protection in improving waste management performance. In this context, several studies identify factors that contribute to raising the awareness towards consequences of waste management. For example, Sidique *et al.* (2010) argue that communication efforts raise the awareness towards benefits of recycling behavior and positively affect attitude towards recycling and consequently behavior.

Wang and Yuan (2011) demonstrate that *work experience*, which represents the number of years of professional experience, is a critical factor affecting workers' attitudes in construction projects. Ajzen (1993) argues that *past experience in a behavior* contributes to the formation of attitude that in turn plays an important role in shaping the behavior, Macey and Brown (1983) report that past experience is the best direct predictor of conservation behavior. Terry *et al.* (1999) indicate that previous recycling experience should be included when considering the prediction of behavior towards recycling.

Studies show that behavior is highly influenced by *social pressure*, also called social norms, especially when the perception of the social group about the behavior is important to the individual performing the behavior (Chan 1998, Shaw 2008). For example, social pressure has the greatest impact on recycling behavior, where an individual has a great intention to perform a certain behavior positively when he/she perceives that it is important what others think he/she should be doing.

3.2 Corporate Factors

Attending *training courses* is one of the best sources of information for gaining knowledge about various waste management strategies, and thus is proved to be an effective way in improving waste minimization and recycling behavior (Udawatta *et al.* 2015). Kulatunga *et al.* (2006) report that training and education about waste minimization methods play an important role in 1) raising the awareness on the importance of waste management as a cost-effective and profitable technique, 2) forming positive attitudes regarding waste management at both corporate and individual levels, and 3) enhancing actual behavior towards waste management.

Adequate *supervision* of waste management activities with clear instructions from the top level to the bottom level of the organization is identified as an important solution for proper implementation of waste management strategies (Kulatunga *et al.* 2006). Tam and Tam (2008) report on a *financial reward* scheme used to motivate construction workers to adopt a waste management strategy, resulting in waste reduction by up to 23%. Therefore, introducing financial incentives for better waste management practices would help to develop and implement waste management applications in the construction industry (Kulatunga *et al.* 2006).

Several statistical methods have been used to examine the effect of personal and corporate factors on behavior towards CWM practices. However, these methods cannot quantify the relationships between causal factors and do not offer practical strategies for improving behavior towards waste management. Hence, this paper advocates the use of Bayesian Networks (BN) to quantify the relationships between related factors. BN allow for the quantification of such relationships through probabilistic relations. What follows is a review of relevant literature on BN demonstrating the effectiveness of this method in constructing a behavioral model that relies on probabilistic analysis.

4 BAYESIAN NETWORKS (BN)

A BN is a mathematical model where each variable is represented by a node, with arrows representing the dependencies between related variables. It is a directed acyclic graph that cannot include a feedback, i.e., a path traced along its links cannot pass through a variable more than once. The first step in constructing a BN requires identifying the important nodes, the states of each node and the relationships between them, which can be done based on existing literature, experts' consultation and/or the analyst's experience and knowledge of the analyzed system (Dambacher *et al.* 2007).

The BN model shows the causal structure in a problem using conditional probability distributions to define relationships between variables, where a conditional probability table (CPT) is generated for each variable after compiling the model. If a variable has no incoming arcs and hence does not depend on any other variable in the model (i.e., conditionally independent), its CPT includes one probability distribution. However, the CPT of a variable that has parents consists of one probability distribution for every combination of possible values of the parents, leading to a large number of combinations in case of complex BNs. In this case, filling in the CPTs may require many conditional probability assignations and can thus be difficult to complete (Ames *et al.* 2005, Uusitalo 2007). For this reason, Cain (2001) recommends simplifying the BN by either combining nodes or reducing the number of states within each node (called "data pre-processing"), which in turn reduces the overall size of the CPTs. Netica, Hugin, GeNIe and B-course are examples of software packages, which have been ported to most commonly used operating systems (Uusitalo 2007).

The BN technique has no minimum sample size requirements. It has the ability to show accurate prediction of outcomes even with small sample sizes (Kontkanen *et al.* 1997). The technique works with variables of categorical data type, which allows capturing non-linear relationships between the variables (Myllymaki *et al.* 2002). Once the BN is compiled and the conditional probabilities are estimated, then the model is complete and the downward and upward propagation of information through the model can be used to view how observed conditions or evidences at certain nodes affect the probable conditions at the other nodes (i.e., states are known with certainty) – the so-called "reasoning" (Zhou *et al.* 2008). The reasoning characteristic is one of the key points of BNs as it allows stakeholders and decision makers to visually interpret a given problem, thus serving as an effective tool in predicting the probable outcomes of management decisions on selected variables.

5 CONCLUSION

The effective implementation of CWM practices requires understanding the influencing factors of behavior towards waste and the quantification of the relationships between these factors as well. As such, this paper represents a comprehensive literature review covering possible influencing factors, limitations of existing studies related to behavior towards waste management, and applicability of BN as a methodology to generate BN model with quantified relationships between causal factors. Based on this literature review, a research model as well as a questionnaire required to collect data about factors affecting the behavior of construction workers are being built, thus serving as seeds for the ongoing project aiming at generating a CWM behavioral BN model. Construction field workers from various construction projects located in Beirut (Lebanon) are currently being interviewed. Data will be incorporated into the research model to generate the behavioral BN model and strategies for improving the behavior regarding waste management will be proposed accordingly.

Acknowledgments

This paper and the ongoing research project have been funded through a grant from the National Center for Scientific Research (CNRS) (No. 102655), Lebanon.

References

Ajzen, I., Attitude theory and the attitude-behavior relation, *New Dir attitude Meas.* , Krebs, D. Schmidt, P. (Eds), 41–57, Walter de Gruyter, Berlin, 1993.

Ames, D. P., Neilson, B. T., Stevens, D. K., Lall, U, Using Bayesian networks to model watershed management decisions : an East Canyon Creek case study, *J Hydroinformatics*, 7(4), 267–282, 2005.

Begum, R. A., Siwar, C., Pereira, J. J., Jaafar, A. H., Attitude and behavioral factors in waste management in the construction industry of Malaysia. *Resour Conserv Recycl.,* 53(6), 321–328, 2009.

Cain, J., Planning improvements in natural resources management: Guidelines for using Bayesian networks to support the planning and management of development programmes in the water sector and beyond, *UK: Centre for Ecology and Hydrology*, 2001.

Chan, K., Mass communication and pro-environmental behaviour: waste recycling in Hong Kong, *J Environ Manage.*, 52(4), 317–325, 1998.

Coe, R., Analyzing ranking and rating data from participatory on farm trials, in *Quantative Anal data from Particip methods plant Breed*, Bellon, M. R., Reeves, J. (Eds.), 44–65, Mexico, DF:CIMMYT, 2002.

Dambacher, J. M., Shenton, W., Hayes, K. R., Qualitative modelling and Bayesian network analysis for risk-based biosecurity decision making in complex systems, *Aust. Cent. Excell. Risk Anal.* 2007. Retrieved from http://www.acera.unimelb.edu.au/materials/endorsed/0601.pdf on August 12, 2015.

Esin, T., Cosgun, N., A study conducted to reduce construction waste generation in Turkey, *Build Environ*, 42(4), 1667–1674, 2007.

Grewal, R., Cote, J.A., Baumgartner, H., Cote, J.A., Multicollinearity and Measurement Error in Structural Equation Models : Implications for Theory Testing, *Mark Sci.,* 23(4), 519–529, 2004.

Kontkanen, P., Myllymaki, P., Silander, T., Tirri, H., Comparing Predictive Inference Methods for Discrete Domains, *Proc sixth Int Work Artif Intell Stat. Ft.*, 311–318, Lauderdale, USA, 1997.

Kulatunga, U., Amaratunga, R., Haigh, R., Rameezdeen, R., Attitudes and perceptions of construction workforce on construction waste in Sri Lanka, *Manag Environ Qual.*, 17(1), 57–72, 2006.

Macey, S., Brown, M., Residential energy conservation: the role of past experience in repetitive household behavior, *Environement Behav.,* 15(2), 123–141, 1983.

Merino, M., Gracia, P., Azevedo, I., Sustainable construction: construction and demolition waste reconsidered, *Waste Manag Res.*, 28(2), 118–129, 2010.

Myllymaki, P., Silander, T., Tirri, H., Uronen, P., B-course: a web-based tool for Bayesian and causal data analysis, *Int J Artif Intell Tools*, 11(3), 369–387, 2002.

Poon, C. S., Yu, A. T. W., Ng, L. H., Comparison of low-waste building technologies adopted in public and private housing projects in Hong Kong, *Eng Constr Archit Manag.,* 10(2), 88–98, 2003.

Ramayah, T., Lee, J. W. C., Lim, S., Sustaining the environment through recycling: An empirical study, *J Environ Manage.,* 102, 141–147, 2012.

Shaw, P. J., Nearest neighbour effects in kerbside household waste recycling, *Resour Conserv Recycl.,* 52(5), 775–784, 2008.

Sidique, S. F., Joshi, S. V., Lupi, F., Factors influencing the rate of recycling: An analysis of Minnesota counties, *Resour Conserv Recycl.,* 54(4), 242–249, 2010.

Srour, I. M., Chehab, G. R., El-Fadel, M., Tamraz, S., Pilot-based assessment of the economics of recycling construction demolition waste, *Waste Manag Res.,* 31(11), 1170–1179, 2013.

Tam, V. W. Y., On the effectiveness in implementing a waste-management-plan method in construction, *Waste Manag.,* 28(6), 1072–1080, 2008.

Tam , V. W. Y., Tam, C. M., Waste reduction through incentives: a case study, *Build Res Inf.,* 36(1), 37–43, 2008.

Terry, D., Hogg, M., White, K., The theory of planned behavior:self-identity, social identity and group norms, *Br J Soc Psychol.*, 38(3), 225–244, 1999.

Torres-reyna, O., Getting Started in Factor Analysis (using Stata 10), 2007a. Retrieved from http://dss.princeton.edu/training/ on August 11, 2015.

Torres-reyna, O., Linear Regression using Stata, 2007b. Retrieved from http://dss.princeton.edu/training/ on August 11, 2015.

Udawatta, N., Zuo, J., Chiveralls, K., Zillante, G., Improving waste management in construction projects: An Australian study, *Resour Conserv Recycl.,* 101, 73–83, 2015.

Uusitalo, L., Advantages and challenges of Bayesian networks in environmental modelling, *Ecol Modell.,* 203(3-4), 312–318, 2007.

Wang, J., Yuan, H., Factors affecting contractors' risk attitudes in construction projects: Case study from China, *Int J Proj Manag.*, 29(2), 209–219, 2011.

Young, W., Davis, M., McNeill, I. M., Malhotra, B., Russell, S., Unsworth, K., Changing behaviour: Successful environmental programmes in the workplace, *Bus Strateg Environ.,* 1–15, 2013.

Yuan, H., Key indicators for assessing the effectiveness of waste management in construction projects, *Ecol Indic.*, 24, 476–484, 2013.

Zhou, Q., Fang, D., Wang, X., A method to identify strategies for the improvement of human safety behavior by considering safety climate and personal experience, *Saf Sci.,* 46(10), 1406–1419, 2008.

Interaction between Theory and Practice in Civil Engineering and Construction
Edited by Komurlu, R., Gurgun, A. P., Singh, A., and Yazdani, S.

COASTAL PLANNING AND THE EVALUATION OF COASTAL STRUCTURES IN TERMS OF SUSTAINABILITY

NECATI GULBAHAR[1], IMAM BAKIR KANLI[2], and DIDEM ODABASI[1]

[1] *Dept of Civil Engineering, Istanbul Gelisim University, Istanbul, Turkey*
[2] *Dept of Public Administration, Marmara University, Istanbul, Turkey*

Coasts are cross-sections of the water and land environments, for thousands of years they formed under the factors of climate, geological, geomorphological, topographic and hydraulic and it has reached a natural balance. But increasingly this natural balance is disrupted due to human impact. Coasts, with their strategic position, are a wealth of resources for people with endless possibilities and mild climate. Therefore, coasts continuously have remained under great pressure just because of residential, transportation, energy, economy and tourism. These pressures can be minimized by applying the principles of science and engineering with a rational urban planning. It will be held after an urban planning at the macro level, as a part of that plan, it should be considered in the design of optimal engineering structures. Therefore, the structure of the planning and design engineering for optimal site selection should be done and compatible structures integrated with topography and bathymetry such as coastal hydrodynamics should be designed. In this paper, after briefly the occurrence and type of coasts, the strategic importance of coastal will be highlighted and coastal planning and engineering structures in Turkey will be examined in terms of sustainability by demonstrating some examples and solutions will be proposed for encountered problems.

Keywords: Urban planning, Natural balance, Shoreline protection, Land use, Zoning.

1 INTRODUCTION

Mankind has been interested in coastal areas since historical times. This is due to coastal zones not only provided many possibilities such as fishing, navigation, transport, trade for coastal zones but also have been source of wealth for beyond the coastal areas. That is why many civilizations were born in coastal areas. Presently, humans are also taking a strong interest in these areas. Therefore, the integrated planning approach is too serious issue about coasts. In fact by Integrated Coastal Zone Planning it is provided to preserve the ecological, natural characteristics of coastal areas, as it is also provided to prevent possible harms to that delicate balance which is formed of the interaction of water, land and air.

The rapid population increase in urban areas started where the economic values of this land increased and speculations were observed in those days. A number of buyers or capitalists have conflict each other in order to possess land from these regions. Hence the unplanned changes in urban areas and coast line got started in this way. As a result of these adverse effects the over loading at the eco-systems and environments such as the excess use of natural sources, the damage coastal areas and pollution of soil. Water and air were observed. Istanbul is typical clear example for these unpleasant changes in Turkey.

In this paper, considering the characteristics and formation of the coastal geomorphology at the specified location, which is Istanbul, the impact of human to the physical environment and the need of integrated planning at especially cities having shores as well as the impact of these adverse changes will be discussed. Therefore, it was thought that Istanbul would be a clear example. For this reason, some observations in the field were made and, related reports, projects, articles were considered.

2 BASIC TERMS AND DEFINITIONS

2.1 Sustainability

Sustainability concept was first mentioned by Burundtland Report which could be defined as meeting the needs of the present without compromising the ability of the future generations to meet their own needs.

2.2 Coastal Planning

The coastal planning is regulating the shore use within the scope of scientific methods and public demands. Coastal cities have been susceptible to any changes.

2.3 Shoreline

Shoreline can be defined as the line where water surface and land meet. Naturally, water level is variable containing maximum and minimum values, therefore, shoreline is also changeable.

3 ENTEGRATED PLANNING

Unplanned and uncontrolled accumulation of overpopulation, which took place in the coastal city, results in much more catastrophic consequences. Deteriorated structures are not only ecological, but also the social and economic structure. This situation has inevitably made the adoption of sustainable management methods necessary by using of coastal areas within taking a sustainable approach to planning in a integrated approach. For example, the principles which are prepared by DAMPO (Research Planning Architecture Consultancy Company) for the Ministry of Public Works and Settlement Republic of Turkey, determined for Iskenderun Gulf Coastal Zone Integrated Planning and Management Project is provided in Table 1.

4 ENGINEERING PLANNING

Planning may be defined as organizing the intending future works in a sensible way and sorted by priority of each work. The coastal planning is regulating the shore use within the scope of scientific methods and public demands. The basic parameters that will be considered in terms of coastal engineering can be expressed as coastal climate, morphology, wave and sediment movements (Figure 1). Coastal engineering generally carries on a work in the following fields (Kabdasli 2015).

- **Protection of Shoreline:** Coastal structures, coastal revetments, seawalls, groins and strakes are considered.
- **Protection of the area behind the shore:** The structures listed above may be designed under this topic too.
- **Harbor Construction:** A suitable design of seawall has great importance.
- **Protection of embouchure:** The structures listed above may be designed under this topic too. Besides, bottom scanning may be performed.
- **Special marine structures:** Headworks, discharges, oil rigs, open sea wave energy power plants, submarine pipelines can be considered.

In terms of coastal engineering, sustainable planning and design of the structures mentioned above contribute to sustainable shore use. For design of the structures, the following data are analyzed.

- **Climate Data:** Wind, precipitation and fog are prominent data.
- **Wave Analysis:** Frequency, period, maximum and minimum wave height and possibility are needed
- **Flow and vortex analysis:** Due to differences density of sea water, or some other factors flow or vortex may occur in the area.
- **Morphological Data:** Geological and geophysical analysis of the plant area and determine of bathymetry.
- **Environment impact Assessment:** Analyzing the plant's adverse impacts to the environment in construction and run phases.
- **Economic Analysis (Feasibility):** Plant must be feasible in technical and economic ways.

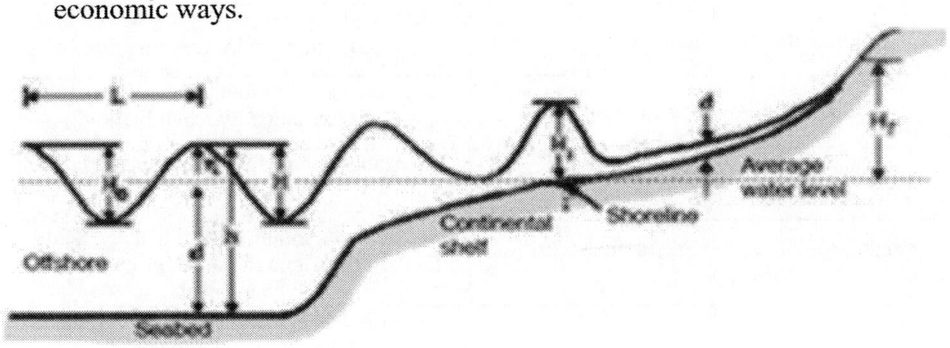

Figure 1. A profile of coastline of sea (Kabdasli 2015).

However, harbors may be divided into different types based on functions. Here, the general criteria for the determination of harbor site location will be given as follows (Yuksel 2000).

- Geographical location should be well chosen.
- Port inlet must be determined in accordance with waves, currents and sand movements.
- Allocating sufficient space to ships to berthing easily to the dock should be considered.
- Convenient locations for the installations should be determined.
- At least a period of 50 years, should be considered for development and expansion of the port.
- Highway, railway and airway connections should be considered.
- Shallow seas and mountainous regions are the most important factors for port site selection.

Table 1. Iskenderun Gulf, Holistic Coastal Planning General Principles (DAMPO 2009).

GENERAL PRINCIPLES	
Sustainable Development Principle	Comprehensive, integrated and rational planning approach should be adopted.
Preventiveness Principle	Environmental Impact Assessment activities that should be carried out before investment, should be given great importance.
Prudence Principle	The competent authorities should use its authority on prohibition of the activities that might cause irreversible damage on the coastal regions or making regulations about these situations.
Estimation Principle	Risk analysis should be performed and necessary measures should be taken according to natural and man-made disaster scenario.
Rehabilitation Principle	Deformed texturesshould be get back to the old formation by rehabilitation.
Polluter/UserPays Principle	Deformed texturesshould be get back to the old formation by rehabilitation.
Appropriate use of Technology and Good Environmental Practices Principle	The extent of available sources, the high-tech should be used.
Informing the public and Public Participation Principle	The accession process to management focused decisions should be provided at the local level
International Cooperation Principle	In the fields of planning and management model develop ment and application, international cooperation opportu nities should be developed to develop a policy or produce a project.

5 REGULATIONS

5.1 International

European Union (EU) conducted some works in this context, Thus, Conference of the Peripheral Maritime Regions (CPMR) was established in1973. It is working to promote a more balanced development of all Europe's regions and to increase regional competitiveness for the CPMR, sustainable development and the conservation of the environment and biodiversity are also priority issues. Its focuses in particular on energy policies and climate change mitigation and adaptation measures.

5.2 National (for Turkey)

The first regulations on the coasts in Turkey were made with the amending Law No. 1605 at No. 6785 Construction Zoning Law in 1956. At which there has been 10 m of a zone from the coast where buildings cannot be built for the individual purposes. According to the regulations in Article 43 of the 1982 Constitution; "The coasts are under the sovereignty and disposal of the State. Public interest is considered first to benefit from the shores of the surrounding coastline of sea, lake and river. Thus, coasts, with the effect of the 1982 Constitution, Law No. 3086 issued in 1984 for the first time. However, the Law No. of 3086 with the entry into force, existing problems could not be solved and they remain. In subsequent years, with the legal arrangements made the width of coast lane, was first 30 m, and then it was 100 m. However, all these arrangement are not enough for the solution of the problems.

6 RESULTS AND CONCLUSION

Since the 1970's, in order to encourage tourism, especially forests and fertile agricultural areas of Marmara, Aegean and Mediterranean coasts of Turkey have been polluted and destroyed. This was due to take benefit from the legal loopholes. Unfortunately, this period is currently in process. Some proposal for the solutions can be given such as; first, integrated coastal management policy should be established in the development plans. Goals and objectives should be clearly defined ın plans. Geomorphological structure of the coast and bathymetry should be examined in detail, Shorelines must be determined in accordance with the scientific principles and coast should be considered as a whole with the impact of behind coast and related areas. Coordination between institutions and data sharing should be provided. Coastal structures must be compatible with the coastal geomorphology. People's demand should also be considered. Finally, while designing of the coastal structures should be taken into account the earthquake effects.

References

DAMPO Ltd..C., Iskenderun Gulf Coastal Zone Integrated Planning and Management Project, 2009.

Kabdasli S. Kıyı ve Liman Yapıları, retrieved from www.google.com.tr, 2015.

Yüksel Y., Ozmen H., Cevik E., Ozguven O., Celikoglu Y., BostanT., Gurer S., Gokoglu F., Doğu Marmara Depreminin Körfezdeki Deniz Yapıları Üzerindeki Etkileri, III.Coastal Engineering Symposium, Canakkale, 2000.

Interaction between Theory and Practice in Civil Engineering and Construction
Edited by Komurlu, R., Gurgun, A. P., Singh, A., and Yazdani, S.

OPTIMIZATION OF FLAT ROOF THERMAL RENOVATION TECHNOLOGIES

RUTA MINIOTAITE

Dept of Civil Engineering Technologies, Kaunas University of Technology, Kaunas, Lithuania

The created methods and the logical scheme of an alternative selection based on it, allow for the adoption of decisions that optimally satisfy the technical parameters of roof repairs and the financial possibilities of a builder in each concrete case. It is here suggested that one selects a rational version of roof repairs in a step-by-step way by assessing the alternatives with the help of multi-criteria decision methods. Each stage has pre-selected assessment criteria, decision methods and created database. The initial stages include the following steps: the assessment of the condition of the roof following the developed methods, the expected price of the work and preliminary selection of alternatives. By using the method of mathematical analysis – efficiency value analysis – the most economically and technologically efficient option of thermal renovation was determined. The reconstruction of non-usable flat roof is optimal in technical, economic and humidity exposure terms. The price of this roof construction is not the minimum price, but duration of construction work, resistance to humidity exposure are optimal.

Keywords: Technology of renovation, Analysis of workability.

1 INTRODUCTION

Recently renovation and modernisation of buildings is a hot topic in Lithuania. Various programmes are implemented, people are encouraged to explore renovation possibilities and renovate their homes. Renovation means the improvement and modernization of buildings and living environment, the creation of new quality. It is a separate part of reconstruction work that involves a complex of construction and installation works aimed at improving the physical, mechanical, aesthetic and other operating properties of the present building in order to achieve better energy. Renovation involves the renewal of building structures, interior, exterior, insulation of the building envelope and other work requested by the customer in order to improve the present properties of the building (Bieksa *et al.* 2011).

Renovation is not defined in technical regulations of construction work; therefore buildings are renovated in accordance with technical regulation of construction "Types of construction of the building" (STR 1.01.08 2002). According from this regulation renovation of a building can be treated as reconstruction of the building.

Depending on the roof structure, the building may lose up to 30 – 35% heat through the roof. Therefore, renovation of the building should start from the roof (Rapceviciene 2010). Roofing technologies and materials used in twenty year old and older building are outdated; the insulation used those days does not meet the current requirements for roof structures. All roofs that were not renovated using new generation roofing materials are in very poor condition: the rood deck is uneven, there are blisters and cracks in bitumen cladding, the seams between the roof cladding and other roof elements are leaky. Building are loosen big amount of energy through the roofs.

Renovated roofs must meet the currently applicable structural requirements according to the Regulation "Structures of buildings. Roofs" (STR 2.05.02 2008) and energy efficiency requirements according to the Regulation "Energy efficiency of buildings. Certification of energy efficiency" (STR 2.01.09 2012).

The main roof repair and heat insulation renovation technologies are described in the recommendations "Flat roofs repair methods and solutions" (R 26-00 2000) approved by the Ministry of Environment of the Republic of Lithuania.

Both newly-built and renovated roofs must meet the minimum requirements set forth in construction regulations. First of all the roofs must be weatherproof and resistant to design loads. They must be designed, build and maintained in accordance with the essential requirements for the building, i.e. they must be mechanically resistant and intact, the structure must meet fire safety, hygiene, health, environmental and operation requirements, roofs must ensure protection from noise, save energy and heat. The roof structure must satisfy the flammability $B_{roof}(t1)$ class, whereas the heat transfer coefficient of a renovated roof must be at least $U_{n.r.}$=0,20W/m^2·K (STR 2.05.01 2013).

Thermal renovation is the best method for renewing a flat roof. Before starting a flat rood renovation, the condition of the roof must be evaluated and appropriate constructional solutions must be selected. The most feasible method of renovation is selected after considering the present condition of the roof, the present conditions of work, financial resources and the desirable additional roof performance properties.

The most common problem of old flat roofs is "confined" humidity. The pressure of accumulated water vapour is confined in the structure and produces blisters in the roof coating. These blisters deteriorate the performance of the roof. Renovation must ensure proper ventilation of roofing layers.

The goal of this paper is to analyze the methods of thermal renovation of flat roofs taking into account the materials used, roof structures and roofing technologies as well as conduct the integrated analysis of selected technological alternatives by means of efficiency value analysis.

2 EVALUATION AND OPTIMIZATION OF FLAT ROOF RECONSTRUCTION

Design solutions of construction processes can be evaluated by different methods. According to the number of criteria used they are divided into single criterion and multiple criteria solutions (Juodis 2005). In order to find the best solution, a multiple criteria optimization of selected structural solution alternatives must be done.

2.1 Initial Data for Multiple Criteria Evaluation of Alternative Solutions for Flat Roof Reconstruction

Five structural alternatives for the thermal renovation of a flat roof are selected. The selection is followed by economic and humidity calculations, workability analysis, and the most rational variant are determined. Structural solutions of reconstructed roofs are presented in Table 1. For building a terrace or a roof garden, additional loads for bearing structures should be considered. The price of such roofs is also higher.

Table 1. Structural solutions of reconstructed roofs.

Alternative	Structural drawing	Structural layer, materials
Alternative 1 Not used flat roof	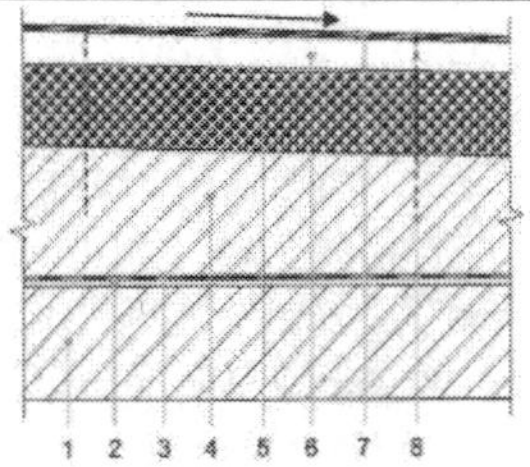	1 Existing reinforced concrete slab 2 Surface levelling layer 3Present sound insulation 3 Present heat insulating porous concrete slab 4 Lower polystyrene foam layer 5 High performance polystyrene foam layer 6 Roof waterproofing with protective layer
Alternative 2 Flat roof in use	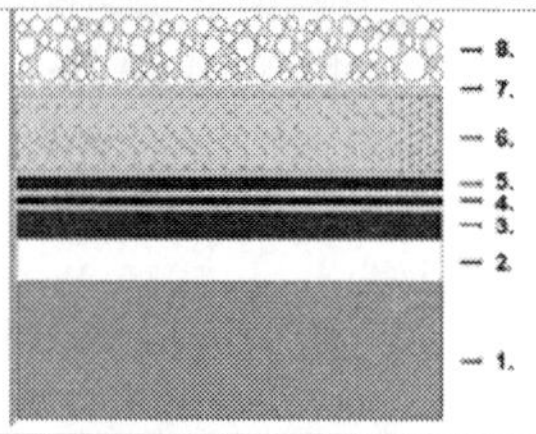	1 Roof base layer 2 Surface levelling layer 3 Old bitumen cladding 4 Lower Superbase Ultratorch layer 5 Upper Superbase Ultratorch layer 6 Thermal insulation – extruded polystyrene 7 Filtering layer from geotextile 8 Gravel or concrete slabs
Alternative 3 Reverse roof	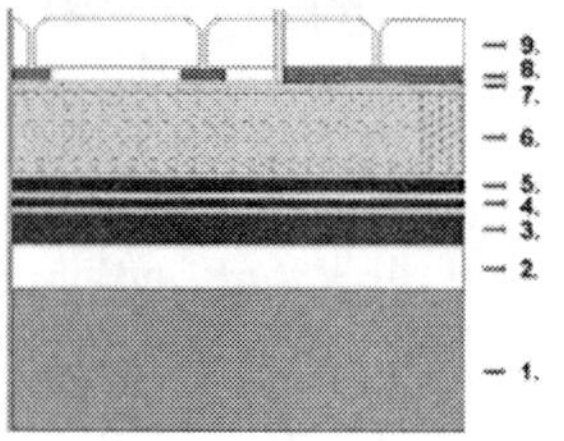	1 Roof base layer 2 Surface levelling layer 3 Old bitumen cladding 4 Lower Superbase Ultratorch layer 5 Upper Superbase Ultratorch layer 6 Thermal insulation – inverse type 7 Filtering layer from geotextile 8 Load distributing layer 9 Concrete slabs
Alternative 4 Not used flat roof	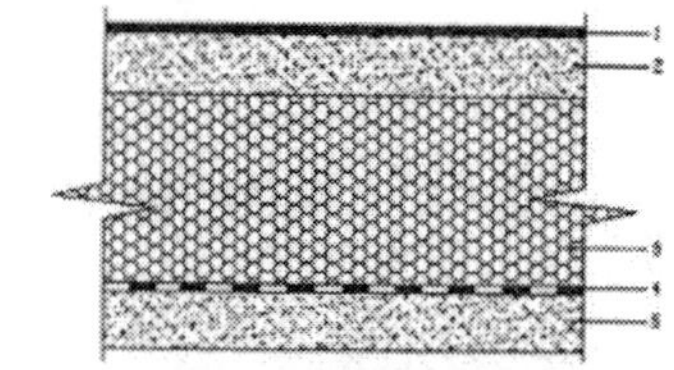	1 Bearing structure 2 Old bitumen cladding 3 Lower thermal insulation layer – polystyrene foam plates (EPS 80, EPS 100) 4 Upper thermal insulation layer – thermal insulation plates 5 Upper layer of reinforced bitumen waterproof plates
Alternative 5 Flat roof in use	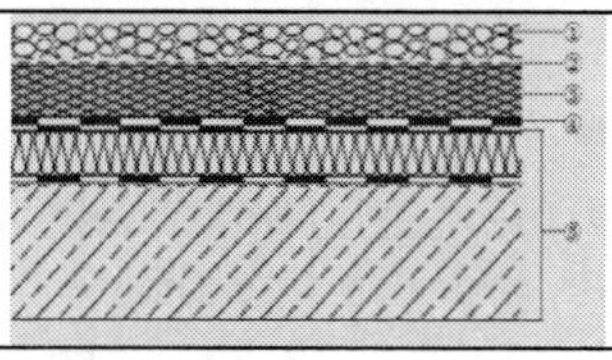	1 Breakstone /gravel Ø 16/ 2 Geotextile filtering layer 3 STYROFOAM 300 4 Final new waterproof layer 5 Old roofing layer

2.2 Using Efficiency Value Analysis to Find the Effective Structural Solution for the Reconstructed Flat Roof

The efficiency value analysis is the assessment of solutions basing on a set of criteria. This method can be applied both with dimensional and non-dimensional values, therefore, the solutions can be evaluated in terms of different criteria: technical, legal, economic, social etc.

Method evaluation algorithm:

- A set of evaluation criteria is formed;
- A relative weight to each criterion is assigned;
- Values of criteria are found;
- The efficiency value from 0 to 10 points is assigned;
- The numerical value of efficiency is determined;
- Total efficiency value of alternative solutions is determined.

Different criteria are not equally important in the evaluation of alternative constructional solutions, therefore the significance of criteria in relation to each other is taken into account during the evaluation.

The following criteria were selected to evaluate the workability of flat roof insulation renovation methods:

Technical – economic criteria (C):

C1 – The price of 1 m^2 of the roof, EUR,

C2 – Fire resistance of roof structure (1 – not resistant, 5 – resistant),

C3 – Complexity of roof structure installation (1 – complicated, 5 – not complicated);

C4 – duration of building 100 m^2 of roof structure by a team of 6 builders (in days);

Criteria for the evaluation of humidity regime of a roof structure:

C5 – vapour resistance factor of thermal insulation;

C6 – amount of condensed vapour per year (g/m^2).

Combinations of evaluation criteria are done to optimize the alternative structural solutions:

Combination 1. Technical – economic criteria (*CR1, CR2, CR3, CR4*).

Combination 2. Criteria evaluating humidity regime of the roof structure (*CR5, CR6*).

Program DOFTHERM has been used to calculate the moisture regime of the roof constructions. It determined the variation of moisture condensation by changing insulation thickness, the type and the roof cover. The analysis determined the changes in the moisture content of the roof structure when changing internal relative humidity and air temperature.

Evaluation of technological alternatives for roof renovation by means of efficiency value analysis using the determined importance of criteria is presented in Table 2. Figure 1 illustrates graphical comparison of alternative roof renovation technologies by technical-economic criteria. Figure 2 illustrates graphical comparison by humidity regime criteria.

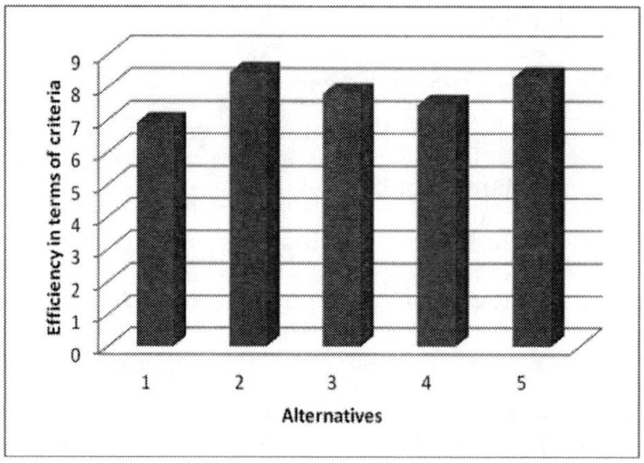

Figure 1. Technical-economic comparison of alternative roof renovation technologies.

After the optimization of structural solutions for flat roof reconstruction by applying the sets of criteria of efficiency value analysis it was found that Alternative 4 was the most optimal (bearing structure, old bitumen cladding, lower thermal insulation layer from polystyrene foam (EPS 80, EPS 100), upper thermal insulation layer from heat insulating slab, top layer made of reinforced bitumen waterproof sheets). This alternative is optimal in technical-economic and humidity regime terms. The cost of this roof structure is not minimum price but the time of installation, resistance to external fire and weather exposure is optimal.

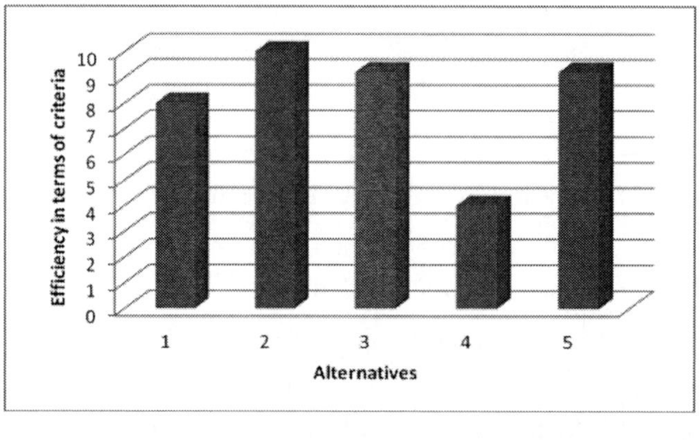

Figure 2. Comparison of alternative roof renovation technologies by humidity regime criteria.

3 CONCLUSIONS

Thermal renovation is the most effective for flat roofs. This renovation involves the laying of additional layer of thermal insulation in order to achieve the thermal resistance value required by current construction standards. New high quality construction materials and technologies ensure the durability and reliability of roofs for few decades.

The effect of moisture must be evaluated in the construction of flat roofs because moisture migration and condensation causes serious problems: moistening of thermal insulation, reduced thermal resistance, destruction of heat insulation structure, moulding due to condensate thawing, damage of rolled cladding structure.

The analysis of the efficiency value of alternative design solutions showed that the reconstruction of not used flat roof (Alternative *4*) is optimal in technical-economic and humidity regime terms. The price of this alternative is not the lowest but the timing of construction work, resistance to external fire and moisture are optimal.

Table 2. Evaluation of alternative design solutions according to the first set of reconstructed flat roofs.

Criteria	**Relative weight, %**	**The value of criteria / Alternatives of roof renovation technologies**				
		1	**2**	**3**	**4**	**5**
C1	0.50	194,48	245,51	242,98	240,99	261,31
C2	0.25	5	4	5	4	5
C3	0.15	3	2	2	3	2
C4	0.10	5	9	8	7	8
C5	0.40	1	2	1	0	1
C6	0.60	5	0	0	2	0

Table 2. Continued.

Criteria	**Efficiency values (0-10)**					**Efficiency in terms of criteria**				
	1	**2**	**3**	**4**	**5**	**1**	**2**	**3**	**4**	**5**
C1	5	10	8	7	9	2.5	5.0	4.0	3.5	4.5
C2	10	9	10	9	10	2.5	2.25	2.5	2.25	2.5
C3	6	4	4	6	4	0.9	0.6	0.6	0.9	0.6
C4	10	6	7	8	7	1.0	0.6	0.7	0.8	0.7
					AMOUNT	6.9	8.45	7.8	7.45	8.3
C5	8	10	8	1	8	3.2	4.0	3.2	0.4	3.2
C6	8	10	10	6	10	4.8	6.0	6.0	3.6	6.0
					AMOUNT	8.0	10.0	9.2	4.0	9.2

References

Bieksa, D., Siupsinskas, G., Martinaitis, V., Jaraminiene, E., Energy efficiency challenges in multi-apartment building renovation in Lithuania: *Journal of Civil Engineering and Management*, 17(4), 467-475, 2011.

Juodis, A., *Mathematical modeling and optimization of construction processes*. Technologija, Kaunas, 2005.

Rapceviciene, D., Evaluation of multi residential house renovation efficiency: *Science – Future of Lithuania,* 2(2), 83-89, 2010.

Recommendations R 26-00, Methods and solutions of repairing flat roof structures, 2000.

Technical regulation of construction STR 2.05.01:2013, Design of the Energy Performance of Buildings, 2013.

Technical regulation of construction STR 2.01.09:2012, Energy efficiency of buildings. Energy efficiency certification, 2012.

Technical regulation of construction STR 2.05.02:2008, Structures of the building. Roofs, 2008.

Technical regulation of construction STR 1.01.08:2002, Types of construction of the building, 2012.

IDENTIFICATION OF BARRIERS THAT HINDER THE EFFECTIVE IMPLEMENTATION OF BUILDING ENERGY CODES

MOHAMMAD DANESH EDALAT[1], MOHAMMAD MEHDI MORTAHEB[2], and HAMED KASHANI[2]

[1]*MSc. Student University, CE Dept, Sharif University, Tehran, Iran*
[2]*Faculty, CE Dept, Sharif University, Tehran, Iran*

In response to the growing energy consumption levels in buildings, many countries have introduced standards and codes aimed at reducing their national building energy consumption and managing the amount of emissions. Title 19 of Iran's National Building Regulation (NBR) was first introduced in 1991. It was projected that Title 19 can lead to a 30 to 50% reduction in building energy consumption in the country. Nevertheless, evidence suggests that the introduction of Title 19 NBR has not led to substantial saving in building energy consumption. Unlike Title 19 NBR, similar standards in other countries have proven to be very effective in reducing the building energy consumption levels. For instance, it is believed that the implementation of Title 24 of California's Building Energy Efficiency Standards has led to $66 billion dollars of energy savings over the past 35 years. The objective of this study is to utilize content analysis and experts' opinion in order to identify and categorize the barriers that hinder the effective implementation of Title 19 NBR in Iran. The contribution of this study to the state of knowledge is the identification of barriers that could hinder the effective implementation of building energy codes.

Keywords: Sustainable construction, Environment impact, Built environment, Construction policy, Standards, Developing countries.

1 INTRODUCTION

The growing problems caused by the use of energy-consuming equipment in the building sector and the increased utilization of electricity and fossil fuels have led to the development of standards and codes that are aimed at optimizing building energy consumption. Increased energy consumption is directly related to the emission of environmental pollutants. These emissions could be reduced through appropriate measures. Building energy efficiency codes are regulatory tools that establish minimum levels of energy efficiency for different building types. They typically cover the design and construction of various building energy systems. They play a fundamental role in achieving energy efficiency objectives, particularly for new construction.

Title 19 of Iran's National Building Regulation (NBR) developed in 1991. It was expected that its implementation can promote energy saving in buildings (National building regulation office 1997). It was projected that Title 19 NBR can lead to a 30 to

50% reduction in the amount of building energy consumption (IFCO 2005). Nevertheless, evidence suggests that the introduction of Title 19 NBR has not led to substantial reduction in building energy consumption. As a result, the average energy consumptions by buildings in Iran are currently 1.9 times the world average (Development Organization of Kashan 2011).

2 BACKGROUND

Unlike Title 19 NBR, similar standards adopted by other countries have proven to be very effective in reducing the building energy consumption levels. China is the second largest energy consumer, after the United States. Buildings account for 28 percent of Chinese energy consumption (DOE 2009). Building energy efficiency issues in China have drawn increasing attention from the government since the mid-1980s (Huang & Joe Deringer 2007). The Ministry of Housing and Urban-Rural Development coordinates and develops China's national building energy codes. It has issued a series of national and industrial codes to promote building energy efficiency. As a result, China has become one of early adopters of national building energy codes (1986) among Asia-Pacific Partnership on Clean Development and Climate countries. In 2004, an ambitious energy conservation plan for Chinese buildings became effective. It required 50% energy conservation for all new buildings. Several major cities such as Beijing and Tianjin shall take a lead in implementing s more strict 65% energy saving standard. Tianjin is one of the most successful Chinese cities in compliance enforcement of building energy efficiency codes with nearly 100 percent compliance. Strong determination of the Chinese government and broad international collaboration are among the key factors contributing to the success of these plans (PNNL 2009).

In United States, see a range of benefits have been realized from the development and implementation of building codes. These benefits include lower energy use, reduced energy costs, reduced pollutant emissions, stronger local economies, improved energy resource reliability and improved health (State Climate and energy program). An analysis by the Department of Energy (DOE) estimates that cumulative energy savings from 1992–2012 were approximately 4.2 quads and cost savings to consumers have been more than $44 billion in USA (PNNL 2014). The use of building energy efficiency code is not just a national endeavor. In 1978 California became the first state in the United States to enact a statewide building energy code (California Title 24 Part 6). Title 24 establishes prescriptive and mandatory guidelines for construction methods, materials, equipment, and controls. These guidelines are used in new construction and major retrofits. It is believed that the implementation of Title 24 of California's Building Energy Efficiency Standards has led to $66 billion dollars of energy savings over the past 35 years. The Energy Commission estimates that energy efficiency investments in California (encouraged by Title 24 building codes and Title 20 appliance standards along with utility efficiency programs) have resulted in an estimated $56 billion in electricity and natural gas savings through 2006. An additional $23 billion in savings anticipated through 2013. These savings have yielded almost $5 billion in net benefits from avoided generation, transmission, and distribution and natural gas usage. More than $900 million in savings has been reported in year 2008. Between 2006 and 2008, utility programs generated savings at a cost of less than $0.03/kWh compared with $0.08/kWh for base-load power (Berkeley National

Laboratory 2012). These savings resulted primarily from DOE supported activities that help upgrade model energy codes, accelerate their adoption by states and localities, and improve code compliance via software tools, training, and technical support (State Climate and energy program).

Therefore, evidence suggest that, compared to similar energy efficiency codes, Title 19 NBR has not been successful in achieving its target saving levels. Research is needed to identify barriers that hinder its effective implementation. The objective of this research is to identify and classify these barriers.

3 METHODOLOGY

To achieve the objective of this research, a comprehensive review of literature was conducted in order to identify the barriers that hinder the effective implementation of Title 19 NBR. The findings of this literature review were verified and enhanced through interviews with several experts. In order to evaluate the impact of each barrier, a survey was distributed among subject matter experts. The responses to questionnaires were compiled and analyzed in order to identify the most prominent barriers.

Through literature studies and numerous interviews with experts, the barriers that reduce the effectiveness of Title 19 NBR, are identified as following:

- Authorities have competing and conflicting interests (Kari 2014)
- Authorities have failed to recognize the need for localized code (Kari 2014)
- Code failed to offer appropriate validation methods (PNNL 2011)
- Authorities have failed to recognize the need to separate the codes for new construction and retrofit (Berkeley National Laboratory 2012)
- Authorities have failed to update and revise the code in a timely fashion (PNNL 2003, Pakravan 2014)
- Municipalities have not participated in the code development process (Eslami 2015)
- Compliance costs are high (IFCO 2005)
- Investors do not consider the implementation of Title 19 as value generating (Pakravan 2014)
- Check lists are complicated (IFCO 2005)
- The code is not accompanied by appropriate software (PNNL 2011, Iwaro & Mwasha 2010)
- Construction materials are sub-standard (Pakravan 2014)
- The current manufacture rating system is ineffective (Pakravan 2014)
- Liability insurance services are inadequate (Pakravan 2014)
- Engineers and inspectors are unfamiliar with technical details (Kari 2014)
- Due diligence is not performed (Eslami 2015)

- Practitioners fail to meet ethical codes and standards (Pakravan 2014)
- Policies imposed by ministries are conflicting (IFCO 2005)
- Penalty system is ineffective (Pakravan 2014)
- Government incentives are inadequate (Pakravan 2014)
- End-users are not willing to pay price premium for buildings that comply with the code (Kari 2014)
- Energy subsidies are high (Eslami 2015)
- A new generation of engineers has failed to thrive (Kari 2014).

Table 1. Barriers hindering the Implementation of Title 19 NBR.

Description of Barriers	Category	Impact
Code failed to offer appropriate validation methods	Adoption	Very High
Authorities have failed to recognize the need to separate the codes for new construction and retrofit	Adoption	
Authorities have failed to update and revise the code in a timely fashion	Adoption	
Municipalities have not participated in the code development process	Adoption	
Investors do not consider the implementation of provided Title 19 as value generating	Implementation	
Check lists are complicated	Implementation	
The code is not accompanied by appropriate software	Implementation	
Construction materials are sub-standard	Implementation	
The current manufacture rating System is ineffective	Implementation	
Due diligence is not performed	Control	
Policies imposed by ministries are Conflicting	External factors	
Penalty system is ineffective	External factors	
End-users are not willing to pay price premium for buildings that comply with the code	External factors	
Energy subsidies are high	External factors	
A new generation of engineers has failed to thrive	External factors	
Authorities have competing and conflicting interests	Adoption	High
Compliance costs are high	Implementation	
Liability insurance services are inadequate	Implementation & Control	
Engineers are unfamiliar with technical details	Control	
Inspectors are unfamiliar with technical details	Control	
Practitioners fail to meet ethical codes and standards	Control	
Government incentives are Inadequate	External factors	
Authorities have failed to recognize the need for localized code	Adoption	Medium

7 RECOMMENDATIONS

Based on a comprehensive content analysis and interviews with experts, a series of strategies that can potentially overcome the abovementioned barriers were identified. The recommended strategies to overcome the barriers to the effective implementation of building energy codes are the followings:

- The content and background calculations of energy codes should be developed considering the capabilities of stakeholders.
- Energy codes should be updated and revised on a regular basis to reflect the changes in technologies and policies.
- The use of modeling and simulation methods in order to calculate and compare the effect of codes and their provisions should be promoted.
- Engineers and inspectors should be trained and familiarized with details of the code on a frequent basis.
- Checklists should be designed with the objective of facilitating effective and efficient monitoring of compliance level.
- An appropriate building rating system should be developed as a supplement to building energy codes.
- An appropriate manufacturer rating system should be developed in order to prevent low quality materials and equipment from entering the market.
- Appropriate regulatory and statutory environment should be created to guarantee the effective implementing and monitoring of building energy codes.
- Ethical standards should be enforced throughout the implementation and monitoring process.
- Providing liability insurance mechanisms should become mandatory as part of the building energy code implementation process.
- Public authorities should offer appropriate incentives to promote the implantation and compliance with regards to building energy codes.
- Public authorities should invest in enhancing the culture of energy conservation.

8 CONCLUSION

Currently, buildings account for about 30% of all energy consumption globally and a significant share of greenhouse gas emissions. In response to the growing energy consumption levels in buildings, many countries have introduced standards and codes aimed at reducing their national building energy consumption and emissions. Title 19 of Iran's National Building Regulation (NBR) was first introduced in 1991 with the aim of promoting energy saving in buildings across the country. Title 19 NBR helps ensure that new buildings use energy efficiently. It is estimated that this code can reduce building energy use by 50% or more compared to buildings designed without energy efficiency in mind. Nevertheless, evidence suggests that so far the introduction of Title

19 NBR has not led to substantial reduction in building energy consumption. The objective of this study was to identify the barriers that hinder the effective implementation of building energy codes such as Title 19 NBR. Using literature review and interviews, with experts, barriers to effective implementation of Title 19 NBR were identified. In addition, through a survey, the level of impact of each barrier was identified. The results of survey suggest that, lack of attention to validation methods, integrate issues of new construction and retrofit, postponing the time of edition, absence of municipality in development process and attitude of investors are the most important barriers in reducing the impact of Title 19 NBR in Iran. A variety of strategy can be implemented by the authorities to promote the adoption and effective implementation of the code.

References

Berkeley National Laboratory, Building Energy-Efficiency Best Practice Policies and Policy Packages, 2012. Retrieved from www.//eaei.lbl.gov/sites/all/files/GBPN_Final.Oct_.2012.pdf on December 8, 2015.

Development organization of Kashan, *The Need for Energy Efficiency in Industry*, 2011.

Eslami, R., Interview conducted on July 5, 2015.

DOE, World Carbon Dioxide Emissions from the Consumption and Flaring of Fossil Fuels, 2009. Retrieved from www.eia.gov/environment/emissions/ghg_report/ghg_carbon.cfm on September 2, 2015.

Huang, J., & Joe Deringer, Energy Efficiency Building Standards in China 2007, 2007. Retrieved from www.asiabusinesscouncil.org/docs/BEE/papers/BEE_Policy_China.pdf on July 18, 2015.

IFCO, *Studies and Performance Planning of Tile 19 NBR*, Iranian Fuel Conservation Company, 2005.

Iwaro, J., & Mwasha, A, A Review of Building Energy Regulation and Policy for Energy Conservation in Developing Countries, *International Journal of Energy Policy*, Elsevier, 38(12), 7744-7755, 2010.

Kari, B., Interview conducted on April 30, 2014.

National Building Regulation Office, *History Books and the Development of National Building Regulations*, Tehran, 1997.

Pakravan, S., Interview conducted on April 29, 2014.

PNNL (Pacific Northwest National Laboratory), Achieving the 30% Goal: Energy & Cost Savings Analysis of ASHRAE Standards 90.1-2010, 2011. Retrieved from www.energycodes.gov/sites/default/files/documents/BECP_Energy_Cost_Savings_STD2010_May2011_v00.pdf on June 2, 2015.

PNNL, Building Energy Codes Program: National Benefits Assessment 1992-2040, 2014. Retrieved fromwww.energycodes.gov/sites/ default/files/ documents/ BenefitsReport_Final_March20142.pdf on August 21, 2015.

PNNL, Country Report on Building Energy Codes in China, 2009. Retrieved from www.eia.doe.gov/emeu/international/carbondioxide.html on Febuary 9, 2015.

PNNL, Understanding Building Energy Codes & Standards, 2003. Retrieved from www.pnl.gov/ main/ publications/external/technical_reports/PNNL-14235.pdf on December 20, 2014.

United States Environmental Protection Agency, Energy Efficiency Policies: Building Code for Energy Efficiency. Retrieved from www3.epa.gov/statelocalclimate/documents/pdf/guide_action_chap4_s3.pdf on September 8, 2015.

CHARACTERISTICS OF MOLTEN SALTS AND RECOMMENDATIONS FOR USE IN SOLAR POWER STATIONS

SAMAAN G. LADKANY, WILLIAM G. CULBRETH, and NATHAN LOYD

HRH College of Engineering, University of Nevada Las Vegas, Las Vegas, NV, USA

Molten salts (MS) in the 580°C range could be used to store excess energy from solar power stations and possibly from nuclear or coal. The energy can be stored up to a week in large containers at elevated temperature to generate eight hours of electricity to be used at night or during peak demand hours. This helps to reduce the fluctuation experienced at thermal solar power stations due to weather conditions. Our research supported by Office of Naval Research (ONR), presents a survey of salts to be used in molten salt technology. The physical characteristics of these salts such as density, melting temperature, viscosity, electric conductivity, surface tension, thermal capacity and cost are discussed. Cost is extremely important given the large volumes of salt required for energy storage at a commercial power station. Formulas are presented showing the amount of salt needed per required megawatts of stored energy depending on the type of salt. The estimated cost and the size of tanks required and the operating temperatures are presented. Recommendations are made regarding the most efficient type of molten salt to use. Commercial thermal solar power stations have been constructed in the US and overseas mainly in Spain for which molten salt is being considered. A field of flat mirrors together with collection towers are used in some designs and parabolic troughs used in others.

Keywords: Commercial electric station, Energy storage, Energy production, Molten salt technology, Solar salts, Thermal solar power.

1 INTRODUCTION

Molten solar salts are a great and effective way to store excess solar energy for future use due to the vast heat storage capacities of solar salts. In order for the solar salts to effectively store heat, the salts must be contained. This is done by storing the solar salts in large insulated tanks in order to keep the molten salts in a closed system.

This project examines the current method of using insulated stainless steel cylindrical shells to store molten salt and presents a preliminary design of real life examples. In addition, this design solution is compared to alternative shell designs that are expected to be more efficient in reducing shell thicknesses and stainless steel using hybrid shell design and shapes other than cylindrical shells.

2 TYPES OF MOLTEN SALTS

There are various kinds of salts, all of which can be melted for use as a molten salt. This report will mostly focus on five salts: sodium nitrate, lithium nitrate, potassium nitrate, sodium chloride, and a mixture of 60% sodium nitrate and 40% potassium nitrate. These salts have been most prominently mentioned in the literature and are being used in experimental thermal sun storage facilities since they are cost effective (Janz 1967). Other salts that can be used in these applications, both alone and in mixture form, include calcium nitrate, potassium chloride, and lithium chloride (Janz 1967).

3 PHYSICAL PROPERTIES OF MOLTEN SALTS

The first aspect of solar salts that must be considered are there physical properties, including melting point, density, viscosity, surface tension, heat capacity and electrical conductance. The density of these solar salts directly affect the loading exhibited by the storage tanks and any piping used. The melting point reflects an approximation of the temperatures these storage tanks will experience, which can be used to determine thermal expansion, ultimate strength and thickness along with heat shielding requirements of the tanks. The viscosity determines the resistance of the molten salt while flowing through any pipes used. Surface tension is the measure of force a liquid exerts on a surface by interacting with the surface. Lastly, the electrical conductance determines the salt's ability to conduct electricity. Table 1 compares the densities and melting points of these various salts.

Table 1. Physical properties of solar salts (Haynes 2012a, Janz *et al.* 1972).

Compound or Mixture	Melting Point (°C)	Density at MP (g/cm3)
Sodium Nitrate – $NaNO_3$	306.5	1.900
Lithium Nitrate – $LiNO_3$	253.0	1.781
Potassium Nitrate – KNO_3	334.0	1.865
Sodium Chloride – NaCl	800.7	1.556
60 % $NaNO_3$ / 40 % KNO_3	225 (approximate)	1.870 (at 625 K)

Comparing the melting points, the 60% sodium nitrate and 40% potassium nitrate mixture has the lowest melting point with an approximate melting point of 225°C (Janz *et al.* 1972). The next lowest melting point is lithium nitrate at 253°C (Haynes 2012a). On the other side of the spectrum, sodium chloride (basic table salt) has the highest melting point considered at 800.7°C (Haynes 2012a). The melting point of a salt is an important consideration for solar salt applications, which means that based on melting point, the best salt, for our applications is the 60% sodium nitrate and 40% potassium nitrate mixture since it has the lowest melting point considered while sodium chloride is the worst salt considered since it has the highest melting point.

Comparing the densities of these salts, the salt with the lowest density considered is sodium chloride with a density of 1.556 g/cm3 (Haynes 2012a). The salt with the next lowest density is lithium nitrate with a density of 1.781 g/cm3 (Haynes 2012a). At the other end, the salt with the highest density considered is sodium nitrate with a density of 1.900 g/cm3 (Haynes 2012a). Unlike melting point, density is not as important of a consideration, especially since the relative difference in densities between these salts is small.

Table 2 compares the viscosities, surface tensions, and electrical conductance of various solar salts.

Table 2. Physical properties of solar salts at melting point (Janz 1967) (Janz *et al.* 1972).

Compound or Mixture	Viscosity (mPa-s)	Surface Tension (mPa)	Electrical Conductance (S/cm)
Sodium Nitrate – $NaNO_3$	3.038	116.35	0.9713
Lithium Nitrate – $LiNO_3$	7.469	115.51	0.3958
Potassium Nitrate – KNO_3	2.965	109.63	0.6324
Sodium Chloride – NaCl	1.459	116.36	0.8709
60 % $NaNO_3$ / 40 % KNO_3	3.172*	121.80 (at 510 K)	0.7448*

Note: Values with a single asterisk (*) have been extrapolated for the 60% $NaNO_3$ mix at 580 K

Comparing the viscosities, the salt with the lowest viscosity is sodium chloride with 1.459 mPa-s (Janz 1967). The next lowest salt is potassium nitrate with 2.965 mPa-s (Janz 1967). Conversely, the salt with the highest viscosity is lithium nitrate with 7.469 mPa-s (Janz 1967). In comparison with other physical properties, considered, viscosity is not the most important property to consider in comparing molten salts. However, it is a property of some importance as the viscosity compares the resistance exerted against the molten salts while flowing through a pipe, which is something the molten salts will have to do in the containment units.

Comparing the surface tension, the salt with the lowest surface tension is potassium nitrate with 109.63 mPa (Janz 1967). The next lowest salt is lithium nitrate with 115.51 mPa (Janz 1967). On the other side, the salt with the highest surface tension is the 60% sodium nitrate and 40% potassium nitrate mixture with 121.80 mPa (Janz *et al.* 1972). In comparison with other properties considered, surface tension is also not one of the most important properties to consider in comparing molten salts to be used in our applications. However, it is a property of some importance because it affects the tanks and piping of the containment units

Comparing the electrical conductance, the salt with the highest electrical conductance is sodium nitrate with 0.9713 S/cm (Janz 1967). The next highest salt is sodium chloride with 0.8709 S/cm (Janz 1967). On the other side, the salt with the lowest electrical conductance is lithium nitrate with 0.3958 S/cm (Janz 1967). Compared to the other physical and thermodynamic properties considered, electrical conductance is a minor consideration when comparing solar salts for energy storage applications.

4 THERMODYNAMIC PROPERTIES OF MOLTEN SALTS

Solar salts are known for their ability to store heat for long periods of time. The heat of fusion measures the required amount of heat needed to convert a substance from a solid state to a liquid state, or simply the amount of heat needed to melt a substance. The specific heat capacity measures a substance's ability to store heat. Lastly, thermal conductivity measure a substance's ability to conduct heat through said substance. All three properties considered are of major importance since these properties compare how the salts conduct and store heat. Table 3 compares the thermodynamic properties of solar salts.

Table 3. Thermodynamic properties of solar salts (Janz 1967) (Cornwell 1970) (Haynes 2012b) (Janz *et al.* 1979).

Compound or Mixture	**Specific Heat Capacity (J/mol/K)**	**Thermal Conductivity (kW/mol/K)**	**Heat of Fusion (kJ/mol)**
Sodium Nitrate – $NaNO_3$	131.8	5.66	15.50
Lithium Nitrate – $LiNO_3$	99.6	5.82	26.70
Potassium Nitrate – KNO_3	115.9	4.31	9.60
Sodium Chloride – NaCl	48.5	8.80	28.16
60 % $NaNO_3$ / 40 % KNO_3	167.4 (at 510 K)	3.80	13.77
Note: Since some values were given in calories in some sources, they were converted into joules for this table (1 cal = 4.184 J or 1 kcal = 4.184 kJ) (IUPAC n.d.).			

Comparing the specific heat capacity, the salt with the highest specific heat capacity is the 60% sodium nitrate and 40% potassium nitrate mixture with 167.4 J/mol/K (Janz *et al.* 1979). The next highest salt is sodium nitrate with 131.8 J/mol/K (Janz 1967). On the other side, the salt with the lowest specific heat capacity is sodium chloride with 48.5 J/mol/K (Janz 1967). Based on this comparison, the best salt to use for energy storage is the 60% sodium nitrate and 40% potassium nitrate mixture since it has the highest heat capacity considered while sodium chloride is the worst salt considered since it has the lowest heat capacity.

Comparing the thermal conductivity, the salt with the highest thermal conductivity is sodium chloride with 8.80 kW/mol/K (Cornwell 1970). The next highest salt is lithium nitrate with 5.82 kW/mol/K (Cornwell 1970). On the other side, the salt with the lowest thermal conductivity is the 60% sodium nitrate and 40% potassium nitrate mixture with 3.80 kW/mol/K (Cornwell 1970).

Comparing the heat of fusion, the salt with the lowest heat of fusion is potassium nitrate with 9.60 kJ/mol (Haynes 2012b). The next lowest salt is the 60% sodium nitrate and 40% potassium nitrate mixture with 13.77 kJ/mol (Janz *et al.* 1979). On the other side, the salt with the highest heat of fusion is sodium chloride with 28.16 kJ/mol (Haynes 2012b). Based on the comparison of salt characteristics presented in Table 1.3, the 60%/40% sodium/potassium nitrates present, for now the most interesting option for molten salt energy storage. However other options will be considered, such as, the addition of Nano silica to the salt mix in order to improve its specific heat capacity by 30% or more.

5 COST OF SOLAR SALTS

Ultimately, compared to the other considered salts, the most promising solar salt to use, so far, in molten salt energy storage, is the 60% Sodium Nitrate and 40% Potassium Nitrate mixture since it compares favorably against other salts in terms of thermodynamic and heating properties, which are the primary factors to consider for use as a solar salt.

However, when considering the use of solar salts, one must consider the costs of various types of salts. Table 4 compares the 60% sodium nitrate and 40% potassium nitrate mixture to various other solar salt substitutes that are available in the marketplace.

Table 4. Costs of solar salts (Kearney *et al.* 2001).

Compound or Mixture	ΔT (°C)	Cost of Salts ($/kg)	Cost of Power ($/kWH)
Hitec XL in 59% Water (42:15:43 Ca:Na:K)	200	1.43	18.20
	200	3.49 (w/o H_2O)	18.20
Hitec (7:53 Na:K: Nitrate, 40 Na Nitrate)	200	0.93	10.70
Solar Salt (60:40 Na:K Nitrate)	200	0.49	5.80
Calcium Nitrate Mixture Dewatered (42:15:43 Ca:Na:K Mixture)	200	1.19	15.20
	150	1.19	20.10
	100	1.19	30.00
Therminal VP-1 (Diphenyl Biphenyl Oxide)	3.96	100.00	57.50

The solar salt mixture (60% NaNO3 – 40% KNO3) is both the least expensive in terms of cost to purchase, which is 49 cents per kilogram, and the costs per kilowatt-hour of power generated, which is $5.80 per kilowatt-hour (IUPAC n.d.). The next best priced mixture in both aspects is the Hitec mixture, which costs 93 cents per kilogram to purchase and has a power cost of kilowatt-hour of $10.70 (IUPAC n.d.). In addition, the mixture the most expensive in both aspects is the Therminal VP-1, which costs $100 per kilogram to purchase and has a power cost of $57.50 per kilowatt-hour (IUPAC n.d.).

6 CORROSION FROM MOLTEN SALTS

In addition to being able to hold large quantities of heat, molten salts can be corrosive. Table 5 shows corrosion properties of stainless steel exposed to various molten salts.

Table 5. Corrosion properties of stainless steel using molten salts (Sohal *et al.* 2010).

Compound or Mixture	Temp (°C)	Corrosion Rate (mm/y) SS 304	SS 316
Sodium Chloride – NaCl	845	7.2	7.2
Hitec Salt	538	0.21	<0.03
	430	------	0.007
	505	------	0.008
	550	------	0.074

The sodium chloride at a temperature of 845°C corrodes both types of stainless steel at 7.2 millimeters per year (Sohal *et al.* 2010). At 538°C, the Hitec Salt corrodes through SS 304 steel at 0.21 millimeters per year, and through the SS 316 steel at less than 0.03 millimeters per year (Sohal *et al.* 2010). In addition, the Hitec Salt corrodes through SS 316 steel 0.007 millimeters per year at 430°C, 0.008 millimeters per year at 505°C, and 0.074 millimeters per year at 550°C (Sohal *et al.* 2010).

7 CONCLUSION

A survey of molten solar salts for use in energy storage shells is presented, to provide electric generation stations with power for eight hours. Tables are shown providing the characteristics of various molten salts to be used in thermal solar energy stations.

Recommendations for the selection of an economical molten salt compound is made using various characteristics, including thermal capacity, availability, melting temperature, and the cost of salts.

References

Cornwell, K., The Thermal Conductivity of Molten Salts, Department of Mechanical Engineering, Heriot-Watt University, 1970. Retrieved from http://iopscience.iop.org/0022-3727/4/3/313/pdf/0022-3727 4 3 313.pdf on Nov. 6, 2014.

Haynes, W., Density of Molten Elements and Representative Salts, CRC Handbook of Chemistry and Physics, 2012. Retrieved from http://www.hbcpnetbase.com//articles/04_07_92.pdf#xml=http://www.hbcpnetbase.com/search/pdfHits.asp?id=04_07_92&DocId=118023&hitCount=10&hits=1984 1980 1969 1965 1386 1382 160 159 141 41 on Nov. 5, 2014.

Haynes, W., Enthalpy of Fusion, CRC Handbook of Chemistry and Physics, 2012. Retrieved from http://www.hbcpnetbase.com//articles/06_26_92.pdf#xml=http://www.hbcpnetbase.com/search/pdfHits.asp?id=06_26_92&DocId=118085&hitCount=4&hits=507 366 365 155 on Nov. 5, 2014.

IUPAC, 1.6 Conversion tables for units, n.d. Retrieved from http://iupac.org/publications/analytical_compendium/Cha01sec6.pdf on Feb. 3, 2015.

Janz, G. J., *Molten Salts Handbook*. Academic Press. New York, NY, 1967.

Janz, G. J., Allen, C. B., Bansal, N. P., Murphy, R. M., and Tomkins, R. P. T., Physical Properties Data Compilations Relevant to Energy Storage. II. Molten Salts: Data on Single and Multi-Component Salt Systems, Molten Salts Data Center, Cogswell Laboratory, Rensselaer Polytechnic Institute, 1979. Retrieved from http://www.nist.gov/data/nsrds/NSRDS-NBS61-II.pdf on Nov. 10, 2014.

Janz, G. J., Krebs, U., Siegenthaler, H. F., and Tompkins, R. P. T. Molten Salts: Volume 3, Nitrates, Nitrites, and Mixtures, Molten Salts Data Center, Department of Chemistry, Rensselaer Polytechnic Institute, 1972. Retrieved from http://www.nist.gov/data/PDFfiles/jpcrd10.pdf on Nov. 18, 2014.

Kearney & Associates and Flabeg Solar International, Engineering Evaluation of a Molten Salt HTF in a Parabolic Trough Solar Field, pg. 7, Nrel.gov, 2001. Retrieved from http://www.nrel.gov/csp/troughnet/pdfs/ulf herrmann salt.pdf on Dec. 13, 2014.

Sohal, M., Ebner, M., Sabharwall, P., and Sharpe, P., Engineering Database of Liquid Salt Thermophysical and Thermochemical Properties, Idaho National Laboratory, 2010. Retrieved from http://www.inl.gov/technicalpublications/Documents/4502650.pdf on Feb. 6, 2015.

Interaction between Theory and Practice in Civil Engineering and Construction
Edited by Komurlu, R., Gurgun, A. P., Singh, A., and Yazdani, S.

MULTI-CRITERIA DECISION ANALYSIS OF UP-TO-DATE CONSTRUCTION TECHNOLOGY

RUTA MINIOTAITE

Dept of Civil Engineering Technologies, Kaunas University of Technology, Kaunas, Lithuania

At the very moment various construction and projection firms offer to use all kinds of new technologies and its systems, structural decisions, proper materials used as well as work implementation methods, without paying any attention to the factors, effecting the choice of selecting the most effective technology of construction process. Having in mind that funds should be realized rationally, it is the matter of vital importance to measure the practical adaptability of technologies being used either they are new or have been used for many years to achieve the best technological decisions. In order to achieve the efficiency of the funds invested and diminishing construction duration as well as increasing its quality, various construction process technologies are developed and improved by using modern technical and informational tools, which are integrated into the stages of projection and construction. The paper analyzes the methods of multi-criteria alternative technological solution evaluation in construction environment. Methodology, which is used, allows to complex evaluate the efficiency of construction designed decisions at the stage of preparation phase. There has been done a practical technological modeling of installation process of concrete floor as well as there have been determined optimal solutions. In order to achieve the results mentioned, a method of proximity to an ideal point was used.

Keywords: Optimization, Method of proximity to an ideal point, Construction technology, Algorithm.

1 INTRODUCTION

The prospective technological processes' decisions and operations, their efficiency and competitiveness are programmed at the stage of construction objects' projection. In order to achieve the efficiency of construction technology it is advisable to use the achievements of fundamentals and applied science to solve the main tasks in various spheres of construction decision modeling and optimization.

While modeling and projecting construction technologies, it is advisable to perform the multi-criteria analysis of technological processes and decisions on the grounds to fully coordinate the goals of the interested parties (customer, projector and contractor). That is why major principles and methods of construction decision technology multi-criteria optimization as well as questions associated with alternative decision technologies and mathematical modeling are analyzed in this paper. As for example, which provides the practical technological modeling of installation process of concrete floor and determining the optimal decisions based on the method of proximity to an ideal point.

2 MAJOR PRINCIPLES OF CONSTRUCTION TECHNOLOGICAL DECISION OPTIMIZATION

Applicable technological and other project decision optimization methods used in projection and construction processes could be divided into two major groups: applied mathematics and systemic-technical analysis methods. A great deal of construction organization tasks could be solved by using mathematic statistics, theory of chances, mathematic programming, 'gambling' theory, multi-criteria optimization and other methods. The selection of optimization method depends on the task character which is solved, the possessed source information and frequently it requires local interpretation. While solving practical construction optimization tasks, most frequently only one of several economic criteria is chosen. The significance of the criterion selected is very important. It shows that one of the criterions mentioned, which is selected by the interested party, is much more important than the other one, for example, total construction price. So to be concrete it shows the significance of one criterion to the interested party in comparison to the rest, which are left as less significant or insignificant. The significance of the criterion could be determined by employing statistical, expert opinion based methods, even comparison, and entropy methods. These methods help to determine various theoretical, subjective, and complex values of great significance that are further used in the decision optimization counting processes. By counting values it is meant that the interested party could choose the criterion of the greatest significance. So any construction technological decisions could be described and optimized according to the following system of criteria evaluation, where the criteria could be expressed by the indices of technological economy and quality characteristics. For this purpose methods of multi-criteria decisions are used (Zavadskas and Kaklauskas 1996, Janusaitis 1998).

We can definitely state that each of the decision optimization method mentioned has got its own advantages and disadvantages. Moreover, each of them could be used to solve the tasks of specific constructions groups. They help to create various optimization models of theoretical objects, technological or work processes like technological net models, mathematic models, expert level systems, decision support systems and many others models.

Researches of modeling and optimization were founded on the basis of applied mathematics method, economics, system theory, cybernetics, and in the sphere of counting technological science and its integration. While optimizing technological construction processes, it is advisable to apply the theoretical principles of system methodology on the grounds that in nowadays the technological projection methods which are used do not correspond to the requirements of effective decision making. It could be noted that one of the major disadvantages are the decisions accepted synonymous, without any preliminary examination and evaluation of the model of construction process technology and many the like multi-criteria evaluation (Zavadskas *et al.* 1999, Rapceviciene 2010). While solving the any technological decision optimization problem, it is necessary to perform 3 major steps of construction process systemic examination (Figure 1).

While modeling the construction composite process technological decisions, it is advisable to accept these main preconditions (Zavadskas *et al.* 1999, Sarka *et al.* 1999):

(1) All possible variations of complex process technologic decisions have to be constructed. Moreover, technological connections and partial variants of the processes (partial alternative decisions) have to be established;

(2) While forming the net model, consisting of partial process technological variants, it is necessary to take a precondition that only one of many other partial process technological variants will be implemented;

(3) Every partial process has got its individual time duration, which is either technologically based or depends on the work expenditure.

So while creating the mathematical model of alternative technological decision making, it is advisable to define the set of the compared alternative decisions and their evaluation criteria. In that case, the source data matrix P is prepared and most often consists of different units of measurement.

That is why the matrix should be normalized, i.e., it has to be transformed into the anti-dimensioned unit or sizes. Knowing the aims of the solution as well as applying the methods of normalization various normalized values of indices are obtained, which play the key role in other stages of solution multi-criteria optimization.

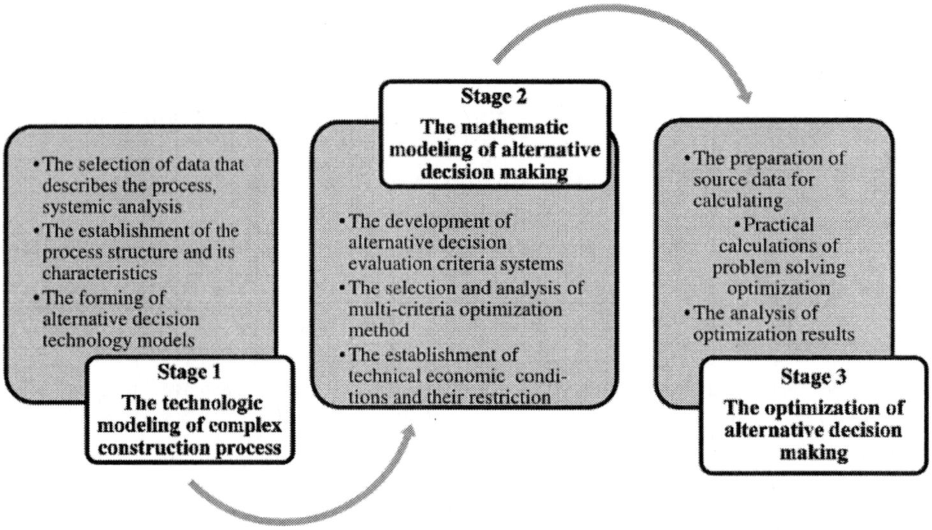

Figure 1. The steps of construction composite process systemic examination.

3 THE MAJOR RUDIMENTS OF APPLYING THE METHOD OF PROXIMITY TO AN IDEAL POINT, USED TO EVALUATE THE TECHNOLOGY

The main essence of the multi-criteria evaluation method is formation of generalized composed criterion. It is based on the comparison deviation of the criteria from so called the ideal criteria, consisting of the best variant criteria being analyzed. By applying the method and K_{bit} criteria, it is advisable to take into account that each variant of the task problem solving utility function has got the tendency to monotonously increase or monotonously decrease, i.e. the larger value of any indices, the better it is or worse for less of the same index value. It depends on the fact whether the utility function increases or decreases. Indices have to be either cardinal or ordinal. If we have got the ordinal indices, they should be quantified. Besides, significance values should be determined, otherwise, they all are accepted as being equals. The application algorithm of the method of proximity to an ideal point, estimating the significance of the criteria, is presented in Figure 2.

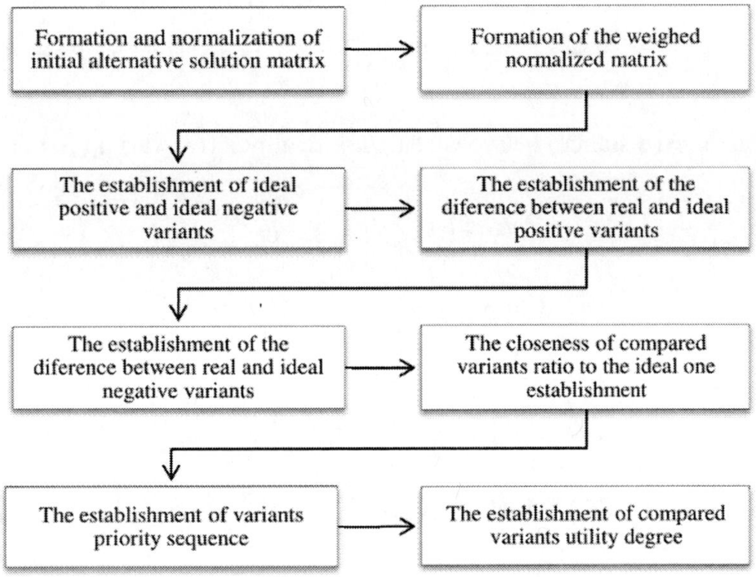

Figure 2. Application algorithm of the method of proximity to an ideal point.

The matrix P of alternative architectural decisions is created. There could also be criteria either grouped or ungrouped. The matrix normalization is being dome according to the formula:

$$\overline{x_{ij}} = \frac{x_{ij}}{\sqrt{\sum_{i=1}^{m} x_{ij}^2}}, \qquad \text{where } i=\overline{1,m};\ j=\overline{1,n} \qquad (1)$$

If the significance of the subjective or theoretical ($\overline{q}$ or q_t) criteria is known, then the vector column multiplied by the normalized matrix corresponding column.

$$\text{We get weighed matrix } \overline{P^*} = [\overline{P}]\cdot[q] \qquad (2)$$

If there are no values of significance, then $\overline{P} = \overline{P^*}$ ($\overline{P}$ matrix is compared to the weighed matrix), i.e. we take the precondition that entire alternative solution criteria are equally important. The ideal positive variant is being established:

$$a^+ = \left\{ \left[\left(\max_i f_{ij} / j \in I \right), \left(\min_j f_{ij} / j \in I' \right) \right] / i = \overline{1,m} \right\} = \left\{ f_1^+, f_2^+, ..., f_n^+ \right\} \quad (3)$$

where I – indices of ratio (maximizing), which possess the highest values.

The ideal negative variant is being established:

$$a^- = \left\{ \left[\left(\min_i f_{ij} / j \in I \right), \left(\max_j f_{ij} / j \in I' \right) \right] / i = \overline{1,m} \right\} = \left\{ f_1^-, f_2^-, ..., f_n^- \right\} \quad (4)$$

The difference (distance) between real and ideal positive variant is being found:

$$L_i^+ = \sqrt{\sum_{j=1}^{n} \left(f_{ij} - f_j^+ \right)^2} \quad (5)$$

where a_i – real variant; a^+ - ideal positive variant; L_i^+ - positive distance.

The difference between real and ideal negative variant is being found:

$$L_i^- = \sqrt{\sum_{j=1}^{n} \left(f_{ij} - f_j^- \right)^2} \quad (6)$$

$K_{bit,\,i}$ calculation of values (each alternative value is found):

$$K_{bit,i} = \frac{L_i^-}{L_i^+ + L_i^-}, \text{ when } \forall_i ; i = \overline{1,m} \quad (7)$$

$0 \le K_{bit} \le 1$, besides,

$$K_{bit,i} = \begin{Bmatrix} 1, jei & a_i = a^+ \\ 0, jei & a_i = a^- \end{Bmatrix} \quad (8)$$

The best (the most rational) architectural solution will become the one, which K_{bit} value will be max ($K_{bit,i}$=max). Using the values we form the priority sequence utility degree establishment. We compare the value of the variants: examined with the ideal.

$$N_i = \frac{K_{bit,i}}{K_{bit,\max}} \cdot 100\% \quad (9)$$

4 THE ESTABLISHMENT OF THE OPTIMAL FLOOR COVER EQUIPMENT TECHNOLOGY, APPLYING THE METHOD OF PROXIMITY TO AN IDEAL POINT

The change appeared in technological systems affect not only the rest systems parts or elements, but also the expected final result. This is why it is of vital importance to deliver systemic structuralization of the construction technological processes.

The installation of the floor cover is only a partial installation process, which could be analyzed in the complex manner. The complex process of the floor cover equipment maximally can form 7 partial processes. Each partial process possesses the original structure, which is distinctive from the rest.

While establishing the combinations of partial processes, the technological combination sequence of partial processes is found. It is important to establish the combinations of technological connections among the separate partial processes, because it could happen so that the variant of the partial process does not possess the connectivity with the other variant of the partial process, i.e. one of them could have no technological connection with another.

Evaluation criteria system: *K1* – the price of floor installation variant; *K2* – work costs; *K3* – the level of work mechanization in percentage, %; *K4* – the floors' resistance to wear and tear; *K5* – the chemical resistance of the floor cover, in points; *K6* – the beginning of using the floor cover, in days; *K7* – comfortability, in points; *K8* – ecology conditions, in points; *K9* – aesthetic view, in points; *K10* – hygiene, in points.

Subjective criteria significance are determined by the even comparison method. While performing the calculation of criteria significance on the basis of even comparison method, we received the significance of subjective criteria. The source matrix is formed in the following stage.

While performing the normalization of the matrix, and evaluating the significance of the criteria, as well as forming weighed normalized matrix, we received calculation results.

5 CONCLUSIONS

At the very moment various construction and projection firms offer to use all kinds of new technological projecting systems as well as work execution technologies, without paying any attention to the circumstances, affecting the choice of selecting the most effective technology of construction process. In order to rationally employ the technological process of the funds, it is advisable to perform the modeling and multi-criteria evaluation of the alternative solution technology.

To perform the selection of the most suitable floor cover installation variant, the algorithm of solution optimization was formed a well as the criteria evaluation system, which describes the goals of interested parties striking for the aim in projection, work execution and exploitation processes.

References

Janusaitis, R., The evaluation of wall heating alternative solutions, applying the method of technological net model of the "component cutting" method, *Civil engineering*, IV, 2, Technika, Vilnius,161-170, 1998.

Rapceviciene, D., Evaluation of multi residential house renovation efficiency: *Science – Future of Lithuania,* 2(2), 83-89, 2010.

Sarka, V., Ustinovicius, E. K., and Zavadskas, E. K., Synthesis of the projects applying the compromised compensating models in construction, *Civil engineering,* V, 6, Technika, Vilnius, 374-385, 1999.

Zavadskas, E. K., Kaklauskas, A., *Construction systemic technical evaluation,* Technika, Vilnius, 1996.

Zavadskas, E. K., Simanauskas, L., and Kaklauskas, A., *Solution support in system construction.* Technika, Vilnius, 1999.

Interaction between Theory and Practice in Civil Engineering and Construction
Edited by Komurlu, R., Gurgun, A. P., Singh, A., and Yazdani, S.

A COMPARATIVE STUDY BETWEEN CHINA AND USA ON THE INFLUENCING PATH OF TRANSACTION COSTS IN CONSTRUCTION PROJECTS

HUIMIN LI[1] and CHENGYI ZHANG[2]

[1]North China University of Water Resources and Electric Power, Zhengzhou, China
[2]Eastern Kentucky University, Richmond, Kentucky, United States

The actual cost of a construction project is composed of not only production costs, but also transaction costs. The presented model includes the uncertainty of the owner's behavior, the uncertainty of the contractor's behavior, the uncertainty in the transaction environment and mechanism, and project management efficiency, to illustrate the influence path of the transaction costs borne by the owner. The model is tested by the structural equation modeling (SEM) using the data collected from construction project owners. The finding indicate that uncertainty in the transaction environment and mechanism appears to be the core construct of the model due to its strong linkages with transaction costs and the owner's behavior, the contractor's behavior, and project management efficiency. The uncertainty of the contractor's behavior is also found to affect project management efficiency positively. A certainty behavior on the part of the owner reduces the uncertainty in the transaction environment and mechanism, increases the efficiency of project management, and reduces the transaction costs of construction project.

Keywords: Construction project, Transaction cost, Influence path, Structural equation modeling.

1 INTRODUCTION

In transaction cost economics, a transaction occurs when a good or service is transferred across a technologically separable interface (Williamson 1987). The transaction cost economics approach provides a useful framework for analyzing the inevitable differences in interest between different contracting parties who are members of the project coalition (Winch 1989). Transaction costs are different from production costs; while production costs are the costs of transforming inputs into outputs, transaction costs arise from economic exchange. The majority of these studies on transaction costs in construction projects have focused on the theoretical and qualitative aspects of this issue. Williamson (1985) argue that transaction costs include the costs of drafting, negotiating and enforcing an agreement, and the costs of governance and bonding to secure commitments. The generation of transaction cost has its path dependence. Not only the amount of transaction cost of construction project is different under different institutional environment, but also the influence path to the transaction cost is different. We first present a review of transaction cost literature as the theoretical background of the study. Then, we discuss a research model and corresponding research hypotheses. After presenting the research method adopted to test our research hypotheses, we report cross-country data analyses and results. The paper concludes with a discussion of findings and implications for theory and practice.

2 TRANSACTION COSTS IN CONSTRUCTION PROJECTS

Hughes *et al.* (2006) classify transaction costs by project phase, namely pre-tendering costs (marketing, forming alliances, and establishing reputations), tendering costs (estimating, bidding, and negotiating) and post-tendering costs (monitoring performance, enforcement of contractual obligations, dispute resolution). In this study, pre-contract transaction costs include the cost of market research, the cost of exploring financing opportunities, the cost of conducting a feasibility study, the cost of bidding/ negotiation, and the cost of day-to-day pre-contract project management. Post-contract transaction costs include the cost of day-to-day contract administration, the cost of administering claims and change orders, the cost of dispute resolution, and incentive payments. Respondents are asked to estimate the approximate cost of pre-contract and post-contract transaction costs with respect to contract value in the last project they completed for their company/agency.

3 RESEARCH HYPOTHESES

In this study, the latent variables that determine transaction costs are considered to be the uncertainty of the owner's behavior, the uncertainty of the contractor's behavior, the uncertainty in the transaction environment, and project management efficiency. Four hypotheses are put forward from four factors that affect transaction costs. On basis of the four hypotheses, and the relationship between them, another five hypotheses are put forward. Finally, there are nine hypotheses in total.

H1: The uncertainty of owner's behavior have a positive effect on the transaction costs; H2: The uncertainty of contractor's behavior have a positive effect on the transaction costs; H3: The efficiency of project management have a negative effect on the project transaction costs; H4: The uncertainty of trading environment and mechanism have a positive effect on the project transaction costs; H5: The uncertainty of owner's behavior have a positive effect on the transaction environment and mechanism; H6: The uncertainty of owner's behavior have a negative effect on the efficiency of project management; H7: The uncertainty of transaction environment and mechanism have a negative effect on the efficiency of project management; H8: The uncertainty of transaction environment and mechanism have a positive effect on the uncertainty of contractor's behavior; H9: The uncertainty of contractor's behavior have a negative effect on the efficiency of project management;

4 RESEARCH METHOD

First, the sample testing of questionnaire was carried out in a small scale. After adjustment, the questionnaire was distributed to the owners who were responsible for construction managing, including public enterprises and institutions, and real estate development companies. Respondents were asked to evaluate the previous completed project of their organizations, using Likert Scale 1-5. Samples of China were mainly from the owners of public enterprises, and real estate development companies. Samples of America were mainly from the Construction Owners Association of America (COAA), Federal Highway Administration (FHWA), American Association of State Highway and Transportation Officials Members (AASHTO), and the list of the top owners in 2009 from Engineering News-Record (ENR). Respondents were asked to evaluate the previous completed project of their organizations, using five Likert Scale to answer. From March 2011 to April 2011, 502 Chinese questionnaires were sent out, and 108 completed responses were returned. The rate of response is 21.6%. 2628 English questionnaires were sent out, and 239 completed responses were returned. The rate of response is 9.09%.

4.1 Comparison of Measurement Models

The data from China and the United States were input into the measurement model, and the indicator of reliability test for each latent variable was received. As shown in Table 1, the Cronbach's α of each latent variable is above 0.7 in samples from China and America. The most standardized loading of factors are above 0.7, and the values of construct reliability and composite reliability are both above 0.6. The factor loads of each latent variable are greater than the acceptance criteria 0.5. The amount of average variance extracted (AVE) of each latent variable are above 0.5, which indicates that the explanatory power of measurement index exceeds the error variance, and the measurement of each constructed variable has sufficient convergent validity.

From data of both China and the United States, as shown in Table 2, all the indices of the measurement model are meet the requirement. The fitting effect is well. This model is suitable for the study of the influencing path of transaction costs in China-US construction projects.

Table 1. The measurement model—confirmatory factor analysis.

Latent variables	**Cronbach's alpha (α)**		**Average variance extracted**		**Composite reliability**	
	China	**USA**	**China**	**USA**	**China**	**USA**
The uncertainty of owner's behavior	0.701	0.714	0.650	0.640	0.944	0.898
The uncertainty of contractor 's behavior	0.724	0.786	0.577	0.646	0.844	0.927
The efficiency of project management	0.917	0.921	0.585	0.707	0.873	0.922
The uncertainty of transaction environment and mechanism	0.720	0.810	0.65	0.579	0.944	0.925
The uncertainty of transaction environment and mechanism	0.720	0.810	0.65	0.579	0.944	0.925
Transaction costs	0.728	0.811	0.714	0.757	0.833	0.862

Table 2: The measurement model—model fit indices.

Goodness of fit indices	China	US
Chi-square (χ^2)	788.8	876.5
Degree of freedom (df)	340	340
χ^2/df	2.320	2.580
Root mean square residual (RMR)	0.048	0.042
Goodness-of-fit index (GFI)	0.903	0.908
Adjusted goodness-of-fit index (AGFI)	0.854	0.900
Parsimonious goodness-of-fit index (PGFI)	0.610	0.620
Comparative fit index (CFI)	0.901	0.932
(Root mean squared error (RMSEA)	0.070	0.072

4.2 Comparison of Structural Models

The data from China and the United States are input into the structural model, the received fitting indexes are shown in Table 3. All of the indices meet the requirement of the model fit. From the above fit indexes, the fitting effect is better, and the model can be accepted. Thus, it can be concluded that data from China and the United States are suitable for this model.

Table 3. The structural model—model fit indices.

Goodness of fit indices	China	US
Chi-square (χ^2)	592.47	740.7
Degree of freedom (df)	331	331
χ^2/df	1.790	2.238
Root mean square residual (RMR)	0.049	0.032
Goodness-of-fit index (GFI)	0.889	0.901
Adjusted goodness-of-fit index (AGFI)	0.805	0.842
Parsimonious goodness-of-fit index (PGFI)	0.673	0.681
Comparative fit index (CFI)	0.935	0.921
Root mean squared error (RMSEA)	0.063	0.071

As seen from the comparison on influencing path of transaction costs in China and US construction projects in Table 4, we found support for all hypotheses except for the path from the uncertainty of owner's behavior to transaction cost (H1), which was not supported by the US sample. It has indirect effect on the transaction cost through the project management efficiency and the uncertainty of project transaction environment. While the other eight hypotheses have been validated. Conclusions about the relationship between latent variables are also similar to Chinese samples. There are different coefficients between Chinese and US samples, is it caused by the different systems of the two countries? There is need to make a multi-group comparison.

4.3 Multi-Group Comparison

In order to test whether the differences in transaction systems of construction project between the two countries have a significant impact on the path coefficients of the model. The multi-group comparison model of the structural equation model is employed. Presupposing a benchmark model A, and model B (assuming the path coefficients of H1 between China and the United States are equal, i.e. $W_C=W_M$) is the one that fixed parameter limit is more than model A, hereby assuming the path coefficients of H2-H9 respectively, with each model comparing with model A, and taking model B as an example, the null hypothesis and the alternative hypothesis under test are:

Null hypothesis: Model A= Model B

Alternative hypothesis: Model A $\neq$Model B

As in Table 5, after being tested, only path hypotheses of H1, H2 and H7 are significant, while H3, H4, H5, H6, H8 and H9 are not significant. It indicates that the influence of the uncertainty of owner's behavior on the transaction cost in China is more than in the US; while the influence of the uncertainty of contractor's behavior on the transaction cost in the US is more than in China. The influence of the uncertainty of the transaction environment on the project management efficiency in China is more than in the US.

Table 4. Comparison of the path coefficient of China and the United States.

Hypotheses	Path coefficient(China)	Supported	Path coefficient(US)	Supported
H1	0.21	Y	0.07	N
H2	0.25	Y	0.42	Y
H3	-0.32	Y	-0.29	Y
H4	0.23	Y	0.35	Y
H5	0.36	Y	0.21	Y
H6	-0.13	Y	-0.14	Y
H7	-0.56	Y	-0.48	Y
H8	0.42	Y	0.25	Y
H9	-0.17	Y	-0.18	Y

Table5. Hypothesis Testing—Mutil-Group Comparison.

Hypothesis	**DF**	**CMIN**	**P**	**NFI Delta-1**	**IFI Delta-2**	**RFI rho-1**	**TLI rho-2**	**Sig.**
H1	1	2.100	0.024	0.000	0.000	-0.001	-0.001	Sig.
H2	1	3.982	0.046	0.001	0.001	0.000	0.001	Sig.
H3	1	0.002	0.967	0.000	0.000	-0.001	-0.001	No
H4	1	1.518	0.218	0.000	0.001	0.000	0.000	No
H5	1	1.523	0.217	0.000	0.001	0.000	0.000	No
H6	1	3.168	0.075	0.001	0.001	0.000	0.000	No
H7	1	15.133	0.000	0.004	0.005	0.004	0.005	Sig.
H8	1	3.132	0.077	0.001	0.001	0.000	0.000	No
H9	1	3.118	0.077	0.001	0.001	0.000	0.000	No

5 CONCLUSION

This study starts from the determinants that influence the project transaction costs: the uncertainty of owner's behavior, the uncertainty of contractor's behavior, the uncertainty of transaction environment and mechanism and the project management efficiency, which construct nine hypothetical models of the influencing path of the construction project transaction cost. From the comparative analysis on the influencing path model of the transaction cost in China-US construction projects, it can be found that the model is suitable for data from both China and the US. In addition, each indicator of the measurement model of American data meets the requirements, while only H1 is not supported in the structural model, which indicates that the uncertainty of owner's behavior in the United States has no direct influence on the transaction cost of construction projects. It is mainly because the US project management market is more developed, if the owner has little engineering experience, he can solve the problem by hiring a professional project management company in the market for the project management contracting, so, the uncertainty of the owner behavior has little effect on the transaction costs. But whether in the United States or China, the same is that the transaction cost in the project transaction process can be reduced by reducing the uncertainty of contractor behavior and the uncertainty of trading environment and mechanism, and it can also be reached by increasing the efficiency of project management. The cohort analysis is carried out in the model, it can be found that the influence of the uncertainty of owner behavior on the transaction cost in China is more than in the US; while the influence of the uncertainty of contractor behavior on the transaction cost in the US is more than in China; the influence of the uncertainty of the trading environment on project management efficiency in China is more than in the US.

References

Hughes, W., Hillebrandt, P., Greenwood, D. and Kwawu, W. Procurement in the Construction Industry: the impact and cost of alternative market and supply processes. Taylor & Francis, London and New York, 2006.

Williamson, O. E. The Economic Institutions of Capitalism, The Free Press, New York, 1987.

Williamson, O. E. The Economic Institutions of Capitalism: Firms, Markets, Relational Contracting, Free Press, New York. 1985.

Winch, G. "The construction firm and the construction project: a transaction cost approach." Constr. Manage. Econom, 7(4), 331-345, 1989.

Interaction between Theory and Practice in Civil Engineering and Construction
Edited by Komurlu, R., Gurgun, A. P., Singh, A., and Yazdani, S.

DIFFERENCES IN CONSTRUCTION COMPANIES' PREFERENCES IN MAKING BID/NO BID DECISIONS

DUYGU AKALP and MEHMET EGEMEN OZBEK

Dept of Construction Management, Colorado State University, Fort Collins, USA

In today's competitive business environment, bid/no bid decision is crucial for a construction company, as poorly made bidding decisions can have bad consequences. Given the fact that longevity of construction companies partly depends on how they make bidding decisions, determining the key factors for bid/no bid decisions has been the topic of research since mid-1950s. However, there is not much literature on how much value/weight different construction companies put on those factors, and more specifically if there are differences between different companies' valuation of those factors. The purpose of this study is to investigate how differently the construction companies in the United States (US) value the key factors that are commonly utilized to make bid/no bid decisions. For this purpose, first, 14 key factors were determined based on the literature. Then, a survey instrument was sent to the US construction companies with different demographics such as size, revenue, sector, etc. to gather data to determine the values/weights they assign to those factors. For this, Analytic Hierarchy Process (AHP) was used. AHP enabled pairwise comparisons to be made between the key factors that affect bidding decisions to deal with the complex relationships among those key factors. As a result of the AHP process, the weights for the key factors for bid/no bid decisions were determined for each company. The AHP results were combined with the demographic data of the construction companies and Kruskal-Wallis test was conducted to identify whether there is a statistically significant difference between the weight of importance of the key factors based on the contractor type, contractor sector and contractor size. The preliminary results showed that there is a statistically significant difference between the general contractors and subcontractors for weighing the importance of the owner identity key factor. Another significant difference was found within the contractor sector in the determination of the weight of importance of the project type key factor.

Keywords: Bidding, Key factors, Multi criteria decision-making, Analytic hierarchy process, AHP, Kruskal-Wallis test.

1 INTRODUCTION

In today's competitive business environment, bid/no bid decision is complex and crucial for construction companies. It is complex due to the consideration of many intangible and tangible factors in the decision-making process (Mohanty 1992). It is crucial because poorly made bidding decisions can have bad consequences. For example, not bidding a favorable project can result in a lost opportunity for a company to make profit and establish a potential long-term relationship with a new client. On the other hand, bidding a project that actually does not fit the company's priorities may result in wasted time/effort (Ahmad 1990, Wanous *et al.* 2003). Moreover, the reputation of a company can be damaged by submitting many non-winning proposals and thus sometimes giving a "no bid" decision could be the right thing for companies (Gido & Clements 2009).

Given the fact that longevity of construction companies partly depends on how they make bidding decisions, determining the key factors for bid/no bid decisions has been the topic of research. However, there is not much literature on how much value/weight different construction companies put on those factors, and more specifically if there are differences between different companies' valuation of those factors. The overall purpose of this study is to investigate how differently the construction companies in the United States (US) value the key factors that are commonly utilized to make bid/no bid decisions.

2 BACKGROUND

The most influential factors for bidding decisions and bidding strategies have been extensively discussed in the literature since mid-1950s (Harris *et al.* 2006). Various key factors have been identified and ranked and numerous decision-making models have been developed to minimize the risks of complex comparison process.

The first known model was proposed by Friedman (1956), which investigated the issues related to the probability of winning and estimating the optimum bid amount by using probabilistic approaches. Since then, several bid/no bid decision support models have been introduced in the literature based on Friedman (1956)'s study (King & Mercer 1987, Whittaker 1981). Ahmad & Minkarah (1988) discussed inapplicability of the probability-based models by asserting the heuristic nature of the bidding environment. The authors conducted a survey among 400 general contrators in the USA and determined 31 factors that affect decision making process. The top three factors were listed as "Type of job", "Need for work" and "Owner". Since then, most of the research has been based on the factors determined in Ahmad & Minkarah's study.

Shash (1993) modified the questionnaire by Ahmad & Minkarah (1988) and identified 55 factors affecting decision making process. Wanous, Boussabaine, & Lewis (2003) implemented a model by using artificial neural network (ANN) based on 157 real-life projects from Syrian construction companies in which 18 key factors were determined. Jarkas, Mubarak, & Kadri (2014) identified 43 factors based on the literature review and a survey conducted in the State of Qatar. Shash (1998) studied subcontractors' bidding decisions. Lowe & Parvar (2004) determined 21 factors based on the literature review and conducted correlation analysis between the factors and decisions to bid. Based on the results, a significant positive linear correlation was found for eight key factors and the contractors' decisions to bid on a project.

In Bageis & Fortune's (2009) study, 87 factors were identified and statistical approaches were used to determine the interrelations between contractor characteristics and the bidding decisions. Egemen & Mohamed (2007) investigated the factors that affect bidding and mark-up decisions of the 80 Northern Cyprus and Turkish construction firms. For the final model, 50 and 44 factors were included into the framework, respectively. The results showed that bidding and mark-up decisions of the small and medium sized companies were significantly different.

3 RESEARCH METHODOLOGY

This study identifies the differences in the contractors' valuation of the bid/no bid decision-making key factors by taking into the consideration the contractor type, contractor sector, and contractor size. For this purpose, first, 14 most-commonly identified and utilized factors were determined based on a comprehensive literature review and grouped under two as firm-related and project-related factors as shown in Table 1.

Table 1. The key factors that affect bid/no bid decision as determined from the literature review.

Firm Related (Internal) Factors	Project Related (External) Factors
1) Current workload	8) Project size
2) Experience in similar projects	9) Project duration
3) Availability of equipment, materials and human resources	10) Location of the project
4) Financial ability	11) Project type
5) Need for work	12) Contract conditions and type of contract
6) Technical knowhow	13) Owner identity
7) Compliance with the business plan	14) Competition

Then, to identify the weights of importance of (i.e., value given to) each of these factors, Analytic Hierarchy Process (AHP) was conducted. AHP is a multi-criteria decision-making method that utilizes pairwise comparison technique by providing a preference scale. It determines the relative importance of the factors based on the subjective preferences of the respondents (Saaty & Vargas 1991). Including the 14 sub-factors and 2 main factors, a pairwise comparison table in Excel format was created resulting in 43 pairwise comparisons. The respondents were asked to identify which factor is more important than the other and how much more important that factor is over the other one by indicating absolute numbers provided in the AHP comparison scale. The pairwise comparison tool was sent to approximately 900 construction professionals (includes different individuals from the same companies) who have a relationship with the Department of Construction Management at Colorado State University and consequently 48 responses were received. The demographics of the companies were also collected to be used in statistical analyses. The results of the one participant who did not provide demographic information and six participants whose consistency ratio (see Saaty & Vargas (1991) for an explanation of consistency ratio) highly exceeded the acceptable limit of 0.10 were eliminated; therefore 41 responses were included in the study.

Table 2. The revenue range and the corresponding company size based on the responses received.

Revenue	Size
<=$39,500,000	Small Size Construction Company
$39,500,000< <=$125,500,000	Small-Medium Size Construction Company
$125,500,000< <=487,500,000	Medium-Large Size Construction Company
>487,500,000	Large Size Construction Company

The weights of the key factors were estimated for each respondent using the AHP methodology as outlined in Saaty & Vargas (1991). Furthermore, the demographic data was sorted based on contractor type (general contractor vs. subcontractor), contractor primary sector (Residential, Commercial, Industrial, and Heavy/Highway), and contractor size which was determined based on the quartiles of revenue (see Table 2).

To determine whether the differences in the identified weights of the bid/no bid decision-making key factors (by taking into the consideration the contractor type, contractor sector, and contractor size) are statistically significant; several statistical approaches were utilized. By considering the non-normality of the collected data, Kruskal-Wallis test which is a non-parametric approach of the one-way Anova test was conducted (Stokes *et al.* 2000). To meet the independency assumption of the Kruskal-Wallis test, each factor was investigated separately. The analyses were conducted with a statistical computing program, SAS (SAS 2015).

4 RESULTS

The hypotheses were constructed to test whether the given importance to the factors by various groups of contractors are significantly different or not. For this purpose, 14 different analyses were conducted for each demographic classification (i.e., Contractor Type, Contractor Sector and Contractor Size). The significance level was set at 0.05 and estimated p-values were compared to that. The estimated p-values as a result of implementing the Kruskal-Wallis test are provided in Table 3.

Table 3. Estimated p-values based on the analysis between the weights of the key factors and contractor classification.

Key Factor\ Contractor classification	Contractor Type	Contractor Sector	Contractor Size
Current workload	0.3513	0.0654	0.3089
Experience in similar projects	0.4449	0.3777	0.7436
Availability of equipment, materials and human resources	0.5681	**0.0124**	0.1595
Financial ability	0.9532	0.1171	0.0583
Need for work	0.5231	0.2155	0.4417
Technical knowhow	0.8839	**0.0376**	0.9177
Compliance with the business plan	0.1823	0.0899	0.1255
Project size	0.8984	**0.0291**	0.9395
Project duration	0.6001	0.3883	0.2151
Location of the project	0.178	**0.0354**	0.3281
Project type	0.5439	**0.0015**	0.644
Contract conditions and type of contract	0.8851	0.5	0.0906
Owner identity	**0.0295**	0.25	0.0849
Competition	0.2649	**0.0227**	0.8736

* *The bold cells show the p-values that are less than 0.05.*

As can be seen in the results, the owner identity was found statistically significant for the "Contractor Type" classification. The p-value (0.0295) gives the information that there is enough evidence that we can conclude the weight of importance given to the owner identity factor by general contractors and subcontractors are statistically significantly different from each other.

The "Contractor Sector" classification includes six subgroups which were determined based on revenues: Residential, Commercial, Industrial, Heavy, Residential-Commercial, Commercial-Industrial. Based on the Kruskal Wallis test results, contractor sectors are significantly different from each other for the weights given to the following factors: availability of equipment, materials and human resources, technical knowhow, project size, location of the project, project type, and competition. To determine specifically which of the groups in the "Contractor Sector" differ from each other; the Bonferroni correction/adjustment multiple testing procedure was conducted. The results showed that Commercial-Industrial vs. Industrial, Commercial vs. Industrial, Heavy vs. Industrial, and Residential vs. Industrial contractor sector groups were found significantly different within each comparison in the valuation of the importance weighing of the project type key factor. It is important to note that even though availability of equipment, materials and human resources, technical knowhow, project size, location of the project, and competition key factors were found to be significantly different for "Contractor Sector" groups as indicated by the Kruskal Wallis test, those factors were not captured in subsequent testing by the adjusted multiple testing procedure because of the conservativeness of the Bonferroni correction method.

The "Contractor Size" was also investigated in the same manner. The respondents were divided into four groups as shown in Table 2. The results didn't provide enough evidence to support the hypothesis, which was at least one of the groups, is significantly different.

5 CONCLUSIONS

The results provide evidence that statistically significant differences exist in the weights of importance given to the bid/no bid decision-making key factors based on the contractor type and contractor sector. However, the results do not provide sufficient information if there is a significant difference based on the contractors' size on the valuation of the key factors. It should be stated that the limited sample size of this study might be influential on the results. Therefore, larger sample size might improve the results. However, even considering the small sample size, two significant relationships (contractor type and contractor sector) were identified that points out the need for further research. This study contributes to the construction engineering and management body of knowledge by advancing the current state of the knowledge on the topic from "factors that affect bid/no bid decisions" to "different weights/values given to the factors that affect bid/no bid decisions by construction companies with different demographics".

References

Ahmad, I., Decision-Support System for Modeling Bid/No-Bid Decision Problem, *Journal of Construction Engineering and Management*, 116(4), 595–608, December, 1990.

Ahmad, I., & Minkarah, I., Questionnaire Survey on Bidding in Construction, *Journal of Management in Engineering*, 4(3), 229–243, July, 1988.

Bageis, A. S., & Fortune, C., Factors Affecting the Bid / No Bid Decision in the Saudi Arabian Construction Contractors, *Construction Management and Economics,* 27(1), 53–71, January, 2009.

Egemen, M., & Mohamed, A. N., A Framework for Contractors to Reach Strategically Correct Bid/No Bid and Mark-Up Size Decisions, *Building and Environment*, 42(3), 1373–1385, March, 2007.

Friedman, L., A Competitive-Bidding Strategy, *Operations Research*, 4(1), 104–112, February, 1956.

Gido, J., & Clements, J. P., *Successful project management*. South-Western, 2009.

Harris, F., McCaffer, R., & Edum-Fotwe, F., *Modern Construction Management*. Wiley, 2006.

Jarkas, A. M., Mubarak, S. A., & Kadri, C. Y., Critical Factors Determining Bid/No Bid Decisions of Contractors in Qatar, *Journal of Management in Engineering*, 30(4), 05014007, July, 2014

King, M., & Mercer, A., Differences in Bidding Strategies, *European Journal of Operational Research*, 28(1), 22–26, January, 1987.

Lowe, D. J., & Parvar, J., A Logistic Regression Approach to Modelling the Contractor' s Decision to Bid, *Construction Management and Economics,* 643–653, July, 2004.

Mohanty, R., Project Selection by a Multiple-Criteria Decision-Making Method: An Example from a Developing Country, *International Journal of Project Management*, 10(1), 31–38, February, 1992.

Saaty, T. L., & Vargas, L. G., *Prediction, Projection and Forecasting: Applications of the Analytic Hierarchy Process in Economics, Finance, Politics, Games and Sports.* Kluwer Academic Publichers. Boston, 1991.

SAS., SAS Studio, University Edition 2.2.9., 2015. Retrieved from http://www.sas.com/en_us/software/university-edition.html on January 15, 2016.

Shash, A. A., Factors Fonsidered in Tendering Decisions by Top UK Contractors, *Construction Management and Economics*, *11*(2), 111–118, 1993.

Shash, A. A., Subcontractors' Bidding Decisions, *Journal of Construction Engineering and Management*, *124*(2), 101–106, March-April, 1998.

Stokes, M. E., Davis, C. S., & Koch, G. G., Categorical Data Analysis Using The SAS System, *SAS Institute Inc.,* 2000.

Wanous, M., Boussabaine, H. A., & Lewis, J., A Neural Network Bid / No Bid Model : The Case for Contractors in Syria, *Construction Management and Economics*, *21*(1), 737-744, October, 2003.

Whittaker, J., Implementing a Bidding Model, *The Journal of the Operational Research Society*, *32*(1), 11–17, January, 1981.

Interaction between Theory and Practice in Civil Engineering and Construction
Edited by Komurlu, R., Gurgun, A. P., Singh, A., and Yazdani, S.

IMPACT OF BIDDING ENVIRONMENT ON CHANGE ORDERS

BAHADIR V. BARBAROSOGLU and DAVID ARDITI

Dept of Civil, Architectural, and Environmental Engineering, Illinois Institute of Technology, Chicago, IL, USA

Change orders occur during construction, and usually increase both construction cost and project duration. This paper investigates how change orders are related to the bidding environment. The data used in this study were collected from 74 public building construction projects awarded by the Public Building Commission of Chicago in the seven-year period 2008-2015. The contract values range between $0.3 - $70 million, and the total worth of the projects is a little over $1.1 billion. While delay may result in liquidated damages for contractors, increase in project cost can cause financial problems for owners. It is in the interest of owners and contractors to study similar past projects, and to observe how change orders are affected by the level of competition in bidding. Owners must budget for contingencies, and contractors must adjust their resources in anticipation of change orders. It would therefore be quite desirable for owners and contractors to be able to predict the dollar amount and duration implications of potential change orders by making use of past historical data. The prediction of the dollar amount and duration implications of potential change orders before the construction starts can alert construction owners and contractors to potential delays in total project completion time and potential overruns in total project cost. Knowledge about potential change orders creates a predictable environment and improves the relationship between owners and contractors. The results of this study indicate that contractor-driven change orders are positively related to the difference between the lowest and the second lowest bids.

Keywords: Contract changes, Public building construction, Bidding climate, Prediction models, Cost overruns, Bid spread.

1 INTRODUCTION

Competitive bidding is the most commonly used procurement method in public works projects, where contracts are awarded to the lowest qualified bidder. However, the total cost of a project often exceeds the lowest bid due to change orders. In a highly competitive bidding environment, bidders may lower their bids, and expect to regain profit by submitting claims and change orders (Wright and Williams 2001). Lower bids create higher cost deviations between the bids. According to Chaovalitwongse *et al.* (2011), there is a trade-off between the average cost deviation and cost overruns. The larger cost deviation between bids may increase the number of change orders, and actual project cost. The difference between the lowest and second lowest bids is commonly used in the literature to measure deviation between bids, also called "spread", "bid-spread", and "money left on the table" (Skitmore *et al.* 2001).

The average number of bidders per project is expected to be high when the number of available construction projects on the market is low. In such situations, bidders lower their mark-ups to increase their chances of winning. However, the lowest bidder may tend to regain their foregone profit by submitting change orders during contraction. The objective of the study presented in this paper is to investigate the impact of the bidding environment on contractor-driven change orders. This paper is organized as follows: the next section briefly reviews the literature on the bidding environment, bid competitiveness measures, and contractor- driven change orders. The succeeding section describes how the data are collected and used in order to test the research hypothesis. This section is followed by the "Conclusion" that presents the results, discusses the limitations of the study, and provides suggestions for future studies.

2 LITERATURE REVIEW

The relationship between bid prices and bid competition was analyzed in several studies. Factors that were used as measures of competition in the literature include, but are not limited to number of bidders, need-for-work, market conditions, availability of other projects, project's public exposure, and prestige.

A statistical analysis conducted by de Neufville and King (1991) found that the relationship between the "need-for-work" and bid mark-up is statistically significant. In other words, bid mark-up decreases as need-for-work increases. The study also showed that the lowest bid decreases as the number of bidders increases. Factors affecting "need-for-work" were identified by Chua and Li (2000) as current workload of projects, need for continuity in employment of key personnel and workforce, relationship with owner, and required rate of return in investment. Similar to the study of de Neufville and King (1991), Carr (2005) examined the statistical relationship between the number of bidders and bid prices. The interaction found in the study was statistically significant; an increase in the number of bids caused a lowering of the lowest bid. Awwad *et al.* (2014) studied the relation between risk attitudes and the mark-up decision, and found that contractors tend to lower their mark-ups as the intensity of the competition increases, regardless of their risk attitudes. Oo *et al.* (2010) found that in addition to the number of bidders, market condition, as well, is a measure of competitiveness that is significantly effective on mark-up decisions. Oo *et al.* (2010) found that contractors lower their mark-ups in recessionary environments; their study also showed that mark-up decisions were related with project size but not project type.

Chua *et al.* (2001) studied the impact of competition on contractor profits. Nine factors that are related to competition in lump sum contracts were listed as availability of other projects, cash flow requirements, degree of technological difficulty, identity of owner/consultant, project public exposure and prestige, project time scale and penalty of non-completion, safety hazards, size of project, and time allowed for bid preparation. The result of the study showed that expected profits are considerably lower in high-competition environments.

The negative effects of low bid mark-ups are emphasized in several studies. Lo *et al.* (2007) state that the excessive number of bidders may cause abnormally low bids, and some of the contractors that submit abnormally low bids try to get extra profit cutting corners to lower their costs, and submitting claims and subsequent change orders. According to Meulen and Money (1984), the number of bidders per bid is high because of high competition during recession, and some contractors purposely submit low bids in such an economic environment. It is suggested that contractors' opportunistic behavior negatively affects the quality of public works. Submitting unrealistically low bids makes it impossible for contractors to complete projects according to plans and specifications. Due to the sacrifice made in profit, contractors submit claims and change orders that lead to schedule delays (Ioannau and Leu 1993). Owners carry a large risk when they select a contractor that submits an unusually low bid and will not complete the project as planned.

3 DATA COLLECTION AND EXPLORATION

This study analyzed information about 74 public building construction projects awarded by the Public Building Commission of Chicago (PBC), including 434 bids and a little over $1.1 billion in construction value over a seven year period 2008-2015. These projects are of eight types, namely elementary schools, high schools, parks and playgrounds, libraries, fire stations, police stations, community centers and field houses, and other facilities. Each project involves either construction of a new structure, or renovation of an existing structure, or construction of an annex/addition to an existing structure. Information was extracted from bid tabulations and meeting minutes available on the PBC's website. This information was used to evaluate the impact of the bidding environment on the number and magnitude of contractor-driven change orders. Table 1 shows a summary of the bid data collected in this study.

Table 1. Bid data summary (2008-2015).

Construction Type	Total Contract Value	Number of Bidders	Number of Contracts Awarded
New Construction	$949,657,312	292	52
Addition/Annex	$135,043,186	84	11
Renovation	$102,912,407	58	11

The amounts and types of change orders were identified for each project by examining the minutes of the meetings of the Board of Commissioners of the PBC. The change orders approved by the Board were grouped in six categories including owner-directed changes, client-directed changes, errors and omissions in contract documents, discovered conditions/differing site conditions, code compliance issues, and other changes. The change orders that were not specified under any of these categories were labeled "non-categorized changes" and then removed from the dataset. Client-directed and owner-directed change orders were removed from the dataset, and the remaining change orders were labeled under the name of contractor-driven change orders. In the 2008-2015 study period, the yearly average of contractor-driven change orders expressed in percent of contract value ranged between 0.54% and 1.92%.

The hypothesis of the study is that the amount of contractor-driven change orders during construction increases with increased competition at the time of bidding. In this study, the indicators of competition at bidding were selected based on the literature cited in Table 2 as annual number of available bids, annual average number of bidders per bid, number of bidders in the bid, money left on the table (the difference between the lowest and the second lowest bids in percent of the lowest bid), number of works under contract, total uncompleted works under contract in percent of contract value, and number of recently completed contracts.

Table 2. Indicators of competition at bidding.

Literature	Factor	Corresponding Indicator Used in the Study
• Chua *et al.* (2001)	Availability of other projects	• Annual number of available bids
• Meulen and Money (1984) • De Neufville and King (1991) • Carr (2005) • Loe et al (2007) • Oo *et al.* (2010) • Chaovalitwongse *et al.* (2011)	Number of bidders	• Annual average number of bidders per bid • Number of bidders in the bid
• Skitmore et al (2001) • Chaovalitwongse *et al.* (2011)	Bid-spread, cost deviation	• Money left on the table
• De Neufville and King (1991) • Chua and Li (2000)	Need for work	• Number of works under contract in percent of contract value • Total uncompleted work under contract • Number of recently completed contracts

A stepwise regression was performed using SPSS to create a predictive model. The bid competition indicators were regressed against the predictor variable of contractor-driven change orders in percent of contract value. As seen in Table 3, the only predictor included in the model is "money left on the table". The rest of the variables are not in the model because some of the information contained in them is already present in the model. The reason why the remaining independent variables did not make it to the model is because of the high inter-correlations between some of these variables. For example, the Pearson correlation coefficients between "money left on the table" and "annual average number of bidders per bid", and between "money left on the table" and "number of bidders in the bid" were respectively -0.321 and -0.266, both significant at $\alpha = 0.05$.

The model accounts for 13.2% of the variance in contractor-driven change orders in percent of contract value. The model is statistically significant $F(1,50) = 8.621$, $p = 0.005 < 0.05$. The coefficient (B) of the independent variable is presented in Table 4. The low coefficient of determination ($R^2 = 13.2\%$) in this study can be due to the fact that the dollar value of change orders may be related to a multitude of other factors than "money left on the table" such as project characteristics, completeness and quality of design, opportunities to cut costs or improve quality, and other reasons that were not investigated in this study. The fact that the model is statistically significant constitutes evidence that competition at the time of bidding (as measured by "money left on the table") is an important predictor of the magnitude of change orders to come in the construction phase.

Table 3. Stepwise regression model to predict the value of change orders in percent of contract value.

Model	*r*	*R* Square	Adjusted *R* Square	Standard Error	Significant *F* Change
1	0.387	0.150	0.132	2.68864	0.005

Table 4. Regression coefficients.

Model	*B*	Standard Error	*β*	*t*	*p*
(Constant)	0.741	0.743		1.567	0.124
Money left on the table	0.229	0.078	0.387	2.936	0.005

The model presented in Table 3 indicates that the dollar amount of contractor-driven change orders increases as the "money left on the table" increases. This finding suggests that bidders who bid low are expected to recover the lost profits through the change orders they initiate. Why would the money left on the table be higher in some circumstances? The "money left on the table" is a measure of competitiveness. It is known that in competitive environments, bidders are likely to bid lower than normal (Meulen and Money 1984; de Neufville and King 1991; Chua *et al.* 2001; Carr 2005; Lo *et al.* 2007; Oo *et. al* 2010; Awwad *et al.* 2014). Also, judging from the correlation coefficient of -0.265 ($\alpha = 0.05$) between "money left on the table" and "number of years the lowest bidder has been active in the industry", the money left on the table appears to be higher if the lowest bidder has fewer years of experience. Indeed, inexperienced contractors lack external legitimacy; i.e., they lack a good understanding of owner requirements, and subcontractor operations (Fu *et al.* 2003).

4 CONCLUSION

The impact of the bidding environment on contractor-driven change orders was investigated in this study. The annual number of available bids, the annual average number of bidders per bid, the number of bidders in the bid, the money left on the table, the number of works under contract, total uncompleted works under contract in percent of contract value, and the number of recently completed contracts were identified as potential independent variables effecting the amount of contractor driven change orders. A stepwise regression analyses was performed, and a statistically significant relationship was found between contractor-driven change orders and "money left on the table". The prediction model showed that as the difference between the lowest and the second lowest bids increase, the amount of change orders goes up. A correlation analysis indicated that "money left on the table", is correlated with "company experience" at $\alpha = 0.05$. In other words, the more experienced the company, the less difference between the lowest and the second lowest bids even in times of economic downturn.

As found in several studies in the literature and in this study, unusually low bids can be a sign of upcoming change orders. Owners must be prepared for potential change orders and adjust their budget contingencies by making use of historical data. They should also be prepared for delays in the work schedule. Contractors must avoid submitting unrealistically low bids that may affect the company's survival in the industry.

The findings of this study suggest that a more detailed future study may investigate the impact of project characteristics such as project type and contract value on contractor-driven change orders. Exploratory research found that the study of project type and contract value in the context of contractor-driven change orders is expected to result in interesting findings.

References

Awwad, R., Asgari, S., and Kandil, A., "Developing a Virtual Laboratory for Construction Bidding Environment Using Agent-Based Modeling." *Journal of Computing in Civil Engineering*, 2014.

Carr, P. G., "Investigation of bid price competition measured through prebid project estimates, actual bid prices, and number of bidders." *Journal of construction engineering and management* (131)11, 1165-1172, 2005.

Chua, D. K. H., and Li, D., "Key factors in bid reasoning model." *Journal of Construction Engineering and Management*, (126)5, 349-357, 2000.

Chua, D. K. H., Li, D. Z., and Chan, W. T., "Case-based reasoning approach in bid decision making." *Journal of construction engineering and* management (127)1, 35-45, 2001.

de Neufville, R. and King, D., "Risk and need-for-work premiums in contractor bidding." *Journal of Construction Engineering and* Management, (117)4, 659-673, 1991.

Fu, W. K., Drew, D. S., and Lo, H. P., "Competitiveness of inexperienced and experienced contractors in bidding." *Journal of Construction Engineering and Management* (129)4, 388-395, 2003.

Ioannou, P. G. and S.-S. Leu. "Average-bid method-competitive bidding strategy." *Journal of Construction Engineering and Management* (119)1, 131-147, 1993.

Lo, W., Lin, C. L., and Yan, M. R., "Contractor's opportunistic bidding behavior and equilibrium price level in the construction market." *Journal of construction engineering and management* (133)6, 409-416, 2007.

Oo, B.-L., Drew, D. S., and Lo, H. P., "Modeling the heterogeneity in contractors' mark-up behavior." *Journal of Construction Engineering and Management* (136)7, 720-729, 2010.

Skitmore, M., Drew, D., and Ngai, S., "Bid-spread." *Journal of Construction Engineering and management* (127)2, 149-153, 2001.

Wright, M. G. and Williams, T. P., "Using bidding statistics to predict completed construction cost." *The Engineering Economist*, (46)2, 114-128, 2001.

Interaction between Theory and Practice in Civil Engineering and Construction
Edited by Komurlu, R., Gurgun, A. P., Singh, A., and Yazdani, S.

VALUE ORIENTED PROJECT DEVELOPMENT – EFFECTS OF AN EARLY INTEGRATION OF STAKEHOLDERS TO THE VALUE OF PROPERTIES

EDWIN HARRER

Institute for construction management and economics, TU Graz, Austria

Since the beginning of the 1990s there can be watched some mayor changes at the European real estate market. One of them is the tendentiously saturated market, which leads to much higher user requirements and have essential influences on the design of real estate properties. In the future, it is necessary that the design of durable asset properties is suitable for short-changing user requirements. As that, the variability of properties, with the aim of a long-term marketability, will also be a very significant factor to use properties as security for loans. The changings above result in increasingly complex buildings. The rapid advancement in technology and with it the adjustments of norms and regulations, require a quantity of professional disciplines and special knowledge. To that, the number of project participants increases rapidly. Financiers and builders, who usually have little or no expertise in the management of construction projects, are mostly overtaxed. Its overarching objective is the completed building, which correspond to the desired user requirements and deliver the anticipated rate of return. Consequently, the question arises whether the conventional workflow at real estate project development, which is marked by a sequentially participation of the stakeholders, can fulfill these requirements. Or is it more purposeful to put some incentives on the stakeholders and gather them to operate parallel in earlier project phases. From this initial position, the hypothesis is deduced, that the application of an integrated workflow in project development affects the value of real estate properties. The research should give an answer to the questions: firstly, whether the model of an early integration of the participants at real estate project development is suitable to increase the value of properties and secondly, what relative difference of the asset value is between a conventional process and an integrated process model.

Keywords: Sustainability, Valuation methods, Market value, Collaborating models, Partnering, Partnership, Simultaneous engineering, Incentives.

1 INTRODUCTION

From the perspective of an investor, and consequently from the viewpoint of a project developer, a long-term and maximized return on investment of their properties is crucial. This means that a permanent income and the return of interests have to be ensured (Brauer 2013). So the economic success of a property directly relates to its life cycle value. In the past, this was shown in many researches as well monetarily and not monetary (e.g., Pelzeter 2006, Eser 2009, Fröch 2012). In contrast to these life cycle based optimizations, the present work additionally implements the integral process of the workflow and investigates its impact on the value of properties (Figure 1).

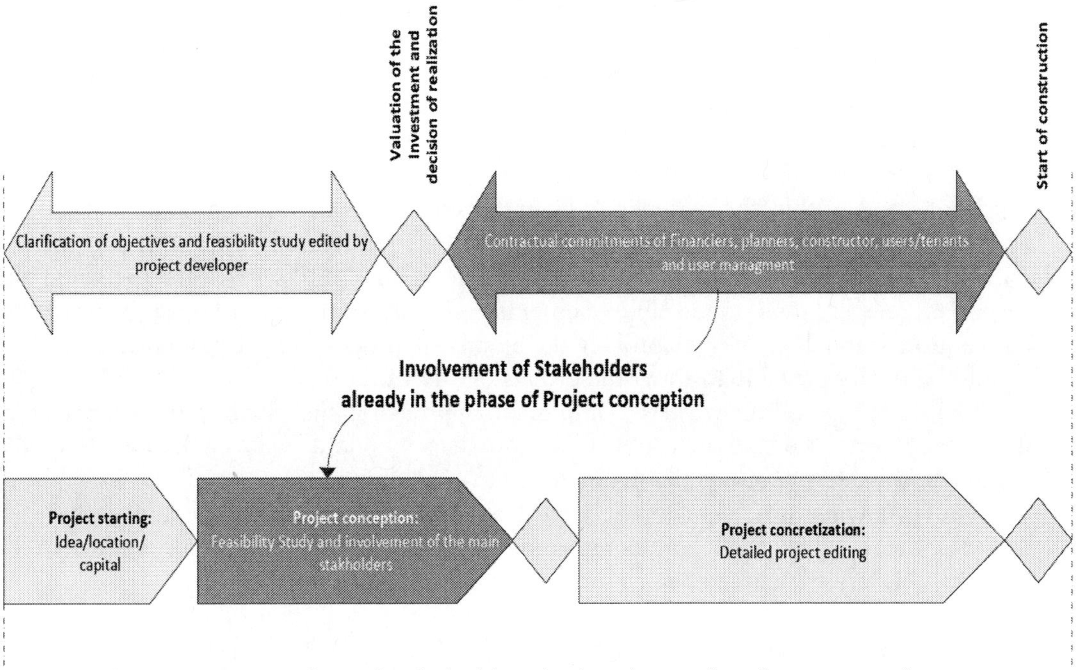

Figure 1. Integration of stakeholders in the phase of project-conception.

By integrating the stakeholders, there are several synergy effects, which have influence on the project development progress (Racky 2006). Some examples are listed below:

- Efficiency of planning and construction
- Quality and functionality of the building
- Building costs
- Renovation cycles
- Costs of operation and maintenance
- Risk relocation
- Vacancy
- etc.

This leads to the assumption that the change over from a sequential project development progress to an integral progress has an impact on the value of real estate properties.

2 METHODOLOGY AND OBJECTIVE

The high heterogeneity of the property market and the uniqueness of construction projects require to set tight constraints. Therefore, in a first step, it is necessary to select an appropriate valuation method and a specific partnering model in relation to the examined property submarket. Collaborating models in the construction process, such as GMP, PPP, etc., are determined by an early involvement of the necessary project participants, which in turn has an impact on the cost-, schedule- and quality-characteristics of projects (Grimscheid 2014). These parameters form the basis of yield expectations and consequently the value of real estate projects.

Regardless of this, the parameters from the selected collaborating model, which are under influence by the project participants, will be investigated. Due to a comparison with the examined parameters of the valuation method, it is possible to match the essential, relevant and modifiable elements. About a comparative analysis of the conventional and the integral project development process in an example project, it should be finally possible to show the effects on the value.

The advantage of this relative comparison approach is that the non-consideration of unchanged parameters, which aren´t able to detect conclusively, will not affect the result and misinterpretations like in an absolute consideration.

The following article is an excerpt from the ongoing research and shows the analysis of an appropriate value investigation method and the identification of its decision relevant parameters.

3 IDENTIFICATION OF DECISION RELEVANT PARAMETERS

Considering the market value (ON 1997) of real estate properties as the basis for its recoverability, it can be assumed that all factors affecting its value are reflected by it. That means, that the willingness of a tenant or investor to pay the appropriate rental or purchase price only will be given if all decision relevant parameters are involved in this price. This approach is also pursuing Fröch (2013) in his work, in which he demonstrates the optimization of project development by integration selected parameters of sustainability.

According to current international methods, the market value of income oriented real estate properties is determined by capitalization of its net income referring to the remaining period of life. For this, a number of different standardized and non-standardized procedures exist, such as *Ertragswertverfahren*, Discounted cash flow method, Residual method, etc. (ON 1997). For deducting the decision relevant parameters, at this work the choice dropped on the DCF method. On the one hand, this method is used internationally and on the other hand, it's also adopted to the standards of the German language speaking regions like Austria, Germany or Switzerland.

The principle of the DCF method is dividing the whole lifetime of the property into a detail-prediction period (explicit growth model) and the period of the remaining lifetime (implicit growth model). All expected cash inflows and outflows are recognized directly in the detail-prediction period of the DCF method and not if considering the remaining lifetime of the property on the capitalization rate. The base of both models is the principle of a present value calculation where all of the expected cash inflows and outflows are discounted to the point of time $t=0$. The residual value at the time $n+1$ is the present value of a finite annuity, calculated on the remaining life of the property and also discounted to the point of time $t=0$. This displays the following equation:

$$K_0 = \sum_{t=1}^{n} \frac{CF_t}{(1+r)^t} + \frac{RW_{n+1}}{(1+r)^{n+1}} \tag{1}$$

CF = Cashflow
RW = Restwert
r = Diskontierungszinssatz
n = Detailprognosezeitraum
K = Barwert

$$RW_{n+1} = CF_{n+1} * V \tag{2}$$

$$V = \frac{(i+1)^{Rd} - 1}{(i+1)^{Rd} * i} \tag{3}$$

i = Kapitalisierungszinssatz
Rd = Restnutzungsdauer

The main parameters of the equation above are:

- Cash flow [*CF*] consisting of the rental income from properties less the operating costs, maintenance and the shortfall of rental income e.g. by the changing of tenants.
- Capitalization rate [*i*] with the implication of rental growth and major risks such as location risk, refurbishment and modernization risk, risk of third party using, risk of rent progression, etc.
- Discounting rate [*r*] with explicit designation of rental growth - so the risk of rent progression is directly shown in the cash flow.
- Residual value [*RW*] consisting in turn of the cash flow and the remaining economic period of life [*Rd*] of the corresponding object.

In this approach, a derivation of the corresponding parameters initially appears without problems. On closer analysis, by which components they are made up, the main difficulties quickly become evident.

The factors mentioned above, like the cash flow, the interest rates and the remaining economic period of life directly are in context to the success factors recoverability, location, market and the property itself (Eser 2009). Furthermore, there are varieties of dynamic correlations between these parameters. Pelzeter (2006) for instance, describes a dependency between the economic life of an object and its location. Accordingly, the restoration cycles for objects in better locations are longer then in worse ones. Also the qualities of properties affect their renovation cycles and in addition have direct influence on incomes and costs. A property, which fulfils the user demands in large measure will earn more revenue at the market and get a lower vacancy rate than anyone with a low performance level. These are only some of the major effects that have to be taken into account.

4 PROSPECTS

An essential part of the following approach is to show the dependencies of the parameters above and to find out the relations to the modifiable parameters of the selected partnership model. The aim is to identify correlations and link qualitative, not directly with costs detectable parameters, to quantifiable factors for making them measurable.

With this method, it should be possible to ensure under the general conditions of the investigation that, in addition to the quantitatively detectable parameters, also all relevant qualitative parameters are included in the calculation model.

References

Brauer, K.-U., *Grundlagen der Immobilienwirtschaft. Recht - Steuern - Marketing - Finanzierung - Bestandsmanagement - Projektentwicklung,* Springer Fachmedien Wiesbaden, 2013.

Eser, B., *Erzielung nachhaltig hoher Büroimmobilienwerte. Ein Entscheidungsmodell für die Planungsoptimierung,* Gabler Verlag / GWV Fachverlage GmbH, Wiesbaden, 2009.

Fröch, G., *Probabilistische Bewertung und systematische Optimierung von Projektentwicklungen unter Berücksichtigung von Nachhaltigkeitskriterien. Bauwirtschaft und Projektmanagement,* Innsbruck University Press, 2013.

Girmscheid, G., *Projektabwicklung in der Bauwirtschaft-prozessorientiert. Wege zur Win-Win-Situation für Auftraggeber und Auftragnehmer,* Springer Berlin Heidelberg, 2014.

Österreichisches Normungsinstitut (ON), *Liegenschaftsbewertung Grundlagen,* Ö-Norm B1802, 1997.

Pelzeter, A., *Lebenszykluskosten von Immobilien. Einfluss von Lage, Gestaltung und Umwelt,* Müller, 2006.

Racky, P. (ed.), *Innovative Abwicklungsformen für Bauprojekte: Partnering und PPP. 15. September 2006 in Kassel,* Kassel Univ. Press, 2006.

Interaction between Theory and Practice in Civil Engineering and Construction
Edited by Komurlu, R., Gurgun, A. P., Singh, A., and Yazdani, S.

FINANCIAL ISSUES IN PUBLIC PRIVATE PARTNERSHIPS

ANTONIO COSIMO DEVITO[1], DAVID ARDITI[2], and ARIE GOTTFRIED[1]

[1]*ABC Dept, Politecnico di Milano, Milan, Italy*

[2]*Dept of Civil, Architectural, and Environmental Engineering, Illinois Institute of Technology, Chicago, IL, USA*

Public-Private Partnerships (PPPs) have gained significant importance in academia, government, and the construction industry over the last three decades. As a result, numerous studies have been published on the topic. These studies investigate various issues associated with PPP projects such as the length of the concession period, the selection of the private consortium, and the laws, rules and regulations that govern PPP implementation. However, there is a lack of systematic overview about issues with particular attention to financial aspects. This paper presents the outcome of a literature review that has been conducted to collect information about financial issues encountered in PPP implementation. This review included a total of 33 papers published in major journals and conference proceedings in the construction field in the last 25 years. At the end of the review, it was possible to identify a set of 19 financial issues that were grouped into five categories, namely (1) financing system, (2) financial market, (3) transaction costs and delays, (4) the public agency, and (5) the private consortium. The choice of focusing on this theme was prompted by the fact that the other issues (i.e., social, legal, and political) are all linked to financial issues. The identification of the salient issues related to financial aspects in PPPs is expected to contribute to the state-of-art in PPP research and implementation as it can pave the way to an empirical investigation that can in turn lead to a streamlining of the PPP processes throughout the world.

Keywords: Project delivery/financing systems, Financial factors, Financial viability, PPP project management, Infrastructure, Construction industry.

1 INTRODUCTION

Public Private Partnerships (PPPs) have emerged as one of the major delivery/financial mechanism for infrastructure projects since their inception. They are born from the idea of letting the private sector into the infrastructure market to alleviate the demands on public money problems (Li and Akintoye 2003). Put simply, thanks to a PPP contract, a public agency, i.e., a local or a central government, gives to a private consortium the provision of a public service or public utility, by entering into an agreement governing the long-term obligations of the parties. In particular, the private party bears the responsibility for financing and building the necessary infrastructure, and operating, managing and maintaining the service. A common mistake is to mix up PPPs and simple concessions and it should be mentioned that much of the literature does not explain the difference between these two mechanisms. In simple concessions, the private party pays the public agency for the right to operate the asset. In this way the transfer of risk to the private partner is generally higher than that of a PPP agreement (OECD 2010).

If a PPP contract is properly built and managed, it can provide many benefits to both sectors. On the one hand, public agencies can alleviate the financial burden on their own funds, transfer risks to the private sector, and obtain a higher value for money thanks to the contribution of the private consortium's expertise. On the other hand, the private sector can enter an expanding market for infrastructure projects where government involvement is reduced and taking advantage of opportunities to earn profits (Algarni *et al.* 2007).

However, since the inception of PPPs, projects have revealed several problems due to the complex aspects of the arrangement, including numerous interrelationships between social, political and legal risks, and unfavorable and uncertain commercial conditions. The objective of the research presented in this paper is to investigate the financial issues that may occur over the entire life-cycle of a PPP project because financial issues constitute a control point that can prevent the occurrence of several problems of different nature. For example, Soomro and Zhang (2015) show that financial problems may trigger a slow and obstructed project progress, create social and legal complications, and cause the concessionaire's insolvency and/or the cancellation of the concession. Therefore, this paper identifies the mostly discussed financial issues in the literature.

2 RESEARCH METHODOLOGY

The academic community has developed a vast body of knowledge about PPPs. A systematic analysis of articles published is undertaken in this paragraph. Following Al-Sharif and Kaka's method (2004), a search was conducted using the "title/abstract/keyword" field on the search engine "ASCE library" considering in the selection the top-ten construction journals listed in Chau's ranking (1997). Search keywords included public–private partnership, private finance initiative, PPP, PFI, financial management. In addition, conference proceedings were reviewed for a total of 33 papers over the last 25 years.

The number of PPP studies that strictly deal with financial issues is limited. Some of the studies focus on legal, economic, and political/social obstacles (Li *et al.* 2005, Chan *et al.* 2010, Hwang *et al.* 2013). Given that the literature offers many studies that investigate critical success and best value contributing factors, the search method was extended to cover articles that mention financial issues as success/failure factors. For example, Zhang (2005) found that a "creative financial package" is a critical success factor for a PPP project. Tiong *et al.* (1997), Zhang (2005), and Yuan *et al.* (2012) identified another critical success factor in a strong financial commitment of the private consortium. The financial issues were thus identified by surveying not only papers where the main focus is financial issues in PPPs, but also papers that explore success/failure factors in PPPs.

3 FINANCIAL ISSUES

A total of nineteen issues were identified in the literature review. Some of these issues share similar themes; hence the five macrocategories were defined in the following way: (a) unattractive financing system, (b) adverse financial market, (c) high transaction costs and delays, (d) financial weakness of the public agency, and (e) financial weakness of the private consortium.

(a) UNATTRACTIVE FINANCING SYSTEM

(1) **Inappropriate financial analysis:** This problem occurs when inputs, such as demand forecasts, technical feasibility, and cost estimates, are not available or detailed enough. If the financial performance of the project over its life is not evaluated properly, the possibility to achieve the best value for money is jeopardized (Zhang 2005, Yuan *et al.* 2012, Tang *et al.* 2013, Xie *et al.* 2013).

(2) **Unappealing financial package:** A financial package should be based on appropriate cost (e.g. tariff, water rate, or energy change costs), minimal financial risks to stakeholders, minimal burden on debt-servicing capacity of project revenue, and creative thinking (e.g., opening up additional markets within the same project). An unappealing package could necessitate lengthy and troublesome negotiations and delays in project progress with a reduction of the lenders' attraction. (Zhang 2005, Jefferis 2006, Yuan *et al.* 2012, Tang *et al.* 2013).

(3) **Weak commercial arrangement:** This issue occurs when the contract payment mechanism, the financial arrangements, the sources of loans, and the sources of stand–by financing facilities for possible eventualities are not formulated in a clear way (Zhang 2005, Tang *et al.* 2013).

(4) **Unfair concession period:** Public agencies want a shorter period in order to take possession of the asset as soon as the debt is repaid. Private consortia want a longer period for earning more revenues. The concession period has to be fair in order to protect the interests of both the public agency and the private consortium (Tiong *et al.* 1997, Zhang 2005, Xie *et al.* 2013).

(5) **Inappropriate equity to debt ratio:** A low equity to debt ratio reveals the vulnerability of the private consortium to manage financial issues. On the other side, a high ratio implies a strong commitment in terms of risks for the private parties (Zhang 2005).

(6) **Level of tolls or tariffs:** For public agencies, high tolls or tariffs means high service/product costs; for private consortia, low tolls or tariffs means low revenues (Tiong *et al.* 1997, Zhang 2005, Xie *et al.* 2013).

(7) **Lack of appropriate toll/tariff adjustment mechanisms:** Given the long-term nature of a PPP contract, changes, such as economic growth, price elasticity and user behavior, could occur over years. Therefore, an inappropriate and inflexible toll/tariff mechanism makes it difficult for a private consortium to recover its investments (Zhang 2005, Yuan *et al.* 2012).

(8) **Net present value of toll/tariff revenues in the concession period:** A high net present value of toll/tariff represents an issue for public agencies because it implies a high total cost for using the services or products of the asset. A low net present value of toll/tariff is an indication of a low cash intake for private consortia (Zhang 2005).

(9) **Inadequate total investment schedule:** Knowing the evolution of total investment over the concession period can help the stakeholders to assess the status and the strength of the financing mechanisms (Zhang 2005, Yuan *et al.* 2012).

(b) ADVERSE FINANCIAL MARKET

(10) **High and variable interest rates:** A high rate of interest increases project costs and financial risks (Tiong *et al.* 1997, Zhang 2005, Hwang *et al.* 2013).

(11) **Low rate of return and restriction on the cap:** A low rate of return and a restriction on the cap indicate low profitability, with the effect of discouraging the investors (Tiong *et al.* 1997, Zhang 2005).

(12) **Weak local financing and high financial service charges:** This factor weighs on the cash flow of the project and its feasibility. Indeed stable local financing could prevent foreign currency risks, e.g., convertibility and fluctuation in interest/exchange rates (Tiong *et al.* 1997, Li *et al.* 2005, Zhang 2005, Chan *et al.* 2010, Yuan *et al.* 2012).

(c) HIGH TRANSACTION COSTS AND DELAYS

(13) **High transaction cost:** The transaction costs of PPPs arise because of complex contract negotiations, legal, financial, and technical advisory, for organizing the procurement phase, and for monitoring and managing the contract over the life-cycle of a project. High transaction cost may damage PPP benefits and not achieve the best value for money. (Li *et al.* 2005, Zhang 2005, Corbett and Smith 2006, Chan *et al.* 2010, Hwang *et al.* 2013).

(14) **Long delay in reaching financial closure and long time in construction period:** Obstacles in reaching financial closure increase costs in the bid phase. Delay in construction duration increases project development costs (Li *et al.* 2005, Zhang 2005, Corbett and Smith 2006, Chan *et al.* 2010, Hwang *et al.* 2013, Tang *et al.* 2013).

(d) FINANCIAL WEAKNESS OF THE PUBLIC AGENCY

(15) **Weak financial commitment of the public agency:** Limited financial participation on the part of a public agency discourages a private consortium because it could be read like a lack of interest (Tiong *et al.* 1997, Li *et al.* 2005, Zhang 2005, Yuan *et al.* 2012, Hwang *et al.* 2013).

(16) **Corruption and bribery:** This factor may result in dangerous compromises with negative effects on the project (Li *et al.* 2005, Hwang *et al.* 2013).

(17) **No government control on tolls/tariffs:** The government has to monitor and regulate the tolls/tariffs in order to prevent public opposition with consequent high political costs (Zhang 2005).

(e) FINANCIAL WEAKNESS OF THE PRIVATE CONSORTIUM

(18) **Weak financial commitment of the private consortium:** Limited financial contribution on the part of the private consortium may damage project progress because the private consortium may not be able to solve possible problems with appropriate funds (Tiong *et al.* 1997, Li *et al.* 2005, Zhang 2005, Yuan *et al.* 2012, Hwang *et al.* 2013).

(19) **Weak financial standing of the private consortium:** A weak financial standing on the part of a private consortium has heavy consequences such as a higher commercial risk, and higher interest rates on borrowed funds (Li *et al.* 2005, Zhang 2005, Corbett and Smith 2006, Jefferies 2006, Hwang *et al.* 2013, Tang *et al.* 2013).

4 CONCLUSION

Since a PPP project involves multiple factors of different nature, it is not possible to investigate all of them in great depth. This paper presents the state-of-art in PPP research about financial issues, a critical factor in the success/failure of a PPP.

The research presented in this paper involves a review of the literature. A total of 19 financial issues were identified and were organized in five macrocategories that include the strength of the financing system, the stability of financial markets, the level of transaction costs and delays, and the financial strength of the stakeholders. Decision–makers in public agencies and private consortia should carefully analyze these issues in order to understand if a PPP project is appealing to both parties.

References

Al-Sharif, F., and Kaka, A., PFI/PPP topic coverage in construction journals." *Proc., 20th Annual ARCOM Conference*, Vol. 1, 711–719, Heriot Watt University, Edinburgh, Scotland, U.K., 2004.

Algarni, A. M., Arditi, D., and Polat, G., Build-Operate-Transfer in Infrastructure Projects in the United States, *Journal of Construction Engineering and Management*, 133(10), 728–735, 2007.

Chan, A. P. C., Lam, P. T. I., Chan, D. W. M., Cheung, E., and Ke, Y., Potential Obstacles to Successful Implementation of Public-Private Partnerships in Beijing and the Hong Kong Special Administrative Region, *Journal of Management in Engineering*, 26(1), 30-40, 2010.

Chau, K. W., The ranking of construction management journals, Construction Management and Economics, 15(4), 387–398, 1997.

Corbett, P., and Smith, R., An analysis of the success of the private finance initiative as the government's preferred procurement route, *Proc., Accelerating Excellence in the Built Environment Conf.*, Birmingham, U.K., 2006.

Hwang, B. G., Zhao, X., and Gay, M. J. S., Public private partnership projects in Singapore: Factors, critical risks and preferred risk allocation from the perspective of contractors, *International Journal of Project Management*, 31(3), 424–433, 2013.

Jefferies, M., Critical success factors of public private sector partnerships: A case study of the Sydney SuperDome, *Engineering, Construction and Architectural Management*, 13(5), 451-462, 2006.

Li, B., Akintoye, A., Edwards, P. J., and Hardcastle, C., Perceptions of positive and negative factors influencing the attractiveness of PPP/PFI procurement for construction projects in the U.K.: Findings from a questionnaire survey, *Engineering Construction and Architectural Management*, 12(2), 125–148, 2005.

Li, B., and Akintoye, A., An Overview of public-private partnership, in *Public-Private Partnerships: Managing risks and opportunities*, A. Akintoye, M. Beck & C. Hardcastle (eds), Oxford, 3-30, 2003.

OECD, *Dedicated Public-Private Partnership Units: a survey of institutional and governance structure*, 32, 2010.

Soomro, M., and Zhang, X., Roles of Private-Sector Partners in Transportation Public-Private Partnership Failures, *Journal of Management in Engineering*, 10.1061/(ASCE)ME.1943-5479.0000263, 2015.

Tang, L., Shen, Q., Skitmore, M., and Cheng, E., Ranked Critical Factors in PPP Briefings, *Journal of Management in Engineering*, 29(2), 164–171, 2013.

Tiong, R. L. K., and Alum, J., Final Negotiation in Competitive BOT Tender, *Journal of Construction Engineering and Management*, 123(1), 6-10, 1997.

Xie, J. and Ng, S. T., Multiobjective Bayesian Network Model for Public-Private Partnership Decision Support, *Journal of Construction Engineering and Management*, 139(9), 1069–1081, 2013.

Yuan, J., Wang, C., Skibniewski, M., and Li, Q., Developing Key Performance Indicators for Public-Private Partnership Projects: Questionnaire Survey and Analysis, *Journal of Management in Engineering*, 28(3), 252–264, 2012.

Zhang, X., Concessionaire's Financial Capability in Developing Build-Operate-Transfer Type Infrastructure Projects, *Journal of Construction Engineering and Management*, 131(10), 1054–1064, 2005.

Interaction between Theory and Practice in Civil Engineering and Construction
Edited by Komurlu, R., Gurgun, A. P., Singh, A., and Yazdani, S.

PUBLIC PRIVATE PARTNERSHIPS FROM THE POINT OF VIEW OF LARGE CONSTRUCION COMPANIES

ANTONIO COSIMO DEVITO[1], DAVID ARDITI[2], and ARIE GOTTFRIED[1]

[1]*ABC Dept, Politecnico di Milano, Milan, Italy*
[2]*Dept of Civil, Architectural, and Environmental Engineering, Illinois Institute of Technology, Chicago, IL, USA*

Stringent budget constraints and the need to build infrastructures have resulted in Public–Private Partnerships (PPP) being acknowledged as a desirable delivery/financing system for public projects. A PPP is an agreement between a public agency and a private consortium that allows the financing, design, construction, and operation of large scale infrastructure. While the public agency can be a single government department (e.g., a highway department), the private consortium is composed of several parties including contractors, designers, financial institutions, and other investors. The party that plays a critical role in the consortium is typically a large construction company. Various researchers have identified some of the critical factors that affect the success of a PPP, but only few studies have undertaken a comprehensive examination of financial issues. This paper aims to fill this research gap by analyzing information collected from large construction companies pertaining to financial issues. A structured questionnaire survey was administered to the largest 190 international construction companies listed by *Engineering News–Record* to investigate the relative significance of 19 financial issues that were identified in a thorough review of the literature. The findings indicate that factors such as inappropriate financial analysis, a low rate of return, and long delays in reaching financial closure have a high impact on the success of PPPs. Moreover, the public agency's tendency for corruption and the private consortium's weak financial standing can also undermine the financial strength of the arrangement. The findings can be used by all participants in PPPs to evaluate the financial issues in such projects.

Keywords: Project delivery/financing systems, Financial factors, Success measures, Management of PPP projects, Large contractors, Private consortia.

1 INTRODUCTION

Different types of public procurement have been used in worldwide infrastructure development over the years. Among these, Public–Private Partnerships (PPPs) is seen as a pragmatic tool for the provision of infrastructure development and delivery of public service when the government does not have the required budget to cover the capital cost (Xie and Ng 2013).

PPPs are agreements between entities in the public and private sectors. The public partner, usually a government agency, establishes project requirements and sets the tender to choose the private consortium. The private partner usually is a consortium of designers, construction companies and lenders. It finances, designs, builds, and operates the public asset. In this venture, construction companies cover an important role because they must have the necessary skills and expertise to deliver the asset according the public agency's requirements. Furthermore, construction companies must enjoy sound finances in order to manage risks that can occur over project progress, and provide cost effective facilities/services (Zhang 2005a).

In light of these considerations, the objective of this paper is to investigate construction companies' perspective about the major financial issues in PPPs by surveying the larger construction companies involved in PPP projects in the past.

2 RESEARCH METHODOLOGY

This study combines a literature review and a questionnaire survey to collect data on the financial aspects of PPPs. The research methodology involved three steps: (1) identifying the financial issues encountered in the literature. These issues are summarized in Table 1 (Devito *et al.* 2016); (2) administering a questionnaire survey to the largest international construction companies; and (3) analyzing the data collected.

Table 1. Financial issues in PPPs.

Macro-Category	Financial Issue
Unattractive financing systems	Inappropriate financial analysis, Unappealing financial package, Weak commercial arrangement, Shorter concession period, High equity to debt ratio, Low tolls or tariffs, Lack of appropriate toll/tariff adjustment mechanisms, Low net present value of toll/tariff revenues in the concession period, inadequate total investment schedule.
Adverse financial market	High and variable interest rates, Low rate of return and restriction on the cap, Weak local financing and high financial service charges.
High transaction costs and delays	High transaction cost, Long delay in reaching financial closure and long time in construction.
Financial weakness of the public agency	Weak financial commitment of the public agency, Corruption and bribery, No government control on tolls/tariffs.
Financial weakness of the private consortium	Weak financial commitment of the private consortium, Weak financial standing of the private consortium.

The questionnaire survey was designed to rate the impact of the financial issues listed in Table 1 on the success of a PPP project as perceived by construction companies. The survey was administered in May 2015 – June 2015 using a data collection platform called SurveyMonkey (2015). The target respondents were selected from the top 225 construction companies listed by *Engineering News-Record* (ENR 2015) that have been part of PPP projects. Of these 225, only 190 had a "Contact Us" or "Board Members" section in their website that made it possible to collect 1,186 email contacts that were sent emails with a link to the questionnaire.

The questionnaire was organized in two sections. Section I was composed of five sets of questions, one for each financial macro-category listed in Table 1. These questions asked to evaluate the impact of financial issues on PPP success according to a

Likert scale (1931) of 1-5, where 1 was " very low impact", and 5 was " very high impact". A "don't know" alternative was also available. Section II asked information about the respondents such as years of experience with PPPs, and the types of PPP projects in which the companies had been involved.

3 SURVEY FINDINGS

Sixty-five respondents responded to the survey, a response rate of response rate of 5.50%. This rate of response is not high but is acceptable for an exploratory survey of this nature. The respondents were from 16 different countries out of the 32 listed by ENR, a rate of 50% (Table 2). All respondents were from construction companies with a good knowledge and experience in PPPs. Indeed, the majority of the respondents reported being involved in PPP projects for more than 10 years. They have mainly been involved in PPP projects relative to transportation and public facilities (Table 3).

Table 2. Countries represented in the survey.

Country	Number of respondents
Australia	2
Belgium	1
Brazil	1
Canada	7
China	1
Egypt	2
Greece	1
India	4
Italy	11
Korea	1
Malaysia	1
Netherlands	1
Spain	6
Turkey	2
United Arab Emirates	2
United States of America	12
Not specified	10
Total	65

Table 3. Types of PPP projects represented in the survey.

Type of PPP project	Number of respondents
Transportation	35
Public Facilities	34
Environment	23

*Some respondent companies were active in more than one type of PPP project

Each respondent expressed a judgment based on experience. Only issues rated 4.00 or higher (i.e., "high impact" and "very high impact") are considered for discussion in the following bullets. The survey findings are the followings:

1) Looking at the responses to the first set of questions about the financing system, the three most important financial issues that received the highest average ratings are: inappropriate financial analysis (4.34), unappealing financial system (4.02), and

weak commercial arrangement (4.02). The average rating of "inappropriate financial analysis" was expected to be high as it agrees with the findings of Zhang's (2005b) and Yuan *et al.*'s (2012) studies which showed that this issue, being related with costs and revenues, is focused on assessing the project from an investment point of view. Therefore, a private company cannot understand the viability of the project if the finances are not analyzed properly. As far as the other two financial issues are concerned, the risk of getting into extensive negotiations is higher when the financial package is not appealing and when the commercial arrangement is weak. These risks scare construction companies because negotiations lead to delays and increased costs.

2) With regard to financial markets, there is only one issue rated to have high impact on PPP success: "low rate of return and restriction on the cap" with an average rating of 4.03. Li *et al.* (2005) underlined the importance of having an attractive market for the PPP project. This condition is feasible only if unstable interest rates and strict restrictions on the cap are prevented. Furthermore, the public agency should not promulgate special regulations that undermine the economic growth of the country because it could have an effect on user behavior and on the rate of return of a project. According to Harris (2004), if the rate of return is low, the private consortium may use its resources and skills on other more appealing projects.

3) Concerning transaction costs and delays "long delay in reaching financial closure and long time in construction" is rated at 3.95 very close to "high impact". As stated by Zhang (2005b), delays may pose a major risk to the private consortium because the actual profitability of the project is reduced. The construction company will be compelled to postpone the operational phase of the facility, hence postponing the possibility to cover the debt from the cash flow of the project.

4) With regard to the critical aspects of the public agency, the issues with a high impact on the PPP success are "corruption and bribery" with an average rating of 4.32 and "weak financial commitment of the public agency" with rating of 4.12. The first finding is consistent with the study conducted by Hwang *et al.* (2013) who showed the importance of sharing the "corruption and bribery" risk among the stakeholders, which implied that the private consortium should play an important role in developing a transparent procurement without corruption. The second finding reveals the construction company's concern about engaging with public agencies with a weak financial commitment. A lack of sound finances in the public agency undermines the willingness of private consortia to take part in PPPs.

5) Looking at the last macro-category, the only issue rated at as having high impact on PPP success is "weak financial standing of the private consortium" with an average rating of 4.11. This result is similar to Zhang's (2005b) finding. Furthermore, another issue in this macro-category, "weak financial commitment of the private consortium" received an average rating of 3.98, very close to "high impact". Actually, the two issues are linked to each other. A private consortium's weak financial standing influences the financial evaluation conducted by banks. Banks have to assess a construction company's financial strength because, having to

protect their investment, banks are focused on the guarantees of the debt repayment. An inappropriate financial standing compromises the likelihood to obtain external financing, hence inevitably reducing the financial commitment of the private consortium.

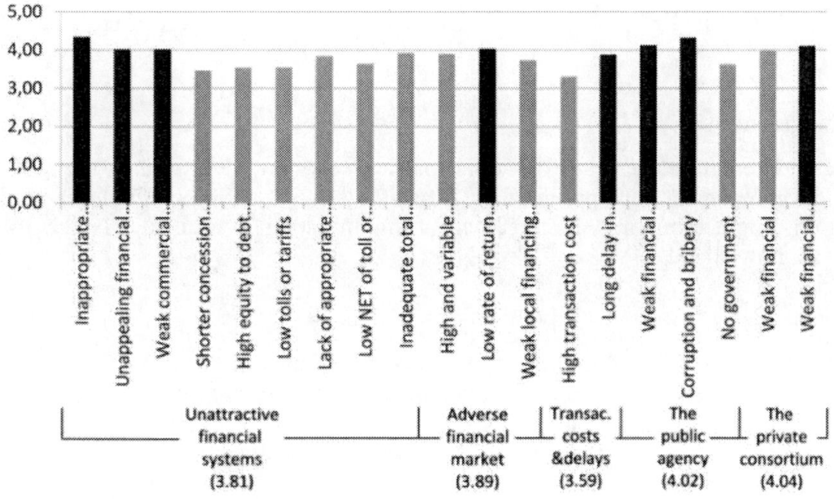

Figure 1. Weighted average impact ratings of financial issues on PPP success.

4 CONCLUSION

This paper presents the findings of a questionnaire survey administered to the largest international construction companies listed by *Engineering News-Record*. The questionnaire dealt with 19 previously identified financial issues in PPP implementation. The relative impact of each financial issue within each macro-category has been determined based on the respondents' judgment and experience.

As seen in Figure 1, the findings show that the financial issues with a high impact are: inappropriate financial analysis, unappealing financial package, weak commercial arrangement, low rate of return and restriction on the cap, long delay in reaching financial closure and long time in construction period, weak financial commitment of the public agency, corruption and bribery, weak financial standing of the private consortium. Overall, these issues are rated at an average level greater than 4.00 (i.e., "high impact" or higher).

According to the information presented in Figure 1, two of the five macro-categories, received average ratings higher than 4.00 (high impact on PPP success): "financial weakness of the private consortium" (4.04), and "financial weakness of the public agency" (4.02). This is no surprise because the strong financial position of the two main actors in PPPs should be conducive to the success of the project. Actually, the remaining macro-categories are quite dependent of the financial standing of public agencies and private consortia. For example, a weak financial standing of either party can create uncertainties that could make financial analysis difficult to perform, resulting in unreliable outcomes. Financial uncertainties could also affect the competition in the marketplace, and increase transaction costs and delays. It can therefore be concluded from this study that an important prerequisite for the success of a PPP is the strong financial standing of the two main parties in the PPP organization, i.e., the public agency and the private consortium.

The financial issues revealed by the questionnaire survey as having high impact on PPP success should be recognized by all stakeholders in PPP projects. It is possible to achieve success in PPP projects if financial issues are recognized and managed effectively.

Acknowledgments

The names and email addresses of potential participants that were used in the survey of large construction companies were provided by Rushit Thakkar, a former graduate student in the Construction Engineering and Management Program at IIT. Rushit Thakkar's support is gratefully acknowledged.

References

Devito, A. C., Arditi, D., and Gottfried, A., Financial Issues in Private Public Partnerships, in *Interaction between Theory and Practice in Civil Engineering and Construction*, Komurlu, R., Gurgun, A., Yazdani, S., and Singh, A. (Eds.), Istanbul, Turkey, 2016.

Engineering News-Record, 2015, retrieved fromwww.enr.construction.com/toplists/sourcebooks on April 1, 2015.

Harris, S., Public Private Partnerships: Delivering better infrastructure services, *Proc. Conf: Investment in Infrastructure in Latin America and the Carribbean: Second Generation Issues and Perspectives*, Iter-American Development Band, Washington, D. C., 2004.

Hwang, B. G., Zhao, X., and Gay, M. J. S., Public private partner- ship projects in Singapore: Factors, critical risks and preferred risk allocation from the perspective of contractors, *International Journal of Project Management*, 31(3), 424–433, 2013.

Li, B., Akintoye, A., Edwards, P. J., and Hardcastle, C., Perceptions of positive and negative factors influencing the attractiveness of PPP/PFI procurement for construction projects in the U.K.: Findings from a questionnaire survey, *Engineering Construction and Architectural Management*, 12(2), 125–148, 2005.

Likert, R., *A technique for the measurement of attitudes. Archives of Psychology*, Columbia University Press, New York, USA, 1931.

SurveyMonkey, 2015, retrieved from www.surveymonkey.com on June 30, 2015.

Xie, J. and Thomas Ng, S., Multiobjective Bayesian Network Model for Public-Private Partnership Decision Support, *Journal of Construction Engineering and Management*, 139(9), 1069–1081, 2013.

Yuan, J., Wang, C., Skibniewski, M., and Li, Q., Developing Key Performance Indicators for Public-Private Partnership Projects: Questionnaire Survey and Analysis, *Journal of Management in Engineering*, 28(3), 252–264, 2012.

Zhang, X., Criteria for Selecting the Private-Sector Partner in Public–Private Partnerships, *Journal of Construction Engineering and Management*, 131(6), 631–644, 2005a.

Zhang, X., Concessionaire's Financial Capability in Developing Build-Operate-Transfer Type Infrastructure Projects, *Journal of Construction Engineering and Management*, 131(10), 1054–1064, 2005b.

Interaction between Theory and Practice in Civil Engineering and Construction
Edited by Komurlu, R., Gurgun, A. P., Singh, A., and Yazdani, S.

UNDERSTANDING THE ENGINEER'S ROLE IN ADMINISTERING THE CONSTRUCTION CONTRACT

MOHAMED-ASEM U. ABDUL-MALAK[1] and FARAH Y. EL MASRI[2]

[1]*Civil and Environmental Engineering, American University of Beirut, Beirut, Lebanon*
[2]*Civil and Environmental Engineering, University of Alberta, Edmonton, Canada*

The role of the Engineer has been carried out by the design consulting firm, traditionally in charge of not just designing and providing technical supervision for the construction work, but also of administrating the construction contract and further assisting in adjudication and arbitration processes in the case of disputes arising along the way. However, due to the increase of the projects' size and complexity, employers nowadays tend to assign other entities for handling the managerial and administrative tasks normally required of the Engineer. Using the FIDIC's 1999 Red Book as a platform, the Engineer's roles have been systematically identified on a clause-by-clause basis, and various statistics as to the roles' frequency are offered. A method has been devised to properly classify these identified roles. The proposed classification is based on parameters and attributes which include: the exact deduced role; the role's timing and frequency of application; the time-barring stipulated for its application; the role's nature, being reactive, proactive, or passive; the role fulfillment capacity; and the role fulfillment exclusivity. The outcome of this work offers employers with a profound understanding of the roles expected of the Engineer in administrating the construction contract, particularly assisting in the apportionment of these roles between or among multiple employer-appointed entities, when such a split in role-assignment is deemed desirable by employers.

Keywords: Administration, Engineer, Contract, Classification, Roles, Project, FIDIC.

1 INTRODUCTION

Traditionally, employers assign the design consulting engineers to undertake the responsibilities of the "Engineer" for administrating the construction contract. Under such an appointment, the Engineer's services during the construction phase include the technical (quality) supervision of the works as well as the administration of the commercial and scheduling issues related to the contract. As such, the Engineer can also be referred to as a decision-maker, a function that requires a certain degree of impartiality and fairness from him. In common law, the role of the Engineer can be understood to involve, on the one hand, actions to be taken as the employer's agent and, on the other hand, actions warranting the rendering of a professional opinion" (Jaeger and Hök 2010). On projects of certain complexity, the ability of managing, predicting and innovating in the contract administration process can present a great challenge (Reynolds 2008). Nowadays, the owners of construction projects are becoming increasingly oriented towards a more value-based and operational management process (Ryd 2003). There is a need for professionals that can act as the owner's connection between business development and technical design and supervise the execution process in a less traditional way (Lindahl and Ryd 2007). As such, the design consulting engineers are no longer the owners' first choice of advisers (Nicol and Pilling 2000), and there is a simple suggestion of changing the profession's tradition of "protectionism" and "exclusivism" in order to deliver a higher quality in the construction project (Siva and London 2011). The question remains: is there a good understanding on the part of the concerned parties as to what the Engineer's traditional roles exactly entail before judging whether any such roles can be the subject of change or not?

2 RESEARCH SCOPE

The objective of the research work reported on in this paper is to examine and highlight the major roles expected of the Engineer, and to investigate a method that can help more rigorously classify them. The desired classification method shall allow practitioners a better chance to differentiate between the two general classes of technical versus administrative and managerial roles and to be able to judge the instances where the Engineer is expected to act as the Employer's agent versus an independent decision maker. The work consisted of scrutinizing all the roles given to the Engineer through a clause-by-clause reading of the 1999 FIDIC Red Book's conditions of the contract. Suitable criteria were then conceptually explored to help better understand and classify each role. Finally, a classification matrix was proposed which differentiates the roles that can be exclusively assigned to the design consulting entity from those other roles that can alternatively be assigned to other equally or better qualified entities.

3 SCREENING OF THE ENGINEER'S ROLES

3.1 Deduction of Roles

The 1999 FIDIC Red Book's conditions are structured under 20 headings summarizing all possible duties, roles and authorities of the three main participants in the project. The word "Engineer" is cited in 99 out of 163 sub-clauses, representing 61% of these conditions. It is to be noted that the scanning of the Engineer's different roles was limited to clauses uniquely citing the word "Engineer" and did not include the roles assigned to the Employer's Personnel.

3.2 Interpretations and Analysis

Out of the 99 sub-clauses mentioning the word "Engineer", 93 sub-clauses express 33 distinct roles for the Engineer. The roles cited in definitions, listings or carrying the same name as their sub-clause remained hidden from this screening to avoid redundancy. It can be seen from Figure 1, that 13 out of the 93 sub-clauses of relevance (14%) are heavy, in terms of their numerous instances (four or more) of making a reference to a role expected of the Engineer. The selected threshold of four has no basis except that which is deduced visually from the scatter.

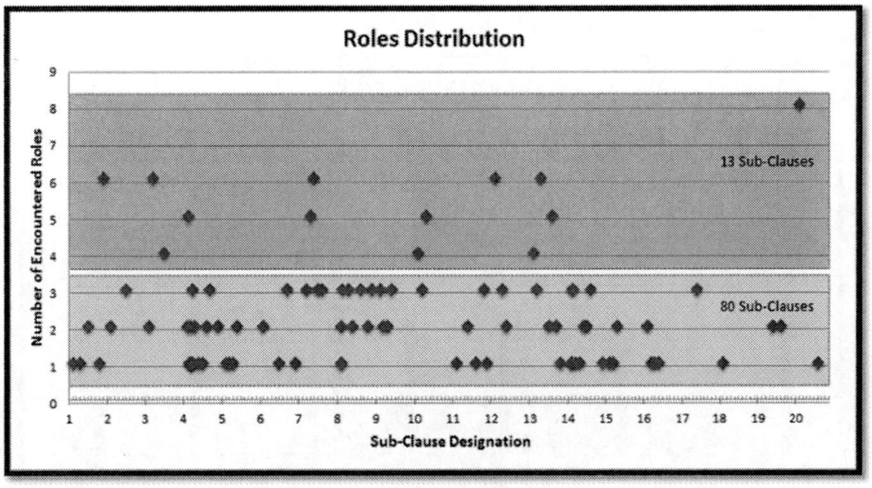

Figure 1. Number and distribution of encountered roles.

Sub-Clause 20.1, concerned with contractor's claims, is ranked with the highest number of roles of eight, as shown in Figure 2. Related sub-clauses, which deal with matters that can be the cause for time and price increases, include variations, work measurement, delayed drawings, differing site conditions, among others. Quality-related sub-clauses include testing, inspection, and tests on completion.

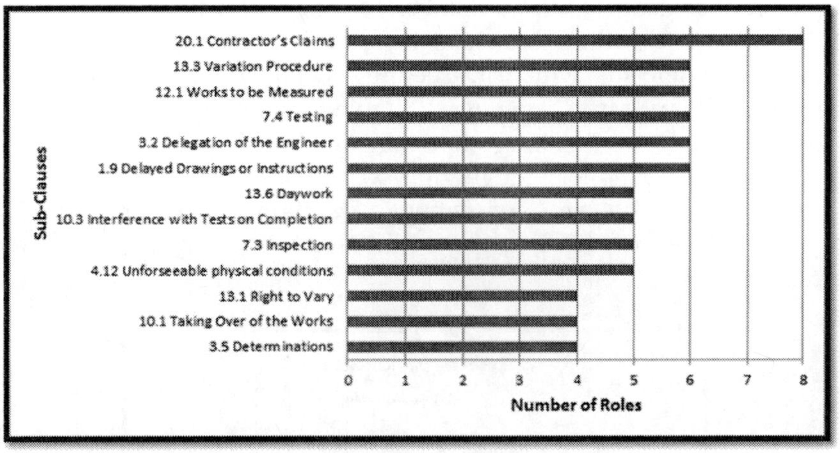

Figure 2. Sub-clauses with role frequencies of four or more.

From another perspective, Figure 3 shows the frequency of all the 33 identified roles, in an attempt to discern the important ones. The most frequent role is that of "making determination", which appears in 33 sub-clauses excluding Sub-Clause 3.5 Determinations. It can be argued that this particular role, of determining an extension of time and/or an additional compensation, is indeed of great importance since it requires of the Engineer to act as a decision-maker at 33 different situations, under which the Engineer is required to defend his identity of an independent professional.

When it comes to determining any matter, the engineer is expected to consult with each of the parties. Even if consultation is hindered, the engineer is obliged to issue the determination out of professional ethics (Bunni 2005).

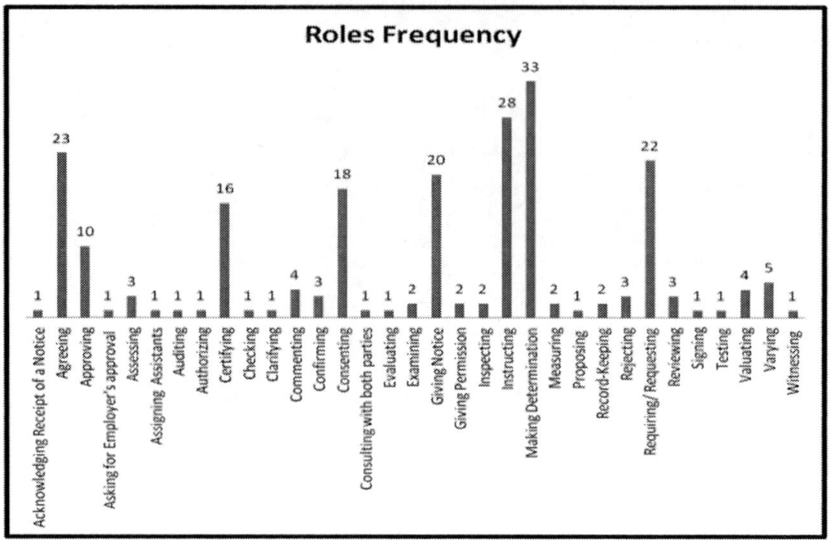

Figure 3. Frequencies of the deduced roles.

It can be further deduced that a number of other roles have a high frequency of occurrence. Figure 4 shows the basket of those roles whose frequencies are equal or greater than five percent, with "instructing" being ranked second to "making determination" followed by "agreeing", a role that is intimately referred to in conjunction with that of "making determination". Two other roles related to "instructing" are also found to be highly ranked; these are "requiring/requesting" and "giving notice". As also shown, "consenting", "certifying", and "approving" are to be regarded as key roles expected to be fulfilled by the Engineer with remarkable frequencies.

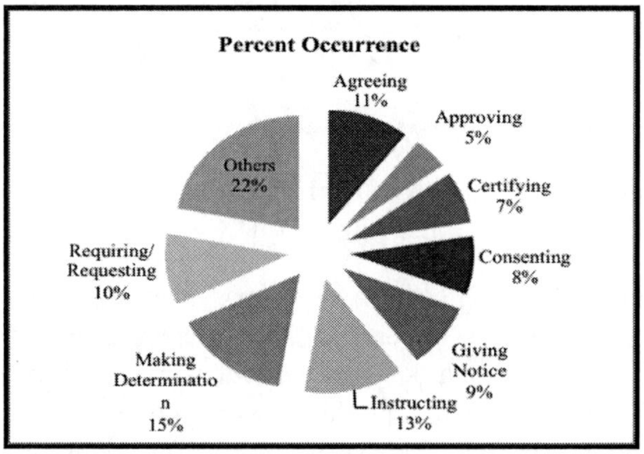

Figure 4. Engineer's roles with 5% occurrence or more.

4 CLASSIFICATION OF ROLES

4.1 Parameters and Attributes

All along the screening of the encountered Engineer's roles, a number of observations were being recorded, according to which six parameters (in addition to the concerned role itself) and their corresponding attributes were selected to help design a system of classification of these roles, as summarized in Figure 5. At the end of the scrutiny process, a total of 218 roles (resulting from the repetitions of the 33 distinct roles that were identified throughout) filled a similar number of rows for building the intended classification matrix.

Classification Designation	Classification Parameter	Parameter Attributes	Underlying Rationale for Assigning Attribute(s)
CD 1	Deduced Main Roles	Agreeing / Determining / Consenting / Approving / Certifying / Assessing / Evaluating / Valuating / Etc.	Descriptive and adopted from the exact wording used in the examined FIDIC clauses
CD 2	Time Constraint(s)	Time Barring	Stipulated period within which the Engineer shall act
		////versus////	
		No Stipulation	No period is stipulated for the role in question
CD 3	Timing and Frequency	Timing of Role Fulfillment	Stage of intervention along the construction contract lifespan
		////and////	
		Role fulfillment frequency	The rate of occurrence of every role over the project lifecycle
CD 4	Role Nature	Proactive	Taking the initiative in the role
		Reactive	The required role is the answer to an act or request by the Others (The Employer and/or the Contractor)
		Passive	The nature of a role where no reaction is necessarily required from the Engineer and where basically the Engineer is only informed in the process
CD 5	Role Type	Administrative	Defines an automated role in a defined process/ paperwork related
		Technical	Requires specialty or expertise in the matter
		Managerial	Includes some decision making within a specific frame of time
CD 6	Role Fulfillment Capacity	Employer's Agent	Where the Engineer is supposed to act on behalf of the Employer
		////versus////	
		Independent	Where the Engineer is supposed to act in a professional objective manner
CD 7	Role Fulfillment Exclusivity	Candidates excluding the Designer	Due to certain conflict of interest or lack in experience
		Candidates including the Designer	Unrestricted to a specific party
		Exclusively the Designer	Purely technical designer expertise required

Figure 5. System for classifying the encountered roles.

4.2 Interpretations and Analysis

In this section, only a very brief summary of general inferences is presented. Following the construction of the classification matrix, it became readily evident that the majority of the roles are:

- Reactive (70%), meaning that the Engineer is largely at the receiving end of submittals, requests, and claims from the contractor;
- Technical (62%), showing that the role of the Engineer does not entail dealing with technical matters at all times, thereby suggesting that someone else other than the design professional could fulfill such roles;
- Intended to be performed by an independent professional (67%), thereby emphasizing the importance of impartiality; and

- Intended for candidates including, but not exclusive to, the design professional (75%), indicating that such roles can also be alternatively assigned to another professional who is equally or better qualified for performing such roles.

Further to the observation stated under the fourth bullet above, it is found that almost 22 percent of the roles should not be assigned to the design professional, whereas the roles that are exclusive to the design professional represent only 4 percent. In reiteration to the observation stated under the second bullet above, the technical roles counted for 62 percent of the total roles, versus 25 percent and 13 percent for the administrative and managerial roles, respectively. Such a deduced distribution of roles could send a misleading message that the designer should take the lead when it comes to administrating the construction contract as opposed to assigning such a lead to a project management professional. Here, it should be noted that the role of making determinations is also considered technical, involving scheduling and cost assessments and valuations, and the majority of these determinations do not suit being handled by the design professional. That alone can revoke the above-hinted misconception.

By way of concluding, the intended classification of the different roles, expected of the Engineer in acting as the administrator of the construction contract, aims at aiding projects' owners/employers in deciding on how best to allocate the contract administration roles among the entities that are lined up to contribute to the process of overseeing the construction phase.

Acknowledgments

The research work reported on in this paper was funded via a graduate assistantship by the Engineering Management Program. The authors are grateful for this support.

References

Bunni, N. G., *FIDIC forms of contract, 3rd Edition* Blackwell, Oxford, 2005.

FIDIC, International Federation of Consulting Engineers, *Conditions of Contracts for Construction*, 2007.

FIDIC Geneva, The Role of Construction Managers, Notes on the use of FIDIC in the Middle East, 2008.

Jaeger, Axel-Volkmar and Hök, Götz-Sebastian, *FIDIC - A Guide for Practitioners,* 2010.

Lindahl, G. and Ryd, N., "Clients' goals and the construction project management process", *Construction Project Management*, 25(3), 147-15, 2007.

Nicol, D., Pilling, S., *Changing Architectural Education: Towards a New Professionalism*, E & F Spon, London, 2000.

Ryd, N., *Exploring construction briefing – from document to process, doctoral thesis*, Chalmers University of Technology, Gothenburg, 2003.

Siva, J. and London, K., Client learning for successful architect-client relationships, *Engineering, Construction and Architectural Management*, 19 (3), 253 – 268, 2012.

Interaction between Theory and Practice in Civil Engineering and Construction
Edited by Komurlu, R., Gurgun, A. P., Singh, A., and Yazdani, S.

BARRIERS OF TRADITIONAL PROJECT DELIVERY SYSTEM IN ACHIEVEMENT OF PROJECT GOALS

IKBAL ERBAS

Akdeniz University, Faculty of Fine Arts, Dept of Architecture, Antalya, Turkey

Although traditional project delivery system is one of the project delivery systems that is preferred by both public and private owners, characteristic of the system constitutes some barriers in achievement of the project goals. In the scope of this study the barriers caused by traditional project delivery system were revealed for achieving the basic project goals which are quality, time and cost. For this purpose impact of the relationships between the participants and workflow process of the system were examined in achievement of the quality, time and cost goals of the project. It is found out that impossibility of contractor's involvement to design process, adversarial relationships among the participants, impossibility of using fast track and uncertainty of final project cost because of the unexpected project changes are the main barriers that have impacts on main project goals.

Keywords: Traditional system, Relationship between participants, Workflow process, Project goals.

1 INTRODUCTION

A project delivery option is defined as a method for procurement by which the owner's assignment of ''delivery'' risk and performance for design and construction has been transferred to another party (parties) (Clough 1981, Mahdi and Alreshaid 2005). Traditional project delivery system is the most common procurement system and it is referred to as "design-bid-build" option (Mahdi and Alreshaid 2005). This system is also known as hard bid or the low bid method. This is still considered the traditional project delivery method for design and construction where the design precedes the construction and the contracts provides either a lump sum or unit price bid to obtain the work (Ohrn and Rogers 2008).

Although traditional project delivery method is one of the main alternative delivery methods that is usually preferred by public and private owners, it is clear that traditional system has some weaknesses in terms of project performance. Project performance can be measured and evaluated using a large number of performance indicators that could be related to various dimensions (groups) such as time, cost, quality, client satisfaction, client changes, business performance, health and safety (Cheung *et al.* 2004, DETR 2000, Enshassi *et al.* 2009). Time, cost and quality are, however, the three predominant performance evaluation dimensions (Enshassi *et al.* 2009). In the scope of this study barriers of traditional project delivery system in achievement of quality, time and cost goals were revealed. Due to the complexity of the construction work, relationships between the participants and workflow process of the system have important influence on achievement of these project goals. Therefore barriers were determined by explanation of these two main problem areas in the scope of the study.

2 RELATIONSHIPS BETWEEN PARTICIPANTS IN TRADITIONAL PROJECT DELIVERY SYSTEM

This is a familiar delivery method to most owners and requires a defined scope prior to bidding (El-Sayegh 2007). The owner, designer (architect) and contractor are three prime players of traditional project delivery system. In traditional arrangement the owner has two separate contracts: one with the designer and one with the contractor (Figure 1). The designer is responsible to the owner for the design of the project and also administers the construction contract as the owner`s representative (Mahdi and Alreshaid 2005). The designer deliverables includes plans and specifications for the construction of the project. These documents are subsequently used by the owner as the basis to make a separate contract with the contractor (Hale *et al.* 2009). The contractor is responsible to the owner for the proper construction of the design and is responsible for methods and procedures of construction. This creates an independent relationship between the designer and the contractor with each directly responsible to the owner (Mahdi and Alreshaid 2005). Two separate contracts, with two separate entities, are utilized by owners to complete one construction project, including two solicitations and procurement steps (Hale *et al.* 2009). The separation of the designer and the contractor in this system creates a system of checks and balances because the designer and the contractor are in a position (Mahdi and Alreshaid 2005).

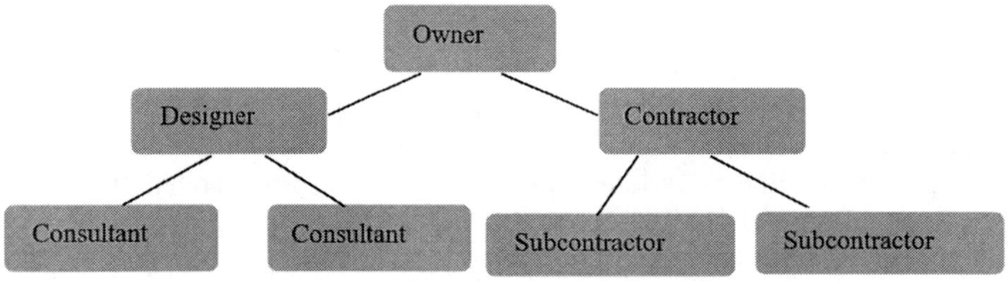

Figure 1. Relationship diagram of participants in traditional project delivery system.

Although many methods are used for awarding this contract, the most common approach is to solicit bids from different construction companies. The company providing the lowest bid will then build the project based on the documents produced by the designer (Hale *et al.* 2009).

This contracting system offers the advantage of being widely applicable, well understood, and with well-established and clearly defined roles for the parties involved. It is the most common approach for public owners having to comply with state procurement statutes. Furthermore, it offers the owner a significant amount of control over the end product, particularly since the facility's features are fully determined and specified prior to selection of the contractor. However, many construction owners have experienced a variety of frustrations using this system, leading to the development of other methods (Titan Reality and Construction 2015).

3 WORKFLOW PROCESS OF TRADITIONAL PROJECT DELIVERY SYSTEM

Traditional project delivery system has a lineer workflow process which consists of decision, design, bidding and construction phases (Figure 2). In decision phase of the system, the design company is hired first to provide design services and develop the contract drawings and specifications (Ohrn and Rogers 2008). In design process, after the approval of preliminary design, the designer uses in-house staff (or alternatively, consultants) to prepare fully completed plans and specifications that are then incorporated into a bid package (Trauner Consulting Services 2007).

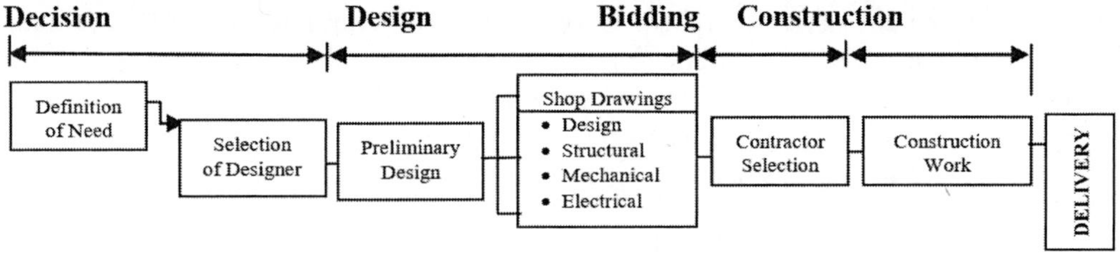

Figure 2. Workflow scheme of traditional project delivery system.

At the end of the design phase, the designer assists the owner in the bidding phase and selecting the contractor (Ohrn and Rogers 2008). The design package is presented to interested general contractors, who prepare bids for the work, and execute contracts with subcontractors to construct various specialty items. In many cases, the contractor submitting the lowest responsive bid is selected to perform the construction. This contractor is then responsible for constructing the facility in accordance with the design. The designer typically maintains limited oversight of the work and responds to questions about the design on behalf of the owner. The designer may also assist the owner in administering the construction contract, including determination of project progress, for interim payments made to the contractor (Titan Reality and Construction 2015).

4 BARRIERS OF TRADITIONAL SYSTEM

At the beginning of the project, main goals of the project participants are to finish the design and construction process as possible as shortest time with a highest quality at a lowest cost. Because of the involvement of multiple different participants, achievement of these goals might be difficult especially for the owners and the owners might face with different barriers. Main sources of these barriers that have influence on project goals in traditional system can be evaluated under two different headings as:

- Barriers caused by relationships between participants
- Barriers caused by workflow process.

4.1 Barriers caused by Relationships between Participants

One of the main barriers related to relationships between the participants is the lack of involvement of the construction professional during the design phase. In this system the contractor can't be part of the design process. The designer's drawings don't necessarily show the assemblies that the contractor chooses. This means that the contractor junks about half of the construction drawings and replaces them with shop drawings. The amount paid to the designer for those wasted drawings amounts to roughly 1-1½ % of the construction cost (Thomsen 2006). Because the quality, price, and completion date of the contract are all established by the contract requirements, there is little incentive to the contractor to provide any expertise beyond what is minimally required to obtain and complete the project within the requirements of the contract (Ohrn and Rogers 2008).

Another barrier related to relationships is that traditional system "tends to create an adversarial relationship among the contracting parties, rather than foster a cooperative atmosphere in which issues can be resolved efficiently and effectively" (Trauner Consulting Services 2007). Characterized in another way what this states is that the traditional project delivery method provides an incentive for the parties to the contract to not create a cooperative project atmosphere (Ohrn and Rogers 2008). As a result of this two barriers, quality goals of the project cannot be succeeded properly.

4.2 Barriers caused by Workflow Process

Another part of the barriers are related to workflow process of the system. Traditional system process takes too long. Because the construction drawings are used for the core of the contract with the contractor, construction work can't start until all the drawings are done (Thomsen 2006). It is a linear sequence during which the owner, procures the designer`s design services separate from the procurement of the construction services. The design must be fully completed prior to the procurement of the construction services. This is due to the fact that the procurement of construction services is typically based upon a hard bid price which cannot be assembled until there is a full set of plans and specifications (Ohrn and Rogers 2008). Nonuse of fast track because of the linear sequence of traditional system`s workflow process constitutes an important barrier in terms of the project time goals.

Another barrier which has influence on achievement of cost goals of the project is the owner's misconception that the bid price is the final price. Designer plans and specifications are rarely if ever perfect and the contractor's interpretation of the plans and specs rarely if ever match the intentions of the designer. As result of this, it is common place in construction process that there will be changes and change orders. This often has the untended consequence of placing a stress on the business relationships between the owner, design professional, and the contractor (Ohrn and Rogers 2008).

5 CONCLUSION

Although traditional project delivery method is seen as the primary project delivery process, when compared with alternative project delivery systems, the traditional system is insufficient in terms of meeting project quality, time, and cost goals.

Nature of the relationships between the participants in traditional system tends to cause more adversarial relationships among the contractor, the designer and the owner. Potential conflicts between the designer and the contractor because of the adversarial relationships and lack of involvement of the contractor to the design process affects the quality of the project in a negative way. Relationships between the participants are directly associated with effectiveness of the design process including selection of the materials, construction means and methods which are closely associated with quality of the end product. On the other hand participation of the contractor to the design process beside the designer will ensure coordination between design and construction phases. This cooperation will help achievement of the project quality goals in some cases.

Workflow process of the system causes arising of barriers that have effects on project time and cost goals. Impossibility of using fast track because of the linear sequence of the workflow process affects time goals of the project. This leads to longer delivery time. On the other hand increase on number of the changes and change orders as a result of being less flexible for changes causes uncertainty of final project cost. Thus cost goals of the project cannot be irretrievable as claimed by the owner. An owner tends to use traditional system should consider the main barriers that he might be encountered during the project process and he should take measures against to potential risks for overcoming these barriers. Only in this case the owner may be avoided from losses associated with time, quality and cost.

References

Cheung, S. O., Suen, H. C. H. and Cheung, K. K. W., PPMS: a Web-based construction project performance monitoring system, *Automation in Construction,* 13, 361–376, 2004.

Clough R. H., *Construction Contracting, 4th ed.*, Wiley, New York, USA, 1981.

Department of the Environment, Transport and the Regions (DETR), KPI Report for the Minister for Construction by the KPI Working Group, January 2000.

El-Sayegh, S. M., Significant Factors Affecting the Selection of the Appropriate Project Delivery Method, *Fifth LACCEI International Latin American and Caribbean Conference for Engineering and Technology (LACCEI'2007),* Tampico, México, 2007.

Enshassi, A., Mohamed, S. and Abushaban S., Factors Affecting The Performance Of Construction Projects In The Gaza Strip, *Journal Of Civil Engineering And Management*, 15(3), 269- 280, 2009.

Hale, D. R., Shrestha, P. P., Gibson G. E. Jr. and Migliaccio, G. C., Empirical Comparison of Design/Build and Design/Bid/Build Project Delivery Methods, *Journal Of Construction Engineering And Management ASCE,* 135(7), 579-587, July, 2009.

Mahdi, I. M. and Alreshaid, K., Decision Support System For Selecting The Proper Project Delivery Method Using Analytical Hierarchy Process (AHP), *International Journal of Project Management,* 23 (7). Pp. 564-572, 2005.

Ohrn, L. G. ve Rogers, T., Defining Project Delivery Methods for Design, Construction, and Other Construction-Related Services in the United States, *International Proceedings of the 44th Annual Conference*, Auburn University - Auburn, Alabama 2008.

Titan Reality and Construction, retrieved from www.titannewyork.com/construction_strategy1.asp on September 22, 2015.

Trauner Consulting Services, Construction Project Delivery Systems and Procurement Practices: Considerations, Alternatives, Advantages, Disadvantages 2007. Retrieved from http://ascpro0.ascweb.org/archives/cd/2008/paper/CPGT293002008.pdf on October 2, 2015.

Interaction between Theory and Practice in Civil Engineering and Construction
Edited by Komurlu, R., Gurgun, A. P., Singh, A., and Yazdani, S.

IN SEARCH OF ALTERNATIVE AWARDING CRITERIA

BERNHARD BAUER, JOHANNES WALL, and DETLEF HECK

Institute of Construction Management and Economics, Graz University of Technology, Austria

In Austria, the public procurement of construction works is mainly based on two different types - the lowest acquisition costs and the most economically advantageous tender. The first one only considers the price, which means, that the contract is awarded to the bidder with the cheapest price. The other principle refers to the most economically advantageous tender, which has to be chosen for projects that exceed a certain overall tender value or are related to projects with higher complexity. This principle focuses on considering other awarding factors besides the price, but it often fails its purpose due to the lower weighting of these issues. Over the last years the market has not expanded very much, therefore there is a lot of competition related to the lowest price, which results in inadequate quality and opens up the field for further claims. In order to end the tense price competition among the competitors and to increase the quality of the projects, new criteria with serious weighting will be needed and to fulfill the most economically advantageous tender and supporting the focus on construction quality. Even though there are acknowledged guidelines dealing with environmental aspects of construction sites, like the RUMBA guideline (*Richtlinien für umweltfreundliche Baustellenabwicklung* - Guidelines for Sustainable Construction Site Management). However aspects of sustainability are no common standard within construction processes. To capture the common standards of sustainable measures on construction sites, and the industries' understanding for environmental topics, research on experts in the field was done. Based on the authors' acknowledgements, the paper targets the aspects of an environmental friendly construction site and introduces new criteria for the awarding process, and mentions some of the findings such as the "technical equipment expertise" or "sustainable construction site management".

Keywords: Public procurement, Construction quality, Most economically advantageous tender, Sustainability.

1 INTRODUCTION

The current situation in the construction industry is characterized by highly competitive markets and a strong focus on tendering for the lowest costs. That means, only the price is considered for awarding the contract. It is related to the principles of public procurement, which tells public clients that they have to take care of public investments and therefore they are forced to base their tendering decisions on something obvious and easy to check and justify by internal audits on how they are spending public funds. However, it is clear, that not only the price is reasonable for awarding a contract.

Related to the current planning and construction conditions, planning services are often not fully terminated when the procurement process starts in order to save time in the construction project. Bauer and Heck (2014) state that this lack of planning depth in the beginning of a project mostly leads to complications in the stages of construction and to changes in the final design.

This often leads to claims and disputes about the scope and extends of the additional works between the client and the contractor. As a consequence those disputes lead to disrupted construction processes and often end up in legal conflicts.

This situation seems to become more and more unsatisfying and the need for other aspects to base the procurement decision on increases.

It also represents the initial basis of the current contribution. In this context several case studies have been investigated (Stabauer 2015) and based on the findings, additional criteria have been developed how the procurement decision can be used for awarding construction works.

2 LEGAL REQUIREMENTS

It is necessary to consider the legal frameworks of public procurement and to keep the main goals of public procurement in mind before thinking about possible alternative awarding criteria. In the following figure 1 the main principles are illustrated. Therefore to be in line with the European procurement law (Directive 2014/24/EU), the principle of non-discrimination as well as the transparency of the contract awarding has to be cared about. Additionally the principle of supporting a competitive market situation should be kept in mind, to foster new and innovative solutions and developments.

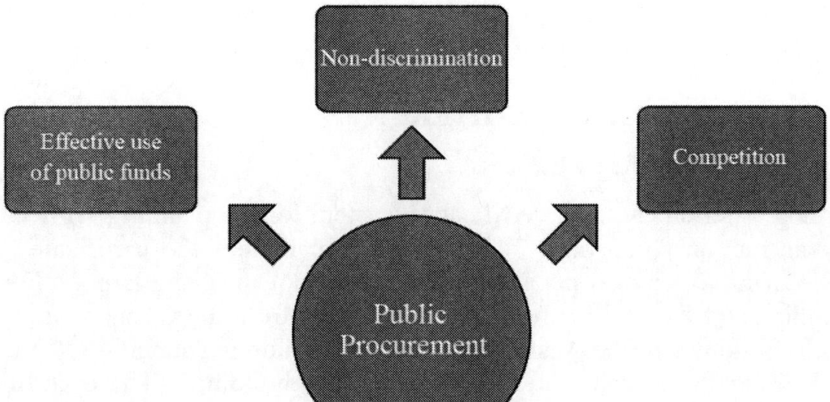

Figure 1. The main principles of public procurement.

Furthermore according to the federal public procurement act in Austria there are several financial boundaries, which are used for defining what kind of procurement procedure can be applied in a certain project. E.g. if the construction costs exceed 5.186.000 EUR, the call for tenders has to be announced on European level according to the Commission Regulation EU No. 1336/2013. This means if the contract increases a certain amount, the most economically advantageous tender needs to be applied. Therefore additional criteria are necessary to be considered during the procurement process. At the moment, according to the state of the art in the Austrian public procurement, there are several awarding criteria used such as

- Extension of warranty
- Reduction/Saving of construction time
- Reduced utilization and maintenance costs

The problems with the mentioned criteria is, that they are just considered to a minimum extend due to the reduced weighting factor. This means overall the price is the only crucial factor. So in terms of thinking about new criteria also the weighting needs to be considered, to guarantee their contribution to the awarding of the contract, otherwise they can only be seen as dummy criteria.

3 METHOD

To gain further understanding for the general issue of the present used criteria for awarding construction contracts, Stabauer (2015) conducted a survey that was taken as a basis to identify potential new awarding criteria and establish quality criteria for construction sites.

The standardized questionnaires were answered by 40 experts, focused on the sense of general awarding criteria and the willingness to accept those criteria for future construction projects. Basing on the results of this research, new criteria were established, that focus on the construction process itself, in order to point out the possibilities to make construction sites more economical and environmental friendly with the introduction of new quality criteria.

4 APPROACHES FOR NEW CRITERIA

4.1 Guaranteed Certificate Success

An awarding criterion can be a certification system for the planning services, but also for the construction companies. The tenderer guarantees the certificate success in acceptance of his offer or a penalty for failure to not meet the target. For example green building certificates like the DGNB certificate from the German Association for Sustainable Buildings or the Austrian Sustainable Building Council – ÖGNI, are often used to indicate the sustainable performance of a building. These certificates are divided into three different performance levels: gold, silver and bronze, with gold for the maximum rating is and thus leads to the maximum number of points for this criterion, indicating the best sustainable performance. Tenderers with guaranteed success certificate silver or bronze will receive proportionately fewer points. The focus of the certification systems is, to focus and improve the entire lifecycle of a building. This criterion should already be included in the project development phase, and conducted in more detail during the planning phase, targeting the whole life cycle of a building.

4.2 Identified Errors in the Tender Documents

Errors within the tender documents are for example quantity changes that represent more than 20% of an item in the bill of quantities, the lack of items, wrong items, but also any faults that were not considered by the client and that would lead to disruptions within the construction process. The criterion "unidentified tendering errors" does not only aim to encourage planners to provide conscientious tender documents, but also to reward the "honesty" of contractors and support a cooperative working climate.

4.3 Technical Machinery Expertise

The criterion "technical equipment expertise" should be applied to increase the quality of construction works by the use of more modern machinery and techniques. Till (2015) mentions that another positive effect is, that the use of state of the art equipment will be more attractive for bidders.

It is intended that this criterion is used in the machine-intensive civil engineering such as road, rail, pipeline, and landfill construction. A table has to be provided by the client where the bidders are able to see how many points they get with the appropriate use of the required equipment.

The site supervision is in charge of reviewing the used equipment. It executes spot checks and imposes a penalty at three times infringement of the treaty.

4.4 Alternative Offers

An alternative offer is an alternate suggestion of the tenderer for improving the way a service is done. Those offers may be allowed by contracting authorities for contracts that are awarded according to the criterion of the technically and economically most advantageous tender. The client must explicitly specify in the tender if and what kind of alternative bids are allowed.

According to Gast (2006), if the client does not specify on the admissibility of alternative bids, variants are not permitted. In the tender documents he has to explain the minimum requirements of alternative offers in terms of their comparability of services and to designate, in what way these bids shall be submitted. In the procurement process, the client must only take those alternative offers into account, which meet the established minimum requirements.

4.4 Environmental-Friendly Execution of Construction Works

This criterion is aimed at the environmentally friendly construction performance. As a tool for awarding the contract, RUMBA (guidelines for Sustainable Construction Site Management) is mandatory. The advantage of using this issue is that the client is able to use a well-known catalogue and does not have to transpose and define own criteria, such as "low-dust construction site".

Following objectives should be pursued to enhance the environmental performance issues of construction sites:

- Reduction of construction site traffic by avoiding and reducing the number of transportation operations or trying to shift transport to rail

- Increasing the recycling rate of construction waste by separated collecting
- Less noise, air pollution, greenhouse gas and light emissions
- Better urban design integration to minimize aesthetic disturbances

4.5 Work Safety and Health Protection

The aim of this criterion is to reduce the number of accidents on construction sites and to protect the life and limb of workers. According to the latest statistics published by the AUVA (Allgemeine Unfallversicherungsanstalt - the Austrian Workers' Compensation Board), one in five work accidents are happening on construction sites, which makes the accident rate in the construction industry approximately two and a half times higher than in other sectors.

In addition to the physical suffering, especially the financial aspect should not be underestimated. So does an average work accident in Austria cost 22.511 EUR. These costs result from the pensions, the accident medical care, as well as costs for businesses and other public areas.

The defined sub-criteria of this criterion form an action plan. The characteristics can be an addition to applicable laws, such as the worker protection and construction site coordination law, and therefore significantly contribute significantly to the reduction of labor accidents at construction sites. The contracting authority will need to define the weighting of the individual criteria, as this mainly depends on the type of the construction site.

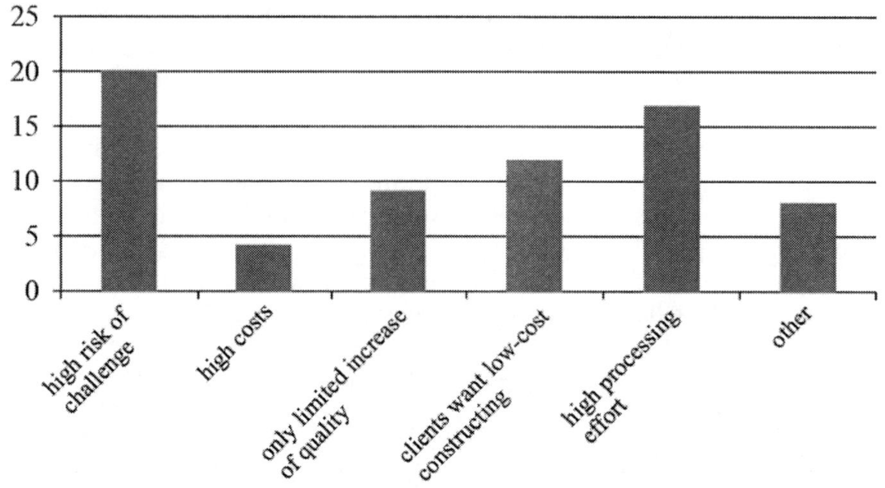

Figure 2. Problems with the implementation of quality criteria (Stabauer 2015).

5 DISCUSSION

Research as well as literature show, that the potential to introduce quality criteria in the tendering process is enormous and generally desired. Even though there are many proponents that require the application of quality criteria in a right weighing, many aspects haven't been clarified so far, so that there are still some uncertainties left.

Figure 2 shows that most of the questioned experts think, that there will be a high risk that tender decisions will be challenged, if awarding criteria are used. Also it is the general understanding that the cost for the procurement procedure will increase as a result of the higher efforts that have to be invested setting up a criteria catalogue, as well as the evaluation of the bids and the monitoring and controlling of implementation on the construction site.

Those facts compete with the facts that quality criteria, like "alternative offers", or "environmental-friendly execution of construction works" will not only improve the quality of building constructions, but also will relax the market situation and the price war between market participants. Also the decreasing price pressure on contractors will benefit the relations between clients and contractors.

6 CONCLUSION

The topic of awarding construction contracts to contractors with the lowest acquisition costs or to the ones with the most economically advantageous tender is heavily discussed. The process of tendering will have to be reconsidered since not only life-cycle-costs of buildings come into consideration, but also the ongoing price war between contractors is a growing issue in the construction industry.

If criteria, as the ones mentioned in this paper, are applied correctly, they can help to establish a tendering culture, that does not only award bids with the lowest acquisition costs, but also improve the general understanding that it is worth to also take economical factors into consideration in order to make construction processes more efficient, environmental-friendly and safer.

References

AUVA, Unfallstatistik Bauwesen 2014. Retrieved from http://www.auva.at/portal27/portal/auvaportal/content/contentWindow?contentid=10008.542535&action=b&cacheability=PAGE&version=1442218376 on October 3, 2015.

Bauer, B., Binder, M., and Heck, D., Preisdruck bei Planerleistungen, *bauaktuell*, Linde Verlag, 178-183, September, 2014.

Commission Regulation EU No. 1336/2013.

Directive 2014/24/EU of the European Parliament and of the Council of 26 February 2014.

Gast, G., *Das öffentliche Vergabewesen in Österreich: Praxishandbuch zur Auftragsvergabe nach dem neuen Bundesvergabegesetz,* Heidelberg, Redline Wirschaft, 2006.

Stabauer, J., *Zuschlagskriterien im Bestbieterverfahren*, Graz, Institute for Construction Management and Economics, Graz University of Technology, 2015.

Till, S., *Zuschlagskriterien im Bestbieterverfahren*, Graz, Institute for Construction Management and Economics, Graz University of Technology, 2015.

Interaction between Theory and Practice in Civil Engineering and Construction
Edited by Komurlu, R., Gurgun, A. P., Singh, A., and Yazdani, S.

OPPORTUNITIES FOR ATTEMPTING AMICABLE SETTLEMENT ALONG THE CLAIM/DISPUTE TIMELINE

MOHAMED-ASEM U. ABDUL-MALAK[1] and SIBA J. JABER[2]

[1]*Civil and Environmental Engineering, American University of Beirut, Beirut, Lebanon*
[2]*Engineering Management Program, American University of Beirut, Beirut, Lebanon*

Claims and disputes are considered an inevitable consequence of the construction process, and project participants are becoming more aware of their costly and lengthy resolution process. The conditions of contract normally include procedures for the submittal, administration and resolution of claims, while it is widely recognized that attempting to resolve claims/disputes through amicable settlement as opposed to arbitration or litigation can save on a lot of time and money. This paper presents a critical analysis of the windows of opportunities offered under both the 1987 and 1999 FIDIC conditions of contract for attempting to resolve claims/disputes amicably. It first focuses on the 56-day period specifically dedicated for attempting an amicable resolution of disputes and highlights the possibilities of effectively stretching this period in more than one way. It further explores the opportunities available at the earlier stages of the claim/dispute timeline, before a claim is escalated to the level of a dispute.

Keywords: Construction, Claims, Disputes, Resolution, Amicable, Settlement.

1 CONSTRUCTION CLAIMS AND DISPUTES

Projects with complex designs and contract conditions are likely to result in extra costs and/or delays for the Contractor, which may cause claims to arise. Contractors resort to construction claims to recuperate the additional costs incurred during any construction project, and tend to argue that owners are not always fair when judging their entitlement to compensation (Fawzy and El-adaway 2012). The costly and lengthy resolution of these claims and disputes has become a matter of concern for the parties involved. If construction conflicts are not effectively addressed and managed, they can evolve into serious disputes, which may result in additional time-delay costs (Seifert 2005). Resolving disputes and managing claims in an optimal and timely manner (in accordance with the contract conditions) can minimize cost, time, and tension, which is why the claim-dispute resolution process is critical to any project and should allow the parties to rightfully achieve their objectives (Cheeks 2003). When submitting a claim, the contractor must closely follow the steps dictated in the contract conditions. The International Federation of Consulting Engineers (FIDIC) has prescribed in its standard contract documents a protocol to be followed by the parties to track and administer claims/disputes. It is expected that, after a claim has been notified and detailed particulars submitted, consultation discussions and various attempts at reaching an agreement on the matter are to take place. At this point in time, the matter at hand is still considered a claim, and has not yet evolved into a dispute. As illustrated in Figure 1, the FIDIC's guidelines explain that a claim develops into a dispute if one or more of the following factors are fulfilled:

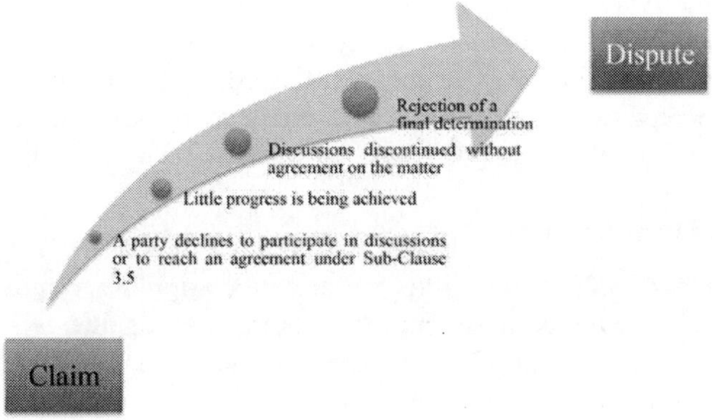

Figure 1. Claim escalating into a dispute.

The completed research work, being reported on in this paper, aimed at the identification of opportunities along the claim/dispute timeline that can be viewed as potential windows for attempting and achieving agreement/settlement of claims/disputes before any such unresolved maters are referred to arbitration/litigation. Five windows have been identified as such, and their candidacy for being explored by the parties for achieving amicable agreements/settlements has been verified through the analysis of claims/disputes administration on nine recent projects undertaken in the Middle East and North African (MENA) region. This paper focus only on the critical analysis part of the research which demarks the windows of opportunities offered under both the 1987 and 1999 FIDIC conditions of contract for attempting to resolve claims/disputes amicably.

2 AMICABLE SETTLEMENT

An amicable settlement of a dispute can be defined as an agreement that is reached to the satisfaction of all parties involved, by way of avoiding what may otherwise be lengthy and costly dispute resolution procedures. Such a settlement can occur at earlier stages along the claim/dispute timeline depending on the parties' attitudes, but it may also be delayed until after the issuance of a determination/decision in regard of any such claim/dispute. According to Sub-Clause 20.5 of the 1999 FIDIC conditions of contract, after a notice of dissatisfaction with a decision by the Engineer or the Dispute Adjudication Board (DAB) is issued, the conflicting parties are allowed a period of time to attempt to resolve the dispute amicably. Here, amicable settlement is required as a condition precedent to the commencement of arbitration, although this sub-clause does not prescribe any method/procedure for pursuing such a settlement. According to Booen (2000), no method is specified in order to give the parties the greatest flexibility in the choice of the procedure, including direct negotiation, conciliation, or mediation.

2.1 Motives for Amicable Settlement

Attempting to resolve disputes through amicable settlement as opposed to arbitration or litigation gives the parties on either end of the dispute the chance to exercise some degree of control over the way their disputes are administered and resolved. A number of factors may motivate a party to settle amicably before resorting to litigation or arbitration. These could include the desire to maintain good relationships with the other party, the concerns for the time and cost that might be spent if the dispute is to be settled arbitration/litigation, the weakness of one's own case, and the concerns about the lack of assets of the opposing party, among others (Panov and Petit 2015).

2.2 The Claim-Dispute Phases

The 1987 and 1999 FIDIC conditions of contract stipulate requirements and their associated time bars for the administration of claims and disputes. As such, the claim-dispute timeline can be understood to involve four major phases, as illustrated in Figure 2. The interests and – correspondingly – the attitudes of the parties are expected to differ between any one phase and the next, depending on how the claim/dispute details and emanating discussions unveil and evolve, respectively. A critical aspect of this timeline is when the determination by the Engineer is challenged by either party, and the matter is subsequently referred as a dispute either back to the Engineer or – alternatively – to the DAB, depending on the contract's stipulation, for the issuance of an Engineer's Decision or a DAB's Recommendation or Determination (as the case warrants), respectively.

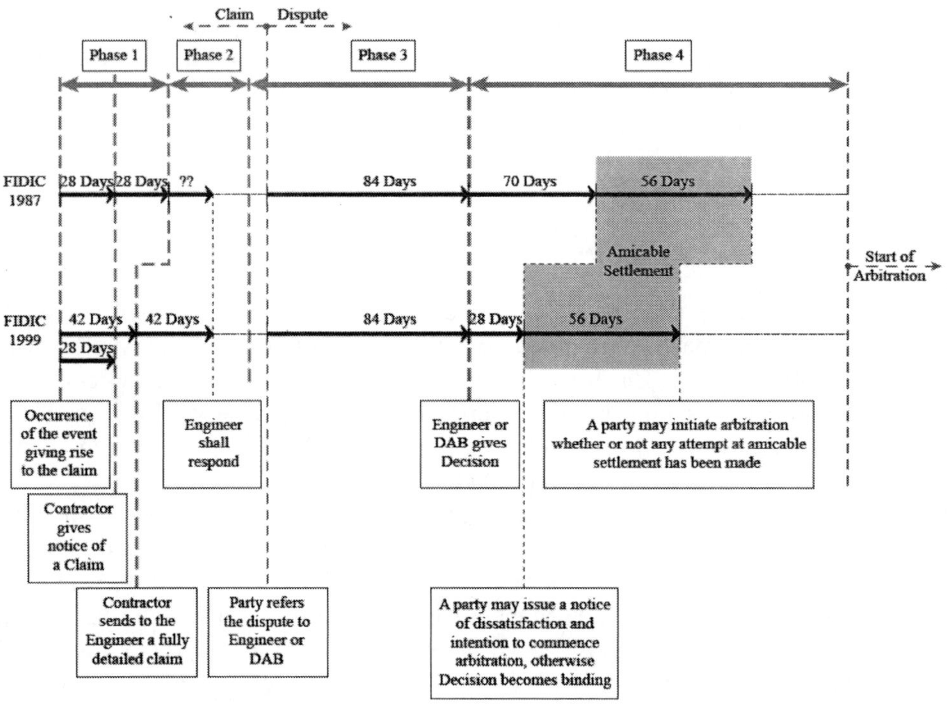

Figure 2. Claim-dispute timeline under the 1987 and 1999 FIDIC's conditions.

2.3 Effective Length of the Amicable Settlement Period

Although Sub-Clause 67.2 of the 1987 and Sub-Clause 20.5 of the 1999 FIDIC conditions of contract both allocate a mandatory period of 56 days along the dispute timeline during which amicable settlement can be attempted and achieved, this period can effectively be construed in several ways, thus resulting in an extension of this stipulated period, as discussed hereunder.

2.3.1 *Issuance of the notice of dissatisfaction*

The disputing parties are allowed a 70-day period (1987 FIDIC) or a 28-day period (1999 FIDIC) following the date when an Engineer's or a DAB's decision is issued to give a notice of dissatisfaction; otherwise, the issued decision becomes contractually binding. This is followed by a 56-day period allowing for amicable settlement. It becomes evident here that the point along the dispute timeline when the notice of dissatisfaction is actually issued can allow for more or less time to attempt amicable settlement. That is, the 70-day or 28-day period given for the disputing parties to consider the issued decision and entertain accepting or rejecting it can be added to the 56-day period specifically stipulated for allowing amicable settlement attempts. It can therefore be argued that indirectly extending the effective period for amicable settlement in such a backward manner, as shown in Figure 3(a), by not rushing the issuance of the notice of dissatisfaction, can be considered advantageous for the disputing parties, as it allows them more time to try and resolve their differences as a last resort prior to arbitration or litigation.

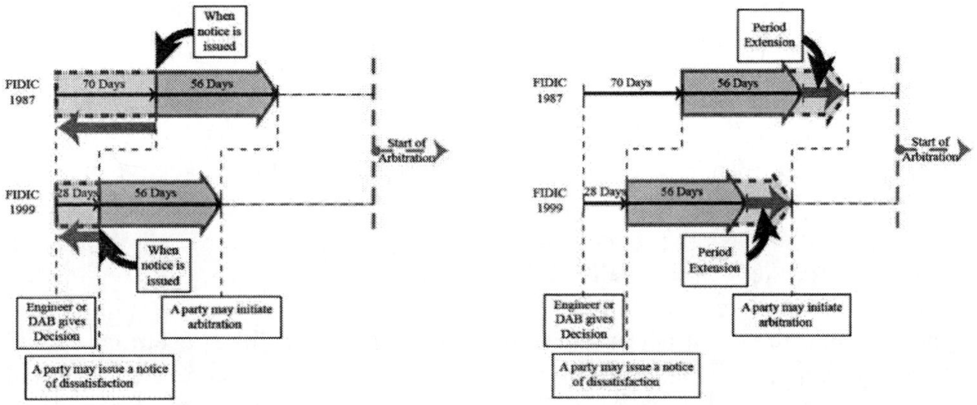

(a) (b)

Figure 3. Extensions to the amicable settlement period.

2.3.2 *Extension by mutual agreement*

Sub-Clause 67.2 of the 1987 FIDIC and Sub-Clause 20.5 of the 1999 FIDIC conditions of contract state that "*unless the parties agree otherwise,* arbitration may be commenced on or after the fifty-sixth day after the day on which notice of dissatisfaction was given, even if no attempt at amicable settlement has been made." It therefore follows that the originally stipulated 56-day period can be extended to allow more time for attempting amicable settlement, provided that both parties mutually agree to any such extension. It is worth noting that the maximum number of days of any such extension can be contemplated at the onset of the project and can therefore be specified in the particular conditions of the contract. Alternatively, the number of days can also be extended at any time before the expiry of the 56-day period, if both parties feel they need more time to attempt settlement. This extension is shown in Figure 3(b).

3 DEMARCATION OF AMICABLE SETTLEMENT WINDOWS

As discussed above and seen in Figure 4, there already exist two windows for attempting amicable settlement following the issuance of a decision on a referred dispute: Window 1, which is demarked by the originally stipulated 56-day period and any extension to it, mutually approved by the parties, and Window 2, which is demarked by extending the originally stipulated 56-day period in a backward manner, to reflect not rushing the issuance of a notice of dissatisfaction by either party. A third window of opportunity can be further identified, which demarks the period extending between the expiry of the 56-day period (or any possible extension to it) and the date of initiation of arbitration. This could also include any further notice period required prior to the commencement of arbitration. Window 3 might extend considerably for various reasons, one of which, as verified by the project cases examined, can be the stipulation of the issuance of the taking-over certificate as a condition precedent to the initiation of arbitration. To be further noted is that while the FIDIC conditions stipulate the expiry of the 56-day period before arbitration may be commenced, an upper bound as to when such an initiation may take place is not prescribed.

The three windows discussed above fall along the dispute part of the claim-dispute timeline. However, additional windows exist on the claim part of the timeline that allow the parties to resolve their differences early on, when the matter has not yet developed into a dispute. To this effect, Window 4 extends from the moment when the Engineer responds with a determination, at least on the principle of the claim, until the matter is referred back to the Engineer, or to the DAB, for a decision. During this period, further particulars may be submitted and the issued determination deliberated. At this stage, the tension between the parties has not escalated, and communication between the parties can usually be attempted without the need for third-party assistance. It was verified through the project cases examined that this period can extend well into the construction duration as well as beyond the substantial completion of the works, due to the claiming party's concerns as to the consequences of potentially having to reject an Engineer's or a DAB's decision and, consequently, having to notify the other party of the intention to commence arbitration. The fifth window represents the first opportunity for reaching agreement on any matter in question. This is inherent in the language stipulated under Sub-Clause 3.5 of the 1999 FIDIC conditions, requiring the Engineer to consult with each party in an endeavor to reach agreement. At this early stage, the role played by the Engineer is expected to have a significant bearing on the way the differences concerning the matter in question may evolve.

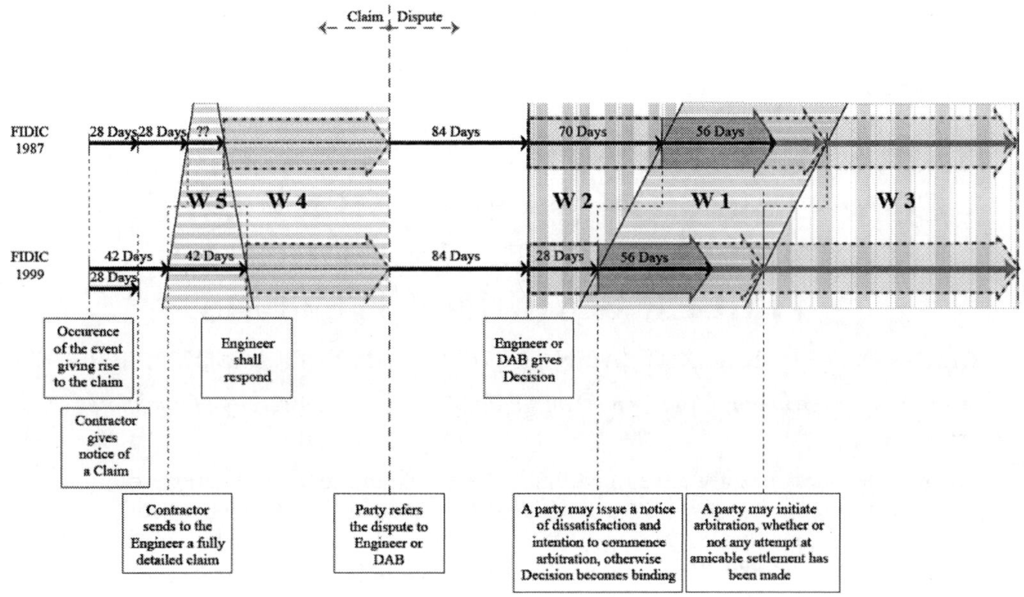

Figure 4. Demarked windows for attempting agreement/settlement.

The above-demarked windows should be considered as opportunities that – if not taken for granted – could help resolve claims/disputes expeditiously, save the parties time and money, and help maintain good relationships among them.

References

Booen, Peter L., The FIDIC Contracts Guide: Conditions of Contract for Construction, Conditions of Contract for Plant and Design-build, Conditions of Contract for EPC/turnkey Projects. Lausanne: Fédération Internationale Des Ingénieurs-Conseils, 2000.

Cheeks, J. Richard, Multistep Dispute Resolution in Design and Construction Industry, *Journal of Professional Issues in Engineering Education and Practice*, 129.2, 84, 2003.

Fawzy, Salwa A. and Islam H. El-adaway, Contract Administration Guidelines for Managing Conflicts, Claims, and Disputes under World Bank Funded Projects, *Journal of Legal Affairs and Dispute Resolution in Engineering and Construction*, 4.4, 101-110, 2012.

Panov, Andrey, and Sherina Petit, Reviews - Arbitration News, Features and Reviews - Global Arbitration Review, globalarbitrationreview.com, Retrieved from www.globalarbitrationreview.com on September 28, 2015.

Seifert, Bryan M. "International Construction Dispute Adjudication under International Federation of Consulting Engineers Conditions of Contract and the Dispute Adjudication Board, *Journal of Professional Issues in Engineering Education and Practice*, 131.2, 149-157, 2005.

Interaction between Theory and Practice in Civil Engineering and Construction
Edited by Komurlu, R., Gurgun, A. P., Singh, A., and Yazdani, S.

INDUSTRIALIZED TIMBER CONSTRUCTION: CONSTRUCTION MANAGEMENT ASPECTS AND INFLUENCES IN MODULAR TIMBER BUILDING SYSTEMS

JOERG KOPPELHUBER, KATHARINA HINTERSTEININGER, and DETLEF HECK

Institute of Construction Management and Economics, Graz University of Technology, AT

Advancements of recently developed timber products and their application within the prefabrication process in construction industry demand more standardization and unification of construction management processes. Within the timber industry the prefabrication of modular building units with an optimised prefabrication grade experienced an ascending development over the last decade. Today's methodology is based on a formal concept of realized projects that provides the methods of the production whereas the basic components behind do not change. Especially modern timber construction is mostly suitable for prefabrication, caused by the material specific properties, possibilities in prefabrication and quick assembly time under dry construction circumstances. According to an expert survey undertaken within a research project, the integration of a consistent data workflow in the production process allows a significant shortened construction period and cost reduction. Following these expert interviews timber construction systems with a high percentage of prefabrication are considered to be less expensive than on-site building methods. High productivity combined with excellent quality in short installation periods delivers arguments for best practice examples in construction management of modular timber building systems.

Keywords: Prefabrication grade, Cost reduction, Potentials, Turn-key system, Standardization, Cross laminated timber.

1 INTRODUCTION

Modern timber construction systems are different to the traditional work of carpenters known over the last centuries. This is mainly caused by technical developments in the recent decade as well as the size, complexity and application was extended significantly. Therefore these building systems require further analysis to ensure confidence and settlement within the building market. Although cross laminated timber (CLT) experienced great success it was hardly examined up to now regarding economic aspects and construction management in general. The presented expert survey gives an insight into the current situation of modern timber building systems with a high variety in the grade of prefabrication (Staib 2008). Highly prefabricated 2D- timber elements and 3D-timber modules have a substantial impact on the performance for large scale timber construction systems and its reliability in future in order to originate a leading appropriate building system.

2 PRINCIPLES OF INDUSTRIALIZED CONSTRUCTION

An industrialized building system is the transfer of various industrialized functions from the stationary industry into the decentralized systems of the construction industry. However the word prefabrication is used generally for building in series, building in systems that include production and construction with 2D-elements and 2D- or 3D-modules. Industrialized construction encompasses a production of building elements in an industry setup that is not subject to changes in weather (Girmscheid 2010).

Building in systems often implements the typical building systems of the 1960's with its unification in repetitive multi-storey concrete buildings in middle Europe. To achieve an economical advantage at that time building components were standardized and produced at a high quantity. In contrast the introduction of the principles of lean production the leading production systems were changed completely (Piller 2006).

2.1 Modular Timber Building Systems

The industrialized timber building section, which differentiates significantly from the traditional carpentries through the implementation of automated production and prefabrication methods has been developed constantly by introducing technical developments on the material level as well as computer based manufacturing process over the last decades (Lennartsson 2009).

The types and systems of the industrialized timber construction division range from simple and unready products to semi-finished building elements and turn-key modules. The variety on the market is high and complex as many producers are selling their own systems especially developed for their needs and markets. However the current state of the art in middle Europe in modular timber buildings with a turn-key prefabrication grade up to 95 % does show the possibilities within this building material and potential for a wide range of use. Especially in the case of buildings with a repetitive room arrangement such as hotels, student accommodation and nursing homes with completely prefabricated room-modules done in timber are indicating the high quality in the building sector in middle Europe nowadays.

2.2 Expert Survey on Industrialized Timber Construction

Accurate data and reflections on industrialized building systems with its advantages, constraints and challenges in future are not well investigated especially when it comes to timber construction methods and systems. To generate considerable data an expert survey was conducted from November 2014 to February 2015 (Hintersteininger 2015). Within these experts of various disciplines were interviewed such as architecture and engineering, builders and SME's, public and private investors as well as experts from R&D. The results from this survey where then compared to existing and comparable investigations on these topics to verify the accuracy of the expert statements.

3 PREFABRICATION IN TIMBER CONSTRUCTION

The characteristic of timber as building material dominates the process, and thus offers a wide range within the prefabrication applications. Partial or entire prefabrication is not clearly defined in-depth and often misused for various types of preassembled elements. Especially during the planning phase, a systematic and consequent coordination is required to develop modules with a clear joining technology. Basic grids form the fundamental set out not only for building physics and structural line up but also for the production process and transportation constraints (Staib 2008).

3.1 Significance of Prefabrication in Timber Construction

Compared to international research also the survey undertaken shows that prefabricated building systems offer a constant growth within the upcoming years. Especially in the field of multi-storey residential and commercial buildings as well as intra-urban and inner-city developments industrialized building systems implement great advantages and a prosperous future is illustrated.

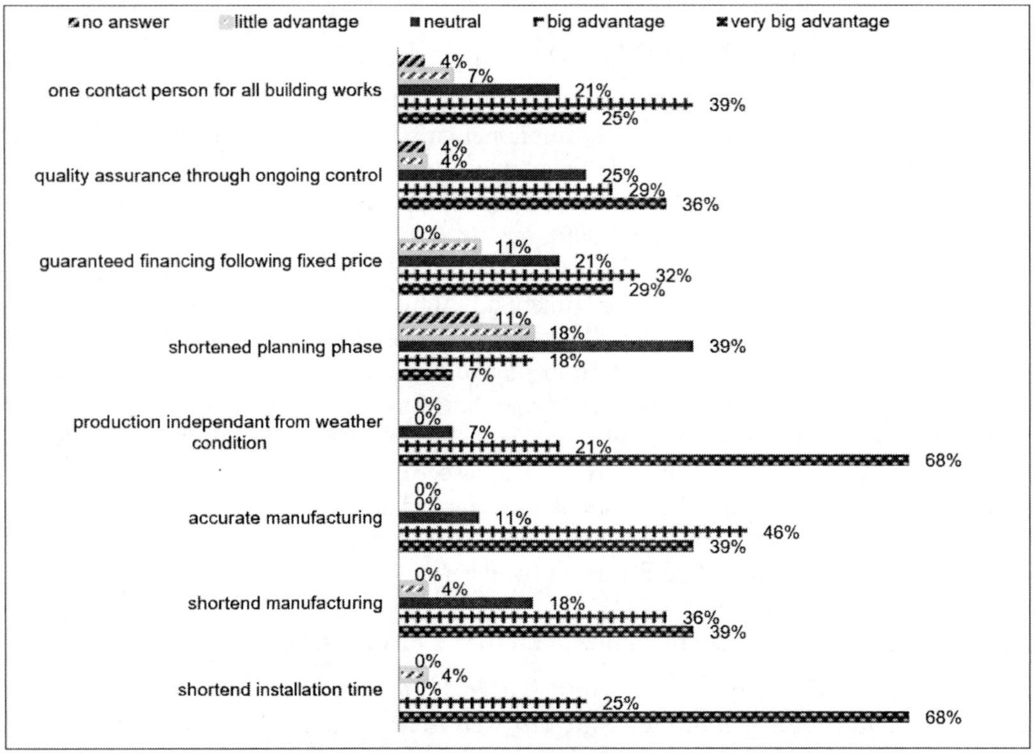

Figure 1. Expert survey advantages of industrialized compared to traditional systems (Hintersteininger 2015).

However the reasons for using industrialized building systems are not related on the material used are diverse. Figure 1 originated through the expert survey shows that especially the quality reasons, lower general building costs on site as well as an easier production under dry and constant conditions do include the greatest advantages for using prefabricated building systems. Accurate production methods as well as fast manufacturing and short installation time on site form the main benefit in implementing more industrialized building systems in the construction industry.

3.2 Criteria Grade of Finishing Works

The grade of prefabrication offers a range of interpretations. However, there is a lack of understanding in literature on the economics of the grade of prefabricated finishing work. Meanwhile, the expert survey features a constant increase of the prefabrication market, especially in timber building systems over the next few years (Schnittich 2012). This includes systems using 3D-modules with a high degree of prefabrication as well as 2D-elements with finished surfaces internally as well as externally.

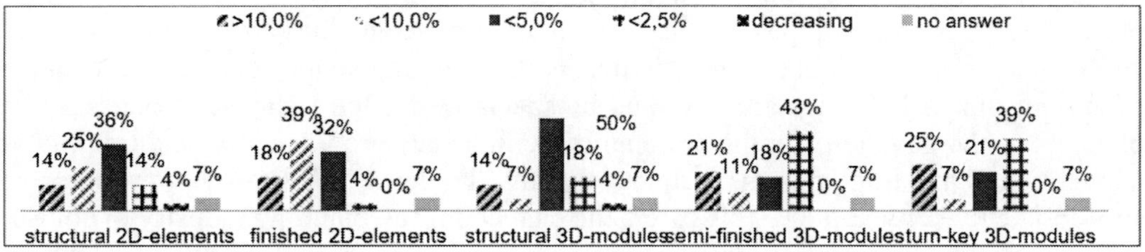

Figure 2. Expert survey market development of industrialized timber building systems depending on the grade of finishing works (Hintersteininger 2015).

4 CONSTRUCTION MANAGEMENT ASPECTS

Within the big variety of products and manufacturing process general construction management aspects such as modularity, turn-key building systems, comparable cost calculation schemes, adequate integral planning tools as well as the ongoing discussion on the topic of interfaces on trades have the advantage to allow the development of timber from a simple product level to a system level within the construction industry.

4.1 Modularity in Timber Construction

Building systems are always generated by a set up and combination of 2D-elements and 3D-modules. Various impact factors have the opportunity to develop traditional building components into industrialized building systems without losing the advantages and significance of the individual tasks but generating a bigger benefit by an intelligent combination of the planning and production factors (Rinas 2011).

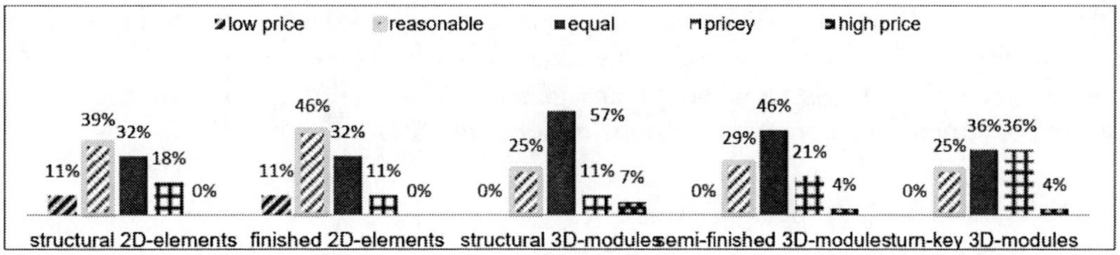

Figure 3. Expert survey cost comparison of traditional to industrialized building systems (Hintersteininger 2015).

This can be spotted in the expert survey by identifying the cost influence and chance using a higher grade of prefabrication within the finishing works. So the expert opinion offers a fast benefit when using prefabricated 2D-elements and 3D-modules; however finalized turn-key 3D-modules are still indicated with higher costs but with a very large opportunity in the coming years.

4.2 Planning in Modular Timber Building Systems

When it comes to the discussion of planning systems, level of detail and simple questions of who is responsible for developing which detail and drawing, not only the timber industry has the discussion on responsibilities. However as various timber products have been developed recently, a lack of knowledge does appear not only when it comes to construction but especially during the planning stage. Therefore a majority of current timber buildings are designed by specialized often in-house timber industry planners as the general planning community with a standard education and training on conventional building systems cannot comply the level of detail required for an appropriate timber solution. This offers the chance for the planning industry to develop new and simple integral planning systems on one hand with a big market development in the future. But on the other hand a large variety of new materials, construction systems, details in connection and building physics and new timber building systems contain the risk of failure by incorporating too much variability for simple planning tasks. More complexity in 3D-systems has an exorbitant higher risk when used. Therefore the implementation of integral planning systems such as building information modeling (BIM) is needed for the entire timber construction industry in future (Girmscheid *et al.* 2010).

4.3 Interfaces – Materials – Building Systems

The increase in complexity in the construction industry has a major impact also on the timber building systems. Especially the fact that recently developed building materials and products have a low level of trial and error and impound a big risk when implemented in today's building systems. The number of interfaces increased over the last decades exorbitant as numerous trades and business urge into the construction market. The level of detail is much higher than 10 years ago where many details could be solved on site when the question of how a new systems should be include in conventional building solution didn't even exist. Nowadays complex CAD-CAM combined systems do offer the chance to reduce the risk of failing by too many interfaces. Just-in-time deliveries of products and systems which need to be implemented in standard structural elements do offer a big chance especially cost wise and the majority of costs nowadays are generated by building services engineering instead of structural components (Koppelhuber *et al.* 2015).

5 POTENTIALS AND TENDENCY

The high potential of timber construction building systems can clearly be spotted in the short period of installation required on site in comparison to conventional building systems. Especially prefabricated 2D-elements and 3D-modules deliver arguments for the wider use of timber as the main structural material even though the cost impacts are not commonly used as arguments as they are hardly quantifiable and ratable.

5.1 Potentials of Prefabrication

Although timber construction has a high potential for pioneering tasks especially when implementing turnkey strategies transforming the advantages into an integral building systems (Rinas 2011), the successful way over the last years of this building material can be extended further as also the expert survey covers these facts. Effort is required not only in the technical and material sector but also in the building economic domains.

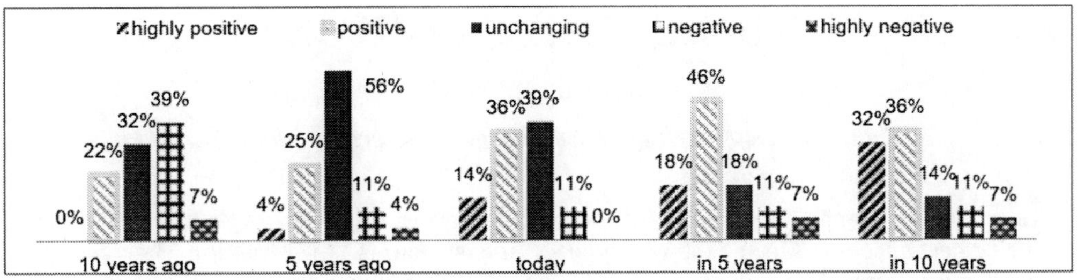

Figure 4. Expert survey potential of modular timber building systems (Hintersteininger 2015).

Particularly for achieving a higher planning reliability and better quality assurance a standardization of building components as well as system components in general are required and the planning and management process needs to be more professionalized.

In addition the rising demand makes the development of a comprehensive prefabricated timber construction system, as well as neutral cost studies of the effects of an industrialized construction and adaptation of building regulations mandatory. It can create another step towards industrialization in the building industry by implementing turn-key concepts to create meaningful construction management tools to allow ecological friendly material to become a widely used standardized building system.

References

Girmscheid, G., Scheublin, Frits., (Eds.) for CIB Task Group, *New Perspective in Industrialisation in Construction – A State of the Art Report*, Zürich, Eigenverlag des IBB an der ETH Zürich, 2010.

Girmscheid, G., *Strategisches Bauunternehmensmanagement*, Berlin, Springer Verlag, Berlin-Heidelberg, 2010.

Hintersteininger, K., *Kennzeichen und Aspekte des industriellen Bauens – Anwendbarkeit im Holzbau*, Masterthesis, Graz University of Technology, 2015.

Koppelhuber, J., Zuegner, D., and Heck, D., *Construction management of multi-storey residential timber buildings – assessment criteria and their impact in the cost calculation*, technical paper and poster, Forum wood building Nordic Växjö, Sweden, 2015.

Lennartsson, M., *Modularity in Industrialised Timber Housing*, Luleå. Universitetstryckeriet, Luleå, 2009.

Piller, F. T.: *Mass Customization; Ein wettbewerbsstrategisches Konzept im Informations-zeitalter*, Wiesbaden, Deutscher Universtitätsverlag, 2006.

Rinas, T., *Kooperationen und innovative Vertriebskonzepte im individuellen Fertigteilbau. Entwicklung eines Geschäftsmodells,* PhD thesis, Zürich, ETH Zürich, 2011.

Schnittich, C., *Vorfertigung – Hightech und Handarbeit*, In: Detail, Zeitschrift für Architektur und Bauteil, Vorfertigung, 6/2012.

Staib, D. R., *Elemente und Systeme, modulares Bauen, Entwurf Konstruktion neue Technologien*, München, Institut für internationale Architektur Dokumentation, 2008.

THE CRUCIAL ROLE OF "CMa" IN PRECONSTRUCTION PHASES: KNOWLEDGE AND SKILL REQUIREMENT

TAYİBE SEYMAN

Architecture Dept, Istanbul Kemerburgaz University, İstanbul, Turkey

Construction Management is one of a significant project delivery system in search for the project success. One of the two types of this system is "Construction Management as Advisor" which involves a professional entity called Construction Manager as Advisor (CMa) which advises the Owner from the initial decisions to the completion of the construction. There has been a growing movement towards the project success factors in the past number of decades. However conducting research on project success factors related to the preconstruction phase is not as common. The aim of this paper is to review the crucial role of CMa in the preconstruction phase and identify the knowledge and skill requirements. By doing so, this study also aims to be a reference to the owners in selecting or hiring a CMa which can be a person or an organization. Since construction management delivery system is relatively a new method in Turkey. In order to define knowledge and skill requirements; the services to be given by a CMa in American Institute of Architects contract documents are analyzed by one by. The research findings showed that 22 different services have to be given by CMa in the preconstruction phase. The skill and knowledge requirements of those services are discussed and identified accordingly.

Keywords: Project success, Pre-construction, Construction management, Building defects.

1 INTRODUCTION

The focus of construction industry for project success in construction industry is mostly on the construction process itself. However a building project is a cascade of phases starting with the decision of owner and the conception of the project of which outcome will be an input fort the next phase. Moreoever most of the decisions are given before the construction starts in preconstruction phases. Since those decisions given will have high impacts on the construction process and the building itself as an end product, this study aims to explore the advantages of a CMa to improve the preconstruction phases. Sommerville (2007) examined more than a thousand design deficiencies and reached a conclusion that the reasons of them are; poorly prepared technical details and misconsidered spesific requirements. Also it is mentioned that if the design team put a clear image for both the contractors and the occupants, problems would be less. According to the study of Reddy and Andrew (2005), the design and the construction of a building is actually an integrated system, however construction professionals are mostly not take into consider the strong relationship. At the construction process, one of a main reason of delays and cost increases is rework. According to Sommerville (2007), ambiguity causes the delays. Additionaly ambiguity rises because of incomplete, uncertain and conflicting type of information which is produced at the preconstruction phases. Preconstruction phases consist of four different stages; Project Conception, Design, Construction Documents and Bidding. Design involves Schematic Design and Design Development (PRM 2005). According to PRM (2005), the Project conception activities which are the responsibility of the owner are feasibility study, programming, site analysis and site selection. Meanwhile the owner is required to produce program, budget and schedule of the Project properly. As an investor and not an expert in building construction, owner will need to coordinate and manage different proffesionals or consultants to provide those activities and documents which are required to include complete, certain and clear information. On the other hand all those first information such as budget, schedule, programming will be the determinants of the quality of the design of the building after all. Additionally, the ability of decisions given in the early stages to influence costs of building project's is higher as seen on the Figure 1. The conception has the highest influence on the cost which will be a performance measurment for the construction and success level for the building.

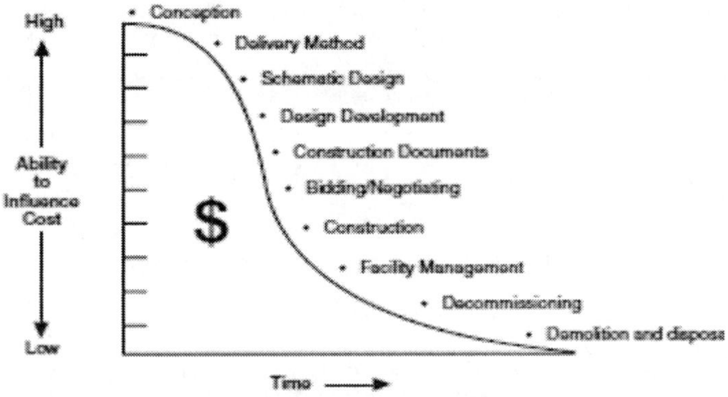

Figure 1. Ability of decisions to influence costs at different stages (PRM 2005).

2 CONSTRUCTION MANAGER AS ADVISOR

Construction management is the process of professional management applied to a construction project from conception to completion for controlling project time, cost, and extent (PRM 2005). One type of the Construction Management Project delivery system is the Construction Manager as Advisor (CMa). Without having any financial risk, the CMa's role is to advise the owner throughout the phases of the building project starting from the initial decisions. When an owner implements a project using a CMa, it allows the owner to make use of advice that is unaffected by any potential conflict of interest (PRM 2005). One of the primary advantages of CMa is utilizing the CMa's construction expertise during the Pre-Construction Phases to advise on design constructability, scheduling, cost and budget control, contract and bid document preparation, and other (CMMA 2015). Therefore in this study, by focusing on the availability of a CMa in preconstruction phases, it is going to be disscused that how crucial is the preconstruction services of a CMa for the success of the building.

3 PRE-CONSTRUCTION PHASES SERVICES OF CMa

The pre-construction services to be given by a CMa are listed as brief descriptions below as Table 1, by the light of C132 2009, the standard contract document of American Institute of Architects (AIA), between CMa and the Owner. Required knowledge areas are noted for each service (Table 1).

Table 1. Knowledge area requirements for CMa pre-construction services.

No	CMa Pre-construction Services (AIA C132,2009)	Required Knowledge Area
3.2.1	*to Review the program furnished by the Owner.*	PROGRAMMING
3.2.2	*to Provide a preliminary evaluation of the Owner's program, schedule and construction budget requirements, each in terms of the other.*	PROGRAMMING COST&BUDGET CONT.
3.2.3	*to Prepare a written Construction Management Plan*	COORDINATION
3.2.4	*to Prepare preliminary estimates of the Cost of the Work or the cost of program requirements.*	COST& BUDGET CONT.
3.2.5	*to Review design documents during their development and advise on proposed site use &improvements, selection of materials,& building systems&equipment. Provide recommendations on constructability, availability of materials&labor, sequencing for phased construction,*	DESIGN REVIEW SITE EVALUATION
3.2.6	*to Prepare and periodically update Project schedule*	SCHEDULING
3.2.7	*to Prepare and update estimates of the Cost of the Work of increasing detail&refinement.*	COST& BUDGET CONT.
3.2.8	*to Make recommendations whenever determines that design details adversely affect constructability, cost or schedules.*	DESIGN REVIEW
3.2.9	*to Provide recommendations and information regarding the assignment of responsibilities for temporary Project facilities and equipment, materials and services.*	CONSTRUCTABILITY REVIEW
3.2.10	*to Provide recommendations&info regarding the responsibilities for safety programs among the Contractors.*	CONSTRUCTABILITY REVIEW
3.2.11	*to Provide recommendations on the division of the Project into individual Contracts for the construction of various categories of Work*	CONSTRUCTABILITY REVIEW
3.2.12	*to Update the Project schedule to include the components of the Work, including phasing of construction, etc.*	SCHEDULING
3.2.13	*to Expedite and coordinate, the ordering and delivery of materials*	SCHEDULING
3.2.14	*to Assist the Owner in selecting, retaining and coordinating the Professional services of surveyors, special consultants, testing laboratories*	CONSTRUCTABILITY REVIEW
3.2.15	*to Provide an analysis of the types&quantities of labor required for the Project&review the availability of labor.*	CONSTRUCTABILITY REVIEW
3.2.16	*to Assist the Owner in obtaining information regarding applicable requirements for equal employment opportunity programs&other programs*	LEGALITIES
3.2.17	*to Update and submit the latest estimate of the Cost of the Work and the Project schedule, after the approval of the Drawings and Specifications,*	COST&BUDGET CONT.
3.2.18	*to Submit the list of prospective bidders.*	BIDDING
3.2.19	*to Develop bidders' interest in the Project and establish bidding schedules. to Issue bidding documents to bidders and conduct pre-bid conferences*	BIDDING CONSTRUCTABILITY R.
3.2.20	*to Receive bids, prepare bid analyses&make recommendations*	BIDDING
3.2.21	*to Assist the Owner in preparing Construction Contracts and advise on the acceptability of Subcontractors and material suppliers.*	CONSTRUCTABILITY REVIEW
3.2.22	*to Assist in obtaining building permits and special permits.*	LEGALITIES

4 KNOWLEDGE AND SKILL REQUIREMENTS OF CMa IN RELATION TO THE PRE-CONSTRUCTION PHASES

It is defined in this study that there are twelve knowledge areas related to the pre-construction phases, which will be explained in detail and discussed in the following sections.

4.1 Architectural Programming

It is the starting point and if the program has deficiencies, design defects will be inevitable. Through the availability of a CMa at the very beginning of the project, the needs of owner will be converted into architectural space requirements, which will be the start of high quality design. In order to give architectural programming service or to consult the owner CMa has to have knowledge of architectural design and space standards in addition to cultural and community context, legal issues, cost and construction methods. Basically, architectural programming requires investigative and information gathering skills, analytical skills in addition to strong communication and management skills.

4.2 Design Review

Design review will be only affective if it is done at the preconstruction phases by considering the architectural program, functional requirements, legalities, site restrictions and construction methods. Without having any experience of building design, to make a design review is not quite possible. Skills related to understanding of architectural and MEP drawings in addition to the ability to use the related computer aided design softwares such as Autocad are required.

4.3 Site Evaluation

The size of the site, its anticipated use, and the programming requirements will have a major effect on the scope of work for site analysis services. Having knowledge on climate, topography, soils, and natural features, site utility distribution systems, evaluating site access, understanding of building siting considerations, ability to analyze multiple factors objectively and also ability to work with related or specialty disciplines are needed (Fasla 2015). Additionally, knowledge of feasibility reviews, ability to use related softwares, investigative and analytical skills are required.

4.4 Legalities

Knowledge related to the local building law,codes and regulations is required to make design and constructablity reviews. Any design not considering the legalities would fail since the building permits will depend on to be allign with the local authorities. Knowlegde of code compliance reviews in addition to communication skills and anticipating the possible future requirements are required for a CMa to be a proper guidance.

4.5 Scheduling

Scheduling of preconstruction phases require design process knowledge. Since estimating the times of approvals from the legislative authorities after the submittals is also critical, the knowledge of legalities is required for CMa. Milestones, key dates and submittals timings would affect the start of construction and the completion of the building. One of an important skill required is risk analysis, making assumptions and foreseen the alternatives to set the milestones.

4.6 Constructability Review

Constructability Review is the process of evaluating the construction documents for clarity, consistency, compeleteness and ease of construction to achieve overall project objectives (PRM 2005). Constructibility seeks to eliminate non-productive work on site, make the production process simpler and provide the opportunity for more efficient site management and safer working (Emmit and Groese 2014). The knowledge of construction means, methods, process and systems are required for an efficient constructability review. Without knowing the labor availability to be used for a specific type of a detailing, it is not possible to incorporate that detail into the design. One of an important required skill is ability to anticipate the design details which will possibly cause problems during construction or in the building as an end product.

4.7 Cost and Budget Control

Making cost estimates at different stages of preconstruction phase requires to use specific information related to each stage. Such as making a cost estimate in the conteption stage is different than to make in construction documents stage. Especially the very early stages are difficult since the need of much more assumption without having concrete desicions. Knowledge required are making cost estimates and budgeting especially for the preconstruction stages. Analytic skills are required.

4.8 Bidding

The most critical part of the process is to prepare bidding documents which will give the same understanding to all parties. Unless it can cause disputes at the time of construction of that work item which would mean delay and additonal cost. Knowledge of building construction is required. CMa requires to have detailed thinking skills and the ability to foresee the possible misunderstandable or unclear explanations for the work item. Also, CMa has to have the knowledge of related softwares to control the process and while evaluating the contractors.

4.9 Management and Coordination

Specifically for the preconstruction phases, the estimates for cost and time will be involved in the construction management plan while the uncertainities are at a high level. Coordination skills by organizing effective meetings, on time communications for confilicts, flexible way of thinking are required.

5 CONCLUSIONS

For the success of building project the main advantage of CMa system is the availability of a building construction expert who has the ability to manage and coordinate. Preconstruction phases involve decisions of which results will be visible at the time of construction or even after the construction.

In the preconstruction phases CMa's key role requires wide knowledge especially on design, constructability, cost management, programming and contract documents. While advising the Owner, CMa's communication, coordination, organization and management skills will affect the preconstruction process critically. Moreover those skills will influence the construction directly by giving the proper decisions and implementing them in all related documents .

Having onsite experience on design and construction is an asset for a CMa in order to manage, coordinate or advise properly not only management skills or technical skills are needed but also at some level of experience is required.

Preconstruction phases requires to associate fluid opinions with the construction process and the building at the end. CMa, have to have responsive actions during the evolvement of architectural design.

As a conclusion, for a succesful construction process and a successful building at the end, pre-construction services of a CMa requires knowledge on "architectural programming", "design review", "site evaluation", "legalities", "scheduling", "constructability review", "cost & budget control", "bidding" and "management & coordination" in addition to strong communication, analytical thinking and investigative skills.

References

AIA C132 "Standard Form of Agreement Owner and Construction Manager, CMa Edition", American Institute of Architects, 2009.

CMMA, The Construction Management Association of America, An Owner's Guide To Construction and Program Management, 2011. Retrieved from https://cmaanet.org/sites/default/files/files/Owners%20Guide%20Ver%209-2011.pdf, on December 2015.

Emmit S, Gorse C., *Barry's Introduction To Construction Of Buildings,* Wiley 3rd Ed, 2014

Fasla, F.Z, Site Analysis, AIA The Architect's Handbook of Professional Practice, 13th edition, 2000. Retrieved from http://www.aia.org/aiaucmp/groups/aia/documents/pdf/aiab089275.pdf, on December 2015

PRM, The Project Resource Manual, CSI Manual Practice Courier, Wesford, 2005.

Reddy, G. and Andrew T.N., Construction Project Failures: Towards Systems Thinking Strategies For Improving The Construction Delivery Process, 2005.

Sommerville, J., Defects and Rework In New Building: An Analysis of the Phenomenon and Drivers. *Structural Survey*, Vol.25,No 25, 391-407, 2007.

Interaction between Theory and Practice in Civil Engineering and Construction
Edited by Komurlu, R., Gurgun, A. P., Singh, A., and Yazdani, S.

USE OF THE DELPHI METHOD TO IDENTIFY AND ANALYZE INCENTIVES FOR US HIGHWAY CONSTRUCTION WORKERS

R. EDWARD MINCHIN JR., GAVIN N. HANCOCK, and JAMES G. SULLIVAN

University of Florida, Gainesville, Florida, USA

Crew leaders and foremen on construction job sites are skilled individuals that are dwindling in numbers due to the push on students throughout their secondary schooling to attend college, in addition to other economic factors. The focus of this research is to determine what incentives construction companies implement in order to retain, motivate and increase the productivity levels of their skilled crew leaders and foremen. This study utilizes the Delphi method, which consists of questioning a panel of experts in a series of rounds—with the summaries of responses from each round made available to each of the experts—until a consensus is reached. The Delphi method allows a researcher to utilize the expertise of a smaller population sample than traditional statistical methods, but has gained worldwide acceptance as a substitute for some of those traditional methods. Each construction company operates in a unique, independent manner as they compete to attract and then retain these valuable workers. The main goal of this research is to investigate this as an economic issue, and determine if offering more, or larger, incentives to the skilled crew leaders and foremen is worth the additional cost. The purpose of this paper is to describe how the Delphi Method was utilized to accomplish this.

Keywords: Employee retention, Contractors, Heavy, Civil.

1 INTRODUCTION

In the construction industry, it is critical that general contractors consistently work well with owners and fulfill their contractual requirements, providing a high quality product by the agreed completion date. In order to meet these demands in such a highly competitive industry, while also turning a profit after project completion, these companies must hire highly skilled workers throughout all career levels of the company and maintain a positive reputation in their respective markets. Having such highly skilled workers throughout the company, especially the hourly workers in the field, makes the company productive and enables these client-owner relationships to form. Forming these relationships attracts repeat and new business for the construction company, earning the company more revenue overall.

Every year, construction companies around the nation contribute a portion of their profits to the hard working employees that enable the company to operate efficiently and effectively in their respective sectors. Most of these incentives are distributed to the salaried individuals within the company and are usually weighted in monetary value based on the specific individual's performance. In addition to these employees, it is also important to recognize the individuals that are compensated on an hourly basis due to their high level of involvement in the field operations throughout the construction process, such as the foremen and skilled crew leaders that self-perform work throughout the different trades. Without highly qualified and skilled field workers, the productivity of the construction process is jeopardized, which could result in excessive costs for the construction company or a breach of contract with the owner.

2 RESEARCH OBJECTIVES

The purpose of this research is to compare the differences between the incentives upper management offers the different levels of employees throughout multiple construction companies, with the goal to determine whether or not the foremen and skilled crew leaders are compensated according to their level of involvement and control of the construction process. Prior research has shown that it is important for highly productive workers to be recognized and rewarded in some manner for their accomplishments in many different industries, which contributes to their overall job satisfaction and financial stability; therefore, this study will focus on how this concept is applied throughout the construction industry, specifically the skilled crew leaders and foremen. These incentives can come in many forms, such as cash bonuses, stock options, health benefits and job-specific training, which are all used to retain, motivate and increase the productivity of the employee. This research will also focus on the views that upper management personnel have towards their foremen and skilled crew leaders, as well as their reasons for offering, or not offering, certain types of incentives.

3 RESEARCH MEHODOLOGY

This research was conducted using a procedure known as the Delphi method, which produces statistically significant results in a manner different from conventional statistical methods. The Delphi method is a highly structured and systematic technique of communication that allows interaction between the participants anonymously in order to accumulate expert judgement on a topic, or phenomenon, where there is a lack of knowledge regarding the problem. This research method has been proven to work well when the end goal for the researcher is to better understand a problem, determine a solution, or develop forecasts of what could happen if the problem is left unsolved (Linstone and Turoff 1975).

To implement the Delphi method most effectively, all identities of the participants are to remain anonymous, which allows them to better express their opinions without feeling any pressure from the group to conform to the other respondents' answers. An identical questionnaire is also sent out to all participants in order to collect data that can be compared and analyzed properly. Once all of the respondents return their questionnaires, a summary of the responses are then returned to each of the experts in order for them to examine all of the different points of view regarding the topic. The experts are then asked to provide feedback pertaining to the responses from the summary in order to determine what was disagreed upon. If a consensus is not reached, a new questionnaire is compiled and sent back out to each of the respondents in order to further research the issue. This process is repeated until a conclusion can be drawn

(Skulmoski *et al.* 2007). These factors were all taken into account throughout this research to provide relevant results that can be used to draw accurate conclusions.

The members of the panel that completed the questionnaire are regarded as experts in the heavy construction industry. Due to the prescribed rules outlined in the Delphi method, a minimum number of six respondents were sought for the expert panel in order to make this research statistically significant, yet eight ended up participating. This expert panel was composed of individuals from a mix of small, medium, and large contracting firms, that all operate in the heavy construction industry, their companies ranging in annual revenue from roughly \$100,000,000 to nearly \$11,000,000,000 in 2014. When determining which companies to target in the medium to large range, the Engineering News-Record's "Top 400 Contractors" list was referenced (2014).

Each of the experts from these companies were required to have a minimum of 15 years working experience and have worked in the field at some point in their career. The panel for this research consisted of two project managers, three superintendents, two presidents, and one business development manager.

4 RESULTS

4.1 Round 1 – Respondent Answers

This study, composed of 11 total questions, focused on the labor trends in the construction industry and the incentives companies provide to retain, motivate, and increase the productivity of their skilled crew leaders and foremen. The results of this study, pertaining to the first round of questions that were given to the panel of eight respondents, are as follows:

4.1.1 *Questions 1-3*

The first three questions of this study pertain to the recent trend in the decline of skilled crew leaders and foremen available, the experiences that each of the respondents have faced finding such workers for their jobs, and the major contributing reasons for these shortages, respectively. All eight respondents "highly agreed" or "agreed," that there is a problem with the construction labor force and the decline of skilled crew leaders and foremen available. Seven of the eight respondents have faced the issues listed below either "often" or "very often" in attracting / retaining workers, while one respondent "rarely" faced these issues. The respondent that rarely saw this type of problem explained in greater detail that their company does a significant amount of in-house training and promoting from within, which helps them counter this problem. The respondent further explained that they move their foremen from project to project and "rarely depend on local talent at the foreman level."

Reasons for the shortage of skilled construction workers is detailed in the literature (Minchin *et al.* 2016). The list of reasons below was derived from the respondents as to why they felt there is currently such a large shortage in the labor force throughout the construction industry.

4.1.2 *Question 4*

The fourth question of this study focuses on whether or not the panel of experts noticed much of a skill/productivity level gap between the younger generations (millennials) of laborers versus the older generations (baby boomers). One of the respondents pointed out that younger individuals may not arrive with the necessary, skills but they "definitely have the ability to learn and acquire them at a quicker rate" than older individuals. Another replied "Blue collar work is portrayed as an 'if you can't do anything else' career." Many school systems have eliminated "trade" or "shop" classes, so most young workers have no experience, unless they had a summer job with a home builder or a landscape company.

4.1.3 *Question 5*

This question, dealing with the percentage of the work in this sector of the industry that is self-performed by individuals working for the general (prime) contractor, was included in order to validate that each of the companies hired a large number of in-house crew leaders and foremen. The smallest of the companies researched reported that all foremen and crew members are in-house. Their foremen are typically responsible for anywhere between one to six laborers, depending on the job size. Larger companies noted that "the work force percentages vary significantly from discipline to discipline," which relates to the specific scopes of work performed. Most civil work is 70-80% self-performed, but can be dictated by the state where the work is being performed. It was also noted that general contractors on the commercial side of the industry tend to self-perform roughly 5% of the work under contract, with the rest being subcontracted out, in order to show the stark difference between the different sectors of the construction industry.

4.1.4 *Question 6*

The sixth question focused mainly on whether or not the companies researched offered any types of incentives to the in-house labor force for self-performed work, since these firms performed the majority of their work themselves. "Best Quality Crew of the Month" or "Safest Crew of the Month" rewards, Lunches or barbeques, T-shirts, hats, or other company merchandise, Group milestones/productivity goals, Profit sharing plans, and Bonus programs are popular ways to reward excellence.

The company that noted utilizing a bonus program was one of the smaller firms that can keep track of these employees easier and "knows who is really contributing to the bottom line and who is just collecting a paycheck," which is reflected in the bonus, if even given. This could be measured by schedule performance or man-hour performance. The incentives given to the in-house workers are typically dispersed on a case-by-case basis, and are typically determined by the project manager on the jobsite.

4.1.5 *Question 7*

The seventh question of this questionnaire pertained to whether or not safety was implemented as an incentive or not for the labor force and gave more insight into some of the responses from the previous question, with a 50/50 split between the respondents' answers. Half of the companies researched gave incentives to their employees for upholding a high level of safety throughout each project, especially if zero accidents were recorded. The safety-based incentives varied from barbeques, or other types of lunches paid for by the company, to a percentage of their year-end bonus based off of the employee's safety record for the entire year, but this was limited to only the foremen and above for the bonus system. The other half of the respondents noted that they do not offer incentives for safety due to the fact that OSHA feels as if "safety incentives offered will only keep the work force from turning in or admitting to any incident in order to receive the reward."

4.1.6 *Questions 8-10*

These questions focused on what types of incentives each company offers their labor force that solely focus on increasing the productivity levels of these workers, as well as the frequency these incentives are given, such as time-related or milestone-based incentives.

Specifically in regards to productivity, three of the respondents offer their labor force incentives that focus solely on productivity, which are predominantly milestone-based or determined by the productivity of a specific operation. One of the respondents, whose company offers these incentives, explained that they are not offered to all employees or on all projects, rather "individual projects can develop incentive programs for particular components of the job to ensure the success of that work" specifically. Another respondent noted that their labor supervision receives annual bonuses based on productivity, which is a cash-based incentive. This respondent noted that their company offers a pay incentive typically when their projects enter their final stages, explaining that the employees are offered "end of job bonuses if they stay until the project is no longer in need of their services," which is directly tied into their "safety and quality performance until the end of the project." The remaining respondents did not offer their labor force productivity-related incentives, or stated that there was "nothing official" for their company.

In regards to the expert's personal preference as to which incentive types they prefer to offer their labor force, in order increase their productivity, the responses were very similar to the incentives that they felt best motivated them. These incentives included bonuses, extra hours, recognition, pay raises, positive working environments and friendly competition between crews for awards.

4.1.7 *Question 11*

The eleventh question of this questionnaire pertained to the different incentives that the experts felt are the best motivators for the labor force. Four of the respondents noted that for their employees, they have noticed that they respond best to monetary, or cash-based, incentives. Of those four respondents, one of them also noted that "enjoying the work you do" is important as well. The other respondents all noted that recognition and respect in the work place are crucial motivators for these employees as well. Recognition was noted as a good motivator because it shows that one "matters and that management cares" about them.

4.2 Round 2 – Respondent Answers

Upon distribution of the Round 1 responses to the panel, the unanimous response from the panel was that they (all panel members) agreed with all the comments of all the other members of the panel. The researchers explained to the panel that such a conclusion meant that all eight of them were "on record," so to speak, that every comment made by every panel member could be attributed to every panel member. All members agreed that this was accurate, so the researchers stopped the research at the second round.

5 CONCLUSIONS

This study specifically focused on what incentives were offered to the skilled crew leaders and foremen in the construction industry that retain, motivate and increase their productivity levels most effectively. The foremen and skilled crew leaders are not offered the same incentives, nor at the same frequency. Crew leaders were offered the fewest incentives, with the foremen being the second-least incentivized position. Although monetary incentives are not as effective to specifically retain an employee, many construction workers will leave for a higher paying job very quickly. Recognition is the single most important incentive, not only for increasing worker's productivity, but also for motivation and retention of workers.

The Delphi method, which was used to conduct this research, proved to be a suitable process for obtaining data throughout this study. Questioning a pool of eight experts—all with a vast amount of knowledge regarding this issue, especially in regards to the field operations and what drives their workers to perform at the highest level—ensured the information collected was relevant, due to the fact that the minimum number of required respondents for the Delphi method needed to be at least six. This method enabled the experts to anonymously provide feedback regarding what each of the other respondents stated for all of the questions, which allowed them to avoid groupthink and look at every question from multiple perspectives and at a high level of detail. This also enabled each of the respondents a significant amount of time to think through each of the questions carefully and critique the other's answers without being pressed for time, which would occur if this was done in a conference type setting.

The process went through two rounds, with all respondents agreeing with all responses given in the first round. As no changes were recommended in the second round, the process was ended by the researchers.

References

Engineering News-Record, 2014, The Top 400 Contractors, Retrieved from http://enr.construction.com/toplists/TopContractors/001-100.asp on January 18, 2016.

Linstone, H. and Turoff, M., *Delphi Method: Techniques and Applications.* Addison-Wesley Reading, 1975.

Minchin, R. E., Close, S., and Flood, I., Public Perceptions of the Construction Industry in Florida, Submitted for publication to Proceedings, *International Structural Engineering and Construction Europe and Mediterranean Conference*, Istanbul, Turkey, 2016.

Skulmoski, G., Hartman, F., and Krahn, J., The Delphi Method for Graduate Research, *Journal of Information Technology Education: Research,* 6(1), pp 1-21, 2007.

Interaction between Theory and Practice in Civil Engineering and Construction
Edited by Komurlu, R., Gurgun, A. P., Singh, A., and Yazdani, S.

ORGANIZATIONAL PROJECT MANAGEMENT MATURITY FROM THE CONSTRUCTION PRACTITIONERS POINT OF VIEW

SEPIDEH MOTAMEDPOOYA and EHSAN ASNAASHARI

Allaodoleh Semnani Institute of Higher Education, Garmsar, Iran

The organizational project management maturity model (OPM3) aims to implement project management principles into project-based organizations. In fact, OPM3 by focusing on organizational level, can help construction companies to integrate their various activities and follow their goals strategically. However, this useful tool has not been recognized by most of the Iranian construction organizations. The purpose of this research is to introduce OPM3 to the experts and collect their views on how it can be implemented in their projects. First, the concept of OPM3 introduced to participants. To reach this aim, 52 questions were selected from PMI self-assessment questionnaire. These questions were categorized under four groups of planning, performance, procurement and prediction. Interviewees answered each question with Yes and No and then scores were added up in each group. At the next step, based on maximum and minimum score of groups a qualitative interview was carried out to understand reasons behind scores attained by interviewees. Identifying detailed elements can enhance the Iranian practitioners' understanding about OPM3 and help them to apply it to their organizations in Iran's working conditions.

Keywords: Iranian projects, Maturity assessment, OPM3, Construction organizations.

1 INTRODUCTION

There are several obstacles in the way for executive managers to review their strategies in order to comply with the expectations of stakeholder and supply the needs of the market. These barriers include the change towards improvement, complexities in economy and worldwide competition. The low rate of efficiency in construction, the overuse of resources and the small amount of output of the working are the reasons to this phenomenon (Guangshe *et al.* 2008). The low profit rate of investments in Iran's construction industry is the consequence of weak management practice. Hence, project managers are in need of improving organizational project management practice. As a result organizations must find a way through a project-oriented view to turn strategy into organizational success (PMI-OPM3 2013).

Organizational Project Management Maturity Model (OPM3) is the tool to execute strategies that converts the project, design and portfolio processes into high-quality, stable, repeatable, and predictable output. OPM3 serves as a milestone to evaluate the maturity of companies in Iran and it presents guidance for the improvement organizational project managing in organizations. This article, after introducing

OPM3, will look over measuring the maturity of organizations and identifying the contributing factors in organizational project management maturity.

2 WHAT IS OPM3?

As a developmental model for organizational project management, this standard is presented by Project Management Institution (PMI) in 2003. This model is flexible and executable for projects of various scales and supports all sorts of institutions regardless of their size, complexity, geographical situation, age, or their level of maturity (PMI-OPM3 2013). Organizational maturity refers to readiness and experience of staff, processes, and technologies which are measured continuously. The improvement element in OPM3 differentiates this model from all others. This element paves the way for the organization to reach maturity by gathering organization information.

3 A LOOK OVER THE PREVIOUS PROJECTS

A series of interviews with experts and then evaluating with the Principal Component Analysis (PCA) system, done by the Qi *et al.*, 2014 research group and their colleagues, fourteen effective factors in project management in construction companies are identified as follows: management by stakeholders, handling the flow of information inside the organization, coordinated decision making in the organization, coordination between portfolios in order to achieve organizational strategies, controlling the timeline schedule, capability in quality management, coordination of the project team, capability of safety management, capability of contract management, capability of understanding the market, managing the financial aspects of the project, dedicating resources, capability of managing the project limitations, and foreseeing and management of risk. The most important effective factor is contract management and the least important factor is market understanding. The above items are categorized in four groups of Planning, Performance, Procurement, and Prediction (Qi *et al.* 2014). Experts were asked to point out the advantages and disadvantages of each of the fourteen factors and the results are as follows. The most important factor in stakeholder management is identifying the stakeholders. Controlling the critical ways are noteworthy in timeline schedule managing. In quality management, controlling the quality is deemed most important. Recognizing the weak and strong points of each staff member is viable to the coordination of the project team, just as reviewing and improvement of the system is an effective matter in safety management. Respondent did not had enough knowledge to answer risk management matter in regards with organizational maturity management, and there was no evidence of any common ground with contract management.

4 RESEARCH METHOD

This research uses OPM3 as an evaluation method for organizational maturity. As suggested by PMI, 151 questions are presented for self-evaluation. The self-evaluation questions in this article are based on the four categories of Planning, Performance, Procurement, and Predicting (Qi *et al.* 2014). After categorizing the questions in these four groups, a minimum and maximum of score was appointed for each category. In the first step, a limited number of questionnaires were distributed between experts.

Before handing the main questionnaire to high level managers, they were piloted, tested and corrected. There needed to be an informative approach towards the interviewee, due to the lack of knowledge about the maturity model. Therefore, prior to the questioning, they were presented with enough information about the model, and then they were asked to thoroughly introduce the organization in order to have an organizational structure in terms of project, program and portfolio which led to having a better understating of the mission, vision and the fundamental values of the organization as well as its strong and weak points. The third step, was asking the interviewees to answer the yes/no self-evaluating questions. In the next step, after calculating the scores, a set of deeper questions were asked in accordance with their scoring of the four categories and the resulting answers were analyzed in terms of quality and briefly presented.

5 INTERVIEW RESULTS

Although the standards of this model are well accepted and they are more known that other forms of maturity model (Ghoddousi *et al.* 2011), there is not enough knowledge about this model and most of the managers provided with the model and having been explained to by a researcher, still feel the need to be presented with a pre-executed plan in other organizations, which unfortunately does not exist. Therefore all interviewees showed great eagerness to find out the result of this model in their own organization.

Table 1. Score of interviewees on each category.

Interviewee code	Experience (year)	Planning	Performance	Procurement	Prediction
I1	30	8	4	8	11
I2	17	8	11	7	8
I3	19	7	6	8	10
I4	11	9	9	9	12
I5	15	7	4	6	9
I6	7	8	9	8	10

5.1.1 *The analysis of the collected data*

The results derived from interview are presented in four groups. Detailed analysis is provided below.

5.1.2 *Planning*

Strategic planning is a type of long-term programming .I4 with the highest point thinks of programming as "doing things today for a better tomorrow to come" because of this belief that "the future belongs to those who are able to make difficult and strategic decisions today."

It may be right to believe that survival in the insecure and competitive environment of today has made organizations try to recruit professionals and experts. Organizations highly depend in their success on those forces who step beyond their occupational duties. There are a lot of role-playing elements leading to this behavior. Those include job promotion, energy, and encouraging a sense of capability which lead to a boost of spirits and efficiency in the staff. Most of the interviewees believed that aside from corruption, promotion is given only based upon one's own talents and capacities. Interviewee no. 4, with regard to their experience working abroad, recommended that there is a need for designing an evaluation system which is capable of studying one's operation for a possible promotion half-yearly, or annually, and in a three-level basis of their self-evaluation.

It should be noted that knowing those on-job learnings bring value to the organization and could help make a change and improve, or avoid making the same errors and mistakes over again. Every organization is to keep a record of their failure and success stories of the past and review them, they as well should learn lessons from them in a systematic fashion, so that all employees can fathom and have access to them. According to the questions raised in this section, interviewees I1 and I5 believed that due to the traditional structure of the company, exchanging experiences is done orally, but the interviewee I3 emphasized on written recording of data in governmental projects. Interviewee I2 believed that the records in their private company, in addition to economic profitability, had an impact on reduction of project time. I6 also emphasized that public and technical information is recorded both in written form and electronically. Interviewee I4 suggested software which helps to record data, based on his past experiences.

In addition to the above-mentioned points, it is important to optimize the communication system in organizations. If there is no proper communication in an organization, it will stop functioning well and everything will be disorganized. I5 believes that open communication depends on the circumstances, the type of project, and its runtime. For instance, hierarchical organizational structure is important in simple projects and open communication is effective in complex ones. Other interviewees supported open communications because they believe it will save time.

5.1.3 *Prediction*

In this section, I2 interviewee scored last because he considered his organization low self-esteemed in the section related to the prediction, which was categorized according to Qi *et al.* (2014). Another factor in this part is the organization's strategy to maintain internal and external sources. An efficient organization is an organization that achieves its aims through meeting the demands and needs of stakeholders, balancing resources and flexibility towards environmental changes and encourages commitment, creativity, shared values and mutual trust. These strategies deal with processes as well as the structures or systems and in a way preserve source in the organization. Given the importance of maintaining internal and external sources for the organization, the management should strive to make huge investments on this strategic source. According to the interviews, senior director I6 believes that the layout of the system and good job promotion maintains internal sources and assessing and rating the performance of contractors, maintains external sources. I2 believes that to protect foreign sources, the best way is to preserve internal sources that the organization, due to economic problems, does not pay much attention to.

5.1.3 *Procurement*

Procurement means goods or services provision from external sources. It can be said that communication with external sources need that new management practices replace traditional ones. According to I4, one of the urgent needs in order to support this successful trend is the implementation of new methods of project management, and creating scientific sources and references which managers need. One of these sources is PMBOK which reflects the knowledge of project managers and presents project management principles in a way that can be used in wide range of projects. Investigations done in this study shows that all the interviewees see their organization in need of project management knowledge guidelines before implementation of the maturity model, while this knowledge is not clear for managers in practice.

5.1.4 *Performance*

One of the most important items from the interviewees' point of view is the direct influence of senior managers of the projects in order to evaluating the performance. Due to cultural matters, the personnel in are not placed in their own position as assorted by the organization. So the roles in the lower levels may have contradiction with the senior managers because of having more experience. According to interviewees I1 and I5, senior managers interferes in lower levels because of the traditional way of thinking and believe that in order to more control of the expenses, there is the need of senior manager's role. However, I6 opposed the role of direct interference of the senior managers with the project management, because he emphasized that in the sessions with project managers, they only have general information about the projects and their interference may prevent the temporary progression of the project which in turn leads to delay in its total progression.

Project control is a process going in the direction of maintaining the project route in order to gain a justified economical balance among three factors of cost, time and quality while project execution. In fact, the control is the complete and exact administration of the program for the project in a way that by recognizing the causes, it can be returned to its original path in a closest possible condition. In this context I6 stated that, all of the project levels are gauged by base lines and standards. I2 and I3 pointed to the project control by an independent unit. I4 believes that control in this company is mostly on the expenses of financial resources.

Quality control is also related to the study of performance and the level of this value is implicitly determined by the agreement between customer and the suppliers. Reaching to a high quality by standardization is one of the effective factors in organizational mature model. According to the interviewees, in Iranian companies there are quality control teams which act separately from the company. I2 believes that because of the friendly relationship of employer and the consultant the matter of quality is frequently ignored. I1 says that only in public projects the quality is checked.

Standard is a criteria for evaluating the function of the devices, workers and managers. It is also a criteria for recognizing skills, educations, specialties, equipment and required technologies. According to the interviewees all of the companies follow the available standards, however, I2 and I6 felt more required to obey standards because of administrating governmental projects. I1, I3, I4, and I5 believed that standards were mostly followed in designing phase up to the confirmation of the relevant authorities, but after that, because of the friendly relationship with the employer and the presence of the difficulties in cultural and economic infrastructures, most of the standards were ignored.

6 CONCLUSIONS

The construction industry plays a basic role in economic development and social relationship. On this basis, OPM3 model is considered as a tool to reach this strategic goal of the organization. According to studies on this subject, the most important barrier against the recognition of factors affecting organizational maturity in Iran, are the lack of information and the avoidance of the organizations in the matter of presenting the organizational internal relationship. Following the effort to recognize more effective factors, cases such as: standardization, quality control after standardization, recording taught lessons from the project, the strategy for maintenance of internal and external resources, upgrading the job level of the employees, the existence of open relationship at all levels, the amount of senior manager's interference in project management, application of PMBOK and the control administration were pointed to. But to administrate the above mentioned cases, obstacles such as the lack of economic and cultural infrastructures, administrative corruption, the existence of too traditional point of views to accept the systems and models of improvement and at the end, the lack of sufficient information to manage the project were recognized. According to this study, it is proposed that, apart from their age, experience and the number of their projects, try to adapt the management maturity model. This can be also applied to new companies with short history and is considered as the secret of their future success in national and international environment.

References

Ghoddousi, P., Amini, Z., and Hosseini, M., A Survey on the Maturity State of Iranian Grade One Construction Companies Utilizing OPM3 Maturity Model, *Technics technologies education Management*, 2011.

Guangshe, J., Li, C., Jianguo, Ch., Shuisen, Z., and Jan, W., International conference on Information Management, Innovation Management and Industrial Engineering, 2008

Project Management Institute, *Organizational Project Management Maturity Model Project Management Institute*, Inc. 2013.

Qi, Sh., Chen, W., Zhang, Y., Wu, J., and Jiazhong, C., Identification of influence factor and Establishment of evaluation index system for OPM3 in mega construction engineering Enterprise, Applied mechanics and materials, 2014.

Interaction between Theory and Practice in Civil Engineering and Construction
Edited by Komurlu, R., Gurgun, A. P., Singh, A., and Yazdani, S.

A BAYESIAN-BASED MODEL FOR ALLOCATING CONTINGENCY TO A PORTFOLIO OF CORRELATED CONSTRUCTION PROJECTS

PAYAM BAKHSHI[1] and ALI TOURAN[2]

[1]Dept of Construction Management, Wentworth Institute of Technology, Boston, MA
[2]Dept of Civil and Environmental Engineering, Northeastern University, Boston, MA

Despite the availability of various probabilistic and non-probabilistic methods for determining budget contingency, still many large capital projects are suffering from cost overrun. Furthermore, most of the effort has been concentrated on calculation of contingency for a single project rather than a group of projects. This paper introduces a new Bayesian-based model for allocating contingency budget to a portfolio of correlated construction projects. The proposed model enables an owner agency to define the individual project confidence level for contingency calculation taking into account the portfolio budget and the desired portfolio confidence level. The model recognizes the correlation between each pair of projects in the portfolio and calculates the required increase in budget in such a way to ensure adequate budgets with respect to individual projects and the portfolio. Using the information from newly completed projects, the Bayesian technique can be used to update the model parameters periodically so that more accurate contingency budget can be established for the portfolio. A numerical example is presented to show the application of the model on a portfolio of transit projects. The proposed model can be employed as an effective tool for the owner agencies in charge of funding a group of projects every year.

Keywords: Probabilistic model, Risk, Monte Carlo, Truncated normal distribution, Pearson correlation.

1 INTRODUCTION

In 2012, Bakhshi and Touran proposed a model for calculation of contingency in a portfolio of construction projects. The model assumes normal distribution for the cost overruns/ underruns and truncated normal distribution for the cost of each project in the portfolio. The promise of the model is to protect a portfolio of projects against cost overrun by adjusting their original budgets. The proposed model helps an agency find the level of confidence needed at the individual project level to ensure that the portfolio budget will meet the minimum level of confidence based on available funding and the agency's policy goals. A Bayesian approach is employed to update the model on regular intervals. As more information becomes available in the future, the required adjustment in portfolio budget will be reduced, because the accuracy of estimating the contingency is improved. However, the limitation of this model is that it is not considering the correlation among the projects in Bayesian approach when updating the model. This current research is a continuation of Bakhshi and Touran (2012) where dependencies (Pearson correlation coefficient) between each pair of projects in the portfolio is recognized. This enhances the efficiency and accuracy of the model.

2 BASIS OF THE MODEL

The model presumes truncated normal distribution for the cost of each project in the portfolio. To form this distribution, it is assumed that the probability of experiencing underrun m is α as the discrete portion of distribution. m is an arbitrary number based on agency's objectives and α can be determined by reviewing the historical cost overruns/underruns. Let us assume that there is a database of construction projects comprising of $i = 1, ..., n$ projects with the initial budget of b_i. It is found through this historical data that there is $\alpha\%$ chance to have m percent underrun and get the project done with $c_i = (1-m).b_i$. Figure 1 depicts the model steps, the parameters, and their definitions as used in the process. Interested readers can refer to Bakhshi and Touran (2012) for details of the model and how the equations were derived.

$$\eta = \Phi\left\{\Phi^{-1}(\alpha) + \left[(1-\alpha).\Phi^{-1}(\alpha) - \frac{e^{\frac{-[\Phi^{-1}(\alpha)]^2}{2}}}{\sqrt{2\pi}}\right]\left[\frac{\varphi.\sqrt{\sum_{i=1}^{n} c_i^2 + 2\sum_{i<j}^{n}\sum_{j}^{n}\rho_{ij}c_ic_j}.\Phi^{-1}(\gamma)}{(1-m).(1-\rho).B} - 1\right]\right\} \tag{1}$$

$$\frac{B^*}{B} = (1-m).\left[1 + \frac{(1-\rho).[\Phi^{-1}(\eta) - \Phi^{-1}(\alpha)]}{(1-\alpha).\Phi^{-1}(\alpha) - \frac{e^{\frac{-[\Phi^{-1}(\alpha)]^2}{2}}}{\sqrt{2\pi}}}\right] \tag{2}$$

3 BAYESIAN APPROACH FOR UPDATING THE MODEL WITH *k* CORRELATED PROJECTS

It is assumed that k recently completed and correlated projects need to be updated. To conduct the analysis with correlated cost overruns/underruns, the joint density function of project cost overruns/underruns is required; the probability distribution of each project's cost overrun/underrun is the marginal distribution. Knowing the marginal distributions of cost overruns/underruns is not sufficient to obtain their joint density function. Multivariate normal distribution is the special case in which the only information required other than marginal distribution of each random variable is the values of covariance among the variables (Rowe 2003). When there is a multivariate normal distribution, each of its marginal variables by itself is normally distributed. The converse, however, is not generally true (Kutner *et al.* 2005). Despite this, a simplifying assumption that the joint density function of the cost overruns/underruns to be a multivariate normal distribution. The multivariate normal PDF is:

$$f(\delta_1, ..., \delta_k) = \frac{|\mathbf{V}^{-1}|^{1/2}}{(2\pi)^{n/2}}.\exp\left[-\frac{1}{2}(\boldsymbol{\delta} - \bar{\boldsymbol{\delta}})'.\mathbf{V}^{-1}.(\boldsymbol{\delta} - \bar{\boldsymbol{\delta}})\right] \tag{3}$$

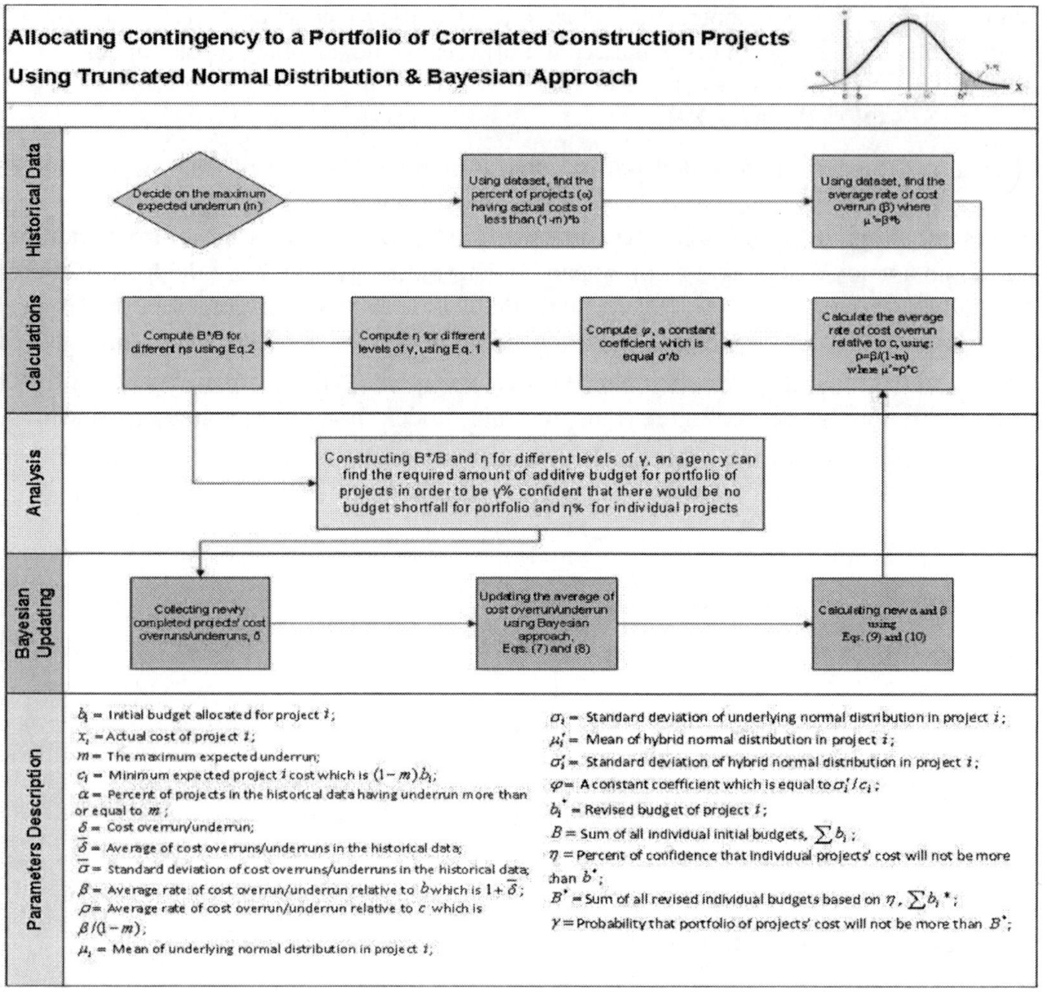

Figure 1. The model steps and parameters.

Where all bold letters represent a matrix/vector and: (1) $\mathbf{\delta} = (\delta_1, ..., \delta_k)^T$ are the cost overruns/underruns of k newly completed projects (T denotes the transpose of matrix). (2) $\bar{\mathbf{\delta}} = (\bar{\delta}_1, ..., \bar{\delta}_k)^T$ are the means of each cost underruns/overruns distribution. Since it was assumed that all these k projects are coming from a normal population with the mean $\bar{\delta}$, $\bar{\mathbf{\delta}}$ can be written as $\bar{\mathbf{\delta}} = (\bar{\delta}, ..., \bar{\delta})^T$. (3) $\mathbf{V}$ is the variance-covariance matrix which is a symmetrical ($k \times k$) matrix. (4) ρ_{mk} is the correlation coefficient between project m and k. (5) σ_j is the standard deviation of cost overrun/underrun of project j. Again, since it was assumed that all k projects are coming from a unique normal population with the standard deviation of $\bar{\sigma}$, all σ_j $(i = 1, ..., k)$ in the matrix $\mathbf{V}$ are equal to $\bar{\sigma}$. Since the standard deviation of the population is not known, $\bar{\sigma}$ is assumed to be the standard deviation of cost overruns/underruns of k observed projects σ. (6) $|\mathbf{V}^{-1}|$

is the determinant of matrix $\mathbf{V}^{-1}$ which is the inverse of matrix $\mathbf{V}$. One should note that Eq. (3) is the joint probability function of project cost overruns. In other words, assuming that $\bar{\boldsymbol{\delta}}$ and $\mathbf{V}$ are known, the probability of observing $\boldsymbol{\delta} = (\delta_1, ..., \delta_k)^T$ is found. With Bayesian updating, to find the likelihood of having $\bar{\boldsymbol{\delta}} = (\bar{\delta}, ..., \bar{\delta})^T$ as the parameter of distribution is needed when $\boldsymbol{\delta} = (\delta_1, ..., \delta_k)^T$ is observed. Thus, Eq. (3) can obtain the likelihood function of $\bar{\boldsymbol{\delta}}$ which is the function of just $\bar{\delta}$ as follows (Rowe 2003):

$$L(\bar{\delta}_1, ..., \bar{\delta}_k) = L(\bar{\delta}, ..., \bar{\delta}) = \frac{|\mathbf{V}^{-1}|^{1/2}}{(2\pi)^{n/2}}.\exp\left[-\frac{1}{2}.(\boldsymbol{\delta} - \bar{\boldsymbol{\delta}})'.\mathbf{V}^{-1}.(\boldsymbol{\delta} - \bar{\boldsymbol{\delta}})\right] \tag{4}$$

Having the likelihood function and assuming a prior distribution $f'(\bar{\delta})$ like Eq. (5), the posterior distribution $f''(\bar{\delta})$ is a normal shape given as Eq. (6).

$$f'(\bar{\delta}) \sim N(\delta', \sigma') = \frac{1}{\sqrt{2\pi}.\sigma'}\exp\left[-\frac{1}{2}.\left(\frac{\bar{\delta} - \delta'}{\sigma'}\right)^2\right] \quad \& \quad f''(\bar{\delta}) = k.L(\bar{\delta}_1, ..., \bar{\delta}_k).f'(\bar{\delta}) \sim N(\delta'', \sigma'') \tag{5 \& 6}$$

In the correlated case, to calculate mean and standard deviation (δ'', σ'') of $f''(\bar{\delta})$, only numerical approach is available and there is no close form formula. To this end, a range of possible $\bar{\delta}$ is selected. To be conservative, the range is assumed to be from -99.99% to 200% with the pace of 0.001. These values are input in Eqs. (4) and (5) to respectively calculate likelihood value, $L(\bar{\delta}_1, ..., \bar{\delta}_k)$ of each possible $\bar{\delta}$ and the correspondence prior PDF value, $f'(\bar{\delta})$. It should be noted that for the values of $\bar{\delta}$ outside the range of [-99.99%, 200%], the $L(\bar{\delta})$ and accordingly $f''(\bar{\delta})$ become too small so that they can be ignored from the analysis without any significant impact. Reviewing Eq. (4), it is found the term in the exponential function is the product of three $(1 \times k)$ and $(k \times k)$ and $(k \times 1)$ matrices which results in a polynomial function of $\bar{\delta}$. It means that Eq. (4), for any $\bar{\delta}$, gives a scalar likelihood value. Having variance-covariance matrix, $L(\bar{\delta}_1, ..., \bar{\delta}_k)$ can be easily calculated by any available mathematical package such as MATLAB. Multiplication of the prior PDF and likelihood values of each $\bar{\delta}$ gives $f''(\bar{\delta})$ which is the posterior PDF value of $\bar{\delta}$ before normalization (the area under the curve is not equal one). Both curves $L(\bar{\delta}_1, ..., \bar{\delta}_k)$ *vs.* $\bar{\delta}$ and $f''(\bar{\delta})$ *vs.* $\bar{\delta}$ need to be normalized. To calculate δ'' and σ'', the area under the curve of $f''(\bar{\delta})$ *vs.* $\bar{\delta}$ after normalizing is assumed to be divided to t narrow rectangles. The area of each rectangle is the probability of having the $\bar{\delta}_{(mid)j}$ (midpoint of the rectangle j). Then:

$$\delta'' = E(\bar{\delta}) = (\bar{\delta} : f''(\bar{\delta}) \; is \; Maximum) \tag{7}$$

$$\sigma''^2 = E(\bar{\delta}^2) - (E(\bar{\delta}))^2 = \sum_{j=1}^{t} \bar{\delta}^2_{(mid)j}.P(\bar{\delta}_{(mid)j}) - \delta''^2 = \frac{1}{\sum_{j=1}^{t}\left[\bar{\delta}_{(mid)j}.f''\left(\bar{\delta}_{(mid)j}\right)\right]}.\left[\sum_{j=1}^{t} \bar{\delta}^2_{(mid)j}.\left(\bar{\delta}_{(mid)j} - \bar{\delta}_{(mid)j-1}\right)f''\left(\bar{\delta}_{(mid)j}\right)\right] - \delta''^2 \tag{8}$$

The term $\sum_{j=1}^{t}\left[\bar{\delta}_{(mid)j}.f''\left(\bar{\delta}_{(mid)j}\right)\right]$ in Eq. (8) is to normalize the posterior distribution and plays the role of k in Eq. (7).

4 UPDATING THE PRIMARY PARAMETERS OF THE MODEL

In Section 3, using Bayesian approach and having the information of newly completed projects the distribution of $\bar{\delta}$, the average of cost overruns/underruns, was updated and posterior distribution $f''(\bar{\delta}) \sim N(\delta'', \sigma'')$ was calculated. The mean δ'' and standard deviation σ'' of the posterior distribution is now used to update the primary parameters of the proposed model α, β, and ρ as follows:

$$\alpha_{new} = P(x < -m) = P(Z < \frac{-m-\delta''}{\sigma''}) = \Phi(\frac{-m-\delta''}{\sigma''}) \quad \& \quad \beta_{new} = 1+\delta'' \Rightarrow \rho_{new} = \frac{\beta_{new}}{1-m} \qquad (9 \,\&\, 10)$$

where Φ is the cumulative function for standard normal distribution. The proposed model is updated by α_{new}, β_{new}, and ρ_{new} values and becomes ready to be applied to any prospective set of projects which are in budget allocation process.

5 APPLICATION OF THE MODEL

The application of the model is demonstrated using cost data from 31 transit projects funded by the Federal Transit Administration (FTA) of the US Department of Transportation. These projects are divided into three groups: (1) a set of 22 project completed before 2004 named Historical Dataset for determining the model parameters of α, $\bar{\delta}$, β and ρ; (2) a set of 5 projects completed in 2004 and named First Dataset to see the effect of the model on cost overruns/ underruns and updating the model with actual costs of this projects; and (3) a set of 4 projects completed in 2005-06 and named Second Dataset to further investigate the effectiveness of the model and the Bayesian updating part. After verifying the normality of Historical Dataset using a test of goodness of fit, $m = 15\%$ was assumed as the maximum expected underrun defined by the FTA. Then from this dataset, it was calculated that $\alpha = 9.1\%$ and the average of cost underruns/overruns is $\bar{\delta} = 8.79\%$; thus $\beta = 1.0879$ and $\rho = 1.2799$. To determine the correlation among each pair of the projects in the First and Second Datasets, an approach called the Proposed Structured Guideline (PSG) was used to elicit correlation (Bakhshi 2011). The results of applying the proposed approach on aforementioned datasets are summarized in Table 1. Column "Actual Cost Overrun/ Underrun" depicts the actual mean and standard deviation of cost overruns/ underruns in three datasets. Column "Adjusted Cost Overrun/ Underrun" shows the mean and standard deviation of cost overrun/ underrun if the model had been applied to the data. The last Column "Updated Cost Overrun/Underrun" presents the mean and standard deviation of cost overruns/underruns after using the Bayesian updating which will prepare the model for the next application. The required adjustment in the value of factor B^*/B and cost overrun/underrun are diminished after each updating. For example, the First Dataset could have ended up with a 6.00% cost underrun instead of the actual 13.84% cost overrun by assigning increasing factor of 1.2111 to the budget and individual risk assessment confidence level of 77%. For the Second Dataset, using the model could have reduced the cost overrun from 21.04% to 7.73% by increasing the budget by a factor of 1.0832 and individual risk assessment confidence level of 77%. Table 1 shows the improvement that can be gained by applying this model over a period of time. Due to the inherent characteristic of Bayesian updating, it is expected that the model gets more accurate as more projects become available and more updating occurs.

Table 1. Summary of the results from applying the proposed model on transit projects.

Data	α	β	For $\gamma = 85\%$		Actual Cost Overrun/ Underrun		Adjusted Cost Overrun/ Underrun		Updated Cost Overrun/ Underrun	
			B'/B	η	$\bar{\delta}$	$\bar{\sigma}$	$\bar{\delta}_{Adj}$	$\bar{\sigma}_{Adj}$	δ^*	σ^*
Historical Dataset	N/A	N/A	N/A	N/A	8.79%	0.2053	N/A	N/A	N/A	N/A
First Dataset	9.10%	1.0879	1.2111	77%	13.84%	0.1844	-6.00%	0.1522	1.30%	0.0965
Second Dataset	4.55%	1.0130	1.0832	77%	21.04%	0.4118	-7.73%	0.3139	1.10%	0.0888
Prospective Dataset	3.50%	1.0110								

6 CONCLUSION

To control the cost overrun, in this paper, a new probabilistic model was introduced for allocating contingency in a portfolio of correlated construction projects. The model assumes hybrid normal distribution for cost of projects and utilizes available historical data. Then, a Bayesian approach is employed to update the model as more projects are completed and new information becomes available. The proposed model first helps an agency to find the required portfolio's budget increase in order to have a certain confidence γ that the budget will be sufficient. Also, the model gives the required confidence level η to conduct risk assessment,at individual project level to insure that the portfolio budget will not overrun with a probability of more than $1-\gamma$. The model can be updated with the information of newly completed projects where the dependencies between the projects are acknowledged and incorporated into the model.

References

Bakhshi, P. and Touran, A., A New Approach for Contingency Determination in a Portfolio of Construction Projects, *J.of Risk Analysis and Crisis Response*, 2 (4), 223-232, Dec, 2012.

Bakhshi, P., A Bayesian Model for Controlling Cost Overrun in a Portfolio of Construction Projects, PhD Dissertation, Northeastern University, Boston, March, 2011.

Kutner, M., Nachtsheim, C., Neter, J., and Li, W., *Applied Linear Statistical Models*, 5th Ed., McGraw-Hill, New York, NY, 2005.

Rowe, D., *Multivariate Bayesian Statistics: Models for Source Separation and Signal Unimixing*, Chapman & Hall/CRC Press, Boca Raton, FL, 2003.

Interaction between Theory and Practice in Civil Engineering and Construction
Edited by Komurlu, R., Gurgun A. P., Singh, A., and Yazdani, S.

DELAY MANAGEMENT IN BUILDING CONSTRUCTION: A COMPARATIVE STUDY

RUVEYDA KOMURLU[1] and DAVID ARDITI[2]

[1]*Dept of Architecture, Kocaeli University, Kocaeli, Turkey*
[2]*Construction Engineering & Management Program, Illinois Institute of Technology, Chicago, USA*

Construction projects are one-time endeavors with three strong constraints i.e., time, cost and quality. Construction management pursues success through the balanced accomplishment of the target levels of these. Since construction management has been and is still being performed to some extent by architects, AIA's guidance has a strong influence on the profession. Thus, AIA's "A201-2007 General Conditions of the Contract for Construction" is widely preferred among owners and contractors in the U.S. In Turkey, on the other hand, the government is the owner of a large number of public building projects with a total contract value that accounts for a sizeable portion of the national construction activity. Thus, the "General Conditions for Construction" that is issued by the Ministry of Environment and Urbanization is widely used in the industry. Delays have an important effect on both the owner and the contractor; while the owner experiences cost overruns as a direct result of delays, contractors suffer from excessive overhead. A201-2007 addresses the issue in "Article 8-Time"; the Turkish General Conditions addresses it in "Clause 30-Job Duration and Time Extension". A literature review is performed in this paper to understand the importance of the deviations from the project duration specified in the contract. The objective of this study is to determine the similarities/differences in the management of delays, as mandated by the general conditions commonly used in the U.S. and Turkey. Recommendations are made in the light of the findings to improve delay management.

Keywords: Work schedules, Contracts, Contract administration, Delays, General conditions.

1 INTRODUCTION

Building construction projects are relatively complex and expensive endeavors that take a long time from start of the investment decision to the delivery of the end product (Chui and Bai 2010). The success of a building construction project depends on three major constraints which are scope, time, and budget. Quality is the general aspect that is pursued through all processes. The three constraints are directly related, which means when one is altered, at least one of the others is affected. The contract documents regulate how these issues are managed among the owner, the architect and the contractor.

One of the most important pieces of the construction contract documents is the General Conditions (Chui and Bai 2010, Fisk 2002). This document defines the roles and liabilities of the parties i.e., the owner, the architect, and the contractor, and expresses the methodology to be used for regulating the relationship between these parties throughout the construction process.

There are several General Conditions preferred by owners in the U.S. for managing the construction process. The most commonly used ones were introduced by the American Institute of Architects (AIA), the U.S. government (Federal Acquisition Regulation), and the Engineers Joint Contract Documents Committee (EJCDC) composed of the American Society of Civil Engineers (ASCE), the National Society of Professional Engineers (NSPE), the American Consulting Engineers Council (ACEC), and the Construction Specifications Institute (CSI) (Chui and Bai 2010). The General Conditions issued by the AIA is the most commonly used document (El-adaway *et al.* 2014) since in the traditional design-bid-build project delivery system, the bidding and construction phases are managed by architects.

The American Institute of Architects (AIA) has a long history of regulating the construction industry in the U.S. publishing the first General Conditions in 1911 (AIA 2007). With the "A201 Family of Documents", AIA regulates the roles of the owner, the architect/engineer, the contractor, and subcontractors. These also set standards for contract types such as stipulated sum, cost plus fee, etc. AIA's A201 – 2007 General Conditions of the Contract for Construction is the general set of rules covering all contracts.

Government is one of the most important investors in building construction projects in Turkey with 42.33 percent (YEM 2015). The bidding process and execution of public investments are regulated the Public Procurement Authority of the Ministry of Finance. Like other public institutions, the Ministry of Environment and Urbanization executes many public building projects, most of which are social mass housing projects. All public investments executed by government agencies are regulated by the General Conditions for Construction issued by the Public Procurement Authority (MEU 2014). This document is also referred to by a considerable number of investors in the private sector since it is used as baseline by courts of law.

2 MANAGEMENT OF TIME IN BUILDING CONSTRUCTION PROJECTS

Being one of the three constraints in building construction projects, time plays an important role in the success of a project. Thus, any variation in the duration set for a project, either affects the other two constraints i.e., scope and/or budget, or threatens the feasibility of the project.

Delay may occur in the planning, the design, and/or the construction phases of a project. The major reason why delays take place during the planning and design phases is changes in the owner's project requirements (Yang and Wei 2010). In the construction phase, delays may be caused by strikes, rework, lack of organization, material shortage, equipment failure, natural disasters, and change orders (Birgonul *et al.* 2015). Sometimes, two or more of these issues may cause a delay making the situation more complicated (Birgonul *et al.* 2015, Alkass *et al.* 1996). According to AlSehaimi *et al.*'s (2013) review of sixteen papers focusing on delays, ineffective planning and control is found to be the primary cause of delays in the construction phase with 87%, problems of supply and procurement of materials second with 69%, and poor site management third with 56% of the studies pointing to these causes of delay. Depending on the nature of the work and the probability of the problems, concurrent delays may sometimes take place, which multiplies the effect of simple delays (Birgonul *et al.* 2015, Alkass *et al.* 1996).

Given that construction is a long term investment initiative, completing a building construction project on or ahead of schedule is important for success. In addition to ensuring a stable scope of work and completing the job within budget, time management is most important in the performance of a project (Kazaz *et al.* 2012, Enshassi *et al.* 2009). Much like the construction industry in many countries around the world, the Turkish construction industry suffers from delayed project completion and considerable financial losses related to these delays (Arditi *et al.* 1985). Because of the importance of time and the negative implications of delay, General Conditions place special emphasis on schedules and procedures to be followed in case of delays.

A201 – 2007 General Conditions of the Contract for Construction addresses time related issues in Article 8 - Time. This article defines the commencement, execution and substantial completion of the work, sets the use of day as the unit for measuring time, expresses the importance of time for the contract, and states the methods to be used to respond to variations from predefined deadlines in three sections. Section 8.3 Delays and Time Extensions states the steps to follow if a delay occurs. According to Section 8.3, if there is a delay caused by the owner or the architect, or other reasons justified by the architect, then the contractor should be awarded an extension of time determined by the architect. The claim for an extension has to be submitted in written form and in a timely manner, stating the additional cost of the delay and the effect on progress, including any additional documentation, as defined in Article 15. Section 8.3 also states that both the owner and the contractor have the right to seek recovery of damages for delay.

Table 1. Articles, sections and provisions of A201 – 2007 General Conditions of the Contract for Construction, and Turkish General Conditions for Construction regarding time of building construction contracts.

A201 – 2007 General Conditions of the Contract for Construction		General Conditions for Construction	
ARTICLE 8	TIME	ARTICLE 5	EXECUTION OF WORK
SECTION 8.3	DELAYS AND EXTENSIONS OF TIME	CLAUSE 30	JOB DURATION AND TIME EXTENSION
8.3.1	Architect's change order for a time extension		
8.3.2	Claims for a time extension		
8.3.3	Recovery of damages for delay		

Article 5 - Execution of Work of the General Conditions for Construction addresses several issues about how the contractor should carry out the work and how the owner should manage problems such as mobilization, labor, equipment, subcontractors, undefined work, change orders, non-conformity, expenses, and time and extensions. Clause 30 - Job Duration and Time Extension defines under what circumstances the contractor should be charged liquidated damages, and under what circumstances the contractor should be awarded a time extension. If a delay is caused by the contractor or by natural disasters that are beyond the control of the contractor or the owner, the contractor should notify the owner about the cause of the delay with supportive documents within 20 days of the occurrence of the incident. The notification should clearly state the effect of the delay on the work schedule. If that is not possible, a time extension will be considered by the owner whenever the delay's impact on the total project duration is determined. It is also stated that a notification is not necessary if the delay is caused by the owner.

3 CONCLUSION

Time management is important in the success of a building construction project. Both the delivery of the product and the financial returns depend on the parties' ability to complete the works within scheduled deadlines. The General Conditions emphasize the importance of time management and clearly state the responsibilities of the parties in case delays occur. Since AIA has a long history of regulating the industry, and architects have a key role in managing the procurement and management of construction services, A201 – 2007 General Conditions of the Contract for Construction is the most commonly used general conditions in the U.S. Although limited by the law of the state in which the project is executed, this document expresses that,

- The contractor has to submit a schedule to the owner and the architect, within the time limit stated in the contract, not for approval but for information, since the contractor is solely responsible for the work.
- If delay is caused by the owner, or a factor not related to the contractor's performance, a time extension needs to be awarded by a change order in the amount determined by the architect.
- For a time extension to be awarded, the contractor should file a claim in a timely manner.
- Recovery of damages for delay is limited for both the owner and the contractor by state law.
- The contract may include a provision for liquidated damages to be applied in case the substantial completion date is not met by the contractor.

Being the biggest owner in Turkey, the government sets the rules and legal procedures to be followed in managing projects in the building construction industry. Since its use is mandatory by law in any kid of public investment, the General Conditions for Construction issued by the Public Procurement Authority is the most common document regulating the relationship between the owner, the architect and the contractor not only in public projects but also in private projects. This document states that:

- The contractor has to submit a schedule to the owner for approval, within the time limit stated in the contract or its addenda.
- The owner has the right to reject the schedule, or accept it with modifications.

- If the contractor fails to meet the substantial completion date recorded in the schedule, the penalty stated in the contract document is applied.
- If the delay is caused by a factor that is beyond the control of the contractor, a time extension needs to be issued by the owner to compensate for the entire delay or part of the delay depending on the cause of the delay.
- The contractor should file a claim within 20 days of the occurrence of the delay.
- The claim should state the reasons and their effect on the schedule.

Turkish construction firms are familiar with the General Conditions for Construction since they commonly perform in the domestic market, but they also compete in international markets such as Russia, Middle Eastern and Arab countries. Turkish companies also operate in Europe and the U.S. (Ugur 2010). Turkish construction companies should have a good understanding of the General Conditions used in other countries.

References

AIA, *A201 – 2007 General Conditions of the Contract for Construction,* American Institute of Architects, 2007, Retrieved from http://www.aia.org/contractdocs/referencematerial/aiab099119 on November, 2015.

Alkass, S., Mozerolle, M., Harris, F., Construction Delay Analysis Techniques, *Construction Management and Economics,* 14(5), 375-394, 1996.

AlSehaimi, A., Koskela, L., Tzortzopoulos, P., Need for Alternative Research Approaches in Construction Management: Case of Delay Studies, *Journal of Management in Engineering,* ASCE, doi: 10.1061/(ASCE)ME.1943-5479.0000148, 407-413, October, 2013.

Arditi, D., Akan, G. T., Gurdamar, S., Reasons for Delays in Public Projects in Turkey, *Construction Management and Economics,* 3(2), 171-181, 1985.

Birgonul, M. T., Dikmen, I., Bektas, S., Integrated Approach to Overcome Shortcomings in Current Delay Analysis Practices, *Journal of Construction Engineering and Management,* ASCE, 141(4):04014088, 1-11, 2015.

Chui, K. W., Bai, Y., Comparison of Contract General Conditions between United States and China, *Journal of Architectural Engineering,* ASCE, doi: 10.1061/(ASCE)AE.1943-5568.0000020, 119-125, December, 2010.

El-adaway, I. H., Fawzy, S. A., Cody, K., Fast, S., Spencer, G., Bond, D., Cushman, D., Stieffel, T., Contract Administration Guidelines for Contractors Working under AIA A201-2007 Contract for Construction, *Journal of Legal Affairs and Dispute Resolution in Engineering and Construction,* ASCE, 6(1):03013002, 1-10, 2014.

Enshassi, A., Arain, F., Al-Raee, S., Causes of Variation Orders in Construction Projects in the Gaza Strip, *Journal of Civil Engineering and Management,* 16(4), 540-551, 2010.

Fisk, E.R., *Construction Project Administration,* 7th Edition, Wiley, Hoboken, N.J., 2002.

Kazaz, A., Ulubeyli, S., Tuncbilekli, N. A., Causes of Delays in Construction Projects in Turkey, *Journal of Civil Engineering and Management,* 18(3), 426-435, 2012.

MEU, *General Conditions for Construction,* Ministry of Environment and Urbanization of Turkey, 2014, Retrieved from https://www.csb.gov.tr/db/bayburt/webmenu/webmenu4604.pdf on November 1, 2015.

Ugur, L. O., Cantitative Comparison of Responsibilities' and Risks' Distribution Between Turkish General Conditions of Construction and FIDIC Red Book General Conditions, *e-Journal of New World Sciences Academy,* 1A0069, 5(2), 104-120, 2010.

Yang, J., Wei, P., Causes of Delay in the Planning and Design Phases for Construction Projects, *Journal of Architectural Engineering,* ASCE, doi: 10.1061/(ASCE)1076-0431(2010)16:2(80), 80-83, June, 2010.

YEM, *Turkish Construction Sector Report 2014*, The Building Information Center, Istanbul, 2015. Retrieved from http://www.yapi.com.tr/TurkYapiSektoruRaporu2014/#1/z on January 1, 2106.

Interaction between Theory and Practice in Civil Engineering and Construction
Edited by Komurlu, R., Gurgun, A. P., Singh, A., and Yazdani, S.

STAFF COMPETENCY ISSUES IN LINEAR SCHEDULING APPLICATIONS

JUNESEOK YANG and DAVID ARDITI

Dept of Civil, Architectural, and Environmental Engineering, Illinois Institute of Technology, Chicago, USA

Linear scheduling method (LSM) is a scheduling method that is based on achieving a continuous flow of resources as well as precedence requirements. Although LSMs have recognized strengths in dealing with resource-intensive projects that are composed of repetitive activities, LSMs are not used extensively. This study proposes a staff-technology fit model that aims to measure the extent to which the features provided by LSM match scheduling staff experience, know-how and capabilities. The aim is to understand why LSM is not used as extensively as expected in construction scheduling. A questionnaire survey was administered to professionals listed in the directory of the Construction Management Association of America (CMAA) to measure staffing-technology fit in LSM applications. Research findings indicate that LSM applications satisfy user requirements in construction scheduling by (1) providing information that is useful to project participants depending on their position and function in the project organizations, (2) presenting graphics that are easy to understand, and (3) providing ease in communication. Research findings also reveal that LSM is not only a scheduling tool but also a control tool. However, the findings also point out that (1) LSM schedules require much time and effort to produce, (2) there are only very few commercially available software programs capable of running LSM schedules, and (3) there are not enough training opportunities for project managers and their staff to learn how to use LSM.

Keywords: Staff-technology fit, Scheduling methods, Linear scheduling, Scheduler competencies, Scheduling software, Visualization.

1 INTRODUCTION

The critical path method (CPM) is by far the most popular method used in larger construction projects. Alternative techniques to CPM were developed in the last 40 years under the generic term, linear scheduling methods (LSMs). However, while the general consensus is that LSMs are well suited to projects composed of activities of a repetitive nature, their usage is not widely accepted (Arditi *et al.* 2002).

Can the fit between the scheduling staff's expertise and the characteristics of LSM explain the reason why LSM is not used as extensively as expected? The objective of this study is to identify the barriers which prevent LSM from being used widely in the industry.

2 THE PROPOSED STAFFING-TECHNOLOGY FIT MODEL

The staffing-technology fit model shown in Figure 1 leads to three general propositions. The first two propositions deal with the technical capabilities of the scheduler and the characteristics of the LSM technology, respectively. The third one is a "fit" relationship between the capabilities of the scheduler and the characteristics of the LSM technology.

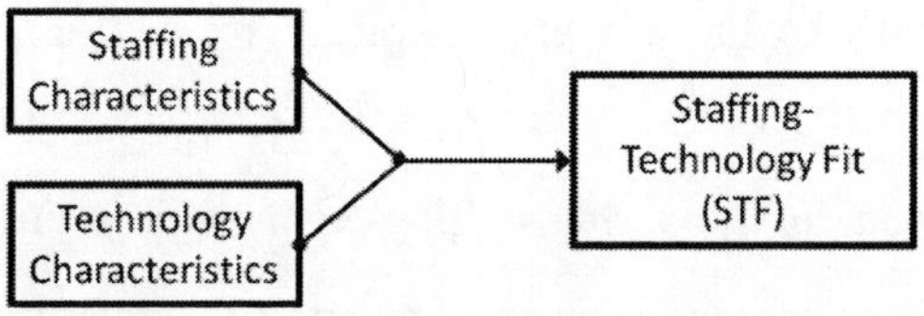

Figure 1. Staffing-technology fit model.

2.1 Characteristics of Staffing

The characteristics of the staff could affect how easily and well schedulers utilize a scheduling method. The difficulty of scheduling may vary depending on the competence of a scheduler. Schedulers, who are more competent, well trained, or quite familiar with a particular scheduling system will be better able to successfully manage their work.

2.2 Characteristics of the Technology

Technologies can be defined as tools that individuals use in carrying out their tasks (Goodhue and Thompson 1995). Scheduling and control can be achieved by using LSM that offers advantages such as maintaining continuity of resources, conciseness, and vividness in linear and repetitive projects.

2.3 Staffing-Technology Fit

Staffing-technology fit can be defined as the degree to which technology functionally matches the individual abilities of schedulers and managers. To be specific, an experienced scheduler can select an optimum level of detail, whereas insufficient knowledge of the scheduler may lead to poor schedule development. Thus, by examining to what degree the LSM scheduling system fits the capabilities of schedulers, the scheduling system can be judged to be acceptable or not.

3 METHODOLOGY

The methodology of the study involved an online survey that was designed by means of a survey tool provided by www.SurveyMonkey.com. The study was targeted to the professionals listed in the directory of the Construction Management Association of America (CMAA).

The first part of the questionnaire required the respondent to indicate agreement or disagreement with eight statements described in the following section on a scale of 1–5 (1=strongly disagree and 5=strongly agree). The second part of the questionnaire included three questions that inquired about demographic characteristics such as the experience of the respondents, and the type and characteristics of projects in which they were involved.

4 STAFFING-TECHNOLOGY FIT IN LSM APPLICATIONS

Eight statements presented in Table 1 were identified as staffing-technology fit measures that affect the linkage between LSM and the capabilities of the staff involved in scheduling and control. They were inspired from the study of Goodhue and Thompson (1995).

Table 1. Statements for measuring staffing-technology fit in LSM applications.

Statement	Clarification	Sources
S1. The schedule provides information that is useful to all project participants.	Participants in the construction project require information that is useful to them depending on their position and function in the project organization.	Galloway (2006)
S2. The schedule generates graphics that are easy to understand by all parties in the project.	The graphical format enables all parties in the project to better visualize the plan of action and to more easily communicate the plan to the group involved with the project.	Duffy *et al.* (2010)
S3. The degree of detail that the schedule provides coincides with the needs of the project manager.	The level of detail varies from low density (e.g., executive summary) for construction owners to high density (e.g., detailed schedule) for contractors.	Ceran and Dorman (1995) Hartley (1993)
S4. Schedulers are familiar with and have access to a software package.	Software packages allow schedulers to define many phases and tasks, and then to assign the labor, materials and resources required to complete each task and phase.	Hartley (1993)
S5. The schedule requires little time and effort to produce.	If a scheduling method is easy to use and user-friendly, the schedule requires less time and effort to produce.	Badiru and Pulat (1995)
S6. There is enough training for project managers and their staff on how to use the system.	Training on how to use the scheduling system improves the productivity of project participants.	Birrell (1980) Martin (1988) Benjamin *et al.* (1990)
S7. Project participants can effectively communicate with each other on the basis of the schedule.	Communicating on the basis of the schedule makes project participants understand the day to day objectives of the project, and, reduce uncertainty and disagreement.	Orlikowski and Yates (1994) Walker (1995) Jergeas and Hartman (1994) Coble and Snow (1996) Mackanzie *et al.* (1999)
S8. Project managers can respond to schedule-related issues promptly.	Project managers should respond promptly to the schedule-related issues such as contractual compliance, reasonability of job logic, and activity durations.	Goodwin (1993) Moosavi and Moselhi (2012) Hartley (1993)

5 DATA ANALYSIS AND DISCUSSION

The questionnaire was designed for response over a web link emailed to the potential respondents in February/March 2015. A total of 251 completed responses were received for data analysis. Of the 251 respondents, 68% indicated that they had experience in building construction (e.g., commercial, residential, educational, etc.), 63% in civil works (e.g., roads, bridges, tunnels, etc.), and 33% in industrial construction (e.g., power plants, refineries, etc.).

Concerning project size, 71% of the respondents had been involved in projects worth over $50 million. Also, the average number of their years of experience in the construction industry was 24.4 years. All respondents stated that they were familiar with LSM. Given their extensive experience, especially in large projects, the respondents appeared to be well qualified to answer the questionnaire administered in this study.

According to the results presented in Figure 2 respondents mostly agree that (1) LSM satisfies the requirements of all project participants who need information that is useful to them depending on their position and function in the project organization(S1); (2) LSM is not only a scheduling tool but also a control tool (S8); (3) LSM allows a good degree of detail in the schedule (S3); (4) LSM provides ease in communication between project participants (S7); and (5) LSM presents graphics that can be of great value to users (S2).

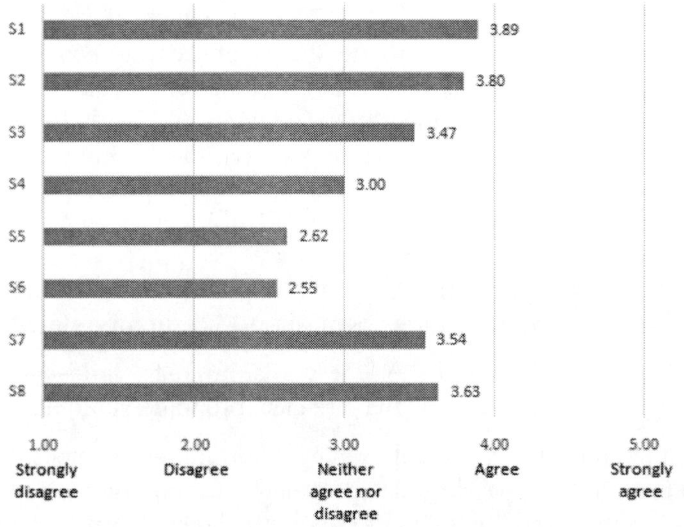

Figure 2. Mean scores of LSM applications.

On the negative side, respondents mostly agree that (1) there are not enough training opportunities for project managers and their staff to learn how to use LSM (S6); (2) LSM schedules require much time and effort to produce (S5); and (3) several well-designed software packages are not commercially available for schedulers' use (S4).

The literature supports these findings. A study conducted by Jongeling and Olofsson (2007) shows that a major reason why the construction industry has been slow to adopt LSM is the lack of supporting software that is user-friendly and easy to use so that the schedule requires less time and effort to produce. In addition, Tavakoli and Riachi`s (1990) work indicated that the main reason behind unsuccessful LSM applications was the lack of support from employees who were presumably not knowledgeable enough about LSM. To accomplish a successful implementation of LSM, all levels of users should be trained well. According to Nageeb and Johnson (2009), most of the limitations and disadvantages are due to limited training opportunities for LSM methods and software because scheduling theory is often lacking at the university level and is usually taught as a two-week seminar.

6 CONCLUSION

This study proposed a staffing-technology fit model to understand why the use of LSM is limited in construction. A questionnaire survey was conducted to collect information about LSM applications. The staffing-technology fit model was used to find out respondents` perceptions of how their capabilities fit the requirements of LSM. Professionals` opinions regarding staffing-technology fit show that respondents mostly agree that LSM requires significant effort to produce and that LSM is not taught at school or at work. Also, schedulers are not familiar with and have not enough access to a software package. Thus, these reasons seem to hinder LSM from being accepted in the industry.

This empirical study considers only the theory of staffing-technology fit. However, LSM use can also be explained by theories of attitudes and behavior. In addition, the fit between LSM and the task performed by schedulers can also help to understand LSM implementation. Although the study has limitations, the findings and implications are significant in that the fit between the technical capabilities of the staff involved in scheduling and control and the characteristics of LSM technology does affect the extent to which LSM is used in the construction industry.

The findings of this study indicate that the performance of LSM is likely to be enhanced if all project participants are cognizant of scheduling-related matters.

- For owners, LSM should monitor work progress. Indeed, LSM applications provide production diagrams that are easier to understand than CPM networks.
- For contractors, LSM should create opportunities to meet deadlines because contractors have to pay liquidated damages for contractor-caused delays. Thus, a scheduling method that can predict bottlenecks that may result in delays will allow contractors to complete the project in the shortest possible time.
- For construction managers, LSM should allow managers to coordinate the activities over the entire project, communicate with other participants, and respond to schedule-related matters.

Future work may involve the examination of other factors than staffing-technology fit, such as task-technology fit, expected consequences of utilization, social norms, habit, and facilitating conditions. Also factor analysis can be used to group the variables for further quantitative analysis.

References

Arditi, D., Tokdemir, O. B., and Suh, K., Challenges in Line-Of-Balance Scheduling, *Journal of Construction Engineering and Management*, 128(6), 545-556, Dec, 2002.

Badiru, A. B., and Pulat, P. S., *Comprehensive Project Management: Integrating Optimization Models, Management Principles, and Computers*, Prentice-Hall, New Jersey, 1995.

Benjamin, C., Babcock, D., Yunus, N., and Kincaid, J., Knowledge-Based Prototype for Improving Scheduling Productivity, *Journal of Computing in Civil Engineering*, 4(2), 124-134, Apr, 1990.

Birrell, G. S., Construction Planning—Beyond the Critical Path, *Journal of the Construction Division*, 106(3), 389-407, May, 1980.

Ceran, T., and Dorman, A. A., The Complete Project Manager, *Journal of Architectural Engineering*, 1(2), 67-72, Jun, 1995.

Coble, R., and Snow, K., *Non Verbal Communication As It Relates To Safety Management*, CIB Working Commission *W99*, Lisbon, 1996.

Duffy, G. A., Oberlender, G. D., and Seok Jeong, D. H., Linear Scheduling Model with Varying Production Rates, *Journal of Construction Engineering and Management*, 137(8), 574-582, Nov, 2010.

Galloway, P. D., Survey of the Construction Industry Relative to the Use of CPM Scheduling for Construction Projects, *Journal of Construction Engineering and Management*, 132(7), 697-711, Jul, 2006.

Goodhue, D. L., and Thompson, R. L., Task-Technology Fit and Individual Performance, *MIS Quarterly*, 213-236, Jun, 1995.

Goodwin, R. S., Skills Required of Effective Project Managers, *Journal of Management in Engineering*, 9(3), 217-226, Jul, 1993.

Hartley, K. O., How to Make Project Schedules Really Work for You, *Journal of Management in Engineering*, 9(2), 167-173, Apr, 1993.

Jergeas, G. F., and Hartman, F. T., Contractors' Construction - Claims Avoidance, *Journal of Construction Engineering and Management*, 120(3), 553-560, Sep, 1994.

Jongeling, R., and Olofsson, T., A Method for Planning of Work-Flow by Combined Use of Location-based Scheduling and 4D CAD, *Automation in Construction*, 16(2), 189-198, Mar, 2007.

Mackanzie, J., Gibb, A., and Bouchlaghem, N., Communication of Safety in the Design Phase, *15th ARCOM Annual Conference*, 2(1), 569-578, Sep, 1999.

Martin, G. L., Training Programs for Engineering Consulting Firms, *Journal of Construction Engineering and Management*, 114(1), 121-132, Mar, 1988.

Moosavi, S. F., and Moselhi, O., Schedule Assessment and Evaluation, *Proceeding of the Construction Research Congress*, 535-544, May, 2012.

Nageeb, M. R., and Johnson, B. T., Line of balance scheduling: Software Enabled Use in the US Construction Industry, *Proceedings of the 45th Annual conference of the Associated Schools of Construction of University Florida*, Apr, 2009.

Orlikowski, W. J., and Yates, J., Genre Repertoire: The Structuring of Communicative Practices in Organizations, *Administrative Science Quarterly*, 541-574, Dec. 1994.

Tavakoli, A., and Riachi, R., CPM Use in ENR Top 400 Contractors." *Journal of Management in Engineering*, 6(3), 282-295, July, 1990.

Walker, D. H., An Investigation into Construction Time Performance." *Construction Management and Economics*, 13(3), 263-274, Jul, 1995.

Interaction between Theory and Practice in Civil Engineering and Construction
Edited by Komurlu, R., Gurgun, A. P., Singh, A., and Yazdani, S.

TASK-TECHNOLOGY FIT IN THE ADOPTION OF LINEAR SCHEDULING METHOD

JUNESEOK YANG and DAVID ARDITI

Dept of Civil, Architectural, and Environmental Engineering, Illinois Institute of Technology, Chicago, USA

Although the general consensus is that linear scheduling methods (LSMs) are quite powerful, their use in construction has been very limited. The linkage between the characteristics of scheduling methods and the requirements of the tasks performed by schedulers has been an on-going concern in the construction industry. This study proposes a "task-technology fit" model to understand why LSMs are not being used as extensively as expected. The model aims to determine whether the characteristics of LSM (technology) satisfy the duties and obligations of construction schedulers (tasks). By scrutinizing the task-technology fit in LSM applications, deficiencies can be detected which hinder the wider use of these methods in the industry. A questionnaire survey was administered to measure task-technology fit in LSM applications. The target population included schedulers, project managers, construction managers, and other professionals listed in the directory of the Construction Management Association of America (CMAA). The findings indicate that LSM is effective in repetitive projects and is able to provide a smooth and efficient flow of resources by adjusting activities' rate of production. In addition, research findings point out that LSM effectively shows activity sequences as well as progress. However, the findings also reveal that LSM is not applicable when reliable resource data are not available. It should also be noted that very few software packages that perform LSM scheduling are commercially available on the market.

Keywords: Task-technology fit, Scheduling methods, Linear scheduling, Scheduling tasks, Scheduling software, Visualization.

1 INTRODUCTION

Even though the general consensus is that linear scheduling methods (LSMs) are quite effective in projects composed of activities of repetitive nature, their use in construction has been limited. In order to understand the reasons why these methods are not used as extensively as expected, an attempt is made in this study to analyze the linkage between the characteristics of LSM (technology) and the requirements of the tasks performed by schedulers. By scrutinizing the "task-technology fit" in LSM applications, deficiencies which deter LSM from being used widely in the industry could be found. Thereby, the ways to increase its level of acceptance could be developed.

2 THE PROPOSED TASK-TECHNOLOGY FIT MODEL

The task-technology fit model leads to three general propositions. The first two propositions deal with the characteristics of the scheduling task and the LSM technology, respectively. The third, and most critical proposition, is that task and technology characteristics interact to define a relationship. Such interaction is the essence of what is meant by a "fit" relationship.

2.1 Characteristics of the Tasks

Activities are identified to describe the project in sufficient detail so as to satisfy the schedule objectives (Hartley 1993). According to Chua and Shen (2005), construction can be viewed as a production line flowing through the activities of a project, and being supported by resources.

Many construction operations in building, industrial and civil works are repetitively performed. These repetitive projects consist of a large number of similar or identical units. Thus, maintaining work crew continuity in projects composed of repetitive units is essential in minimizing disruption that makes schedules difficult to develop as well as maintain (El-Rayes and Moselhi 2001).

Construction time and resources should be considered simultaneously for proper project scheduling. Time often takes precedence over resource utilization in construction projects. To be specific, when the interrelationships between project participants and activities are critical to the project or when the time constraints assume contractual significance, time has a higher priority than resource utilization (Hartley 1993). In contrast, optimal resource utilization is recognized as the key to meeting a repetitive construction project schedule (El-Rayes and Moselhi 2001). To be specific, construction managers need to develop a schedule for directing and controlling resources of manpower, machinery, and materials, which play a significant role in making work plans reliable, in a coordinated and timely fashion in order to deliver a project within the limited time available (Halpin and Woodhead 1976). Thus, construction project scheduling should be performed under resource constraints by considering flexibility for time through proper resource leveling.

2.2 Characteristics of the Technology

Technologies can be defined as tools that individuals use in carrying out their tasks (Goodhue and Thompson 1995). In the context of construction scheduling, schedulers utilize LSM to perform their tasks in projects that exhibit repetitive characteristics. LSM is based on a continuous flow of resources. The general consensus in the literature is that LSM is better suited in situations that involve repetitive activities.

2.3 Task-Technology Fit

Task-technology fit can be interpreted as the extent to which a scheduling method is supportive of the tasks generally performed by schedulers. A scheduling system that does not have a good fit is considered failed or unacceptable. Thus, examining the fit between the characteristics of a scheduling method and the tasks usually performed by schedulers may help to identify the barriers to extensive implementation.

3 METHODOLOGY

The methodology of the study is presented in Figure 1. An exhaustive literature review was conducted to understand the characteristics of LSM scheduling. Similarly, technology acceptance theories were reviewed to propose a "task-technology fit" model to investigate the current situation. The questionnaire survey method was chosen for data collection because the unit of analysis is users of LSM. The study was confined to the professionals listed in the directory of the Construction Management Association of America (CMAA). The selection of the respondents was based on their experience in construction scheduling. A cover letter was emailed to the recipients, which emphasizes the intent of the study and acknowledges the confidentiality of the information that is requested. This letter also included a link to the questionnaire.

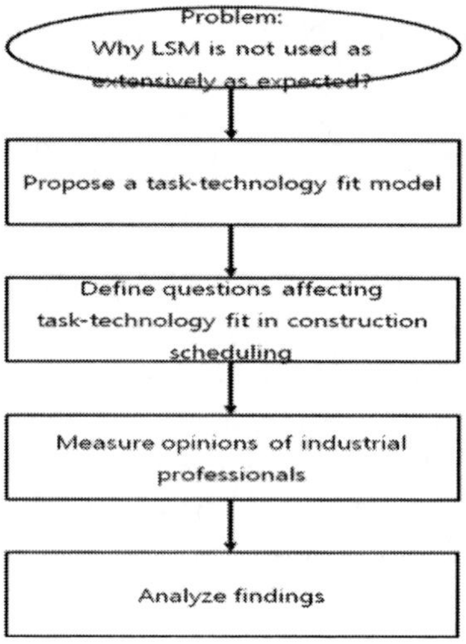

Figure 1. Research methodology.

The eight statements described in the following section were identified as task-technology fit measures that affect the linkage between LSM and the tasks of the staff involved in scheduling and control. The first part of the questionnaire required the respondent to indicate agreement or disagreement with these eight statements on a scale of 1–5 (1=strongly disagree and 5=strongly agree). The second part of the questionnaire included three questions that inquired about demographic characteristics such as the type and size of projects in which they were involved and the respondents` years of experience in construction industry.

The online survey was designed on a web-based platform www.SurveyMonkey.com. The reason for selecting an online survey tool was to obtain a wider sample of respondents and to reduce cost and time (Rubin and Babbie 2013).

4 TASK-TECHNOLOGY FIT IN LSM APPLICATIONS

The eight statements presented in Table 1 were designed to measure the relationship between scheduling-related tasks and LSM technology. The statements are inspired from the study of Goodhue and Thompson (1995).

Table 1. Statements for measuring task-technology fit in LSM applications.

Statement	Clarification	Sources
T1. The schedule clearly shows activity sequences in my projects well.	For a schedule to be realistic, the physical relationships that exist among the different construction components must be considered.	Moosavi and Moselhi (2012) Echeverry *et al.* (1991) Tommelein and Ballard (1997) Hartley (1993)
T2. Software exists that does all the tedious calculations instantaneously.	Commercial scheduling software packages are used to extract, read, and analyze the data with accuracy and high speed.	Badiru and Pulat (1995) Lee (2005) Mattila and Bowman (2004)
T3. Activities' rates of production can be adjusted for efficient performance.	Production rates should be updated and adjusted on a regular basis to assure they accurately reflect the site conditions in the construction area, allowing a smooth flow of resources and working continuity.	Hartley (1993) Moosavi and Moselhi (2014)
T4. It is easy to schedule projects that are composed of repetitive activities.	Although most construction projects are dominated by non-repetitive activities, some construction projects such as highways, railways, pipelines, and tunnels are characterized by a series of successive and repetitive activities.	Harris and Ioannou (1998)
T5. A realistic schedule can be developed even if reliable resource data are not available.	Time constraints sometimes have a higher priority than resource constraints. For example, reliable resource data are desirable but not required in network-based scheduling systems.	Zheng *et al.* (2005) Harris and Ioannou (1998)
T6. The schedule provides a smooth and efficient flow of resources.	In order to control the flow of resources smoothly, equipment and labor should be utilized in the most efficient way possible and the total cost of resources should be minimized.	Karaa and Nasr (1986)
T7. The schedule defines all contractual interfaces clearly.	A scheduling method defining all contractual interfaces clearly provides the legal basis for the administration of construction disputes and claims.	Moosavi and Moselhi (2012) Hartley (1993)
T8. The schedule can measure progress compared to a baseline schedule.	A scheduling method that provides progress measurement compared to a baseline schedule ensures the fitness of the schedule.	Moosavi and Moselhi (2014)

5 DATA ANALYSIS AND DISCUSSION

A cover letter including a link to the questionnaire was emailed to the recipients in February/March 2015. The letter included the intent of the study and an acknowledgement of the confidentiality of the information that was requested. A total of 251 completed responses were received for data analysis. Of the 251 respondents, more than half indicated that they had experience in building construction (e.g., commercial, residential, educational, etc.) (68%) and civil works (e.g., roads, bridges, tunnels, etc.) (63%), while fewer had experience in industrial construction (e.g., power plants, refineries, etc.) (33%).

Concerning project size, 71 % of the respondents had been involved in projects over $50 million. Also, the average number of their years of experience in the construction industry was 24.4 years. All respondents stated that they were familiar with LSM. Based on their extensive experience in large projects, the respondents appeared to be well qualified to answer the questionnaire administered in this study.

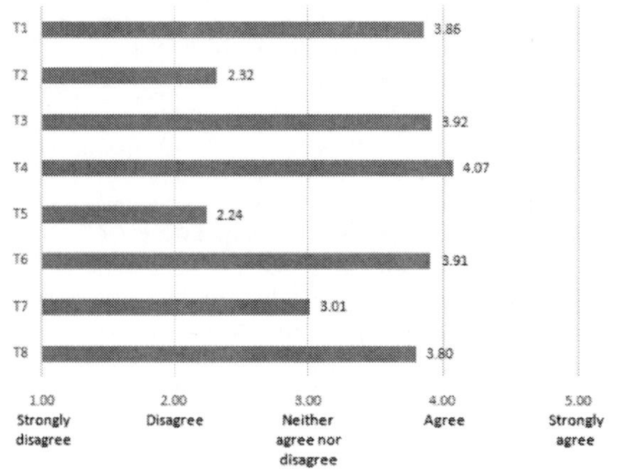

Figure 2. Mean scores of LSM applications.

The results presented in Figure 2 reveal that professionals` view concerning task-technology fit tended to "agree" in five of the eight statements that (1) LSM is effective in repetitive projects (T4); (2) LSM adjusts activities` rate of production efficiently (T3); (3) LSM provides a smooth and efficient flow of resources (T6); (4) LSM shows activity sequences well (T1); and (5) LSM can show progress well (T8).

On the negative side, respondents mostly disagree that (1) LSM is applicable when reliable resource data are not available (T5); (2) Software packages for LSM are not sufficient (T2); and (3) LSM defines all contractual interfaces clearly (T7). There is enough evidence in the literature to support these findings. The LSM scheduling methods' criticalness is based on time and resources, unlike network scheduling where it is based only on time. To be specific, the LSM diagram can be developed once the number of crews and the expected rate of output have been computed for each activity and then, the number of units to be produced is plotted against time (Arditi *et al.* 2001). Therefore, LSM is indeed not applicable when reliable resource data are not available.

Also, studies such as the one conducted by Jongeling and Olofsson (2007) show that a major reason why the construction industry has been slow to adopt LSM is the lack of supporting software packages that help to produce LSM schedules.

6 CONCLUSION

Despite the obvious strength of LSM, its use in construction has been limited. This study proposed a task-technology fit model to understand why LSM is not used as extensively as expected in construction scheduling. A questionnaire survey was conducted to collect information about LSM applications.

The findings of this study regarding task-technology fit in LSM applications indicate that LSM is superior in repetitive projects by its very nature, and that LSM should not be used if resource data are not available. Also, software packages that help to produce LSM schedules have to be developed for wider acceptance of LSM applications. LSM requires significant expertise and effort to produce but LSM is seldom taught to schedulers at school or at work.

Although this empirical study is limited only to an investigation of task-technology fit, the findings and implications are significant in that the fit between task and technology does affect positively the use of scheduling technology. Future work may involve the examination of additional factors such the level of the schedulers` expertise and their attitude toward an unusual technology such as LSM.

References

Arditi, D., Tokdemir, O. B., and Suh, K., Effect of Learning on Line-Of-Balance Scheduling, *International Journal of Project Management*, 19(5), 265-277, Jul, 2001.

Badiru, A. B., and Pulat, P. S., *Comprehensive Project Management: Integrating Optimization Models, Management Principles, And Computers*, Prentice-Hall, New Jersy, 1995.

Chua, D., and Shen, L., Key Constraints Analysis With Integrated Production Scheduler, *Journal Of Construction Engineering And Management*, 131(7), 753-764, Jul, 2005.

Echeverry, D., Ibbs, C. W., and Kim, S., Sequencing Knowledge For Construction Scheduling, *Journal Of Construction Engineering And Management*, 117(1), 118-130, Mar, 1991.

El-Rayes, K., and Moselhi, O., Optimizing Resource Utilization For Repetitive Construction Projects, *Journal Of Construction Engineering And Management*, 127(1), 18-27, Feb, 2001.

Goodhue, D. L., and Thompson, R. L., Task-Technology Fit And Individual Performance, *MIS Quarterly*, 213-236, Jun, 1995.

Halpin, D. W., and Woodhead, R. W., *Design Of Construction And Process Operations*, John Wiley & Sons, Inc, New York, 1976.

Harris, R. B., and Ioannou, P. G., Scheduling Projects With Repeating Activities, *Journal Of Construction Engineering And Management*, 124(4), 269-278, Jul, 1998.

Hartley, K. O., How to Make Project Schedules Really Work for You, *Journal of Management in Engineering*, 9(2), 167-173, Apr, 1993.

Jongeling, R., and Olofsson, T., A Method for Planning of Work-Flow by Combined Use of Location-Based Scheduling and 4D CAD, *Automation In Construction*, 16(2), 189-198, Mar, 2007.

Karaa, F. A., and Nasr, A. Y., Resource Management In Construction, *Journal of Construction Engineering and Management*, 112(3), 346-357, Sep, 1986.

Lee, D.-E., Probability of Project Completion Using Stochastic Project Scheduling Simulation, *Journal of Construction Engineering and Management*, 131(3), 310-318, Mar, 2005.

Mattila, K. G., and Bowman, M. R., Accuracy of Highway Contractor's Schedules, *Journal of Construction Engineering and Management*, 130(5), 647-655, Oct, 2004.

Moosavi, S., and Moselhi, O., Review of Detailed Schedules in Building Construction, *Journal of Legal Affairs and Dispute Resolution in Engineering and Construction*, 6(3), Aug, 2014.

Moosavi, S., and Moselhi, O., Schedule Assessment and Evaluation, *Proceedings of the Construction Research Congress*, 535-544, May, 2012.

Rubin, A., and Babbie, E., *Research Methods For Social Work*, Cengage Leanring, Belmont, 2013.

Tommelein, I. D., and Ballard, G., Look-Ahead Planning: Screening and Pulling, *Civil Environmental Engineering Department of University of California, Berkeley*, Technical Report No. 97-9, Oct, 1997.

Zheng, D. X., Ng, S. T., and Kumaraswamy, M. M., Applying Pareto Ranking and Niche Formation to Genetic Algorithm-Based Multiobjective Time–Cost Optimization *Journal of Construction Engineering and Management*, 131(1), 81-91, Jan, 2005.

Interaction between Theory and Practice in Civil Engineering and Construction
Edited by Komurlu, R., Gurgun A. P., and Singh, A.

DESIGN AND PROJECT PLANNING OF THE RENOVATION OF A RAILWAY LINE

AMELIE OUTTIER[1,2], BART DE PAUW[1,2], and HANS DE BACKER[2]

[1]*TUC Rail NV, Brussels, Belgium*
[2]*Dept of Civil Engineering, Ghent University, Ghent, Belgium*

The railway line L50A is one of the most important lines in Belgium, connecting the capital Brussels with Ghent. The line originally consisted out of two tracks. Because this formed a constricting bottleneck for the entire railway network in Belgium, the existing capacity was expanded. Between 2008 and 2015 a third and fourth track was constructed, one on each side of the original 2 tracks. Now that the expansion works have ended, the 2 original tracks have to be refurbished. The necessary works include: renewal of ballast and rails, changing all wooden sleepers by pretensioned concrete ones, strengthening of the subgrade and renewal and improvement of the waterproofing layers at the bridges and viaducts. This article focuses on the first 10 kilometers of the trajectory, including 10 locations where the waterproofing has to be renewed. At each location, the existing ballast, sleepers and tracks have to be removed for the middle two tracks, so a new polyethylene-waterproofing layer can be installed. Afterwards the tracks have to be restored. All of this has to be performed while the 2 outer tracks remain in operation and without disturbing the electrification cables above the tracks. Because of this, the project will necessitate a combination of creative planning, use of modified and small-scale equipment and a judicious use of night-time labor. All of these items will be discussed in the article.

Keywords: Refurbishment, Expansion, Waterproofing, Labor planning, Management.

1 INTRODUCTION

Railway line L50A was expanded from two existing track to 4 tracks with an additional track on each side of the original construction. After completion of this expansion project, the two central tracks have to be refurbished. This includes:

- Renewal of the actual rails and ballast layer;
- Wooden sleepers have to be exchanged by concrete ones;
- Sanitation of the rack bed;
- Renewal of the waterproofing layers and adaptation of the ballast walls because of the additional tracks.

This section includes 10 bridge structures within a length of 10 km. The overall project planning is that in first step, the track bed is renewed at the location of all the bridge structures, including a renewal of the waterproofing and new ballast. Afterwards, the entire ballast layer will be renewed once more, but now along the entire length of the section, using a ballast work train.

The use of a thin polyester layer is preferred as waterproofing for all bridges, since they can be installed using smaller equipment, when compared with the installation of mastic asphalt. It is the intention to work at all bridges separately, and well before the actual track renovation. By planning this way, some of the bridges can be combined and the overall cost of the project should be reduced considerably.

2 SCOPE AND SAFETY

This entire project has to be constructed working in a narrow area, in between two railway tracks that remain in operation during the majority of the construction period. During the construction of the additional third and fourth track, the choice was made to use a very wide cross-section. Because of this, the new work area will always respect a safety distance of 1.5 m, measured from the outer rail of each parallel track. Pede viaduct has the narrowest safety zone and is shown in Figure 1 (De Backer *et al.* 2008).

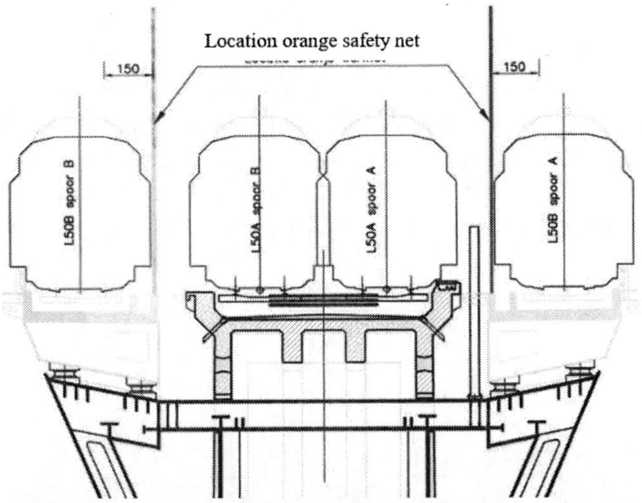

Figure 1. Safety zone vs. work area on the Pede Viaduct.

Since the work program for this location includes working with machinery, including moving cranes, safety guards will have to be present. The use of a railway crane will also be considered.

3 BRIDGE RENOVATIONS

3.1 Borrestraat Bridge

The bridge at the Borrestraat will be suppressed completely. During earlier works, the bridge opening under the railway has already been filled up, using stabilized sand and foam concrete. Since the height in length profile of the new tracks will be quite considerable, an additional filler layer will be installed on the old bridge.

The existing ballast stabilizing structures, shown in Figure 2, will additionally have to be expanded to allow for easy maintenance in the future.

Figure 2. Ballast walls for the Borrestraat bridge.

3.2 Herdebeekstraat Bridge

The bridge deck at the Herdebeekstraat has recently been renewed, which implies that the waterproofing layer should still be in good condition. However, a slight repositioning of the tracks, because of the use of a widened profile, will necessitate a repositioning of the concrete through decks. The decks are shown in Figure 3, already in their new and final position. Because of this, all of the pathways, etc. will have to be adapted as well.

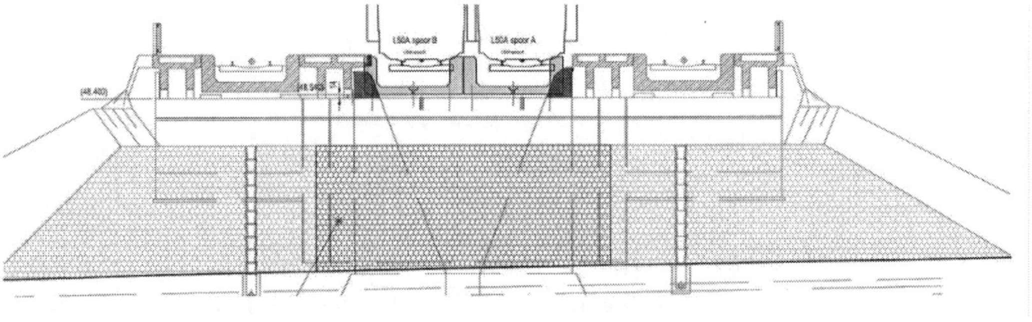

Figure 3. Cross-section of the Herdebeekstraat bridge, abutment closest to Brussels.

3.3 Pede Viaduct

The line L50A crosses the valley of the river Pede by a 523 m long historic viaduct, built in the 1930's. In those days the viaduct was chosen over large backfills in the valley due to poor soil conditions. The structure consists of 16 three-hinged reinforced concrete arches with a span of 32 m and a maximum height of 20 m. Four arches form an independent group, as they are separated from the rest of the viaduct by double pier structures, allowing for compensation of the thrust force of each group.

The viaduct is a benchmark in the rural environment through its dominance midst the gentle slopes of the Pede valley. Extension of the railway facilities to 4 tracks needed to widen the existing structure by two additional lateral viaducts, with respect to the protected work of art. Therefore, the arch structure and the existing piers were left apparent as much as possible. The arches kept their function and continued to behave as a four-span group. Important guidelines in making the new design were the contrast of old and new technology, keeping the four-span static behaviour with the repetition of 32 m spans and leaving the characteristic view of the hollow piers.

After consideration of various alternatives during the pre-design stage, the final design consists of a steel superstructure with variable hollow sections supported by integrated cantilever pier structures, as shown in Figure 4.

Figure 4. Earlier expansion works on the Pede viaduct.

The steel box girders of the superstructure are continuous over 4 spans, in accordance with the existing arches. The box section is characterised by waving patterns, both in the plan view as in the cross-sections. The upper flange of the box section is constant and stays horizontal along the structure's length. The lower flange however has a variable width of minimum 3,65 m at the piers and maximum 5,15 m at the span centre. The lower flange rises according to a sine wave from the supports towards the span centre obtaining less height and is twisted about a horizontal axis as it becomes wider. As a result, the vertical box web near the existing concrete arches has variable height, whereas the outer web plate shows torsion along the bridge axis. This creates a waving pattern of the steel structure complying with the existing arches, both in horizontal plane as in the front view, as illustrated in Figure 4. The new superstructure is supported by steel cantilever structures, fixed to the existing piers. Each vertical pier has a rectangular box section in a conical shape and fades into the lower part of the concrete pier. Two cantilever piers are joined by a transverse internal steel framework located in the hollow parts of the existing piers to ensure horizontal stability in transverse direction.

A large opening exists at the arch crown of the original concrete viaduct, as can be seen on Figure 5. This was originally intended as a drainage chamber, so it was equipped with drainage pipes. However, this greatly complicates the installation of the new waterproofing layer on the bridge. Two possible methods have been considered:

- The drainage chamber can be filled using a light concrete, after which a waterproofing layer in polyester (available in rolls) can be installed on the flat bridge surface. Because of weight restrictions the light concrete should actually be lighter than the ballast layer that is present within the drainage chamber. In addition, an arch joint is present at the arch crown, so a joint will have to be included within the filler, allowing for some rotation. This solution has the advantage that it is faster, allows for easier maintenance in the future, but the execution is slightly more complicated;
- If the drainage chamber remains in place, another waterproofing layer is necessary because of the uneven bridge surface. Acrylic resins can then be considered, because they allow for spraying on the irregular, but thoroughly cleaned, surface of the bridge. However, this method needs a considerably longer work period and the use of at least two cranes.

Since the expansion works on the viaduct assuming that the waterproofing layers would be renewed consecutively, these works will have to be carried out as soon as possible. After all, the expansion works have already resulted in some water damage.

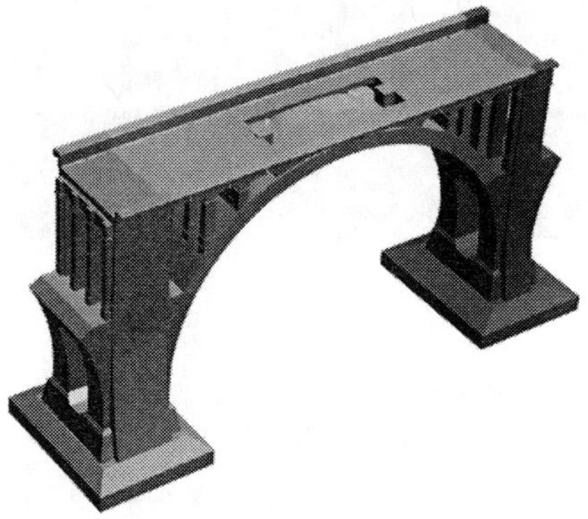

Figure 5. Original construction of the middle track supporting structure of the Pede viaduct.

3.4 Concrete and Masonry Arch Bridges: Kouterstraat, Bullenbergstraat

Both bridges, shown in Figure 6, are identical masonry/concrete arches where the most important works include the renewal and expansion of the pathways.

3.5 Concrete Culvert Bridge

For the bridges shown in Figure 7, the scope is quite similar. Special for this location is the fact that older abutments are still present from ancient arch bridges in the area behind the new culvert bridges. Since the new tracks will be installed at a lower position, these older abutments will have to be removed, working from supported trenches.

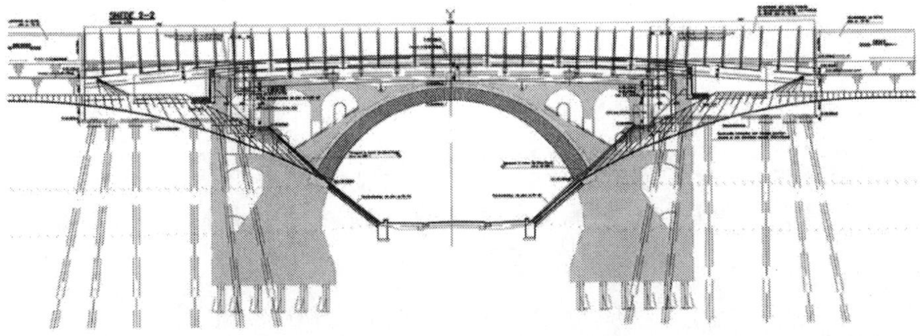

Figure 6. Masonry arch bridges at the Kouterstraat and Bullenbergstraat.

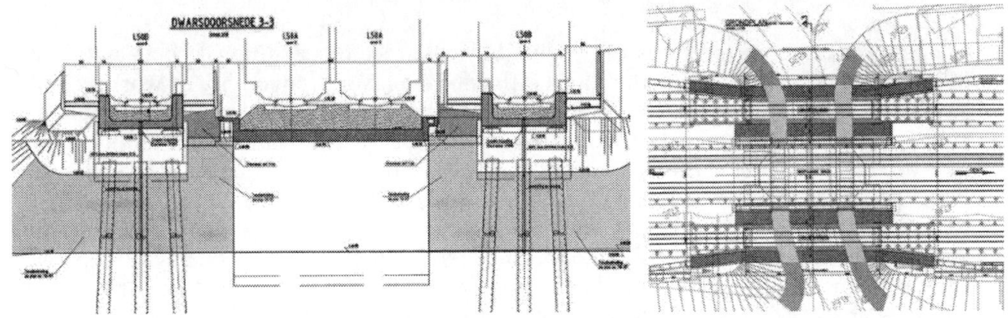

Figure 7. Concrete culvert bridges, cross-section (left) and plan view (right).

Table 1. Comparison of the project times for each bridge.

Bridge	Necessary time (weeks)
Borrestraat	10
Herdebeekstraat	15
Pede viaduct	23-27
Bullenbergstraat	15
Kouterstraat	14
Culvert bridges	12-14

4 CONCLUSION

Since the central tracks of Line 50A will be out of service for a period of 2 years because renewal of the track switches in a neighboring section, time is not really an issue, as is illustrated in Table 1. This will allow for a relaxed planning of the execution, where elements can be combined when they offer an opportunity for cost reduction.

References

De Backer, H., and Outtier, A., Quantification of thermal loads in steel box girders, *11th East Asia-Pacific Conference on Structural Engineering and Construction (EASEC-11): Building a sustainable environment*, Yang, Y., Leu, L., Chen, C., and Su, P. (eds.), 178-179, Taipei, Taiwan, 2008.

Interaction between Theory and Practice in Civil Engineering and Construction
Edited by Komurlu, R., Gurgun, A. P., Singh, A., and Yazdani, S.

FREQUENTLY OCCURRING LOSSES OF PRODUCTIVITY IN THE SHELL CONSTRUCTION PHASE

MARKUS KUMMER and CHRISTIAN HOFSTADLER

Institute of Construction Management and Economics, Graz University of Technology, Graz, Austria

In construction projects, losses of productivity often lead to additional costs and time delays. Suboptimal working conditions influence the labor productivity of workers and have an effect on the number of working hours required, for instance, to produce one cubic meter of reinforced concrete. Costs can be reduced and total productivity increased if the sources of potential losses of productivity are identified at the process planning stage or in early project phases. An expert survey conducted in 2014/2015 at Graz University of Technology, Austria, collected data on the losses of productivity that most frequently occurred in conjunction with reinforced concrete works in the shell construction phase. A standardized questionnaire was used for this survey. In addition, an interview was held with each of the experts to discuss the questionnaire in order to improve the quality of the survey and to gather important background information that prompted the responses provided by the experts. For this survey, reinforced concrete works were divided into ramp-up, main construction and final phases. Experts from Austria and Germany were able to select from a range of predefined losses of productivity and to specify other losses separately. This paper reports on the progress of the survey and refers to the related analytical methods. Furthermore, findings of the expert survey are represented in bar diagrams and interpreted. The survey revealed that the main construction phase was generally associated with the greatest potential for losses of productivity. Furthermore, most frequently stated losses of productivity varied depending on the specific construction phase. Future construction projects should particularly focus on the losses of productivity that were most frequently mentioned in the expert survey. The benefit of this approach is that it will provide the opportunity to implement early compensatory measures and to minimize the potential for losses of productivity.

Keywords: Expert survey, Reinforced concrete works, Ramp-up, Main construction, Final phase.

1 INTRODUCTION

In construction projects, losses of productivity often incur additional costs and construction time delays that give rise to disputes and legal proceedings as well as to the commissioning of expert opinions (Kummer 2015a).

To ensure a timely response to losses of productivity and to implement related compensatory measures, it is of particular interest to ascertain those losses of productivity that typically occur in the individual shell construction phases. An expert survey was conducted by means of standardized questionnaires because site observations or analyses of final costing exercises are usually difficult to perform. This survey determined the losses of productivity that occurred most frequently in the individual shell construction phases (reinforced concrete works).

An idealized trapezoidal model that comprises ramp-up, main and final construction phases is assumed for the utilization of production factors. In the main construction phase, most resources, such as the maximum number of workers, are utilized on-site (see Figure 1 – idealized representation). This is also the period in which the highest output is achieved (Hofstadler 2014).

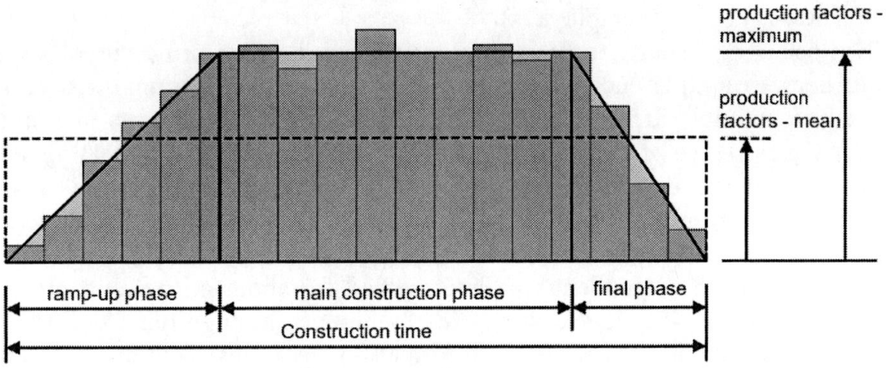

Figure 1. Development of production factors over construction time.

2 SELECTION OF EXPERTS

It is usually impossible to interview all experts in a given field, with the exception of a few, highly specialized disciplines. This is why sampling is necessary to arrive at conclusions as to the overall population on the basis of received responses. The selection of such experts relies on the following fundamental questions (Gorden 1969):

- Who is in possession of relevant information?
- Which of these experts are available?
- Whose willingness to provide information is the highest?
- Who is most likely in a position to provide accurate information?

In this context, experts are defined as individuals who possess specific knowledge in a clearly delineated field and who serve as a source of specific knowledge for the purpose of a survey (Gläser and Laudel 2010).

According to this definition, they possess knowledge and intellectual skills and competencies in a given field that clearly surpass the average level. Expert knowledge usually comprises exceedingly large amounts of information, including simplifications, lesser-known facts, rules of thumb and smart practices (heuristics) that enable efficient problem solving (in the relevant field) (Springer and Gabler 2014).

As part of a preselection process, a total of about 130 experts with experience in the fields of costing, process planning, construction, final costing and invoicing were contacted in writing and asked to participate in the survey. These were individuals with the related expert knowledge and sufficient professional experience who were added to this "pool" of potentially competent and willing experts either due to the fact that they were personally acquainted with those conducting the survey or by establishing contacts through third parties.

In total, 27 experts from Germany and Austria were recruited for the survey. 26 responses were submitted to the question regarding losses of productivity that occur most frequently during the shell construction phase. The majority of respondents (approx. 63%) worked for large companies with more than 250 employees, about 26% worked for medium-sized businesses (50 to 249 employees), and about 11% came from small businesses (10 to 49 employees).

When selecting experts, one of the key criteria is their professional experience. Only this degree of experience will enable sound responses regarding the occurrence of losses of productivity. In the survey presented in this paper, experts had an average professional experience of 17.7 years; this experience ranged from 5 to 41 years.

Questions were designed and developed together with social researchers (sociologists) in several revision steps, applying the principles of simplicity, clarity, impartiality, and specificity.

On average, each respondent was interviewed for about 45 minutes either on the phone or in a face-to-face session to overcome ambiguities, collect missing information, or obtain background information and justifications of responses (Kummer 2015b).

3 ANALYSIS OF SURVEY

The expert survey determined the types of productivity losses that occurred in the individual project phases. Respondents were able to tick boxes to indicate the project phases in which predefined productivity losses (see Figure 2) occurred particularly frequently according to their experience. Multiple choices were permitted.

Results are represented and described as bar diagrams for each of the three shell construction phases (see Figure 1). Experts had the opportunity to specify other losses of productivity, including the following additional categories: losses upon handover during multi-shift operation, changes to plans, limited available storage space, unskilled workers, increased material utilization and extended periods of material provision.

Losses of productivity were reported to occur most frequently in the main construction phase (134 mentions), followed by the ramp-up phase (87 mentions). Respondents identified the lowest potential for losses of productivity in the final phase (69 mentions) (Kummer 2015b).

For the ramp-up phase (see Figure 2), a particularly large number of respondents indicated losses of productivity occurring due to exceedingly short lead times in the planning process and poor planning quality. This opinion was justified by the fact that, in most cases, the planning process had not yet been finalized upon commencement of the construction phase, and that plans were completed only shortly before the actual construction process began. The factor of "changes to plans" additionally mentioned by one of the respondents can also be grouped under the above-mentioned, most frequently occurring type of productivity loss identified in the survey.

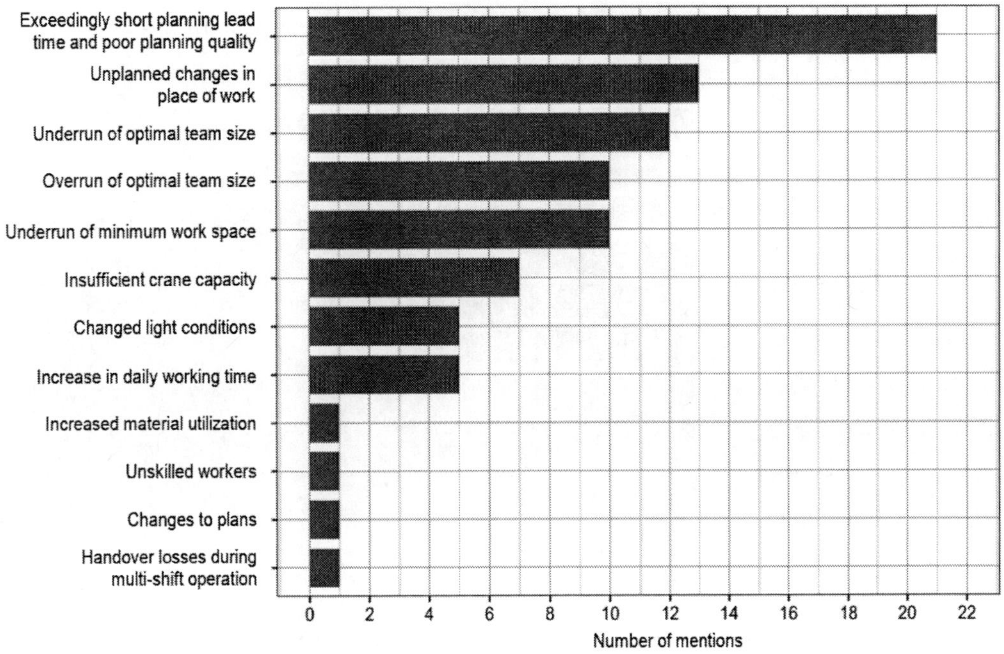

Figure 2. Losses of productivity – ramp-up phase (total number of mentions: 87).

Furthermore, unplanned changes in the place of work were frequently mentioned as a reason because on-site processes and workflows often need to be aligned with each other at the initial stage, and employees may also have to perform residual work on other construction sites, which prevents them from being available for an entire day or week on the affected site. Changed lighting conditions were mentioned relatively rarely as a cause of loss of productivity in the ramp-up phase because extended daily working times are rather infrequent at this stage. According to the experts, the same applied to insufficient crane capacity, which is a relatively rare phenomenon in the ramp-up phase because the maximum number of workers has not been reached and the amount of resources used in this phase is smaller than at the main construction stage.

Figure 3 shows the number of mentions of losses of productivity that occurred during the main construction phase. Overall, this diagram revealed a marked increase in the number of mentions compared to the ramp-up phase. This phase was considered to be associated with the highest potential for losses of productivity.

In this phase, the most frequently mentioned losses of productivity were increases in daily working time, changed lighting conditions, and insufficient crane capacity. Moreover, the experts identified exceedingly short planning lead times and poor planning quality as one of the other key factors that promoted losses of productivity in the main construction phase (as previously stated for the ramp-up phase).

Less significant factors that led to productivity losses in the main construction phase included unplanned changes in the place of work and underrun of the optimal team size. Since this phase is likely to be associated with the maximum number of workers present on-site, overruns of the optimal team size or underruns of the minimum work space are more likely to occur at this stage.

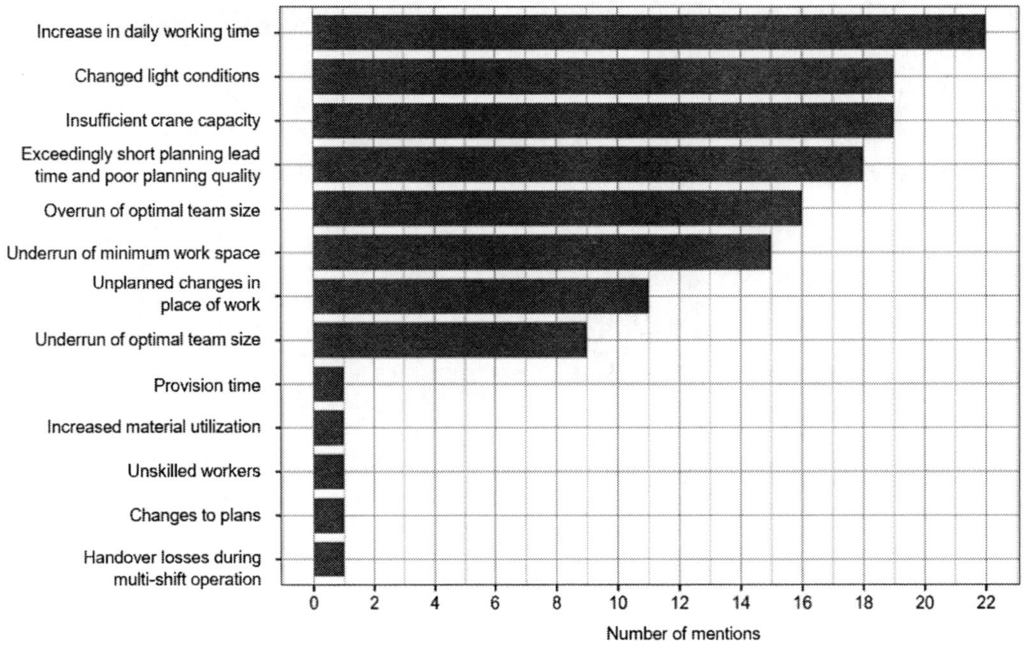

Figure 3. Losses of productivity – main construction phase (total number of mentions: 134).

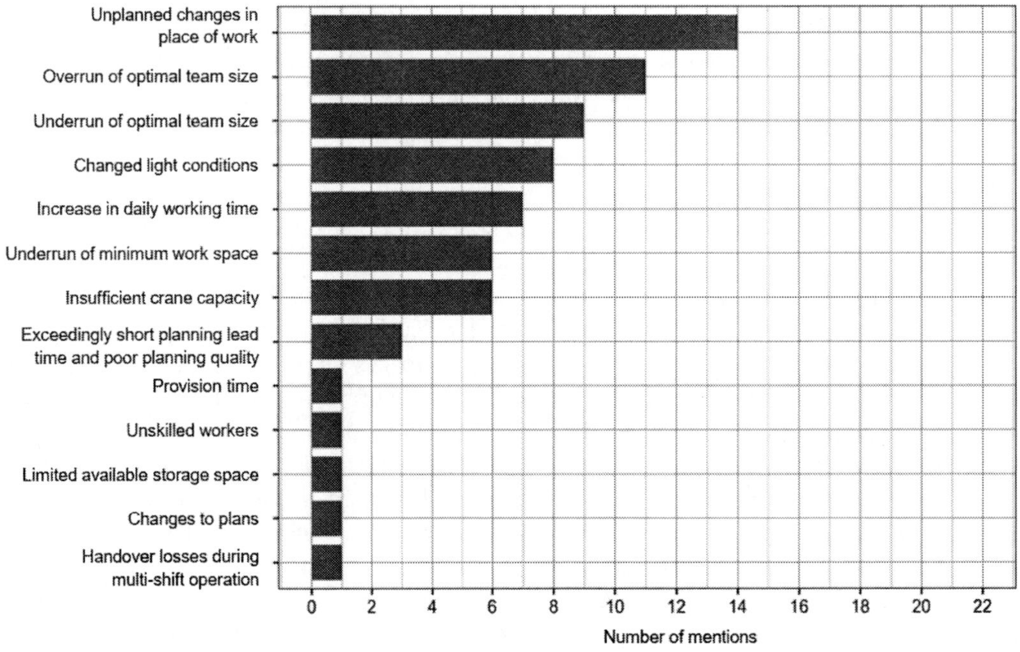

Figure 4. Losses of productivity – final phase (total number of mentions: 69).

Overall, the number of mentions of expected losses of productivity became smaller again in the final phase compared to the main construction phase (see Figure 4), and generally decreased to a level below the ramp-up phase.

In the final phase, the most frequently mentioned loss of productivity resulted from unplanned changes in the place of work. Experts justified this statement by workers often either having to perform residual work in several different places when the project is nearing completion or working on other construction sites already. Respondents stated that overruns and, similarly, underruns of the optimal team size may occur at this stage. Underruns should be expected whenever the completion date is not at risk and workers are removed from the construction site for this reason. Overruns should be expected whenever the final project phase is under high time pressure, and work intensity increases accordingly to comply with the scheduled completion date. In the final phase, experts rarely indicated losses of productivity due to short planning lead times and poor planning quality because, at this stage, the detailed design exercise will usually have been completed, and related plans are available early enough in the process.

4 SUMMARY

This paper identified the specific shell construction phase in which losses of productivity are most likely to occur. The main construction phase is generally associated with the highest potential for losses of productivity. This is where the number of mentions was the highest across all cited causes of losses of productivity. Experts justified this observation by the fact that the maximum amount of resources is utilized on-site at this stage, which may result in mutual obstructions between trades and longer working times as well as insufficient equipment capacity. Future construction projects should particularly focus on the losses of productivity that were most frequently mentioned in the expert survey. Its results should primarily raise awareness of frequently occurring losses of productivity among all parties involved in the construction process to prevent such losses in future and to increase productivity during the shell construction process.

References

Gläser, J., Laudel, G., Experteninterviews und qualitative Inhaltsanalyse als Instrumente rekonstruierender Untersuchungen, Springer Fachmedien GmbH, Wiesbaden, 2010.

Gorden, R., Interviewing – Strategy, Techniques, and Tactics, The Dorsey Press, Homewood, Illinois, 1969.

Hofstadler, C., Produktivität im Baubetrieb – Bauablaufstörungen und Produktivitätsverluste, Springer Vieweg, Berlin Heidelberg, 2014.

Kummer, M., Non-Linearities in the Calculation of Construction Costs, ISEC Press, Fargo, North Dakota, 2015a.

Kummer, M., Aggregierte Berücksichtigung von Produktivitätsverlusten bei der Ermittlung von Baukosten und Bauzeiten – deterministische und probabilistische Betrachtungen, Doctoral Thesis, Graz University of Technology, Graz, 2015b.

Springer Gabler Verlag (eds.), Gabler Wirtschaftslexikon, Stichwort: Expertenwissen, retrieved from http://wirtschaftslexikon.gabler.de/Archiv/55229/expertenwissen-v9.html on October 20, 2014.

Interaction between Theory and Practice in Civil Engineering and Construction
Edited by Komurlu, R., Gurgun, A. P., Singh, A., and Yazdani, S.

BARRIERS TO NEW TECHNOLOGY DISSEMINATION IN THE CONSTRUCTION INDUSTRY

MASOUD ZAVARI, MOHAMMAD MEHDI MORTAHEB, and HAMED KASHANI

Dept of Civil Engineering, Sharif University of Technology, Tehran, Iran

Despite the common belief, there is a considerable amount of innovation that occurs in the construction industry. However, the construction industry has been relatively slow to embrace the full potential of many new technologies. Evidence suggests that there is a lag in the uptake and effective implementation of several new technologies such as mechanical rebar splicing system. The objective of this study is to identify the impediments or barriers to the implementation of innovative technologies such as mechanical rebar splicing systems in the construction industry. A comprehensive literature review is conducted to identify and classify the barriers to new technologies dissemination. The findings of this literature review are validated using interviews with the construction industry experts. This study contributes to the state of knowledge by identifying and classifying the barriers that generally hinder the dissemination of new technologies in the construction industry. The practitioners can utilize the findings of this study to design appropriate strategies to overcome the barriers for implementation of new construction technologies. As part of this study, the extent to which the identified barriers have hindered the adoption of rebar splices, a technology that has not been adapted to its full potential. A questionnaire was designed, which was categorized the identified barrier in five major categories of cost, society and regulation, knowledge, marketing, and technical matters. The findings of the survey it was concluded that the lack of knowledge has the highest level of impact.

Keywords: Innovation, Impediment, Coupler splices, Mechanical splices, Implementation, Technology adoption, Diffusion of innovation.

1 INTRODUCTION

Innovative products, means and methods, technologies, services and management strategies are essential to economic growth, sustained competitive advantage and the achievement of landmark projects (Schumpeter 1934). There is a common perception among some professionals that construction innovation is rare and that the industry is slow to change (Slaughter 1998). This perception sometimes stems from comparisons of the construction industry to other industries, in which the introduction and adoption of new technologies and products have been extensive over the past several decades. Despite this perception, innovation in construction does occur and has been documented by researchers (Slaughter 1998).

The present research aims identify the impediments or barriers to the implementation of innovative technologies such as mechanical rebar splicing systems in the construction industry. A literature review was conducted to identify and classify the barriers to new technologies dissemination. The findings of this literature review were validated using interviews with the construction industry professionals. The knowledge presented in this paper can be used by aspiring entrepreneurs and existing organizations in the construction industry to identify the barriers that hinder the adoption innovative technologies and design strategies that increase the likelihood of success of innovative method and technologies.

2 LITERATURE REVIEW

Several studies on how innovation could be implemented in construction projects have been undertaken (Slaughter 2000, 1999, 1998, 1993a, 1993b, Winch 1998, Tatum 1987). These studies usually focus on how innovation is managed within one firm. There are also a variety of barriers that can hinder the dissemination of innovation in a filed. Fundamentally for technology dissemination to take place a number of obstacles such as uncertainty surrounding the costs and benefits of adoption, asymmetric information on the value of the innovation, financial and skill requirements, externalities, and regulatory barriers must be overcome (Johnson and Lybecker 2009). Bowley (1966, 1960) found several barriers that hinder progress of innovation: form of contract, cost of carrying out research, lack of information on cost savings arising from the innovation and restrictions imposed by regulations. Gann (2000) identified other impediments to construction innovation: contractors and consultants are isolated from one another, contractors are often of small size and fragmented, there is absence of competitive pressures on non-innovative firms and contractors may be unable to specify their products. However, there is no academic research in construction industry to find and categorize all probable barriers. Finding different barriers by studying other fields' literature such as environment, agriculture and energy is helpful to determine effective factors for technology dissemination in construction industry (Johnson and Lybecker 2009, Mercado *et al.* 1998, Gan 1998).

3 RESEARCH METHODOLOGY

A hybrid research methodology is applied to identify a set of significant factors in dissemination of construction innovation. A preliminary list of factors from the literature is presented through semi-structured interviews with researchers and industry practitioners who were asked to provide their comments. A final list of variables is obtained from these interviews and is presented in Table 1.

A questionnaire is designed and distributed among construction professionals to determine the significance of each barrier in dissemination of mechanical splices' implementation. In-depth interviews were conducted to gather further information to supplement the data gathered from the questionnaires.

3.1 Identifying Barriers in Dissemination of Construction Technologies

Twenty five barriers in dissemination of construction technologies were identified through an intensive literature review using important databases. To ensure the adequacy and comprehensiveness of these variables, five construction professionals from Tehran were invited for a preliminary study that consists of semi-structured interviews. The five respondents were two structural designers, two contractors and one construction owner. The professionals that were selected had at least 20 years of experience in the construction industry in order to ensure valuable discussions on factor that hinder the adoption of innovation in construction. After removing two controversial factor, 23 preliminary factors were determined. The resulting factors are shown in Table 1.

Table 1. Barriers of construction technologies dissemination.

Cost effectiveness	Society	Regulations	Knowledge	Providers	Technical issues
Increase in labor cost	Lack of infrastructures	Challenge existing legal systems	Limited knowledge of clients	Inappropriate marketing	Existence of technical limitation
Limited added value	Incorrect governmental policy	No incentive regulations	Inappropriate teaching and prerequisite knowledge	Limited providers	Inappropriate supporting equipment
Uncertainty regarding to the qualities of innovation as well as future prices	Society informal regulations		Limited knowledge of designers	Inappropriate ongoing support and training	Lack of expert labor
Disadvantage on stakeholders' income			Limited knowledge of scientists	Unsuitable assistance with maintenance	
Increase in cost claims of stakeholders			Limited knowledge about providers		
Increase in cost of related matters to innovation					

3.2 Questionnaire Survey

The initially identified 23 factors vary in their significance for every innovation construction technology or method. Therefore, a questionnaire was designed based on the table 1 to determine the significance of each barrier for couplers' dissemination. In this study, the questionnaire is adopted as an appropriate quantitative data collection method. The main participants selected for the survey come from top companies. They were invited to rank the significance of each variable under five categories, namely, cost, society and regulation, knowledge, marketing and technical matters, on a five-point Likert scale, with 5 indicating the most significant and 1 indicating the least significant. Likert scales are widely used and considered suitable to measure the importance of factors (Wang *et al.* 2010).

The full survey was conducted in Tehran over a two-month period. A total of 50 copies of the questionnaire were distributed between construction professionals. The demography of the respondents is summarized in Table 2.

3.3 Data Processing

A mean importance rating was used to calculate the importance of each variable, Eq. (1):

$$u = \frac{n_1 + 2n_2 + 3n_3 + 4n_4 + 5n_5}{n_1 + n_2 + n_3 + n_4 + n_5} \tag{1}$$

where u is the mean importance rating of a variable, and n_1, n_2,n_3, n_4, and n_5 represent the number of respondents who rated the variable as 1, 2, 3, 4 and 5, respectively. A statistical test was also conducted on the mean importance rating of each factor to determine whether the respondents consider the variables significant or otherwise. The null hypothesis H_0: $u \le u_o$ was tested against the opposite hypothesis and H_1: $u > u_0$, where u_0 is the population mean. The decision rule was to reject H_0 when the calculated t value was larger than $t_{(n-1,\, a)}$, as shown in Eq. (2):

$$t = \frac{x - \mu_0}{S_x/\sqrt{n}} > t_{(n-1,\alpha)} \tag{2}$$

where the random variable $t_{(n-1,\, a)}$ follows a t-distribution with (n-1) degrees of freedom, x is the sample mean, S_x is the sample standard deviation, n is the sample size, which is 50 in this study, and u_0 is the critical rating garnered by the factor considered the most significant, which is fixed at 3.

Table 2. Demography of the respondents.

Educational background	Percentages of respondents (%)
B.S.	56
M.S.	38
Ph.D.	6
Age	**Percentages of respondents (%)**
20-33	66
34-47	25
48-61	9
Position	**Percentages of respondents (%)**
Structural designer	21
Construction contractors	26
Construction owners	16
Construction supervisors	23
Researchers	14

4 IN-DEPTH INTERVIEWS

After the barriers were identified through the questionnaire survey, in-depth interviews were conducted to help interpret and elaborate the collected data. A total of eight respondents expressed interest in participating in the in-depth interviews. Among the interviewees, three were structural designers, four were construction contractors and one was an owner. The interviews were conducted with an average duration of 25 min. These interviews mainly aimed to gather further elaboration and explanation to the critical barriers.

5 ANALYSIS

The mean importance ratings, standard ratings, and t values are calculated from the survey results. The factors are ranked according to the order of importance shown in Table 3.

Table 3. Survey results on mean importance ratings and hypothesis testing.

Variables	U	S	t value	Sig.	Rank
Cost					
Increase in cost	3.28	1.054	1.509*	.141	14
Limited added value	3.25	1.016	1.392*	.174	15
Price uncertainty	3.94	0.619	8.569*	.000	2
Cost of quality uncertainty	3.66	0.971	3.824*	.001	9
Increase in cost claim	3.22	1.039	1.191*	.243	16
Increase in cost of rebar contractors	2.78	1.039	-1.191	.243	20
Increase in cost of transportation	3.09	1.058	.501	.620	17
Society and regulation					
Informal society regulation	4.06	0.878	6.849*	.000	3
Challenge to legal system	2.84	1.019	-.867	.393	19
Lack of legal incentive	3.91	0.777	6.597*	.000	5
Knowledge					
Limited knowledge of structural designers	3.84	0.884	5.400*	.000	6
Inappropriate teaching and prerequisite knowledge	3.94	0.801	6.623*	.000	4
Lack of knowledge about performance	3.41	0.979	2.347*	.025	11
Limited knowledge of construction owners	4.19	0.644	10.424*	.000	1
Limited knowledge about providers	3.69	0.859	4.527*	.000	7
Marketing					
Inappropriate marketing	3.63	0.793	4.458*	.000	8
Limited providers	3.41	0.979	2.347*	.025	12
Inappropriate ongoing support, training and assistance with maintenance	3.31	0.859	2.058*	.048	13
Technical matters					
Oppose technical standards	2.69	0.738	-2.396	.023	22
Harder performance compared to other alternatives	2.97	1.332	-.133	.895	18
Inadequate expert labor	3.59	1.103	3.045*	.005	10
Inappropriate supporting equipment	2.56	0.914	-2.709	.011	23
Increase in conflict of construction teams	2.66	0.827	-2.350	.025	21

* t-value is larger than $t_{(49,0.95)}$, which is 1.6765.

According to survey results, fifteen barriers affect mechanical splices dissemination. As the conclusion of in-depth interview, all of the barriers were specified by this method and construction industry has to eliminate them if diffusion of mechanical splices is needed.

6 CONCLUSION

In its simplest form, innovation is positive change that results from the implementation of new ideas. There is a perception that innovation in the construction industry is lacking or occurs at a very slow pace. Innovations in the construction industry may take place at a lower rate compared to other industries, but it does, and must, occur in a competitive market. The literature suffers from the lack of research on critical factors that can be used to guide construction industries to transfer technologies. In this study, through an examination of the literature and through semi-structured interviews with various professionals, 23 barriers under five categories were identified. These factors can potentially affect new technologies dissemination in construction industry and they were ranked for mechanical splices according to their importance ratings based on a questionnaire survey. Fifteen factors with a t value above 1.6765 were perceived as the most significant in this field. These critical factors have been shown to be reasonable and reliable through a series of statistical analysis. The identified factors can be adopted to inform the development of effective dissemination strategies and reduce barriers.

Notably, this study is conducted in line with the particular context of Iran, especially Tehran. Nevertheless, the findings can serve as useful references for similar research attempts in other cities.

References

Bowley, M., *Innovation in Building Materials*, Gerald Duckworth, London, 1960.

Bowley, M., *The British Building Industry*, Cambridge University Press, Cambridge, 1966.

Gan, Lin., Wind energy development and dissemination in China, *CICERO Working Paper*, 1998.

Gann, D.M., *Building Innovation: Complex Constructs in a Changing World*, Thomas Telford, London, 2000.

Johnson, Daniel K.N., and Kristina M. Lybecker, Challenges to technology transfer: A literature review of the constraints on environmental technology dissemination, *Colorado College Working Paper*, 2009.

Mercado, Agustin, JR., Marco Stark, and Dennis P. Garrity, Enhancing sloping land management technology adoption and dissemination, *IBSRAM Proceedings*, 25-45, 1998.

Schumpeter, J., *The Theory of Economic Development*, Harvard University Press, Cambridge, MA, 1934.

Slaughter, E.S., Builders as sources of construction innovation, *Construction Engineering and Management*, 119(3), 532–549, 1993a.

Slaughter, E.S., Innovation and learning during implementation: a comparison of user and manufacturer innovations. *Research Policy*, 22, 81–95, 1993b.

Slaughter, E.S., Models of construction innovation, *Construction Engineering and Management*, 124(3), 226–31, 1998.

Slaughter, S., Assessment of construction processes and innovations through simulation, *Construction Management and Economics*, 17, 341–50, 1999.

Slaughter, E.S., Implementation of construction innovations, *Building Research and Information*, 28(1), 2–17, 2000.

Tatum, C.B., Process of innovation in construction firm, *Engineering and Management*, 113(4), 648–63, 1987.

Wang JY, Yuan HP, Kang XP, Lu WS., Critical success factors for on-site sorting of construction waste: a China study, *Resources, Conservation and Recycling*, 54:931–6, 2010.

Winch, G., Zephyrs of creative destruction: understanding the management of innovation in construction, *Building Research and Information*, 26(4), 268–79, 1998.

Interaction between Theory and Practice in Civil Engineering and Construction
Edited by Komurlu, R., Gurgun, A. P., Singh, A., and Yazdani, S.

PERCEPTIONS OF THE CONSTRUCTION INDUSTRY

R. EDWARD MINCHIN JR., SHANNON M. CLOSE, and IAN FLOOD

University of Florida, Gainesville, Florida, USA

Despite the recent recession, the construction industry remains a vital part of the nation's economy. As the current generation of workers grows older, the industry is beginning to face a lack of skilled workers. In addition to the growing need for workers, the industry has traditionally had a difficult time attracting younger workers to enlist in its trades due to stereotypes such as low wages and opportunities and the work taking place in a dirty and dangerous environment. The aim of this study was to determine and document the attitudes of post-high school adults about the construction industry. Understanding these attitudes could help construction companies and trade schools develop recruitment and training programs that attract younger workers to the industry. A survey was conducted and the respondents' answers were analyzed statistically in order to determine general attitudes, differences in attitudes within the population, and differences between their attitudes and opinions and those of high school students. The results of the study show that many of the traditional stereotypes connected with construction, such as a dirty work environment, still exist. However, post-high school adults are increasingly aware of the opportunities and competitive wages available in construction-related jobs. Significant differences in responses were also found within the populations when looking at gender, age, and education level. Lastly, post-high school adults were found to have more favorable opinions of the construction industry than students in high school.

Keywords: Construction, Image, Perceptions, Shortage, Career.

1 INTRODUCTION

The construction industry touches every aspect of a person's life, from the home they live in to the roads they drive on to the utilities that fuel daily life. The construction industry employs around nine million people directly (salaried, waged and self-employed), about 6% of the total workforce of the U.S. Most people are not aware of the breadth of the industry and its importance to the country's economic development and health.

Skilled labor and craft professionals are the backbone of the industry. These jobs are vital to every construction project completed and ensure that critical components of infrastructure are completed safely and successfully. Management positions, at both the project and office level, also greatly influence the industry. Their expertise and supervision are key to ensuring that a project meets all its objectives, including those of cost, time, quality and safety performance. Recently, however, as the baby boomer generation begins to retire, a substantial shortage of workers is developing that threatens to put an ever-increasing strain on the industry.

1.1 Statement of the Problem

The oncoming shortage of workers in the construction industry has the potential to not only harm the construction industry long-term, but also to severely cripple the current economic recovery. The potential for this shortage of workers was noted by the industry as far back as the 1980s (Whyte and Green 2006). However, the recession of 2008 has made the problem worse. Construction unemployment has been as high as 15.8% (Goldman 2009) since the beginning of the recession. As the economy rebuilds, it may be difficult to fill the number of jobs that will hopefully be created.

High school students are an ideal demographic to target for new workers, but young adults a few years removed from high school can also be vital to filling the expected void. Many young adults that choose not to attend college or a trade school would be ideal for entering the industry as trained construction workers. Moreover, many students currently in college are not aware of opportunities in the field available to those with a degree. It is important to determine how the job market may appear to people as the country pulls out of the recession and industry recovers, as well as how to attract and retain workers in a rebound period.

In recent years, the industry has had a difficult time attracting and retaining new workers. Much of this problem is caused by the negative image of the construction industry. Construction jobs are often considered low paying, dangerous, and low in prestige. It will be important to develop recruitment and retention programs that address these issues as the need for workers increases.

1.2 Aim of Study

The purpose of this study is to discover and document young people's opinions of the construction industry. The study focuses on their knowledge and perception of the industry as they are about to enter the work force. These insights will be helpful in developing potential training and/or recruitment programs that attract more young people to enter the industry.

1.3 Hypotheses

This study aims to analyze several hypotheses and a general observation. The first hypothesis is that the image of the construction industry is still viewed negatively by post-high school adults. The second is that there are significant differences in the attitudes of males and females toward the construction industry. The third hypothesis is that there are differences in attitudes of different age groups. The fourth is that there are differences in the attitudes of those with different education levels, and the fifth is that there are differences in the attitudes of those with different employment statuses. These will be tested by statistically comparing the responses of survey participants. The last focus of the research was to observe and measure the differences in perceptions of the construction industry among post-high school adults and high school students. It is believed that post-high school adults will have a more favorable opinion of the industry. This is discussed as an observation because of lower raw high school response data than is needed to make a statistical comparison.

2 LITERATURE REVIEW

The construction industry is an integral part of the US economy. The industry accounted for about 4.8% of the national $14 trillion Gross Domestic Product (GDP) in 2007, according to the US Department of Commerce's Bureau of Economic Analysis. However, construction also helps fuel other industries such as manufacturing, finance and insurance, and transportation. The US Census Bureau reported that the industry provided 7.2 million jobs (not including the self-employed) ranging from day laborers to skilled labor and construction management, and had about 730,000 employers in 2007. The largest group of these workers, 1.76 million, is laborers followed by metalworkers, carpenters, supervisors, electricians, painters, and plumbers (Caulfield 2008).

The future lack of workers first became apparent in the construction industry in the 1980s (Whyte and Greene 2006). According to Whyte and Green, one of the large problems contributing to the worker shortage is the aging population. It was estimated that over half of the US population is now over the age of 50 and the youth population has a lower growth rate than the elderly (Whyte and Greene 2006). Up to 73% of workers in the construction industry within craft trades plan to leave the industry after the age of 46 (Rowings *et al.* 1996). This older population increase will lead to retirees outnumbering workers entering the workforce. Competition from other industries will also increase this problem, as the entire country will be searching for workers in a smaller worker population. In 2006, it was predicted that the industry would need to bring in and retain up to 275,000 workers a year over 10 years in order to keep up with growth projections at the time (Whyte and Greene 2006).

Working in the construction industry is often seen as an inferior job choice. As more students begin to enroll and complete college, this stigma may hurt efforts to recruit young people into the industry. In a 2002 list of job rankings, bricklayer, carpenter, and ironworker were ranked in the bottom ten of the 250 vocations listed (Kantz 2000). A 2006 survey of Alachua County (Florida) high school students showed that the majority believed the industry is dangerous and involves physical labor. The survey also showed that 50% of the students with a friend or family member in the industry would not choose to pursue a construction career, and almost 80% without such a connection would not work in the industry (Diavolistis 2006).

Despite the fact that the recession lowered the number of construction jobs, the job opportunities that exist in construction today are extremely promising. The 2012-2018 jobs outlook found in Occupational Outlook Quarterly released by the Bureau of Labor Statistics examines the trends of jobs for over 300 occupations and determines growth potential (Occupational Outlook Quarterly 2010).

3 DATA GATHERING

After completion of the literature review, a survey was developed based on a previous survey that focused on high school students, in order to facilitate comparison of the responses. All of the questions/statements in the original survey that were applicable to post high school adults were used in this survey. The survey was administered to consenting participants on the University of Florida campus. The results of the survey were analyzed arithmetically and statistically.

3.1 Survey Information

The survey as administered contained two sections. The first section gathered demographic information about each participant. The information on gender, age, race, education, and employment status allowed for the development of statistical differences between groups. The second section was a list of 20 statements which the participant judged using a 7 point Likert Scale labeled as follows: 1-Strongly Disagree, 2-Disagree, 3- Tend to Disagree, 4-Average, 5- Tend to Agree, 6- Agree, 7-Strongly Agree. In order to have similar data for an adult-to-high school comparison, the statements were taken from the previous study on the attitudes of high school students. Some statements on the high school student survey were not used on the adult survey because they were developed to examine the motivations of students or relations to the construction industry, which were beyond the scope of this study.

The survey was conducted over three days at the University of Florida. People passing by two locations were given an information sheet and asked to take the survey. At both locations, surveys were administered to those willing to participate between the hours of 10 am to 2 pm.

4 ANALYSIS

4.1 Demographics

The demographics collected in the survey were analyzed to gain a perspective of the survey participants and to examine differences between demographic groups. The total number of survey participants was 52. All participants chose to fully answer the demographic information.

The first demographic recorded was the gender of the participant. Of the 52 responses, 36 (69%) were male while 16 (31%) were female. Comparing these two populations was done to identify any differences in the perceptions of males and females when considering construction as a career.

The age of the participant was asked, as well as how many years post high school they were. The longer someone has been out of high school, the more experience in the workforce they should have and therefore, theoretically, the more knowledge they'd have about any industry. For analysis, the ages were broken down into five ranges: 18-20, 21-23, 24-26, 27-28, and Above 28. The number of participants for each group was 18 (34%), 12 (23%), five (10%), four (8%), and 13 (25%) respectively. Because of the disparity in number of participants in each range, two groups were used for comparison tests: "18-23" and "24 and above".

Participants were also asked to indicate their race. This information could be useful in seeing what groups of people are knowledgeable about the industry and which could be targeted for recruitment programs. The racial demographics of the survey takers was 41 Caucasian (78%), two African American (4%), four Hispanic (8%), four Asian (8%), and one Other (2%). Because of the low numbers of races other than Caucasian, this demographic was not used further.

The level of education was also examined in order to determine trends within different education levels. Of the survey population, one (2%) had some high school education, two (4%) had completed high school, 29 (56%) had some college education, and 20 (38%) had at least a bachelor's degree.

The last demographic examined was employment. The survey participants were comprised of 41 (79%) employed people and 11 (21%) unemployed. A full-time student was considered employed.

4.2 Statistical Analysis

The population was next analyzed using the Mann-Whitney Rank Sum Test to determine significant differences between groups. The groups analyzed were as listed earlier. The statements were classified as statistically significant if the p-value was 0.05 or less.

Four statements showed significant differences in the attitudes and perceptions of male and female respondents (all p-values were 0.026 or less). In three of these four statements, women rated the construction industry more favorably than men. The statement to which females answered less favorably than males was in thinking the occupation was dangerous. These results indicate that women may be open to careers in construction more so than in the past when the career field had been more male dominant.

Five statements were found to yield significant differences when the results of the population by people aged 18-23 and Above 23 were compared, and the older group answered them all more favorably (all p-values were 0.047 or less). Statements dealt with opportunities in the field, opportunities for women and the belief that construction is a dirty job. These results show that more education may be needed for younger people outside of high school in order for them to understand the opportunities available and to also help shed the negative reputation that the industry may still hold.

The responses to three statements were found to have significant differences when analyzing the population as people who had completed college and those who had not (all p-values were 0.046 or less). College graduates answered significantly more positively on the statement dealing with construction being a dirty job. The other statements are about the training and education necessary for a job in construction and the opportunities available therein. College graduates were more inclined to agree with these statements also.

No significant differences were found between the answers of employed and unemployed participants. All p-values were well over 0.05.

4.4 High School vs. Post-High School

The last analysis compared the responses of high school students to those of post-high school adults. Since the individual response data of the high schoolers were not available, the means were compared between the two groups.

Post-high school students had mean responses more favorable to the construction industry than did the high school students for all twenty statements. The more favorable answers indicate that construction seems a more attractive opportunity to post-high school respondents than to high school students.

Results show that after high school, construction industry opportunities may become more evident than in high school. Job training and recruitment programs would be much more effective at targeting these people if the targeted people knew that there are career opportunities requiring college degrees available in construction and that the industry pays well.

5 CONCLUSIONS

The survey has shown that the industry is viewed more favorably than the research team originally anticipated based on the earlier study involving high school students. An important result was that the sampled population showed an understanding of the opportunities available. As people begin to see construction as a well-paying and respectable career, more people may be willing to take jobs as both management and craft professionals. However, a glance at the means showed that the traditional belief that construction is a dirty and dangerous job still exists.

It was originally believed that differences would exist in many of the statement responses because of the cultural and knowledge differences in the groups created. When looking at the population split up by gender, significant differences were found. When looking at the population divided into people aged 18-23 versus Above 23, significant differences were found when scrutinizing based on education level; and no significant differences were found when comparing the responses of employed participants and non-employed participants.

Outside of seeing construction as an outdoors job, the post-high school population had favorable responses to all the statements dealing with traditional construction stereotypes. The post-high school population also had higher rankings for the opportunities in construction, particularly in regard to education and earnings.

In fact, post-high school adults answered more favorably to the construction industry on all 20 questions in the survey. While this strongly suggests that "post-high school adults view the industry more favorably than high school students," the low volume of raw high school response data makes a meaningful statistical analysis impossible. Therefore this statement cannot be supported statistically.

References

Bureau of Labor Statistics, The 2008-2018 job outlook in brief, *Occupational Outlook Quarterly*, (54)1, 2010.

Bureau of Labor Statistics, 2010, U.S. Department of Labor, Retrieved from http://www.bls.gov.

Caulfield, J., 2008, Construction Industry Employed 11 Million People in 2007, *Builder*. Retrieved from http://www.builderonline.com on June 6, 2010.

Diavolistis, A., *Absence of Craft Professionals: Alachua County High School Students' Perceptions of the Construction Industry*, M.S. thesis, University of Florida, Gainesville, FL, 2006.

Goldman, D., 2009, *Worst year for jobs since '45*, CNNMoney. Retrieved from http://money.cnn.com on June 6, 2010.

Kantz, L., *Jobs Rated Almanac: the best and worst jobs - 250 in all – ranked by more than a dozen vital factors, including salary, stress, benefits, and more*, St. Martins Griffin, New York, 2000.

Rowings, J., Federle, M., and Birkland, S., Characteristics of the Craft Workforce., *Journal of Construction Engineering and Management*. 122(1), 83-90, 1996.

Whyte, D., and Greene, S., 2010, The Skilled Workforce Crisis, *National Center for Construction Education and Research*. Retrieved from http://www.nccer.org.

Interaction between Theory and Practice in Civil Engineering and Construction
Edited by Komurlu, R., Gurgun, A. P., Singh, A., and Yazdani, S.

DEMOTIVATING FACTORS FOR ARCHITECTS IN THE CONSTRUCTION INDUSTRY

GÜLDEN GÜMÜŞBURUN AYALP[1] and FATMA ARSLAN[2]

[1]*Architecture Department, Zirve University, Gaziantep, Turkey*
[2]*Evşen Konukoğlu Architecture Firm, Gaziantep, Turkey*

Motivation is one of the factors that influence the productivity of people. Architects play a vital role in the management of projects in the construction industry. The success of construction projects relies heavily on their active involvement and effective performance. Hence it is important to assess the impact of motivation on the performance these architects. On the other hand, architects are frequently confronted with problems that could lead them to demotivation. Demotivation is caused not simply by a lack of motivators but the existence of certain situations that cause dissatisfaction and discourage individuals from pursuing desired goals and aims. From this point of view, the present study conducted with the aim of identifying factors affecting demotivation of Turkish architects. After a review of extant literatures in construction management, and design management 69 demotivating criteria was produced and used in questionnaire survey, and data were collected from architects who are working at construction industry, especially at design firms. The questionnaires were administered to architects via e-mail, and 71 participants responded to the survey. Utilizing the particular statistical analyzes, the factors were identified for detailed analysis and discussion. From the findings of the present study, ten underlying demotivating factors have been identified for architects in Turkish construction industry. This study would help managers of design firms to develop healthy workforce through eradication of the identified demotivating factors using some of suggested solutions.

Keywords: De-motivation, Design firms, Construction management, Architects, Turkey.

1 INTRODUCTION

The construction industry is complex, dynamic and uncertain, and requires highly motivated employees. The issue of employee motivation is important as it establishes a substantial foundation for high performance levels and less unproductive time. Therefore, improving productivity or the employees on a project can have a significant improvement on the project outcome (Zakari, et al., 1997; Kazaz, et al., 2008).

There is an implicit assumption that lack of motivation is the same as demotivation as conceptualized by Gorham and Christophel (1992), while other researches (Ng *et al.* 2004) argued that the two are different. Lack of motivation means lack of inner or social stimulus or impetus for an action on the part an individual (also refers to as "no motivation" or "zero motivation"). Demotivation on the other hand entails dampening of morale or spirit in carrying out a particular action. It generally causes dissatisfaction, and in actual fact, it is even worse than lack of motivation as it leads to feeling of being downcast, dispirited, depressed and despondent on the part of employee.

Motivation and de-motivation are an important aspect in heavy industries such as construction due to the project-based nature of the sector. Construction industry relies heavily on human resources who work in teams over a period of time, to deliver client needs and requirements to desired quality and within budget. During the projects, individual project team members are confronted with many problems, such as poor team interaction, low-quality workmanship, material unavailability, co-workers' incompetence and the project itself among others, which could lead to demotivation and low morale (Ng *et al.* 2004).

Studies on motivation in the construction industry have been done since the early 1970s. Nearly all of these studies have focused on the motivating factors of workers and foremen. However, there have been only a few studies focusing on the demotivation of architects and civil engineers. Although, architects play an important role in the management of construction projects, there has been only a one study, which belongs to Oyedele (2013) focusing on the demotivation of architects. The main area of interest for that study is the examination of architects' demotivation that is involved in the design phase of construction projects. The difference between our study and other one is Oyedele (2013) determined 43 demotivating factors for UK architectural practices, whereas 69 factors are defined in this study. The other difference is this study is the first and unique study, which investigates the Turkish architects' demotivation factors.

2 RESEARCH METHOD

After a thorough review of extant literature particularly in the knowledge areas of project management, organizational behavior and design management and practice, which are deemed important to the subject area, 69 demotivating factors, were identified. The motive for this was to deduce whether the 69 identified criteria influence architects demotivation or not.

3 FINDINGS

3.1 Participants

The sample comprised of architects who are currently employed in the construction sector. Therefore, the participants' experience (in years) is important for evaluating the significance of de-motivational matters. The distribution of the respondents' experience (in years) and working area are presented in Table 1 as crosstab.

Table 1. Demographics of survey respondents.

Years of experience	***Type of Working Area***			**Total**
	Office	*Site*	*Office+Site*	
1-5 years	22	3	16	41
6-10 years	9	0	6	15
11-15 years	3	0	3	6
16-20 years	1	0	1	2
21 years and above	2	1	4	7
Total	37	4	30	71

3.2 Reliability and Non-Parametric Tests

Using the SPSS (Statistical Package for Social Sciences) software tool, the overall Cronbach's alpha coefficient for this study was 0.976, demonstrating a very good reliability and internal consistency of majority of the criteria. To confirm that all the criteria are actually contributing to this internal consistency. According to Field (2005), if a criterion were not contributing to the overall reliability and therefore not a good measure of the construct, its associated Cronbach's alpha coefficient would be higher than the overall coefficient (0.976). This higher value implies that if the criterion were deleted, the overall reliability of the whole data would improve (Field 2005). Based on this rule, three criterions (DC5, DC11, DC13) were identified having a value of 0.977. It means that the criterions – "incompetence among design teams/co-workers", "isolation by opposite gender" and "inadequate social activities out of working hours" are therefore unreliable and not a good construct in measuring architect demotivation based on the population surveyed, and were consequently dropped from the list, leaving only 66 criteria.

After establishing a statically reliable list of architects demotivating criteria, it was essential for this study to examine whether the criteria were perceived similarly or differently by the respondents based on their demographics. These include the architects' years of experience as an architect in the construction industry and the type of working area that the respondents are currently located. Since the data were not drawn from a particular probability distribution and normal distribution is not assumed, non-parametric tests of Kruskal-Wallis were used for both demographics respectively.

For the differences in perception based on years of experience and working area type, Kruskal-Wallis test was used since there are more than two samples that are independent. The null hypothesis is that there are no differences in the perception of the respondents among the groups with regards to their years of experience and working area type, respectively.

3.3 Demotivation Severity Index and Ranking of Architects' Demotivating Criteria

In order to measure respondents' perception on the level of severity of each demotivating criteria, a demotivation severity index formula was computed using Eq. (1) below. The equation was derived from similar formula computed by Kometa et al. (1994) and Chan and Kumaraswamy (2002).

Demotivation severity index:

$$(DS) = \left(\frac{\sum_{i=1}^{N} (s_i)}{NS} \right) \times 100\% \tag{1}$$

From the overall ranking, the top five demotivating criteria in descending order include: DC49 – "Inept leadership behaviors of managers (e.g. dictorial, bullying, intimidation, etc.)", *DC6* – "Inadequate commitment among design team members", *DC8* – "Distrust and dishonesty among design team members", *DC60* – "Disrespectful manager", *DC48* – "Poor co-ordination between manager and employee". The three factors (DC49, DC60, DC48) are demotivating criteria arising from manager or supervisor. The next two demotivating factors belong to design team / co-workers related group.

3.4 Underlying Demotivating Factors – Exploratory Factor Analysis

Using exploratory factor analysis, the 66 reliable criteria (identified from the reliability analysis) were inputted to the SPSS software tool. The results indicated a ten-factor solution, which accounted for 84.18% of total variance as shown in Table 2.

Table 2. Exploratory factor analysis.

		Eigen value	% of variance	Factor loadings
Factor 1	***Poor tangible and intangible conditions of organization***	27.50	41.67	
DC31	Lack of complimentary lunch, beverages (coffee, tea, etc.) at work			0.773
DC33	Lack of on the job training activities			0.759
DC69	Long distance between work and city center			0.710
DC34	Inadequate opportunity for career development			0.682
DC68	Transportation problem- Long distances between accommodation and work			0.682
DC29	Poor working environment (overcrowded, poor lighting, noise, etc)			0.628
DC42	Organizational politics hinders performance and success			0.583
DC27	The work is unattractive and dull for me			0.577
DC30	Inadequate project resources (software, computer, etc.,)			0.552
DC32	Inadequate holiday entitlement			0.521
DC40	Lack of giving adequate information to employees about their performance at work by managers			0.509
DC41	Unequal authority and responsibility among co-workers at same position			0.455
Factor 2	***Poor interpersonal relationship***	4.82	7.31	
DC2	Poor co-ordination of design teams			0.820
DC7	Inadequate co-operation among design team members			0.785
DC1	Poor communication within design teams			0.768
DC6	Inadequate commitment among design team members			0.699
DC4	Unhealthy competition among co-workers			0.617
DC12	Lack of helping each other among co-workers			0.566
DC14	Team members dislike their work			0.564
DC3	Incompatibility of design team members			0.448
Factor 3	***Project induced stress***	3.19	6.84	
DC64	Dissatisfaction of clients			0.751
DC24	Excessive work load pressure			0.682
DC23	Unfair duty distribution among co-workers			0.679
DC38	Display of lack of interests in subordinate's work/ideas by supervisors			0.638
DC22	Lack of employee's winning recognition by supervisors			0.595
DC19	Unfairness distribution of promotion and bounty			0.566
DC17	Unfairness and unequal salaries and promotion among co-workers who work at same position			0.542
DC39	Inadequate freedom in day-to-day conduct of work			0.524
DC25	Working excessively long hours			0.514

Table 2. Exploratory factor analysis (Continued).

		Eigen value	% of variance	Factor loadings
Factor 4	***Poor managerial relationship***	2.59	5.93	
DC46	Co-workers cannot use initiative about work			0.696
DC37	Having inadequate authority for doing this work			0.595
DC47	Poor communication between manager and employee			0.701
DC48	Poor co-ordination between manager and employee			0.625
DC58	Lack of synergy between organizational goals and leadership behavior			0.574
DC45	Inadequate planning and control in organization			0.628
DC36	Unclear / Undefined job description			0.425
DC66	Lack of labor discipline			0.443
Factor 5	***Negative leadership behavior***	2.48	4.76	
DC49	Inept leadership behaviors of managers (e.g. dictorial, bullying, intimidation, etc.)			0.696
DC50	Inadequate leadership support			0.658
DC44	Employees do not feel that they belong to that firm			0.604
DC51	The working environment is focused on negative criticism			0.534
DC52	Frequent changes of project priorities by supervisors			0.511
DC60	Disrespectful manager			0.480
Factor 6	***Organizational injustice***	1.99	4.01	
DC15	Inadequate social insurance rights			0.749
DC18	Lack of monetary prizes (e.g. promotion and bounty) expect from salary			0.678
DC20	Late salary and promotion payment			0.666
DC16	Inadequate salaries			0.592
DC53	Clients demand for radical and high-degree of innovation in design			0.576
DC21	Lack of rise opportunity			0.463
Factor 7	***Dysfunctional design team***	1.97	3.57	
DC54	Design decisions are dictated by cost and not quality factors			0.810
DC55	Unrealistic project demands from clients			0.702
DC26	Low participation in decision making			0.567
DC59	Restrictive and negative effects of legal legislations to architectural designs			0.546
DC61	High ratio of failures at work			0.433
Factor 8	***Poor organizational culture***	1.87	3.20	
DC67	Cultural differences among co-workers			0.740
DC65	Compulsion for attending long distance business travel			0.734
DC10	Lack of liveness among co-workers			0.709
DC62	The jobs that are given by clients are not continuous and balanced throughout the year (some months are very busy, whereas others are not)			0.586
DC63	Chaos environment at the firm			0.478
Factor 9	***Perceived career decline***	1.57	2.68	
DC9	Perception of lack of respect among co-workers			0.782
DC8	Distrust and dishonesty among design team members			0.674
DC28	Work assignments do not always match my skills and interests			0.522
DC35	Poor safe and healthy conditions			0.518
Factor 10	***Lack of concurrent engineering***	1.05	2.00	
DC56	Uncooperative behaviors of clients/other project stakeholders			0.715
DC57	Difficulty in understanding idiosyncratic and tacit needs of clients			0.682
DC43	Lack of creativity easement of co-workers			0.522
			81.97	

The KMO value and the Bartlett tests of sphericity were 0.760 (above 0.5) and 0.000 (less than 0.05) respectively, indicating that the data set is suitable for factor analysis (Pallant 2005).

4 CONCLUSIONS

The focus of this study is on factors causing demotivation to Turkish architects within architectural organizations. These include factors originating from organizational behavior, project processes and design team/co-worker related activities. From the findings of this study, ten underlying demotivating factors have been identified, and include "poor tangible and intangible conditions of organization", "poor interpersonal relationship", "project induced stress", "poor managerial relationship", "negative leadership behavior", "organizational injustice", "dysfunctional design team", "poor organizational culture", "perceived career decline", "lack of concurrent engineering".

Of importance to this subject area is the comparison of the demotivating factors identified in this study with motivational factors identified in previous studies; specifically, the questions whether lack of motivation means presence of demotivation or vice versa.

References

Chan, D. M. W., Kumaraswamy, M. M., Compressing Construction Durations: Lessons Learned from Hong Kong Building Projects. *International Journal of Project Management,* 20 (1), 23–35, January 2002.

Gorham, J., Christophel, D. M., Students' Perception of Teacher Behaviors as Motivating and Demotivating Factors in College Classes. *Communication Quarterly,* 40 (3), 239–252, 1992.

Kometa, S. T., Olomolaiye, P. O., Harris, F.C., Attribute of UK construction clients Influencing Project Consultants'. *Construction Management and Economics* 12 (5), 433–443, March,1994.

Ng, S. T., Skitmore, M. R., Lam, K. C., and Poon, A. W. C., Demotivating Factors Influencing the Productivity of Civil Engineering Projects. *International Journal of Project Management,* 22, 139–146, February 2004.

Oyedele, L. O., Analysis of architects' Demotivating Factors in Design Firms. *International Journal of Project Management,* 31, 342-354, April 2013.

Pallant, J., *SPSS Survival Manual, 2nd Edition.* Open University Press, Berkshire, 2005.

Zakari, M., Olomolaiye, P., Holt, G. D., and Harris, F. C., Factors Affecting the Motivation of Iranian Construction Operatives, *Building and Environment,* 32(3) 161- 6, March,1997.

CURRENT STATE OF PRACTICE IN DISASTER RISK MANAGEMENT

EHAB MLYBARI, IMTIAZ AHMED, and MUHAMMAD KHALIL

Dept of Civil Engineering, Umm Al-Qura University, Makkah, Saudi Arabia

The Kingdom of Saudi Arabia has historic profile of multifaceted disasters, including, floods, earthquakes, volcanoes, cyclones, sand storms, rock falls, collapse of structures, epidemics, fire, terrorist acts, stampedes and other complex emergencies. The recurring enormous human and material losses emphasize the need for pursuing modern concept and approach which entail paradigm shift in handling disasters, from 'response centric' to 'proactive' disaster risk management (DRM) leading to disaster-resilient infrastructure, building integrated response capability and achieving enhanced awareness in the societal context. The DRM encompasses all activities related to disaster mitigation, preparedness, emergency response, recovery, rehabilitation and reconstruction. The study synthesizes information obtained on DRM from published literature, personal experiences and interaction with the stakeholders. Besides, questionnaire survey was also conducted to assess the community awareness and preparedness of concerned departments. The study undertakes comprehensive review of KSA disaster profile, current practices, response capability and human resource development. It has been identified that KSA is confronting multifaceted challenges including, lack of formal training and professional expertise; lack of understanding about conceptual aspects of disasters and coordination amongst various stakeholders; and absence of hazard risk assessment, mapping or micro-zoning, building codes, etc. This paper gives KSA sate of practices, institutional mechanism and capacity building needs to manage disasters on integrated and sustainable basis.

Keywords: Integrated, Sustainable, Infrastructure, Community, Awareness, Stakeholders.

1 INTRODUCTION

Over the past few decades, the devastating impacts of disasters have aggravated manifold, adversely affecting almost all countries across the globe. The increase in population, changed land use patterns, unplanned development and climate change were mostly the underlying causes of many catastrophic events. To minimize losses, the concept of disaster risk management (DRM) is evolved to evaluate and mitigate the damages through proactive risk management. The Kingdom of Saudi Arabia (KSA) also suffered enormous losses due to recurring multifaceted disasters. The disaster profile of KSA indicates that its north western region is prone to earthquakes and volcanic hazards, the central and western region to floods, the south west mountainous region to landslides and the central and eastern regions to dust storms (Al-Bassam *et al.* 2014). This paper, based on synthesis study and questionnaire survey, gives state of the practice, institutional mechanism and capacity building needs to manage disasters in KSA on integrated and sustainable basis.

2 DISASTER RISK MANAGEMENT - CONCEPTUAL ASPECTS

DRM is defined by UN International Strategy for Disaster Reduction (UNISDR 2009) as "The systematic process of using administrative directives, organizations, and operational skills and capacities to implement strategies, policies and improved coping capacities in order to lessen the adverse impacts of hazards and the possibility of disaster". The DRM mainly encompass three phases; prior, during and post disaster. The pre-disaster phase comprises four steps: a) risk identification and assessment: b) risk prevention, mitigation and reduction; c) risk transfer; and d) preparedness. The during-disaster phase encompass emergency response, to include rescue, relief and recovery. Whereas, post-disaster phase comprises rehabilitation and reconstruction. The primary objectives of DRM include: develop resilience in society against disasters; minimize damages through sustainable development; reduce disaster risks and vulnerabilities; and develop institutional framework, defining roles and responsibilities of all the stakeholders.

3 DRM PRACTICES IN KSA

The history of emergency management in KSA can be traced back to late 1920s when the first emergency management body, i.e., fire brigade, was raised in Makkah, primarily aimed at serving pilgrims. In 1948, the fire brigade was merged into a new department, General Security and Fire Services, which was replaced in 1965 by General Directorate of Civil Defense (GDCD). However, the structure, goals and responsibilities of GDCD were later reformed and it is now the main organization responsible for DRM (Alamri 2010).

The current framework of the GDCD is organized into three levels: Board of GDCD, Executive Committee, and volunteers. The board consists of Minister of Interior as Chairman, Assistant Minister of Interior as Deputy-Chairman and a number of members from various divisions of GDCD such as fire services, police, emergency medical services, etc. The Executive Committee manages the daily affairs and operational matters. Whereas, volunteers are citizens willing to assist GDCD during any calamity. GDCD is also assisted by a number of related departments and services, like Hajj and Umrah Special Forces, Makkah's Disaster Management Center, Red Crescent, Traffic, General Police and Highway Forces, etc.

Table 1. Disasters causing human losses in KSA.

Disaster Type	Date	No. of Affected	No. of Killed	Reference
Fire during Hajj	Apr, 1997	> 1,500	343	(Pararas-
Rift Valley Fever	Sep, 2000	500	87	Carayannis
Earthquake in Harrat Lunayyir	May, 2009	40,000	-	2013)
Jeddah Floods	Nov, 2009	>10,000	161	(Alamri 2010)
Crane Collapse in Makkah	Sep, 2015	394	111	(Batrawy 2015)

3.1 Significant Disasters in KSA

KSA has experienced numerous devastating events, suffering enormous human and material losses. Table 1 gives inventory of significant disastrous events, whereas salient aspects of various disasters are briefly summarized in succeeding sub paragraphs.

3.1.1 *Flood*

Floods remained the most frequent hazard in KSA during 1982-2005, with an average return period of 7 times per years along with average economic losses of USD 19 million (Al-Saud 2010). Also, floods account for 7th of the 10 most damaging disasters during 1900-2010. In 2009, Jeddah experienced the worst floods, resulting in 161 deaths and monetary losses of US$ 900 Million (Alamri 2010).

3.1.2 *Earthquake*

KSA is relatively less vulnerable to earthquakes since it is not located along the collision boundaries of the Arabian Plate (Pararas-Carayannis 2013). However, its north western region experienced numerous shocks during April-June 2009, with maximum recorded earthquake of magnitude 5.4 (Pallister *et al.* 2010).

3.1.3 *Tsunami*

The Arabian Peninsula has been affected by tsunamis in the past, with most of the recorded tsunamis occurring along its eastern and southern edge. The Tsunami formed by the 1883 Karakatau volcano eruption, the 1945 Makran earthquake or the 2004 Sumatra event, were extremely destructive. Two, localized tsunamis have been recorded in the Red Sea and another in the Persian Gulf (Pararas-Carayannis 2013).

3.1.4 *Fire*

The worst fire incident occurred during Hajj in 1997 that claimed 343 lives and wounded more than 1,500 pilgrims (Alamri 2010). Fire remains potentially the most often occurring hazard that has caused enormous losses.

3.1.5 *Volcanoes*

Historically, the area has witnessed volcanic eruption, with last recorded around 1400 years ago (Grainger 2007). In 2009, the lava field of 'Harrat Lunayyir' experienced numerous shocks, highlighting its potential vulnerability (Pararas-Carayannis 2013).

3.1.6 *Collapse of structures*

In September 2015, a crawler crane toppled over the Grand Mosque in Mecca that killed 111 people and injured 394 (Batrawy 2015). Earlier, in 2006, a hostel near the Holy Mosque in Mecca collapsed after a fire had spread in lower floors of the building that killed 76 people and injured 64 (Alamri 2010). The old infrastructure and services in the ancient cities remain potentially vulnerable.

3.1.7 *Epidemics*

The worst epidemic disaster that has occurred was 2000 Rift Valley Fever outbreak which killed at least 87 people and affected more than 500 people (Alamri 2010). The two holy cities are visited by millions from across the world and, despite all preventive measures, epidemics remain potential source of disaster.

4 QUESTIONNAIRE SURVEY

Two survey questionnaires were developed to determine the state of community awareness and the preparedness of DRM organizations. The community awareness questionnaire mainly covered five areas and it was responded by 147 persons. Whereas, questionnaire pertaining to DRM organizations had 19 questions, which was responded by 14 stakeholders. The results and analysis of the survey is presented below.

4.1 Community Awareness Survey

The results of community awareness survey are given in Table 2.

Table 2. Results of community awareness survey.

DRM Aspects	Summary of Question	Percent
Disaster related knowledge	Knows about Disasters	5
	Disasters cannot be totally prevented	83
	Have basic education of DRM	84
	Attended training on DRM	20
	Listened to experts on DRM	27
Disaster preparedness and readiness	Preparation for disasters is important	90
	Importance of disseminating knowledge about disaster	92
	Government and its organizations can manage disasters	75
	Government will provide relief facilities	89
	Government will undertake reconstruction after disaster	88
Disaster Adaptation	Lacks awareness about evacuation areas	77
	Has no knowledge about organization to be contacted	50
	Knowledge about disaster prone areas	70
	Lacks awareness about community participation.	47
Disaster Awareness	Never participated in voluntary activities	84
	Lacks awareness about the retrofitting of building	73
	Lacks knowledge about emergency aid	76
	Giving priority to disaster awareness at local, regional and national level	74
	Early recovery after disaster is crucial.	83
Disaster risk perception	Large-scale disasters could occur	39
	Locality is not safe against disasters	79
	Buildings are not well designed to withstand disasters.	71

4.2 DRM Stakeholder Organizations Survey

The results of DRM stakeholder organizations survey are briefly summarized in the following sections.

4.2.1 *Probability of occurrence and damage potential of disasters*

Disasters are most likely or likely to occur in the following order: fire, collapse of structures, terrorist acts, epidemics, cyclones, landslides / rockfalls, sand storms, floods and industrial accidents. Whereas the damage potential of disasters is reported in the following order: floods, fire, epidemics, cyclone, collapse of structure, terrorist acts, earthquake, tsunamis, rockfalls, industrial accidents and sand storms. It has also been reported that existing infrastructure lacks resilience against disasters.

4.2.2 *State of preparedness to manage disasters*

Organizations are either not prepared or the state of preparedness is below satisfactory against almost all disasters. Also, they are not properly equipped or state of equipment held is below satisfactory. It has also been reported that either no warning arrangements exist or the state of multi-hazard early warning systems is below satisfactory.

4.2.3 *State of training & community awareness*

Organizations are not well trained to efficiently manage disasters. The state of community awareness and training is also reported as below satisfactory.

4.2.4 *Need for DRM plans & implementation*

The planning to respond to disaster is reported unsatisfactory and serious problems have been reported in DRM implementation. DRM organizations strongly endorsed the need for the following: hazard mapping and urban planning, education and awareness, shelter and foods, medical kits, transportation of victims, community participation and early warning mechanism.

5 FINDINGS OF THE STUDY

Saudi Arabia is vulnerable to multifaceted natural and manmade disasters, whereas floods, sand storms, fires, collapse of structure and epidemics are considered recurring and most hazardous. The community is generally aware of the nature and consequences of these disasters, but lacks formal education, training and exposure to any kind of community awareness or participation programs. The DRM organizations are also deficient of requisite training, professional expertise and modern equipment to manage various disasters efficiently. The DRM organizations generally pursue response oriented event specific approach, whereas modern concepts emphasize proactive integrated DRM. Besides, KSA lacks framework for coordinated response, hazard risk assessment, mapping or micro-zoning, building / services guidelines, specifications, codes and regulatory mechanism, which are essential for sustainable development.

6 WAY FORWARD

The study concluded that a paradigm shift in concept and approach to handle disasters in KSA is essentially required. The enormous challenges relating to community awareness, education, training and capacity building have to be addressed on urgent basis. Following is recommended in this respect:

- Need to formulate DRM institutional framework outlining roles and responsibilities of various stakeholders, along with addressing training and capacity building needs, with focus on innovation, technology and integration.
- Need DRM strategy, plans and manuals, containing specifications, standards and guidelines for developing well planned, technically sound, environmentally compatible and socially acceptable sustainable urban infrastructure.
- Need to establish an institution for DRM education, training and awareness programs, besides introducing academic and research activities, like degree program, short certification courses, seminars and workshops.
- KSA being signatory of the UNISDR's Hyogo Framework for Action 2005-2015: Building the resilience of nations and communities to disaster, entailing paradigm shift from reactive to proactive approach, is re-emphasized for compliance. A national level effective body needs to be established to develop framework, ensure phased implementation and enforce monitoring mechanism.

REFERENCES

Alamri, Y. A., Emergency Management in Saudi Arabia: Past, Present and Future. *University of Christchurch Report*, New Zealand, 2010.

Al-Bassam, A. M., Zaidi, F. K., Hussein, M. T., Natural Hazards in Saudi Arabia, *Geo-Information for Disaster Management*, 243(1), Antalya, Turkey, 2011.

Al-Saud, M., Assessment of Flood Hazard of Jeddah Area 2009, *JWRP*, Scientific Research, 2(9), 839-847, July, 2010.

Batrawy, A., Pilgrims Traumatized, asking how Mecca Crane could collapse, Thebigstory.ap.org, September 13, 2015. Retrieved from http://bigstory.ap.org/article/8e9be9edf5664802b232a40d7196cdd5/saudi-king-top-imam-visit-those-injured-crane-collapse on November 9, 2015.

Grainger, D.J., *The Geologic Evolution of Saudi Arabia*, Global Mapping, UK, 2007.

Pararas-Carayannis, G., Critical Assessment of Disaster Vulnerabilities in the KSA-Strategies for Mitigating Impacts and Managing Future Crises. *The First Saudi International Conference on Crisis and Disaster Management*, Riyadh, KSA, 2013.

Pallister, J.S., McCausland, W.A., Jónsson, S., Lu, Z., Zahran, H.M., Hadidy, S.E., Aburukbah, A. Stewart, I.C.F., Lundgren, P.R., White, R.A. & Moufti, M.R.H., Broad accommodation of rift-related extension recorded by dyke intrusion in Saudi Arabia, *JNG*, 3, 705–712, September, 2010.

UNISDR, Terminology on DRR, unisdr.org, August 30, 2007. Retrieved from https://www.unisdr.org/we/inform/terminology#letter-d on November 9, 2015.

Interaction between Theory and Practice in Civil Engineering and Construction
Edited by Komurlu, R., Gurgun, A. P., Singh, A., and Yazdani, S.

DESIGN CONSIDERATIONS FOR MAINTENANCE: FACILITY MANAGERS' PERSPECTIVE

ARMIN JALALI SOHI[1], DAVID ARDITI[1], and AFSHIN JALALI SOHI[2]

[1]*Civil, Architectural and Environmental Engineering, Illinois Institute of Technology, Chicago, USA*

[2]*Civil Engineering and Geosciences, Delft University of Technology, Delft, Netherlands*

Maintainability is one of the considerations designers take into account in the design stage of building construction projects. In other words, maintainability is a design consideration that may improve ease of maintenance during operation. The designer is expected to include maintenance knowledge and experience into the design, including planning for specific tools to facilitate implementation. Maintainability issues appear during the occupancy stage and have a direct impact on the cost of owning and operating a facility. Annual operation and maintenance expenses during the life cycle of a building could amount to many times the initial construction cost. Design deficiencies constitute one of the main reasons for high annual maintenance cost. Facility management involves activities to maintain and operate a constructed facility. It is essential to evaluate facility managers' perspective on how maintainability characteristics should be considered at design phase of facilities. A survey of 168 facility managers in the mid-west region of the US is uncovered that (1) maintainability should be top priority in design, especially for mechanical systems, (2) deferred maintenance should be avoided because it increases costs over the life-cycle of buildings, and (3) constructed facilities should be handed over with a maintenance schedule provided by the designer. Paying special attention to maintainability in the design phase is expected to help reducing the cost of operation.

Keywords: Facility management, Maintainability, Design priorities, Design deficiencies, Maintenance methods, Operation phase.

1 INTRODUCTION

After a building is constructed, facility managers are usually in charge of maintaining the building. Different maintenance methods are available to keep various systems and equipment at the required functional level. The selection of the appropriate method is based on the architect's recommended maintenance schedule. In case there is no maintenance schedule provided by the architect, facility managers choose an appropriate method for each component based on their experience.

In addition to operating a building's systems, facility managers take care of the maintenance of these systems. Occupants' demands evolve during the life cycle of a facility, which requires different maintenance strategies for different situations. Generally, maintenance can be preventive or corrective. Preventive maintenance includes a routine schedule for different components of each system to be checked, serviced or replaced. Corrective maintenance happens once a system breaks down.

There are several considerations an owner and a designer take into account in the design of a building Ogburn and El-adaway 2014). Owners and designers attach priorities to these considerations. Paying special attention to maintainability in the design phase helps to reduce the cost of owning and operating because taken over the life of a facility, maintenance may cost more than the initial cost of construction.

The main objective of this research is to collect and evaluate the opinions of facility managers about the extent to which maintainability is considered in design, the impact of different maintenance methods on facility costs, and the effect of a designer-provided maintenance schedule on facility management.

2 MAINTENANCE AND MAINTAINABILITY

In order to facilitate maintenance, during the life of a building, designers should consider maintainability as a serious design parameter. Injecting a combination of maintenance knowledge and experience into the project delivery process is a formal method of easily including maintainability into the process. For a better prediction of malfunctions and breakdowns in building systems, the building must be planned and designed for maintainability and a pre-defined maintenance program with a structured process should be in place and should be supported by corporate resources.

Facility managers are typically responsible for fixing the defects discovered after handover (Tay and Ooi 2001), but there are notable liabilities for inferior design and construction (Chong and Low 2006, Lam 2000). Design error-related defects often cause a chain effect, restrain performance, and lead to almost half of the maintenance cost (Arditi and Nawakorawit 1999a, 1999b, Josephson and Hammarlund 1999, Ilozor *et al.* 2004).

The life cycle cost analysis of a facility begins in the preliminary design phase and ends when the facility is decommissioned. In an effort to minimize the cost of construction, design decisions made during the preliminary design and construction phases typically consider only the construction costs. Deficiencies in the use of constructability standards at the design phase often lead to difficulties in maintenance. The concept of maintainability addresses this concern and can become an extension of the constructability process. As such, the formal constructability process may also incorporate a maintainability review (Dunston and Williamson 1999).

According to Chong and Low (2006), there are many maintenance issues that designers are not able to prevent. However, by using property managers' experience in regards to repeated maintenance issues; designers can improve overall building quality if they concentrate their efforts on a few major maintenance issues.

One of the main concerns of the construction industry is construction defects that require excessive maintenance, repair, or replacement. The performance of a building component depends on the function, system type, and materials used. Systems and materials are the main subjects of most research studies about defects.

The conventional construction management focus on saving money during design and construction is only a small percentage of the potential savings during the life cycle of facilities. Significant cost savings could be obtained by delivering facilities' complete construction information to facility managers. There are many approaches chosen by owners to transfer construction information to facility managers. Paper manuals are typically stored in a room on site at each facility (East *et al.* 2013).

3 METHODOLOGY

The investigation was conducted in two phases. The first phase involved a literature review. The maintainability considerations that should be accounted for in the design of a building were identified in this phase. In the second phase, a questionnaire was developed, using the considerations identified in the literature review. A survey tool was designed to collect facility managers' points of view. In addition to demographic information about annual revenue, years of experience in the industry, and the type of building designed, the questions explored (1) the efficiency of commissioning, (2) the existence of maintenance schedules provided by architects, (3) the reasons for high maintenance cost, (4) the most important considerations taken into account by architects in designing a building, (5) the systems that generate the highest maintenance expenditures, (6) the most important factors considered by architects in selecting equipment/materials, (7) the method of maintenance used, and (8) the impact of BIM and sustainability considerations on maintenance decisions. The findings were tabulated and organized in charts and bar diagrams. Justification for the three most pertinent factors is presented in the following subsections. The findings are discussed in some detail in Section 4.

3.1 Annual Maintenance Cost

The annual cost of maintenance depends on different factors but maintenance issues constitute the predominant factor. Maintenance issues arise when there is a shortcoming in the design concept, defective workmanship, or inappropriate material used. In order to avoid inappropriate decisions in the design phase, early contractor and facility management involvement is suggested (El-Haram and Agapiou 2002, Song *et al.* 2009). Facility managers were asked to state their opinions about the impact of inappropriate maintenance practices, deferred maintenance and inadequate design on annual maintenance cost.

3.2 Design Priorities

In the design phase, architects prioritize different considerations. Studies show that cost, time and quality are the most important considerations for architects. Barnes (1988) refers to them as the triangle of project objectives. According to Meng (2013), designers keep a balance between considerations such as aesthetics, reliability, durability, conformance, functionality and practicability. The lack of maintainability among these considerations is driven by the separation of design from the post-design process. Facility managers were asked about their preference about design considerations in addition to time, cost, and quality.

3.3 Maintenance Budget

Each building consists of different systems, including the structural, sanitary, mechanical and electrical systems. Each system requires a separate maintenance plan as well as a separate annual budget for maintenance activities. One of the responsibilities of a facility manager is to identify the budget for each system. Facility managers were asked to identify the causes for high annual maintenance costs.

4 FINDINGS AND DISCUSSION

The survey was electronically sent to 3,845 facility managers who are members of the International Facility Management Association (IFMA) in twelve mid-west states as defined by the U.S. Census Bureau, namely Illinois, Indiana, Iowa, Kansas, Michigan, Minnesota, Missouri, Nebraska, North Dakota, Ohio, South Dakota and Wisconsin. A total of 168 facility managers participated in the study, which amounts to a response rate of 4.4 %. This is a low rate of response, but it is not unexpected in online surveys. It should nevertheless be noted that even though the number of respondents is as large as 168, the responses may not necessarily reflect the opinions of the much larger facility manager population at large. This was an exploratory study that aimed to detect trends in the industry, and as such, 168 responses is a respectable sample size.

The results indicate that only few buildings are handed over to facility managers with a maintenance schedule provided by the architect. Indeed, almost three quarters of facility managers (73%) stated that the handover documents include a maintenance schedule in less than 25% of buildings. Whereas facility managers believe they receive most completed facilities without a maintenance schedule, in a parallel survey administered to architects (Jalali Sohi 2015), most architects stated that they deliver facilities mostly with a maintenance schedule at handover. This discrepancy in opinions is controversial. If it is true that only few buildings are received by facility managers with a maintenance schedule at handover, this may be an explanation for the high costs of maintenance throughout the life of these buildings. Scarponcini (1996) demonstrated that savings in design and construction (e.g., by avoiding the expense of generating a maintenance schedule) is only a fraction of the probable savings (e.g., in maintenance costs) throughout the life of a facility. According to East *et al.* (2013), who has collected information from facility managers over a decade, owners are trying to transfer construction information to facility managers by different methods. Although electronic data storage and management has gained ground, storing paper data manuals is still the most common method of retaining construction specifications. Shifting to computerized data storage practices is expected to reduce maintenance costs in the long run.

Forty five percent of facility managers think deferred maintenance is the main reason for high annual maintenance cost. Inappropriate design practices scored 32% and inadequate design received 23% as the second and third factors, respectively. Deferred maintenance may be a result of limited budgets, incapable staff, lack of planning, and deficient understanding of consequences. In the ASCE's 2015 Report Card, the reason why the infrastructure and structural facilities obtained low ratings was mainly because of deferred maintenance. Moving away from deferred maintenance is expected to have a positive impact on maintenance costs. Judging from the responses of 55% of the facility managers, the impact of design problems on maintenance has to be recognized too. A study conducted in 2013 of the curricula of all architecture programs in the U.S. showed that the word "maintenance" did not appear in any course description. There seems to be a disconnect between the wishes of facility managers and the education of architects relative to building maintenance.

Cost, duration, and quality of construction are considered to be important factors in designing a building. Other considerations, some of which are as important, include constructability, maintainability, aesthetics, occupants' comfort, and functionality. In addition to cost, time, and quality, 40% of facility managers' top two concerns were found in this study to be about maintainability and functionality. Aesthetics is considered to be the least important concern as it was marked by less than 2% of facility managers. On the other hand, architects usually seem more interested in aesthetics and constructability (Jalali-Sohi 2015). Again, there seems to be a disconnect between facility managers' needs and the priorities of architects.

A building involves four major systems, including structural (walls, slab, frames, etc.), sanitary (plumbing, fire protection, etc.), mechanical (HVAC), and electrical (elevators, circuits, cables, etc.) systems. The findings show that the large majority of facility managers (86%) think that mechanical systems have the highest annual maintenance cost, whereas only 8% think that annual maintenance cost is driven by structural or sanitary systems. Electrical systems were rated to have an impact on maintenance cost by only 6% of the respondents. These findings agree with the findings of an earlier study by Arditi and Nawakorawit (1999) who found that the mechanical system is the most troublesome component in maintenance activities. Mechanical systems deserve special consideration during design.

5 CONCLUSION AND RECOMMENDATIONS

Typically, an owner's goal is to minimize the cost of design, construction and operation, and to maximize occupant comfort and environmental quality over the life of a building. Maintenance is an important part of this equation. A variety of factors affect maintenance issues in the design, construction and operation of buildings. These factors include but are not limited to maintenance schedules, methods of maintenance, maintenance cost, maintenance considerations during design, and building systems that need special attention for maintenance. The factors that affect maintenance in the operation of buildings were explored by means of a survey administered to facility managers. The results highlight interesting facts that may have long lasting impact on the research literature and on actual practice.

The findings show that about three quarters of facility managers provide or receive a maintenance schedule for less than 25% of buildings at handover. Computerized management practices during design and construction are expected to alleviate this problem.

The findings also show that deferred maintenance and design problems are ranked by facility managers as the most important reasons for high annual costs of maintenance. Shifting to preventive maintenance practices is a good solution to this problem. Additionally, recognition of the importance of maintenance by educators in architecture appears to be an important part of the cost reduction equation. Design checklists such as the one developed by Hassanain *et al.* (2015) for water supply and drainage systems could help.

While this study shows that facility managers' priority rests with maintainability, architects are usually preoccupied by aesthetics and constructability. Injecting maintainability considerations to constructability reviews is most desirable.

Finally, facility managers believe that mechanical systems cause the highest annual cost of maintenance compared to other building systems. This finding reconfirms the outcome of previous studies and draws the attention of architects and facility managers alike that mechanical systems have to receive special attention during design and during operation.

References

Arditi, D. and Nawakorawit, M. (1999). "Designing Buildings for Maintenance: Designers' Perspective." *J. Archit. Eng.*, 5(4), 107–116.

Arditi, D. and Nawakorawit, M. (1999). "Issues in Building Maintenance: Property Managers' Perspective." *J. Archit. Eng.*, 5(4), 117–132.

Barnes, M. (1988). ''Construction Project Management.'' Int. *J. Proj. Manage.*, 6, 69–79.

Chong, W. and Low, S. (2006). "Latent Building Defects: Causes and Design Strategies to Prevent Them." *J. Perform. Constr. Facil.*, 20(3), 213–221.

Dunston, P. and Williamson, C. (1999). "Incorporating Maintainability in Constructability Review Process." *J. Manage. Eng.*, 15(5), 56–60.

East, E. W., Nisbet, N., and Liebich, T. (2013). "Facility Management Handover Model View." *J. Comput. Civ. Eng.*, 27(1), 61–67.

El-Haram, M. A., and Agapiou, A. (2002). "The Role of the Facility Manager in New Procurement Routes." *J. Qual. Mainten. Eng.*, 8(2), 124–134.

Hassanain, M., Fatayer, F., and Al-Hammad, A. (2015). "Design Phase Maintenance Checklist for Water Supply and Drainage Systems." *J. Perform. Constr. Facil.*, 29(3), 04014082.

Ilozor, B. D., Okoroh, M. I., and Egbu, C. E. (2004). "Understanding Residential House Defects in Australia from the State of Victoria." *Build. Environ.*, 39(3), 327–337.

Jalali Sohi, A. (2015). "Maintainability Issues In Building Design and Operation". Thesis presented to Illinois Institute of Technology at Chicago, IL, in partial fulfillment of the requirements for the degree of Master of Science, Chicago, IL.

Josephson, P. E., and Hammarlund, Y. (1999). "The Causes and Costs of Defects in Construction: A study of seven building projects." *Autom. Constr.*, 8(6), 681–687.

Lam, K. C. (2000). "Quality Assurance in Management of Building Services Maintenance." Southeast Asia Facility Management, Building Services Engineering Dept., The Hong Kong Polytechnic Univ.

Meng, X. (2013). "Involvement of Facilities Management Specialists in Building Design: United Kingdom Experience." *J. Perform. Constr. Facil.*, 27(5), 500–507.

Ogburn, M. and El-adaway, I. (2014). "Biddability, Constructability, Operability, and Environmental Checklist: Potential Role in Reducing Conflicts, Claims, and Disputes." *J. Leg. Aff. Dispute Resolut. Eng. Constr.*, 6(1), 05013001.

Scarponcini, P. (1996). "Time for an integrated approach to facility management." *J. Comput. Civ. Eng.*, 10(1), 3.

Song, L., Mohamed, Y., and AbouRizk, S. M. (2009). "Early Contractor Involvement in Design and its Impact on Construction Schedule Performance." *J. Manage. Eng.*, 25(1), 12–20.

Tay, L., and Ooi, J. T. L. (2001). "Facilities Management: A 'Jack of All Trades.' " *Facilities*, 19(10), 357–362.

Interaction between Theory and Practice in Civil Engineering and Construction
Edited by Komurlu, R., Gurgun, A. P., Singh, A., and Yazdani, S.

INTEGRATED DESIGN APPROACH FOR BUILDING MAINTENANCE ISSUES

AYŞE MÜGE ÖZ and SEMA ERGÖNÜL

Architecture Faculty, Mimar Sinan Fine Arts University, İstanbul, Turkey

Maintenance and repair is a principal task in building operation in terms of time spent and cost. Therefore, improving maintenance performance can result in remarkable operational benefits. Traditionally, maintainability is not fully considered when designing a building, whereas many decisions to influence building maintenance are taken in design phase. Because of this, there is a need for a new approach to handle design and maintainability as a whole. Integrating design with maintenance eliminates design oriented maintenance issues and promotes knowledge transfer between designers and facility managers. This paper aims to identify design requirements for maintainability and present possible acquisitions of maintenance integration in design. Design parameters which are important for maintainability of a facility management are considered as accessibility, flexibility, standardization, modularization and selection of materials and equipment. Designers should give more attention to these parameters in order to create more valuable buildings for the users and to prolong the life cycle of the facility. A facility manager's involvement in design steps would also improve maintainability of the building because of the specific technical details that a designer cannot fully designate. With a collaborative, new design approach considering maintainability, buildings profitability would be enhanced with less cost, time and labor.

Keywords: Building design, Maintainability, Facility management, Operation, Repair, Life cycle.

1 INTRODUCTION

Facility maintenance and repair is an important topic of Facility Management (FM) with intent to prolong building life and have maximum profit from the operation. Regrettably, it is not sufficiently considered by designers because of the limited knowledge and interest about operational details and maintenance requirements. However, design decisions are of vital importance for operating a facility. Thompson (1994) asserts: 'Designers should be aware of capital costs, plant reliability, operational economies, and maintenance, as the building operator will be interested in the lifetime costs of the building and not just initial purchase and installation'. Thus, early integration of facility management is needed to combine design parameters with operational needs. Otherwise, design originated operational problems that harder to solve would arise. Kuda and Berankova (2014) express this issue as: 'When architects design constructed facilities they commit errors in design which are not endangering as to statical or engineering aspects, but these will ensue during exploitation of the facilities'.

Integrating design and operation requires better communication between design and FM professionals. Nutt (2000) points out that: 'Knowledge transfer between designers and facility managers ensures that; facility objectives support organizational objectives, facility provisions are aligned with end-user needs, and design concepts are compatible with property strategy overall'. As an ongoing attempt, the communication between designers and facility managers is getting more frequent. 'On average, design firms communicate with the property managers during the design process between sometimes and often' (Arditi and Nawakorawit 1999). But it is not always possible to meet designers and facility managers together, just because operational team is set up after the design (Liu and Issa 2013). For this reason, designers ought to be aware of core operational issues and design requirements for life cycle.

For high performance, high qualified and properly operated facilities; there are numerous matters that should be handled in planning and design which have significant impact on future costs and quality. As a great expenditure and also intense field of operation; maintainability comprises remarkable part of these considerations. Operational phase constitutes approximately 60% of the total lifecycle cost of a facility and the main activities during operations are related to maintenance and repair (M&R) (Liu and Issa 2013). This great expense requires more focus on decreasing and expediting maintenance. Integrating facility maintenance with design is able to reduce the cost of maintenance and improve operating profitability. 'If all the maintenance unfriendly design issues can be solved in the design phase, the maintenance cost would be reduced and the life cycle cost of a facility would be much lower' (Liu and Issa 2014).

Considering all of these, there is a need to determine design needs for maintenance. The purpose of this paper is to set a framework for design originated maintenance issues. First of all, maintainability and its context are explained. Then, important points to consider during planning and design are discussed. It is concluded with the possible acquisitions by integrating design and maintenance.

2 BUILDING MAINTAINABILITY

Maintenance and repair directly contributes to physical condition, performance, reliability and comfort of the facility. Eva and Katerina (2013) define maintenance as 'all works carried out to keep and restore an item at a current acceptable standard'. It is not possible to avoid a component to wear off, but is possible to minimize breakdowns, extend useful life and ease maintenance. Primary factors to lead maintenance are poor design, poor workmanship, detective materials and components, usage and age, climatic or environmental factors and nature of users (Olanrewaju and Abdul-Aziz 2015). Most of these factors are related to decisions taken in design phase. That means; design steps are so valuable to maximize physical performance, decrease component inefficiency and minimize maintainability issues. Fitzgerald (2001) argues that 'In the early phases of a project, flexibility is high and design change cost is low. Thus, product maintainability can eliminate maintenance cost, reduce downtime and improve safety'. Blanchard *et al.* (1995) also define maintainability as follows:

- Probability that an item will be retained in or restored to a specified condition within a given period of time, when maintenance is performed in accordance with a prescribed procedures and resources
- Probability that maintenance will not be required more than x times in a given period, when the system is operated in accordance with prescribed procedures by personnel with proper skills.
- Probability that the maintenance cost for a system or product will not exceed y dollars per a designated period of time, when the system is operated and maintained in accordance with prescribed procedures.

It is clear that, maintainability is quite related with facility reliability, quality, performance and economy. A design approach considering all these factors is needed to enrich the building for operational efficiency.

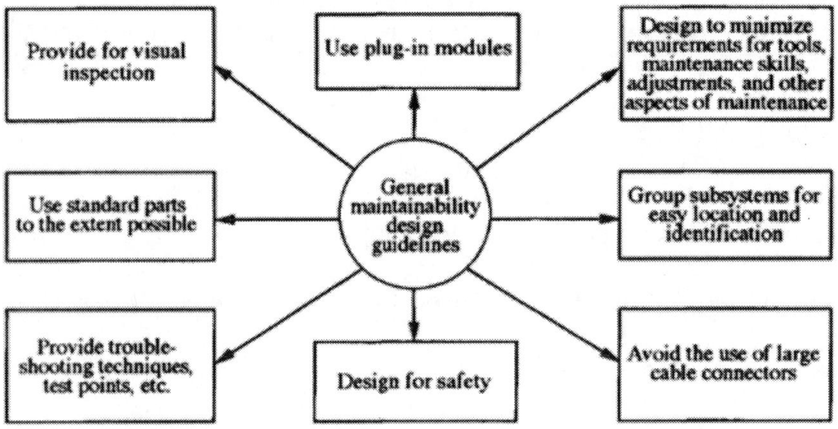

Figure 1. Design for Maintaibiltity Guideline (Dhillon 1999).

3 HOW TO INTEGRATE DESIGN AND MAINTENANCE

This paper aims to remark important points for design oriented maintenance issues and emphasize the importance of collaboration between designers and facility managers. For this purpose, there is a need to change the way facilities are designed. Even there are some basic design points for maintainability (Fig.1); a designer cannot fully designate the maintenance needs because of specific technical details. Integration of facility professional would result in more effective solutions for maintainability. A facility professional can involve in design by attending design meetings or reporting his opinions. But usually, their knowledge about building design is restricted and designers are disposed to exclude them from the process when designing the facility (Jensen 2008). Because of this, bi-directional collaboration is needed to share knowledge about both facility design and operation. That means, a facility manager need to have core competences and knowledge about building design, as well as designer has operational knowledge. Another way to integrate FM in design is the use of specialist consultants and/or involving specialists from FM providers (Jensen 2008).

Developing technologies also contribute to integrating maintenance to design. As a life cycle tool and promising technology, Building Information Modeling (BIM) Technology is also able to meet designers and facility managers on the same 3D model. BIM model not only visualizes the design defects that make maintenance activities impossible to perform but also acts as a bridge between design and operation phases.

3.1 Design Issues

Liu and Issa (2014) states that: 'While there is no cost or very low cost to correct such design defects in the design phase, it would lead to a much higher maintenance cost if the non-accessible equipment breaks down in operation'. Generally, facility managers are not involved in early phases of project and design considerations don't include operational details. These details include organizational requirements, technological requirements, HVAC requirements, lighting, security, signage and accommodations, and many other factors. If the facility manager is engaged to early design, an integrated project team is naturally developed and many problems like lack of constructability, operability, maintainability, and serviceability can be resolved (Meng 2013). If we look from the viewpoint of maintainability, there are some design points to consider.

Another study by Liu and Issa (2013) investigated facility personnel's problems about maintainability with a couple of interviews. Results indicate that, the lack of space designed in the ceiling to contain MEP systems is the most frequently occurring problem, followed by lack of adequate space for the mechanical room, limited space for AHU filter access, poor design of equipment layout, and lack of equipment accessibility (Liu and Issa 2013).

Main design requirements for maintainability include parameters like accessibility, flexibility, standardization, modularization and selection of materials and equipment, which are briefly explained in following sections.

3.1.1 *Accessibility*

All components must be easily accessible to perform maintenance and repair in minimum time. Moreover, the space that the maintenance operator will need must have been designated. Thompson (1994) asserts: 'The designer has to provide adequate access to plant and equipment, such as working space within ducts and the plant room, as well as working platforms for ladder access to equipment at high and low levels'. However, a designer can hardly designate the accessibility of all components, especially in high detailed technical equipment like HVAC. For example, on-end installed components cannot be installed without knowing the life expectancy and maintenance requirement of each. Maintenance frequency must also be considered; otherwise rarely maintained components are dissembled and assembled in every maintenance job unnecessarily (Liu and Issa 2014). Access openings are also important for maintainability. Operator size and the required actions of operator like turning, pushing and pulling in space must be designated (Dhillon 1999).

3.1.2 *Flexibility*

At the present time, business conditions are quite changeable. Besides, organizations are under pressure of market competition. As a result, business itself and the way of handling the business changes by time. Naturally, facility that the business is operated in also must be changed not to become useless or underperformer. This issue is a part of maintainability; building must proceed to be used against any changes. Nutt (2000) defines five design criteria for future flexibility as use flexibility, operational flexibility, operational flexibility, physical flexibility, property flexibility and market flexibility.

3.1.3 *Standardization*

Standardization is an important parameter for maintainability which can be defined as: 'Minimizing the variety of parts and components that a product or system will need' (Dhillon 1999). By this way, different products can use common and interchangeable parts. Moreover maintenance process is standardized and carried out faster and more reliable way. Interchangeability is result of standardization, which means any item can be replaced within a product by any similar item (Dhillon 2006). Both standardization and interchangeability remarkably minimize the effort for maintenance and repair.

3.1.4 *Modularization*

Modularization separates the component into units to quickly assemble, dissemble, maintain and repair. By this way, operator can work on only the problematic part of the component and minimizes the effort, cost and time consumed. Modularization also helps dividing up maintenance responsibilities, less costly and less time consuming training of maintenance personnel, lower levels of skills and tools needed to maintenance and reduced equipment downtime (Dhillon 2006).

3.1.5 *Selection of materials and equipment*

In the selection of materials and equipment; designers cannot easily make operationally right decision. This is because of the wide variety of products, furthermore pressure from dealers can make designer to choose the wrong product (Arditi and Nawakorawit 1999). This decision process is so valuable for the rest of the building life, the quality, reliability and efficiency must be in accordance with operational needs. In terms of maintainability, materials and equipment mustn't need maintenance for a designated period of time, or mustn't cost more than designated maintenance cost. A study carried out by Arditi and Nawakorawit (1999) shows that, when selecting materials, maintenance is considered as the third important factor, after cost and aesthetics; however, while selecting equipment, it is the most important one.

4 CONCLUSION

In the discussion of building maintenance issues, it is seen that maintainability of facilities needs wider recognition by designers. Decisions that determine maintainability are taken in design phase; thereby a facility manager's guidance in design would improve operational efficiency. Main design considerations for maintainability are considered as accessibility, flexibility, standardization, modularization and selection of materials and equipment. If these parameters are thoroughly embedded by designers, design originated maintainability issues can be averted. Furthermore, building performance would be enhanced with minimum breakdowns and maintenance related complaints. In terms of cost, time and labor, designing buildings for maintainability is not an option but a requisite to increase life cycle value.

References

Arditi, D., & Nawakorawit, M., Designing buildings for maintenance: designers' perspective. *Journal of Architectural Engineering*, 5(4), 107-116, 1999.

Blanchard, B. S., Verma, D., & Peterson, E. L., *Maintainability: a key to effective serviceability and maintenance management (Vol. 13).* John Wiley & Sons, 1995.

Dhillon, B. S., *Engineering maintainability: how to design for reliability and easy maintenance.* Gulf Professional Publishing, 1999.

Dhillon, B. S., *Maintainability, maintenance, and reliability for engineers.* CRC Press, 2006.

Eva, Š., & Katerina, M., The Effect of Implementing a Maintenance Information System on the Efficiency of Production Facilities, *Journal of Competitiveness*, 5(3), 60-75, 2013.

Fitzgerald, A., "Design for maintainability", *START*, Vol. 8 No. 4, pp. 1-4, 2001.

Jensen, P. A., Integration of considerations for facilities management in design, *Design Management in the Architectural Engineering and Construction Sector: CIB W096 Architectural Management & TG49 Architectural Engineering*, 191-199, Rotterdam, Holland, 2008.

Kuda, F., & Berankova, E., Extending the life cycle of buildings using Project and Facility managements, *Applied Mechanics and Materials*, Vol. 584, 2291-2296, 2014.

Liu, R., & Issa, R. R. A., BIM for facility management: Design for maintainability with BIM tools, *Proc., 30th Int. Symp. of Automation and Robotics in Construction and Mining (ISARC 2013),* Montreal,Canada 2013.

Liu, R., & RA Issa, R., Design for maintenance accessibility using BIM tools. *Facilities*, 32(3/4), 153-159, 2014.

Meng, X., Involvement of facilities management specialists in building design: United Kingdom experience, *Journal of Performance of Constructed Facilities,* 27(5), 500-507, 2013.

Nutt, B., Four competing futures for facility management, *Facilities*, 18(3/4), 124-132, 2000.

Thompson, P., The maintenance factor in facilities management, *Facilities,* 12(6), 13-16, 1994.

Interaction between Theory and Practice in Civil Engineering and Construction
Edited by Komurlu, R., Gurgun, A. P., Singh, A., and Yazdani, S.

BUILDING CONSTRUCTION CULTURE IN GERMANY

ALBERT BEN-DAVID and AVIAD SHAPIRA

Faculty of Civil and Environmental Engineering, Technion, Haifa, Israel

In the course of a study of mega building construction projects in Germany, the need arose to identify and analyze the characteristics of the local construction culture, particularly upon realization that these characteristics may constitute significant contextual factors in equipment and site logistics planning. The study of the German building construction culture was based on (1) a focused literature review; (2) case studies of multicrane construction sites; and (3) structured interviews with experts of various professions from the German construction industry. The findings highlight the role of the following factors: (1) the typical German professional character; (2) various systems of contractual relationships; (3) state and city regulations; (4) technology and industrialization; (5) workforce sources; (6) rooted safety culture; (7) protection of the environment; (8) the crane market; (9) relationships between planning, engineering, and construction; (10) logistics concepts; and (11) additional unique features of Berlin as a major construction hub. Some of the German culture characteristics detected were further supported by a comparison with other advanced European construction cultures.

Keywords: Equipment planning, Contextual factors, Multicrane sites.

1 INTRODUCTION

An ongoing study of equipment and site logistics planning (ELP) for mega building construction projects, conducted in Germany, called for the identification and analysis of the characteristics of the local construction culture, given the understanding that these characteristics may constitute significant contextual factors in ELP. Such characteristics should also be considered factors that may limit the generalizability of the study's results. Hence, this paper aims to provide an answer to the question, what are the unique characteristics of the building construction culture in Germany? The answer was sought mainly through 15 case studies of multicrane building construction sites in Germany, located in urban surroundings. The paper first presents a concise review of the relevant literature. Then, the data collection method is presented, followed by a presentation of the findings. The paper closes with some concluding remarks.

2 LITERATURE REVIEW

German construction companies have traditionally relied on skilled and trained workers. Qualified individuals are required to attend various professional courses (Syben 1998) in order to obtain the degree of *Meister* (Richter 1998). Due to the flow of guest workers (*Gastarbeiters*) into Germany, based legally on bilateral agreements between Germany and countries that are not part of the EU (Rudolph 1996, Rixler 2003), the competence of the construction workforce has declined sharply in recent years. Construction companies have had to change their policy of employing workers, hiring less trained (but cheaper) workers, on the one hand, and tightening control and management, on the other. The issue of employing guest workers in big numbers has caused a chain of conflicts and structural changes in the German construction industry (Faist 2003). Nowadays, construction activities are increasingly executed by general contractors under turnkey contracts (Rixler 2003), which are highly regulated by the German construction contract procedures (Johansen and Walter 2007).

The German construction process is traditionally controlled by the owner or client and their representative, the architect. While this contractual mode places emphasis on maintaining success measures such as cost and quality, it also causes various problems for the construction companies (Syben 2000). Companies are hindered from taking part in the construction process until all important decisions have been made by the owner, ignoring the other players' potentially relevant abilities and resources. The dominant character of the owner also influences project concepts for dealing with uncertainties, often lacking a much desirable holistic view of the overall uncertainties (Bryde and Volm 2009).

The German construction industry is considered advanced and modern, using innovative techniques mainly in residential projects (Thuesen and Hvam 2011) and office buildings (Hegger 1996). Construction systems are based, among other things, on optimization of cost and derived value, control over the predicted project duration (Stoy *et al.* 2007a), construction speed (Stoy *et al.* 2007b), and implementation of the principles of lean construction (Johansen and Walter 2007).

3 RESEARCH METHOD

Following a focused literature review, data were collected through (1) case studies of complex building projects, which included on-site visits and interviews to expose the influence of local characteristics; and (2) a series of structured interviews with experts of various professions (engineering, management, finance, legal, authorities) from the German construction industry.

3.1 Case Studies and Interviews

Each of the 15 case studies consisted of one to three in-depth structured interviews with the contractor's project manager, site visits, and an inspection of exclusive project materials provided by the site management. The interviews contained both open-ended and structured questions and addressed, among other issues, topics that characterize the German construction culture. Selection criteria for the case study mega building projects included the following (mean values are given in parentheses): (1) gross built area $\geq$ 40,000 m^2 (80,800 m^2); (2) construction budget $\geq$ €100M (€125); (3) number of tower cranes on site $\geq$ 5 (6.5, with max. of 12); (4) distinct horizontal dimension; (5) single building; and (6) busy urban location. Seven of the projects were located in Berlin, and the rest in Frankfurt, Hamburg, Leipzig, and Stuttgart. All 97 cranes used on the projects were of the top-slewing type (of which 92 had a trolley jib and 5 had a luffing jib). The 15 projects were executed by nine different construction companies known in Germany (and several also out of Germany) as leading contractors. The projects included various uses such as shopping centers, office buildings, residential, and a variety of mixed uses. The number of workers per project at peak time ranged between 400 and 1200 and site administration staffs numbered between 40 and 80 persons.

3.2 Interviews with External Experts

Structured interviews were also conducted with senior representatives of several independent firms and disciplines, mainly to gain a deeper insight and understanding of the issues addressed in the current study. The interviewees had professional affinity to the German construction culture with regard to ELP, project management, engineering, finance, law, and regulations. The total number of these interviewees was 26, and their mean length was about two hours per interview.

4 FINDINGS: CHARACTERISTICS OF GERMAN BUILDING CULTURE

The study's field findings constitute first-hand evidence and highlight the role of various characteristics of the German construction culture as significant contextual factors. Main findings are as follows (the number of case study projects exhibiting the finding is given in brackets):

- Contractual relationships: In response to the dominant role played by the owner in project planning and control, leading construction contractors now tend to execute their project using a turnkey delivery system [9]. Exceptional proactive involvement of owners in equipment and site logistics planning was revealed in several projects [4].
- Productivity: High productivity rates are reflected by the low mean number of workers per crane – 12 to 18. These numbers are much lower than those observed in other developed countries (Shapira *et al.* 2012) [8].
- Regulation: Strong involvement of state and city regulatory agencies is visible with respect to construction and planning codes, workplace safety, environment and wildlife protection, underground water contamination, etc. [12].
- Technology and industrialization: Contractors are using the most advanced technologies and construction methods; innovative solutions are encouraged by company managements; and the high level of mechanization and industrialization is indicated by the number of cranes and the high rate of prefabricated elements, respectively [10].
- Workforce: Teams of guest workers are supervised by local senior foremen who are responsible for training and quality control [7]. Most of the crane operators are outsourced together with the teams of guest workers; to a lesser extent, subcontractors employ their own crane operators. No cases of operators who are on the permanent payroll of the construction company were observed [8].

- Crane market: A clear preference for locally manufactured cranes was observed: 80% of the 97 cranes were German made. A near balance between cranes owned by the construction firms (43%) and rentals (57%) was maintained. Major rental companies also report a state of supply-and-demand balance in the current construction climate and volumes.
- Berlin as a major construction hub: (1) Berliners have become used to "living in a construction site" and have developed a compliant attitude toward noise, traffic interruptions, and oversailing cranes in the vicinity of their homes; (2) project planning, cost and schedule are affected by engineering issues such as high levels of underground water, proximity of construction sites to each other (requiring the coordination of logistics and work envelopes of overlapping cranes), and the vast number of unexpected obstacles (such as old bombs from the war).

5 COMPARISON WITH OTHER CONSTRUCTION CULTURES

A comparison with other construction cultures – an unavoidable byproduct of the current study – was also assisted by case studies conducted earlier in four mega building projects in the United Kingdom and France as part of a pilot study (Shapira *et al.* 2015).

Germany is positioned high compared with other construction industries in advanced European countries. The German construction industry, among other characteristics, is more highly regulated (Mahapatra *et al.* 2012) and more productive (Clarke and Hermann 2004) than countries such as The United Kingdom, and is based more on professionality and less on contractual formal procedures (Syben 2000). Several studies that compared productivity rates indicated clearly that the construction industry in Germany is more efficient than in other European countries such as France and The United Kingdom (Proverbs *et al.* 1998a, Proverbs *et al.* 1998b, Proverbs *et al.* 1999). In recognition of the virtues of the German construction culture, a study was conducted by the British Government, Department of Trade and Industry, in order to acquire knowledge regarding innovative off-site manufacturing technologies in Germany for implementation in the United Kingdom (Venables *et al.* 2004). The advanced level of construction systems in Germany compared with those in the United Kingdom, France and other European countries was the topic of various other studies (Stoy *et al.* 2007a, Thuesen and Hvam 2011) that indicated the superiority of the German construction industrialization

Findings of the case studies conducted in the United Kingdom show similarity to Germany with regard to state and city regulations – a strict attitude toward meeting the legal construction regulations concerning the environment, safety, and public order. Regulations concerning noise, dust, pollution and contamination produced by construction sites in urban areas were also well enforced by the authorities.

In terms of workforce, the greater emphasis placed on training is rooted in the German tradition. Richter (1998) found that, compared with other European countries such as the United Kingdom, France, and the Netherlands, German workers – albeit mostly foreigners – are more highly trained and professional, which in part explains the high quality and productivity achieved by the German construction industry.

6 CONCLUSION

Based on the relevant existing literature, on 26 interviews with experts from various disciplines and sectors of the German construction industry, and mainly on 15 case studies of mega building construction projects in Germany, this study exposed the characteristics of the local construction culture. The comparison with other cultures, which was limited in its scope, was assisted by four case studies conducted in the United Kingdom and France. Although most findings are in favor of the German construction culture, the local industry certainly faces difficulties, including, among others, the opposing views of owners and contractors regarding uncertainties and risk management, the involvement of owners in typical construction planning issues commonly addressed by the project's management, and the structural changes the industry underwent following the influx of cheap foreign and less trained workforce. Full exposure of the strengths and weaknesses of the German construction industry requires a comparative investigation of "regular" and not only mega building projects, which is indeed the next phase of the current study. In addition to gaining deeper insights, this will also allow to accord statistical significance to the findings. The current study was conducted within a broader research aimed at identifying and quantifying the influence of equipment and site logistics planning on the overall success of construction projects. Recognizing the role played by contextual factors in construction planning in general, and in ELP in particular, the findings will help pinpoint those characteristics of the local culture that should be given greater attention, as opposed to those of lesser effect.

References

Bryde, D. J., and Volm, J. M., Perceptions of Owners in German Construction Projects: Congruence with Project Risk Theory, *Construction Management and Economics,* 27(11), 1059–1071, 2009.

Clarke, L., and Hermann, G., Cost vs. Production: Labour Deployment and Productivity in Social Housing Construction in England, Scotland, Denmark and Germany, *Construction Management and Economics*, 22(10), 1057–1066, 2004.

Faist, T., Protecting Domestic vs. Foreign Workers: The German Experience during the 1990s, *Workshop on Temporary Migration,* International Labour Organization, Geneva, 2003.

Hegger, J., High-Strength Concrete for a 186m High Office Building in Frankfurt, Germany, *Engineering Structures,* 18(11), 850–854, 1996.

Johansen, E., and Walter, L., Lean Construction: Prospects for the German Construction Industry, *Lean Construction Journal,* 3(1), 19–32, 2007.

Mahapatra, K., Gustavsson, L., and Hemstroem, K., Multi-Storey Wood-Frame Buildings in Germany, Sweden and the UK, *Construction Innovation*, 12(1), 62–85, 2012.

Proverbs, D. G., Holt, G. D., and Olomolaiye, P. O., A Comparative Evaluation of Concrete Placing Productivity Rated amongst French, German and UK Construction Contractors, *Engineering, Construction and Architectural Management*, 5(2), 174–181, 1998a.

Proverbs, D. G., Holt, G. D., and Olomolaiye, P. O., A Comparative Evaluation of Reinforcement Fixing Productivity Rates amongst French, German and UK Construction Contractors, *Engineering, Construction and Architectural Management,* 5(4), 350–358, 1998b.

Proverbs, D. G., Holt, G. D., and Olomolaiye, P. O., Productivity Rates and Construction Methods for High Rise Concrete Construction: A Comparative Evaluation of UK, German and French Contractors, *Construction Management and Economics*, 17(1), 45–52, 1999.

Richter, A., Qualifications in the German Construction Industry: Stocks, Flows, and Comparisons with the British Construction Sector, *Construction Management and Economics*, 16(5), 581–592, 1998.

Rixler, S., *The Effect of Foreign Subcontractors on the German Building Industry*, GRIN Publishing GmbH, Munich, 2003.

Rudolph, H., The New Gastarbeiter System in Germany, *Journal of Ethnic and Migration Studies*, 22(2), 287–300, 1996.

Shapira, A., Ben-David, A., and Steffen, P., Involved Parties and Plan Formats in Equipment Planning for Mega Building Projects," *8th International Conference on Construction in the 21st Century (CITC-8)*, S. M. Ahmed *et al.* (Eds.), Thessaloniki, Greece, 7–14, 2015.

Shapira, A., Simcha, M., and Goldenberg, M., Integrative Model for Quantitative Evaluation of Safety on Construction Sites with Tower Cranes, *Journal of Construction Engineering and Management*, 138(11), 1281–1293, 2012.

Stoy, C., Dreier, F., and Schalcher, H. R., Construction Duration of Residential Building Projects in Germany, *Engineering, Construction and Architectural Management*, 14(1), 52–64, 2007a.

Stoy, C., Pollalis, S., and Schalcher, H. R., Early Estimation of Building Construction Speed in Germany, *International Journal of Project Management*, 25(3), 283–289, 2007b.

Syben, G., A Qualifications Trap in the German Construction Industry: Changing the Production Model and the Consequences for the Training System in the German Construction Industry, *Construction Management and Economics*, 16(5), 593–601, 1998.

Syben, G., Contractors Take Command: From a Demand-Based towards a Producer Oriented Model in German Construction, *Building Research and Information*, 28(2), 119–130, 2000.

Thuesen, C. and Hvam, L., Efficient On-site Construction: Learning Points from a German Platform for Housing, *Construction Innovation*, 11(3), 338–355, 2011.

Venables, T., Courtney, R. G., and Stockerl, K., Modern Methods of Construction in Germany: Playing the Off-site Rule, *Report of a DTI Global Watch Mission,* Pera Innovation, Leicestershire, U.K., 2004.

Interaction between Theory and Practice in Civil Engineering and Construction
Edited by Komurlu, R., Gurgun, A. P., Singh, A., and Yazdani, S.

PROFESSIONAL ETHICS IN THE DAILY WORKING PRACTICE OF CIVIL AND STRUCTURAL ENGINEERS

ULRIKE QUAPP and KLAUS HOLSCHEMACHER

Faculty of Civil Engineering, Leipzig University of Applied Sciences, Leipzig, Germany

Sooner or later in their daily working practice, engineers come in contact with issues of professional ethics. Codes of professional conduct are well developed and popular in a number of countries all over the world. Also in Germany, codes of professional conduct were developed for various professions such like medicine, civil and structural engineering as well as architecture over the past decades. As a planner, consultant or specialized expert, civil and structural engineers are obligated to obey the rules of proven professional conduct and modern professional ethics. Their professional practice requires the observation of secured technical knowledge and the obligation for ongoing professional development. The paper provides an overview about professional ethics and engineers' professional law. Furthermore, it is dealing with the aspects of professional ethics in the civil and structural engineers' daily working practice and it discusses the question, if rules of professional conduct help to establish clients' trust in the competence and the moral attitude of civil and structural engineers. The author explains why it is necessary to implement ethical education in universities' civil and structural engineering programs and to continue the education in the courses for ongoing professional development. Finally, the paper mentions relevant case law. The conclusion is: Codes of ethics may not have the same importance in all countries all over the world. But they can be inspirational and motivating as well as instructive. Codes of conduct help structural engineers to take decisions and are guiding principles in their daily working practice. Therefore, it is necessary to distribute knowledge of professional ethics and to enforce codes of professional conduct.

Keywords: Codes of ethics, Professional conduct, Engineers' professional law, Ethical education, Client's trust.

1 INTRODUCTION

The construction industry develops rapidly in a globalized world. New technology, increasingly complex construction projects as well as more energy efficiency and environmental requirements bear an enormous burden of responsibility on civil and structural engineers. Every day in their working life engineers are at risk to be confronted with ethical difficult situations. The pressure to perform and the need to reduce construction costs do not make it easier to decide in an ethical correct way. Furthermore, not everything that is technically feasible also is ethically acceptable. For example, within the current intensive debate concerning Hydraulic Fracturing - also called Fracking - ethical aspects of technological progress were discussed as well. Codes of professional ethics can support civil and structural engineers to manage their everyday working life. Furthermore, they are able to increase clients' confidence in the engineering profession. But, it is to state that codes of ethics do not create ethics (Strahlendorf 2004). They are only guidance for ethically correct professional behavior (Quapp 2015).

2 CODES OF ETHICS

Professional ethics is written down in codes of ethics. In Germany, professional ethics are part of the professional law of engineers and enforceable. Codes of ethics express the rights, duties, and obligations of the members of the profession. Obviously, the codes of ethics are not comprehensive enough to cover all possible ethical dilemmas that an engineer might encounter in his or her career. The codes serve as starting points for making ethical decisions (University of Missouri 2015).

The first codes of ethics were developed in United States of America in the year 1912 (Hubig and Reidel 2003). In the beginning, they only consisted of rules concerning the conduct towards clients and competitors. It took years, until after the Second World War, before principles for the orientation of the engineering practice at the safety, health, and welfare of the public were added (Mitcham 2009), as for instance by regulations concerning the environmental protection. Today, in United States codes of professional ethics are well developed and can be easily followed to avoid problematic situations. In Germany, private professional associations set the first codes of ethics. Later, after the Second World War, these regulations were included in the binding codes of professional conduct (Taupitz 1991) set by professional chambers which were established by law and are so-called indirect public administration at federal level as well as an instrument of professional self-administration. Beside these binding and enforceable rules of professional conduct also codes of ethics developed by private professional associations exist in Germany.

Rules for civil and structural engineers' professional ethics can be found nearly all over the world. Thus, e.g. the American Society of Civil Engineers, defined Ethics Guidelines for Professional Conduct for Civil Engineers as well as the *Fédération Ingénieurs et Scientifiques de France*, the Nigerian Institution of Civil Engineers, and the Japan Society of Civil Engineers, just to mention a few. The guiding principles for professional conduct of engineers are comparable all over the world, whereby it is to note that in Germany a distinction must be made between professional law and self-imposed standards of professional associations for instance set by the *Verein deutscher Ingenieure* (2002). Mostly, avoiding conflicts of interest, a negative attitude to bribes and kickbacks, ensuring legal compliance, refusing benefits as attempts to influence the engineer's performance or judgment as well as protecting the environment etc. are content of such ethical guidelines.

Normally, codes of ethics are set by engineers' organizations or professional associations of engineers and contain:

- General orientation (principles and mission statements) characterizing the engineer's profession
- Professional rules for the engineer's everyday work life concerning relationship to clients, supervisors, colleagues (esp. competitors) as well as regarding profession and public

- Procedural rules for behavior in difficult moral situations (Hubig and Reidel 2003).

Not all of them are binding or enforceable. Furthermore, most of the codes of ethics do not contain all of the aforementioned elements. Having internal and external effects, addressees of codes of ethics are the public, employers, clients and fellow professionals.

3 PROFESSIONAL ETHICS IN PRACTICE

Nearly every day in their working life engineers have to take ethical difficult decisions. When faced with a problem at work that raises a moral issue, rules of professional ethics can help civil and structural engineers to decide how to act. They can assist in making choices in difficult professional situations (Strahlendorf 2004). Rules for professional conduct help clarify values and form a framework for discipline as well as support to build up a group identity and collegiality (Strahlendorf 2004).

One further advantage is the impact on public confidence in the profession of civil and structural engineers. By codes of ethics, clients get an idea about the principles civil and structural engineers try to follow during their work. This is a sign of quality which is the source of the good reputation of engineers and the basis of the professional pride.

The more detailed the professional code is the more guidance it offers for the engineer. Codes of ethics do not only give orientation for professional conduct, they relieve the engineers of some responsibility as well. Thus, every engineer can invoke the codes of ethics considering the other societies' expectations which means a protecting effect for civil and structural engineers. Furthermore, codes of ethics relieve the state and the legislation which can build upon the professional regulations (Hubig and Reidel 2003).

Consequences for violation of professional codes of ethics are heterogeneous. There is a big difference between countries where engineering only can be practiced with a professional license such as in Germany and other countries where a professional license is not necessary. In Germany, violating the codes of contact can result in serious consequences such as a fine or the removal from the chamber of engineers and in fact to a dismissal from the profession. In other countries (e.g. USA) violating the code of ethics may result in expulsion from a professional society but not in a prohibition to practice engineering (University of Missouri 2015).

Professional societies' ethical guidance falls into three basic categories depending on consequences (Gardenier 1995). First, a "weaker" type of code of ethics which is more a declaration of intent without any consequences is the type of "Aspirational guidelines". They encourage competent and moral practice in general, but do not provide specific rules of conduct (Gardenier 1995). Mostly, they consist of a few advices for desirable ethical conduct. But today, the second type "Educational Guidelines" is usual which articulates more clearly what represents ethical practice, often with interpretive commentary or case illustrations (American Society of Civil Engineers 2008). Their intention is an educational effect on the institution members. However, this category of codes does not imply or impose sanctions for deviations from the guiding principles (Gardenier 1995). "Regulatory guidelines", as the Code of

Ethics of the Institution of Engineers Australia (2010), spell out adjudicable guidelines and informs about consequences, such as fines or expulsion from the society, for failures to comply. Furthermore, ethics are covered by statutory codes of conduct in state or federal law (e.g. the US Foreign Corrupt Practices Act or the German Act Against International Corruption)

The relationship between law and professional ethics is quite interesting. Not everything that is unethical is at the same time illegal too. In the majority of the countries, codes of ethics are not legally binding. But, most of the regulations are in compliance with the national legislation. However, in some aspects professional ethics cover far more issues than the law does. Nevertheless, ethical issues often play a role in cases of professional liability, such as the Kansas City Hyatt Regency walkway collapse or the Bad Reichenhall Ice Rink Collapse (Quapp 2015).

4 PROFESSIONAL ETHICS IN EDUCATION

Significant for a positive effect of professional ethics on society is a wide distribution of codes of ethics among not only engineers but also the public. Knowledge about the code contents is a must for civil and structural engineers. Therefore, a lifelong ethical education, starting at universities and continuing in further training should be offered.

4.1 Professional Ethics in Study Programs

On the basis of the arguments above the question arises if, in civil and structural engineering programs, there is a need for special courses in professional ethics. In the discussion a distinction must be made between extra courses or the inclusion of ethics in subject specific courses. Certainly, higher education institutions must take on the task to educate their students in professional ethics. But it seems to be much more successful to integrate this subject in a teaching concept where practical aspects play a major role (Kledtke 2000). Thus, questions of ethically impeccable professional conduct can be included e.g. in courses in building project management or law as well as in all other subject specific lectures and should be discussed at practical examples. Lecturers should teach their subjects practice-oriented and discuss application in construction industry always in the light of environment and society. Although accreditation organizations such as the Accreditation Board of Engineering and Technology (ABET 2014) have mandated that engineering programs include ethics in their curriculum, additional ethic courses are not to prefer because they would further reduce time for specific technical subjects in civil and structural engineering programs.

4.2 Professional Ethics and Continuing Professional Education

Continuing professional education is an essential element of professionalism, and it is often referred to directly in codes of ethics. In Germany, further training is obligatory for members of state chambers of engineers and publicly certified experts to continuously improve professional knowledge and skills. The legislature in Germany is authorized to impose this obligation for ongoing professional development to civil and structural engineers (Higher Administrative Court of North Rhine-Westphalia 2012). If civil or structural engineer members of the German engineers' chambers are not able to prove continuing education courses in sufficient quantity it will have consequences such as imposing a fine or withdrawal of recognition as government-recognized expert (Vock 2013).

Furthermore, professional ethics should be an educational subject in continuing professional education. Only if engineers know about the codes' contents they can apply. Of course, nearly all civil and structural engineers are informed about the codes of professional conduct but may be not in detail. Furthermore, legislation can change and new court decisions can be published with effects on ethical issues.

5 CONCLUSION

Because of the more complex demands on civil and structural engineers in globalized everyday work life, importance of codes of ethics increased in the last years. Codes of professional conduct are the conscience of the engineering profession. They may not have the same importance in all countries all over the world but codes of ethics can be inspirational and motivating as well as instructive. Day by day, codes of professional ethics help structural engineers to take difficult decisions and are guiding principles in their working practice. Because of high requirements on engineers such as standing responsibility for society as well as liability for construction deficiencies and damages, civil and structural engineers need regulations of ethically impeccable professional conduct which are independent of stakeholders' interests.

The codes are able to convey an image of engineers' competence and reliability in public. Rules of professional conduct for engineers are both a promise and an obligation. They help to establish clients' trust in the competences and the moral attitude of civil and structural engineers. But, it must be noted that codes of ethics do not create ethics. They only define ethically wrong or right professional conduct.

Therefore, it is necessary to distribute knowledge of professional ethics and to enforce codes of professional conduct. Collateral education is required to bring a code of ethics alive, starting in the education at the universities and continuing in the courses for continuing professional development. Furthermore, infringement of regulation about professional conduct must be sentenced without any exception in order to preserve public confidence in the work of civil and structural engineers.

References

Accreditation Board of Engineering and Technology, November 1, 2014. Criteria for Accrediting Engineering Programs. Retrieved from www.abet.org/wp-content/uploads/2015/05/E001-15-16-EAC-Criteria-03-10-15.pdf on October 11, 2015.

American Society of Civil Engineers, Ethics guidelines for professional conduct for civil engineers, 2008. Retrieved from www.asce.org/uploadedFiles/Ethics_-_New/ethics_guidelines010308v2.pdf on July 9, 2014.

Gardenier, J., Hearing, Commission on Research Integrity, January 5, 1995: Summary of Selected Testimony, *Professional Ethics Report*, VIII(2), 3-4, 2015. Retrieved from www.aaas.org/sites/default/files/migrate/uploads/per1.pdf on October 15, 2015.

Higher Administrative Court of North Rhine-Westphalia, decision of 26th of April 2012, court case file 6s A 689/10.S.

Hubig, C. and Reidel, J. (eds.), *Ethische Ingenieurverantwortung,* Verein Deutscher Ingenieure (VDI), Berlin, 2003.

Institution of Engineers Australia, Code of Ethics, 2010. Retrieved from www.engineersaustralia.org.au//sites/default/files/shado/About%20Us/Overview/Governance/codeofethics2010.pdf, with reference to the Royal Charter and By-Laws of 2011. Retrieved from www.engineersaustralia.org.au/sites/default/files/shado/About%20Us/royal_charter_bylaws.pdf on July 29, 2014.

Kledtke, H., *Zum Problem von Technik und Ethik am Beispiel der Ausbildung zum Ingenieur*, doctoral dissertation, University of Dortmund, 2000.

Mitcham, C., A historico-ethical perspective on engineering education: from use and convenience to policy engagement, *Engineering Studies*, I(1), Routledge, London, 2009.

Quapp, U., Professional Ethics for Civil and Structural Engineers under the German Law Governing the Profession of Engineers, *Journal of Legal Affairs and Dispute Resolution in Engineering and Construction*, 10.1061/(ASCE)LA.1943-4170.0000173, B4515001. Retrieved from www.dx.doi.org/10.1061/(ASCE)LA.1943-4170.0000173 on October 14, 2015.

Strahlendorf, P., Professional ethics, *Proc., ASSE Professional Development Conference*, ASSE, 2004. Retrieved from www.bcsp.org/Portals/0/Assets/PDF/Articles_PPT/ProfessionalEthics.pdf on July 20, 2014.

Taupitz, J., *Die Standesordnungen der freien Berufe*, de Gruyter, Berlin, 2002.

University of Missouri, Engineering Ethics, 2015. Retrieved from www.engineering.missouri.edu/mae/files/engineering_ethics1.pdf on October 11, 2015.

Verein Deutscher Ingenieure, Fundamentals of Engineering Ethics, 2002. Retrieved from www.vdi.de/fileadmin/media/content/hg/17.pdf on July 29, 2014.

Vock, W., *Das Recht der Ingenieure*, 2nd edition, Boorberg, Stuttgart, 2013.

LEAVING CHANCE TO CHANCE ON ACADEMIC ARCHITECTURAL DESIGN-BUILD PROCESSES

ENRIQUE VILALCIS[1], LORENA RODRIGUEZ[2], and CYNTHIA AYARZA[2]

[1]*Arquitectura Diseño y Artes, Pontificia Universidad Católica del Ecuador, Quito, Ecuador*
[2]*Research, Ensusitio Arq., Quito, Ecuador*

Once a student asked me before class: what are we going to do tomorrow? I said: I don't know, we will see tomorrow how it goes. The objective of this paper is to confront two different ways of learning, the class versus on site. We usually think that planning for planning might be the solution, but what happens when planning leaves so little chance to chance? Sometimes approaching design problems without a plan, but with tools and scenarios, works better for a student and community. We will confront these two systems for the same design problem to see which deliver better results for the community and the student's learning process. Which system will make us ready for the unexpected, for the un-planned, for a future scenario where resources will be limited and non-expected, global warming, oil production decrease?

Keywords: Teaching process, Planning.

1 INTRODUCTION

Sometimes I tell my students that architecture is an act of witchcraft, almost a form of fortune telling. In architecture we as architects' project, meaning that we plan, envision something that is not there yet. We project, a reality that someday may or may not happen. This is an act of prediction and guessing. But how can we make the most accurate prediction? It is obvious that we would have to know a lot about the topic which we would try to project. This means understanding the environment in which the projection will live. There are infinite ways of approaching the craft of the design learning process, but we will focus on two: a programmatic and a problematic method. The student, during his or her academic life, will have to learn to design and project based on a program method or based on solving problems.

The following paper deals with which would be an adequate way of learning within the field of architecture based on what the future will be: energy crisis and global warming, confronting two alternatives of "teaching", the problematic versus the programmatic. After describing and analyzing this possible reality and these two different approaches of learning, the case study raises the questions on which would be a more adequate alternative based on a specific circumstance. This case study, that takes place in an academic environment, will illustrate these two teaching technics and in a qualitative analysis will determine the relevance of each alternative.

In order to better guide our students it is important for us professors to try to understand the environment where the student will develop his architectural craft. See Figure 1. Our future environment is uncertain, but here are some facts that might let us provide guidelines, global warming and energy crisis.

(1) Global warming and climate control, "Currently, energy constraint and global warming are the biggest challenges confronting the planet. (...) The analysis carried out by the US Energy Information Administration (EIA) estimates that, by 2030, global energy consumption will have grown by over 70% (EIA 2007)." Foruzanmehr (2008) and "the building sector is one of the major energy consumers in the world. The proportion of total energy use attributable to buildings generally ranges from 10 - 15% in undeveloped countries to more than 40% in the developed countries (Robertson 1992, p.129)." Foruzanmehr (2008). Meaning that whatever we, as professors, are able to do in order to create consiusnes within this topic might be a most.

(2) Energy crisis and decreased oil production, "as natural resources become harder to obtain, capital is diverted to extracting more of them. This leaves less capital for investment in industrial output. The result is industrial decline, which forces declines in the service and agricultural sectors. About the year 2030, population peaks and begins to decrease as the death rate is driven upward by lack of food and health services." (D.H. Meadows 2004). But ideally we should not see this as a catastrophic future but an opportunity projection, "Heinberg shows how oil peak, peak water, peak food, etc. lead not only to the end of growth, and also to the beginning of a new era of progress without growth." Heinberg (2012).

For our postmodern time the economists, scientists, and politicians make control more difficult. Some facts remain firm and others, like the behaviors or the society are uncontrollable. What is needed is to know how to react, project, plan based on a specific reality and culture where needed.

These tow facts will lead us to the question: endogenous versus exogenous architecture production. The endogenous prediction involves the creation of new economies and the ideal of using and working with what is available, both in resources and technologies. The exogenous ideal is to follow a unified way of thinking-thought process, for example the international movement within the modern movement, which tried to homogenize a system which resulted on disastrous consequences on local architecture and negating local cultures, economies and process.

Which would be the most adequate way of learning for this endogenous or exogenous projection?

The programmatic technique for learning in architecture is a method that may guide the student to solve a design problem based on the modern movement of scientific facts solution, "according to the model of Technical Racionality...-professional activity consist in instrumental problem solving made rigorous by application of scientific theory and technique." Schon (1987). Normally this process is engaged in the classroom, a controlled environment where the paper or the computer can hold everything and anything, making the stakes low for both the professors and students.

On the other hand the problematic problem solving design process is based on the practical, experimental approach to solve the problem, "in practice, Schon sugest, The profesionals work in a different way. In stead of dwelling in the "hills" of profesional certanty, they have to work in the "sludgy valley"of daily life, facing complex and confusing situations, that do not acept simple technical solutions". Atkinson (2002). The student in this case is brutally confronted by reality in order to solve the problem. The student then has to, or needs to, use his knowledge and research in order to find a way or a solution to solve the problem. When the student applies his knowledge to a practical problem then the information finally gets imprinted. The stakes in this case are very high, because time and resources are limited and also because a failure might have consequences that both the computer and the paper could detain, but not reality. As Michael 2007 states, "is within the workshop where the theoretical comprehension and the social location of a problem is held. With the edification of projects where the evidence of the theoretical knowledge is applied, special in the early steps where the student start his learning by practice but with a deficient theoretical foundation. The learning process in the Architectural Workshop most have a strong experimental focus." Michael (2007). The theoretical knowledge without the practical use would not be a complete learning experience.

The problematic way of teaching seeks opportunities rather than solutions, opportunities for learning rather than the opportunity of creation; it emphasizes the process of learning rather than the result. And later, when the future architect deals with clients or communities will see opportunities for an infrastructure based on their needs and desire rather than an opportunity to design an infrastructure. The difference resides in the way of seeing architecture, not as a product but as a process itself, just as Cesar Pelli says "The practice of an art is an end in itself. We work toward a necessary result, perhaps a sublime result, but the creative act is in the practice." Pelli (1999). Within this process the support of the professor or the guide is essential because he/she has to guide the student and also to ensure that the problem will be solved in an adequate way. If not, the solution might affect the community negatively.

The programmatic teaching is based on the final project; the idea is to construct, to build. The design process is almost scientific leaving no chance to chance, which means almost necessarily that this architecture may not leave chance to chance, and becomes an opportunity for show off the learned or acquired skills but as Schon states "Shein´s use of the term "skill" is of more than passing interest. From the point of view of the model of Technical Rationality institutionalized I the proffesional curriculum, real knowledge lies in the theories and techniques of basis and applied science. Hence, these disciplines should come first. "Skills" in the use of theory and technique to solve concrete problems should come later on, when the student has learned the relevant science-first, because he cannot learn skills of application until he has learned applicable knowledge; and secondly, because skills are ambiguous, secondary kind of knowledge. There is something disturbing about calling them "knowledge" at all." Schon (1987).

During the problematic process the relationship between the student and the "client" or community is highly important because it will create the necessary links and will raise the stakes. The architecture produced by this process in which the process itself is architecture (understanding or thinking about environment) leaves chance to chance. It is an architecture that resides in the use and its subjectivity, rather than in aesthetics or preconceptions or misconceptions of the architect or the professor. Architecture that leaves chance to chance then is not a product but a process of constant construction and life. When should the student or the architect leave? When the architecture itself starts to formulate its own answers.

Within the programmatic structure the set of tools are taught to the student beforehand so he/she might have them in mind in order to apply them to reality at a later day. With the problematic design process the student discovers those tools during the process itself. This means that he/she discovers what is needed or not needed, meaning that sometimes he would have the chance to discover something that he would not need or apply to the specific problem but this knowledge is already there.

In order to illustrate these two ways of design teaching we will analyze two groups of students solving an academic problem within the same environment: After analyzing and experiencing our university campus; find, acknowledge or mitigate a certain problematic within the use of public space, applying the basic design principles and building the 1:1 scale design.

2 GROUP A - PROGRAMMATIC

Students are given all the design tools beforehand and their focus is product oriented, meaning the product that will express this concepts, the design tools, is the final meaning of the project. See Figure 1. During the process the final objective is the product and how it reflects the basic principles of design. The end is to show the composition leaving the site and it's problematic in second or third layer of understanding.

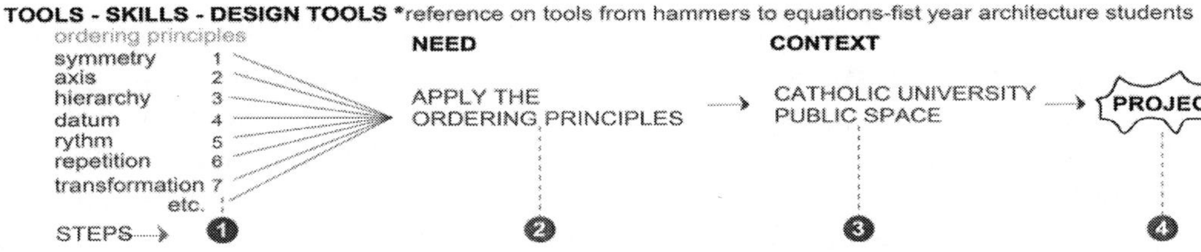

Figure 1. Programmatic approach.

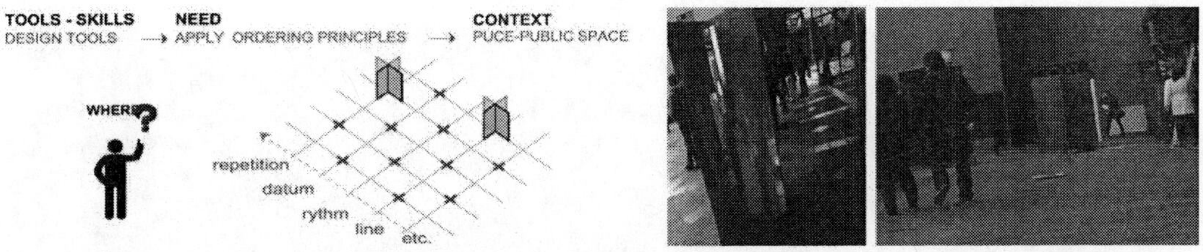

Figure 2. Programmatic approach: proposal and finished product.

The final result of this project was excellent, by the standard point of view in which all the basic principles were present and they made their point in a sculpturesque piece which became something that the people tried to avoid circulating and observe rather than use. See Figure 2. The understanding of the project became almost exclusively for the students and professor. Since the academic exercise was project oriented then the students built their building, installation, and the exercise was evaluated and done.

3 GROUP B - PROBLEMATIC

Students where given the question: what problematic do you would find within the campus? See Figure 3. "Shein's use of the term 'skill' is of more than passing interest. From the point of view of the model of Technical Rationality institutionalized I the proffesional curriculum, real knowledge lies in the theories and techniques of basis and applied science. Hence, these disciplines should come first. 'Skills' in the use of theory and technique to solve concrete problems should come later on, when the student has learned the relevant science-first, because he cannot learn skills of application until he has learned applicable knowledge; and secondly, because skills are ambiguous, secondary kind of knowledge. There is something disturbing about calling them 'knowledge' at all." Schon (1987).

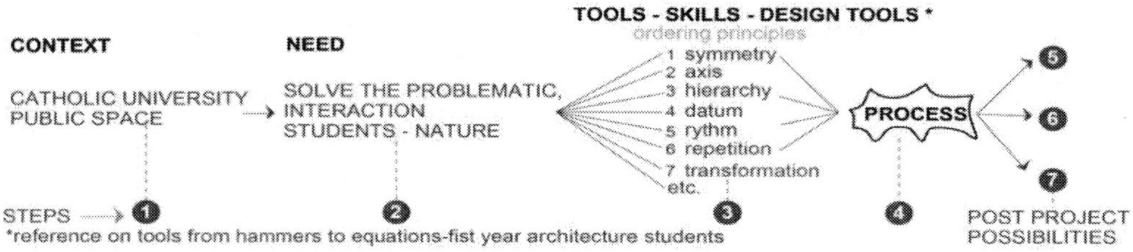

Figure 3. Problematic approach.

During the process the final objective is to try to solve or acknowledge the existing problem using the basic design principles, the means for an end not the end by itself. The final result in this case was a piece that started to be used on the fly and that also used the basic principles of composition but not as an end but as a mean to solve the problem of not acknowledging nature. See Figure 4. Since the exercise was problem oriented the students after the evaluation consider the idea that their project was able to be located in some other place in order to have a more everlasting life, so they install and reuse some of their materials in a park somewhere else.

Figure 4. Problematic approach: proposal, finished product, and reuse in new location.

4 CONCLUSION

For us as professors and professionals it is important to understand the real necessity of learning: learning for pleasure, learning for necessity, or learning by obligation? If necessity and pleasure are combined, then we would not have students, but restless seekers of both knowledge and chances to chances the reality in which they will live.

For a possible future, where energy crisis and global warming, (Foruzanmehr 2008) will be determinant for new generations lifestyles, two probable economic systems where brought: endogenous and exogenous (Lietaer 2005). Based on the academic exercise mentioned before, if we choose to act or project endogenously, then group B might succeed because we would have students that might base their action on solving problems and researching the best ways of doing it based on the study or their physical and cultural environment, and if there are no specific tools to do that they might be able to generate their own. The process is collaborative and pertinent to the culture of the place, acting endogenously. The work is necessary and uses the necessary tools in order to solve the problem, neither more nor less; the same applies for the use of resources. The student then is prepared to not be prepared, leaving chance to chance. Then he/she would be able to acknowledge-recognize problems rather to solve them. The idea is to know what to look for and not look for the answer.

If exogenous (Fernandez 2014) is an option, then probably group A might be a better option because they might have been acting in order to have an object rather than a process. This means attracting attention to the object itself because of its beauty of scientific characteristics but not because of its territorial or cultural pertinence. The student might project his wishes and display his knowledge rather than use this knowledge in order to solve problems, leaving no chance to chance within the projects. Group A developed a high level of competitive spirit due to a need to show off. Based on this observation then the use of tools and resources might be limited only by the need to showoff and not related to the needs inherent in the problem not by the needs of the problem.

References

Atkinson , T., G. C, The Intuitive Practitioner. Barcelona: Octaedro, S.L., 2002.

Fernandez, R., L. G., En la espiral de la energía, Colapso del capitalismo global y civilizatorio. Madrid: Libros en Acción, 2014.

Foruzanmehr, A., F. N., Towards new approaches for integrating vernacular passive-cooling systems into modern buildings in warm-dry climates of Iran. Air Conditioning and the Low Carbon Cooling Challenge. London: Windsor, 2008.

Heinberg, R., The End of Growth. Adapting to Our New Economic Reality. Canada: New Society Publishers, 2012.

Lietaer, B., El futuro del dinero. Como crear nueva riqueza, trabajo y un mundo más sensato. Buenos Aires: Errepar, Longseller, 2005.

Meadows, D.H. , D. M., Limits to growth scenario. United States: Chelsea Green, 2004.

Michael, Y. F., Hacia otro enfoque en la enseñanza del proyecto de arquitectura, Arquitectura y Urbanismo (Vol. XXVIII). La Habana, Cuba: Instituto Superior Politécnico José Antonio Echeverrñia, 2007

Pelli, C., Observations For Young Architects. New York: The Monacelli, 1999.

Schon, D., The Reflective Practitioner , Educating the Reflective Practitioner. San Francisco: CA: Jossey Bass, 1987.

Interaction between Theory and Practice in Civil Engineering and Construction
Edited by Komurlu, R., Gurgun, A. P., Singh, A., and Yazdani, S.

WIND TUNNEL TEST OF SCAFFOLDS SET WITH WALL CONNECTERS WITH BASEBOARD HEIGHT AS A PARAMETER

HIROKI TAKAHASHI and KATSUTOSHI OHDO

Construction Safety Research Group/National Institute of Occupational Safety and Health, Kiyose, Japan

The Japanese Industrial Safety and Health Law were revised in March 2009 to introduce new measures by which to prevent accidental falls in the construction industry. As part of this revision, authorities established regulations on the provision of guard rails, toe boards, mesh sheets, and other components in appropriate positions on scaffolds. When scaffolds are set in construction sites, their strength against wind force needs to be calculated. Japanese design guidelines regulate the strength of scaffolds against wind force; however, the design guidelines were written with old-style scaffolds in mind. It is not known whether the design guidelines are appropriate for new-style scaffolds. At the construction sites, the scaffolds connect to structures through the use of wall connecter, to keep scaffolds from falling down. The wind load that acts on the scaffolds was supported by the wall connecter at the construction sites. On the other hand, in conventional designs, a baseboard is used on construction sites. In this study, to set scaffolds at construction sites while using baseboard height as a parameter, we performed a wind tunnel test to examine the wind load that acts on scaffolds that have been set with wall connecters. The wind tunnel device has a total length 74,900 mm, while the device interior is 2,300 mm wide and 2,000 mm high. The load sell to set the wall connecter was used to measure wind load. The models, each of which was 1/10 in size, were used on scaffolds at general construction sites. The scaffolds were three stories high and one span wide. A baseboard was situated on one side of the long face of the scaffolds. The wind speed was set at a uniform flow of 10 m/s, because the wind force coefficient of a cylinder is stable at this speed. The characteristic length was positioned 5 mm along the diameter of a leg member. The Reynolds number was approximately 3.5 × 103. From our results, the wind force coefficient was found to increase as the baseboard height increased. With regard to efficient scaffold design, calculations of the wind force coefficient should therefore consider baseboard height.

Keywords: Fall accident, Wind force coefficient, Projected area.

1 INTRODUCTION

The Japanese Industrial Safety and Health Law were revised in March 2009 to introduce new preventive measures with regard to accidental falls in the construction industry (Japan Construction Occupational Safety and Health Association 2009). As part of this revision, the authorities have established regulations regarding the provision of guard rails, toe boards, mesh sheets, and other components in appropriate positions on scaffolds. Additionally, the installation of leading handrails requires that handrails be erected before construction work commences, to protect against falls (Ministry of Health, Labor, and Welfare 2003). For this purpose, a special structure called a handrail frame is used. Figure 1 shows examples of modern designs for scaffolds, and Figure 2 shows an example of modern scaffolds used at construction sites.

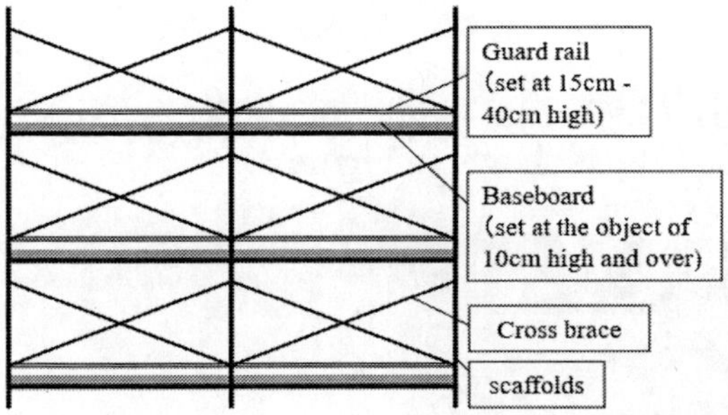

(a) Cross brace, guardrail and toe board.

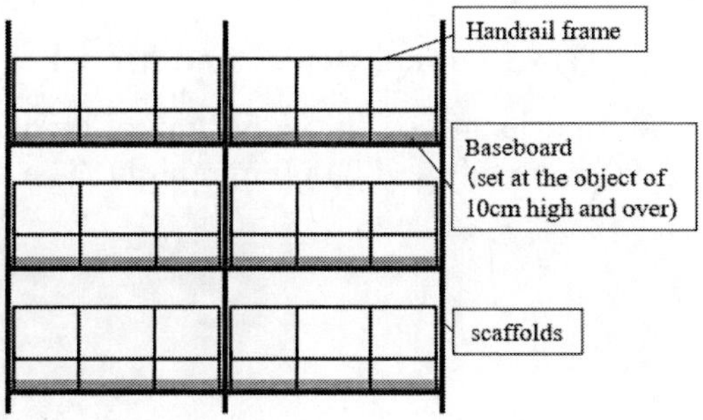

(b) Handrail frame and toe board.

Figure 1. Examples of modern scaffolds.

Figure 2. Modern scaffolds used at construction sites.

When scaffolds are set at construction sites, their strength against wind force needs to be calculated (Scaffolding and Construction Equipment Association of Japan 2004). Japanese design guidelines regulate the strength of scaffolds against wind force; however, those design guidelines were written with old-style scaffolds in mind. It is not known whether the design guidelines are appropriate for new-style scaffolds. Additionally, at construction sites, scaffolds connect to the structure through the use of some kind of wall connecter, to keep the scaffolds themselves from falling down. The wind load that acts on the scaffolds is supported by the wall connecter at the construction site; on the other hand, in conventional designs, baseboards are used at construction sites.

In this study – with reference to the method by which scaffolds are set at construction sites with the baseboard height as a parameter – we performed a wind tunnel test to examine the wind load that acts on scaffolds that are set with some kind of wall connecter.

2 OVERVIEW OF WIND TUNNEL TEST

2.1 Wind Tunnel Device and Model

The total length of the wind tunnel device is 74,900 mm, while the device interior is 2,300 mm wide and 2,000 mm high. The test setup is shown in Figure 3.

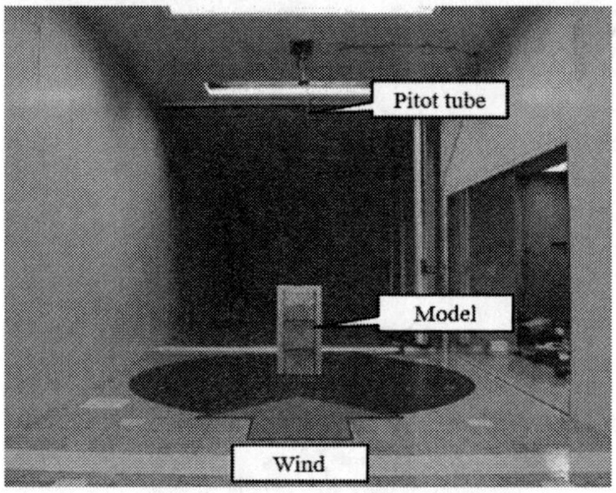

Figure 3. Wind tunnel test.

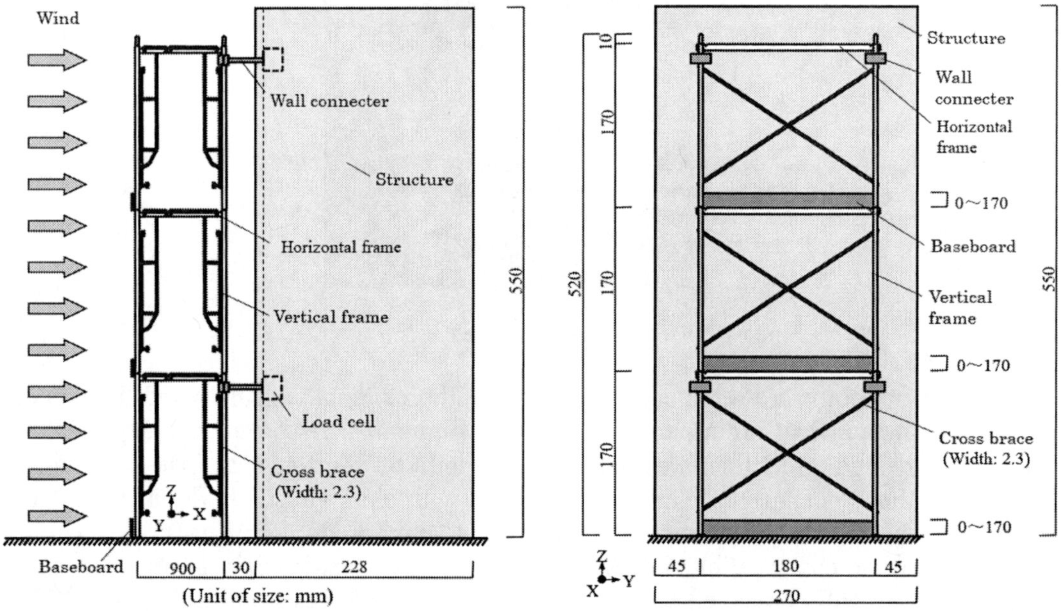

(a) X–Z face (b) Y–Z face

Figure 4. The scaffolds and the structure used for the wind tunnel test.

The model reflects scaffolds used at general construction sites, and it is 1/10 in size. The scaffolds are three stories high and one span wide. The scaffolds and the structure used for the wind tunnel test are shown in Figure 4. The vertical frame direction of the scaffolds is denoted by X, while the cross-brace direction of the scaffolds is represented by Y; the height of the scaffolds is Z. The scaffold model was put into the wind tunnel, and the structure was placed near the scaffold model. The scaffolds were separated from the floor, and the structure was connected to the floor. Wall connecters were set at four places on the scaffolds. Some load cells were set to the structure, and the wall connecters were connected to the load sells. The load sells were used to measure wind force. A baseboard was situated on one side of the Y–Z face of the scaffolds.

2.2 Projected Area and Reynolds Number

The ratios of the baseboard height and the projected area of the scaffolds are expressed as dimensionless coefficients, as follows:

$$\tau = h_b / h_V \quad (1)$$

$$\eta = A_b / A_0 \quad (2)$$

where τ is the ratio of the baseboard height, h_b is the baseboard height, h_V is the height of the vertical frame, η denotes the ratio of the projected area of the scaffolds, A_b is the projected area of the scaffolds with a baseboard, and A_0 is the projected area of the scaffolds with a 170 mm baseboard.

Figure 5 shows the relationship between the ratio of the projected area of scaffolds and the ratio of baseboard height. The ratio of the projected area of the scaffolds η was found to increase as τ increased.

The pitot tube was positioned 550 mm from the ceiling of the wind tunnel, and thereafter, the wind speed was measured. The wind speed was set at a uniform flow of 10 m/s. The characteristic length D was positioned 5 mm along the diameter of a leg member. The Reynolds number Re was approximately 3.5×10^3, and was determined as follows:

$$\mathrm{Re} = \frac{UD}{\nu} \quad (3)$$

where U is the wind speed (in m/s), D is the characteristic length (in mm), and ν denotes the coefficient of kinematic viscosity [$\nu = \mu/\rho$, where μ is the coefficient of viscosity ($\mu = 1.82 \times 105$(N s/m^2) and ρ is the air density].

The scaffolds are made of ring-shaped steel pipes whose lengths are proportional to their diameters. These features require the calculation of the wind force coefficient of two-dimensional cylinders (Scaffolding and Construction Equipment Association of Japan 2004). The wind force coefficient of each cylinder changes in accordance with the Reynolds number.

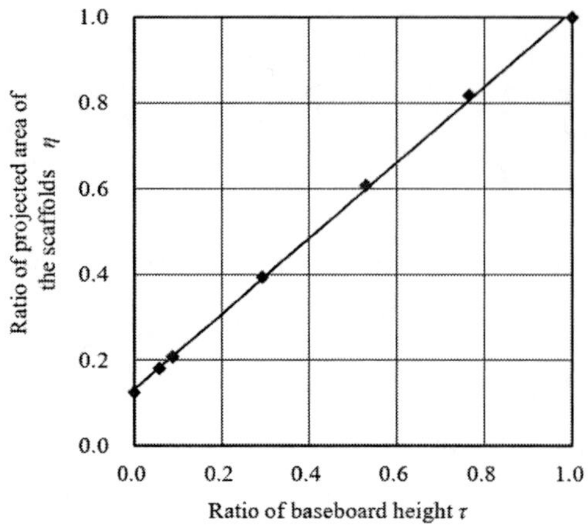

Figure 5. The relationship between the ratio of the projected area of the scaffolds and the ratio of baseboard height.

3 RESULTS AND DISCUSSION

Figure 6 shows the wind tunnel test results. The vertical axis in this figure pertains to the wind force coefficient of the scaffolds C, while the horizontal axis pertains to the ratio of baseboard height τ. The wind force coefficient of the scaffolds in the X direction was calculated as follows:

$$C = \frac{F}{q_F A} \quad (4)$$

where F is the force on the scaffolds, q_F is the reference speed pressure ($1/2\rho V_H^2$, where ρ denotes the air density, and V_H is the wind speed), and A represents the reference area (projected area by the Y–Z directions of the scaffolds).

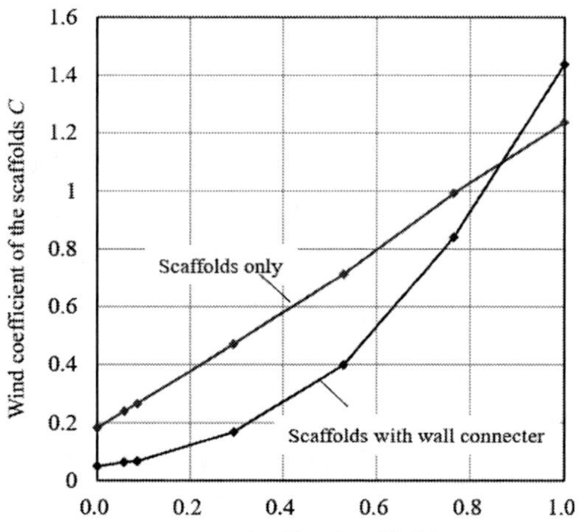

Figure 6. Relationship between the wind force coefficient and the ratio of baseboard height.

Only in the scaffolds was the relationship between *C* and τ proportional (Takahashi *et al.* 2013). However, in scaffolds with a wall connecter, the relationship between C andτ became steep as τ increased. Negative pressure acted on the scaffolds as a consequence of the structure downwind. We think that this has an influence on the distance between the scaffolds and the structure, and so we need to examine this distance. Whenever the scaffolds were set near the structure, we needed to revise the wind force coefficient of the scaffolds. We need to examine ways of making revisions while bearing in mind the wind force coefficient of scaffolds.

4 CONCLUSION

The wind force acting on scaffolds that feature a wall connecter equipped with baseboards was investigated, through the use of wind tunnel testing.

The relationship between the wind force coefficient and the baseboard height, when plotted, became steep as the baseboard height increased. The wind force coefficient differed only with respect to the scaffolds: when the scaffolds were set near a downwind structure, we needed to revise their wind force coefficient. Our future work will examine ways of making revisions while keeping in mind the wind force coefficient of scaffolds.

Acknowledgments

This work was supported by the JSPS Grant-in-Aid for Scientific Research C (grant number 15K01238). We are grateful for the financial aid provided.

References

Japan Construction Occupational Safety and Health Association, *Yearbook of Safety and Health in Construction 2009* (in Japanese), Tokyo, 75–110, 2009.

Ministry of Health, Labor, and Welfare, *Guidelines on the Handrail Preceding Method* (in Japanese), Tokyo, 2003.

Scaffolding and Construction Equipment Association of Japan, *Recommendations for Safety Technology of Scaffolds to Wind Loads* (in Japanese), Tokyo, 2004.

Takahashi, H., Ohdo, K., and Takanashi, S., Fundamental Study on Estimated Method of Wind Force for Fall Protection Method of Scaffolds, *Journal of Japan Society of Civil Engineers, Ser. F6 (Safety Problem)*, 69(2), I_153–I_158, Jan, 2014.

Interaction between Theory and Practice in Civil Engineering and Construction
Edited by Komurlu, R., Gurgun, A. P., Singh, A., and Yazdani, S.

BUILDING INFORMATION MODELING AND FORENSIC ANALYSIS OF DELAY AND DISRUPTION

AGELIKI VALAVANOGLOU and DETLEF HECK

Institute of Construction Management and Economics, Graz University of Technology, Graz, Austria

Delay and Disruption is a common occurrence in construction projects. The challenges of forensic schedule analysis and the evaluation of the extent of project delay increase with the degree of complexity of a project. The occurrence of multiple concurrent delays, secondary effects and the cumulative impact of delay and disruption events can render the establishment of entitlement to extension of time and reimbursement a difficult task. In order for the claiming party to receive compensation for project delay and disruption, causation, liability and quantum have to be demonstrated and proven. Advances in technology have made a shift from conventional delay analysis methods towards delay and disruption analysis supported by Building Information Models possible. Research and application of Building Information Models has increased in recent years, exploring information coordination on multiple dimensions. Linking the fourth dimension of time to a 3D model enables the user to visualize a representation of the construction process. The application of 4D simulation in forensic schedule analysis is a great tool for the visualization of delay events and their effects on the project schedule and the construction process. Although 4D Building Information Models are able to assist forensic schedule analysis, the identification of the cause and effect relationship of delay events of complex construction works, requires an expert who is not only familiar with the software tools but has also experience in delay analysis and is able to clearly determine the accuracy of the produced data. 4D Building Information Models can simulate a high level of project performance, producing great quantities of data. The role of the delay analyst is to identify the relevant facts from the great quantities of data simulated in the 4D model, in order to support his findings of entitlement, causation and resulting damages. The purpose of this paper is to investigate the use of 4D Building Information Modelling in delay and disruption claims and outline the expertise required to perform the forensic schedule analysis.

Keywords: Construction, 4D BIM, Scheduling, Dispute, Expert.

1 INTRODUCTION

Characteristics of construction projects range from simple, low cost programs with short schedules to highly complex with extensive budgets and complicated schedules. The magnitude of a project relates to the complexity of the delay and disruption analysis and therefore to the difficulty of the forensic schedule analysis in the case of dispute. There are various triggers than can cause delay and disruption, some attributable to the owner, other times attributable to the client or in case of true concurrency to both (Nguyen 2007). A contractor can claim excusable or non-excusable delays. Excusable delays can be further categorized into compensable or non-compensable delays. There has to be a distinction between delay and disruption claims, as their investigation requires separate analysis methodologies (Braimah 2008).

Delay occurs when the contractual completion date is not identical to the actual completion date of the construction project (Lee *et al.* 2005, Doughery 2015). Disruption occurs when an event leads to the loss of efficiency of the planned productivity, leading to increased difficulty of performance, cost and possibly delay (Finke 1997, Keane *et al.* 2008). Effects of delay and disruption can be diverse and must be individually identified and analyzed. Compensation claims may include increased home office overhead, additional cost of labor, material and equipment (Schwarzkopf 1995, Haese *et al.* 2001). A disruption that leads to loss of productivity does not necessarily lead to delay, but may still incur extra cost (Braimah 2008). The challenge of delay and disruption analysis does not only lie within the proof of causation, quantification of the delay and damages incurred (Williams *et al.* 2002) but also the distinct separation and identification of the root causes and their primary and secondary effects as well as direct and indirect damages (Doughery 2015).

Apart from conventional methods used internationally in forensic schedule analysis, advances in research and technology as well as the increased application of Building Information Models (BIM) in construction projects in recent years, has revealed a potential implementation in forensic schedule analysis.

2 CONVENTIONAL FORENSIC ANALYSIS METHODS

Forensic analysis methods do not have standardized titles and can be found under various terms (Braimah 2008, Barry 2009). One of the fundamental requirements common to all methods for a correct analysis is that all techniques must be based on accurate data, it is therefore important to use realistic baseline schedules, where resources, linked activities and durations are realistically and accurately calculated (Trauner *et al.* 2009). The baseline schedules must be continuously updated during construction in order to incorporate discrepancies between the planned and the built schedule.

Five commonly used conventional schedule delay analysis methods are the Impacted As-Planned Analysis, the Time Impact Analysis, the Collapsed As-built Method, the Windows Analysis and the As-planned vs. As-Built method (Arditi *et al.* 2008, Streckel 2011).

Schedule delay analysis can be conducted prospectively, contemporaneous with the delay event i.e., during the construction process or retrospective after the delay event has occurred. In highly complex delay claims a combination of prospective and retrospective methods is required, in order to verify the results of the analysis (Keane *et al.* 2008).

In the prospective methods (Impacted As-Planned, Time Impact, Windows Analysis) the baseline schedule is updated and activity relationships, resource allocation and float calculation are reevaluated in order to estimate the likely impact of the delay event on the completion date. Retrospective analysis methods (As-Planned vs. As-Built, Collapsed As-Built) aim to reconstruct historic events to account for actual progress and completion dates (Barry 2008).

Like delay analysis methods, there are a number of methods to calculate damages resulting from disruption. Some of the most widely used methods are the Total Cost Method, Modified Total Cost Method, Industry Studies and Guidelines and the Measured Mile Method. Similar to the delay methods they are described in the literature under various titles (Braimah 2008). All disruption analysis methods aim to identify inefficiencies or loss in productivity, the responsible party or in case of cumulative impact, the apportionment of responsibility, based on different sources of information and data (Braimah 2008).

A number of researchers have pointed out the difficulty for judges and tribunals to interpret technical drawings of great detail as well as gaining an understanding of the project and the events that transpired to cause delay and disruption from the numerous data and information presented in claims based on traditional forensic analysis methods (Keane *et al.* 2008, Gibbs *et al.* 2012).

In recent years the construction industry has experienced a shift from 2D CAD models to the implementation of BIM services, UK Government for example requires all public sector projects to use a fully collaborative 3D BIM as a minimum by 2016. BIM can be used as a collaborative planning and construction tool as well as for the coordination of construction work sequence and the visualization of the construction process (Kessoudis *et al.* 2015).

3 FORENSIC ANALYSIS WITH BIM

Building Information Models have several levels of development commonly described in terms of dimensions, linking the three-dimensional (3D) model to additional information and data of the project, producing additional management parameters to n-dimensional extensions simulating the project life cycle (Azhar 2011). Four-dimensional (4D) BIM links geometrical data from the three-dimensional (3D) model to the construction schedule simulating construction sequences with integrated dependencies of processes and resources (Kessoudis *et al.* 2015). 4D BIM has rendered the identification of time-space conflicts through visual inspection relatively easy in comparison to traditional 2D drawings and their shortcomings in relation to optimal use of workspace allocation (Jongeling *et al.* 2008). There is a variety of commercial software providing 4D schedule solutions with integrated object recognition with assigned categories and materials which offer visualizations of the construction sequence, workspace conflicts and detection of hidden flaws of the planned schedule (Dang *et al.* 2015). Furthermore, BIM tools are able to detect inconsistencies on level of detail amongst activities, omissions and errors in the schedule logic as well as identify potential accessibility problems on construction site (Koo and Fischer 2000).

Researchers have already identified the potential of BIM in dispute prevention and dispute resolution (Greenwald 2012) as well as forensic schedule analysis. Some have concentrated on the visualization option which can aid the demonstration of actual versus planned construction process in standalone or preferably comparative 4D models, highlighting the discrepancies resulting in delay and disruption (Pickavance

2007, Gibbs *et al.* 2013), as well as outline the integration of 4D Modeling tools into typical schedule analysis steps (Coyne 2008).
Apart from the benefits gained through the implementation of BIM, research has also highlighted the shortcomings that need to be further addressed and developed in order to generate automated and dynamic 4D models.

Currently 4D schedules are limited to visualizations, automated calculation of quantities and recourse capacities to establish work flow (Hartman *et al.* 2008, Dang *et al.* 2015), but there is an increasing demand for automated dynamic features in order to use 4D schedules throughout the construction process, especially in cases of changes where schedules, activity logic, relationships and associated resources have to be manually reestablished.

4 FORENSIC ANALYSIS EXPERT

BIM is an innovative tool that can be implemented throughout the lifecycle of a project with great advantages, but the efficient use of the tools lies upon the experts that use it.

Forensic schedule analysis can benefit through the implementation of BIM but it has also aroused the need for a new hybrid model of schedule analyst. Analysis of delay and disruption with BIM requires an expert who has experience in forensic analysis and is familiar with the software, in order to be able to determine the accuracy of the data produced.

The challenge of the analyst is to use all tools provided to identify the cause of the delay or the disruption, appoint the liability to the responsible party and calculate the damages. 4D BIM can simulate a high level of project performance, producing great quantities of data. The role of the delay analyst is to identify the relevant facts from the great quantities of data simulated in the 4D model, in order to support his findings. Visualizations have proven to be a great tool not only for communication inside the project team but also for providing a chronological narrative of the project and the events that led to a deviation from the as-planned to the as-built schedule. Apart from the challenges posed in forensic analysis of delay and disruption using conventional methods, the new challenge of the analyst is to sort through the infinite amount of data, and detect the relevant information, form it into factual evidence and decide upon the level of detail that will form the basis of the claim.

5 CONCLUSION

As technology has progressed and evolved, the application of BIM in construction projects has increased and opened new areas of potential implementation for 4D tools. 4D scheduling shows great promise not only during the planning and construction phase for the prevention of delay and disruption but also for the analysis of events during the forensic schedule analysis. The most important tool identified is the visualization of the chronological events and the comparative 4D models simulating the as-planned versus the as built construction process.

The limitation of automated generation of dynamic 4D schedules renders the use and updating of the schedule an onerous task, especially in case of changes where schedules, activity logic, relationships and associated resources have to be manually reestablished.

Further research is necessary to decrease or remove these limitations in order to be able to fully employ the capacities of 4D BIM during the lifecycle of a construction project. Technological advances and increased implementation of BIM have aroused the need for hybrid experts, who understand the technical part of the construction accompanied by a familiarity with BIM software tools. Forensic schedule analysis will have to find a balance between the two disciplines in order to obtain maximum benefit out of forensic schedule analysis with BIM.

References

Arditi, D. and Pattanakitchamroon, T., Analysis Methods in Time-Based Claims, *Journal of Construction Engineering and Management*, 134(4), 242–252, 2008.

Azhar, S., Building Information Modeling (BIM): Trends, Benefits, Risks, and Challenges for the AEC Industry, *ASCE*, 11, 241-252, 2011.

Barry, D., Beware of the Dark Arts! Delay analysis and the problems with reliance on technology, Society of Construction Law, 2009. Retrieved from http://www.scl.org.uk/papers/beware-dark-arts-delay-analysis-and-problems-reliance-technology on August 11, 2014.

Braimah, N., *An Investigation Into The Use Of Construction Delay and Disruption Analysis Methodologies*, University of Wolverhampton, Wolverhampton, 2008.

Coyne, K., BIM.03, Leveraging the Power of 4D Models for Analyzing and Presenting CPM Schedule Delay Analyses, *AACE International Transactions*, 2nd AACE International Annual Meeting, Canada, 2008.

Dang, T., and Bargstädt, H., 4D Relationships: The Missing Link in 4D Scheduling, *Journal of Construction Engineering and Management*, ASCE, 2015.

Dougherty, J. M., *Claims, Disputes and Litigation Involving BIM*, Routledge, Taylor & Francis Group, London, 2015.

Finke, M. R., Claims for Construction Productivity Losses, *Public Contract Law Journal*, 26(3), 311-338, 1997.

Gibbs, D., Emmit, S., Ruikar, K., and Lord, W., An Investigation into whether Building Information Modelling (BIM) can Assist with Construction Delay Claims, *First UK Academic Conference on BIM*, Greenwood, D. (Ed.), 36-44, BIM Academy, England, 2012.

Gibbs, D., Emmit, S., Ruikar, K., and Lord, W., A Case Study Investigation into the use of Computer Generated Visualisations to Assist with Construction Delay Claims, *CIB World Building Congress*, Kajewski, S., Manley K. and Hampton, K. (Eds.), Queensland University of Technology, Brisbane, 2013.

Greenwald, N., A Creative Proposal for Dispute Systems Design for Construction Projects Employing BIM, *Journal of Legal Affairs and Dispute Resolution in Engineering and Construction,* ASCE, 5, 2-5, 2013.

Haese, G., and Dragelin, T., Types of claims, *Proving and Pricing Construction Claims*, Cushman, R. F., Carter, J. D., Gorman, P. J. and Coppi, D. F. (Ed.), Aspen Law & Business, New York, 2001

Hartmann, T., Gao, J., and Fischer, M., Areas of Application for 3D and 4D Models on Construction Projects, *Journal of Construction Engineering and Management*, ASCE, 134(10), 776-785, 2008.

Jongeling, R., Kim, J., Fischer, M., Mourgues, C., and Olofsson, T., Quantitative Analysis of Workflow, Termorary Structure Usage, and Productivity using 4D Models, *Automation in Construction*, Elsevier, 17, 780-791, 2008.

Keane, P., and Caletka, A., *Delay Analysis in Construction Contracts*, Wiley-Blackwell, Cornwall, 2008.

Kessoudis, K., Teizer, J., Schley, F., Blickle, A., Hiel, L., Früh, N., Biesinger, M., Wachinger, M., Marx, A., and Paulitsch A., BIM bei STRABAG SE, *Building Information Modeling*, Borrmann, A., König, A., Koch, C., and Beetz, J. (Ed.), 541-554, Springer Fachmedien, Wiesbaden, 2015.

Koo, B., and Fischer, M., Feasibility Study of 4D CAD in Commercial Construction, *Journal of Construction Engineering and Management*, 126(4), 251-260, 2000.

Lee, H., Ryu, H., Yu, J., and Kim, J., Method for Calculating Schedule Delay Considering Lost Productivity, *Journal of Construction Management*, 131(11), 1147-1154, November 2005.

Nguyen, L. D., *The Dynamics of Float, Logic, Resource Allocation, and Delay Timing in Forensic Schedule Analysis and Construction Delay Claims*, University of California, Berkley, 2007.

Pickavance, K., Using advanced forensic animations to resolve complex disruption claims. Society of Construction Law, 2007. Retrieved from http://www.scl.org.uk/papers/using-advanced-forensic-animations-resolve-complex-disruption-claims on October 28, 2015.

Streckel, S., *Analyse der Auswirkungen gestörter Bauabläufe und der Anteile ihrer Verursachung durch Auftraggeber, Auftragnehmer und Dritte*, DVP, Berlin, 2012.

Trauner, T., J., *Construction Delays: documenting causes, winning claims, and recovering costs*, 2nd Ed., Elsevier, USA, 2009.

Williams, T., Ackermann, F., Eden, C., Structuring a delay and disruption claim: An application of cause-mapping and system dynamics, *European Journal of Operational Research*, Elsevier, 148(1), 192–204, July 2003.

ADOPTING BIM IN ARCHITECTURAL DESIGN OFFICES

YELIZ TULUBAS-GOKUC[1] and DAVID ARDITI[2]

[1] *Dept of Architecture, Balikesir University, Balikesir, Turkey*
[2] *Dept of Civil, Architectural and Environmental Engineering, Illinois Institute of Technology, Chicago, USA*

Developments in technology such as Building Information Modeling (BIM) have recently created serious competitive pressures on architectural design firms and building construction companies. Design firms and construction companies need to adopt these new technologies to survive and thrive in this environment. Despite the advantages of BIM in building design, its adoption by architectural design firms has been slow due to obstacles such as lack of familiarity with BIM, lack of training in BIM, and lack of supporting resources such as hardware and software. The problems faced in adopting BIM were investigated by only few researchers who focused on the acceptance or penetration of BIM into construction firms. The objective of this research is to develop an integrated model to understand the adoption of BIM in architectural design firms. The model involves (1) the fit between the tasks to be performed by design professionals and the technology provided by BIM, (2) the fit between the organizational competency in the design firm and the capabilities offered by BIM, and (3) the fit between the designers' knowledge/skills in BIM and the sophistication of the BIM product. After a thorough review and synthesis of technology adoption models and theories, this paper proposes an integrated model to understand the adoption of BIM by architectural design firms. The model argues that using BIM depends on BIM's properties as well as the tasks at hand, the organizational competency of the company, and the perceived ease of use.

Keywords: Building information modeling (BIM), Technology acceptance model, Architectural design firms, BIM adoption, BIM technology fit.

1 INTRODUCTION

Design firms need to adopt new technologies to survive in the current environment. For example, Eastman *et al.* (2011) think that individual firms are motivated to adopt BIM, not only to improve their design and construction practices, but also to gain a competitive advantage in the marketplace. Building Information Modeling (BIM) is such a technology.

Despite the advantages of BIM in building design, its adoption has been slow due to some obstacles such as lack of familiarity with adopting BIM, lack of supporting education and training for use of BIM, and lack of supporting resources such as hardware and software (Lee *et al.* 2013). The problems faced in adopting BIM are treated only in a few studies that focus on the acceptance or penetration of BIM in construction firms rather than in design firms (Lee *et al.* 2013, Wang *et al.* 2013, and

Enegbuma *et al.* 2014). The variables that are used in these studies include organizational intention and support of leadership (Lee *et al.* 2013, Wang *et al.* 2013).

This research aims to develop an integrated model to understand the adoption of BIM in architectural design firms. The model involves (1) the fit between the tasks to be performed by design professionals and the technology provided by BIM, (2) the fit between the organizational competency in the design firm and the features offered by BIM, and (3) the fit between the designers' knowledge/skills in BIM and the sophistication of the BIM product.

2 INNOVATION ADOPTION MODELS

User acceptance of technology has been an important field of study. Many models have been proposed to predict and explain the acceptance or rejection of information technologies. Researchers have emphasized task-technology fit (Goodhue and Thompson 1995) and culture-technology fit (Lee *et al.* 2007). Information systems investigators have suggested intention models inspired from social psychology as potential theoretical foundations for research on the determinants of user behavior (Fishbein and Ajzen 1975, Ajzen and Fishbein 1980). Some of these studies were inspired from the Theory of Reasoned Action of Fishbein and Ajzen (1975) and the Technology Acceptance Model of Davis (1989). Some researchers fully implemented these models while others expanded these models by adding new variables. It must be noted however that the majority of the current literature concentrates on the user's acceptance of technology in different industries (health, telecommunications, tourism, banking, etc.), but not in the construction industry except for a small group of studies (e.g., Son *et al.* 2012, Lee *et al.* 2013, Park *et al.* 2012).

The first behavioral models are the theory of reasoned action (TRA) and the theory of planned behavior (TPB). Another behavioral model is the technology acceptance model (TAM) that was derived from the theory of reasoned action (TRA). These models are briefly described in the following subsections.

2.1 Theory of Reasoned Action (TRA)

Fishbein and Ajzen's (1975) theory of reasoned action is an especially well-researched intention model. It has proven successful in predicting and explaining behavior across a wide variety of domains (Davis *et al.* 1989). According to the theory of reasoned action, a person's performance of a specified behavior is determined by the person's "behavioral intention" to perform the behavior, and "behavioral intention" is jointly determined by the "person's attitude" and "subjective norms". The theory of reasoned action provides a model that has potential benefits for predicting the intention to perform a behavior based on an individual's attitudinal and normative beliefs. This model was extended to accommodate developments in the variables and the resulting model was named the theory of planned behavior (Southey 2011).

2.2 Theory of Planned Behavior (TPB)

The theory of planned behavior is an extension of the theory of reasoned action and was proposed by Ajzen (1991) to improve the predictive power of the theory of reasoned action by adding perceived behavioral control. Perceived behavioral control refers to people's perceptions of their ability to perform a given behavior.

2.3 Technology Acceptance Model (TAM)

The technology acceptance model was introduced by Davis (1989). According to Davis *et al.* (1989), the goal of the technology acceptance model is to provide an explanation of the determinants of computer acceptance that is general, capable of explaining user behavior across a broad range of end-user computing technologies and user populations, while at the same time being both parsimonious and theoretically justified. The technology acceptance model uses the theory of reasoned action as a theoretical basis for specifying the causal linkages between two key beliefs: perceived usefulness and perceived ease of use, and users' attitudes, intentions and actual computer adoption behavior. The technology acceptance model is considerably less general than the theory of reasoned action, designed to apply only to computer usage behavior. However, because it incorporates findings accumulated from over a decade of information systems research, it may be especially well-suited for modeling computer acceptance (Davis *et al.* 1989). To determine behavioral intentions, the technology acceptance model depends on two beliefs, perceived usefulness and perceived ease of use.

Perceived usefulness is defined by Davis *et al.* (1989) as the prospective user's subjective probability that using a specific application system will increase his or her job performance within an organizational context. Perceived ease of use is the degree to which the prospective user expects the target system to be free of effort (Davis *et al.* 1989).

3 PROPOSED MODEL

This research proposes an integrated model that was inspired by Venkatesh and Davis's (2000) technology acceptance model, Goodhue and Thompson's (1995) task-technology fit model, and various other research studies such as the work of Lee *et al.* (2013). Figure 1 shows the proposed model. The fit between design tasks and BIM-provided features, the fit between organizational competency and BIM implementation, and the fit between staff competency and BIM requirements are expected to explain the adoption of BIM in design firms.

3.1 Design Task - BIM Technology Fit

The task-technology fit model considers how tasks (i.e., actions carried out by individuals in turning inputs to outputs) affect an individual's use of the technology. Task-technology fit is the degree to which a technology assists an individual in performing his or her portfolio of tasks (Goodhue and Thompson 1995). More specifically, task-technology fit is the correspondence between task requirements and the functionality of the technology. Eight factors were developed by Goodhue and Thompson (1995) to measure task-technology fit: data quality, locatability, authorization, compatibility, production timeliness, systems reliability, ease of use/training, and relationship with users.

The model used in this study, "design task - BIM technology fit" consists of the fit between the tasks normally performed by designers and BIM-provided features. The statements that were used to assess "design task - BIM technology fit" aim to determine to what extent the features provided by BIM overlap with the design tasks normally performed in architectural design firms.

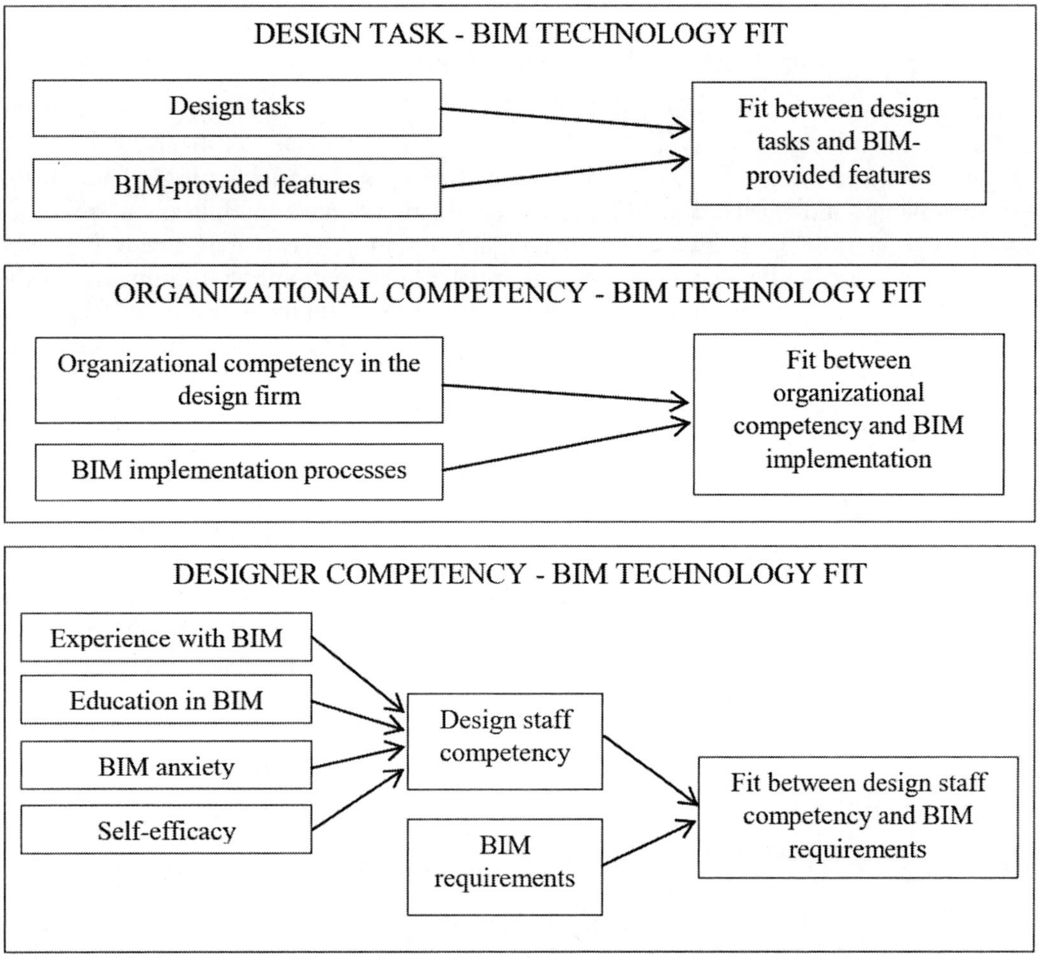

Figure 1. The proposed model.

3.2 Organizational Competency - BIM Technology Fit

Organization-technology fit involves the fit between organizational characteristics and technology implementation processes.

The model used in this study, "organizational competency – BIM technology fit" is defined as the degree of congruence between the know-how accumulated in the firm over the years and the implementation of BIM. In this study, the statements about the fit between organizational competency and BIM implementation processes were formulated by adapting to the BIM environment the findings of a research study conducted by Lee *et al.* 2007 where organizational competency was analyzed with respect to three dimensions: (1) collective efficacy, (2) organizational innovativeness, and (3) top management support. Collective efficacy refers to established beliefs in the organization, whereas organizational innovativeness is the willingness of an organization to try out a new technology. According to Lee *et al.* (2007), the decision by an organization to adopt a new technology may be risky unless there is a firm commitment from top management.

3.3 Designer Competency - BIM Technology Fit

User-technology fit can be defined as the degree to which a technology matches the skills and capabilities of individuals. If users have the necessary skills and knowledge to use a new technology, there is a good 'fit' between user and technology. According to Mohamadali and Garibaldi (2012), not only organizational capabilities but also user capabilities need to be evaluated to investigate the adoption of new technology. A user's knowledge, skills, expertise, and experience must match the requirements of the new technology. User-technology fit is an important factor in the adoption of BIM.

The model used in this study, "designer competency – BIM technology fit" measures the fit between design staff competency and BIM requirements by looking into user characteristics such as experience with BIM, education in BIM, BIM anxiety, and self-efficacy.

4 CONCLUSION

After a thorough review and synthesis of technology adoption theories, this paper proposes an integrated model to understand the adoption of BIM by architectural design firms. The model argues that the adoption of BIM depends not only on BIM-provided features, but also on the design tasks at hand, the organizational competency of the company, and the ease of use as perceived by design staff. An empirical study to justify the components of this model is underway. Once the model is validated, it is expected that it will shed light on the effect of the adoption of BIM on the performance of a design office.

References

Ajzen, I., The theory of planned behavior, *Organizational Behaviour and Human Decision Processes*, 50, 179-211, 1991.

Ajzen, I. and Fishbein, M, *Understanding attitudes and predicting social behavior*, Englewood Cliffs, NJ: Prentice-Hall,1-278, 1980.

Davis, F. D., Perceived usefulness, perceived ease of use, and user acceptance of information technologies, *MIS Quarterly*, 13(3), 319–340, 1989.

Davis, F. D., Bagozzi, R.P. and Warshaw, P.R, User Acceptance of Computer Technology: A Comparison of Two Theoretical Models., *Management Science* (35:8), 982-1002, 1989.

Eastman, C., Teicholz, P., Sacks, R., and Liston, K., *BIM handbook: a guide to building information modeling for owners, managers, designers, engineers and contractors*, Second Edition, John Wiley & Sons, 1-648, 2011.

Enegbuma, W. I., Dodo, Y. A., and Ali, K. N., Building information modelling penetration factors in Malaysia, *International Journal of Advances in Applied Sciences (IJAAS)*, 3(1), pp. 47-56, 2014.

Fishbein, M., and Ajzen, I., *Belief, attitude, intention and behaviour, an introduction to theory and research*, Addison Wesley Publishing Company, 1-573, 1975.

Goodhue, D. L. and Thompson, R. L., Task-technology fit and individual performance, *MIS Quarterly*, 19(2), 213-236, 1995.

Lee, S., Yu, J., and Jeong, D., BIM Acceptance Model in Construction Organizations, *Journal of Management Engineering*, 31(3), 2013.

Lee, I., Choi, B., Kim, J., and Hong, S. J., Culture-technology fit: effects of cultural characteristics on the post-adoption beliefs of mobile internet users, *International Journal of Electronic Commerce*, 11(4), 11-51, 2007.

Mohamadali, N. A. and Garibaldi, J. M., Understanding and addressing the 'fit' between user, technology and organization in evaluating user acceptance of healthcare technology, *5th International Conference on Health Informatics*, February, Vilamoura, Algarve, Portugal, 2012.

Park, Y., Son, H. & Kim, C., Investigating the determinants of construction professionals' acceptance of web-based training: an extension of the technology acceptance model, *Automation in Construction*, 22, 377-386, 2012.

Son, H., Park, Y., Kim C., and Chou, J. , Toward an understanding of construction professionals' acceptance of mobile computing devices in South Korea: An extension of the technology acceptance model, *Automation in Construction*, 28, 82-90, 2012.

Southey, G., The theories of reasoned action and planned behavior applied to business decisions: a selective annotated bibliography, *Journal of New Business Ideas and Trends*, 9(1), 43-50, 2011.

Venkatesh, V., and Davis, F. D., A theoretical extension of the technology acceptance model: four longitudinal field studies, *Management Science*, 46(2), 186–204, 2000.

Wang, Y., Xue, X., and Lia Y., Critical review on the impact factors of BIM application, *International Journal of Digital Content Technology and Its Applications (JDCTA)*, 7(8), 616-624, 2013.

Interaction between Theory and Practice in Civil Engineering and Construction
Edited by Komurlu, R., Gurgun, A. P., Singh, A., and Yazdani, S.

BIM DESIGN FLOW FOR CONSTRUCTION SITE

MARCO L. TRANI, MANUELE CASSANO, and DAVIDE TODARO

A.B.C. Dept, Politecnico di Milano, Milano, Italy

Construction site planning is an important task in a design team whereas a proper design of construction activities has a great influence in time and cost saving as well as in workers' safety. Designers need to focus on site planning since the preliminary phase of building design in order to point out construction criticalities of the project that can affect execution phase. On the contrary, a scarce attention to operational issues during design phase is detectable in industry. Despite of the rapid diffusion of Building Information Modeling (BIM), a lack of BIM tools specifically studied for site design is detectable. Then the purpose of this paper is to describe a workflow developed with the aim to integrate site planning in a BIM design process, gaining thus operational advantages. The development of the workflow follows designers' needs collected through the active participation of the authors in several design teams. A strong coordination and information sharing with other disciplines is the main task of the workflow, so that design choices can take into consideration operational issues as well. In order to simplify and accelerate the tasks of the site designer, the system is based on the use of some tools (such as databases of construction site elements and automatic checking tools), specifically developed for the purpose. The satisfaction of the designers involved demonstrates the efficiency of the system, in particular concerning the decrease of design time and the assistance of design choices thanks to operational information management.

Keywords: Building information modeling, Construction safety, Design optimization, Operational information, Construction simulation.

1 INTRODUCTION

The planning of construction site activities since the early design phases has a great impact on the success of the subsequent execution phase (Hare *et al.* 2006). In particular a preliminary construction site design has the main task to point out construction criticalities of the project that can affect execution phase in term of time, cost and safety loss. If studied during the whole design phase, these criticalities may be solved before affecting construction performances or, if not possible, faced with the right awareness during works. In order to reach this goal a construction site designer has to join the design team from the early stages of the project. In fact all designers have to strongly collaborate, not only to avoid design errors but also to guarantee the real feasibility of the designed works. Only a deep information exchange (between the designers of the building elements and the designer of the works to put in place these elements) can guarantee the fulfillment of such an objective. Regarding in particular safety issues, this collaboration is underlined also by legislation about safety (e.g., European directive 92/57/EEC and related national versions). On the contrary, a scarce attention to operational issues during design phase is diffused in the construction industry. Nowadays Building Information Modeling (BIM) is certainly the most popular method to assure collaborative design. Its quick diffusion and its well-known advantages make it a powerful tool also for construction site design, since the possibility to collect and share information in a unique platform. Nevertheless, a lack of BIM tools specifically studied for construction site design is detectable among BIM software. For this reason, the ongoing research aims to develop a precise workflow to integrate site design in a BIM design process. The site design becomes thus a part of design workflow since the preliminary phase by the development of a proper Construction site information Model (CoSIM).

2 BIM AND CONSTRUCTION SITE LAYOUT PLANNING

Since many years BIM has been studied as a collaborative tool able to optimize the building design process. Lot of research developed BIM tools in different disciplines and some studies concern particularly the site layout planning. Vimonsatit *et al.* (2014) underline the possibility to enhance site planning thanks to simulations and BIM that shows available spaces throughout the construction stage. Lot of the researches that concerns construction sites are indeed focused on construction scheduling (with the implementation of 4D models) and safety planning. In particular from many years researches on 4D scheduling have been carried out starting from 4D CAD tools (Zhou *et al.* 2012) to 4D BIM and use of data to generate schedules (Kim *et al.* 2013). In general the main goal is to facilitate safety planning production (Azhar *et al.* 2012) ensuring a detail level suitable for hazard identification before starting construction. The VTT Technical Research Centre of Finland studied deeply BIM tools for construction site layout and safety planning focusing on the visualization potentiality of BIM for the managing and communicating safety plans by BIM (Kiviniemi *et al.* 2011). Such a research focused also the need of a creation of objects libraries of temporary structures and equipment (Sulankivi *et al.* 2009). The researches on the field focused on single aspects of construction planning (safety, scheduling, costs, etc.). The presented research aims to go ahead single applications and give an overview of the design process by showing design steps that a site designer need to follow according to the development of the project.

3 CONSTRUCTION SITE INFORMATION MODELING

The development of the method focused on two main research environments. The first one concerns the deep study of the actual standards for building design. Furthermore, in order to guarantee the usefulness of the method, the authors collaborated in several design team and got the needs specifically expressed by the designers involved. Moreover, the created workflow follows precise design steps in order to guarantee the integration with other disciplines during the whole design development. In particular the method takes into consideration three main phases of the design to be carried on before the tender phase (Draft, Developed and Detailed design). Such a simple and schematized model is taken into account for the development of the CoSIM workflow in order to comply as much as possible both the design tendencies of those countries that still not have BIM regulations and some of available BIM standards (e.g., UK PAS

1192-2:2013). Each phase has the aim of complying, as much as possible, the needs of all designers in order to gain an integrated model. A particular attention is given to the information contained in the construction model and the relationships between these information and the design information contained in the building model. The result of the whole workflow is a model ready for the tender and, then, for the chosen contractor that can properly implement it in the execution design.

3.1 Concept Pre-Design

The concept design starts with the preliminary draft of the intervention and shows, in particular, the first ideas of the construction by the implementation of surfaces, volumes and shapes. In this first design step the site designers has in particular the main tasks of the modeling of the Construction site master plan around the volumes. Then is possible to define the first shape of the construction site and to introduce a first brief of the logistic plan in order to verify the available spaces and the surrounding situation. So the concept "Site & Surrounding model" has the principal aims of underline the needs in term of construction spaces and evaluate the possible impact of the adjacent constraints. The constraints founded need to be underlined and to be subject to a proper risk assessment in order to evaluate the possible solution to be carried on (e.g., elimination of the constraints, protection, bypassing etc.). Concept phase, as the following phases, has to be concluded by a final checking between the different models created by different actors in order to avoid possible interferences.

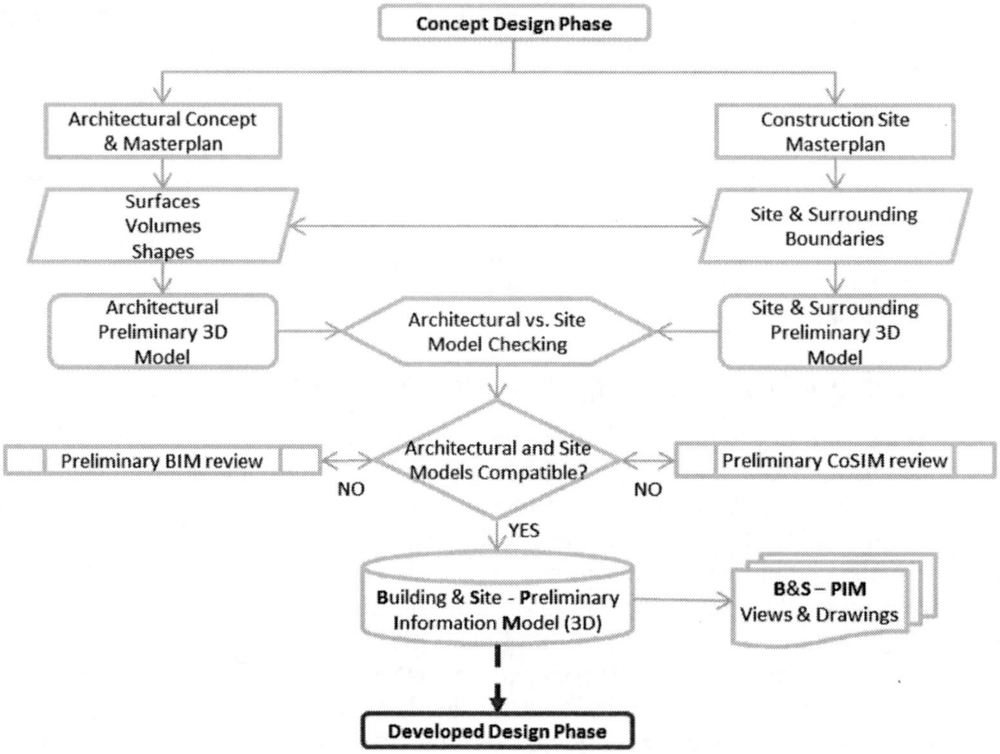

Figure 1. Concept pre-design workflow.

3.2 Developed Pre-Design

In the developed design phase the main task is the design of the construction site logistic plan. In this phase the designers insert information about the singular building systems and their technical solutions. This way the intervention becomes clear and the site designer can focus on these elements by developing a proper construction breakdown structure. Through this he recognize the works to be carried out and it is now able to check the building operational information (e.g., the weight of a structural element) with the site information (the load of the crane) in order to choose the right element to put in place. Hence, the building system technical solutions are placed, side by side, near the Site System technical solutions in order to build the Developed Construction site Information Model. Substantially it represents the construction site layout in the different main phases of the construction process. The developed model is thus characterized by the insertion of an adequate number of site elements completed with their information useful to be matched with building operation information for the choice of the site element itself as well as its position in the site logistic. This issue underlines, in particular in this phase and in the next, the need of a proper library of site elements, not yet so diffused and complete in term of graphical and information details. As in the concept phase, also in developed phase the CoSIM needs to be checked with the others in order to not compromise the rightness of the project. The Building & Site – System Information Model is the result of the developed design and represents a dimensionally correct and coordinated model (PAS 1192-2:2013) able to be submitted for authority permissions.

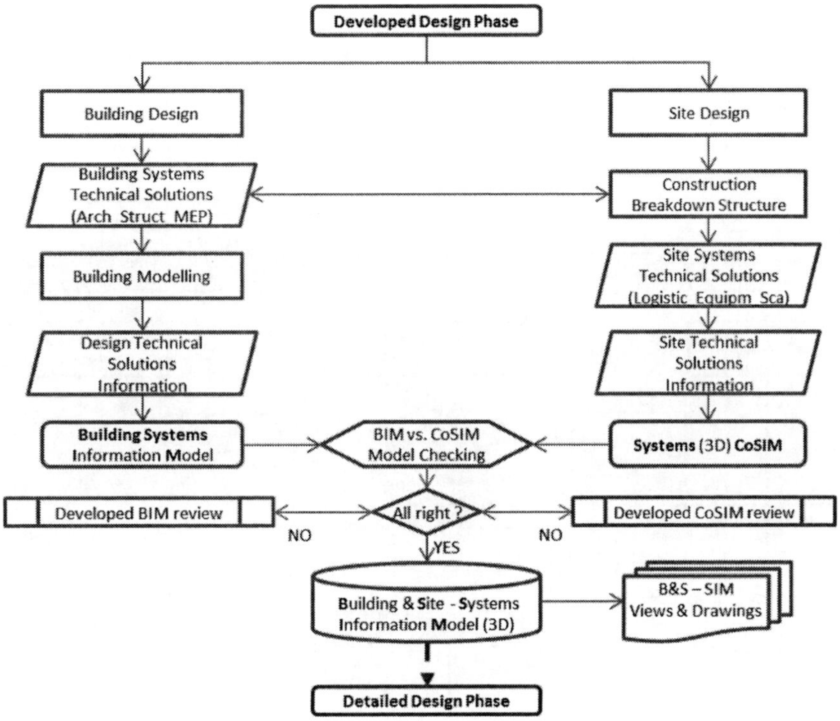

Figure 2. Developed pre-design workflow.

3.3 Detailed Pre-Design

The detailed design phase has the aim to create the final model to be used for the tender. The main task for the site designer in this phase is to have a deep information exchange (through the model) with the other designers about the detailed technological choices. Thanks to these information about the building, the site designer has to manage the operational information in order to evaluate each single work to be carried out and produce a 4D work plan. To do this he has to add to the model further data concerning resources such as work force and unit times. All these information about the single works permit to design each work-zone by defining minimum spaces, temporary structures and possible equipment needed. Therefore the 4D Work plan permits to verify the need of some contemporary works and to assess the risk of interference between these works. The risk assessment permits to solve previously the interferences between works by changing the site design solution (e.g., adding different safety devices) or modifying the work plan with a time shift of the interferential works. Through this, the site designer is able to previously verify the feasibility of the works in a precise context (environment, expected time and costs, safety, etc.) and to give back to the design team information for changing the design solutions if it can create problems to the success of the project. After the tender the model is transmitted to the contractor that becomes the principal actor of design and has to produce the construction model. It represent the final model able to manage construction and has to contain all the elements (both of the building and of the construction site) that will really be used for construction.

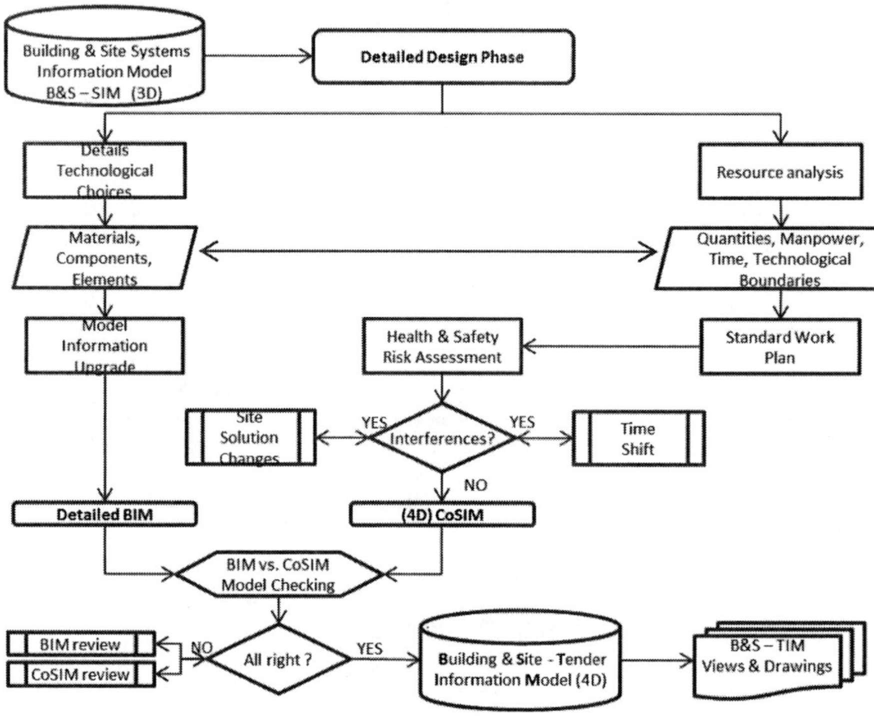

Figure 3. Detailed pre-design workflow.

4 DISCUSSION AND CONCLUSIONS

The development of the method follows the needs of some designers involved in real design contexts. Thanks to these experiences we made many tests about the efficiency of BIM tools for construction site design in this environment, finding both strength points and weaknesses. Obviously the availability of information in a model means efficient information exchange decreasing design time. However, especially regarding site design, BIM tools are not yet properly developed. In fact, in order to efficiently manage the site design, we created a library of BIM elements typical of construction site (equipment, temporary structures etc.) containing the needed info for design choices (Trani *et al.* 2014). Furthermore we store all the studied elements in a database equipped of information searching and checking tools in order to make more efficient and fast information exchange (Trani *et al.* 2015). However there are many elements to be developed in order to make the method more efficient. In fact the development of the library and the database is still in progress. Both the research among the datasheets and the modeling of the elements require an amount of time not compatible with design time. For this reason, finished libraries and databases are required in order to manage faster the whole process. Anyway the results encourage the task group to continue the improvements of the CoSIM method in order better manage construction projects thanks to an increased consciousness of operational issues in the design phase. The path traced is still long and must take into consideration the principal need to introduce BIM among construction professionals and companies. The CoSIM should be an efficient method in a BIM context with the aim of better managing and automating some choice, with the think that must be used not as the solution of any problem but as a design support for decision making that remains, anyway, a designer's task.

References

Azhar, S., Behringer, A., Sattineni, A., Maqsood, T., 2012. BIM for Facilitating Construction Safety Planning and Management at Jobsites, in: *CIB W099 International Conference on Modelling and Building Health and Safety*. pp. 82–92.

Hare, B., Cameron, I., Roy Duff, a., 2006. Exploring the integration of health and safety with pre-construction planning. *Eng. Constr. Archit. Manag. 13*.

Kim, H., Anderson, K., Lee, S., Hildreth, J., 2013. Generating construction schedules through automatic data extraction using open BIM technology. *Autom. In Constr. 35*, 285–295.

Kiviniemi, M., Kähkönen, K., Mäkelä, T., Merivirta, M.-L., Sulankivi, K., 2011. BIM-based Safety Management and Communication for Building Construction, *Vtt Tiedotteita – Research Notes 2597*.

Sulankivi, K., Mäkelä, T., Kiviniemi, M., 2009. BIM-based site layout and safety planning, in: *VTT Symposium (Valtion Teknillinen Tutkimuskeskus)*. pp. 125–140.

Trani, M.L., Cassano, M., Minotti, M., Todaro, D., 2014. Construction site BIM requirements. *Proc. 30th Annu. Assoc. Res. Constr. Manag. Conf. ARCOM 2014*, 663–672.

Trani, M., Bossi, B., Cassano, M., Todaro, D., 2015. BIM oriented equipment choice on construction site, in: *ISEC 2015, Sust. Sol. In Struct. Eng. And Constr.* pp. *1139*–1144.

Vimonsatit, V., Lim, M., 2014. Use of BIM tools for site layout planning, in: *ISEC 2015, Sustainable Solutions in Structural Engineering and Construction*. pp. 755–760.

Zhou, W., Georgakis, P., Heesom, D., Feng, X., 2012. Model-based groupware solution for distributed real-time collaborative 4D planning via teamwork, *Joutnal of Computing in Civil Engineering*. 26 (5). Pp 597-611

Interaction between Theory and Practice in Civil Engineering and Construction
Edited by Komurlu, R., Gurgun, A. P., Singh, A., and Yazdani, S.

MOBILE APPLICATIONS FOR THE CONSTRUCTION INDUSTRY

BAHADIR V. BARBAROSOGLU and DAVID ARDITI

Dept of Civil, Architectural, and Environmental Engineering, Illinois Institute of Technology, Chicago, IL, USA

The construction industry has been criticized as an "old-school" industry, because of being a slow adopter of mobile technologies. This has changed recently. A 2014 survey of 1,048 participants found that 72% of construction practitioners have smartphones, and use smartphone applications for work purposes. While there are thousands of smartphone applications advertised as "construction apps", the most popular smartphone applications offered by software providers to the construction industry are for field data collection, project management, bidding, building information modeling (BIM), accounting, customer relationship management, and estimating. This paper discusses the current state of smartphone applications available to the construction industry, and examines the apps' functions. Due to the rapid growth in the smartphone applications market, new applications become available every day for use in different industries. Given the large number of choices, both companies and individuals in the construction industry must beware when selecting and purchasing smartphone applications. The business needs of the potential users and the expectations from the applications must be well identified, and the selection must be made accordingly. Properly selecting and deploying smartphone applications for construction-related tasks is expected to improve communication, enhance workflow with real time information, and increase productivity.

Keywords: Mobile apps, Smartphone apps, Construction tasks, Computerization, Construction productivity.

1 INTRODUCTION

Smartphones and mobile applications that run on them have changed the world and have become a part of everyday life. As of July 2015, the total number of mobile applications available in leading mobile application stores (e.g., Google Play, Apple App Store, Amazon Appstore, Windows Phone Store, and BlackBerry World) is almost four million. The applications cover many fields and can execute a variety of tasks, but they have one main purpose, and that is to deliver a service (Krouse 2012).

According to *Engineering News Record* (ENR) (2011), mobile devices are also changing construction management. Construction professionals use tablets and smartphones to increase job-site efficiency. Both tablets and smartphones combine mobile accessibility with mobile construction management applications, with the added advantage for smartphones that they can fit in a pocket and provide the same benefits. The main advantage gained from the use of such mobile devices is that they enable construction professionals to work interactively and dynamically; data collected from the site is shared in real time among the project participants with visual attachments, and site reports are generated with more accurate and up-to-date information.

It is apparent that construction companies need to go mobile in order to stay competitive in the industry. However, it is essential to establish a mobile device strategy before selecting a mobile device and applications. Since carrying multiple devices is not ideal for construction professionals, the selected applications must be compatible for the device used for work purposes. Different mobile devices with different operating systems limit the choice of applications. In selecting a mobile device, the advantages and disadvantages of its operating systems must be taken into consideration. For example, Apple's iOS is easy to use and learned quickly by many people, whereas no device that uses iOS has an actual keyboard. Most people are already familiar with Windows' operating system, but application support between its different versions is lacking. Android has the biggest market share, operates on many devices available in different sizes, and with native keyboards. However, Android's information security is a concern. It follows that first, companies should prioritize their business needs and expectations from mobile technologies, and then select a mobile device and applications to implement. A successful implementation requires a combination of good applications and sufficient training, regardless of the chosen device (Bulley 2013).

Chen and Kamara (2011) developed a framework to examine how mobile computing technologies can be used in construction sites to exchange on-site information. Three key issues about mobile devices, mobile applications, and wireless networks were stated in the study. First, mobile devices used on site should be equipped with the hardware that allows users to process site information effectively and efficiently. This hardware includes but is not limited to keyboard, touch screen, voice recorder, and camera. Second, mobile applications used for onsite information exchange should have specific functions for different users that have different roles in construction projects. For example, a project manager needs to review drawings, monitor progress, update schedule, and distribute records, whereas a site engineer needs to review drawings and run necessary calculations. Third, the wireless network that the mobile device uses should provide coverage for large geographic areas especially if users work on several construction projects simultaneously.

Currently, construction information exchange is done through traditional information and communication methods that are non-automatic and paper-based. However, having construction information digitalized by using a mobile computing technology that runs automatic information management activities is ideal and much desired. There are three strategies that can be implemented to improve on-site information management from the current state to the desired level; utilization of a commercially available mobile application that can be synchronized with the main information system, utilization of a commercially available mobile application that can transfer data to the main information system wirelessly, and design/development of a specific mobile application that exchanges data with the main information system wirelessly.

2 MOBILE APPLICATIONS IN THE CONSTRUCTION INDUSTRY

Lu *et al.* (2014) investigated the current state of information and communication technologies (ICT) in the architecture, engineering, and construction (AEC) industry through a review of 145 papers that were published in selected journals between 1998 and 2012. Application development was the second most popular research method in the reviewed papers after empirical analysis. Even though the use of mobile applications is gaining popularity in the construction industry, its adoption is a vital process that must be conducted properly. The study showed that there are five elements that affect adoption: the user, the organization, the technology, the project, and the environment. A successful implementation of new technologies requires training and participation of users. Top management executives' involvement in new technologies facilitates the adoption process. The technology adopted by organizations must be easy to use, compatible, reliable, and provide information security. Project type, cost, duration, specifications, and location must also be considered in selecting and adopting technologies. The environment affects the use of mobile technologies indirectly. For example, government regulations, competitive forces, market demand, vendor strategies, professional norms, current status of technologies, and socio-economic condition can affect organizational behavior in selecting and adopting new mobile technologies.

Studies of mobile applications in the construction industry can be grouped into three categories; general construction management, defect/damage management, and equipment management. This section reviews the studies conducted under these three categories.

In recent years, construction companies have become aware of the potential in using mobile applications. Some companies are developing their own mobile applications. The reason behind this shift is that a large amount of information is produced during construction, but this information is not managed efficiently. Chen (2013) claimed that two thirds of the problems encountered in construction projects is related to information and communication problems, and proposed a mobile application prototype that aims to solve these problems in construction management. Five main functions were provided within the application: contract administration, progress control, equipment management, cost control, and quality management. With the given capabilities, it was suggested that using mobile applications in construction management can minimize time-space constraints, reduce time spent in decisions making, provide real-time information, and strengthen competitive advantage.

Construction professionals with different roles require different types of information to manage the construction activities that they are responsible for. Nourbakhsh *et al.* (2012) investigated the information requirements for mobile application use in construction management. The importance of different categories of information was evaluated from the perspective of consultants, contractors, and owners. It was found that owners need information regarding delay records, site instructions, schedule updates, productivity performance, change orders, variation orders, daily reports, and QC/QA problems, whereas contractors need to access schedule updates, violation reports, QC/QA problems, accident reports, productivity information, inspection results, progress photos, and change orders. On the other hand, consultants need design intent and clarification information. In response to these diverse needs,

Nourbakhsh *et al.* (2012) developed a mobile application called Construction Mobile Application (CMA) that was evaluated by a number of construction practitioners, and was found to be efficient, user-friendly, and easy to use.

Kim *et al.* (2013) presented a mobile application for on-site management that allows construction managers and site engineers to monitor construction sites, manage activities, and exchange real-time information. The application was tested in a hospital building project to validate its usefulness. The test showed that the application has great potential to improve the overall performance of on-site management by improving the level of data sharing and communication, reducing travel time, minimizing rework, and improving quality.

Most of the information and communication related activities in the construction industry are time consuming when conducted manually. Chen *et al.* (2013) stated that it takes time to obtain an organized and complete report of visual inspection of damaged structures because engineers capture photos and gather information on site individually, and because the detection and quantification of the damage is measured through cumbersome manual approaches. Chen *et al.* (2013) developed a mobile application called Collaborative Mobile-Cloud Computing (CMCC) to improve the current practice in team-based visual inspection of civil structures. In a field based experiment, it was found that CMCC can be used for post-disaster and routine inspection of civil structures in an environment that requires collaboration, imaging, and real time information. In a parallel effort, Walker *et al.* (2013) developed a mobile application called Mobile Information Collection Application (MICA) whose purpose was to provide data collection and organization support in infrastructure assessment, and to eliminate unorganized and paper-based documentation. The application was used by sixty inspection teams that investigated the Mississippi River flood in 2011. It was found that the teams that used the application operated the inspection more efficiently and effectively than other teams. The application allowed the user to capture visual data and location information and share it with the command center in almost real time. The study showed that the use of MICA saves travel hours, increases productivity and accuracy of information, and improves safety.

A smartphone provides a constant feed of information among people that interact with each other on a daily basis. Using a smartphone for work purposes is a decision that many contractors are considering today. The potential benefits of using smartphones on site outweigh the costs of user training, purchase of mobile devices and applications. One of the potential benefits is that a construction manager can monitor job-site productivity, and take necessary actions to maximize efficiency. Built-in cameras allow project participants to exchange site data with photos and videos which, in some cases, are much more detailed and informing than a conversation. Real time job site data with visual attachments can even increase a contractor's credibility with the owner. With a variety of selections, construction companies and professionals can find commercially available applications that can assist with day-to-day operations, increase productivity, credibility, and cost efficiency (Yunorich 2011).

As stated, development of a specific mobile application is not the only method that can be used to improve productivity and efficiency in the construction industry. There are several mobile applications that are commercially available and already being used by construction companies. Table 1 shows the most commonly used mobile applications in the construction industry that are commercially available in Google Play and Apple App store.

Table 1. Commercially available mobile applications for the construction industry.

Specific Use	Application Name	Platform	Cost
Bid Management	SmartBidNet	iOS	Subscription Required
BIM	Autodesk BIM360 Field	iOS	Subscription Required
	BIMx	iOS/Android	Free
CAD, Design & Drawings	CAD Touch	iOS	Free
	Finger CAD	iOS	$5.99
	REVITKeys	iOS	$0.99
	iRhino 3D	iOS	$3.99
	AutoCAD 360	iOS/Andorid	Free
	DAKO PRO Civil Eng.	Android	$9.99
	Rilievo	Android	$4.08
	AndCAD	Android	$34.99
Calculator	Construction Master Pro	iOS/Android	$24.99
Contract Management	Contract Maker Pro	iOS	$4.99
	Contract Maker Elite	iOS	$19.99
Document Viewing	Drawvis	iOS	Free
	iBlueprint	iOS	$0.99
Estimating	Quick Service Estimates	iOS	Free
	Contractors Estimating	iOS	Free
	Construction Cost Estimator	iOS	$19.99
	Ultimate Estimator	Android	$4.99
	A Estimate All Pro	Android	$3.99
LEED	Leed BDC Flashcards	iOS	$19.99
Project Management	Procore	iOS	Subscription Required
	OnSite Punchlist	iOS	Free
	Construction Manager	iOS	Free
	ArchiReport	iOS	Free
	Safety	iOS	Free
	Safety Meeting	iOS/Android	Free
	Wrike	Android	Free
Scheduling	Project Plan 365	Android	Free
	Project Planning Pro	iOS/Android	Free

The studies cited in this paper demonstrate the important role of mobile applications in the construction industry, and validate that using smartphones and mobile applications can increase productivity, efficiency and information accuracy, decrease travel time and costs due to rework.

3 CONCLUSION

The construction industry has realized the potential in using mobile devices and applications for work purposes. Several studies conducted recently validated that using smartphone applications for construction-related tasks can improve the level of communication, enhance workflow with real time information and increase productivity. There are two methods that are most commonly used in adopting mobile technologies in terms of mobile applications; development of a specific application for identified needs, and utilization of a commercially available application corresponding to the identified needs. However, regardless of the method chosen, establishing a mobile strategy is of vital importance for a successful implementation. Before making the decision of investing in a mobile application, the roles of the construction practitioners that are going to use the application must be identified and the expectations from the mobile application regarding its functions must be determined accordingly.

References

Bulley, D. Finding Your Mobile Device Strategy. *Contractor Magazine*, 60(2), 34, 2013.

Chen, X. Web Service Based Mobile Worker Supporting System for Construction Industry Applications. *Telecommunication Systems*, 54(3), 277-286, 2013.

Chen, Y., & Kamara, J. M. A framework for using mobile computing for information management on construction sites. *Automation in Construction*, 20(7), 776-788, 2011.

Chen, Z., Chen, J., Shen, F., & Lee, Y. Collaborative mobile-cloud computing for civil infrastructure condition inspection. *Journal of Computing in Civil Engineering*, 04014066, 2013.

ENR: Engineering News-Record. How Tablets Are Changing Construction Management, 267(9), 6, 2011.

Kim, C., Park, T., Lim, H., & Kim, H. On-site construction management using mobile computing technology. *Automation in construction*, 35, 415-423, 2013

Krouse, A. iPads, iPhones, Androids, and Smartphones: FDA Regulation of Mobile Phone Applications as Medical Devices. *Ind. Health L. Rev.*, 9, 731, 2012.

Lu, Y., Li, Y., Skibniewski, M., Wu, Z., Wang, R., & Le, Y. Information and Communication Technology Applications in Architecture, Engineering, and Construction Organizations: A 15-Year Review. *Journal of Management in Engineering*, 2014.

Nourbakhsh, M., Mohamad Zin, R., Irizarry, J., Zolfagharian, S., & Gheisari, M. Mobile application prototype for on-site information management in construction industry. Engineering, Construction and Architectural Management, 19(5), 474-494, 2012.

Walker, R. S., Pettitt, J. A., Scruggs, K. T., and Mlakar, P. F. Data collection and organization by smartphone for infrastructure assessment. *Journal of Infrastructure Systems*, 20(1), 06013001, 2013.

Interaction between Theory and Practice in Civil Engineering and Construction
Edited by Komurlu, R., Gurgun, A. P., Singh, A., and Yazdani, S.

THE TECHNICAL AND ECONOMIC FEASIBILITY STUDY OF USING SMART PHONES IN CONSTRUCTION SITE MANAGEMENT

HAMID REZA SAFA[1] and MEHDI RAVANSHADNIA[2]

[1]*Dept of Civil Engineering, Islamic Azad University-Science and Research Branch-Arak, Arak, Iran*

[2]*Construction Engineering and Management Group, Islamic Azad University-Science and Research Branch-Tehran, Tehran, Iran*

Construction project management is the application of knowledge, skills, tools and techniques related to project activities achieving the project requirements. Accurate and timely acquisition of the site data and information could improve the project management decision processes. However, obtaining the data and information in minimum possible time is the complex process due to nature of the construction site condition. There are several limitations with regards to the site data acquisition including, monitoring construction site, and sharing a daily status report. This study investigates the potential of using smartphone as a new technology in construction sites: (1) to improve monitoring and controlling the site activities; (2) to accelerate decisions making process by improving the speed and accuracy of information exchange process. This study provides an overview of the feasibility of using smartphones in three areas of project management knowledge, discussed in a Guide to the Project Management Body of Knowledge (PMBOK), including time management, quality management and human resources management. Validation was conducted through the real case study based on the proposed scenarios. Return of investment and sensitivity analysis of the two scenarios were also compared to find the efficient scenario. The results of this study can be applied to improve the monitoring and understanding of the site status and can potentially facilitate the decision making process.

Keywords: Construction projects, Project management, Quality management, Time management.

1 INTRODUCTION AND RESEARCH REVIEW

On-site construction management is an important component for the execution of successful construction megaprojects. Accurate and timely acquisition of on-site data and information about construction activities and resources facilitate decision making process, which entails productivity improvement. It is always challenging for site engineers/managers to collect and analyze the site information such as locations of materials, labor, and equipment along with the conventional status of progress. These challenges necessitate the development of methodology and using new technologies, with suitable sensing and communication capabilities to acquire and exchange construction information.

In recent years, several research studies have been conducted to investigate the feasibility of using mobile computing. For example, Son *et al.* (2002) extended the concept of technology acceptance model to investigate the factors that influence successful application of mobile computing devices in the construction industry. Another study by Chen and Kamara (2011) introduced a framework for the implementation of mobile computing on construction sites, comprising an application of a technical model.

Other studies investigated the process improvement in construction site. For instance, Pena-Mora and Dwivedi (2002) presented a collaborative management platform that enables project participants in different locations to share project information using Smartphones. Bowden *et al.* (2006) developed a vision to illustrate to industry professionals how the use of mobile information technology can improve construction processes. Wang (2008) proposed an RFID-based quality management system for monitoring and sharing quality data.

There are several related studies in other industries that can be applied in construction management such as location tracking, sharing data, and facilitating disaster response. For example, Pena-Mora *et al.* (2010) presented an information-technology-based collaboration framework to facilitate disaster response operations. The framework incorporated a web collaboration service, RFID tags, a building black box system, a geo-database, and a Geographic Information System (GIS). Behzadan *et al.* (2008) argued that a mobile worker's spatial context must be continuously tracked in both outdoor and indoor environments for a location tracking system to effectively support construction projects. Song *et al.* (2005) use of RFID and GPS to determine the location of RFID tags to track materials at the construction site. Ko (2005) also used an algorithm to develop a concept of trilateration with 4 readers and tags for three-dimension positions.

The previous studies showed that Smartphone-technology have great potential to considerably improve construction activities, including labor and material tracking, defect management, and progress monitoring.

This study attempted to improve the on-site management system by using smartphone-technology. Through the integration of location and construction site information, the proposed system would enable site engineers and managers to efficiently understand and find the location of work tasks and resources. The proposed methodology has a potential to improve the work efficiency and reduce cost and time of information transfer. Major features of the system are discussed in this paper. A real case study is also presented in order to validate the applicability.

2 METHODOLOGY DEVELOPMENT

Smartphones are the backbones of the proposed model. Several challenges should be considered in developing this model including, transferring, sharing, and collecting data. Hence, three scenarios have been defined; first scenario includes the conventional practice. The advantage of using smartphone and the potential of the technology has been investigated in the second scenario. The use of smartphones technology and other technologies that accompanying with the smartphones technology is investigated in the third scenario. Figure 1 illustrates the information flow in construction site. In the following sections, three scenarios will be discussed.

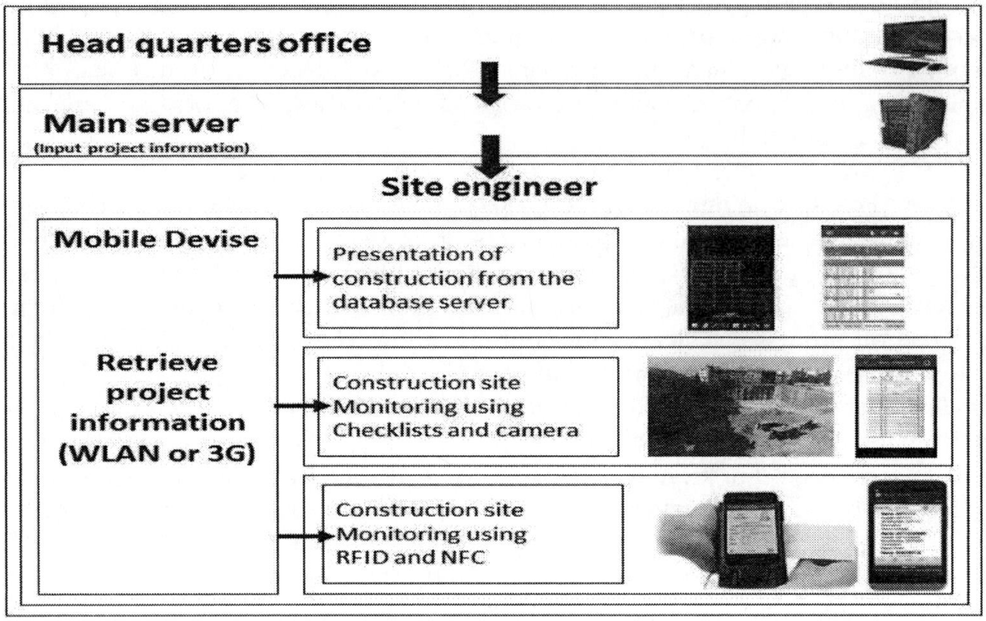

Figure 1. Data flow of the site monitoring.

2.1 Scenario 1: Conventional Practice

Understanding the current status of construction project is quite essential for construction engineers to accomplish successful on-site management. In this scenario all of information will be collected through visual surveying process and visual estimates. The information is recorded on paper, which is sometimes incorporate, time consuming, inaccurate, and difficult for data transferring.

2.2 Scenario 2: Using Smartphones Technology

In the second Scenario, Smart phones are the main part of the model; the sensors and applications of smartphones have been investigated in order to obtain, record, and transfer information in the site. In this scenario, maps, checklists, and timing plans are kept in smart phones. Therefore, it is not necessary to carry paper in order to gather, record, and control information. In addition, it is possible to control human forces, and also to do positioning through GPS. In this scenario, The speed of control process is higher than Scenario 1.

2.3 Scenario 3: Using Smartphone and the other Technologies that they operate With

In third scenario, smart phones are also the main part of the model. In this scenario other technologies, which can cooperate with smart phones are utilized, NFC and RFID are good examples. In Scenario 3, many processes can be done in a semi-automatic manner. These processes include: control of the project's physical progression, internal and external positioning of the site, and warning employees and engineers about risks.

3 CASE STUDY

The case study is an official-commercial project in the city of Tehran. Constructing of this project will take 5 years, the cost estimation is about 100,000,000 USD. We are in the third year of this project. Specifications of Smartphones that use are at a price of 399 dollars, which are capable of 3G and 4G communications, with 13-mega-pixel cameras, are used in this model. The other capabilities of these smart phones include the ability to support Microsoft Office files, and the ability to run PDF files. These phones are equipped with Wi-Fi, GPRS, GPS, and bluetooth.

In this section the cost of each scenario has been estimated. In the first scenario, the costs included the cost of human resource and the fund of the project. In the second scenario, the costs included the cost of smartphone, software, maintenance, and human resource. And lastly, costs of third scenario are all of the costs combined from the first two scenarios (Table 1). As mentioned, this study defined first scenario to clarify the cost of scenarios, therefore the second and third scenario can be compared to the first.

Table 1. Annual cost of scenarios.

	The initial cost $ US	Year 1 $ US	Year 2 $ US	Year 3 $ US	Year 4 $ US	Year 5 $ US
Scenario1	0	25235	33176	39882	46588	52941
Scenario2	4560	18940	25205	30235	35264	40294
Scenario3	10530	19530	25795	30823	35852	40882

After estimating costs, this study has initially determined the internal rate of return for state 1 (the comparison of scenarios 1 and 2 is observable in table 2) and state 2 (the comparison of scenarios 1 and 3 is observable in table 3) in order to assess the design economically. The reason for comparison scenarios 2 and 3 with scenario 1 is to figure out how much benefit, in proportion to the previous state (scenario 1), is gained for the project though the implementation of scenarios 2 and 3. This rate in state 1 is 150 percent, and in state 2, it is 62 percent. The initial Investments and annual Costs in state 1 is less that state 2. In addition, revenues from the implementation of Scenario 2 is more than Scenario 3. These factors cause of IRR differences in state 1 and state 2.

Table 2. State 1.

	The initial cost $ US	Year 1	Year 2	Year 3	Year 4	Year 5
Investment	4560	0	0	0	0	0
Cost	0	411	500	588	676	764
Benefit	0	6300	7970	9650	11320	12650

Table 3. State 2.

	The initial cost $ US	Year 1	Year 2	Year 3	Year 4	Year 5
Investment	10530	0	0	0	0	0
Cost	0	1000	1088	1176	1265	1352
Benefit	0	5700	7382	9060	10735	12060

Then the sensitivity of the internal rate of return of the three items of "Invest, Cost, and Benefit" is examined within changes from 20 percent to – 20 percent, in order to figure out what changes in percentages will be seen in each item of the internal rate of return. The analysis of the sensitivity of the internal rate of return in states 1 and 2 is observable in Fig 2.

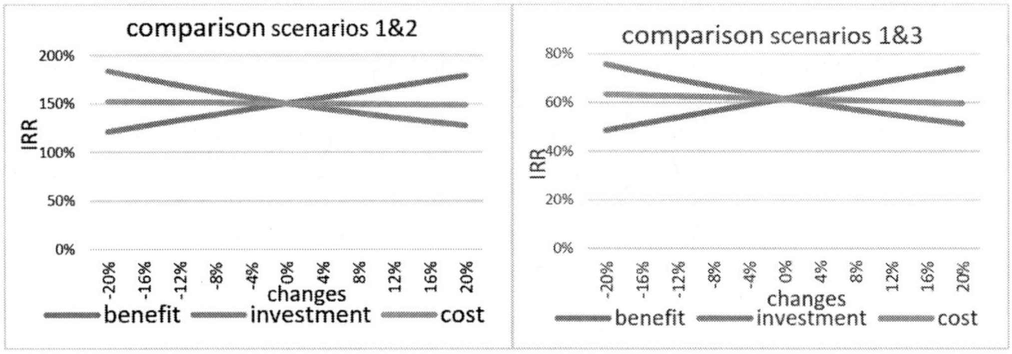

Figure 2. Sensitivity analysis regarding investment, cost, and benefit.

Delays in civil projects are common happenings. Therefore, the internal rate of return has been examined considering the construction period. The internal rates of return have been presented in Figure 3 according to different construction period between 5 to 10 years. Another item which is important in this project is Smartphone costs. Considering the characteristics of different projects, different mobile phones with different prices might be needed. Thus, the internal rates of return have been presented in Figure 4 according to different smart phone costs.

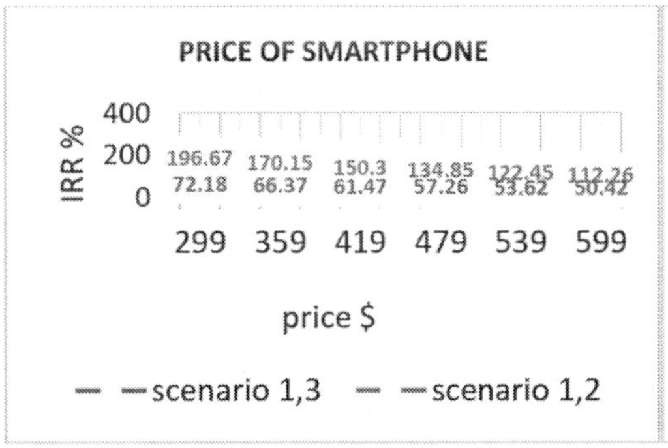

Figure 3. Sensitivity analysis regarding years of operation.

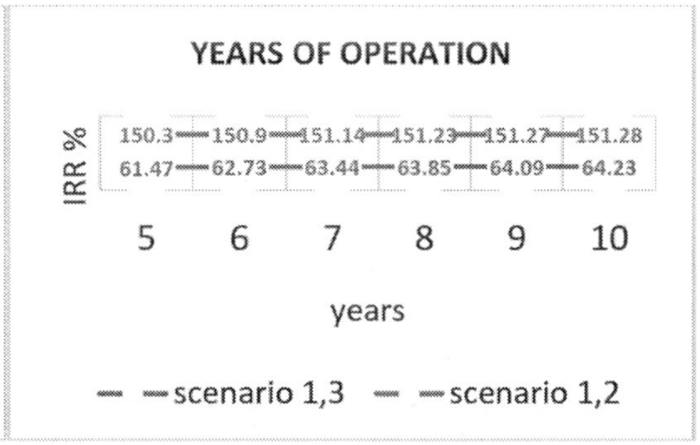

Figure 4. Sensitivity analysis regarding price of smartphone.

Considering the percentage of the internal rate of return, the plan is applicable. But the plan that must be selected is the one which is effective both economically and technically. Second scenario is more economic comparison with the third one; however, there is more capability in the third scenario. Also more money should be spent in scenario 3. Regarding Above tables the changes of IRR sharply increase or decrease because of the role of smartphones, in the event that the costs of the other technologies effect on the third scenario.

4 CONCLUSIONS AND DISCUSSION

Using smartphone-technology has provided a flexible and powerful environment for project managers and experts by facilitating on-site construction management processes. This study investigate feasibility of using smartphones in three areas of project management knowledge, discussed in PMBOK, (1) time management; (2) quality management, and (3) human resources management. The first developed scenario includes the current practice to on-site construction management. The potential of using smartphone has been discussed in second scenario. The third scenario investigated the use of smartphones and some related technologies that can be employed with this technology.

A case study, a commercial construction building, was investigated to validate the proposed model. The results of this study shows that the reduction in travel time in visiting a field office, for acquiring project information, is one of the benefits of this model. The results illustrates that the quality of a construction project can be improved using the proposed system. This study also investigated sensitivity analysis regarding investment, cost, benefit, sensitivity analysis regarding years of operation, and sensitivity analysis regarding prices of smartphones.

References

Behzadan, A. H., Aziz, Z., Anumba, C. J., and Kamat, V. R., Ubiquitous location tracking for context-specific information delivery on construction sites, *Automation in construction*, 737-748, 2008.

Bowden, S., Dorr, A., Thorpe, T., and Anumba, C., Mobile ICT support for construction process improvement, *automation in construction*, 664-676, 2006.

Chen, Y. and Kamara, J. M., A framework for using mobile computing for information management on construction sites, *Automation in construction*, 776-788, 2011.

Hyun, S. and Feniosky, P., Visualization of Construction Progress monitoring, *international Conference*, Montréal, Canada, June 14-16, 2006.

Ko, C. H., "RFID-3D location sensing algorithms." Automation in Construction. Building Information Modeling and Collaborative Working Environments, 19(5), 588–595, 2010.

Ohmori, N., GPS Mobile Phone-Based Activity Diary Survey, *Proceedings of Eastern Asia Society for Transportation Studies*, Vol. 5, pp. 1104 - 1115, 2005.

Pena-Mora, F. and Dwivedi, G., Multiple device collaborative and real time analysis system for project management in civil engineering, *Journal of Computing in civil Engrg.*,23-38, 2002.

Pena-Mora, F., Chen, Y., Aziz, Z., Soibelman, L., Liu, L.Y., Ei-Rayes, K., and Arboleda, C.A., collaboration framework supporting civil engineering emergency response operation, *journal of Computing in civil engineering*, 302-312, 2010.

Slaughter, S., Models of Construction Innovation, *Journal of Construction Engineering and Management*, ASCE, Reston, Virginia, May/June, Vol. 124, No. 3, pp. 226-231, 1998.

Son, H., Park, Y., Kim, C., and Chou, S., Toward and understanding of construction professionals` acceptance of mobile computing devices in South Korea: an extension of the technology acceptance model, *automation in construction*, 82-90, 2002.

Song, J., Haas, C. T., Caldas, C., and Liapi, K. "Locating materials on construction site using proximity techniques." Proc., *Construction Research Congress,* ASCE, San Diego, 552–557, 2005.

Wang, L. C., Enhancing construction quality inspection and management using RFID technology, *Automation in construction*, 467-479, 2008.

Interaction between Theory and Practice in Civil Engineering and Construction
Edited by Komurlu, R., Gurgun, A. P., Singh, A., and Yazdani, S.

DESIGN OF MOLTEN SALT SHELLS FOR USE IN ENERGY STORAGE AT SOLAR POWER PLANTS

SAMAAN G. LADKANY, WILLIAM G. CULBRETH, and NATHAN LOYD

HRH College of Engineering, University of Nevada Las Vegas, Las Vegas, NV, USA

Design of a steel tank for the storage of excess energy from thermal solar power plants using molten salts (MS) at 580°C is presented. Energy can be stored up to a week in large containers to generate eight hours of electricity for use at night or to reduce weather related fluctuation at solar thermal energy plants. Our research supported by Office of Naval Research (ONR) presents a detailed design of a cylindrical shell for the storage of high temperature molten salts. The storage shell consists of an inner stainless steel layer designed to resist corrosion and an external steel structural layer to contain the large pressures resulting from the molten salt. The cylindrical tank is 54 feet (16.459 meters) high and has an 80 feet (48.768 meters) diameter, with the salt level at a height of 42 feet (12.802 meters). Given the heat of the molten salt and the size of the tank, the design includes a flat shell cover supported on stainless steel columns and a semispherical utility access dome at the center. Considerations are made for the reduction of strength of steel at elevated temperatures. Layers of external insulation materials are used to reduce heat loss in the storage shell. The design presents a posttensioned concrete foundation analysis for the storage tank, which sits on a layer of sand to allow for thermal expansion.

Keywords: Commercial electric station, Energy production, Molten salt tanks, Post tensioned concrete slabs, Solar salts, Steel cylindrical shells.

1 INTRODUCTION

Molten solar salts are a great and effective way to store excess solar energy for future use due to the vast heat storage capacities of solar salts. These solar salts are contained in large insulated tanks in order to keep the molten salts in a closed system. This project examines the current method of using insulated hybrid steel cylindrical shells to store molten salt and presents a preliminary design of real life examples.

2 DESIGN METHODS FOR MS STORAGE TANKS

Currently, molten salt (MS) storage shells are usually cylindrical tanks made of stainless steel. The MS steel tanks has a hybrid design of A588 Carbon Steel and an inner layer of 316 Stainless Steel to protect against corrosion, varying in thickness from one inch (25 mm) for a fifty year plant life span to 0.6 in (15 mm) for a thirty year plant life span.

3 TANK REQUIREMENTS

For this stage of the project research, the tanks need to store enough molten solar salt, which is a 60:40 sodium nitrate ($NaNO_3$) and potassium nitrate (KNO_3) mix, to provide power for a 300 megawatt power plant for eight hours each night. Calculations determined that in order to satisfy these requirements, the two tanks need to be able to store 12,048 cubic meters of salt or 425.5 x 10^3 cubic feet.

In order to determine the total mass of salt required to operate the power plant, one must start with the basic energy equation, which is shown in Equation 1 (Holman 1986).

$$E = P_{thermal} * \Delta t_{storage} = m * c_p * \Delta T \quad (1)$$

In Equation 1 above, E represents the total energy in the system. The power generated by the power plant is $P_{thermal}$, which as stated earlier is 300 megawatts. The required time of storage is $\Delta t_{storage}$, which is 8 hours or 28,800 seconds. The required amount of solar salt needed for the power plant is represented by m. The specific heat capacity of the salt is c_p, which is 1540 joules per kilogram of salt per degree Kelvin. The temperature range of the salt in the system is ΔT, which is calculated using Equation 2 below.

$$\Delta T = T_{salt,max} - (T_{sat} - 20\text{ K}) \quad (2)$$

In Equation 2 above, the maximum temperature of salt in the system, or $T_{salt,max}$, is 853.15 degrees Kelvin. The temperature of the Rankine cycle, or T_{sat}, is 620.55 degrees Kelvin. Equation 2 determined that the temperature range for the salt is 252.6 degrees Kelvin.

In order to determine the required mass of salt, Equation 1 is rearranged into Equation 3 as shown.

$$m = \frac{P_{thermal} * \Delta t_{storage}}{c_p * \Delta T} \quad (3)$$

This determined that the power plant requires 22.88 x 10^6 kilograms of salt, or 50.44 x 10^6 pounds (25,220 tons).

Equation 4 is used to determine the volume of solid salt required.

$$V_{salt} = \frac{m}{\rho_{salt}} \quad (4)$$

Equation 4 determined that the volume of solid salt required is 12,048 cubic meters of salt, or 425.5 x 10^6 cubic feet (12,048 cubic meters). This volume will be divided over two tanks, requiring 212.7 x 10^6 cubic feet (6,024 cubic meters) for each tank. However, a third and fourth tanks, all of carbon steel, are recommended for the storage of cooled MS after power generation and for safety and continued operations during maintenance of the other tanks.

All structural steel used is A588 Grade 50 steel. The cylindrical tank designed with a 40 feet (12.192 meters) radius at the base. This results in a height of salt of 42 feet (12.802 meters) and a salt height for the cylindrical tank.

4 STEEL CYLINDRICAL TANKS

The steel structural design was divided into five elements for individual analysis and design, which are the shell wall, the top cover with a central 10 feet (3.048 meters) diameter steel access dome, support columns, a steel bottom, and the concrete slab below a layer of sand. All of these structural elements are made of structural and stainless steel except the concrete slab. Shell theory was used to perform the structural analysis of the cylindrical tank and central access dome.

The first design performed was for the shell wall. Based on shell theory, axial bending in a cylindrical shell occurs mainly at the base of the wall, at the junction with the ring and base plate, before dissipating further up the wall (Urugal 2009). Further analysis determined that axial bending dissipates nine feet above ground. The first step was to determine the bending in the shell wall. The maximum positive axial bending moment is 4.085 kip-foot/foot, at the bottom of the shell and the maximum negative bending moment is 1.225 kip-foot/foot at a height 2.7 feet above the bottom of the shell. Circumferential moments are equal to the Poisson ratio multiplied by the axial moments. The bottom of the wall contains the maximum circumferential tensile force, which is 153.05 kips per linear foot (klf), which is 2234 kN/m. Tensile membrane force is determined by Equation 5b (Urugal 2009). While maximum axial compressive force, N_x, in the wall at the bottom of the shell is equal to the total dead weight of the shell, top slab, live load and service dome, which is the total weight (W), divided by the circumference of the shell. Equations 5c through 5h are used to determine the bending in the shell wall (Urugal 2009).

$$p = \gamma z \tag{5a}$$

$$N_\theta = pr \tag{5b}$$

$$D = \frac{Et}{12(1-v)} \tag{5c}$$

$$\beta = \sqrt{\frac{\sqrt{1-v^2}}{rt}} \tag{5d}$$

$$C_1 = \frac{\gamma h r^2}{Et} \tag{5e}$$

$$C_2 = \frac{\gamma r^2}{Et}\left(h - \frac{1}{\beta}\right) \tag{5f}$$

$$w = e^{-\beta x}(C_1 \cos \beta x + C_2 \sin \beta x) + \frac{\gamma(h-x)r^2}{Et} \tag{5g}$$

$$M_x = D\frac{d^2 w}{dx^2} \tag{5h}$$

$$M_\theta = vM_x \tag{5i}$$

$$N_x = \frac{W_x}{C} \tag{5j}$$

In determining the applied pressure on the tank from Equation 5a, it is the product of the salt unit weight (γ) and the depth of salt (z) at the specified point. In Equation 5b, p is the applied pressure on the wall and r is the radius of the wall (Urugal 2009).

In Equations 5c through 5h, D, β, C_1, and C_2 are coefficients, E is the Young's Modulus of the shell material, t is thickness of the shell wall, ν is the Poisson's ratio of the shell material, h is the total height of molten salt, w is shell wall deflection at a height of x above ground, and the second derivative of w is used to determine the moment at that point (Urugal 2009). M_x is the axial moment at a height of x above ground, W_x is the weight of the shell including dead and live loads on its top at level above x (Urugal 2009). Figure 1 details the design of the cylindrical shell and the top dome.

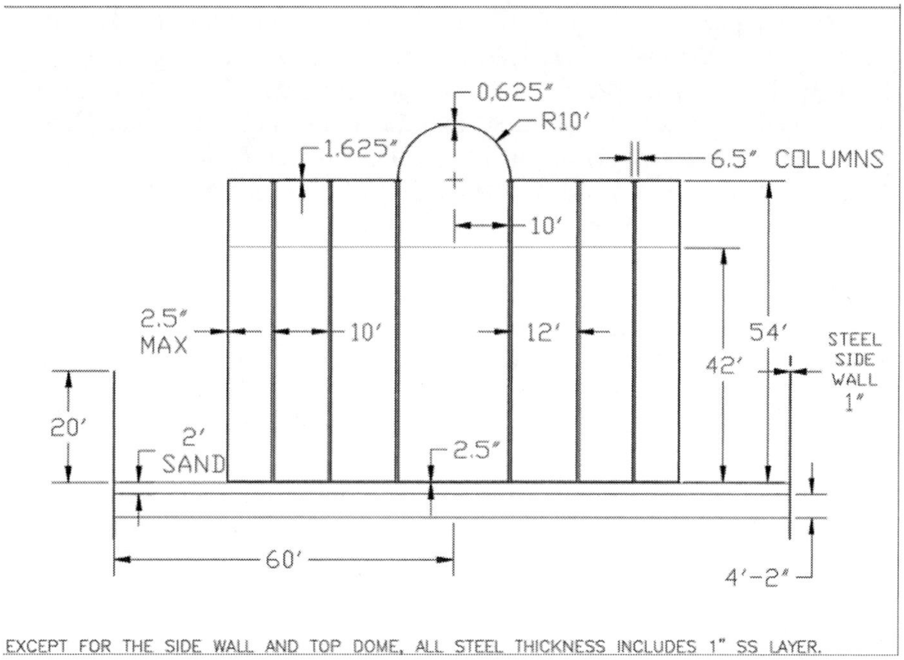

Figure 1. Steel cylindrical shell model including top dome, supporting rows of columns, 2' sand layer, 50" posttension slab, and safety steel walls at the edge.

The shell was designed in sections of varying thickness based on the loading. The bottom nine feet of the shell wall was designed to accommodate excess bending, require 1.5 inches of structural steel thickness due to the combined axial membrane and bending stresses. The next section of the wall, from 9 to 15 feet (2.734 to 4.572 meters) above ground, requires 0.625 inches (15.9 mm) of steel thickness. Starting from 15 feet above ground, the thickness of the shell wall is decreased by 0.125 inches (3.2 mm) every seven feet until a thickness of 0.125 inches (3.2 mm) remain. This results in the wall being 0.5 inches (12.7 mm) thick between 15 and 22 feet (4.572 to 6.706 meters), 0.375 inches (9.5 mm) between 22 and 29 feet (6.706 to 8.839 meters), 0.25 inches (6.4 mm) between 29 and 36 feet (8.839 to 10.973 meters) above ground, and 0.125 inches (3.2 mm) for the remaining portion of the wall above 36 feet (10.973 meters). Due to corrosion effects, a one inch liner of 316 Stainless Steel covers the steel wall.

The next design was for both the top steel plate and the columns supporting it in the cylindrical tank. The top plate is 0.625 inches (15.9 mm) thick and is supported by three circular rows of columns. One row of columns is located ten feet (3.048 meters) away from the center of the tank, at the tip of the opening and the 0.625 inches (15.9) thick service dome. It contains eight equally spaced columns. The second row of columns is located 22 feet (6.706 meters) away from the center of the tank and contains eight equally spaced columns. Lastly, the third row of columns is located 32 feet (9.754 meters) away from center and contains 16 equally spaced columns. These columns are made of carbon steel covered with a layer of stainless steel because of the heat and corrosion from MS. When designing the columns and shell walls, an extra factor of safety is used due to the expected heat of the molten salt. At 580 degrees Celsius, steel is expected to only maintain 60% of its nominal yield strength (Salmon 2009). As a result, the final design load for the first row of columns is 6.5 kips (28.9 kN), 19.6 kips (87.2 kN) for the second row, and 11.7 kips (52.0 kN) for the third row. Ultimately, it is determined that the first row of columns be designed as HSS 4½ x 4½ x 1/8" columns, the second row as HSS 4½ x 4½ x ¼" columns, and the third row as HSS 4½ x 4½ x 1/8" columns (AISC 2012). Due to corrosion effects, a one inch (25.4 mm) liner of 316 Stainless Steel covers the steel column. In addition, the column will be connected to the top steel shell with a 14 inch by 14 inch (356 mm) plate that is two inches thick (50.8 mm).

In order to design for bending in the top steel flat slab, Timoshenko's method was used to design the top plate as a continuous simply supported plate over the edge of the shell and supported by rows of columns as discussed earlier. Moments at the supporting columns are found from the column pattern of annular arrays normalized as rectangular arrays. Based on Timoshenko's (1959) theory, the maximum negative bending moment in each direction is located at the column. With the maximum positive moments, being the radial moments occurring at the center of the normalized annulus, and the maximum circumferential moment, occurring directly halfway between columns. For this shell, the maximum negative moment is 1.785 kip-foot/foot and the maximum positive radial moment is 1.040 kip-foot/foot.

In addition, an opening with a 10 feet (3.048 meters) radius is carved out of the top shell so that a removable steel dome with the same radius can be placed on top of the steel plate. This opening is to allow pipes into the shell and service access into the tank.

5 FOUNDATION DESIGN

A complete design was performed on the concrete slab sitting over dense sand. Included in the foundation design is a 2 feet (610 mm) layer of sand between the tank and the concrete slab as shown in Figure 1 to allow for thermal expansion of the shell. Figure 2 details the radial posttensioning cable layout, the steel ring, and circumferential reinforcement in the concrete slab. The steel ring is necessary because the posttensioning cables cannot intersect with each at the center of the 50 inch concrete slab.

For the slab, 96 radial posttensioning 55/0.5 WG cables that connect to the inner steel ring are required as shown in Figure 2. In addition, six #14 circumferential bars per foot are required under the MS tank, with number of bars decreasing toward the free edge. In addition, the minimum radial posttensioning cables depth is 12.75 inches (324 mm), the maximum radial posttensioning cables depth is 38.75 inches (984 mm), and the circumferential reinforcement depth is 44.125 inches (1.121 meters). This requires a slab thickness of 50 inches (1.270 meters) as shown in Figure 1.

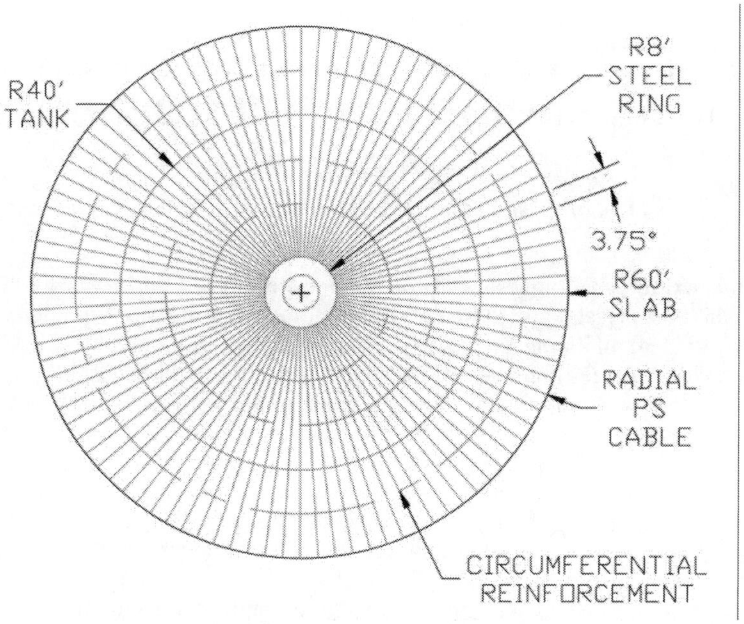

Figure 2. Posttensioning cable and circumferential reinforcement layout for concrete slab including inner steel ring.

6 CONCLUSION

The design of a cylindrical A588 Grade 50 steel shell, having a diameter of 80 feet (24.384 meters), for the storage of molten salts is presented. The shell is 54 feet (16.459 meters) high, has a height of salt of 42 feet (12.802 meters), and has a top access dome with a radius of 10 feet (3.048 meters). The two tank system is designed to store enough molten salt to provide 300 megawatts of power for eight hours. The shell has a one inch (25.4 mm) stainless steel liner to protect against corrosion for a 50 year design life. Also shown is a 120 feet (36.576 meters) diameter concrete foundation with posttensioning, which has a 50 inch (1.270 meters) thickness and steel side walls that are 20 feet (6.048 meters) high for safety in case of an accident.

References

American Institute of Steel Construction (AISC), *Steel Construction Manual*, 14th Ed., 2012.

Holman, J. P., *Heat Transfer, Sixth Edition*. McGraw-Hill. New York, NY, 1986.

Salmon, C. G., Johnson, J. E., and Malhas, F. A., *Steel Structures: Design and Behavior*. 5th Ed. Pearson Prentice Hall, Upper Saddle River, NJ, 2009.

Timoshenko, S., *Theory of Plates and Shells*. McGraw-Hill, New York, 1959.

Urugal, A. C., *Theory of Beams, Plates, and Shells*. 4th Ed. CRC Press, Boca Raton, FL, 2009.

Interaction between Theory and Practice in Civil Engineering and Construction
Edited by Komurlu, R., Gurgun, A. P., Singh, A., and Yazdani, S.

REFUGEE CAMPS AND GATED COMMUNITIES AS A SOLUTION FOR HOUSING

BÜŞRA ÖZAYDIN ÇAT

Dept of Architecture, Kocaeli University, Kocaeli, Turkey

Today the World has a biggest crisis of refugee since The World War II. Refugee is a person who is depressed due to his/her religion, race and ideas or who defect to another country with fear of being oppressed. The refugee camps are high intensity places which provide refugees housing and other social and physical needs. On the other hand today in the capitalist and global cities the most important places for housing are gated communities. The scope of this study is to examine the social and physical similarities of refugee camps and gated communities. Within this framework when we look at some definitions of the concept of gated community, we can see the imitation of refugee camp. In this study, firstly the concept of housing/dwelling and the concept of security which is the most important reason of emerging of gated communities and refugee camps will be analyzed. Then physical and social resemblances of gated communities and refugee camps will be examined. For identifying physical similarities being surrounded by wall or fence, location of the gated communities and refugee camps in the city, their outbuildings like market, pharmacy and their intensity will be analyzed. For social similarities the sense of belonging of refugees and residents and their relations with city will be examined. The results of these will be summarized and evaluated.

Keywords: Security, Intensity, Outbuildings, Sense of belonging.

1 GENERAL APPEARANCE

The largest refugee crisis since World War II is taking place in today's world. Most of the experts consider this era as a "refugee era". Refugee is someone who defected to another country because of persecution or fear of persecution in his home country due to religion, race, and membership of social groups or political thoughts. According to the last report published by The United Nations High Commissioner for Refugees, about 60 million people across the world are forced to move away from their homes. While the number of displaced people was 37.5 million in 2003, in 2013, 51.2 million and in 2014 , 59.5 million (UNCHR 2016). The increase between 2014 and 2015 is the highest per one single year. The most important reason of this increase is the war and crisis in Syria. According to UNCHR, the number of displaced people in Syria is 4,597,436. Together with displaced people within the country, this value reaches to 7.6 million. Due to the close proximity to Syria, there are 2,503,549 registered Syrian people in Turkey (UNCHR 2016). Turkey has signed the Geneva Convention with a geographical limitation in 1951. Due to the number of asylum seekers/refugee, Turkey has an important place in this regard.

Refugee camp is the living space of people who are displaced from their homes as a result of political and social conflict according to the international refugee law. Refugee camps are established as a tent city or a container city across the world. Tan (2015), refugee camps in Turkey are investigated in two parts: One of them is official camps which are managed by AFAD and hosted people from Syria and Kobane. The other one is the empty spaces which are used as camps such as schools, bus terminals, indoor sport halls, factories or warehouses. In general, designed refugee camps are surrounded by barbed wire or wall. Refugee camps meet the requirements of education, health, hygiene and some social needs of the residents in addition to the requirement of shelter. Refugee camps, being are taking placed inside or outside a city, have different types of relationships with the city. Refugee camps, which are economic burden for the country, are mostly founded on narrow lands which causes increased of density. On the other hand, refugees can go outside of the camp only under certain circumstances. This situation causes limited interaction between refugees and the city.

Agamben (2002) describes camp concept as an exception. Tan (2015) said that, this exception status is seen in close sites, dissociated neighborhoods named as ghetto and urban renewal neighborhoods. Therefore, these parts of cities can be compared with the refugee camps. In this context, gated communities can be considered.

Gated communities are the public fields that customize as residential area and have controlled entrance (Blakely and Snyder 1997, Low 2003). According to Ozgur's definition (2006), gated communities are the residential areas which are formatted as a predefined life style and offer some recreational facilities. In this context, it can be said that gated communities cause a security-based kind of decomposition.

The most important definition, which is the starting point of this study, is that gated communities are the modern and desirable refugee camps (Diken 2004). Tan's (2015) definition for refugee camps is also valid for gated communities which is that a designed village, city or neighborhood in small land in order to meet the needs of a community. The housing need is in the center of these two types of construction. In this context, the scope of the study is to examine primarily the concept design of housing.

It is known that there is a danger of legitimate override about refugee camps by producing a discourse about them. However the aim of this study is not analyzing the legitimacy the refugee camps, but to question the constitutionalism of gated communities by comparing them with the refugee camps. On the other hand, the transformation of the concept of "refugee camp" should be kept in mind.

2 THE CONCEPT of HOUSING/DWELLING and SECURITY

The dictionary definition of housing has two important points,

- To take refuge in the closed area to protect from natural effect.
- To settle in somewhere that is suitable to live (TDK 2016).

To better understand we can use dwelling instead of housing. Housing/dwelling is a concept that defining the existence of person. It is all about the physical, social and psychological relations of people that people interact with space (Ersoy 2002). According to Ersoy (2002) tell from Saegert (1985) dwelling provides the most closed relation between the people and environment or place. The act of protection that is in the dictionary definition of the housing derives from security need of person. So the concept of security that is effective on emerging refugee camps and gated communities should be analyzed.

Regarding the definition of the concept of refugee, it can be said that the most important reason of the refuging is unsecured conditions and environment that people live in and people who want to live in a safer place. In this context, it is possible to say that refugee camps are the places that make it possible to house the people have a risk of death. But on the other hand, this delicate space is not a free space.

Gated communities are the spaces that people want to live in to protect themselves from urban crimes and insecure environment based on urban crimes. Especially in big cities the spaces that have uncontrolled risk areas cause the people to tend to the low risk and protected spaces. But this secure area only made possible with monitoring. Çoban ve Özarslan, (2008) report that gated communities are the panoptic places that includes monitoring to be secure. So it can be said that like in refugee camps, people are not totally free in gated communities. Moreover, in gated communities people accept the monitoring voluntarily for the sake security. Refugee camps and gated communities have some physical and social similarities like reason of arising. So the concept of security that is effective on emerging refugee camps and gated communities should be analyzed.

3 PHYSICAL SIMILARITIES

3.1 Walls and the other Monitoring Means

Refugee camps are the spaces that are surrounded by a wall or fence, a restricted zone and having a watch tower for the providing security and prevent the escapes. But these spaces that emerge being surrounded by a wall or fence are not only limiting the movement of the refugees inside, but also border the other people living in the city. About this, Ömer Faruk Gönenç (2015) says that Israel creates a prison in its land by settle the Palestine camps. It can be said that the idea of enclosing by bordering comes from the concept of nation-state. On the other hand, there is a similar situation in the gated communities. Yönet (2011) reports that considerable numbers of the researchers define the gated communities as, separate worlds, isolated communities and social island. According to Landman (2000), protected placements are not only residence but also commercial and residential areas which are separated by walls or balustrades from their surroundings that have controlled entrances using doors and barriers. Even if these walls are built for the sake of security, they also represent one other thing. Luymes, (1997) say that the walls are the symbols of the power on economy and control. Also in both the refugee camps and gated communities, being surrounded by a wall or fence means to be separated from the public space and but the use instead of common field instead.

3.2 Location in the City

In general refugee camps are built away from the center of the city. This situation causes both some problems during both the construction process of the camp to provide infrastructure services, and the use as weakening the relations of refugee with the city center. Also providing health, education and infrastructure services cause economic burden for the cities. However, there are some examples for refugee camps in the city center. For example, some refugee camps in Jordon are in the city center and they work as a part of the city. The relations between refugee camp in the center and city reduce of the effect of border. Like refugee camps, gated communities are built in the near vicinity of the city. This situation prevents the city's problems to affect the residents of the gated communities. On the other hand, to build these buildings in the far places, make them known. Luymes (1997) defined the gated communities by a developing form of the territorial control. Also it can be claimed that these are a form of privatization.

3.3 Outbuildings

In addition to the housing opportunity, refugee camps provide some outbuildings to refugees like personal hygiene spaces, sports areas, television rooms and mosques, places of education and health and market. They also include hairdressers, barbers and tailors. They both give job educations about this job and services to the refugee. But the outbuilding in the camps both provide some services to the refugees and legitimize the reality of the camps. The outbuildings are the indicators of the possibility of living in the camps conditions. Gönenç (2015) exemplify this with the refugee camps in the Palestine and report that people living in these camps don't want these outbuildings. On the other hand gated communities enable lots of activities in the area. Gated communities contain some services like; recreation and sports areas, green areas and walkways, cultural activity areas, swimming pools, parking, education, health and shopping center. Variety and quality of this change as regards to people who live and use these places (Yönet 2011). Also these outbuildings that provide people some social and cultural opportunities are generally used as a marketing instrument.

3.4 Intensity

Even if legally designed refugee camps are built as regards to the estimated number of people, the increase in the number of refugees causes intensity in the camps rise. For example, although Zaatari refugee camps in Jordon have only 60.000 people capacity, according to the last data of UNCHR, 79.457 live in this camp (UNCHR 2016). We can give another example for the intensity of the camps from Turkey. Suleymansah Accommodation Area is in the Akcakale, Sanliurfa. Although population of Akcakale is 27.000, the number of the refugees in the Suleymansah Accommodation Area is nearly 35.000. When we look at the intensity of the gated communities we can see the same situation. For gated communities the concept of intensity can be analyzed under the two titles. One of them is higher land price in big cities. It causes more people to live in a smaller area. The other one is changing master plans within the scope of urban transformations. To build the high rises instead of the low rises in the same area cause the intensity to increase.

4 SOCIAL SIMILARITIES

4.1 Refugees and Resident's Relations with the City

For people living in the refugee camps, it is difficult to form a relationship with the city because of both their political statue and border of the camps. The effects of this problematical relation increase in refugee camps that are far away from center of the city. So, for a considerable number of people that live in the camps, the city is no longer a part of land. But there are some exceptions. For instance, Amman New Camp (Wihdat) is in the center of the capital Amman in Jordon. The camp is the representative and leader of the social life by producing the actor of the daily and social life (Gönenç 2015). On the other hand, the relation of people who live in the gated communities with the city is similar to the refugees. There are two reasons for this problematic relation. One of them is walls that surround the gated communities and the other one is outbuildings in the area. There is no need for residents to get out, because all of the things that they need are in the area. They don't contact the other part of the cities. They even interact with nature to the extent permitted by security staff or wall (Hook and Vrdoljak 2002). In gated communities there are inside and outside areas. But in camp the area is not inside or outside. It descripts third area (Gönenç 2015).

4.2 Sense of Belonging

It is possible to say that people living in refugee camps cannot contact the city, so they don't have a sense of belonging to the area. On the other hand, the housing spaces in the camps are the same; none of them are special for anyone. So people cannot relate personally or profoundly. The weakness of the sense of belonging of people living in gated communities can be estimated to be a result of the metropolis. Metropolitan person have belonging problem (Simmel 2003). Gated communities are suitable for the metropolitan person. People don't feel themselves belong to the city or space they live in. They feel themselves belong to images that are produced by space or city.

5 CONCLUSION

In this study the social and physical similarities of refugee camps and gated communities was examined. Firstly the concept of housing/dwelling and the concept of security which are the most important reasons of emerging of gated communities and refugee camps were analyzed. Then physical and social resemblances of gated communities and refugee camps were examined. For physical similarity it was analyzed that being surrounded by wall or fence, the location of the gated communities and refugee camps in the city, their outbuildings and their intensity. The sense of belonging of refugees and residents and their relations with city was examined to be the social similarities.

In this context it was determined that there are some similarities between refugee camps and gated communities. In this sense, it can be claimed that people living in the gated communities are the refugees of the metropolis. On the other hand, it was determined that refugee camps have a sense changing in the framework of their existence. This change requires estimating the camps in a new context. So a new terminology is needed. The concept of camp-villes that are produced by Agier (2002) can be estimated as a part of the new terminology.

References

Agamben, G., Kutsal İnsan - Egemen İktidar ve Çıplak Hayat, trans.: İsmail Türkmen, Ayrıntı, 226-227, 2002.

Agier, M., "Between War and City: Towards an Urban Anthropology of Refugee Camps", Ethnography, 3, 322. 2002.

Blakely, E. J., Snyder, M. G., Fortress America: Gated Communities in the United States, Washington DC and Cambridge, MA: Brookings Institution Press and Lincoln Institute of Land Policy, 1997.

Çoban, B., Özarslan, Z., Panoptikon: Gözün İktidarı, Su Yayınevi: İstanbul, 2008.

Ersoy, A., Z., Konut ve Ev Kavramlarinin Karsilastirmali Analizi, Unpublished PHD Thesis, Dokuz Eylul University, 2002.

Diken, B., From Refugee Camps to Gated Communities: Biopolitics and the End of the City, Citizenship Studies, 8, 1, 3-106, 2004.

Gönenç, Ö., F., Kamp/Mülteci: Çatışma Mekanlarında Sömürgesizleştirme Mimarlığı, Kentlilik ve Eve Dönüş: Batı Şeria Mülteci Kampları, Arredamento Mimarlik, Boyut, 3, 68-76, March, 2015. (2015)

Hook, D., Vrdoljak, M., Gated Communities, Heterotopia and a "Rights" of Privilege: a "Heterotopology" of the South African Security-Park, Geoforum, 33, 195-219, 2002.

Landman, K., Gated Communities and Urban Sustainability: Taking a Closer Look at the Future, 2nd Southern African Conference on Sustainable Development in the Built Environment (Strategies for a Sustainable Built Environment), Pretoria, South Africa, 2000.

Low, S., Behind the Gates: Life, Security and the Pursuit of Happiness in Fortress America, New York: Routledge, 2003.

Luymes, D., The Fortification of Suburbia: Investigating the Rise of Enclave Communities, Landscape and Urban Planning, 39, 187-203, 1997.

Özgür, F. E., Sosyal ve Mekansal Ayrısma Çerçevesinde Yeni Konutlasma Eğilimleri: Kapalı Siteler, İstanbul, Çekmeköy Örneği, Planlama Dergisi, 4, 79–95, 2006.

Saegert. S., Home Environemt. In I. Altman & C. Werner (Eds), The Role of Housing in the Experience of Dwelling, 1985.

Simmel., G., Modern Kulturde Catisma, trans. Gen, E., Kalayci, N., Bora, T., Iletisim Yayinlari, Istanbul, 2003

Tan P., Kamp/Mülteci: Çatışma Mekanlarında Sömürgesizleştirme Mimarlığı, Müştereklеşme Pratiği Olarak Mülteci Kampı, Arredamento Mimarlik, Boyut, 3, 77-84, March, 2015.

TDK retrieved from http://tdk.gov.tr on January, 2016.

UNCHR retrieved from http://www.unhcr.org/turkey/home.php?content=640 on January, 2016.

UNCHR retrieved from http://data.unhcr.org/syrianrefugees/regional.php on January, 2016.

UNCHR retrieved from http://data.unhcr.org/syrianrefugees/region.php?id=77&country=107 on January, 2016.

Yönet., A.,N., Korumali Yerleşimlerin Tercih Sebeplerinde Etkili Olan Faktörleri Açiklamaya Yönelik Kavramsal Bir Model, Unpublished PHD Thesis, Istanbul Technical University, 2011.

Interaction between Theory and Practice in Civil Engineering and Construction
Edited by Komurlu, R., Gurgun, A. P., Singh, A., and Yazdani, S.

PERFORMANCE ANALYSIS OF EARTHQUAKE RESISTANT BASED URBAN TRANSFORMATION PROJECTS

TUGBA SARSILMAZ and DENİZ GÜNEY

Dept of Architecture, Yıldız Technical University, Istanbul, Turkey

Earthquakes have often produced economic and social disaster. The last destructive earthquake on west side of the NAFZ is 1999 İzmit (Mw = 7.4). On August 17, 1999, a major earthquake occurred in the Marmara region, western Turkey, with a magnitude Mw = 7.4. The total observable length of the rupture was about 100km. The number of condemned buildings after the earthquakes amounted 23,400. There were 18,373 accounted deaths. First reason of this damage is inadequate design and low construction quality of buildings without control and supervision. The second important result is wrong urbanization policies of local and central government. Among the cities of the well developed and developing countries, it is possible to create urban surroundings resistant to earthquakes by comprehensive mitigation activities. Turkish Government has decided to perform many urban transformation projects in Istanbul in order prevent future seismic damages. However these projects have been started without detailed design and analysis. The aim of this study is to analysis performance of urban transformation projects in Istanbul. The feedbacks are collected from the sides of urban transformation projects such as landowners, constructors, officials, non governal organizations (NGO). Especially Fikirtepe Urban Transformation Projects are analyzed with details. Based on studies on site and later analysis, the performances of urban transformation projects in Istanbul are evaluated.

Keywords: Urban, Transformation, Construction, Earthquake, Buildings, Performance, Regeneration.

1 INTRODUCTION

Turkey has been experiencing a great deal of changes in its urban structures over the last two decades, like in many other developing countries under rapid change in demographics. Rapid population increase combined with further industrialization caused massive immigration from rural to the metropolitan areas, which have resulted in increasing urbanization rates in Turkey. The urbanization rate in Turkey was around 25% until the early 1950s, increasing to 40% in 1980 and 65% in 2000. Turkey ranked third among countries with the highest urbanization rate in the world between 1980 and 2000. As of 2012, according to TURKSTAT, Turkey's urbanization rate has reached 77% as shown in Figure 1.

Not only population growth, but also migration from rural areas led to high demand for housing in the metropolitan cities. Burdened with all the problems of urbanization, cities became the subject of dramatic changes as squatter settlements, environmental degradation and deterioration of existing infrastructure came into play. Turkey has experienced very destructive earthquakes in the past 70 years as shown in Table 1 listed by AFAD.

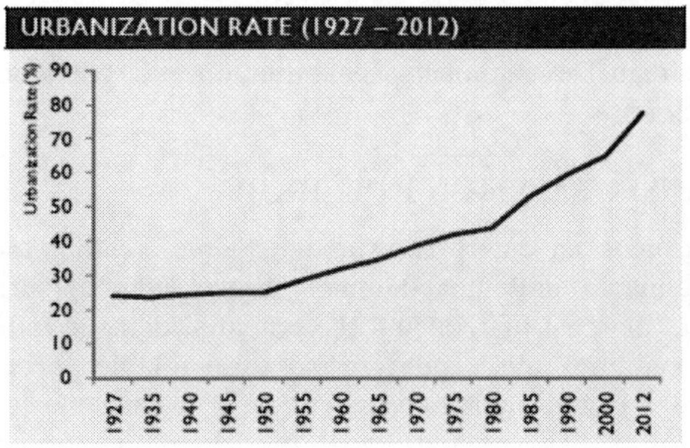

Figure 1. Urbanization rate of Turkey.

Table 1. Largest earthquakes in Turkey.

Year	Location	Magnitude	Life Losses
1939	Erzincan	7,8	32.700
1942	Tokat,Erbaa	7	3.000
1943	Samsun, Lâdik	7,4	4.000
1944	Bolu, Gerede	7,5	3.959
1966	Muş, Varto	6,7	2.396
1975	Diyarbakır, Lice	6,6	2.385
1976	Van, Muradiye	7,5	3.840
1999	İzmit	7,6	17.118
2011	Van	7,2	688

Earthquake performance of cities can be improved by changing the functional characteristics through urban transformation, land-use planning and increasing the quality and redundancy of the infrastructure. Almost all of these risk management measures are being implemented in Istanbul to prepare the city for a large earthquake that has an annual probability of occurrence of about 2%, one of the largest in the world. According to Istanbul Scenario Earthquake (Mw = 7.5) Losses (2009 Estimates), about 16,000 buildings were found to be damaged beyond repair (2%). About 40,000 (5%) buildings would be extensively and 150,000 (19%) buildings would be moderately damaged. The financial losses due to building damages caused by the scenario earthquake are estimated to be in the range of USD 12 Billion. Total Direct Losses will be about USD 40 Billion. The deaths were found to vary between 20,000 and 30,000. The Istanbul Metropolitan Municipality has commissioned the services of four leading Turkish universities for the preparation of the Earthquake Master Plan for Istanbul. The master plan aims for the planning of the activities in these fields, preparation of implementation programs, and identification of the responsibilities and responsible authorities for earthquake disaster mitigation activities. According to this study, new (post 2000) buildings in Istanbul are in general being built much better than the existing building stock. That's why; the Istanbul Metropolitan Municipality of Istanbul pursues programs on Urban Development, Rehabilitation and Transformation, where the focus is transformation projects for highly vulnerable building districts as well as master projects for rehabilitation and transformation of most risky zones. In this paper properties of the Urban Development, Rehabilitation and Transformation projects are analyzed and discussed.

2 URBAN TRANSFORMATION IN TURKEY

Urban transformation is an enterprising and resource-creating tool that enables the realization of sustainable strategic planning. Since Istanbul suffers a high risk of earthquake damage, the cost of reducing the potential damage will be high. Turkey's economic conditions limit the capital resources available for activities to reduce earthquake damage, so public institutions will need to take the lead in such a major urban transformation. The primary goal of urban transformation projects is to eliminate infrastructural and superstructure deficiencies by cleaning up urban spaces from illegal settlements. Nevertheless, urban transformation projects have a social dimension as well. In such projects, informing the local people about the issue concerned, cooperating with the local people and participation of the people are extremely essential for the project to result in success (Karadağ and Mirioğlu 2012). The requirement for urban transformation in less developed and developing countries is based on processes of the unplanned growth of cities, squatting, illegal building, urban sprawl, and clarification of the social problems caused by them (Özden 2008). The most fundamental factor requiring urban transformation in Turkey is squatting. The squatting areas which increased with the urbanization movements that particularly commenced as of the 1950s have now reached dimensions to threaten both the planned urban fabric of the cities in which they are located and urban livability. The ongoing migrations to cities and the fact that squats and the squatting population.

2.1 Urbanism Properties of Istanbul

At present, Istanbul is more modern than many major cities of the world in many respects. This is due to the new investment in the latest technology, and to the concentrations of wealth and affluence that made it possible. Since the cities' rapid progress into modernity only began a few years ago, it is no wonder that most modern available technologies were brought into the city. However excessive, unplanned, uncontrolled construction without protecting green areas threatens future of the city.
The total area of Istanbul City is 5343 km^2. According to last census, Istanbul's population is 14,160,000 inhabitants (2013), its population grows at a rate of 3.45% mostly accounting for the countryside migrations shown in Figure 2. The presence of squatters, mass transportation and solid waste disposal are some of the most critical problem of the city administration phases, further challenged by the increasing irregular and illegal urbanization.

Istanbul is identified as Turkey's economic capital. Main sector of the economy relates to provision of services and utilities (60%), manufacturing accounts for the 35% and the primary sector only for the 5%. When 1999 Marmara earthquake hit the

Marmara Region, in Istanbul, Avcilar (in south-west) and Tuzla (in south-east) were the most affected districts with collapsed buildings. In Istanbul 1-2% of the buildings were damaged, 454 people were killed and 3600 people were injured because of uncontrolled and illegal settlements (Erdik *et al.* 2000).

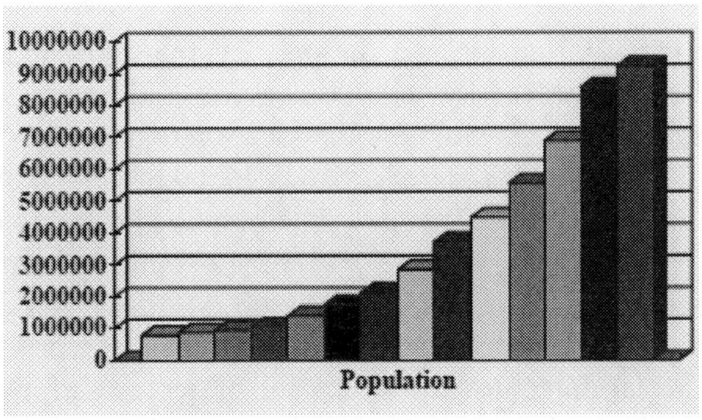

Figure 2. Population chance of Istanbul for each 5 years between 1930-2000.

2.2 Relationship between Earthquakes and Urban Transformation

The following factors play a major role in reducing earthquake damage in a country (Köktürk 2007): 1) A high level of economic development; 2) A balanced distribution of income; 3) Minimal inter-regional differences in economic development and growth; 4) A nationally balanced population distribution; and 5) Implementation of a suitable urbanization policy.

Developing countries like Turkey suffer from a fundamental problem: urbanization and building construction that are contrary to urbanization rules and zoning plans. In addition to this supervising (by municipality or ministry of construction) quality of construction is unacceptable. It is tempting to take the resulting unplanned and unhealthy shanty areas that are at high risk of earthquake damage and merely reinforce the buildings as needed. However, this approach is inadequate because: 1) The lack of a legal basis for reinforcing those buildings constructed contrary to zoning legislation and plans; 2) The difficulties of reinforcing buildings constructed in great density and adjacent to one another; and 3) The possibility of reinforced buildings then suffering earthquake damage, which is a cost increasing factor.

Therefore, it is recommended (IBB 2003) that activities to reduce earthquake damage should cover many interlinked activities such as strategic planning, administrative planning, legal arrangements and financial modelling. Urban transformation serves as one of the most important action plan models in this process. In this context, the demolition or reconstruction of dilapidated or unsound buildings is called "urban renewal" while the strengthening and improving of existing buildings to preserve them is called "sanitation" or "revitalization" (Tekeli 2003). An action plan that enables the realization of projects to improve urban spaces in accordance with real-estate policies and their ownership arrangements, with the participation of the public and private sectors, civil organizations and local people, and leads to the harmonious integration of all urban functions (Bilsel *et al.* 2003). Urban transformation is one of the most important ways of solving issues arising where significant risk is already present (IBB 2003).

3 FİKİRTEPE: URBAN TRANSFORMATION PROJECTS EXAMPLE IN ISTANBUL

Fikirtepe urban transformation project is one of the most controversial projects. Fikirtepe is located at the most risky seismic zone. In addition to seismic risk, soil has liquefaction capacity. Fikirtepe is situated in a location that was once in the periphery of city but became an integrated part of urban settlement and gained increasing land ratings. The area is about 909.253 m^2. There are 17.728 dwellings with 47.655 populations. As shown in Table 2, only %3 of the buildings are in good condition (satisfy the latest code requirement). The %65 of those buildings are reinforced concrete, %34 of them is masonry. Many of those building are illegally constructed buildings without any architectural or structural project. Unfortunately those illegal buildings got residential permits before elections as election promise by politicians without any technical consideration. So their status became semi-legal or semi-illegal. %95.5 of the buildings are private, %4.5 is public ownership. The location of Fikirtepe is very attractive. Because location is very close to Metrobus and metro (main transportation) lines which make this place valuable.

Table 2. Building conditions in Fikirtepe.

Building Condition	Number	Percentage %
Good	208	3,28
Middle	2342	36,93
Bad	3759	59,28
Damaged	32	0,5
Total	**6341**	**100**

The population concentration in Fikirtepe is 364 person/ha as gross and 445 person/ha as net value. Average habitation duration is between 40-60 years. Today, complains are rising from people living in Fikirtepe. According to our survey results, their main complains are: 1) Unclear points in their contracts between them and construction companies; 2) Lack of supervision by ministry and municipality; 3) Speed of project; 4) Expensive monthly revenues after completion of project; 5) Insufficient rent support; 6) Projects lead to the separation of community according to their socio-economic classes, ethnic backgrounds and cultural choices; 7) Unbelievable prices of new projects.

4 CONCLUSION

Urban regeneration and transformation in Turkey, which is a nationwide issue considering the scale and required financing, is expected to support growth in the construction and real estate sectors in Turkey. The earthquake that is expected to hit Istanbul and its attendant potential for loss and damage are a significant issue. Regional planning and macro-level solutions need to be developed to counter this threat. To achieve this, the utilization of resources must be extremely well. However transformed areas must be designed with social and green areas. If just concrete quality is increased, we will lose future because next generations will have to live without well planned and designed city opportunities. Those transformation project managers have to consider habitants' complaints and objections to the project in order to protect social peace. Instead of applying same model and project to everywhere, a unique project which considers all technical and social properties of the region must be designed with participating with inhabitants. In addition to this laws and legislations must be revised according to existing experiences. Municipality and ministry have to supervise projects and protect inhabitants against strong and well organized construction companies.

References

Bilsel, S. Polat, G., Yılmaz, E, 2003, Değişim-Dönüşüm Sürecinde "Kimlik Arayışları" ve "Kentsel Yenileme" Kavramı („Searching for Identity" and the Concept of „Urban Renewal" in the Process of Change-Transformation), *Urban Transformation Symposium* , TMMOB Chamber of Urban Planners, Istanbul Branch Publ., ISBN: 975-395-602-9, Istanbul, pp: 53-64, 11-13 June 2003.

Erdik, M., Durukal, E., Biro, Y. ve Birgören, G.. Istanbul'da Binalar için Deprem Riski ve Risk Azaltımına Yönelik Somut bir Öneri, *II. İstanbul ve Earthquake Symposium*, TMMOB İnsaat Mühendisleri Odası, İstanbul Subesi, İstanbul Technical University, pp. 131-149, 2000.

IBB, 2003, *İstanbul İçin Deprem Master Plan (Earthquake Master Plan for Istanbul),* Metropolitan Municipality Planning and Zoning Department, Soil and Earthquake Examination Directorate, ix+1334, 7 July 2003.

Karadağ, A., Mirioğlu, G., . Bayraklı Kentsel Dönüşüm Projesi Üzerine Coğrafi Değerlendirmeler, *Türk Coğrafya Dergisi*, 57, pp. 21-32, 2012.

Köktürk, E., Deprem ve Kentsel Dönüşüm İlişkileri, Jeodezi, Jeoinformasyon ve *Arazi Yönetimi Dergisi*, Vol. 97 (2), 2007.

Özden, P. P. *Kentsel Yenileme*, Ankara: İmge Kitabevi, 2008.

Tekeli, İ., 2003, Kentleri Dönüşüm Mekanı Olarak Düşünmek (Considering Cities as a Space for Transformation), *Urban Transformation Symposium*, ISBN: 975-395-602-9, Istanbul, pp: 2-7, 11-13 June 2003.

Interaction between Theory and Practice in Civil Engineering and Construction
Edited by Komurlu, R., Gurgun, A. P., Singh, A., and Yazdani, S.

STRUCTURAL CHARACTERISTICS OF WOODEN FIVE-STORIED BUDDHIST PAGODAS AND APPLICATION TO THE MODERN "TOKYO SKY TREE" ARCHITECTURE

KEIMI HARADA

Visiting Professor, International Academy of Architecture, Bulgaria

Definition of Five Storied Pagoda: A five-storied pagoda is a symbolic tomb of Buddha. When Buddhism was introduced into Japan in 538, a five-storied pagoda became one of the important components of temple complexes. Present Situations of Wooden Five-storied Pagoda: There are 22 wooden five-storied pagodas in Japan. The oldest is the one in Horyuji Temple built in 710. The tallest is the one in Toji Temple with the height of 54.8 m. The shortest is the one in Murooji Temple with height of 16 m. Structural Characteristics: There is no historical record that five-storied pagoda was destroyed by an earthquake. A contributory factor is that no nail is used but carved holes or grooves in each of wooden pieces interdigitate precisely to form a structure. Another key factor is "A Central column (Shin-bashira)", an important wooden structural component almost 30 m long. There was no theoretical principle. A Central column is effective against an earthquake and it has been regarded to absorb the earthquake energy. Thirdly, five-storied pagoda interior space is effective to control structural deformation and shake of pagoda. The first theoretical analysis was done by 1/5 scale model in 2006 that proved the role of "Central column". Application to the Modern Technology: There are two cases presently. One is Maru-biru Office building 178.5 m high built in 2002 and another is Tokyo Sky Tree, a telecommunication tower 634 m high built in 2012. Those structural design was influenced by a pagoda's central column as a vibration damping tool.

Keywords: Central column, Shin-bashira, Buddhism temple, Earthquake proof flexible structure, Vibration damping tool.

1 PURPOSE OF THE PAPER

The purpose of this paper is to introduce structural characteristics of earthquake proof five storied pagodas and two cases of modern architecture where traditional technology is applied.

2 DESIGN CHARACTERISTICS OF 22 WOODEN FIVE-STORIED PAGODAS

There are 22 wooden five storied pagodas built before the Meiji Period (1868) in Japan. 9 are designated as "National Treasure", and 13 are designated as "Important Cultural Treasure". There is no record that a five storied pagoda was destroyed by an earthquake in spite of occasional earthquakes in Japan.

2.1 The Oldest Pagoda, Horyuji Temple, Nara Prefecture

The oldest pagoda is Horyuji Temple Pagoda with the height of 31.5 m built in 710 in Nara Prefecture, an ancient capital. Those complex consists of the oldest wooden buildings, registered as a World Heritage Site. The width of the first floor is 6.42 m. (Figure 1)

2.2 The Tallest Pagoda, Toji Temple, Kyoto Prefecture

The tallest pagoda is Toji Temple Pagoda in Kyoto City, built in 1644 with the height of 54.8 m. The width of the first floor is 9.48 m.

2.3 The Shortest Pagoda, Murooji Temple, Nara Prefecture

The smallest pagoda is Murooji Temple Pagoda in Nara Prefecture built in early 800s with the height of 16 m. The width of the first floor is 2.45 m. (Figure 1)

2.4 Spatial Characteristics

The first floor is the place for the Buddha's bone case. Space between the second floor and fifth floor is regarded as a loft. The plan is a square shape of three-span structure, there are twelve columns along the outer wall line and there are four columns inside. A Central column is erected in the center.

2.5 Elevation

The proportion of a main part height to a total height of the Horyu-ji Temple of the year 710 is 0.701. It is 0.717 for the Muroo-ji Temple of the year 800, 0.730 for the Myo-oin Temple of year 1348, and 0.730 for the Honmonji Temple of the year 1607. The proportion of main part increased from the early period to the later period. (Figure 1)

The proportion of a main part height to a width of 1st floor of the Horyuji is 3.57. It is 4.73 for the Murooji, and 4.65 for the Honmonji. The proportion of a width of the fifth floor to that of the first floor of the Horyuji is 0.503. It is 0.594 for the Muooji, and 0.675 for the Honmonnji. Horyuji Temple, built in 710, looks stable and Honmonji Temple, built in 1607, looks slender. It is getting slender in later periods.

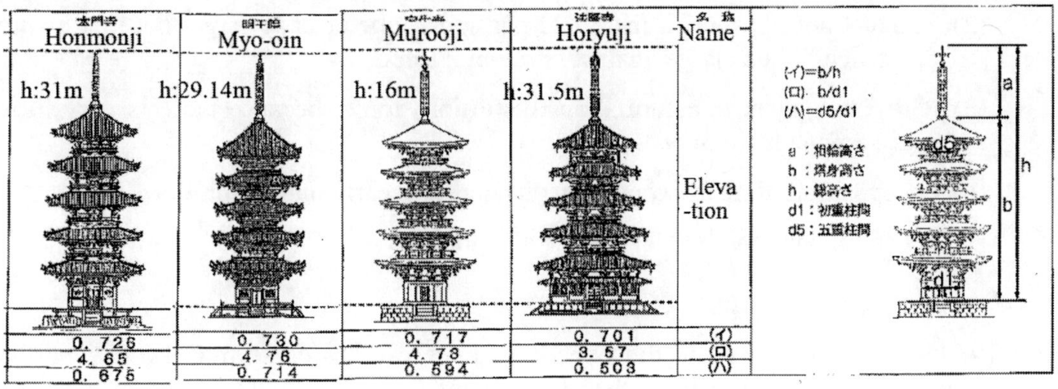

Figure 1. Proportion of five-storied pagodas.

2.6 Structural Features

Structural feature of a pagoda is a Shin-bashira erected in the center supposedly playing a role of earthquake proof. "The weight of a wooden pagoda is about 150 tons, there are 750 types of wooden parts, and 18,800 wooden parts, based on a recent case of a wooden five storied pagoda" (Hamashima and Sakamoto 2010).

3 STRUCTURAL CHARACTERISTICS -EARTHQUAKE PROOF

The five storied pagoda is an earthquake proof with a flexible structural system. Theoretical analyses have been done recently, but reality has not been found yet. There are a few theories of an earthquake proof structure of a pagoda.

3.1 Prof. Ohmori's Theory

In 1910, Prof. Fusakichi Ohmori did a research of vibration measurement in the Horyuji Temple to find that Central column has a role to control a shake of a pagoda. He pointed out "A Central column is an important key factor" like a pendulum (Hamashima and Sakamoto 2010). There was no theoretical principle then, but empirical principles by master carpenters. A central column is effective against an earthquake. Its pendulum isolation has been regarded to absorb the earthquake energy. Composition of each wooden part has high ability to absorb earthquake energy.

3.2 Prof. Riki Sano's Theory

In 1916, Prof. Sano reported that "resonance will not always occur, because five storied pagoda is tall, the natural period is longer especially, in case that the ground is solid, it is longer than the predominant period" (Hamashima and Sakamoto 2010).

3.3 Prof. Makoto Tanahashi's Theory

In 1960's, Prof Tanahashi reported in the Second World Earthquake Engineering Conference as follows,

(1) A pagoda shakes very slowly compared with an ordinary building. The natural period of a pagoda is between 1 and 1.5 seconds).

(2) The amount of wooden materials per square meter is so large that a structure resistance force is large against horizontal force.

(3) It resists against transformation (distortion) force, because there is very little space among many wooden parts.

(4) Composition of many wooden parts decreases earthquake vibration.

(5) A central column plays a role like a pendulum.

3.4 Prof. Kiyoshi Muto's Theory

In 1960's Prof. Kiyoshi Muto did experimental researches using model and reported that transformation performance is large in a pagoda.

(1) Small wooden parts absorb earthquake energy

Because of many wooden parts, a pagoda has a character of damping. Contributory factor is that no nails is used but carved holes or grooves in each of wooden pieces interdigitate precisely to form a structure. Each part will move little by little.

(2) Because of big columns, a pagoda has stability.

(3) Each floor of a five storied pagoda shakes in different direction.

(4) A five storied pagoda's interior space is not used for daily activities so that its structure is effective to control structural deformation and shake of a pagoda.

(5) Because of 17 columns including Central column, a number of columns is enough for its floor area (Hamashima and Sakamoto 2010).

3.5 Recent Research

Wooden Architecture Forum and Research Institute of Disaster Science Engineering executed the first theoretical research of structural behavior analysis of earthquake vibration using a 1/5 scale precise model of the Asuka Period wooden pagoda (592 till 710) from 2004 till 2008. The input data of earthquake energy is the same as the Hanshin Awaji Great Earthquake in 1995. Its magnitude is 7.3, maximum acceleration is 848 Gal, and maximum speed is 105 kine. The results are as follows:

(1) "It curves and its distortion of the second floor is s maximum point. (Figure 2)

(2) Maximum value of deformation is 1/33 radian, but there is no serious damage of test pieces.

(3) Composition of wooden parts has a much ability of damping.

(4) Central column has effects to control distortion and shake, but it does not enough contribute enough against earthquake proof."(Hamashima and Sakamoto 2010).

(5) But theoretical outputs on Central column structural features are not clear yet.

4 APPLICATION TO THE MODERN TECHNOLOGY

At present, there are two cases in which concept of Central column is applied, although design and planning are entirely different from an original five storied pagoda.

4.1 The Tokyo Sky Tree

The Tokyo Sky Tree (Figure 3) is a telecommunication tower of 634 m built in 2012. In consequence of earthquake safety study, engineers decided to set a large cylinder of reinforced concrete like Central column of old wooden pagoda's idea, as a vibration damping tool. (Figure 4)

The size of a central column set in an inner part between the lower part and the upper part is 375 m in height, 8 m in diameter and 60 cm in thickness. There is a stairway inside. "Tower's natural period is as long enough as 10 seconds. 40% of the response shear stress will be reduced compared with the case of no-cylinder" (Konishi 2011). The proportion of the main part height to the lower part width is 7.5 that is higher than the traditional pagodas, because of its peculiarity as a telecommunication tower. (Figure 5)

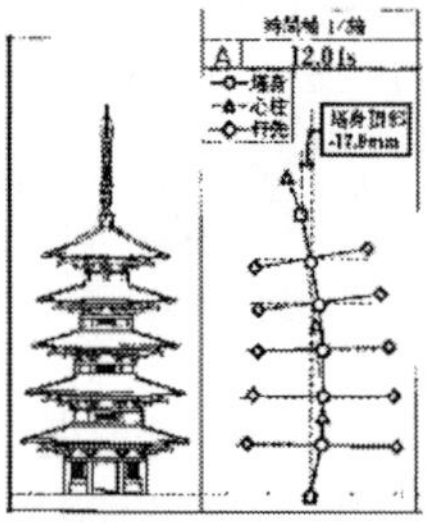

Figure 2. Distortion experiment by 1/5 model.

Figure 3. Tokyo Sky Tree.

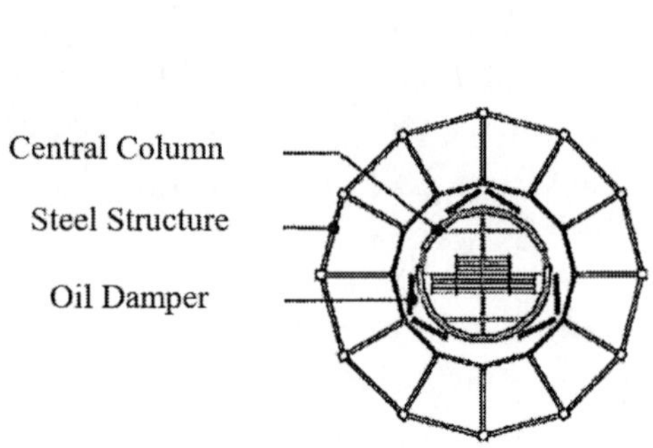

Figure 4. Central Column.

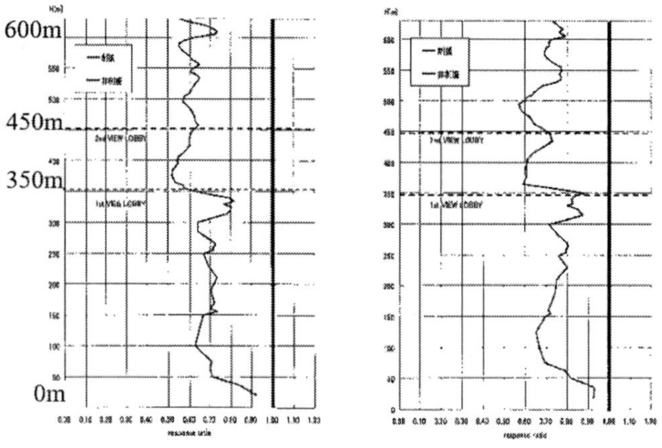

Figure 5. Effect of earthquake energy control by central column.

4.2 Marubiru Office Building, Tokyo

Marubiru (Figure 6) built in 2002 is located in front of the Tokyo Station. It is 37 storied and 178.5 m in height, with total floor area of 159,907.74 m^2. The philosophy of the structural design comes from Central column of a five storied pagoda. The earthquake proof shaft from the ground to the top floor was set in the center of floors. The shaft is connected with each floor by absorbing equipment. Experiment data show as follows (Figure 8):

Case A: Without Central column: It would collapse.

Case B: With Central column, it resists. "The result of structural calculation with the maximum energy of the once-in-a-millennium earthquake, such as the Hanshin-Awaji Great Earthquake in 1995, shows that the earthquake damages disperse and the structure would survive. Maximum Layer distortion decreases by 60%" (Ogawa 2006). The proportion of height to width is 3.39 that is almost the same as the ratio of traditional pagodas.

Figure 6. Maru-biru.

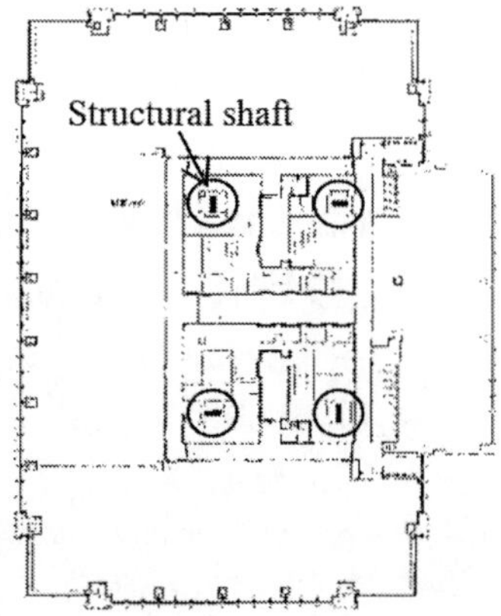

Figure 7. Floor Plan of Maru-biru.

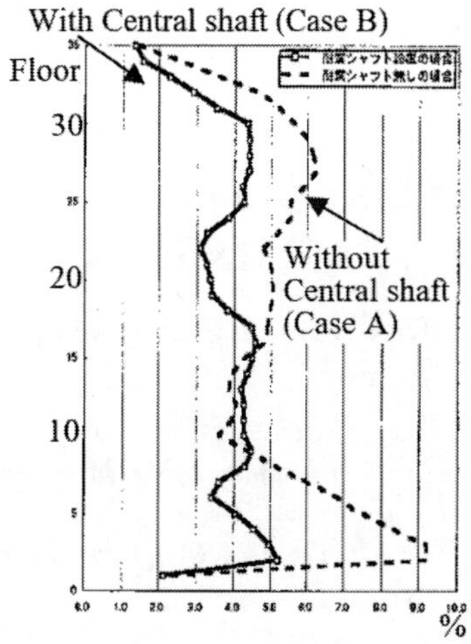

Figure 8. Max. distortion of floor.

5 CONCLUSION

(1) All 22 wooden five storied pagodas (built from 8th century till 19th century) have survived great earthquakes. The fact shows that they are earthquake proof buildings built by master carpenters empirically.

(2) Central column contributes to an earthquake proof structure. But details are not clear yet. Theoretical research has just started, and further interdisciplinary studies including structural engineering, traditional wooden architecture, earthquake science and etc. will be required to clarify the facts.

(3) The idea of Central column is applied to the recent structural design of remarkable modern architecture. In case of Tokyo Sky Tree, 40% of the response shear stress will be reduced. In case of Maru-biru, maximum layer distortion will decrease by 60% compared with the type of without Central column.

References

Hamashima, M. and Sakamoto, I., *Story of Five Storied Pagoda*, Tokyo, 2010.
Konishi, A., Earthquake and wind resistant design of the Tokyo Sky Tree, 2011.
Ogawa, I., High Rise Office, Structural Design, Mitsubishi Sekkei, Tokyo, 2006

Interaction between Theory and Practice in Civil Engineering and Construction
Edited by Komurlu, R., Gurgun, A. P., Singh, A., and Yazdani, S.

STRUCTURAL ENGINEERING PERSPECTIVE ON HISTORIC BUILDING RESTORATION

BURCU BALABAN OKTEN[1], MEHMET SELIM OKTEN[2], CEM HAYDAROGLU[3], and GIANMARIO BENZONI[4]

[1] *Faculty of Architecture, Fatih Sultan Mehmet Waqf University, Istanbul, Turkey*
[2]*Faculty of Architecture, Mimar Sinan Fine Arts University, Istanbul, Turkey*
[3]*Arup, Istanbul, Turkey*
[4]*Dept of Structural Engineering, University of California San Diego, La Jolla,*

The restoration of historic buildings requires a comprehensive approach through a collaborative effort, as well as multi-disciplinary professionals. According to the general conservative guidelines of the Venice Charter of 1964 and the ICOMOS/ISCARSAH Recommendations for the Analysis and Restoration of Structures of Architectural Heritage of 2003, structural engineers are responsible for repair and strengthening of these structures on the basis of principles of authenticity, reliability, compatibility, durability, reversibility, and monitorability. In restoration projects, engineers often face significant issues due to the intrinsic challenges of the projects themselves, as well as the complex level of interaction with other professionals involved. The main goal of this research project is the identification and evaluation of the most significant challenges encountered by structural engineers in restoration applications in Turkey. Data were obtained from semi-structured interviews with six professionals involved in restoration projects in Turkey. The sample group was chosen based on specific competences and involvement in different phases of restoration projects in order to gain different professional perspectives. The results revealed several issues from the early stages of the projects, such as a lack of records about previous interventions, a lack of standards and guidelines with respect to historic structures. The identification and evaluation of these issues is intended to raise awareness by the community as well as by the legislative authorities. In the future, dedicated legislative tools could facilitate the execution as well as the efficiency of these interventions, thus limiting subjective approaches.

Keywords: Historic heritage, Conservation, Project management, Repair and strengthening, Survey, Codes.

1 INTRODUCTION

Restoration projects have increasingly come to take part in building projects in Turkey. According to the T.R. Prime Ministry Directorate General of Foundations, 19,825 cultural properties of various foundations were digitally recorded in archives. Although the number of completed restoration projects was only 46 between 1998 and 2002, 3,363 projects were completed between 2002 and 2008.

Numerous applications of restoration projects reveal significant issues for professionals, generally composed of architects, art historians, restoration technicians, material chemists and engineers. Although all professionals contribute to these projects, structural engineers are the only group who are responsible for repair and strengthening applications of historical structures in view of structural sufficiency and safety guidelines as determined by the legislative authorities.

Lourenço (2006) stresses that "conservation engineering requires a different approach and different skills from those employed in designing new construction" (p. KP05-2). According to Giannopoulos and Asteris (2011), "For an engineer, taking part in the restoration process of a historical structure, through the analysis of its structural system, means mainly to face the demanding task of equipping the historical structure with the capability to withstand future actions with the minimum possible amount of damage, while bearing in mind the characteristics and values which make this structure unique and worthy of special attention" (p. 82). As with many participants in restoration projects, structural engineers need to take into account general conservative guidelines such as the Venice Charter of 1964 and the ICOMOS/ISCARSAH Recommendations for the Analysis and Restoration of Structures of Architectural Heritage of 2003 as related to the principles of authenticity, reliability, compatibility, durability, reversibility and monitorability (D'Ayala 2014). Therefore, structural engineers need to perceive all these aspects, and it therefore appears critical to understand the challenges encountered by structural engineers in restoration applications.

The purpose of our paper is to identify and evaluate the common problems in restoration applications in Turkey from the perspective of structural engineers. The paper is organized as the following three sections. In Section 2, the methodology and characteristics of the interviewees are presented. More specifically, six professionals involved in restoration projects in Turkey were interviewed to obtain data. Then, the findings are identified and a brief discussion is reviewed in Section 3. Finally, the conclusion is summarized.

2 METHODOLOGY

A qualitative research method has been chosen for this study. The interview method provides a closer look into subjects and allows the interviewers to interact with professionals deeply to understand and define research problem.

The professionals selected for the study include five civil engineers specializing in the structural engineering field and one architect working on a conservation committee. The architect was chosen because of his performance duty on a conservation committee. In Turkey, conservation committees do not appoint civil engineers as committee members. Convenience sampling was identified as a suitable sampling method for reaching professionals in different stages of restoration projects. The professionals were recruited through the researchers' informal and formal contacts. Semi-structured interviews were conducted with civil engineers in different stages of restoration projects. The interviews were conducted by telephone between May and August 2015, in accordance with the professionals' busy schedule. The interviews lasted between thirty minutes and one hour. All interviews were recorded and later transcribed. All data was kept in QSR NVivo 10 for collaborative study. Data were coded several times to determine common problems and solution recommendations. The characteristics of the interviewees are summarized in Table 1.

Table 1. Characteristics of the interviewees.

Interviewee	Position	Occupation	Int. Date
I-1	Professor at Faculty of Architecture Structural and Earthquake Engineering	Ph.D. Civil Engineer	21.05.2015
I-2	Founder and Director of a structural engineering firm	M.S. Civil Engineer	28.05.2015
I-3	Assistant Prof. Faculty of Architecture	Ph.D. Architect	17.06.2015
I-4	President of Center for Protecting Cultural Heritages, Application and Research	M.S. Civil Engineer and Architect	18.06.2015
I-5	Site manager	Civil Engineer	30.07.2015
I-6	Site manager	M.S. Civil Engineer	28.08.2015

3 FINDINGS AND DISCUSSION

Interviews were conducted with six selected professionals in different stages of restoration projects to understand the major problems encountered by structural engineers in restoration projects in Turkey. According to the interviews, four significant problems were outlined; they include inadequate restoration courses in civil engineering education, absence of records of previous interventions on historic buildings, lack of standards and guidelines with respect to historic structures, and finally, ineffective participation of civil engineers. The major problems encountered by interviewees are listed on Table 2.

Table 2. Major problems defined from interviews.

Major Problems	I-1	I-2	I-3	I-4	I-5	I-6
Inadequate restoration courses in civil engineering education	YES	NO	YES	YES	YES	YES
Absence of records about previous interventions	YES	YES	YES	NO	YES	YES
Lack of standards and guidelines	NO	NO	NO	YES	NO	YES
Ineffective participation to projects	YES	YES	NO	NO	YES	NO

Almost all interviewees stated that restoration principles and concepts are taught in too limited a manner in civil engineering schools. Dewoolkar *et al.* (2011) emphasizes the importance of integration between the field of historic preservation and civil engineering on the undergraduate level. Although the assistance of structural engineers is of great importance to historical heritage conservation in the way of giving critical structural decisions, there is no course given about conservation and restoration processes in undergraduate civil engineering education in Turkey. This situation causes a lack in the understanding of conservation principles among numerous structural engineering graduates. Graduate architecture programs do offer courses about restoration and conservation, but most of the time enrollment for these courses is limited to architecture students, not engineers. I-6 mentioned that, as a result of absence of education in this field, structural engineers perceive conservation projects like new buildings and this approach causes incorrect implementations in historical structures that include intrinsic problems. According to I-4, civil engineering education related to restoration field should be given in undergraduate programs. I-4 also underlined that lectures in undergraduate civil engineering programs would help to construct an understanding of conservation concepts among civil engineers. Similarly, I-1 pointed out that the studies and theses researched for the strengthening of historical buildings are becoming popular in graduate programs but that there is still a need for courses at the undergraduate level. More specifically, I-3 emphasized that strengthening methods of historical buildings required experimental studies and that more research should be done in civil engineering departments.

Another problem defined by the interviews is the lack of information about past interventions applied in historical buildings. I-1 mentioned that if they had the resources to research information about the history of construction or intervention in historical buildings, they could have started projects with more accurate knowledge, which in turn would provide the option of more efficient, better and applicable restoration projects. Moreover, past interventions might be critical for the structural system, and lack of information could easily cause bad decisions in the design process. According to I-5, half of the restoration projects on site were not useful or applicable because of the uncertainties related to the structure. After construction starts on site, many revisions are revealed within a project due to the discovery of past interventions. Also, I-2 and I-1 mentioned that it was very hard to make appropriate decisions without enough information, and this lack of information caused several unexpected changes in construction steps. In addition, I-3 said that the contractors did not welcome changes in projects and applications because new information causes delays and extra costs during the process. I-3 suggested a governmental institution be established to keep renovation, restoration and conservation information of the historical buildings in an archive. I-6 claimed that it is very difficult for project owners to understand the current condition of historical structures and participate in the restoration projects without knowing these structures' historical background.

Lack of standards and guidelines with respect to historic structures affects all restoration processes from the beginnings of the projects. I-4 suggested that well-designed procedures should be prepared specifically for the field of historic building restoration to prevent unnecessary and redundant applications, such as repetitive material experiments and analyses.

Three interviewees also mentioned ineffective participation in projects. I-1 emphasized that civil engineers ought to join projects during the early stages. Thus, they could have an opportunity to affect structural decisions from the very beginning. This participation could also provide efficient schedules and accurate budgets in restoration projects.

4 CONCLUSION

This paper presented an identification and evaluation of the significant challenges encountered by structural engineers in restoration applications in Turkey. This qualitative study was conducted with six interviewers in order to obtain data. Four major problems were determined, including inadequate restoration courses in civil engineering education, absence of records of previous interventions in historic buildings, lack of standards and guidelines with respect to historic structures, and finally, ineffective participation of structural engineers in restoration projects.

The limitation of the present study is the small number of interviews, and the reliability of our findings is therefore limited. However, it appears also to make an important contribution to the literature for future studies.

References

D'Ayala, D., Perspectives on European Earthquake Engineering and Seismology. In A. Ansal, ed. Geotechnical, Geological and Earthquake Engineering 34. pp. 489–513, 2014.

Dewoolkar, M.M., Porter, D. & Hayden, N.J., Service-Learning in Engineering Education and Heritage Preservation. International Journal of Architectural Heritage, 5(6), pp.613–628, 2011.

Giannopoulos, I. P. and Asteris, P. G., Earthquake Resistant Design of Structures. In M. Papadrakakis, M. Fragiadakis, & V. Plevris, eds. III ECCOMAS Thematic Conference on Computational Methods in Structural Dynamics and Earthquake Engineering M. Greece, pp. 1–12, 2011.

Lourenço, P. B., Structural restoration of monuments: recommendations and advances in research and practice. In 1st International Conference on Restoration of Heritage Masonry Structures. Cairo, Egypt, pp. 1–16, 2011.

Interaction between Theory and Practice in Civil Engineering and Construction
Edited by Komurlu, R., Gurgun, A. P., Singh, A., and Yazdani, S.

COLLAPSE TEST ON RECTANGULAR STEEL BEAM WITH DIFFERENT SUPPORT CONDITION

SUNG-HYUN PARK[1], KYUNG JAE SHIN[1], SWOO-HEON LEE[1], HEE-DU LEE[1], and WOO BUM KIM[2]

[1]*Architectural Engineering, Kyungpook National University, Daegu, Korea*
[2]*Dept of Architecture, Kongju National University, Kongju, Chungnam, Korea*

The behavior of beams with different support condition was investigated through three-point bending test for the purpose of studying difference of maximum load and catenary action of a beam. Beams are made of SPSR400 which has 245MPa of nominal yield strength and 400MPa of nominal tensile strength. The parameters of the supports condition are simple support and rotationally semi-rigid connection with different horizontal reaction strength applied by anchors. The support boundary conditions of beams were classified into three types; (A) simply supported with no anchors, (B) embedded anchors with 50mm depth of ϕ10mm, and (C) embedded anchors with 80mm depth of ϕ10mm. Both ends of the beams were connected by fillet welded angles and supported on a rigid concrete wall through anchors. The test result shows how much the load carrying capacity is increased by catenary action after large deflection at the center of the beam. First peak loads from each types are the loads when first plastic hinge occurs at the mid-span. There were no significant differences among the first peak loads measured from three types. After that, tensile force at anchors due to catenary action increased the load carrying capacity by approximately 55%, which is called as the second peak load. However, second peak load happens when the anchors at a support fully resist a tensile force, therefore it doesn't happen with type (A) and (B). In conclusion, support boundary conditions of a beam don't have an effect on the first collapse load, but the second collapse load is increased as embedded length of an anchor becomes deeper.

Keywords: Bending test, Catenary action, Boundary condition, Rectangular hollow section, Anchor, SPSR400.

1 INTRODUCTION

Beams have different ultimate strength depending on support conditions. General bending tests are conducted in conditions of simple support. However, in this study, the tests were conducted in conditions of a pin-ended beam and moment-fixed beams. This experiment was undertaken for the purpose of studying increment of maximum load due to catenary action of beams, depending on support conditions. Catenary action is the behavior which resists progressive collapse to the last extremity (Khandelwal and El-Tawil 2005). However, precisely how catenary action will behavior is not yet clear (Kim and An 2008). Ahn (2013) explained the characteristics of pullout behavior for anchor. Through this research, the minimum pullout strength that provides a catenary action will be obtained. It shows that the anchor system that satisfies the minimum pullout strength is practical or not in the field.

2 EXPERIMENT

The beam section is a rectangular of 100 (mm) × 50 (mm) × 2T (mm). The span length of the beam is 3,600mm and concrete wall designed as support is relatively very rigid to the horizontal direction so as to generate the catenary action of the beam. The parameters of the supports condition are simple support and rotationally semi-rigid connection with different horizontal reaction strength applied by anchors. The support boundary conditions of beams were classified into three types; (A) simply supported with no anchors, (B) embedded anchors with 50mm depth of ϕ10mm, and (C) embedded anchors with 80mm depth of ϕ10mm. Both ends of the beams were connected by fillet welded angles and supported on a rigid concrete wall through anchors. Concentrated load was applied on the center of beams.

Figure 1 is a sectional view of type A, and Figure 2 is a sectional view of type B and C. In the Figure 1, the beam is placed on the concrete as a simple beam. In Figure 2, the ends of the beam are supported by using anchors. Table 1 shows the details of support conditions of the test types.

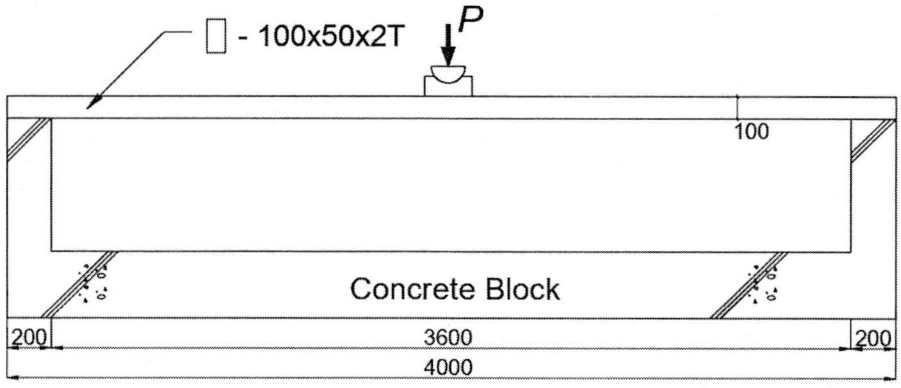

Figure 1. Sectional view of type A.

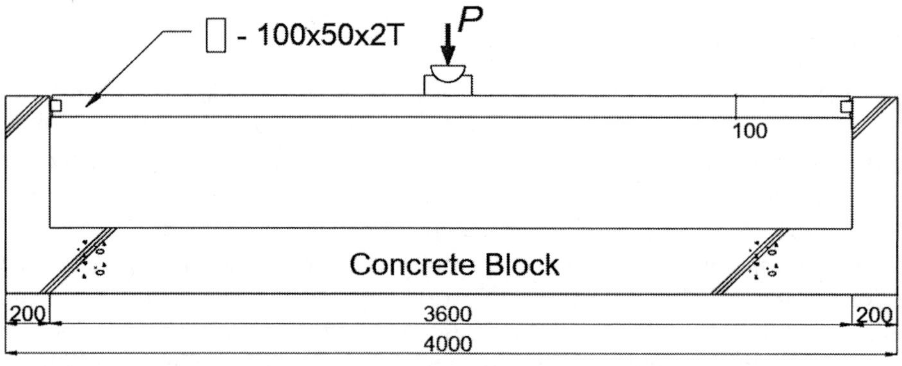

Figure 2. Sectional view of type B and C.

Table 1. Support conditions depending on types.

Type	Support Condition
A	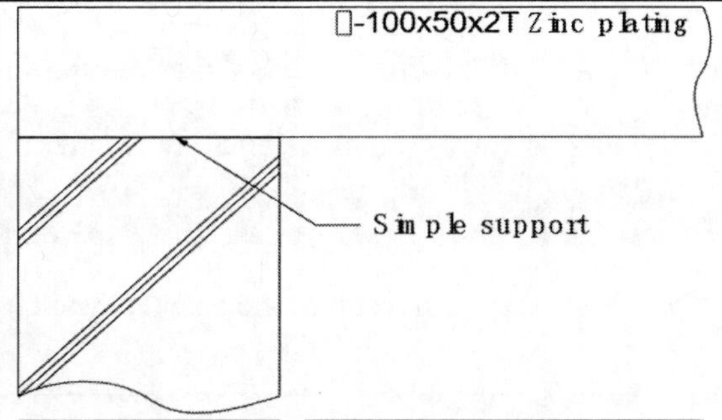
B	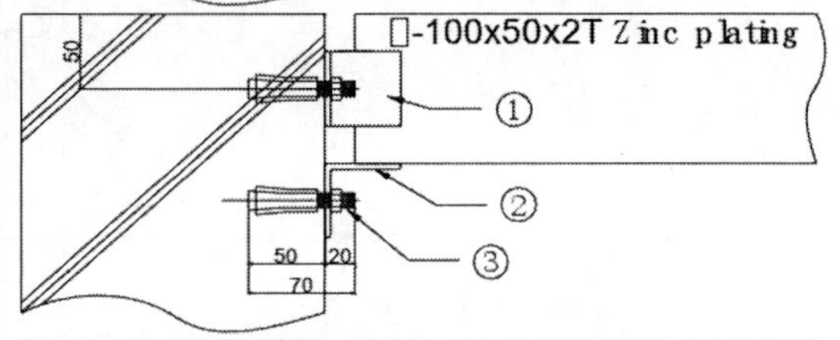
C	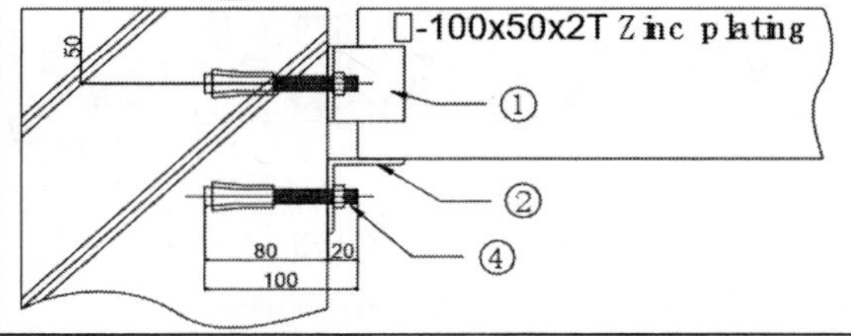
Note: ① 2L - 50×50×4T (Set anchor ϕ10mm -2EA) ② L - 50×50×4T (Set anchor ϕ10mm -1EA) ③ Embedded length 50mm (+20mm) ④ Embedded length 80mm (+20mm)	

3 EXPERIMENTAL RESULTS

Figure 3 shows the test result of the beam. The beam was yielded at the center when the load reached to the first maximum load. After then, when the load reached to the second maximum load, the concrete which was connected to the ends of the beam was fractured as shown in Figure 3.

Figure 4 shows the load-displacement graphs of experimental results depending on support conditions. The first maximum loads from each types are the loads when the first plastic hinge occurs at the center of the beams. After that, tensile force at the anchor due to catenary action results in increasing of load carrying capacity by approximately 55%. This is the second maximum load. The second maximum load happens when the anchors at a support fully resist the tensile force, therefore it doesn't happen with type A 'simply supported' and type B '50mm embedded anchor'.

Figure 3. Test result of the beam. (Type C)

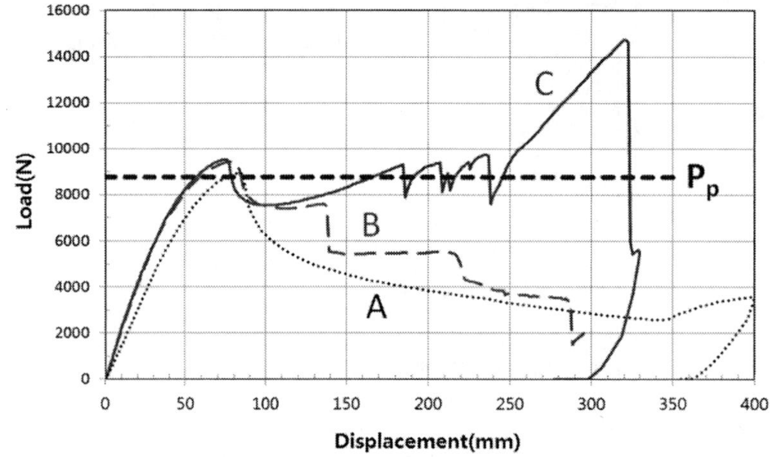

Figure 4. Load-displacement depending on support conditions.

Table 2 shows the first and the second maximum loads according to specimen types. There are no significant differences among the first maximum loads measured from 3 types. However, maximum load increased when catenary action happens with type C '80mm embedded anchor'.

Table 2. 1st and 2nd maximum load.

Type	A	B	C
1st Maximum Load (N)	8,959	9,433	9,531
2nd Maximum Load (N)	-	-	14,733

4 CONCLUSIONS

Support boundary conditions of a beam don't have an effect on the first collapse load, but the second collapse load is increased by 55% as embedded length of the anchor becomes deeper. If the embedded length is short, the catenary action may not expected.

Acknowledgements

This work was supported by the Human Resource Training Program for Regional Innovation and Creativity through the Ministry of Education and National Research Foundation of Korea (NRF-2014H1C1A1067008) and by National Research Foundation of Korea NRF-2013R1A2A2A05006637.

References

Ahn, Y. C., *Characteristics of Pullout Behavior for Expansion Anchor*, Master's thesis, Chungbuk National University, Korea, 2013. (in Korean)

Khandelwal, K. and El-Tawil, S., *Progressive Collapse of Moment Resisting Steel Frame Buildings*, Structures Congress 2005, ASCE, 1-11, 2005.

Kim, J. K. and An, D. W., *Progressive Collapse Analysis of Steel Moment Frames Considering Catenary Action*, Architectural Institute of Korea, 24(1), 55-62, Jan, 2008. (in Korean)

Interaction between Theory and Practice in Civil Engineering and Construction
Edited by Komurlu, R., Gurgun, A. P., Singh, A., and Yazdani, S.

TAP WATER AND INDOOR AIR CONTAMINATION DUE TO AN UNINTENTIONAL CHEMICAL SPILL IN SOURCE WATER

PINAR OMUR-OZBEK[1], DUYGU AKALP[2], and ANDREW WHELTON[3]

[1]*Dept of Civil and Environmental Engineering, Colorado State University, Fort Collins, USA*
[2]*Dept of Construction Management, Colorado State University, Fort Collins, USA*
[3]*Dept of Civil, Environmental and Ecological Engineering, Purdue University, Lafayette, USA*

A chemical storage tank was found leaking into the Elk River, West Virginia, on January 9, 2014. The tank held ~10,000 gallons (38,000 L) of 4-methylcyclohexanemethanol (MCHM). The chemical spilled 2.4 km away from the West Virginia American Water's (WVAW) Kanawha Valley Water Treatment Plant, traveling downstream and entering the treatment plant (50 MGD). West Virginia Poison Center started to get phone calls from the public, concerned about nausea, vomiting, diarrhea, rashes, and other symptoms. Simultaneously, the emergency departments observed an increase in the visit rates to emergency rooms, which were related to the above-mentioned symptoms. It was determined that the chemical spill ended up in the tap water and started affecting the people who used the tap water and inhaled the contaminated indoor air due to tap water use. This study employed available shower, tap water use at the sink, washing machine and dishwasher models and the reported tap water concentrations for the MCHM to determine the indoor air concentrations of MCHM. The results indicated that the exposure to hazardous levels of MCHM is most likely through showering and while during flushing out of the home plumbing. This study also considered natural and forced ventilation to provide guidance while flushing the home plumbing as it plays a critical role in indoor air concentrations of MCHM.

Keywords: 4-methylcyclohexane methanol (MCHM), Drinking water, Air quality modeling, West Virginia USA, Elk River spill

1 INTRODUCTION

In the morning of January 9, the West Virginia Department of Environmental Protection (WVDEP) received complaints of a strong licorice smell near from the Freedom Industries, Inc.'s storage tank farm (Manuel 2014). It was found that, a chemical storage tank was leaking into the Elk River, West Virginia. The tank held ~10,000 gallons (38,000 L) of 4-methylcyclohexanemethanol (MCHM), which is used for coal processing (Manuel 2014, McGuire *et al.* 2014). The chemical spilled 2.4 km away from the West Virginia American Water's (WVAW) Kanawha Valley Water Treatment Plant, traveling downstream and entering into the treatment plant (50 MGD) (Foreman *et al.* 2014). By 4:00 PM, MCHM was detected at the treatment plant. Right after the detection, the do-not-use order was issued at 5:50 PM to inform the residents that the tap water is not safe for drinking, cooking, bathing and washing (Cooper 2014,

Manuel 2014). By that time, West Virginia Poison Center started to get phone calls from the public about nausea, vomiting, diarrhea, rashes, and other symptoms. Simultaneously, the emergency departments (EDs) observed an increase in the visit rates, which were related to the above-mentioned symptoms. To investigate the relationship between increased ED visits and MCHM side effects; West Virginia Bureau for Public Health (WVBPH) and the Agency for Toxic Substances Disease Registry (ATSDR) investigated 369 patients' records of whom visited the EDs (WVBPH and ATSDR 2014). The most common exposures were identified through showering/bathing (52%) followed by drinking (44%) as well as inhalation (15%). To determine a safe consumption level of MCHM in the water, the Center for Disease Control and Prevention (CDC) became involved with local and state health authorities in the investigation. Despite limited available data of MCHM on human health, 1 ppm was established as short-term screening level (CDC 2014).

Previous related studies have focused on the dermal and oral contact/consumption of the contaminated water, while inhalation studies were not the topic of research. The purpose of this study was to estimate indoor air concentrations of MCHM associated with contaminated water use in a dishwasher, washing machine, shower, and kitchen faucet. Additionally, the recommended flushing procedure for West Virginia incident was simulated for kitchen, bathroom and half-bathroom faucets (WVAW 2014).

2 METHODOLOGY

In this study, experimentally supported, two-phase mass balance models were utilized to estimate the chemical volatilization from each source based upon the study of Howard-Reed and Corsi (2000). The models were programmed in a computing program, MATLAB R2015a. The aqueous concentrations of 4-MCHM modeled were based on the levels determined in the tap water between January 18 to 26, when the "do not use order" was lifted for the entire area (see Table 2). To evaluate the exposure under "Do not use order", the highest known concentration in the water distribution system, 3.77 mg/L, was also included in the study (Casteloes *et al.* 2015). 4-MCHM aqueous concentrations were also used for the kitchen, bathroom, half-bathroom sink models flushing procedure under various air exchange rates ranging from 0.53 ac/hr to 10.61ac/hr.

Table 2. The sampled initial liquid concentrations of 4-MCHM at the distribution center.

Date	Highest	1/18	1/19	1/20	1/21	1/22	1/23	1/24	1/25	1/26
4-MCHM (mg/L)	3.77	0.32	0.17	0.02	0.18	0.12	0.02	0.06	0.04	0.27

2.1 Model Development

The volatilization rates of the chemicals are explained by the two-resistance mass transfer theory with the overall resistance, which is the function of gas, and liquid phase resistances (Ömür-Özbek *et al.* 2011). In accordance with the developed models by Howard-Reed and Corsi (2000), the ratio of gas and liquid phase mass transfer coefficients (K_G / K_L) and overall mass transfer coefficients ($K_L A$) of toluene were used to estimate 4-MCHM overall mass transfer coefficients in this study. Toluene was denoted as a surrogate compound. To estimate dimensionless Henry's law constant

(HLC) values of 4-Methylcyclohexanemethanol, the experimentally estimated HLC values at 7°C, 25°C, 40°C, 80°C were included from the study of Sain *et al.* (2015) and linear regression models were applied. The HLC value at 21°C was regressed from the HLC values at 7°C and 25°C. Similarly, the HLC values at 47.7°C, 60°C, 66°C were estimated from the values at 25°C, 40°C, 80°C. HLC values at 25°C and 40°C were assumed equal to the reported values.

2.1.1 *Dishwasher model*

The liquid phase in the dishwasher model was modeled as a well-mixed batch reactor while the gas phase was considered as a continuous-flow stirred-tank reactor (CFSTR) (Howard-Reed *et al.* 1999, Howard-Reed and Corsi 2000). The sequence of a typical dishwasher operation was simulated as: pre-rinse, wash, rinse and final rinse cycles. Dishwasher operation conditions were set as: system headspace volume (181 L), system ventilation rate (5.7L/min), system liquid volume (7.4 L) values and cycle times were taken from Howard-Reed and Corsi (2000). The operation temperature was selected as 66°C based on the requirement of the NSF/ANSI standard 184.

2.1.2 *Washing machine model*

The mass balance of the liquid phase in fill cycle was modeled as continuous-flow stirred-tank reactor (CFSTR) model, while the rinse and wash cycles were modeled as a well-mixed batch reactor (Howard-Reed and Corsi 2000). Typical residential washing machine operations were modeled as fill, wash, spin, rinse and spin cycles. The liquid flow rate (13.8 L), system ventilation rate (55 L/min, 53 L/min), system liquid volume (46 L), total volume of system (150 L) values and cycle times were taken from the study of Howard-Reed and Corsi (2000), and an operation temperature of 21°C.

2.1.3 *Shower model*

In order to determine the rate of mass transfer coefficients of the liquid droplets; the liquid phase was modeled as plug-flow reactor system (Howard-Reed and Corsi 2000) The gas concentrations were calculated for two different shower durations; 10 and 25 min. The water flow rate (9.1 L/Min), the ventilation rate (379 L/Min) and the volume (1745 L) were obtained from Howard-Reed and Corsi (2000). The initial chemical gas concentration in the stall was assumed to be zero.

2.1.4 *Kitchen sink model*

The kitchen sink model was developed by Corsi (1996) was utilized to determine gaseous contaminant concentrations. The water flow rate (0.0079 m^3/min), the room ventilation rate 2 ac/hr the room volume (64 m^3) and the kitchen sink use time (5 minutes) were obtained from Howard-Reed and Corsi, (2000). The water temperature was set as 25°C.

2.1.5 *Flushing procedure*

In West Virginia incident, the residents were required to flush their systems by a two-step procedure; flushing all hot water taps for 15 minutes and flushing cold water taps for 5 minutes (WVAW 2014). In this study, West Virginia flushing procedure was simulated for a kitchen sink, a bathroom sink and a half-bathroom sink for various aqueous concentrations of 4-MCHM. To estimate the gas concentrations under different scenarios: various natural and mechanical air exchange rates were included into the study: 0.53 ac/hr for all windows and doors closed, 0.92 ac/hr for single window-open, 1.7 ac/hr for multiple windows open (Howard-Reed *et al.* 2002). Additionally, the minimum required ventilation rates for the kitchen and bathroom area were determined from ANSI/ASHRAE Standard 62-2001 requirements (Persily *et al.* 2004). The 15 minutes hot flushing water temperature was assumed to be 47.7°C while the cold flushing water temperature was set to 21°C (Whelton *et al.* 2014). The models were also run for the hot water temperature of 60°C, to see gas concentration behaviors at higher temperatures.

3 RESULTS AND DISCUSSION

3.1 Dishwasher Model

The kitchen volume was modeled as 64 m^3 based on the information given for a typical residential kitchen (Corsi 1996). To estimate the gaseous 4-MCHM concentrations in the air, it was assumed that there is no gas contaminant exchange with the other areas or loses in the pollutant level. Once the total mass of MCHM emitted over the cycles within the kitchen was determined, it was divided by the volume of the area to get the concentration. Based on our results, none of the predicted gas concentrations for different aqueous concentrations exceeded the EPA's extrapolated short-term inhalation screening Level (0.01 ppm) (U.S. Environmental Protection Agency 2014). However, when the kitchen volume is equal or less than 46 m^3, the aqueous concentration of 3.77 mg/L led to higher gas concentration than EPA's extrapolated short-term inhalation screening level.

3.2 Washing Machine Model

Based on the model, none of the predicted gas concentrations for the given aqueous concentrations exceeded the EPA's extrapolated short-term inhalation screening level (0.01 ppm). However, when the initial aqueous concentration was 3.77 mg/L, the screening level was approximately 1.09 times more than the estimated gas concentration in the laundry room (3.4 m^3), which is very close to the predicted level.

3.3 Shower Model

Disregarding any loses in the shower stall, to estimate an average 4-MCHM concentration in the shower stall air, the total mass of emitted 4-MCHM was divided by the volume of the shower stall (1.7 m^3). For a 10 min shower model, the results showed that all of the estimated gas concentrations exceeded EPA's extrapolated short-term inhalation screening level, except 0.02 mg/L, which is the lowest liquid concentration sampled in the distribution system and the estimated gas concentrations ranged between 7.6E-03 ppm to 1.7E+00 ppm. For a 25 min shower event, all of the estimated gas concentrations exceeded 0.01-ppm screening level and ranged between 1.60E-02 ppm to 3.5E+00 ppm.

3.4 Kitchen Sink

The gas phase concentrations of 4-MCHM after a 5 min kitchen sink use were predicted. From the calculations, no gas concentrations exceeded the screening level.

3.5 Flushing Simulation

3.5.1 *Kitchen sink*

When the aqueous concentration was 3.77 mg/L, the gas concentrations at 47.7°C and 60°C exceeded EPA's extrapolated short-term inhalation screening level. On the contrary, none of estimated gas concentrations of cold flushing procedure exceed 0.01 ppm for other aqueous concentrations modeled. The gas concentrations at 47.7°C were ~ 27 times higher than the gas concentrations at 21°C. The gas concentrations were ~ 37 times more for 60°C versus 21°C water temperature for the kitchen sink model.

3.5.2 *Bathroom sink*

The results showed that when the aqueous concentration was 3.77 mg/L, the predicted gas concentrations at 47.7°C and 60°C exceeded the EPA's extrapolated short-term inhalation screening level. When the air exchange rate was 0.53 ac/hr, predicted gas concentrations of 0.32 mg/L and 0.27 mg/L were still greater than the guideline. However, when the air exchange rate was modeled as 0.92 ac/hr, the gas concentrations at 47.7°C didn't exceed 0.01 ppm screening limit.

3.5.3 *Half bathroom sink*

For the aqueous concentration of 3.77 mg/L and the air exchange rate of 0.53 ac/hr, all of the gas concentrations exceeded the screening level of 0.01 ppm including the gas concentrations estimated after cold flushing procedure. Similarly when the aqueous concentrations were 0.32 mg/L, 0.18 mg/L and 0.27 mg/L, the gas concentrations after the hot flushing procedure also exceeded the screening level. The cold flushing procedure did not show the same pattern for these initial aqueous concentrations.

4 CONCLUSIONS

Considering the researchers had started to screen the water samples on January 13th, the exact aqueous concentrations of 4-MCHM remained unknown for four days after the chemical spill, hence, the extent of this incident is still unknown. To prevent similar consequences, the lessons learned from this incident should be put into action. For example, the water resource departments and water treatment facilities across the country and around the world should come up with a comprehensive immediate action plan which includes "do not use" news releases and warnings on social media and other outlets. Moreover, consumer complaint tracking should be investigated with hospital emergency data, so that the officials may make more informed decisions and protect public health. Also a backup plan should be developed to obtain water from other resources or cities. The chemical companies and storage tanks should be inspected more frequently and the construction permits should be given when the risks are minimized, and when it is proven that there are barriers to prevent transport of such chemicals in case of an unintended spill. Also, locations for such storage facilities should be significantly away from surface waters to minimize contamination risks.

References

Casteloes, K., Brazeau, R. H., and Whelton, A. J., Flushing Chemically Contaminated Residential Plumbing Systems: A critical analysis, *Environmental Science:Water Research* and *Technology*, 1(3), 787-799, 2015.

Center for Disease Control and Prevention (CDC), Information about MCHM, 2014 . Retrieved from http://emergency.cdc.gov/chemical/MCHM/westvirginia2014/mchm.asp on May 26, 2015.

Cooper, W. J., Responding to Crisis: The West Virginia Chemical Spill, *Environmental Science* and *Technology*, 48(6), 3095–3095, 2014.

Corsi, C. H. and R. L., Volatilization of Chemicals from Drinking Water to Indoor Air: Role of the Kitchen Sink. *Journal of the Air* and *Waste Management Association*, 46(9), 830–837, 1996.

Foreman, W. T., Rose, D. L., Chambers, D. B., Crain, A. S., Murtagh, L. K., Thakellapalli, H., and Wang, K. K., Determination of (4-methylcyclohexyl)methanol isomers by heated purge-and-trap GC/MS in water samples from the 2014 Elk River, West Virginia, chemical spill. *Chemosphere*, 131, 217-224, July, 2015.

Howard-Reed, C., and Corsi, R. L., Volatilization Rates From Water To Indoor Air Phase II. Report to U.S. Environmental Protection Agency, Office of Research and Development: Washington, DC. 2000.

Howard-Reed, C., Corsi, R. L., and Moya, J., Mass Transfer of Volatile Organic Compounds from Drinking Water to Indoor Air: The Role of Residential Dishwashers. *Environmental Science* and *Technology*, 33(13), 2266–2272, May, 1999.

Howard-Reed, C., Wallace, L. A., and Ott, W. R., The Effect of Opening Windows on Air Change Rates in Two Homes, *J. of the Air* and *Waste Management Association*, 52(2), 147–159, 2002.

Manuel, J., Crisis and emergency risk communication. Lessons from the Elk River Spill, *Environmental Health Perspectives*, 122(8), August, 2014.

McGuire, M., Suffet, I. H., and Jeffrey, R., Consumer panel estimates of odor thresholds for crude 4-MCHM, *Journal of the American Water Works Association,* 106(10), 2014.

Ömür-Özbek, P., Gallagher, D. L., and Dietrich, A. M.,Determining Human Exposure and Sensory Detection of Odorous Compounds Released During Showering, *Environmental Science* and *Technology*, 45(2), 468–473, 2011.

Persily, A. K., Bulter, D. S., Beaton, M., Bellenger, L. G., Coggins, J. L., Feldman, E., Borges, D. S., ANSI/ASHRAE Standard 62-2001-Ventilation for Acceptable Indoor Air Quality, 2004.

Sain, A. E., Dietrich, A. M., Smiley, E., and Gallagher, D. L., Assessing human exposure and odor detection during showering with crude 4-(methylcyclohexyl)methanol (MCHM) contaminated drinking water, *The Science of the Total Environment*, 538, 298–305, 2015.

U.S. Environmental Protection Agency, Derivation of an Extrapolated Short-term Inhalation Screening Level for 4-Methylcyclohexanemethanol (MCHM – CAS# 34885-03-5), 2014

Whelton, A. J., Rosen, J. S., Clancy, J. L., Clancy, T. P., and Ayhan, E., The Crude MCHM Chemical Spill 10-Home Study: Resident Behaviors, Perceptions, and Residence Characteristics, 2014, Retrieved from http://www.dhsem.wv.gov/WVTAP/testresults/ Documents/POSTED 10 Home Study Interview Report_FINAL.pdf on July 2, 2015.

West Virginia American Water (WVAW), How to flush your plumbing system and How to flush plumbing faucets and appliances, 2014.

West Virginia Bureau for Public Health WVBPH and the Agency for Toxic Substances Disease Registry ATSDR, *Elk River Chemical Spill Health Effects. Findings of Emergency Dept Record Review*, 2014, Retrieved from http://www.dhhr.wv.gov/News/chemical-spill/Documents/ElkRiverMedicalRecordSummary.pdf

Interaction between Theory and Practice in Civil Engineering and Construction
Edited by Komurlu, R., Gurgun, A. P., Singh, A., and Yazdani, S.

HOW CAN WE STOP THE SEWAGE?

J. M. SÁEZ[1], M. J. M. DAVIS[1,2], M. L. GUTIÉRREZ[1,2], A. VALLEJO[1], and L. AYALA[1]

[1]*Faculty of Architecture, Design and Arts, Pontificia Universidad Católica del Ecuador, Quito, Ecuador*

[2]*Evolution Engineering, Design and Energy Systems Ltd., Exeter, UK*

In Ecuador 46.4% of households lack access to sewerage systems, and then much wastewater is discharged from sewerage into the environment untreated. On the one hand dry toilet systems present themselves as an optimal solution against this backdrop; they mitigate the production of black water, whilst having fertilizer as the end product. On the other hand, they tend to have a low acceptance level by potential end users. This problem of social acceptance is then further exacerbated in urban areas. In order to overcome this, the design of an ecological dry/semi-dry toilet was carried out in two phases. The first phase consisted of purely sociological research, where the aim was to determine the basic design parameters that should be used for ecological toilet prototype. The second phase is the development of ecological toilet prototypes, which are then tested in terms of their level of acceptance/rejection by potential end users. The overall aim is to develop toilet prototypes that on the one hand mitigate the production of black water, whilst on the other are accepted with open arms by end users, and even preferred over the conventional toilet common to urban areas.

Keywords: Dry toilet, Suction toilet, Manual suction, Wastewater treatment, Black water, Grey water, Water recycling.

1 INTRODUCTION

1.1 The Problem of Water Treatment in Quito, Ecuador

In Ecuador wastewater treatment is a serious problem. According to the census of the Ecuadorian National Institute of Statistics and Census (INEC 2010), only 53.6% of households have a connection to public sewerage systems. In addition, wastewater management makes up 53.3% of the average expenditure of national municipalities' budget, totalling 179 142 858 USD (INEC 2011). In 2012 49.8% of national municipalities carried out wastewater treatment, where a total of 128 000 000 m^3 of wastewater was processed. Despite this, 225 000 000 m^3 of wastewater was discharged untreated into the environment. The problem of wastewater treatment could be approached either at a macro or micro scale. A macro scale would involve large, expensive infrastructure works over a long time period. On the other hand, there is a possibility to consider the problem at a micro, or individual scale. In this case, it is suggested that every household would opt to change its toilet installation, to one that avoids the production of black water. The objective of this research project is to determine whether or not it is possible to mitigate the production of black water from

Ecuadorian households, whilst at the same time upholding a maximum level of comfort and lifestyle for the end users.

1.2 The Importance of Participative Design

In this research a participative design approach was adopted in accordance with Reason and Bradbury (2008), which is both considered a methodology for investigation and an approach to applied projects. In this sense, the researchers and end users worked together in order to reach solutions that are innovative and sustainable. It is considered to be a proven methodology in the generation of knowledge (Bergold 2007). In practise, to achieve this it is necessary to involve the end user at all stages in the development of the new product. In this manner it is guaranteed that the end result will be in accordance with the users' needs, given that they were involved in the entire design process. This is reinforced by Liener and Larsen (2009), who mentions how in current society innovative technologies are achieved by being as close to the end user as possible. In is worth mentioning however, that honesty from the end users can only be achieved by creating a secure setting in which the users feel comfortable in expressing their true opinions (Bergold and Thomas 2012).

2 METHODOLOGY

For this research, a series of sociological investigations were carried out: eight site visits to study homes with a dry toilet installed; twenty-eight semi-structured interviews with hypothetical end users of the prototypes to be developed; twelve focus groups with hypothetical end users of the prototypes to be developed. The results of the sociological investigations were used for the parameters of design for an initial prototype, which was given an initial evaluation through a preliminary social study with end users.

3 RESULTS AND DISCUSSION

3.1 The Series of Sociological Investigations

As was mentioned in section 1.2, in order to achieve participatory design it is important to carry out social research in such a manner that the persons being studied feel comfortable and able to express their feelings and opinions. For this reason the site visits were conducted in the homes of dry-toilet users. In addition, the interviews were semi-structured so that they could be adapted to the flow of conversation. Finally, the focus groups were carried out in an informal and familiar environment. The results were as follows:

Eight site visits: eight locations in Ecuador were visited that each had differing climatic conditions. Dry toilets with urine separation or with mixed faeces and urine were both used. A number of installations suffered from ventilation problems, which led to unfavourable smells and proliferation of flies. Sawdust was often used as a drying material and used toilet paper was thrown into the toilet (a practise that is not common in Ecuador). The mix would be left to decompose for a number of months for it to be converted into compost. The users of the dry toilets generally felt comfortable using dry toilet technologies, where the resulting fertilizer was used for ornamental plants and fruit trees. It is also important to note that every end user had had the dry toilet installed under his/her own initiative. In this sense they also had no qualms about mixing the residues of faeces/sawdust/paper/urine in order to produce an optimum compost and fertilizer.

Twenty-eight semi-structured interviews: the results showed the importance of cleanliness for both private and public toilets. On average the interviewees used the toilet four times per day, and tended to change their toilet installation every five to ten years. There was a consistent preference for white toilets, which were felt to reflect cleanliness. With regard to negative aspects, special mention was given to the need for constant cleaning of the toilet, bad smells, the noise made when flushing and the excessive water needed to do so.

Twelve focus groups: various conclusions were reached. First, that the toilet is a basic tool of our everyday lives, and as such should always be in optimum working conditions (with little need for maintenance). Second, cleanliness was seen to be of fundamental importance, and to this extent white, modern toilets were preferred. Third, it was considered that the area in which the toilet is found must be clean, with a light and clear colour scheme to make the user feel at ease. Fourth, there needed to be an effective system to get rid of the waste products, given that there existed an aversion to any contact with them (even visual). Finally, in general there was little interest in knowing about the final destination of the waste products.

3.2 The Initial Prototype

Conventional toilets that are commonly available in the Ecuadorian market tend to be dual flush, using 4.1 litres for flushing urine, and 6 for solids (Viegener 2014). Given the results of the sociological investigations, the parameters used for the design of the initial prototype were to a) maintain the appearance of a conventional toilet, and b) use the least water possible to flush away the faeces, urine and paper. It is important to note that the initial prototype developed to date has drastically reduced the water demand for flushing to but one litre, thanks to a manual suction system. This in turn opens possibilities to a) recycle water from the hand basin used for washing hands to recharge for the next toilet flush, and b) reduce the size of the biodigestor that would be needed in future research, in order to pre-treat the wastewater prior to discharge to public sewerage systems.

Additionally, a further objective was to minimise the economic impact of changing toilet systems, and as such enable it to be easily replicated. To this extent it was of great importance to make use of the existing infrastructures of conventional toilet installations. The prototype was therefore developed using elements that were easily accessible in local stores, and which could be installed by the end user him/herself if necessary.

In Figure 1(a) the first prototype developed is shown, in which a bypass has been installed that through manual suction generates a vacuum to flush away the waste products. The system was tested using various fake faeces and toilet paper mixes. PVC tubing of a 3-inch diameter was used, with a range of heights tested from 90 cm (in accordance to someone standing) to 54 cm (in accordance with the end user sitting). The system was successful in flushing away all the waste products. The next developments of the prototype made were in relation to the water recharge system. The objective was that in the same action the toilet would flush, whilst also recharging the toilet bowl and water deposit in same action. In this manner the toilet was cleaned and left ready for the next user. This was all achieved whilst maintaining a water demand per flush of 1 litre. In Figure 1(b) the subsequent prototype developed is shown. In this case the conventional water cistern is replaced by a compact element that still uses the water float from the original toilet cistern. A pipe makes use of the air compression produced above the manual suction piston, pushing water to recharge the toilet bowl. Finally, in Figure 1(c) a further development of the prototype is shown where water is flushed directly from above the toilet bowl and the original cistern flapper is used, but where the cistern is replaced by a transparent container. The advantage of this option is that the working elements of the toilet flushing system can be clearly seen, which enables the end users to be conscious of how the toilet functions. For all of the prototypes, there is the possibility for them to be connected to either the household water mains or to use recycled water, given that they work independent of water pressure at the recharge point. To this extent successful trails were made in connecting the prototypes to the washbasin, thus mitigating the potable water demand from a toilet.

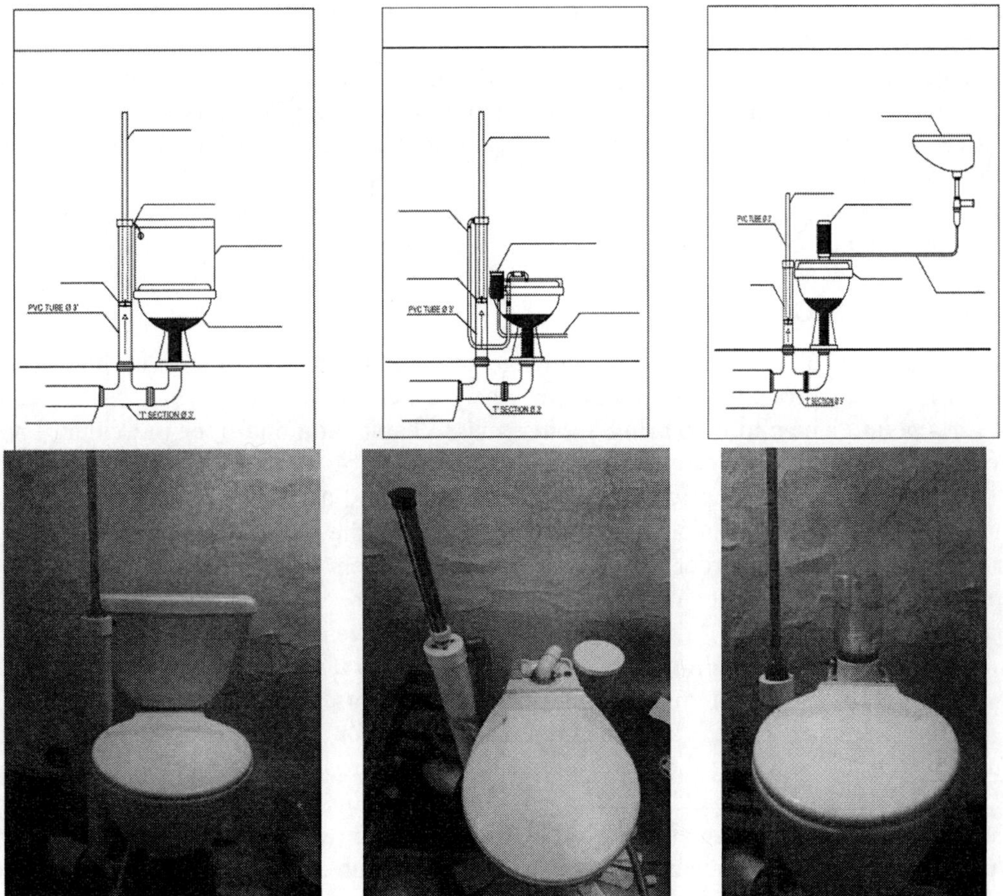

(a) (b) (c)

Figure 1. Manual suction toilet prototypes a, b, and c.

3.3 Initial Testing of the Prototype

This research is in the early stages of carrying out sociological studies to determine the level of acceptance from the end users. As an initial study to gauge this level of acceptance, five semi-structured interviews were carried out with members of the public from a middle-class background. Each person interviewed had previously had the opportunity to use the first prototype (Figure 1a). In general, the first reaction was one of some confusion when confronted with the vertical manual suction appendage, and where it was thought a didactic label might be of use. Otherwise, the prototype was considered to be a normal toilet. It was natural to pull the manual suction handle with force, and as such no discharge problems were encountered. It was also clear that the water demand for toilet flushing had been drastically reduced, which was perceived to be the main advantage. A disadvantage that was encountered was the noise made when flushing, due to the sound of the suction mechanism. Finally, it was noted that used toilet paper could be flushed with the faeces and urine. It was considered unusual however to do so, given that in in Ecuador it is common practise to throw the used toilet paper in a specially allocated dustbin adjacent to the toilet. To conclude, it can be said that the first prototype achieved to: a) drastically reduce the water demand for flushing; b) maintain the appearance of a conventional toilet; c) use readily available materials; and d) have an initial positive level of acceptance overall by the end users. For future research and prototype development, amongst the improvements to be made it is recommended to: a) make the design easier to comprehend regarding the operation of the vertical manual suction unit; b) reduce the level of contact the end user has with the flush mechanism (such as through the incorporation of a pedal-operated system) which was found to be of importance in the focus groups from Section 3.1.

4 CONCLUSIONS

This research project took on the challenge of looking at how to stop sewage (more specifically the production of domestic black water), taking the urban sector as its starting point. The aim is to bring about a mass-change at a micro, or individual level. To this extent it was put forward to design a toilet prototype in such a manner that the water demand for flushing was reduced drastically, and as such paving the way for a micro-biodigestor to be incorporated in the future for the wastewater to be pre-treated prior to discharge. In order to guarantee that end users accepted the prototype with open arms, a series of sociological investigations were carried out to determine the parameters for design. An initial prototype was developed, which used a manual suction system that brought the water demand down to a mere litre. Additionally, the prototype is made from a conventional toilet, making use of elements that are readily available in local stores and without the need for major construction works. Furthermore, by reducing the water demand to one litre, the doors are opened to use recycled water from the hand washbasin for toilet flushing. The initial reactions of end users of the first manual suction prototype were positive overall, where the initial social study shed light on improvements to be made in the future. Finally, the later prototypes that were developed did away with the water tank of conventional toilets. This also reinforced the didactic elements of the toilet, clearly showing the reduced water demand.

5 RECOMMENDATIONS FOR FURTHER RESEARCH

Amongst the suggestions for further research, it is recommended to: i) carry out further development of prototypes, included versions with a mechanized suction system; ii) develop a branding of the prototypes that leads to a positive first reaction and less confusion by the end users; iii) carry out an in-depth sociological study, where the prototypes are installed in a public area such as the Pontificia Universidad Católica of Ecuador; iv) develop a low-cost, micro biodigestor that can be easily incorporated into the toilet prototypes, in order to pre-treat the wastewater prior to discharge. Technical measurements can then be taken of the quality of the water being discharged in order to guarantee that sewage is indeed eliminated.

Acknowledgments

Special thanks needs to be given to Amelia Tapia, student of Architecture at the Pontificia Universidad Católica del Ecuador (PUCE), whose work and dedication was invaluable in the research for this paper. In addition, special acknowledgement needs to be given to Jamie Radford, Civil Engineer at Mott MacDonald, whose wealth of expertise and availability to resolve both technical and academic queries was of great importance.

References

Bergold, J. (2007). Participatory strategies in community psychology research—a short survey. In A. Bokszczanin (Ed.), *Poland welcomes community psychology: Proceedings from the 6th European Conference on Community Psychology* (pp.57-66). Opole: Opole University Press.

Bergold, J. & Thomas, S. (2012) Participatory Research Methods: A Methodological Approach in Motion. Forum: *Qualitative Social Research*. V (13), n.1, art. 30.

INEC, 2010. Resultados del Censo 2010 de Poblacion y Vivienda en el Ecuador. Fascículo Nacional. Retrieved from http://www.ecuadorencifras.gob.ec/censo-de-poblacion-y-vivienda/ on July 3, 2014.

INEC, 2011. Censo de Información Ambiental Económica en Municipios 2010-2011. . Retrieved from http://www.ecuadorencifras.com on July 3, 2014.

Liener, J. & Larsen, T. (2009) High Acceptance of Urine Source Separation in Seven European Countries: A Review. *Environmental Science Technology*. 44, 556 – 566.

Reason, P. & Bradbury, H. (2008). Introduction. In Peter Reason & Hilary Bradbury (Eds.), *The Sage handbook of action research. Participative inquiry and practice* (2nd ed., pp.1-10). London: Sage.

Viegener, F., Catálogo de Productos Institucionales 2014. Retrieved from http://www.franzviegener.com/fvecuador/images/stories/fv/catalogosyrevistas/FV-catalogo-linea-institucional-2015/2015-catalogo-institucional-index.html#/16/ on October, 2015.

Interaction between Theory and Practice in Civil Engineering and Construction
Edited by Komurlu, R., Gurgun, A. P., Singh, A., and Yazdani, S.

THE POTENTIAL FOR GREEN ROOFS IN SUSTAINABLE URBAN DRAINAGE SYSTEMS

M. J. M. DAVIS[1,2] and A. C. TAPIA[2]

[1]*Evolution Engineering, Design and Energy Systems Ltd., Exeter, UK*
[2]*Pontificia Universidad Católica de Ecuador, Dept of Architecture, Quito, Ecuador*

Urban environments tend to lack the capacity to absorb water from precipitation. This is due to vegetated surfaces being replaced by impermeable ones, such as concrete, bitumen or similar. As a result problems can occur, where a period of heavy rainfall coincides with sudden increases in surface water runoff. This in turn can lead to a city's sewerage system becoming overloaded. Sustainable Urban Drainage Systems (SUDS), are recognized worldwide as a successful manner by which to mitigate this phenomenon. One of the principal components of SUDS are permeable areas in an urban setting, which have the ability to absorb and retain rainfall that would otherwise flow as surface runoff. To date there has been little research into what the effect of a massive increase in green roofs would have for cities in Ecuador. As a developing country, it is not uncommon for cities' sewerage systems to suffer collapse when faced with sudden rainfall peak loads. It is suggested in this paper, that instead of looking to implement costly sewerage expansion programs, it would be more cost effective to implement city scale green roof systems. The paper sets out to quantify the theoretical effect of such an initiative.

Keywords: Permeable surfaces, Sewerage, Stormwater, Runoff, Urban planning.

1 INTRODUCTION

The Metropolitan District of Quito (DMQ) has a total surface area of 423,000 ha, where the city itself makes up 19,014 ha (EPMAPS 2011). The main rivers of the city are Machángara, Monjas and Pita (EPMAPS 2011). As an additional point of interest, Quito overlies more than 33 gorges, the majority of which have over time been filled in and built over (EPMAPS 2011). According to the Public Metropolitan Company of Potable Water and Sanitation (EPMAPS) of Quito (2011), the sewerage discharge rates to the main rivers of the city are:

- Manchángara river = 3.48 m^3/s.
- Monjas river = 0.64 m^3/s.
- San Pedro river = 0.69 m^3/s.

In addition to the daily sewerage discharges from the city however, rainwater runoff also plays an important role in determining the urban wastewater distribution system capacity. As the urban area of Quito expands, like any city the amount of permeable ground of that previously unoccupied area, is replaced by hard surfaces that lead to increased surface runoff (Berndtsson 2009). This is turn causes a sharp increase in the flow rate that is needed to be absorbed by the city's sewerage system. One manner in which this can be mitigated is through the separation of rainwater and black water systems. However, in Quito this is not the case and the two flows are combined into one sewerage distribution system (EPMAPS 2011). In this paper it is argued that to better manage Quito's urban drainage systems, three types of intervention are possible:

(1) Increasing the capacity of the sewerage system.

(2) Increasing permeable (green) spaces at the ground level of the city.

(3) Increasing permeable (green) spaces at the level of the roofs of the city.

As a starting point, an annual peak rainfall in the wet season is taken into account. According to the 2014 Annual Meteorological Report of the National Institute of Meteorology and Hydrology (INAMHI 2011), the peak rainfall in Quito over 24 hours was 39.9 mm. This is equivalent to 39.9 l/m^2 in a 24-hour period.

2 INCREASING THE CAPACITY OF THE SEWERAGE SYSTEM

To begin with, an estimate can be made of the surface runoff by a) taking into account the total hard surfaces of the DMQ, and b) adopting an appropriate surface runoff coefficient. In terms of hard surfaces, 10% of the total urban area of Quito is made up of green spaces (Rivadeneira 2014.). This leaves 90% hard surfaces, which for the purposes of the present exercise it is assumed is made up primarily of concrete, paving, asphalt, tarmac, roof tiles and bitumen materials. Table 1 shows the surface runoff coefficients for these material types.

Table 1. Surface runoff coefficients for typical materials in a city's hard surfaces (Pande and Telang 2014).

Material	Surface runoff coefficient
Concrete	0.6-0.8
Paving	0.5-0.6
PVC geomembrane	0.85-0.9
Roof tiles	0.8-0.9
Corrugated metal	0.7-0.9

For this study an average value for the surface runoff coefficient of 0.7 is adopted. Then, if Quito's urban area is 19,014 ha (EPMAPS 2011) and 90% of this can be said to be hard surfaces (Rivadeneira 2014.), it can be deduced that the total surface runoff area is 17,112.6 ha. From this area the peak rainfall over 24 hours is 39.9 l/m^2, with an average surface runoff coefficient of 0.7. As such, the total peak load on the sewerage system is given by: 17,112.6 (ha) * 10,000 (m^2/ha) * 39.9 (l/m2/24hr) * 0.7 = 4,779,549,180 litres over 24 hours, or 55.32 m3 per second average over 24 hours. This is nearly 12 times greater than the total discharges of Quito's residual waters into the Machángara, Monjas and San Pedro rivers combined. It can therefore be concluded that rainwater runoff peak loads, are key in determining whether to increase the capacity of Quito's wastewater discharge systems. This was indeed the case in the Integrated Masterplan for Water and Sewerage for the DMQ, issued by the Metropolitan Public Company of Water and Sanitation (EPMAPS) in 2011. Amongst the suggestions, it was put forward to increase the number of drains to collect surface runoff flows, in addition to improving and increasing the infrastructure links of these drains to the main water network. The estimated cost of works for the Integrated Masterplan came to 161 292,262.25 USD (EPMAPS 2011). It can be said therefore, that increasing the capacity of the city sewerage infrastructure to accommodate peak rainwater runoff flows implies great costs. Additionally, the construction works are likely to be intrusive and over a lengthy time scale.

3 PERMEABLE (GREEN) SPACES AT THE GROUND LEVEL OF THE CITY

Given the lack of SUDS in Ecuador an international case study has been looked into, which has been selected as it was felt to represents an application of SUDS that could potentially be applied to Quito.

3.1 Vibrant Syracuse Spaces and Bioretention Area

The objective of the project was to transform an unused space into a parking lot for the company Gear Factory Inc, but in a manner whereby permeability of the ground was maintained (Mahoney 2011). This was achieved through a mix of using urban vegetation and permeable paving. The areas of vegetation have levels of rainwater filtration in the region of 95% (Berghage *et al.* 2009), and the permeable paving of 84% (Hernández and Martínez 2014). In the state of New York precipitation levels reach 107 mm per month (CustomWeather 2015). By implementing the strategies outlined above practically all of the otherwise 99.51 m^3 of rainwater runoff, from what would have been 930 m^2 of impermeable parking space, was avoided (Mahoney 2011).

4 PERMEABLE (GREEN) SPACES AT THE LEVEL OF THE ROOFS OF THE CITY

In Ecuador it is rare to see a green roof. Nevertheless, there are a number of examples can be drawn on. In this section two case studies of green roofs in Quito are presented. The first is the Ministry of the Environment building in Quito. The third is a private apartment building located in the North of Quito, called vivAlto. In order to understand these green roofs, the basics of green roof design needs to be outlined. A green roof can either be extensive, with a thin substrate and small plants, or intensive, with a thick substrate, large plants and significant subsequent loads on the building structure (Newton *et al.* 2007). For this paper, simple extensive green roofs are considered, where the aim would be that they are applied en-masse without too many complications being incurred on the building structure Two case studies are considered, both to be found in Quito, Ecuador: i) The Ministry of the Environment (MAE) building in Quito, and ii) the Vivalto apartment building. Regarding the MAE building, it was found out from the user of the building conducting the site visit, that the vertical garden was installed as a retrofit to the building (Ministry Technician 2015). A point of interest is that accessible green space was added to the building, without having to sacrifice any floor space of the building itself. Added to this, the garden furniture was made from recycled materials such as tyres, which proved to be not only comfortable, but also resistant to the elements over time. With relation to the Vivalto building, this was built by a private construction firm in Quito. The firm representative highlighted two point of interest in the application of a green roof (Soria, F., Personal Communication, Oct 12, 2015). First it was pointed out how great care needed to be taken over the waterproofing layer, where they had found there to be complications due to water infiltration into the building below. Second, it was pointed out how the green roofs were appreciated by the building residents, where despite being a high-rise, there were nevertheless green areas that children could play in. The roof is shown in Figure 1 (a) and (b).

(a) (b)

Figure 1. a) Green roof at the Ministry of the Environment building in Quito; b) one of the green roofs at the vivAlto building.

5 THE POTENTIAL FOR GREEN ROOFS IN SUSTAINABLE URBAN DRAINAGE SYSTEMS (SUDS) IN QUITO

It was estimated in Section 2 of this paper, that the total urban surface runoff from Quito was 55.32 m^3/sec over a 24 hour period average. This was with the city having 10% green spaces and 90% hard surfaces. In addition, Quito has a total of 763,719 households (INEC 2010). These are then divided into the following typologies:

- House: 57% (435.320)
- Apartment: 29% (221.478)
- Room: 8% (61.097)
- Temporary housing: 6% (45.824)

The Quito Municipality (DMQ 2011) classifies the surface area the household occupies as:

- Small household ≤ 65 m^2
- Large household = between 65 and 120 m^2 (average = 92.5 m^2)
- Extra large household > 120 m^2 (taken for this study as = 120 m^2

By putting together the house typologies with appropriate surface area, the values shown in Table 2 can be calculated:

Table 2. Estimated roof surface areas for Quito.

Household Type	Total No. Households	Surface Area Classification (m^2)	Total Estimated Roof space (m^2)
Rooms and temporary housing	106,921	65	6,949,865
Apartment	221,478	92.5	20,486,715
House	435,320	120	52,238,400
Total			79,674,980

Overall, it can therefore be estimated that Quito's roof area amounts to approximately 7,967.5 ha. In order to assess the theoretical impact of green roofs in

SUDS for Quito, it is put forward that *all* the roofs have a green roof installed. This would amount to 43% of the 19 014 ha of the city's surface area.

In Section 4, the surface runoff coefficient from green roofs was given as ≤ 0.2. However, it also needs to be taken into account that in the event of peak precipitation loads, the capacity of a green roof to retain water will decrease as the substrate becomes saturated. This can especially be the case in winter periods (Berghage *et al.* 2009). According to Moran *et al.,* (2005, cited in Newton, Gedge, Earlyn & Wilson, 2007), in a peak rainfall of 36 mm/h, a 75 mm thick green roof in North Carolina had a peak flow reduction of 87% (surface runoff coefficient of 0.13). Over a six month testing period however, the average retention was 63% (average surface runoff coefficient of 0.37). Nevertheless, the weather of Quito is distinct to North Carolina, where a typical day in the rainy season consists of a morning with clear skies and sunshine, followed by an afternoon of heavy precipitation and a clear night. Given this, it seems reasonable to assume that the surface runoff coefficient from green roofs will be higher for peak rainfall loads that 0.2, but lower that the average in North Carolina of 0.37. For this paper an average has been taken between the peak flow reduction and average retention of Moran *et al.* (2005, cited in Newton, Gedge, Early & Wilson, 2007), giving a surface runoff coefficient of *0.25*. Let it be assumed that the peak rainfall is 39.9 ltr/m^2/24hr (INAMHI 2011), where instead of but 10% of the 19,014 ha of the city as green space, an additional 43% of the cityscape are green roofs with a surface runoff coefficient of 0.25 (given the period of heavy rainfall). The total peak load on the sewerage system would be given by:

Hard surfaces: (17,112.6 – 7,967.5) (ha) * 10,000 (m^2/ha) * 39.9 (l/m^2/24hr) * 0.7 = *2,554,226,430 litres over 24 hours*, or *29.56 m^3 per second average over 24 hours*

Green roofs: 7,967.5 (ha) * 10,000 (m^2/ha) * 39.9 (l/m^2/24hr) * 0.25 = *794,758,125 litres over 24 hours*, or *9.20 m^3 per second average over 24 hours*

Thus giving a total of *38.76 m^3 per second average over 24 hours*. Compared to the original surface runoff to the sewerage system of 53.32 m^3 this represents a reduction of 30%, which could be said to mitigate the need for the expansions proposed by EPMAPS in Section 2.

6 CONCLUSIONS

This paper set out to examine the effect of the widespread application of green roofs in Quito, Ecuador, as a possible alternative to costly sewerage expansion works. Wastewater discharges were studied, and it was shown that peak rainfall runoff was key in justifying expansion works. National case studies of green roof applications were looked into. International case studies of permeable surfaces were also studied. Overall, it was found that in theory should all of the roof space of Quito be harnessed as green roofs, there would be a 30% reduction in peak flows to the sewerage systems. This is likely to have a positive impact in mitigating the needs for costly sewerage expansion works. Additionally, there would be further room to implement permeable surfaces at the ground level of the city.

7 RECOMMENDATIONS FOR FURTHER RESEARCH

The following points are highlighted as of interest for further research. First, further study into the amount of reduction in surface runoff that would be necessary to completely abandon plans for sewerage expansion, coupled with research into possible locations of permeable spaces at the ground level of Quito. Second, study into building typologies and their structures, in order to assess whether green roofs are feasible, in conjunction with social research into the acceptability of green roofs and a study of financial mechanisms to promote their application en-masse.

Acknowledgements

Special thanks needs to be made to Marlon Rios, a student of Architecture at the Pontificia Universidad of Ecuador. His background research in quantifying the total roof space of houses in Quito was invaluable in the development of this paper. In addition, the site visits for the case studies would not have been possible without the support and collaboration of Francisco Soria, Director of the Corprancagua construction company, as well as the Subsecretary of Environmental Quality of the Ministry of the Environment of Ecuador.

References

Berghage, R. D., Beattie, D., Jarret, A. R., Thuring, C., and Razaei, F., 2009, Green Roofs for Stormwater Runoff Control. Retrieved from http://nepis.epa.gov/Adobe/PDF/P1003704.PDF.

Berndtsson, J. C., Green roof performance towards management of runoff water quantity and quality: A review. ELSEVIER (Ed.), 3-4, 2009.

CustomWeather, 2015, My Forecast. Retrieved from http://www.myforecast.com/bin/climate.m?city=24384&metric=true&selectedMonthNum=7.

DMQ, Distrito Metropolitano de Quito, 2011, Ordenanza Municipal No 3746. Retrieved from http://www.departamentos.com.ec/blog/39-municipio-regulaciones/80-ordenanza-municipal-3746.html.

EPMAPS, Empres Pública Metropolitana de Agua Potable y Saneamiento, 2011, Programa de Saneamiento Ambiental para el Distrito Metropolitano de Quito. Retrieved from http://www.aguaquito.gob.ec/sites/default/files/documentos/plan_maestro_alcantarillado.pdf.

Hernández, B., and Martínez, O., 2014, Diseño de un campo de prueba piloto de pavimentos permeables en la ciudad de Cartagena. Retrieved from http://190.25.234.130:8080/jspui/bitstream/11227/1116/1/Dise%C3%B1o%20de%20un%20campo%20de%20prueba%20piloto%20de%20pavimentos%20permeables%20en%20la%20ciudad%20de%20Cartagena%20%5BBrian%20Hernandez%20-%20Omar%20Mart%C3%ADnez%5D.pdf.

INAMHI, 2011, Instituto Nacional de Metereoligía e Hidrologia. Retrieved from http://www.serviciometeorologico.gob.ec/wpcontent/uploads/anuarios/meteorologicos/Am%202011.pdf.

INEC, 2010, Instituto Nacional de Estadística y Censos, Censo de Poblacion y Vivienda Quito. Retrieved from http://www.ecuadorencifras.gob.ec/wp-content/descargas/Manu-lateral/Resultados-provinciales/pichincha.pdf.

Mahoney, J. M., 2011, Quito, una ciudad de grandes parques. Retrieved from http://savetherain.us/wpcontent/uploads/2011/12/GIFVibrantSyrSpaces_FS1.pdf;

Newton, J., Gedge, D., Early, P., and Wilson, S., Building Greener Guidance on the use of green roofs, green walls and complementary features on buildings. CIRIA, London, 2007.

Pande, P, and Telang, S., Calculation of Rainwater Harvesting Potential by Using Mean Annual Rainfall, Surface Runoff and Catchment area. GARJ (Ed.), 3-4, 2014.

Rivadeneira, J., 2014, La función ecológica de las áreas verdes en Quito: el caso del parque la Carolina, Quito, una ciudad de grandes parques. Retrieved from: http://flacsoandes.org/vicisda/presentaciones/Mesas/2rivadeneira.pdf.

Interaction between Theory and Practice in Civil Engineering and Construction
Edited by Komurlu, R., Gurgun, A. P., Singh, A., and Yazdani, S.

COMPARATIVE STUDY OF ENERGY CONSUMPTION FOR AIR CONDITIONING BETWEEN RESIDENTIAL BUILDINGS WITH TWO DIFFERENT SUBFLOORING SYSTEMS

H. ALBAYYAA, D. HAGARE, and S. SAHA

School of Computing, Engineering and Mathematics, University of Western Sydney, Penrith, Australia.

Energy consumed by heating, ventilation and air conditioning account for about 40% of the total energy used in an average Australian home. The main feature that categorizes the construction systems is the thermal mass as it contributes directly to the thermal performance of the entire house. High thermal mass flooring and walls are most appropriate in climate with high diurnal (day-night) temperature ranges. High thermal mass construction system has higher embodied energy but this can offset by reducing heating and cooling energy consumption over the life span of the house. The optimum design, in terms of desirable heat gain or loss, can be achieved by considering the building orientation, thermal mass and careful design of the building envelope including roof, walls, windows and floor systems. To demonstrate relative advantage in terms of energy conservation between houses with different construction systems and thermal mass, two model houses which are detached dwelling with a floor area of 200 sqm and with two levels and four bedrooms were selected in this study. One of the model houses represented modern house with brick veneer walls and concrete slab-on-ground flooring (high thermal mass Model). The second model house represented old house with fibro walls and raised subfloor (low thermal mass Model). The analysis has been carried out using computer software (IDA ICE). The energy performance of the buildings were computed and compared. The results show that the modern house consumed 53% less energy compared to old house and hence the former is significantly cost effective over the long run.

Keywords: Subflooring system, Thermal mass, Heating and cooling of residential buildings, Stack effect.

1 INTRODUCTION AND BACKGROUND

Households account for 26% of the total energy consumed in Australia and is increasing by 2.6% per year (ABS 2006). In NSW, 88% of residential dwellings use some form of heating systems. The principle systems implemented involved reverse air-conditioning (27%), electric resistance heating (25%) and gas heating (23%). Similarly, different types of air conditioning are in use in approximately half of NSW houses for cooling (ABS 2006).

Design of the house with correct orientation is a fundamental step as passive solar design can reduce the energy demand for heating and cooling. Northern oriented dwelling in Australia could use the solar radiation during winter to reduce heating requirement, while in summer solar radiation can be blocked to prevent unwanted heat. Along with passive solar design, building optimization can be the most important step in providing the house with thermal comfort. Increasing insulations, better performing windows design, heat recovery from ventilation and internal air movement are parts of building optimization.

Many studies have been exploring the energy requirements for the residential buildings, through either computer simulation or field studies considering different parameters. In previous studies different scenarios have been investigated in attempt to reduce the energy consumption through building optimization (Hasan *et al.* 2008, Bambrook 2009, Bambrook 2011). Hasan *et al.* (2008) considered five design variables in both building and HVAC system to minimize the life cycle cost. Three of the variables were continuous which the additional insulation is in the external wall, roof and floor, while the other two were discrete variables which are the U-value of the windows and the type of heat recovery. Bambrook (2009) and Bambrook *et al.* (2011) optimized the building materials using different scenarios to the point where simulated energy requirement for the air conditioning is at the minimum value. However, the models for the studies previously mentioned considered the entire building as a single zone. This will affect the final results of the simulation as the internal air movement between the different zones and levels where not included in the analysis. Also, these studies did not include occupants' number and behaviors in the analysis.

The purpose of this paper is to identify the difference in energy consumption between a modern house with brick veneer walls and concrete slab-on ground flooring and an old house with fibro walls and floating timber floor which is raised 1 m above the ground level. This paper uses a holistic system approach whereby the dwelling is considered as a complete unit with multi-zones.

Comparison between the two different house scenarios was investigated using computer software (IDA ICE). IDA ICE is a whole year detailed dynamic multi-zone simulation program for the study of indoor climate of individual zones as well as energy requirements of an entire building. IDA ICE can simultaneously perform assessment of all fundamental issues related to the design of a building: envelope, glazing, HVAC system, windows and shadings, occupant's behaviour, etc...

Table 1. Models specification.

Parameters	Modern House	Old House
Floor	Concrete slab on ground	Timber raised floor
Roof	Terracotta roof	Fibro Roof sheeting
Walls	Brick veneer	Fibro walls
Windows	Double glazing	Single glazing

IDA ICE is delivered in two editions: standard and expert edition. In this study, expert edition has been used to perform the energy loads. The accuracy of the IDA ICE was validated according to CEN13791 (Kropf and Zweifel 2001). This simulation program (IDA ICE) is used to measure the space heating and cooling requirements of a 200 m^2 house. The Sydney weather data for the simulation was obtained from

American Society of Heating, Refrigerating and Air-conditioning Engineers (ASHRAE) database (IDA ICE 2014). Both houses were oriented towards the north to maximize the heat gain during the cold season. This applies to countries located to the south of the equator otherwise houses should face south if they are located above the equator The main differences between the two model houses are given in Table 1.

2 ODERN HOUSE (BRICK VENEER WALLS- CONCRETE SLAB ON GROUND FLOOR)

The parameters on this model have considered using standard building materials (Wilkie 2011) used widely in the industry. Brick veneer-concrete slab on ground was used during the simulation. Main living room with high window to wall ratio was oriented towards the north. This allows maximum heat gain during winter and avoids the heat gain during summer. Gyprock internal walls with light insulation were utilized. Double glazing windows were considered and finally the roof materials considered were terracotta tiles.

Modern House was built as a multi-zones model. With two different floors connected through a stair case, it is very important to simulate the stack effect between the two floors. This was also considered during the simulation. Occupants' behaviors were also set during the whole week throughout the year (occupants are out of the house from 8:00 to 15:00 then half of the occupants were considered inside the house from 15:00 to 17:00 during weekdays and all occupants are considered inside the house during weekends). The simulation was done for both heating and cooling requirements.

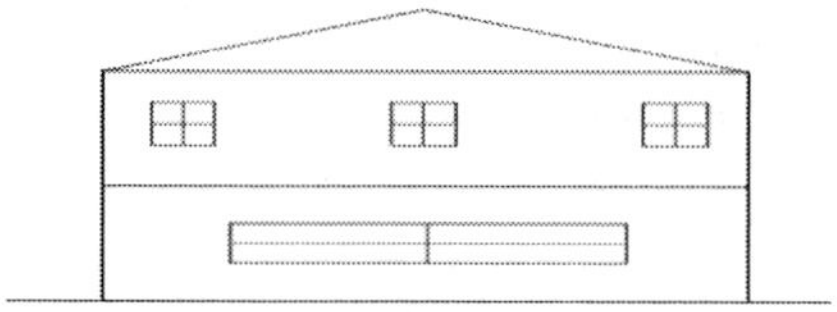

Figure 1. Model house 1.

Air conditioning energy requirement for the Modern House shown in Figure 1 was obtained using the computer model (IDA ICE). The total energy requirement for heating and cooling are also presented in Table 2.

3 OLD HOUSE (FIRBRO HOUSE- RAISED SUBFLOOR)

Standard building materials were used to simulate the model shown in Figure 2. The main key factors were the floating ground floor of the house and the materials used for walls and roof as identified in Table 1. The floor is raised 1 m above the ground. The space between ground level and the ground floor of the house was considered as an open zone in the simulation, which is open to atmosphere. The temperature in this zone was always maintained at ambient levels. This will enable the software to simulate the effect of the outside air movement into the house through the floor and analyze the effect of this air movement on the energy consumption of the house. Timber flooring system was implemented for the floors. Fibro flat sheets are used for models walls and ceilings with thickness of 6 mm and 4.5 mm, respectively (NSW Government 2015). Main living room with high window to wall ratio was oriented towards the north to gain heat in the winter and control solar radiation in summer. Ground and first floors were connected with a staircase. Buoyancy effect was also considered during the simulation of the model to ensure realistic conditions. Occupants' behaviors were also set during the whole week as described in Section 2.

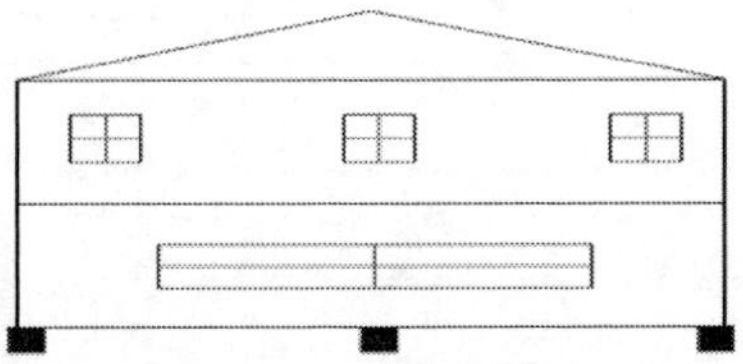

Figure 2. Model house 2.

Energy requirements, as generated by the computer software IDA ICE, for heating and cooling of the model house are shown in Table 2.

4 DISCUSSION OF RESULTS

The results of the annual energy consumption for heating and cooling have been produced in kWh/yr and kWh/(yr.m^2) as shown in Table2. The modeling results indicate that the energy required for heating and cooling of Modern House was 53% less than the energy required for Old House. The annual total energy for Modern House was 10,658 kWh/yr while it was 19,956 kWh/yr for the Old House as shown in Figure 3.

The above results show the considerable differences in the energy requirement for heating and cooling between the modern and old houses. The old houses, in Australia, were built about 40 to 50 years ago. Currently, almost all of the houses built are modern houses. Thus, it can be said that the current houses are far superior to the old houses in terms of energy conservation.

Table 2. Total delivered energy.

	Modern House		**Old House**	
System description	kWh/yr	kWh/(yr/m^2)	kWh/yr	kWh/(yr/m^2)
Electric cooling	9866	24.66	15508	38.77
Electric heating	792	1.98	4448	11.12
Total	10658	26.64	19956	49.89

The results indicate that the modern houses with brick veneer walls and concrete slab on ground flooring are more efficient in energy saving compared to the old house with fibro walls and timber subfloor system. Energy savings over the typical life span of the building of 50 yr period can be significantly high for the modern house. This energy saving works out to be 464,900kWh Based on the average energy tariffs in

NSW, which is AUS $0.278/kWh, the net present value (NPV) of the cost savings over 50 yr works out to be $47,482 . This amount is calculated using an interest rate of 5%. In addition to this cost saving, the energy saving achieved in the case of Modern House will have positive environmental effects, particularly, in terms of reduced greenhouse gas emissions.

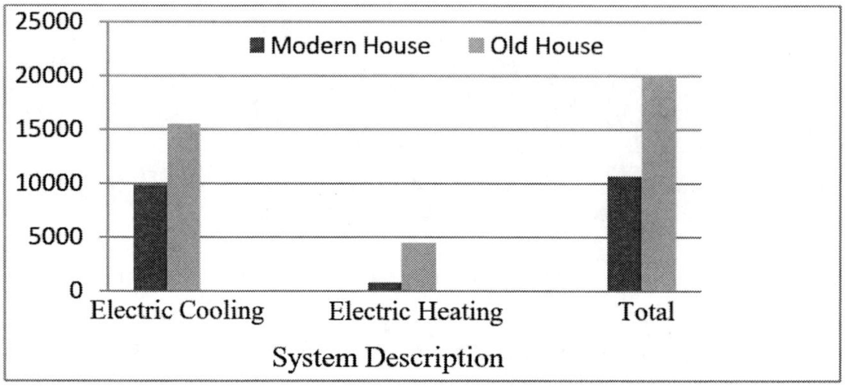

Figure 3. Total energy consumption.

Acknowledgements

Authors express sincere gratitude to School of computing, Engineering and Mathematics IT team, particularly Mr. Keith Rynott and Mr. Rocqueforte O'Leary, for their assistance in installing the software. Also, our thanks are due to EQUA Simulation AB, Sweden, specifically, Ms. Jenny Sjostrom for providing technical support in relation to use of the software.

References

Australian Bureau of Statistics, 2006, Domestic water and energy use in NSW. Australian government. Available online: http://www.abs.gov.au. Accessed on April 8, 2015.

Bambrook, S., Sproul, A., and Jacob, D. *Exploring The Zero Energy House Concept For Sydney*. In Solar09 47th Australian and New Zealand Solar Energy Society Annual Conference, ANZSES, Townsville, Queensland, 2009.

Bambrook, S. M., Sproul, A. B., & Jacob, D. Design Optimisation For A Low Energy Home In Sydney. *Energy and Buildings*, Elsevier,*43*(7), 1702-1711, March, 2011.

Hasan, A., Vuolle, M., and Sirén, K. ,Minimisation of Life Cycle Cost of A Detached House Using Combined Simulation And Optimisation. *Building and Environment*, Elsevier, 43(12), 2022-2034, 2008.

IDA-ICE, 2014, EQUA Simulation, Stockholm, Sweden.

Kropf, S., & Zweifel, G. (2001). Validation of the Building Simulation Program IDA-ICE According to CEN 13791 "*Thermal Performance of Buildings–Calculation of Internal Temperatures of a Room in Summer Without Mechanical Cooling–General Criteria and Validation Procedures*". *Hochschule Technik+ Architektur Luzern. HLK Engineering.*

New South Wales Government (2015). Retrieved from www.nsw.gov.au/fibro/ on August 11, 2015.

Wilkie, G., *A comprehensive guide for owner- builders, revised edition with green supplement*, building your own home Sydney, 2011.

Interaction between Theory and Practice in Civil Engineering and Construction
Edited by Komurlu, R., Gurgun, A.P., Singh, A., and Yazdani, S.

COMPARISON OF LIFE CYCLE COST OF CENTRALIZED AND SPLIT AIR CONDITIONING SYSTEMS

DHARMA HAGARE, JASON HO, and SWAPAN SAHA

School of Computing, Engineering and Mathematics, Western Sydney University, Penrith, Australia

Central and split systems are the two most common air conditioning (AC) systems used in residential applications. Central system employs one large unit to produce and distribute conditioned air through a system of ductwork. On the other hand, the split system, employs several small ACs. Each AC consisted of outdoor and indoor units to produce conditioned air directly to the designed area. Each system has distinct strengths and weaknesses. Depending on the structure of cooling area and operating schedule, the performance of each system will be different. The aim of this paper is to examine the impact of various parameters such as operating schedule and building characteristics to the performance of central and split AC systems over the 25 years of their operation. The life cycle analysis (LCA) considered essential factors which have significant impact on the energy consumption and both initial and operating costs of the two systems. All required sections of life cycle analysis are included according to the relevant Australian Standards. The results indicated that under standard operating conditions, central system is more economical and energy efficient than split system. However, when the flexibility in operation of split system is considered, there was a significant reduction in its operating cost, which was below that of central system. Overall, total life cycle cost of split system was slightly lower than central system. Also, considering the usage flexibility and the comfort of users, it appears that the split system is more suitable than the central AC system for residential buildings.

Keywords: Heating, Cooling, Residential buildings, Sustainability, Initial cost, Operating cost, Life cycle cost.

1 INTRODUCTION

Air conditioning is an essential part of both residential and commercial buildings. This would be the case especially in the light of impending climate changes. Climate change is expected to increase extreme weather conditions in the future. This makes all the more important for buildings to have air conditioning systems which can provide warm or cold air when required. The objective of the air conditioning is to maintain the inside the building air temperature at around 24°C.

This study mainly focuses on the residential buildings air conditioning. Air conditioning requirements for residential buildings is generally met either by centralised or split system. Both systems have their own strengths and weaknesses. Central system was proven in many previous studies (Gheewala and Nielsen 2003,

Yang *et al.* 2001) to have better efficiency in operation than split systems. In central system one single AC unit is employed to provide conditioned air. The unit is connected with a system of ductwork to deliver conditioned air to the designated areas (rooms). The position where the unit is installed greatly affects the design of duct system. The centre of the building is normally taken as the best position to locate the central unit where minimal length of duct is required to run the duct to designated rooms. The unit is usually placed on the ceiling area and hence it is called 'rooftop unit'. On the other hand, a split system can cover a smaller area than the central system, normally one unit per room. A split system consisted of an outdoor and indoor unit. Refrigerant is transferred between these two units to receive and release heat into or out of a building depending on the operating mode.

A building model was developed using Archicad software (https://myarchicad.com/). The model was assigned with a construction set comprising a particular types of external wall, roof, ground floor and windows. More details are given in Ho (2013). Operating cost of the AC systems was simulated for the building using Bentley Hevacomp software (https://www.bentley.com/en/products/brands/hevacomp).

2 INSTALLATION, MAINTENANCE AND DISPOSAL COSTS

The initial installation costs for both central and split systems were estimated by taking into account the market value of various components, which are given in Table 1. Central system was found to be slightly expensive due to the high installation cost associated with the ducting. The assumptions involved in estimating the maintenance cost and the estimated costs for both central and split systems are given in Table 2. Also, Table 2 shows the estimated disposal costs after the life period of 25 years.

Table 1. Initial installation cost of both the systems.

Items	Central system	Split system
Initial installation cost	$9,237 (this included installation of the main air conditioning unit plus the associated ducting)	$8,792 (this included installation of 7 split systems)

3 OPERATION COST

To estimate the operation costs for both the air-conditioning systems two scenarios were developed reflecting the operation of AC systems in residential building in real life (Ho 2013):

- Scenario 1 was aimed to represent typical AC usage schedule in a general family, for example: from 9am to 5pm all members have to leave the house for work and school, hence the system is off. When all members are at home, ACs in selected areas are operated to suit particular activities such as having dinner or go to bed.
- Scenario 2 was developed to reflect the operation flexibility of split system to increase the overall efficiency of HVAC operation. In scenario 2 it was assumed that one bedroom is used as guest room and hence, requires no

cooling need during the week. The guest room will be utilised in the weekend and will be operated on the same schedule with other bedrooms. Operation of AC in other room was also modified. This scenario particularly applied for split system as each system could be operated individually.

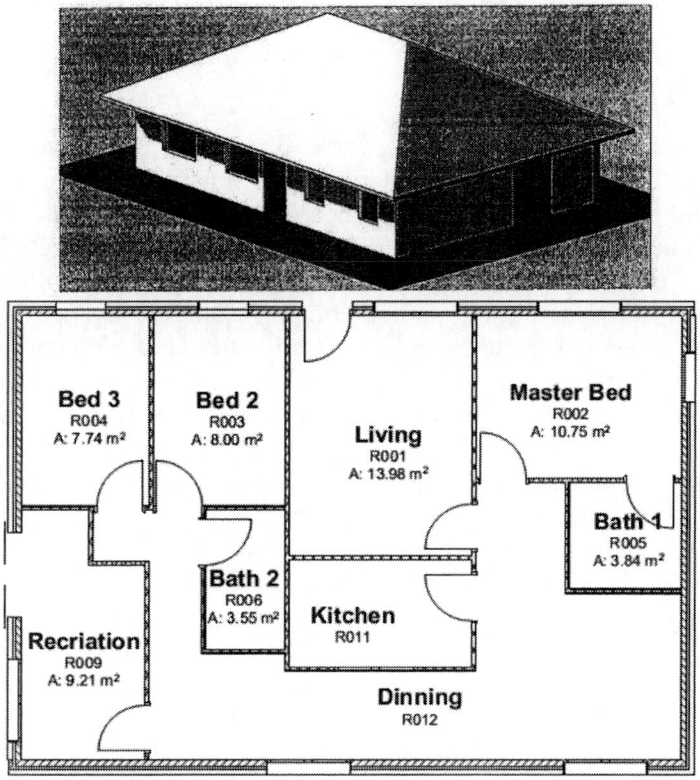

Figure 1. General plan and 3D view of the building considered for simulation.

Table 2. Maintenance and disposal cost of central and split system.

Maintenance	
General tasks	
Refrigerant check; Temperature check; Function check	
Central system	**Split system**

• Return filter replacing
• External condenser coil cleaning
	• Outdoor unit cleaning
	• Indoor unit cleaning
	  o Filters cleaning
	  o Indoor evaporator coils cleaning

Unit cost: $ 250 per system Total annual cost: $ 250	Unit cost: $ 80 per unit Total annual cost: $ 560
Disposal	
Central system, including • Duct disconnection • Refrigerant removal • Unit removal Unit cost: $ 450 per system Total cost: $ 450	Split system, including • Pipe disconnection • Refrigerant removal • Unit removal Unit cost: $ 120 per unit Total cost: $ 840

Annual operating cost for both air conditioning systems were estimated using Bentley Hevacomp software. The results are shown in Figure 2. As explained earlier, while the Scenario 1 was applied for both central and split air conditioning systems, Scenario 2 was applied only for split air conditioning system. An inflation rate of 3 per cent was used to calculate future operating cost base on the first year cost (Trading Economics 2013).

4 NET PRESENT VALUE (NPV)

Using the economic parameters listed Table 3, net present values over 25 years operating life of the air-conditioning systems were computed. The results are shown in Figure 3. As shown in the figure, under standard operating conditions (Scenario 1), the central air conditioning system works out to be cheaper over the 25 years of the operating life. However, when the flexibility of the operating split air conditioner (Scenario 2) is considered, the split air conditioning system works out to be cheaper. In addition to the economic benefits associated with the split air conditioner over the operating life of 25 years, it provides superior comfort for the users as the level of air conditioning can be set individually for each room. On the other hand the temperature settings under the central air conditioning system can only be set as per the zones which may not provide the ability to adjust the temperature at room level. Thus the comfort at the room level is not guaranteed.

Table 3. Input data for life cycle analysis simulation.

Parameter	Value
Annual inflation rate	3 %
Period of analysis	25 years
Electricity price	\$0.30/ kWh
NPV discounted rate	5 %

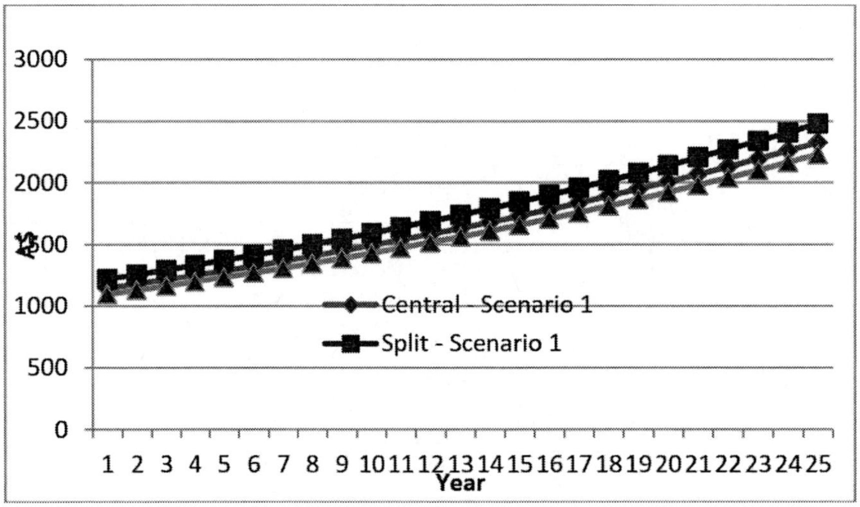

Figure 2. Annual operating cost of central and split systems over 25 years of life.

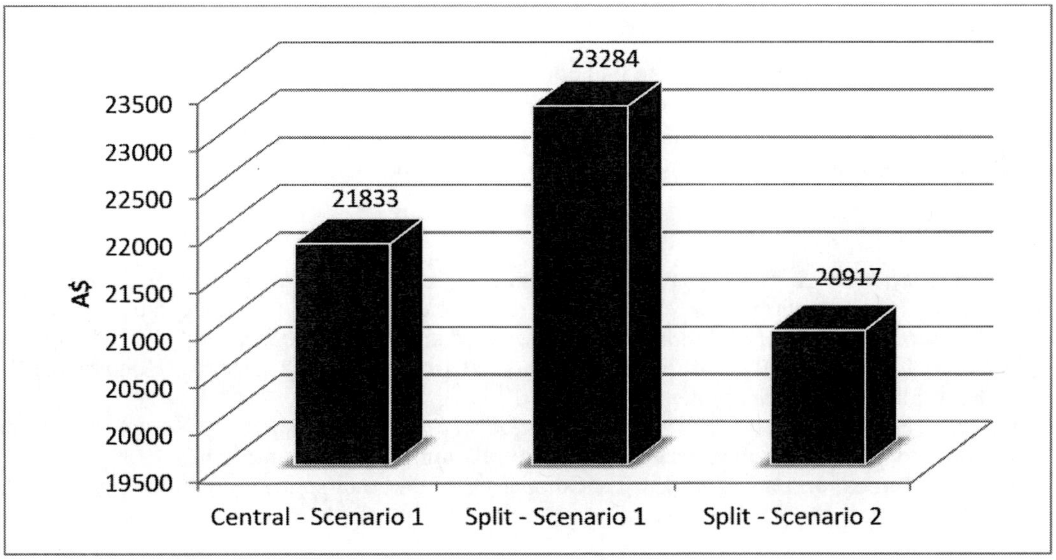

Figure 3. Total operating NPV in 25 years life cycle of central and split system.

5 CONCLUSIONS

This paper compares the overall cost over 25 years period between central and split air conditioning (AC) systems. Central system provides conditioned air by zone and appears to have effective airflow management. However, this type of AC system requires long installation time and higher initial cost. Moreover, zoning does not allow control of conditioned air at room level which can become a source of energy wastage if there is any room belong to the zone but is not required to be conditioned. Also, central AC systems do not provide for user flexibility. Split system has the ability to control conditioned air at room level and hence, can achieve better energy efficiency. However, single split requires one set of indoor and outdoor units for each room to be conditioned which may be time-consuming and costly for maintenance. Operation of split system is less efficient than central system.

Both central and split systems have distinct characteristics that can affect the associated cost in various stages in their life cycle. Central system which the requirement of ductwork for supply and return system has its initial cost varies significantly according to the number of zones, the building structure and the size of cooling area. On the other hand, initial cost of split system seems to be immune with change of building structure and the size of cooling area. Initial cost of split system only changes at a moderate rate when there is change in the number of rooms which usually requires purchase of additional AC units. Operational stage is where central system shows its strength over split system. Simulations in annual and life cycle perspectives show that central system operates more effective and economical than split system in the same scenario.

A 25 years life cycle cost analysis shows that in a long term consideration, split system is a more suitable HVAC system to be chosen. The inefficiency in operation is outweighed by the low initial cost and flexibility in operation which makes life cycle cost of split system more economical than central system.

Acknowledgements

Authors express sincere gratitude to School of computing, Engineering and Mathematics IT team, particularly Mr. Keith Rynott and Mr. Rocqueforte O'Leary, for their assistance in installing the software. Also, our thanks are due to EQUA Simulation AB, Sweden, for providing technical support in relation to use of the software.

References

Gheewala, S. H. & Nielsen, P. H., Central and individual air-conditioning systems - A comparison of environmental impacts and resource consumption in a life cycle perspective. *International Journal of Sustainable Development and World Ecology,* 10, 149-155, 2003.

Ho, K. D., Life cycle analysis of different residential air-conditioning systems. Honours thesis, University of Western Sydney, Austrlia, 2013.

Trading Economics, *Australia Inflation Rate* [Online]. Retrieved from http://www.tradingeconomics.com/australia/inflation-cpi on September 13, 2013.

Yang, H., Burnett, J., Lau, K. & Lu, L., Comparing central and split air-conditioning systems. *ASHRAE Journal,* 43, 36, 2001.

A CONCEPTUAL MODEL FOR EVALUATING THE RELATIONSHIP BETWEEN THE PROFESSIONAL FIRM SERVICE QUALITY AND CLIENT SATISFACTION IN BUILDING PROJECTS

ALUKO OLUSOLA RAPHAEL

Dept of Architecture, College of Engineering & Environmental Studies, Olabisi Onabanjo University, Ibogun Campus, Ogun State, Nigeria

Conceptual model helps the researcher to explore variables of study and their clear interactions. Therefore this study aims at proposing a model of a relationship between the professional firm service quality and client satisfaction in building development projects. This is because client satisfaction remains an important factor in the development of engineering and building consulting services. In today's competitive service market, attaining client satisfaction is very crucial for the continued survival and economic stability of consulting firms. The model is built on the concept of service quality and its effect on client satisfaction. The fundamentals of Servqual theory in addition to competency profile and communication within the consulting arm of building projects constitute the independent variables of the model. The Attribution theory constructs constitute the moderating variables of the study. The proposed model focuses on the relationship between soft service qualities, hard service qualities and the client satisfaction variables of architectural quality, technical quality, functional quality, timeliness of service delivery, cost of services and supervision at implementation stage. The findings will be useful to assist consulting firms in enhancing competitiveness and improve project deliveries in the present global challenges.

Keywords: Competitiveness, Consulting services, Service industry, Knowledge economy, Building consultants, Nigeria.

1 INTRODUCTION

In the knowledge economy, professional services are important drivers of the economic activities and to achieve client satisfaction they face the challenges of organizing and managing their resources in a competitive manner. This is in line with the study of de Jong (2007) who describes *"professional service firms as organization whose core business consists of independent, highly educated and skilled people performing knowledge work and providing non-routine services to other firms or institution in close interaction with client representatives on a temporary basis."*

Professional services in building projects are interconnected and complement each other (Brandl 2010). The services are provided by architectural, engineering and quantity surveying firms. Their activities are largely based on knowledge acquired by the human capital of the firms. The success of both public and private projects is largely dependent on the quality and capabilities of the human resources deployed to perform these technical and design related activities for the clients (Sporrong 2014).

Therefore, there is a need for continuous development and training of the competencies and expertise. The project organization presents opportunity for this through cross project planning and innovations (Koch and Bendixen 2005). The projects determine what they learn and at what proportion. Therefore the type of project and clients become very vital. Critical decision of such clients becomes a pivotal for the concept of client satisfaction itself. The service practitioners provide information flow consisting drawings, specifications, contract conditions, explanations and clarifications which is the basis of all activities in a project. All construction works originate from the flow of this information, and any delay or inefficiencies there from slow down decision-making ultimately causing delay in project delivery. Thus it can be seen that the service quality of these professionals cannot be over-emphasized.

Trend in construction and service industry is moving towards higher quality. With rising expectation, business transformation, there is awareness of the need by building clients to demand for service providers to raise standard. Quality is now the cornerstone for service providers to widen their client base.

ServQual model is the most commonly used theory of customer satisfaction. The model defines service quality (customer satisfaction) as the differences between the customer's expectations and experiences (Parasuraman *et al.* 1988). It is most well known models of perceived service quality which are based on the disconfirmation paradigm. It assumes that customers have certain preconceived expectations of a product or service before actually consuming it. Customers compare the perceived performance of a product (service, goods) with some performance standard. Customers are satisfied when the perceived performance is greater than standard (positively disconfirmed). Dissatisfaction is perceived when the performance falls short of the standard (negatively disconfirmed) (Forsythe 2007) .Dissatisfaction may also be described as the level in which the quality observed by the customer no longer corresponds with the customers' expectations.

Ismail *et al.* (2012) establish three underlying theories of service quality and are: Service quality is more difficult for the customers to evaluate than goods quality; quality evaluations are not made solely on the outcome of a service, but also involve evaluations of the process of service delivery; service quality perceptions result from a comparisons of customers' expectations with the actual service performance. A measure of service quality is usually referred to as customer satisfaction (Vincent *et al.* 2008). Hence measuring customer satisfaction leads to identifying ways to improve customer service quality.

2 SERVICE QUALITY

As perceived by clients, service quality is the extent of discrepancy between expectations and performance. Expectations are the desires and the needs of the client. The performance refers to the client's evaluation of the service provider. Ensuring service quality is meeting or exceeding the expectations from the service (Ismail *et al.* 2012). To manage service quality a clear understanding of what service means to the client is very vital. This measure of service quality is referred to as a client satisfaction.

The European School of thought led by the work of Gronroos (1984) is the pivot on which the concept of service quality revolves. In the concept, service quality is viewed from the technical and functional quality. The technical quality considers if service meets client expectations while the functional quality measures the perception of the client about the production and delivery of the service (Razavi *et al.* 2012)

The American School of thought is built upon the fundamentals of the work of Parasuraman *et al.* (1994). This view regarded as the most prominent has been adopted by a number of researchers in the field of service quality. The study conceptualized service quality as the overall assessment of the difference between perception and expectation of service delivery.

In the model, ten dimensions of service delivery that are generic and relevant to services in general were identified. These dimensions are tangibles, reliability, responsiveness, communication, credibility, security, competence, courtesy, understanding the customer and access. Factor analysis was used in later studies to condense the dimensions into tangibles, reliability, responsibility, assurance and empathy (Parasuraman *et al.* 1991).

Tangible refers to the physical facilities, equipment and appearance of personnel while reliability is the ability to perform the promised service dependably and accurately. Responsiveness is the willingness to help customers and provide prompt service. Assurance refers to courtesy of employees and their ability to inspire confidence, while empathy is the individualized care the firm provides its customers.

Competence is the skill, expertise and professionalism with which the service is executed. Communication is the ability of the service provider to inform the stake holders appropriately in a way he or she would understand.

Professional services are required on an irregular basis. In most cases client experience is very little if any, and these types of services possess a unique set of characteristics (Sonne 1999). Therefore, two elements of the service production processes are categorized as perceived soft process quality and perceived hard process quality.

According to Sonne (1999), soft quality refers to the interaction process between the client and the service providers. Examples are communication and interaction which corresponds to client treatment during service production and includes functional reliability. The hard quality covers non-interactive issues (e.g., professionalism, skills, and physical resources) used in the process of printing technical solution.

The above division of soft and hard quality is consistent with authors who contested that interactive and non-interactive functions should be considered separately in professional services. This is because the nature of consultancy service is characterized by a high degree of complexity about the technical level. This means that the outcome quality of professional services is influenced by his technical ability and partly by the interaction between the consultant and the client. The perceived outcome quality of the service determines the level of client satisfaction (Sonne 1999).

2.1 Service Quality Theory

Technical and Functional Quality Model of Gronroos (1984) emphasizes that an understanding of the client perception of the quality and the way quality is influenced is important for firm to remain competitive. To this end the author indentified three components of service quality which are technical quality, functional quality and image. The Spreng and Mackoy (1996) model tries to enhance a better understanding of perceived service quality and client satisfaction. It highlights the effect of expectations, perceived performance desires, desired suitability and expectation disconfirmation on overall service quality and client satisfaction. Oh (1999) proposes his model as an integration of service quality, customer value and customer satisfaction. Forsythe (2007) on the Disconfirmation of Expectation Model demonstrates how customer satisfaction is affected by the combination of the performer's level of expectation. It holds that satisfaction increases where the performance perception is deemed to be greater than expectations. This is referred to as positive disconfirmation. On the other hand, when a performance perception is lower than the customer's level of expectation, it results into decrease in satisfaction called negative disconfirmation.

3 ATTRIBUTION THEORY

The concept of attribution theory deals with the idea of assigning causality to either external or internal factors in carrying out a responsibility also forms part of the framework development. External and internal attribution assigns causality to an outside force or to an internal force, respectfully. It gives reasons for people to understand their successes and failures. The theory locates attributions in dimensions of internal/external, stable/ unstable and controllable/uncontrollable (AWE 2005).

4 RELATIONSHIP BETWEEN SERVICE QUALITY AND CLIENT SATISFACTION

Amos-Abanyie *et al.* (2014) in their study concluded that in today's sophisticated, large-scale, risky and adversarial construction project environment, the evaluation of a relationship between the services of professionals and client satisfaction is very crucial. It was deduced empirically that the concept of perceived quality is the bedrock of client satisfaction and acceptance of product, and the service outcome. This means service quality is significantly related to customer satisfaction. This demonstrates that a linear relationship exist, meaning that higher level of service quality is related to customer satisfaction. That is to say that the practice of satisfaction is grounded on meeting the needs of the client.

5 METHODOLOGY

This study is part of an ongoing PhD research work, and it uses explanatory method in seeking explanations of the observed phenomenon of interest. It tends to identify causal factors and its outcome on the concept of client satisfaction in the consulting sub-sector of the building projects. The model is being developed from an extensive review of past researches using service quality theory and attribution theory.

6 THE CONCEPTUAL MODEL FOR THE STUDY

Conceptual model helps the researcher to explore variables of study and their clear interactions. In this model, client satisfaction is conceptualized as a function of the professional services quality of the consulting firms. The enabling conditions are the internal controllable factors and external uncontrollable factors influencing the performance of the firms in a given environment (as shown in Figure 1) drawn from the concept of attribution theory.

These influence the quality of both hard and soft quality services and are operationalized to form the set of independent variables. The interaction process between the client and the service providers covering areas of communication, cooperation, ability to keep contracts, deadlines is categorized as soft process. The hard process covers the skills, physical resources and technical qualities deployed towards providing technical solution. The outcome quality is the final document or service the client receives.

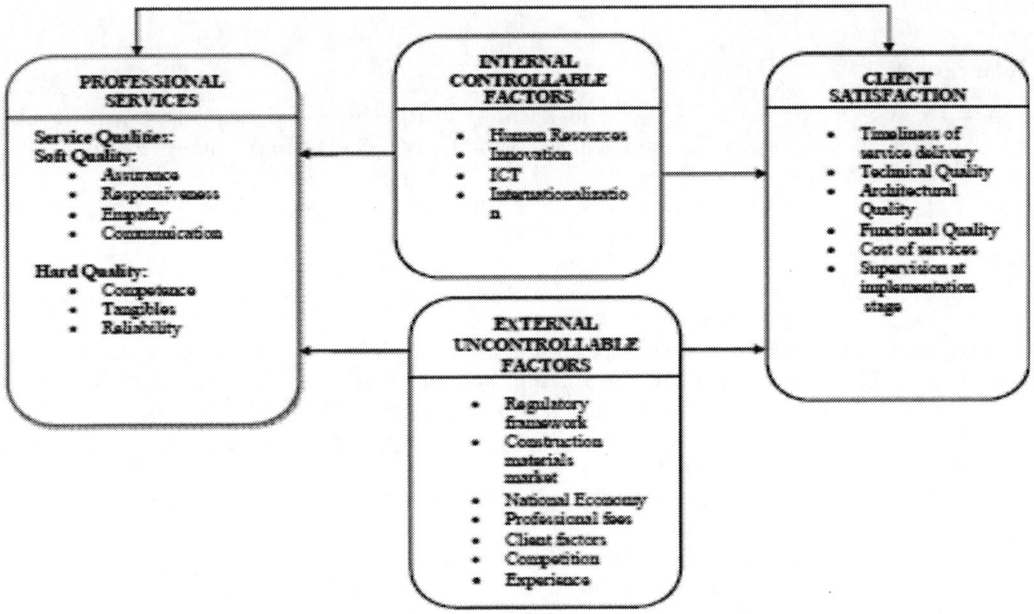

Figure 1. Proposed conceptual model for evaluating the relationship between professional firm service quality and client satisfaction in building projects.

The framework also draws from the conclusions of other researchers (Sonne 1999, Amos-Abanyie *et al.* 2014) that in todays sophisticated, large-scale, risky and adversarial construction project environment, client regards competencies and experience of service providers as very significant factors that relate to the actual service quality in meeting their satisfaction. This often influences clients' decision in team selection for future projects. The perceived outcome which represents a part of the model is operationalized as the client satisfaction measures. The following are the client satisfaction measures: Architectural quality measuring the aesthetic appearance of the completed building project; timeliness of service delivery which relates to the deliverables being timely , detailed and comprehensive ; technical quality, functional quality relating to flexibility in design and fitness for purpose of the project; cost of services which deals with comparing budgeted cost with actual, price of service relative to quality, quality relative to price of service. Supervision at implementation stage involves control of progress of work, control of quality of material and workmanship incorporated in the work, and completeness of site records (Figure 1).

7 CONCLUSION

A conceptual model is a theoretical construct from validated scientific research representing an explicit situation in the real world being investigated. Therefore, this descriptive model represents a position through which the variables of the study can be visualized. It also suggests the relationship of the variables of the study. The idea is that the model can be used to evaluate the concept that professional services can be related to the service outcome using causality internal controllable factors and external uncontrollable factors as moderating variables. These factors are deemed to influence the service quality dimensions as independent variables and equally influence the client satisfaction as dependent variables.

References

Amos-Abanyie, S., Botchway, E. A., and Kwofie, T. E., The Relationship Between Level of Architect's Professional Competencies and Client Satisfaction Level. *Journal of Engineering Management Research. 3(2), 10-19.2014*

Assessing Women in Engineering (AWE) Project. Attribution Theory: An overview. www.aweonline.org, 23/12/2015. 1-9. 2005.

Brandl, K., Perspectives on Globalization and the Danish Architecture and Engineering Industry. *Research Paper for Centre for Strategic Management and Globalization, Copenhagen Business School.* 2010.

de Jong, G. R., (editor) Professional Services Firms; Vrije University, Amsterdam, 2007.

Forsythe, P. J., A Conceptual Framework for Studying Customer Satisfaction in Residential Construction *Construction Management and Economics.* 25(2) 171-182. 2007

Gronroos,C ., A Service quality model and its marketing implications. *European Journal of Marketing*, (18)4, 36-44, 1984.

Ismail, S., Othman, M. H., and Amat, S. C., Measuring Refurbishment Contractors' Service Quality and Client satisfaction :A Case Study at Public Institutions of Higher Education. *Pertanika Journal of Social Science and Humanities .20(1), 107 – 120. 2012.*

Koch, C. and Bendixen., Multiple Perspectives on Organizing; Projects Between Tyranny and Perforation. *Building Research and Information 33(6), 536-546, 2005.*

Oh, H., Service Quality, Customer Satisfaction and Customer Value: A Holistic Perspective. *International Journal of Hospitality Management.* (18), 67-82, 1999.

Parasuraman, A., Zeithaml, V. A., and Berry, L. L., SERVQUAL: A Multiple – Item Scale for Measuring Consumer Perceptions of Service. *Journal of Retailing 64(Spring), 12-40, 1988.*

Parasuraman, A., Berry, L. L., and Zeithaml, V. A., Refinement and Reassessment of the SERVQUAL Scale *Journal of Retailing 67(4), 420-450, 1991.*

Parasuraman, A., Zeithaml, V. A., and Berry, L. L., Reassessment of Expectations as a Comparison Standard in Measuring Service Quality: Implications for Further Research. *Journal of Marketing* .58(1) ,111-124, 1994.

Razavi. S. M., Safari. H., Shafie. H., and Khoram. K., Relationships Among Service Quality, Customer Satisfaction and Customer Perceived Value : Evidence from Iran's Software Industry. *Journal of Management and Strategy 3(3),28-37, 2012.*

Sonne, A., Determinants of Customer Satisfaction with Professional Services-A Study of Consultant Services. *Okonomisk Fiskeriforskning Journal.* 9(2), 97-108.1999

Sporrong, J., Selecting Architectural and Engineering Consultants. *PhD Thesis, Chalmers University of Technology, Sweden, 2014.*

Spreng, R. A. and Mackoy, R. D., An Empirical Examination of a Model of Perceived Service Quality and Satisfaction. Journal of Retailing . (722), 201-214, 1996.

Interaction between Theory and Practice in Civil Engineering and Construction
Edited by Komurlu, R., Gurgun, A. P., Singh, A., and Yazdani, S.

FLEXIBLE SYSTEMS DESIGN

CRISTIANA CELLUCCI[1] and MICHELE DI SIVO[2]

[1]*Architecture Dept, University "G. d'Annunzio" of Chieti and Pescara, Italy*
[2]*Architecture Dept, IDEA Dept, University "G. d'Annunzio" of Chieti and Pescara, Italy*

The research explores the relationship between uncertainty, flexibility and Life Cycle Design in the design of complex systems, in general, and in the particular case of the design of building systems. In the architectural-engineering works, conceived as activities aimed at generating new systems, the comparison with the uncertainty is inevitable. There is a wide range of types of uncertainty; a possible simplification is to consider two types of causes of uncertainty, the presence of the internal or external uncertainties to the system. That is the variability of demand (unknowns about the social and economic context) or the variability of technological market (unknowns about the performance of the system), in other words the functional obsolescence and the technological obsolescence. If flexibility is the ability of a system to be easily modified and to respond to changes in the environment in a timely and convenient, then the flexibility can be considered the antidote to obsolescence, or the characteristic of the system that guarantees slippage over time. It's that property that makes the system resilient able to absorb the shock and/or disturbance without undergoing major alterations in its functional organization, in its structure and in its identity characteristics. In the paper, the flexibility is therefore seen as a fundamental property for designing a generally complex system, and particularly in architectural design, through the identification of design criteria for the implementation of this requirement, that influence on architecture (form more technology) of the system.

Keywords: Uncertainty, Flexibility, Technological flexibility, Spatial flexibility, Life cycle design.

1 INTRODUCTION

One of the main problems affecting the architectural design in the last decades has been the disposal and renovation of these inherited buildings, as well as the risk of becoming technically or functionally obsolete in the short term, because the performance of the buildings no longer has a competitive edge in the housing market nor do they satisfy the needs of the user. This inability to manage the uncertainties of the socio-economic context and the various needs that arise from different types of housing usage tends to render the housing system inadequate and reduce its shelf life. This suggests the need to rethink the concepts of obsolescence, life span and flexibility during the design phase. If flexibility is defined as the ability of a system to be easily modified and to respond to user needs in a timely and effective manner, then it can be considered an antidote to obsolescence and a characteristic of the system that ensures the extension of its life cycle over time. In the paper, the flexibility is therefore seen as a fundamental property for designing a generally complex system, and particularly in architectural design, through the identification of design criteria for the implementation of this requirement, that influence on architecture (form more technology) of the system (Cellucci 2014).

2 UNCERTAINTY: CONTEXT VARIABLES AND SYSTEM VARIABLES

Uncertainty, as "unpredictability" is a fundamental condition in which natural and man-made systems are compared and generally all complex systems. One way to deal with uncertainty is to incorporate flexibility in the initial design, so as to ensure the possibility of choice in the future and be able to tackle successfully the changes that may occur during the life of these systems. Flexibility reduces the exposure of a project uncertainty, provides useful solutions to mitigate risks related to changes and market constraints, and the risks associated with technological obsolescence; is so, that property that makes the system resilient, that is, able to "absorb the shock and/or disturbance without undergoing major alterations in its functional organization, in its structure and in its identity features" (UNEP 2005). A system, in this sense is resilient if retains the idea of its shape, which is maintained despite the subsequent metamorphosis, but it is not permanent for eternity, caged in an unchanging order. There is a wide range of types of uncertainty; a possible simplification is to consider two types of causes of uncertainty, the presence of the internal or external uncertainties to the system. That is the variability of demand or the variability of technological market, in other words the functional obsolescence and the obsolete technological (Sethi and Sethi 1990; Shi and Daniels 2003).

The presence of the internal or external uncertainties to the system (context variables and system variables) involves the application of different theoretical approaches to mitigate the uncertainty of the context and of the system (Fig. 1).

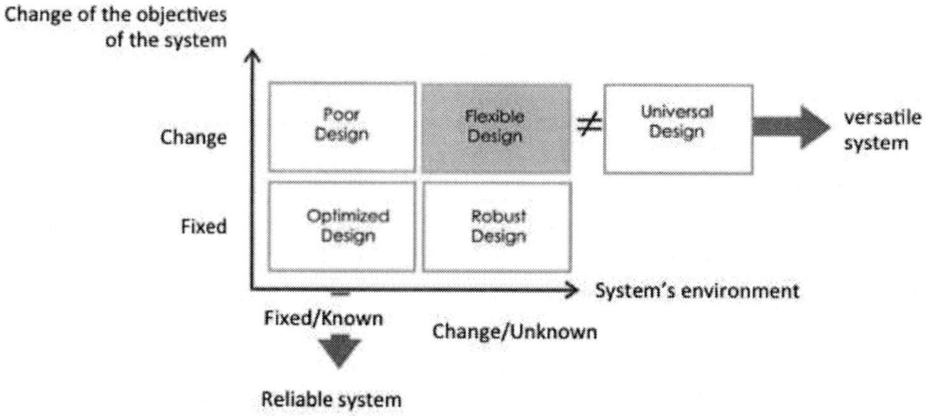

Figure 1. Different theoretical approaches to mitigate the uncertainty of the environment and the system.

Thus, from the foregoing considerations, it is clear that flexibility is required if:

- The uncertainty in the system's environment is such that there is a need to mitigate risks arising from exposure of the system to such uncertainty;

- The system is subject to one or more lens exchange during its life cycle, following the mutability of the user's needs, thus requiring a solution of mitigating risks associated with functional obsolescence;
- The system's technology base evolves on a time scale considerably shorter than system's design lifetime, thus requiring a solution of mitigating risks associated with technological obsolescence (Saleh *et al.* 2003).

3 DEVELOPMENT OF A METHOD FOR IMPLEMENTING FLEXIBILITY IN DESIGNING A SYSTEM

In summary, the flexibility is required if there are internal or external uncertainties to the system, that is if the system is subject to functional and technological obsolescence. In conclusion, we can point two strategies should ensure that extension of the life of the building as well as the same level of performance during the life cycle design (Cellucci and Di Sivo 2016).

UNCERTAINTY	VULNERABILITY	PROPERTIES
Variability of the environment (the uncertainties concerning the social and economic context)	Functional obsolescence of the system	**Spatial flexibility**: The ability of a system to adapt to different uses and functions to respond to the variability of users needs. Spatial flexibility is obtained from the relationship between the requirements that affect the form of the system, the versatility, the convertibility and the modularity of the system.
Variability of the system (the uncertainties regarding the system performance)	Technological obsolescence of the system	**Technological Flexibility**: The ability to work easily on the technological apparatus that governs space. The spatial flexibility is obtained from the relationship between the requirements that act on the structure of the system, which is the maintainability and reversibility.

The requirements of spatial flexibility are:

- *Versatility*. The multiple use of a system within an area with shapes and sizes that is altogether unchanged over time. We say that a versatile system is Universal. This requirement affects a system's internal configuration.
- *Convertibility*. The system's ability to adapt to different physical configurations through a transformation that alters its internal and external configuration, in order to meet the different needs and requirements that arise after the system has been made operational. This requirement affects the system dimension.
- *Modularity*. The Organization of the system into parts which can be subtracted or added to the system according to your needs.
- *Open System*. This requirement affects the system's dimension, its ability to expand over time.

The requirements of Technological flexibility are:

- *Maintainability*. The probability to repair a system at a given time when maintenance operations are implemented in accordance with the prescribed procedures and resources. The implementation of the requirement of maintainability can promote any redevelopment work which is necessary when there is an imbalance between the performance of the technical element and changing levels of need in the users, allowing you to quickly make those adjustment operations to new levels of quality (Di Sivo 2004).
- *Reversibility*. The Organization of the system in sub-systems and separable components, with particular reference to the "features and the" status "of connections. The reversibility of the system makes it possible to decrease the impacts resulting from the disposal of the system, such as for large systems such as buildings; the implementation of this requirement provides the demolition through the disassembly and consequently the separation of constituent parts and materials in order of their possible reuse or recycling.

The spatial flexibility implies a high degree of organizational complexity, which is indeed a fundamental attribute of resilient systems. It is evident that the multiplicity of solutions increases the resilience of the system, the possibility of evolution, of change and adaptation. The concept of technological flexibility implies the concept of simplicity of implementation, intended as quickness and easiness with which they can complete the maintenance operations and reversibility, undertaken in order to correct the gap between requirements and performance in order that the system won't cease its usefulness. The concept of flexibility implies that maintainability and reversibility operations can be performed without the need to undertake other unforeseen or unpredictable collateral activities. Wanting to try a list of the main moments of construction and control of the flexibility of the systems we have reached by developing a list of general criteria of flexibility to be used in reference to the design features of the system (Cellucci and Di Sivo 2016):

LIST OF GENERAL CRITERIA OF FLEXIBILITY	
Programmatic Complexity	Space solutions able to ensure easiness of modifiability over time relative to the variability of users needs and the satisfaction of psychological and functional needs.
Simplicity in the interface:	Choice of technical solutions capable of guaranteeing the possibility of intervening with easiness on system components, that is ensuring the dismantling, substitutability, reparability of system components.
Structural simplicity	Choice of reversible construction techniques, with a particular reference to the characteristics and status of the connections.
Optimization of components	Possibility of substitutability of components, through a modular organization and their optimization towards specific objectives in order to hinder the technological obsolescence.
Optimizing plant integration	Verification of coherence between technical-building solutions and inspection of plant equipment, ensuring the maintainability of the networks without affecting the functionality of other components or subsystems.

4 CONCLUSION

In the paper, the flexibility is therefore concerned as a basis for designing a complex system, through the identification of design criteria for its implementation, that influence on architecture (form more technology) of the system, through the relationship between mitigated requirements of specific sources of uncertainty. This approach can be useful: the uncertainty of the context is such that there is a need to reduce the risks arising from exposure of the system in this uncertainty; the system is subject to one or more target changes, due to variability of requirements of users, requiring a solution to mitigate the risks associated with functional obsolescence, the basic technology of the system evolves in a time shorter than the lifecycle of the system, requesting a solution to mitigate the risks associated with technological obsolescence.

References

Cellucci C., *Tempo e Resilienza: nuove prospettive per la flessibilità spaziale e tecnologica della casa*, PhD thesis, Dipartimento di Architettura, University "G. d'Annunzio" of Chieti and Pescara, 2014.

Cellucci C., Di Sivo M., *Habitat Contemporaneo.Flessibilità tecnologica e spaziale*, FrancoAngeli, Milan, 2016.

Di Sivo M., *Manutenzione Urbana*, Alinea, Firenze, 2004.

Saleh J.H., Lamassoure E. S., Hastings D.E., Newman D. J.,"Flexibility and the Value of On Orbit Servicing: New Customer-Centric Perspective,"in *Journal of Spacecraft and rockets,* Vol.40, No.2, 2003, pp. 279- 291.

Sethi A.K., Sethi S.P., "Flexibility in manufacturing: a survey", *International Journal of Flexible Manufacturing Systems*, Vol. 2 No. 4, 1990, pp. 289–328.

Shi D. and Daniels R.L., "A survey of manufacturing flexibility", *IBM Systems Journal*, Vol.2 No.3, 2003, pp. 414–427.

UNEP, *Climate Change. The Role of Cities*, Nairobi, 2005.

Interaction between Theory and Practice in Civil Engineering and Construction
Edited by Komurlu, R., Gurgun, A. P., Singh, A., and Yazdani, S.

IMPACT OF BUILDING ENVELOPE DESIGN ON ENERGY CONSUMPTION OF LIGHT STRUCTURE SCHOOL BUILDING

BIN SU

Dept of Architecture, Unitec Institute of Technology, Auckland, New Zealand

According to the local climate, an Auckland school building normally does not need air conditioning for cooling during the summer and only needs space heating during the winter. Previous study shows that the mean winter energy consumption is about 38% of the mean annual energy consumption of Auckland school buildings. The mean winter extra energy, which mainly includes space heating, water heating and other appliances related to winter indoor thermal conditions, is about 44% of winter energy consumptions. Extra winter energy of a school building is closely related to and impacted by building thermal design and thermal performance. Although different building design factors related to the main architectural feature, building elements and materials can affect the school energy consumption differently and simultaneously, the relationship between building design data and school energy consumption data can still be identified. With a larger number of sample school buildings, this study focuses on impact strength of building design factors on school winter extra energy consumptions and identifies quantitative relationships between the building design data and the winter extra energy consumption data, which can be used to proximately estimate the amount of saving winter extra energy consumption associated with the change of a design datum for the future school development. In Auckland, there are 425 schools including primary, intermediate and high schools. Real energy consumption and building design data of 57 local schools (13.4% of the total number of Auckland schools) are randomly selected for this study.

Keywords: Building elements, Building energy, Building energy efficiency, Building envelope design, Building thermal design, School building.

1 INTRODUCTION

The World Health Organisation recommends a minimum indoor temperature for houses of 18°C; and 20-21°C for more vulnerable occupants, such as older people and young children (WHO 1987). Current New Zealand Building Codes do not have a general requirement for the minimum indoor air temperature, although it has a requirement of 16°C for more vulnerable occupants, such as older people and young children (SANZ 1990, DBH 2001). There are a number of previous studies on the impact of different building design factors on energy efficiency. These design factors are mainly related to building orientation, geometry and envelope. Some studies focus on building orientation, which is one of most important design factors for building energy efficiency, impacting on solar radiation received (Gupta 2004, Morrissey 2011) and shading (Capeluto 2003). Other studies focus on the impact of building shape (Marks

1997, Florides *et al.* 2002, Mingfang 2002, Adamski 2007) with different orientations (Marks 1997, Florides *et al.* 2002, Aksoy 2006, Adamski 2007); or on energy consumptions under different climates (Depecter *et al.* 2001). All heat exchanges between indoor space and outdoor space are through the building envelope, which has the greatest impact on building energy consumption (Manioglu *et al.* 2006, Radhi 2008). Those studies mainly based on mathematical models and computer simulations. It is difficult to establish universal building passive design guides to achieve energy efficiency for different local buildings and climates. This study uses the difference between mean daily energy usage in the winter months (the likely heating months: June, July and August) and the other months of the year (unlikely heating months) to roughly represent the extra winter energy consumption, which mainly comprises space heating and extra energy for hot water heating, and can also include extra energy for appliances, which are impacted by the winter internal thermal conditions of a school. Smaller extra winter energy usage indicates better indoor thermal conditions and building thermal performance in response to the winter climatic conditions. Actual monthly energy consumption data for a whole year, which can be converted into daily energy usage per unit volume of indoor space (kWh/m^3/day), and building design data derived from real building plans supplied by the Auckland Council of 57 Auckland sample schools are used for this study. This study uses the gradient of the trend line to evaluate the impact strength of a design datum on winter extra energy consumption per cubic meter of indoor space of school building (kWh/m^3/day), and estimate the decrease of winter extra energy consumption when a design datum is changed within a range when other design data also impact the winter extra energy consumption differently and simultaneously for future school development. Table 1 and Table 2 show general information and energy data of the sample school and buildings.

Table 1. General information of sample schools.

General information	**Mean**	**Range**
Number of students	626	109 - 2600
Number of isolated buildings	9.4	1 - 38
Building height (stories)	1.4	1 – 3
Number of classrooms	30.5	5 - 135
Student number per classroom	22	12 - 33
School floor area (m^2)	5257	905 – 22680
Classroom / building floor area	48%	19% - 80%
Floor area per classroom	77m^2	51 – 141m^2

Table 2. Energy consumption data of sample school buildings.

Energy(kWh)	**Mean**	**Range**
Annual	242633	16376 – 1498621
Winter	92951	5662 – 570679
Winter/annual	38%	29% - 49%
Summer	33199	2783 – 164123
Winter/summer	2.8	2.0 – 3.5
Summer/annual	14%	7% - 20%
Heating months	144892	8699 – 900417
Heating months/annual	58%	48%-76%

2 BUILDING DESIGN FACTORS

2.1 Ratio of Building Surface to Volume

The ratios of building surface to volume of the sample school buildings are 0.31 to 0.74 with a mean ratio of 0.48. Over 90% of Auckland schools only have four or fewer classrooms in each isolated school building (see Figure 1). An increase in the ratios of building surface to volume of the sample schools is associated with an upward trend in winter energy consumption (see Figure 2). A school building with a high ratio of building surface to volume has a large external surface area, which results more heat-loss and more energy consumption for space heating and other appliances during the winter. Based on the gradient (0.023) of the trend line in Figure 2, the mean winter extra energy (0.0234kWh/m^3/day) and mean ratio of building surface to volume of the sample schools, Eq. (1) can be used to proximately estimate the amount of increasing or decreasing mean winter extra energy consumption (ΔE_{WE}) associated with the decrease or increase of mean ratio of building surface to volume (ΔR_{SV}) of the future school development under the local climate. For example, decreasing the mean ratio from 0.48 to 0.28, input data into the Eq. (1): 0.0234 - E_{future} = 0.023 × (0.48 – 0.28), the future mean winter extra energy consumption E_{future} = 0.0234 - 0.023 × (0.48 - 0.28) = 0.0234 - 0.0046 = 0.0188kWh/m^3/day. Table 3 shows relationships between reduction (from 3.9% to 19.7%) of mean winter extra energy related to decrease (from 0.48 to 0.32) of mean ratio of building surface to volume for future Auckland school development.

$$\Delta E_{WE} = 0.023 \Delta R_{SV} \tag{1}$$

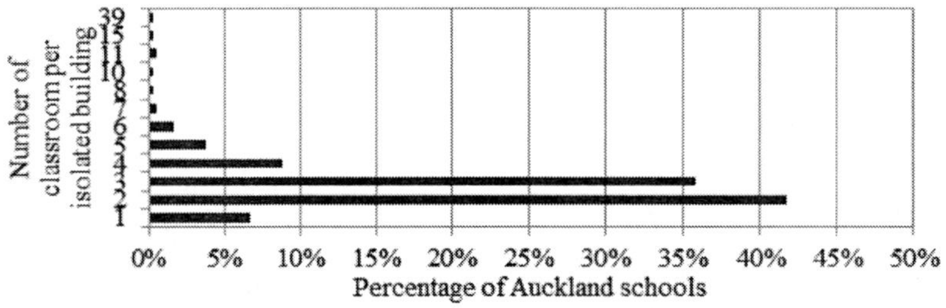

Figure 1. Number of classrooms per school building in Auckland.

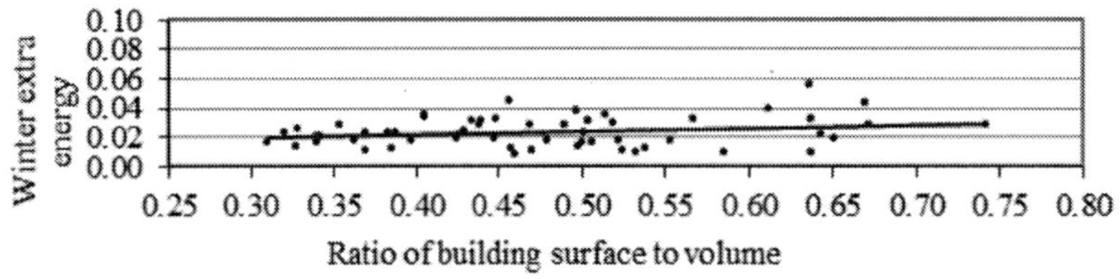

Figure 2. Winter extra energy and ratios of building surface to volume of sample schools.

2.2 Ratios of Roof Surface Area and Wall Surface Area to Building Volume

The ratio of roof surface area to building volume of the sample school buildings are 0.12 to 0.42 with a mean ratio of 0.26 and the ratios of wall surface area to building volume are 0.40 to 0.09 with a mean ratio of 0.22. An increase in the ratio of roof surface area to building volume or the ratio of wall surface area to building volume is associated with an upward trend in winter energy consumption (Figure 3, 4). The gradient of the trend line of roof (0.0426) is higher than wall (0.0141). A low rise building such as a house with 1 to 2 stories loses more heat through roof than wall. Auckland houses with 1 to 2 stories lose about 40% of their heat through ceiling and roof and about 20% through walls during the winter. The ratio of roof surface area to building volume of Auckland houses are 0.12 to 0.40 with a mean ratio of 0.29 (Su 2011). The ratio of roof surface area to building volume of the sample school buildings are very close to the local houses. Roof of Auckland school buildings should be a more important building element or portion of building envelope than the walls for the local school buildings. Decreasing mean ratio of roof surface area to building volume can cause a more positive impact on winter extra energy of Auckland school buildings than the mean ratio of wall surface area to building volume. Eq. (2) and Eq. (3) can be used to proximately estimate the amount of increasing or decreasing mean winter extra energy consumption (ΔE_{WE}) associated with the increasing or the decrease of the mean ratio of roof surface to building volume (ΔR_{RV}) and the mean ratio of wall surface to building volume (ΔR_{WV}) of the future school development under the local climate.

$$\Delta E_{WE} = 0.0426 \Delta R_{RV} \quad (2)$$

$$\Delta E_{WE} = 0.0141 \Delta R_{WV} \quad (3)$$

Table 3. Reduction of mean winter extra energy (kWh/m^{3}day) associated with decrease of mean ratio of building surface to volume of future school building development.

Ratios of building surface to volume	Decrease of mean winter extra energy	Mean winter extra energy	Energy saving
0.44	0.00092kWh/m^{3}day	0.02248kWh/m^{3}day	3.9%
0.40	0.00184kWh/m^{3}day	0.02156 kWh/m^{3}day	7.9%
0.36	0.00276kWh/m^{3}day	0.02064 kWh/m^{3}day	11.8%
0.32	0.00368 kWh/m^{3}day	0.01972 kWh/m^{3}day	15.7%
0.28	0.00460 kWh/m^{3}day	0.01880 kWh/m^{3}day	19.7%

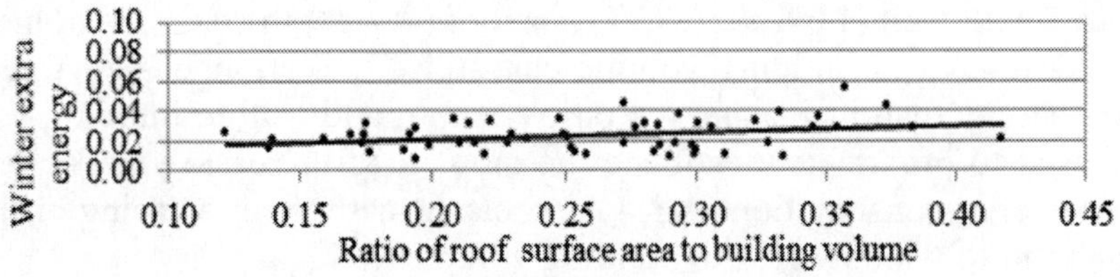

Figure 3. Winter extra energy and ratio of building roof surface area to building volume

2.3 Ratio of Roof Space Volume to Building Volume

The ratios of roof space volume to building volume of the sample school buildings are 0.02 to 0.36 with a mean ratio of 0.13. An increase in the ratios of roof space volume to building volume is associated with a downward trend in winter extra energy consumptions of the sample school buildings (see Figure 5). Increasing roof space volume can positively impact on saving winter extra energy consumption for the low rise school buildings. A roof space can be buffering space for heat-loss during the winter and increase R-value of roof. Eq. (4) can be used to proximately estimate the amount of increasing or decreasing mean winter extra energy consumption (ΔE_{WE}) associated with the decrease or the increasing of the mean ratio of roof space volume to building volume (ΔR_{RSV}) of the future school development under the local climate.

$$\Delta E_{WE} = 0.0171 \Delta R_{RSV} \quad (4)$$

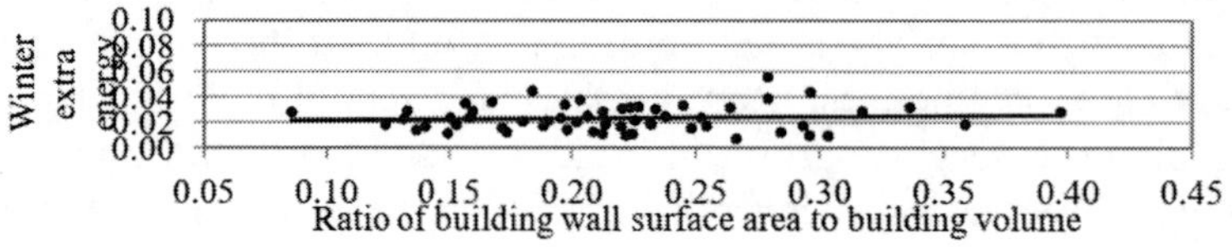

Figure 4. Winter extra energy and ratio of building wall surface area to building volume

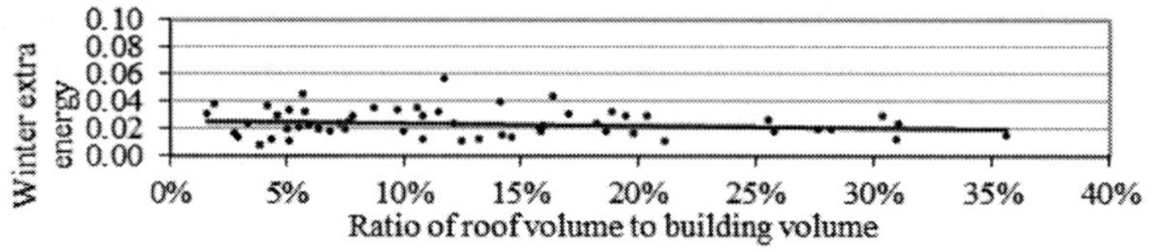

Figure 5. Winter extra energy and ratio of roof volume to building volume.

2.4 Ratio of Window Area to Building Volume

The ratios of window to building volume of the sample school buildings are in the range of 0.02 to 0.1 with a mean ratio of 0.05. An increase in the ratio of window area to building volume is associated with an upward trend in winter energy consumption (See Figure 6). The gradient of the trend line of the ratio of window area to building volume is 0.0762, which is much higher than roof (0.0426) and wall (0.0141). Window area are generally weak part of building envelope with low R-value for building thermal design. Even using double glazed windows, the R-value (0.26 $m^2 °C/W$) is still very low compared with wall (1.5-1.9 $m^2 °C/W$) or roof (1.9-2.9 $m^2 °C/W$), and increasing the ratio of window area to building volume can still cause stronger negative impact on winter extra energy than increasing the ratios of roof and wall to building volume. Eq. (5) can be used to proximately estimate the amount of increasing or decreasing mean winter extra energy consumption (ΔE_{WE}) associated with the increasing or the decrease of the mean ratio of window area to building volume (ΔR_{WV}).

$$\Delta E_{WE} = 0.0762 \Delta R_{WV} \quad (5)$$

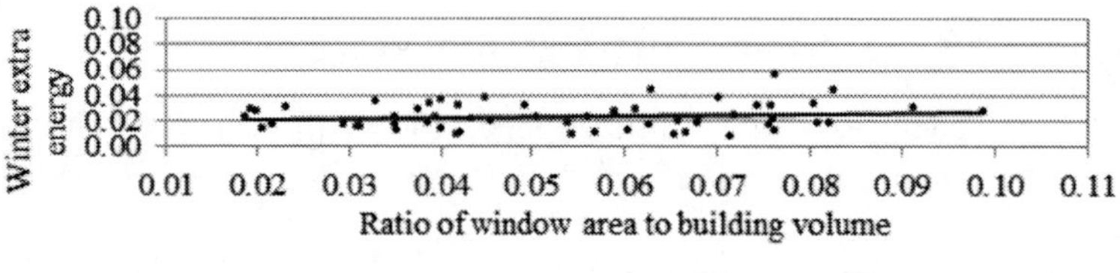

Figure 6. Winter extra energy and ratio of window area to building volume.

3 CONCLUSIONS

From the energy efficiency point of view, Auckland school conventional design with a high ratio of building surface to volume is not good for energy efficiency. Changing school building design conventions from low rise to multi-stories building not only can reduce ratio of building surface to volume but also reduce winter extra energy. Based on energy consumption and design data of the 57 Auckland sample schools, reducing mean ratio of building surface to volume from 0.48 to 0.28 can potentially reduce about 20% of winter extra energy consumption for the future school building development. There are thousands of old classrooms need to be retrofitted for improving their thermal performance and energy efficiency in New Zealand. According to envelope design data for the sample school buildings, the window area is the weakest portion of the envelope, having the lowest R-value, and more negative impact on winter extra energy than roof and wall. The first thing to do for retrofitting an old school building is to improve the thermal performance of the window area to reduce heat loss; and secondly, the roof area, which has more negative impact on winter extra energy than the wall area.

References

Adamski, M., Optimization of the form of a building on an oval base, *Building and Environment*, 42(4), 1632-1643, 2007.

Aksoy, U. T., and Inalli, M., Impacts of some building passive design parameters on heating demand for a cold region, *Building and* Environment, 41(12), 1742-1754, December, 2006.

Capeluto, I. G., Energy performance of the self-shading building envelope, *Energy and Buildings*, 35(3), 327-336, 2003.WHO, *Health Impact of Low Indoor Temperatures*, World Health Organization, Copenhagen, Denmark, 1987.

DBH, *Compliance Document for New Zealand Building Code – Clause G5 Interior Environment*, Department of Building and Housing, Wellington, New Zealand, 2001.

Depecter, P., Menezo, C., Virgone, J., and Lepers, S., Design of building shape and energetic consumption, *Building and Environment*, 36(5), 627-635, June 2001.

Florides, G. A., Tassou, S. A., Kalogirou, S. A., and Wrobel, L. C., Measures used to lower building energy consumption and their cost effectiveness, *Applied Energy*, 73(3-4), 299-328, November-December, 2002.

Gupta, R., and Ralegaonkar, R. B., Estimation of beam radiation for optimal orientation and shape decision of buildings in India, *Architectural Journal of Institution of Engineers. India,* 85, 27-32, 2004.

Manioglu, G., and Yilmaz, Z., Economic evaluation of the building envelope and operation period of heating system in terms of thermal comfort, *Energy and Buildings*, 38 (3), 266-272, 2006.

Marks, W., Multicriteria optimisation of shape of energy-saving buildings, *Building and Environment*, 32(4), 331-339, 1997.

Mingfang, T., Solar control for buildings, *Building and Environment*, 37(7), 659-664, 2002.

Morrissey, J., Moore, T., and Horne, R. E., Affordable passive solar design in a temperate climate: an experiment in residential building orientation, *Renewable Energy*, 36(2), 568-577, 2011.

Radhi, H., A systematic methodology for optimising the energy performance of buildings in Bahrain, *Energy and Buildings*, 40(7), 1297-1303, 2008.

SANZ, *New Zealand Standard 4303-1990 Ventilation for acceptable indoor air quality*, Standards Association of New Zealand, Wellington, New Zealand, 1990.

Su, B., The impact strength of building passive design on housing energy efficiency, *Architectural Science Review*, 54(4), 270-276, 2011.

IMPACT OF BUILDING DESIGN ON ENERGY USAGE OF A PUBLIC LIBRARY

BIN SU

Dept of Architecture, Unitec Institute of Technology, Auckland, New Zealand

This pilot study is to investigate relationships between Auckland library building design and energy consumption. In Auckland, about sixty public libraries belong to the Auckland Council. A number of Auckland public libraries (32% of the total number of public libraries in Auckland) are used for this study. A whole year monthly energy consumption data and library building plans of the sample libraries are recorded and supplied by the Auckland Council. This study explores impact of building design factors on energy consumption, which is based on real energy consumption data and building design data of Auckland public libraries. This study converted monthly energy consumption data into annual, winter and summer daily mean energy consumptions per unit volume of library building indoor space (kWh/m^3/day), which is closely related to building design, indoor thermal conditions and building thermal performance under the local climatic conditions. According to relationships between daily mean energy consumptions per unit volume of library building indoor space (kWh/m^3/day) and library building design data, the study presents energy profiles of local libraries and identify major design factors impacting energy consumption of Auckland public libraries.

Keywords: Building energy, Building energy efficiency, Building thermal design, Library building design.

1 INTRODUCTION

There are a number of previous studies related to impacts of different building design factors on energy efficiency. These design factors are mainly related to building orientation, geometry and envelope. Some studies focus on building orientation, which impacts on solar radiation received (Gupta *et al.* 2004, Morrissey *et al.* 2011) and shading (Capeluto 2003). Other studies focus on impacts of building shape (Marks 1997, Mingfang 2002, Aksoy *et al.* 2006, Adamski 2007) with different orientations (Marks 1997, Florides *et al.* 2002, Aksoy 2006, Adamski 2007) on energy consumptions under different climates (Depecter *et al.* 2001). All heat exchanges between indoor space and outdoor space are through the building envelope, which has the greatest impact on building energy consumption (Manioglu *et al.* 2006, Radhi 2008). Those studies mainly based on mathematical models and computer simulations can be used to compare different building designs for energy efficiency. Some studies used real building energy consumption data and building design data to investigate relationships between them and impact of building design factors on energy efficiency under local climatic conditions (Su 2011, Su 2013). This study explores impact of library building design factors on energy consumption, which is based on real energy consumption data and building design data of Auckland public libraries. This study converted monthly energy consumption data into annual, winter and summer daily mean energy consumptions per unit volume of library building indoor space (kWh/m3/day), which is closely related to building thermal design, indoor thermal conditions and building thermal performance under the local climatic conditions.

2 ENERGY CONSUMPTION AND BUILDING DESIGN FACTORS

2.1 Energy Profiles

Table 1 shows energy consumption data for the sample libraries. Winter energy consumptions of all sample libraries are more than 25% of annual energy consumptions during the 25% time of a year. The mean winter energy consumption is 30.6% of annual energy consumption in the range between 25.4% and 41.2% (see Figure 1). The mean summer energy consumption is 22% of annual energy consumption in the range between 14.1% and 26.3% (see Figure 1). The sample libraries generally use more energy during the winter than summer.

Figure 2 shows 12-months daily mean energy consumptions per unit volume of building indoor space (kWh/m^3/day) of the eighteen sample libraries. During the winter months from June to August or heating months from May to September, daily mean energy consumptions are higher than other months. During the summer, some libraries' daily mean energy consumptions are slightly higher than the months (March, April, October and November) without heating as some libraries use air-conditioning for space cooling during the summer. Figure 3 shows annual, winter and summer daily mean energy consumptions per unit volume of building indoor space (kWh/m^3/day). Some sample libraries' monthly and annual energy consumptions are "outstanding" and significantly higher than the other libraries (see Figure 2, 3). The libraries No. 2, 3, 4, 8, 15 are the top five annual energy consumption libraries (see Figure 3). For the top five annual energy consumption libraries, the mean floor area is 707m^2 and significantly smaller than the other libraries (1479m^2). Mean ratios of winter to annual energy consumptions of the top five annual energy consumption libraries and the other libraries are 29.9% and 30.9% respectively. Mean ratios of summer to annual energy consumptions of the top five annual energy consumption libraries and the other libraries are 23.3% and 21.6% respectively. They are close to mean ratios of winter (30.6%) and summer (22.0%) of all sample libraries. The mean ratio of building surface to volume of the top five annual energy consumption libraries is 0.45 and much higher than the other libraries (0.32). Annual, winter and summer daily mean energy consumptions per unit volume of building indoor space, and annual, winter and summer energy consumptions per unit area of building floor of the top five annual energy consumption libraries are significantly higher than the other libraries (see Table 2). Small library buildings potentially use more energy than large libraries as smaller library buildings have higher ratios of building surface to volume. A library building with a higher ratio has larger building surface area to lose more heat during the winter and get more heat gain during the summer per unit volume of building indoor space, which is not good for energy efficiency.

The mean ratio of roof area to building volume of the top five annual energy consumption libraries is 0.28 and higher than the other libraries (0.2). The top five annual energy consumption library buildings are one or two stories buildings and the mean floor area is only 707m^2. The mean ratio of window to wall area of the top five annual energy consumption libraries is 0.39 and higher than the other libraries (0.35). Windows are commonly weak elements for building thermal design. The thermal resistance (R-value) of a single glazed window (0.15 m^2 °C/W for aluminum window frame and 0.19m^2°C/W for wooden or PVC window frame) is very low compared with walls (1.9-2.0m^2°C/W) and roofs (2.9-3.5m^2°C/W) in accordance with the current standard (SNZ 2009). A library building with a high ratio of building surface to volume potentially increases ratios of roof area to building volume and window to wall area, which is not good for building energy efficiency.

Table 1. Energy consumption data (kWh) of sample libraries.

Energy (kWh)	Mean	Range
Annual	182265	75048 - 369953
Winter: Jun, Jul & Aug.	56278	19629 - 138504
Winter/annual	30.6%	25.4% - 41.2%
Summer: Dec, Jan & Feb.	39469	13198 - 65581
Summer/annual	22.0%	14.1% - 26.3%
Winter/summer	1.5	1.0 – 2.9
Heating months: May-Sep.	87118	31076 - 202805
Heating months/annual	47.3%	40.8%-59.6%

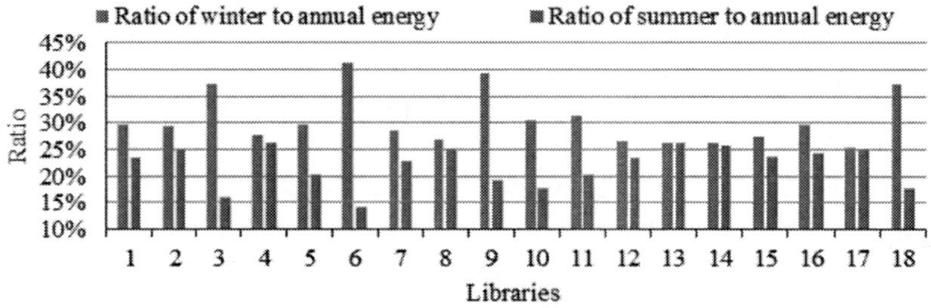

Figure 1. Ratios of winter or summer to annual energy.

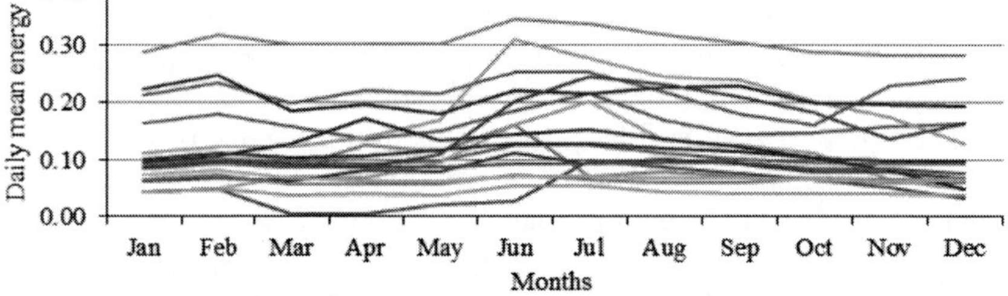

Figure 2. Daily mean energy consumptions (kWh/m^3/day) for 12 months.

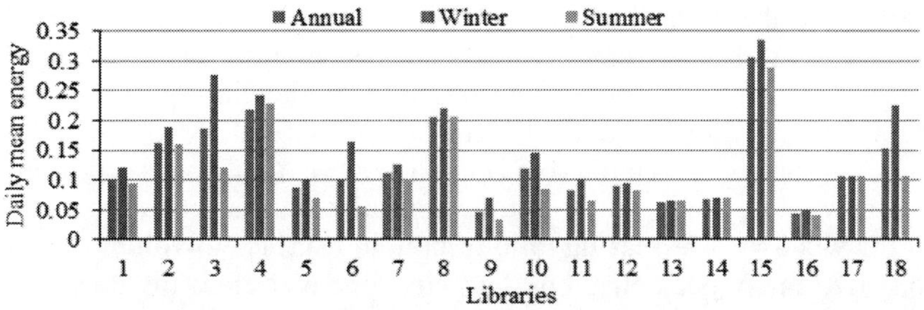

Figure 3. Daily mean energy consumptions (kWh/m^3/day) for annual, winter and summer.

Table 2. Energy consumption data (kWh/m^3/day) of top five energy consumption libraries.

Energy	5 libraries	Other libraries	Difference
Annual(kWh/m^{3}day)	0.2162	0.0897	2.4 times
Winter(kWh/m^{3}day)	0.2525	0.1109	2.3 times
Summer(kWh/m^{3}day)	0.2008	0.0752	2.7 times
Annual(kWh/m^2)	261	130	2.0 times
Winter(kWh/m^2)	77	40	1.9 times
Summer(kWh/m^2)	61	28	2.2 times

2.2 Building Design Factors

The ratio of building surface to volume for a multi-story building with the permanent heating should be less than 0.3 for saving space heating energy (Liu 2000). Ratios of building surface to volume of the four libraries No.9, 11, 12, 16 are less than 0.3. The mean ratio of building surface to volume of the four libraries is 0.24 and much lower than other libraries (0.39). Annual, winter and summer daily mean energy consumptions per unit volume of building indoor space, and annual, winter and summer energy consumptions per unit are of building floor of the four libraries are significantly lower than the other libraries (see Table 3). Mean ratios of winter to annual energy consumptions of the four libraries and the other libraries are 31.7% and 30.3%. Mean ratios of summer to annual energy consumptions of the four libraries and the other libraries are 21.8% and 22.1%. They are close to mean ratios of winter (30.6%) and summer (22.0%) of the all sample libraries. Annual daily mean energy consumptions of the 4 libraries No. 9, 11, 12, 16, are significantly lower than other libraries (see Figure 7, 8) as their ratios of building surface to volume are much lower than others and less than 03. Annual daily mean energy consumption of the library No. 16 is the lowest one in all sample libraries and marginally lower than the other three libraries No. 9, No. 11 and No. 12. The library No. 16 has insulation and double-glazed windows. The other three libraries have insulation and signal-glazed windows.

Annual daily mean energy consumption per unit volume of building indoor space (0.2162kWh/m^{3}day) of the top five annual energy consumption libraries No. 2, No. 3, No, 4, No. 8, No. 15 is 2.4 times of Annual daily mean energy consumption (0.0897kWh/m^{3}day) of the other libraries as the mean ratio of building surface to volume (0.45) of the top five annual energy consumption libraries is significantly higher than other libraries (0.32). The top five annual energy consumption libraries have single glazed windows. Without insulation, annual energy consumption of the library No.15 is higher than the other four libraries No. 2, No. 3, No. 4, No. 8. The library No. 8 is built in 2003 and has sufficient insulation in its envelope according to the building code. Annual energy consumption of the library No.8 with sufficient insulation is the third highest annual energy consumption because it has the highest ratio of building surface to volume (0.55) in all sample libraries. According the current standard, a new library will have sufficient insulation in the building envelope and double glazed window. Decreasing ratio of building surface to volume through designs of building size, sharp, geometry, building envelope and elements is crucial to reduce library building energy consumption. Ratio of building surface to volume should be considered as the first level design factor for future library building energy efficiency.

Table 3. Energy and different ratios of building surface to volume.

Energy	4 Libraries	Other libraries	Difference
Annual(kWh/m^3day)	0.0644	0.1421	45.3%
Winter(kWh/m^3day)	0.0789	0.1706	46.2%
Summer(kWh/m^3day)	0.0559	0.1207	46.3%
Annual(kWh/m^2)	96	186	51.6%
Winter(kWh/m^2)	29	56	51.8%
Summer(kWh/m^2)	21	41	51.2%

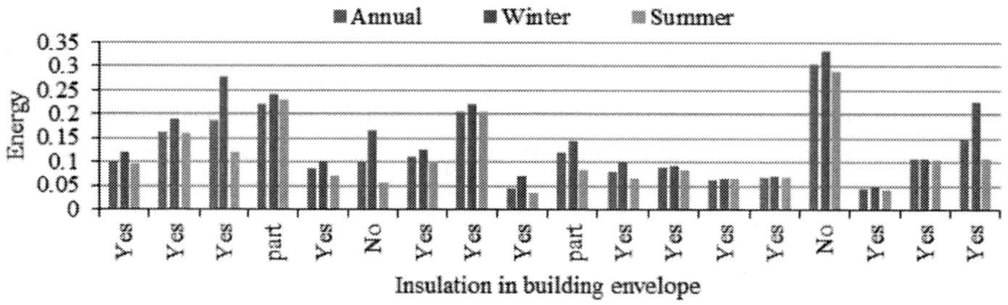

Figure 7. Energy consumptions and insulation.

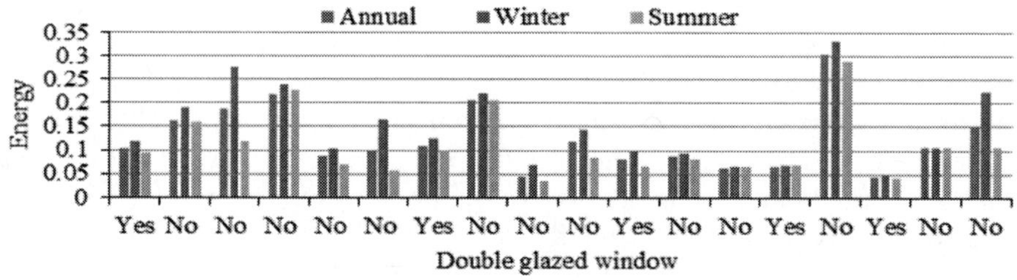

Figure 8. Energy consumptions and double glazed window.

3 CONCLUSIONS

Based on energy data of the eighteen sample libraries, the mean winter energy consumption of Auckland public libraries is about 30% of annual energy consumptions and the mean summer energy consumption of Auckland public libraries is about 22% of annual energy consumptions during the 25% time of a year. Auckland public libraries use more energy during the winter than other seasons. Auckland library design should more focus on winter thermal performance for building energy efficiency.

Small library buildings use much more annual, winter and summer mean energy consumptions per unit area of building floor, and annual, winter and summer daily mean energy consumptions per unit volume of building indoor space than the large libraries as the smaller library buildings have higher ratios of building surface to volume. The top five annual energy consumption libraries with a mean floor area of 707m^2 and the mean ratio of 0.45 use two times more annual, winter and summer energy than other libraries with the mean floor area of 1479m^2 and a mean ratio of 0.32.

Four sample libraries' ratios of building surface to volume are less than 0.3. Annual, winter and summer mean energy consumptions of the four libraries with the mean ratio of 0.24 are less than or about 50% of other libraries with the mean ratio of 0.39. When the ratio of building surface to volume is very low (less than 0.3), double glazed windows can positively impact energy consumption and marginally reduce energy consumption.

References

Adamski, M., Optimization of the form of a building on an oval base, *Building and Environment*, 42(4), 1632-1643, 2007.

Aksoy, U. T., and Inalli, M., Impacts of some building passive design parameters on heating demand for a cold region, *Building and* Environment, 41(12), 1742-1754, December, 2006.

Capeluto, I. G., Energy performance of the self-shading building envelope, *Energy and Buildings*, 35(3), 327-336, 2003.

Depecter, P., Menezo, C., Virgone, J., and Lepers, S., Design of building shape and energetic consumption, *Building and Environment*, 36(5), 627-635, June 2001.

Florides, G. A., Tassou, S. A., Kalogirou, S. A., and Wrobel, L. C., Measures used to lower building energy consumption and their cost effectiveness, *Applied Energy*, 73(3-4), 299-328, November-December, 2002.

Gupta, R., and Ralegaonkar, R. B., Estimation of beam radiation for optimal orientation and shape decision of buildings in India, *Architectural Journal of Institution of Engineers India,* 85, 27-32, 2004.

Liu, J. P, *Architectural Physicals*, China Construction Industry Publication, Beijing, 2000.

Manioglu, G., and Yilmaz, Z., Economic evaluation of the building envelope and operation period of heating system in terms of thermal comfort, *Energy and Buildings*, 38 (3), 266-272, 2006.

Marks, W., Multicriteria optimisation of shape of energy-saving buildings, *Building and Environment*, 32(4), 331-339, 1997.

Mingfang, T., Solar control for buildings, *Building and Environment*, 37(7), 659-664, 2002.

Morrissey, J., Moore, T., and Horne, R.E., Affordable passive solar design in a temperate climate: an experiment in residential building orientation, *Renewable Energy*, 36(2), 568-577, 2011.

Radhi, H., A systematic methodology for optimising the energy performance of buildings in Bahrain, *Energy and Buildings*, 40(7), 1297-1303, 2008.

SNZ, *New Zealand Standard 4218-2009 Thermal insulation: housing and small buildings*, Standards New Zealand, Wellington, New Zealand, 2009.

Su, B., The impact strength of building passive design on housing energy efficiency, *Architectural Science Review*, 54(4), 270-276, 2011.

Su, B., Impacts of Building Design Factors on Auckland School Energy Consumptions, *World Academy of Science, Engineering and Technology. International Journal of Civil, Architectural Science and Engineering*, 7(12), 265-270, 2013.

Interaction between Theory and Practice in Civil Engineering and Construction
Edited by Komurlu, R., Gurgun, A. P., Singh, A., and Yazdani, S.

MORE THAN JUST A GREEN FACADE: VERTICAL GARDENS FOR SOUND ABSORPTION AND ARCHITECTURAL ACOUSTICS

M. J. M. DAVIS[1, 2], M. E. PÉREZ[3], M. J. TENPIERIK[4], and F. R. RAMÍREZ[1]

[1]*Faculty of Architecture, Design and Arts, Pontificia Universidad Católica of Ecuador, Quito, Ecuador*
[2]*Evolution Engineering, Design and Energy Systems Ltd., Exeter, UK*
[3]*Independent Researcher and Vertical Garden Designer, Quito, Ecuador*
[4]*Faculty of Architecture and the Built Environment, TUDelft, Delft, the Netherlands*

Noise can become uncomfortable for us in many situations both indoors and outdoors. External noise consists of activities (airplanes flying overhead, traffic on the road, etc.) that are either loud enough to be considered uncomfortable when outdoors, or are of an elevated volume to the extent that they infiltrate buildings at levels considered uncomfortable. In the case of internal uncomfortable noise, this can either stem from noisy activities that occur inside the building (people speaking loudly, printers, etc.), or when an unexpected sound suddenly permeates an area that has a very low level of background noise. The most common manner by which to mitigate excess noise is through the use of certain materials, which either insulate against noise passing through the material, or absorb the noise wavelengths. In the case of the latter, vertical gardens present themselves as not only an aesthetic element in architecture, but also as a potential acoustic control tool in building design. For this work 10 m^2 of vertical garden substrate modules was tested in a full size reverberation chamber. The objective was to open the doors for vertical gardens to be used in architectural acoustic design.

Keywords: Green wall, Sustainable design, Architecture, Sound engineering.

1 INTRODUCTION

Renterghem and Botteldooren (2008) found that in 2000 up to 44% of residents of the EU were exposed to noise levels that were detrimental to health. Typically noise pollution in cities comes from traffic, in addition to indoor mechanical installations such as HVAC systems. It is also argued that sustainable building design should incorporate noise control systems (Khaleghi *et al.* 2008; Field 2008). There are numerous benefits in the widespread application of green roofs and vertical gardens in cities, where advantages are related to water retention, urban heat island effect reduction, insulation and biodiversity. Wong *et al.* (2010) have put forward that vegetation is also important in fighting acoustic contamination in urban environments. For example, Connelly and Hodgson (2008) put forward that green roofs could be designed for sound absorption in relation to their mass, density and moisture content. In Posada *et al.*'s research (2009), vertical gardens were found to have a beneficial effect in mitigating traffic noise. Furthermore, Restrepo and González (2009) found vertical gardens able to significantly reduce sound levels. More recently, Asdrubali and Mencarelli (2014) sought to determine the absorption coefficient of tropical plants in a porous substrate used in hydroponics cultivation. First an impedance tube was used to determine the normal incidence absorption coefficient. Second, a reverberation chamber was used to determine the diffuse field absorption coefficient. Unfortunately, the chamber dimensions were smaller than those required. Both experiments showed that the main absorber material was the substrate soil, where plants only had a beneficial effect when a large number were planted. Overall, most research to date has been in outdoor vegetation to mitigate the infiltration of uncomfortable sounds into the building interior. However, this paper seeks to build on Asdrubali and Mencarelli's work (2014), where a vertical garden design developed at the *Pontificia Universidad Católica of Ecuador* (PUCE) was tested for interior acoustic design.

2 METHODOLOGY

The main research question of this study was: "Which values does the random incidence sound absorption coefficient of the vertical garden modules filled with substrate have over the frequency range from 100 to 5000 Hz?". These vertical garden modules can be applied in different configurations: connected to form one big garden, or dispersed to form patches of smaller gardens, and directly connected to a wall versus placed in front of a wall. As such, two sub-research questions were formulated:

(1) To what extent will the random incidence sound absorption coefficient of the garden modules be different between a connected and dispersed configuration?

(2) To what extent will the random incidence sound absorption coefficient of the garden modules be different when the modules are directly connected to a wall, placed in front of a wall with a 5 cm air cavity or with a 10 cm air cavity?

In order to answer these questions, measurements were conducted in the reverberation chamber of the Delft University of Technology (TUDelft), which has a size of 200 m^3 and complies with the requirements of ISO 354: 2003. The research followed the procedures of the interrupted noise measurement, as described in this same standard using white noise as sound. The vertical garden modules each had a size of 0.45x0.45 m^2 with a thickness of 10 cm. The modules consisted of a steel mesh with 5 cm apertures, filled with a substrate mixture of potting soil, coco chips and sphagnum moss. The substrate was covered in a fine mesh inside the steel mesh, to avoid loss of fine grains. A total of 50 modules were used, covering a total surface area of 10.125 m^2. The modules were placed in two basic configurations on the floor of the reverberation room: connected and dispersed.

Connected: The modules were connected forming a grid of 7x7 modules plus one module attached to one of the sides (Figure 1a). The total area of the sides in this configuration was 30x0.45x0.1 = 1.35 m^2.

Dispersed: The configuration formed five groups of 2x5 modules (Figure 1b). These five groups of modules were distributed over the floor of the reverberation room. The total area of the sides in this configuration was 70x0.45x0.1 = 3.15 m^2.

(a) (b)

Figure 1. a) Modules in connected configuration, b) Modules in dispersed configuration.

Besides, these two configurations were tested in three different positions in relation to the floor: directly on top of the floor, and with a 5 or 10 cm air cavity between the floor and the modules. To create these cavities, the modules were placed on blocks of 5 cm thick Styrofoam (1 or 2 layers) with dimensions of 10x20 cm. In all cases, the sides of the configuration of modules were open, meaning that the air cavities and the sides of the panel were accessible for the sound. The main reason for choosing the sides to be open was because this more closely resembles the way the modules are used in practice.

Measurements were performed in one-third octave bands with center frequencies starting from 100 Hz up until 5000 Hz. For calculating the random incidence sound absorption coefficient of the modules, the total exposed surface area of the modules was considered, including the top and side surface area. For assessing the sound absorption by the air in the reverberation chamber ISO 9613-1: 1993 was used.

Six positions were marked on the floor of the reverberation chamber. These six positions were used for positioning the microphone. The height of the microphone was roughly 1.3 m above the floor. Positions 5 and 6, also denoted B and A, were also used for positioning the sound source. The center of the sound source was 1.5 m above the floor. As a result, a total of 10 source-microphone positions were used. Besides, each measurement was repeated at least three times.

The equipment used for the measurements was: (1) A Norsonic Nor140 class 1 calibrated sound analyzer; (2) A Norsonic Nor276 dodecahedron omni-directional loudspeaker; (3) A Norsonic Nor280 power amplifier, connected to the Nor276.

Table 1. Weighted (α_w), low frequency (α_L; average of 100 – 315 Hz), mid frequency (α_M; average of 400 – 1250 Hz) and high frequency (α_H; average of 1600 – 5000 Hz) random incidence sound absorption coefficient of the measured configurations.

Configuration	α_w	α_L	α_M	α_H
Connected - on floor	1.00	0.76	1.00	0.94
Connected - 5 cm air gap	1.00	0.82	1.00	0.98
Connected - 10 cm air gap	1.00	0.80	1.00	0.99
Dispersed - on floor	1.00	0.73	1.00	0.93
Dispersed - 5 cm air gap	1.00	0.74	1.00	0.95
Dispersed - 10 cm air gap	1.00	0.73	1.00	0.93

3 RESULTS

Table 1 and Figures 2 (a and b) present the results of the measurement for all of the configurations. The error bars shown in the figure represent the imprecision or uncertainty of the measurements based on the recommendations of ISO 354: 2003. A result of edge scattering yields values of the absorption coefficient higher than 1.

(a)

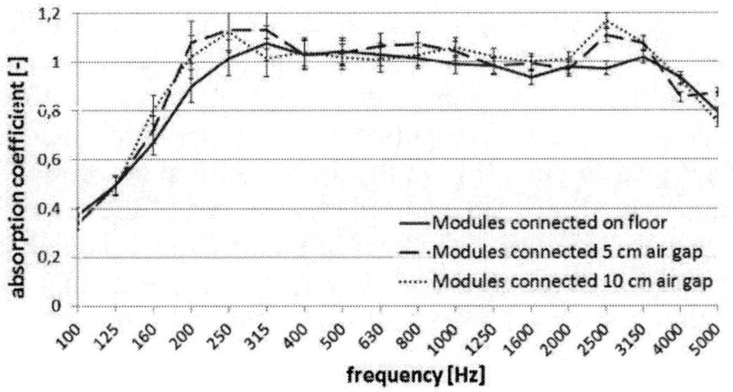

(b)

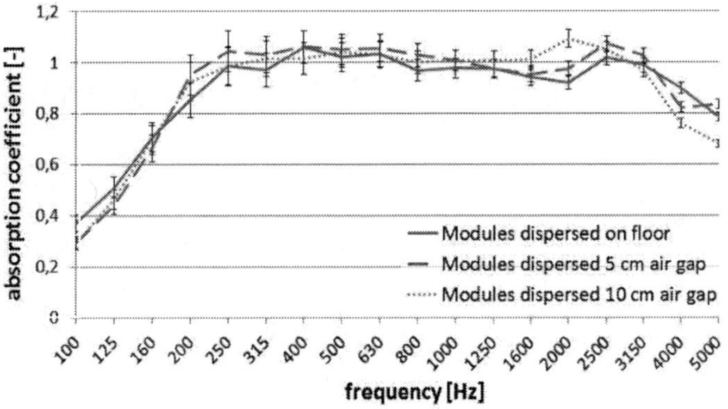

Figure 2. Random incidence sound absorption coefficient, α_s, of the a) connected configurations, b) of the dispersed configurations.

4 DISCUSSION OF RESULTS

4.1 On Floor versus Raised

As can be seen from Figure 2, the substrate follows the behavior of porous absorbers: low sound absorption in the lower frequencies, and high absorption in the higher frequencies. Between 250 Hz and 3150 Hz, the sound absorption coefficient is around 1 (the slightly larger than 1 value is a result of edge scattering). Above 3150 Hz a slight decrease in the sound absorption is visible. This decrease generally is attributable to the measuring equipment and not so much to the material under investigation. Moreover, the difference between the modules directly on top of the floor and the raised modules (5 and 10 cm air gap) is negligible. A peak around 2500 Hz may be attributed to the onset of standing waves inside the modules. Such standing waves arise when the thickness of the material under investigation corresponds to (a random integer times) half a wavelength of the sound. The first frequency where this is expected to happen is

343/2/0.1 = 1715 Hz. Assuming that the thickness of the substrate in many places was less than 10 cm as a result raising this frequency, this value is fairly close to 2500 Hz.

4.2 Connected versus Dispersed

Figure 3 shows the sound absorption coefficient of the modules placed directly on the ground, both in connected and in dispersed configuration. These coefficients are based on the assumption that all exposed surface areas are considered in the calculation. As can be seen, there is no significant difference between the connected and dispersed configuration. The difference between these configurations can be found if not the total exposed surface area is used for computing the absorption coefficient (Figure 3a) but only the top surface area (Figure 3b). In that case, the total absorption is normalized on the top surface area only. Since the exposed surface area of the dispersed configuration is bigger – there is more side area – the total sound absorption is higher, and as a result, the absorption coefficient is higher. However, for characterizing the substrate as a material, it is important to consider the total area exposed to sound.

(a)
(b)

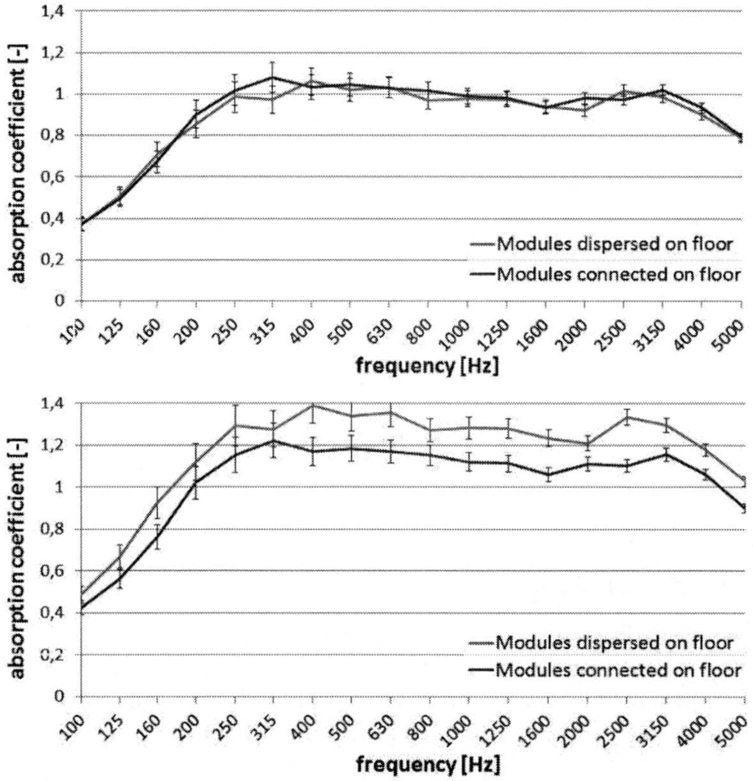

Figure 3. Random incidence sound absorption coefficient, α_s, of the 'on the floor' configurations in case the total exposed surface area (top and side area) (Figure 3a) or only the top area (Figure 3b) is used for calculating the absorption coefficient.

5 CONCLUSION

During this study the random incidence sound absorption coefficient of the vertical garden modules developed at the PUCE were measured in the reverberation room of the TUDelft. A total of 50 modules making up a total floor area of 10.125 m^2 was used for the measurements. Six configurations were measured: connected versus dispersed and directly on the floor versus with an air cavity of 5 or 10 cm. In general, no difference was found between the six configurations: neither air gaps, nor dispersing the modules had an effect on the absorption coefficient. In general, the sound absorption follows the pattern of porous absorbers: low absorption in the lower frequencies, high absorption in the higher frequencies. The weighted sound absorption coefficient of the modules with substrate, for all configurations tested, equals 1.00. This makes this type of substrate highly suitable for applications were sound needs to be attenuated, paving the way for applying vertical gardens for the acoustics of indoor spaces or urban squares.

6 RECOMMENDATIONS FOR FURTHER RESEARCH

First, it is recommended to repeat the study with the same modules, filled with the same substrate, but heavily populated with plants such as ferns or baby tears. Second, certain other characteristics of the substrate, like flow resistivity, porosity and density, were not measured. As a result, a comparison of the measured results to theory is difficult.

Acknowledgments

Special thanks needs to be given to Andrea Vallejo, student of Architecture and at PUCE, whose work and dedication was invaluable in the research for paper. In addition, sincere thanks are due to Arq. MSc. Tatiana Armijos Moya, who was of great help during the work at TUDelft.

References

Asdrubali, F., D'Alessandro, F., and Mencarelli, N., Sound absorption properties of tropical plants for indoor applications in *Proceedings of the 21st international congress on sound and vibration ICSV 21*, Horoshenkov, K.(Ed.), 1-8, Beijing, China , 2014.

Connelly, M. and Hodgson, M., Thermal and Acoustical Performance of Green Roofs. Sound transmission Loss of Green Roofs. *Sixth Annual Greening Rooftops for Sustainable Communities. Conference, Awards & Trade Show*, 1-11, Baltimore, USA, 2008.

Field, C., Acoustic Design in Green Buildings. *ASHRAE Journal*, 60-70, Sep, 2008.

ISO 354: 2003, Acoustics – Measurement of sound absorption in a reverberation room, ISO.

ISO 9613-1: 1993, Acoustics - Attenuation of sound during propagation outdoors - Part 1: Calculation of the absorption of sound by the atmosphere, ISO.

Khaleghi, A., Bartlett, K. and Hodgson, M., Relationship between Ventilation, Air Quality and Acoustics in "Green" and "Brown"" Buildings, *Journal of the Canadian Acoustical Association*, Canadian Acousticas, 36(3), 76-77, 2008.

Posada, M.I., Arroyave, M.d.P., and Fernández, C., Influencia de la Vegetación en los Niveles de Ruido Urbano, *Revista EIA*, 12, 79-89, Dec, 2009.

Renterghem, T.V. and Botteldooren, D., Reducing the acoustical facade load from road traffic with green roofs, *Building and Environment*, Elsevier, 44(2009), 1081-1087, Jul, 2008.

Restrepo, A.C. and González, A., Jardines Verticales para Absorción Acústica de Ruido Urbano. *V Encontro Nacional e III Encontro Latino-Americano sobre Edificações e Comunidades Suntentáveis,* ELECS, Recife, Brazil, 2009.

Wong, N.H. *et al.*, Acoustics evaluation of vertical greenery systems for building walls. *Academic Journal. Building and Environment*, Elsevier, 45(2), 411-420, Feb, 2010.

Interaction between Theory and Practice in Civil Engineering and Construction
Edited by Komurlu, R., Gurgun, A. P., Singh, A., and Yazdani, S.

MULTI CRITERIA MANAGEMENT OF SEWERAGE SYSTEM: ANALYSIS OF THE FAILURES AND THE CRITICALITY OF STRUCTURES

MARZOUK CHERRARED

Faculty of Civil Engineering, University of Sciences and Technology Houari Boumediene, Algiers, Algeria

The FMEA (Failure Modes and Effects Analysis) was used for management of failure risks of the sewerage system of Algiers. It's based on a multi criteria analysis of failures risks of principal structures, their effects and their criticality. The implementation of criteria is based on the knowledge gained in previous phases of investigations. A criticality index "I_c" has been estimated to highlight the most sensitive structures to previously identified risks. This index is calculated based on three parameters: a "risk", a "severity", and a "detectability period" of the failure. Multi criteria analysis was performed to select and rank the criteria on the basis of scores assigned to each of them (from 0 to 7 according to the defined performance states). Four performance classes were defined according to the values of I_c. Classification scale and graphs representing criticality indices were established. A case study is presented. The results highlight the "black points" of the sewerage system. The most sensitive structures have been identified and the targeted interventions in the medium and long terms have been proposed. The weak points of FMEA were also highlighted.

Keywords: Risks management, Criticality index, Multi Criteria method, Sewerage performance, Severity, Detectability time.

1 INTRODUCTION

The FMEA method is designed to study industrial processes, and on which were particularly available accurate data regarding the history of past failures, their consequences and deadlines of return to service of system after a failure. We adapted this approach to data and available knowledge on the sewerage system studied. The study must be described as sensitivity analysis of works to the risks but not a classical FMEA, although the essential basic principles of FMEA were kept. The principle is to "estimate" a criticality index "I_C" that is based on a set of criteria to highlight the most sensitive works to previously identified risks. The adopted methodology consists in:

- Cut out the Algiers's main system into several entities;
- Define, for each entity, the branches of the network and the particular structures as the storm overflow structures, the pumping stations, etc. (the

elements that are ongoing renovation are not integrated into the study that relates to a current state of risks);

- Define, for each work of each entity, the failure criteria and the associated indicators of risk and severity;
- Define a criticality index, denoted I_c, with associating a "failure risk", a "severity" of the considered failure and a "detectability time".

The criticality index is calculated according to the above three parameters.

As part of the methodological principle defined above, a multi criteria analysis was performed to select and rank the criteria on the basis of scores assigned to each of them (from 0 to 7 according to the defined performance states). The aim is to highlight the most sensitive structures and to propose targeted interventions in the medium and long terms. Four performance classes were defined according to the values of I_c: *little sensitive* structures ($I_c \leq 0.2$), *mildly sensitive* ($0.2 < I_c \leq 0.35$), *sensitive* ($0.35 < I_c \leq 0.5$) and *highly sensitive* ($I_c > 0.5$). Classification grids and graphs representing criticality indices were established. A case study is presented in this paper.

2 METHODOLOGICAL APPROACH

The adopted approach is based on the following essential steps:

- Divide the main structure of the sewerage system into twelve units corresponding to the main sub-catchments of the study area and each comprising several hydraulic structures (collectors, storm overflows, pump stations ...),
- Identification of the main types of failures that can disrupt the smooth functioning of the various structures (clogging or pipe rupture, discharges in dry periods or clogging of the overflow pipes of CSOs which happens at rainy weather, stopping of a pumping station ...). Table 1 gives an example.

Table 1. Example failures.

Element	Failure 1	Failure 2
Pipe element	structural failure	Fouling
CSO	discharges in dry periods	Overflows in rainy weather
Pumping station	Stopping	------

- For each element of the system, evaluation of a criticality index (I_c) based on multiple criteria to highlight the most sensitive pipes or structures to the risks previously identified.

2.1 Criteria Classification and Assessment of Retained Parameters

Risk assessment criteria were defined for each type of hydraulic structure. For this, we first defined three evaluation parameters:

- A "risk" of failure: for example, an approved break about collector may be a sign of poor quality of surrounding soil or excessive surface charge. The pipe portion concerned by a past break will present a "risk" of structural failure more important than another with respect to such a criterion.

- A "severity" of failure: consider and quantify the consequences of this failure. For example, the rupture of a pipe, can it cause damages to property and to persons? And if so, what degree of severity could we attributing to it?
- A "detection period": assign to each structure a period of detectability depending on its location, the impact of the failure or its management system. For example, a default in a pumping station remotely managed will be identified more quickly than in a post unattended.

The combination of these three parameters allow to define a criticality index of a failure by the equation (1) (Stamatis 2003, Diadem Press 2003):

$$I_c = Risk\ x\ Severity\ x\ detectability\ period \tag{1}$$

2.2 Assessment and Performance Scales

For each structure, for each type of failure and for each evaluation parameter, we have defined classification criteria for which we have associated an evaluation scale including a performance and notation scale. Table 2 shows the scale adopted for evaluating pipes failure.

Two aspects were considered:

- *The structural failure*: this criterion represents the elements that lead to think that the section concerned could undergo a break causing a collapse.
- *Fouling:* this criterion includes the silting up of networks near the outlet in sea and the areas of decanting of materials carried during rain events.

3 CASE STUDY

In the context of this paper, we present an application on only a part of the sewerage network of Algiers and by dealing only the pipe sections. (See table 1).

3.1 The Part of Studied Network

The part of the studied combined network is located in the center catchment of Algiers with an area of 1327 ha and consists of 849 m of visitable underground pipes. Figure 1 presents the block diagram.

3.2 Classification Scales - Criticality Indexes

By applying the evaluation criteria of table 2 and equation (1), the values of I_c were obtained. Table 3 provides an example for the "structural failure". The performances scale has been chosen as shown in Figure 2.

4 INTERPRETATION OF RESULTS - DISCUSIONS

The sections most critical with respect to the risk "structural failure" are often characterized by a large size of the upstream catchment.

Table 2. Assessment scale of the pipes failure.

Failure type	Parameter	Criterion N°	Criterion Definition	Performance scale	S	S_{max}
Structural	Risk	A	General diagnosis DHW	Good	1	13
				Average	2	
				Bad	3	
				Undiagnosed	2	
		B	SAFEGE diagnosis and SEAAL surveys	Good	1	
				Average	2	
				Bad	3	
				Very bad	4	
				Undiagnosed	2	
		C	observation of past breaking	No	0	
				Yes	1	
		D	Particular surface load	No load	0	
				Normal load	1	
				Temporary significant load	3	
				Permanent significant load	4	
		E	Age of the pipe	After 1962	0	
				Before 1962	1	
	Severity	A	Upstream catchment size	Small (≤200 ha)	1	15
				Average (201 to 500 ha)	2	
				Big (>500 ha)	3	
		B	Ease of replacement (according Φ)	Easy (≤Φ2000)	1	
				No easy	3	
		C	Ease of works	Somewhat favorable	1	
				Moderately favorable	2	
				Very favorable	4	
		D	Potential Impact	Average	1	
				Strong	3	
				Very strong	4	
		E	Possibility of meshing	Yes	-1	
				No	1	
	Detectability period	---	---	Fast	1	3
				Slow	2	
				Undetectable	3	
Fouling	Risk	A	Slope	High	1	11
				Low	3	
		B	Fouling already observed	No	1	
				Yes	2	
		C	Downstream influence	Any	1	
				Temporary	2	
				Permanent	3	
		D	Special flows upstream	Yes	1	
				No	3	
	Severity	---	Upstream catchment size	Small	1	3
				Average	2	
				Big	3	
	Detectability period	---	---	fast	1	3
				Slow	2	
				Undetectable	3	

S : score, S_{max} : maximum score

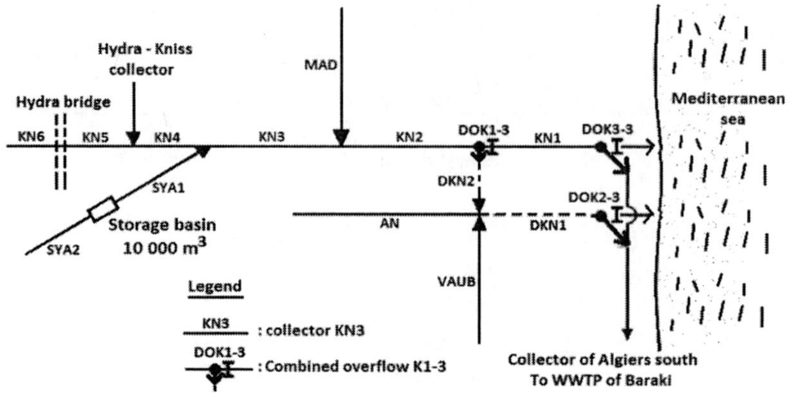

Figure 1. Studied case.

Table 3. Classification scale of the structural failure of pipes.

Pipe	Risk					Severity					DP	I_C
	A	B	C	D	E	A	B	C	D	E		
KN1	2	2	0	3	1	3	3	4	4	-1	3	0,57
KN2	2	2	0	1	1	3	3	2	4	1	3	0,37
KN3	2	2	0	1	1	3	3	2	4	1	3	0,37
KN4	2	2	0	0	1	2	3	1	3	1	3	0,23
KN5	2	2	0	0	1	2	3	1	1	1	3	0,18
KN6	2	2	0	1	1	1	3	4	1	1	3	0,28
DKN1	2	1	0	3	0	3	3	4	3	-1	3	0,40
DKN2	2	1	0	1	0	3	3	4	3	-1	3	0,27
VAUB	2	2	0	1	1	1	1	2	1	1	3	0,15
AN	2	2	0	3	1	1	1	2	1	1	3	0,21
MAB	2	2	0	4	1	2	3	4	3	1	3	0,55
SYA1	2	2	0	1	1	2	3	4	3	1	3	0,37
SYA2	2	2	0	1	0	1	1	4	1	1	3	0,18
Hydra - Kniss	2	2	0	1	1	3	1	1	3	1	3	0,25

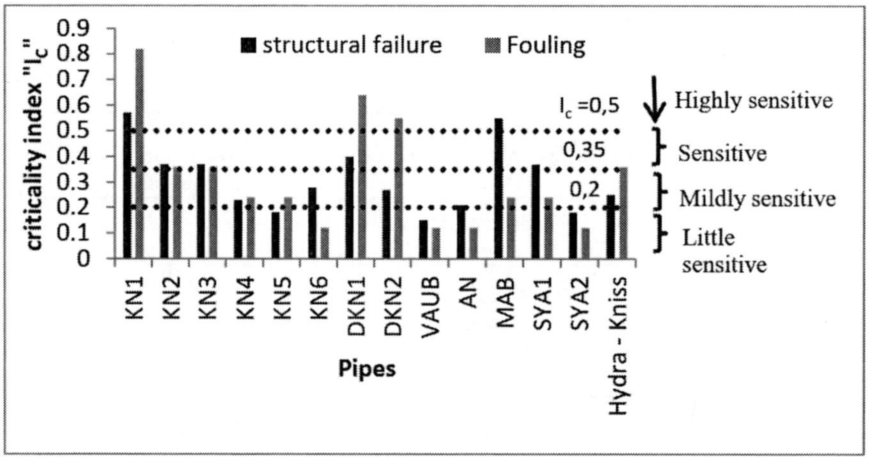

Figure 2. Criticality index and performance classes.

They are located in densely populated areas where the necessary interventions after a collector collapse would be very difficult to achieve. Important factors of increased criticality are also the presence of high loads (passage under building, under railway tracks or high traffic road), the age of the collector or the fact that collapses of the structure have already been occurred in the past. About the fouling risk, the more sensitive pipes are those located in the areas with low slopes, which undergo an downstream hydraulic influence (tide level, threshold of downstream discharge, pumping station), which receive waters containing sands from wadis and of course those whose the silting up has already been observed in the past. The criticality index is also obviously influenced by the upstream catchment size.

Rehabilitation actions of pipes were proposed according to the nature of failures.

5 CONCLUSION

The used methodology has allowed us to establish a credible performance scale of the principal works of sewerage system of Algiers. However it requires an important quantity of data essentially based on surveys and field visits.

The structures whose index of criticality is high can enjoy a relatively optimum maintenance schedule taking into account the priorities established on the basis of performance values or of evaluated failure risk. The Analysis of sensitivities to the risks of the sewerage system allows, as part of sustainability plane and network maintenance, to prioritize interventions and to organize monitoring.

The classic FMEA is a risk assessment tool that allows establishing priorities for intervention. Thus, it allows to reduce and to eliminate the risks of failure. However, this tool could present difficulties such as subjective description of the evaluation criteria, the relative importance among the risk scores, the difference in the representation of risk among the same scores, and the shared knowledge between members of the team of FMEA. We could improve this aspect by using a method less subjective (Kwai-Sang *et al.* 2009, Pillay and Wang 2003, Rajiv *et al.* 2007) like the fuzzy theory (Yeh 2007).

References

Diadem Press, *Guidelines for Failure Mode and Effects Analysis for Automotive, Aerospace and General Manufacturing Industries*, Chemical Rubber Company (CRC) Press, USA, 2003.

Kwai-Sang, C., Ying-Ming, W., Gary Ka Kwai, P., and Jian-Bo Y., Failure mode and effects analysis using a group-based evidential reasoning approach, *Computers & Operations Research*, 36(6), 1768-1779, June, 2009.

Pillay, A. and Wang, J., Modified failure mode and effects analysis using approximate reasoning, *Reliability Engineering & System Safety*, Elsevier, 79(1), 69-85, Jan, 2003.

Rajiv, K. S., Dinesh, K., and Pradeep, K., Modeling and analysing system failure behaviour using RCA, FMEA and NHPPP models, *International Journal of Quality & Reliability Management*, 24(5), 525-546, 2007.

Stamatis, D. H., *Failure Modes and Effects Analysis: FMEA from theory to execution, 2nd Ed.*, American Society for Quality, Quality press, Milwaukee, 2003.

Yeh, R. H. and Hsieh, M. H., Fuzzy assessment of FMEA for a sewage plant - *Journal of the Chinese institute of industrial*, 24(6), 505-512, Jun 2007

THEORY AND PRACTICE IN PROJECT MANAGEMENT: A MULTIDISCIPLINARY APPROACH

SILVIA MAZZETTO

Dept of Architecture and Urban Planning, College of Engineering Qatar University, Doha, Qatar

This paper introduces an approach currently being implemented in the Department of Architecture and Urban Planning at College of Engineering in Qatar University. It describes the primary components of the theoretical research including the course program in Construction and Project Management, its integration with other courses, the cooperation required to support the interdisciplinary emphasis, and the establishment of an innovative academic/industry partnership to provide a state-of-the-art physical and technological infrastructure to support the program goals. The justification and benefits of the step from implicit to explicit theory are discussed based on the influential work of recent studies and the developments of the research approach of the multidisciplinary collaborative project. It is contended that much, although surely not all, of the present implicit theory, can already be pinpointed, and the associated with counter-productive impacts, explained and empirically recognized. Despite the fact that previous studies on interaction between theory and practice offered various degrees of value to the research field, some limitations and debates remain awaiting further investigation, as do the significant changes associated with the construction industry and project management. The research concludes that the efficiency of interaction between theory and practice in the construction industry needs to be further examined, and a possible research method is also proposed for carrying out future studies.

Keywords: Collaboration, Partnership Awareness, Road safety.

1 INTRODUCTION

1.1 Project Management and a Multidisciplinary Research Approach

Projects often comprise of people and organizations from different nationalities and disciplines such as architects, engineers, clients, contractors, suppliers, and banks which generates complexity in the managerial procedures, due to different multidisciplinary approaches inside multicultural teams, foreign managers, and international partners. In the literature it is easy to see that different disciplines can influence a variety of project management issues such as: teams (Binder 2007) leadership (De Bony 2010), communication, performance (Laufer and Tucker 1987), risk assessment (Zwikael and Ahn 2011), business negotiations (Hurn 2007), international project management (Lane *et al.* 2005) and planning. In Qatar there are many complex projects performed in collaboration with foreign companies and local partners such us Doha Metro, Lusail City, Hamad International Airport, Education City and many others. Project management is important to the success of these initiatives, and project managers face numerous challenges, which include multi-disciplinary factors such as managing mixed teams pertaining to different disciplines and additionally many cultural differences within the same team. Specifically, how does the multidisciplinary approach influence project managers' perceptions and utilization of managerial tools? Professionalism starts at the first steps of a student's education in university. Different disciplines, from their initial stages of training, can model and affect the student's attitude and behavior in collaborative work, and the way they act and perform as the new Professional Managers (PMs) of the future. In Qatar University, for example, a multidisciplinary research project has been proposed and developed to involve students and instructors specializing in diverse disciplines and aspects of Architecture and Urban Planning (AUP) and Industrial and Systems Engineering (ISE). The aim is to get them to focus on a common assignment as part of their courses (a class project in this case), using the tools of project management to investigate and develop the students' capabilities and attitudes when working in a multidisciplinary project. The extent that project managers and their teams interact effectively within the project environment is not exclusively an issue of multi-disciplinarily; it is influenced by different national cultures, and past experiences, and can affect a variety of managerial roles and activities, for instance, leadership schemas (the criteria of a good or a bad leader) and behavior scripts (greetings) vary between those of different cultural backgrounds.

2 A MULTIDISCIPLINARY PROJECT

2.1 Multidisciplinary Project Objectives

The multidisciplinary research project was developed at Qatar University, involving a total number of 76 students both form Architecture and Urban Planning (DAUP course ARCT 530 Construction and Project Management) and Industrial and Systems Engineering (ISE course IENG 481 Project Engineering) for the fall of 2015. The project topic was linked to the active competition of the Qatar Road Safety Studies Center (QRSSC), which is a prominent research center in Qatar University, in the form of prize money to be provided to the winning team. The multidisciplinary research project title is "Project Management tools and techniques for the development of a multidisciplinary project on Road Safety in Qatar". The two batches of students derived from various disciplines were requested to use the tools and techniques of PM, through different disciplinary approaches, for the development of the project deliverables, working in groups and demonstrating their collaboration. The aim of the project was to increase road safety awareness in Qatar. As the youth are the basis of the Qatari society, the competition target the youth of Qatar University to engage them in road safety awareness and its importance.

2.2 Project Goal, Deliverables and Proposals

In the last few years, the number of people who have started to drive, of both genders, has increased in Qatar and this has led to an increase in traffic jams and in the number of accidents (Qatar National Road Safety Strategy NRSS 2013-2022). Many factors can contribute to the occurrence of accidents such as lack of road safety awareness, irresponsible and misbehaving drivers and other issues related to the design of the road. The students started their project with collaborative research to monitor the current status of road safety awareness in Qatar and about how road safety in Qatar is a major issue.

The students were working towards producing the project deliverables requested, including reviewed pedestrian safety and road regulations. In addition, their work aims to propose a re-design of some dangerous areas, improving the quality of the road layout in Qatar in order to meet the required road safety measures. It is highly important to design in accordance with the local codes and regulations. However, when it comes to road safety, several additions needs to be considered to ensure safety and that take into account the variations of behavior that one observes while just taking a stroll along the road. For example, pedestrians' safety is quite an issue, since many families visit pedestrian areas with their children, who are at a very high risk of getting into an accident because of the fast cars passing through the street.

With the proposed students' guidelines for re-design, Qatar's road users will benefit from improved road safety, reduced traffic congestion and more environmentally friendly driving, all enabled by better developed roads. Moreover, the research project strives to develop awareness-raising messages regarding road safety to help prevent all road fatalities. The aim of the multidisciplinary project is to promote the development of Qatar road safety educational materials and target road safety publicity campaigns. The active competition of Qatar Road Safety Studies Center (QRSSC) requested the students to produce an awareness video which will develop, diffuse, collate, and disseminate information on good practice in road safety education, training, and publicity throughout Qatar (not only in Doha but also in other areas- Al Khor, Lusail, Dukhan).

2.3 The Methodology for Interaction / Collaboration

A structured research method and templates were selected to assess differences in PM strategies of the multidisciplinary research project team, in using PM tools (Gray and Larson 2002). A convenience sample was used to develop the project and collect data from the students, and a report template, a minute of meeting template, a poster and video template.

The research items covered specific areas of project planning and were then analyzed to compare Architecture and Engineering students in their approach to the multidisciplinary project and in the proper application of managerial tools (Meredith and Mantel 2006) of planning scope, time, cost, risk, quality, integration, innovation and technology, and communication. Due to the impracticality of retaining a large number of managerial tools, and developing a comprehensive research, due to the course duration restrictions (i.e., from September to December 2015) only the planning, risk assessment, time scheduling and project controls were included as mandatory managerial tools of the research. The environmental dimensions examining human nature, responsibility, space, and specific/diffuse were omitted.

Instructors provided course delivery to both batches of students through integrated lectures during the course (Gray and Larson 2002) and through readings, educational audio and video materials, and parts of course assignments. From the initial phase of site analysis, students developed their report, selecting one specific area in Doha, the project title, the group name, and arranging some meetings to define the multidisciplinary collaboration between batches such us project objectives, project methodology and confirmation of each member's role.

Students met in their group approximately twice in a week. They developed a proper agenda each time and discussed it during the meeting. Weekly meeting minutes prepared by the groups described the events of the meeting, starting with a list of attendees, a statement of the issues considered by the participants, and related responses or decisions for the issues, what was discussed and achieved since the previous week, and what were the outstanding items be completed before the next meeting. All meeting minutes were signed by each of the participants before the start of the next meeting. The meeting minutes were collected in a report to monitor the progress of the work; the report was checked against a rubric on written reports.

During the entire semester three main cycles of development and review in the students' proposals and projects were arranged:

- Initial proposal for collaboration between students to identify their role within the group and the new proposed ideas within the selected area of study.
- First project submission including the report meeting minutes; collected data from surveys and interviews; the progress of their project related to urban fabric analysis; and studies to investigate the level of road safety in different zones of Doha, Qatar.
- The progress project's submission and the final submission, including the requested deliverables such as the final report, poster and video for competition, conclusions and final recommendations.

The progress of the submissions phases were discussed within the multidisciplinary groups of students, based on the influential work of each group member, and many samples of projects under construction in Doha, Qatar. Linking multidisciplinary to project outcomes was problematic since many changes in definitions, assessment and numerous factors concerned rose up during the project between groups (Turner 1999).

2.4 Success Criteria

There were two types of success criteria. The first one was the project management success criterion that was related to the students' capability in running the project using the available managerial tools, which meant to finish on time, be able to overcome any delays that might occur during the project and get approval for initiation document. The second type was the project deliverable success criteria that were related to quality in the delivered outcomes as a result of the final project. All the deliverables showed the students' ability to understand the basic principles of project and practice management (Milosevic 1999), the initial evaluation of project costs and budgets, the proper use of activity scheduling, ability to allocate resources to the project, and to demonstrate understanding and appropriate use of methods for controlling projects, including cost and time.

3 RESULTS

Starting with the financial considerations, regarding the fundamentals of building costs, such as financial feasibility, estimating costs with an emphasis on life-cycle cost accounting, students demonstrated their ability to use the basic principles of architectural practice management such as project planning, time management, risk management, mediation and arbitration, and recognizing trends that affect practice. All the students submitted their project report respecting the deadlines.

Many groups of students interacted collaboratively for the whole duration of the project, not only demonstrating their ability to work in interdisciplinary teams and with other multidisciplinary groups to successfully complete their projects. Particularly for the QRSSC competition, the submitted videos addressed objectives such as raising awareness about the danger of speeding on the road, the importance of wearing seat belts, the importance of using baby car seats (booster) and other measures relevant and useful to road safety in Qatar. All the submitted videos were original, not including pictures or videos from any other sources and not containing any interview and completely shot in Qatar and they can be promote by QRSSC as message for advertising, publicity, and promotional purposes without compensation.

Students provided conclusions and recommendations, demonstrating their ability to recognize the dialectic relationship between Project Management and the multidisciplinary approach to recognize the diversity of needs, values, behavioral norms, and social patterns as they relate to the creation of the safe built environment of roads in Qatar. Some specific recommendations regarding guidelines and standards for road design that arose from the students' projects include:

- Proposing homogenous road sections in the redesign of the road to avoid conflicts with oncoming traffic,
- Proposing street furniture and signs that will avoid unpredictable behavior, will separate vehicle types and will reduce speed at potential conflict points
- Limiting the number of design elements in each category and making them uniform.
- Creating a sustainable safe road network having flow function generated by uninterrupted roads having a distributor function – for the distribution and collection of traffic to and from different directions; and slip roads having an access function which need to facilitate appropriate entrances.

4 CONCLUSION

In this paper we have set out to show how the multidisciplinary collaboration between students can enhance research practice in a field dominated by the theoretical approaches of different disciplines. We have also described the methodology used and some of the issues which have arisen in developing this approach and how the research is still facing an initial phase of development, to be further investigated. In giving this partial view of a complex, long-term project it is not possible to describe the full process of evaluation and the outcomes of every single group project in more detail. Only some aspects of evaluation have been referred to briefly and the principle outcomes, but they need to be implemented and investigated through further ongoing multidisciplinary collaborations between students and disciplines in other courses. The investigation included a wide range of research activities, detailed review of the state of the awareness regarding the road safety in Qatar, and site survey and analysis. However, most of the deliverables and outcomes are still in their preliminary stages.

The managerial tools, deliverables and outcomes are similar to those which might be encountered in normal professional practice. The concept of re-designing for the purpose of road safety as a research activity may be difficult to recognize given this similarity, which establishes the relationship, analogies and differences between theory and practice in project management. As already pointed out, the students, working as researchers, rather than professional practitioners, had the opportunity to use their skills to represent and develop a hypothetical proposition in a form which is widely accessible and suitable for many forms of evaluation.

Despite the fact that previous studies on interaction between theory and practice offered various degrees of value to the research field, some limitations and debates still await further investigation. In addition, the significant changes associated with the construction industry and project management in theory also provide opportunities for future research. All the outcomes need to be further investigated and compared with other similar outcomes from other multidisciplinary projects.

References

Binder, J., 2007, *Global PM: Communication, Collaboration and Management*. Gower Publishing, Hampshire.

De Bony, J. D., 2010. *PM and NC: a Dutch–French case study*. International Journal of Project Management 28, 173–182.

Gray, C. F. and Larson, E.W., 2002, *Project Management: The Managerial Process*. McGraw Hill.

Hurn, B. J., 2007, T*he influence of culture on international business negotiations*. Industrial and Commercial Training 39 (7), 354–360.

Lane, H. W., Di Stefano, J. J., and Maznevski, M. L., 2005, *International Management Behavior: Text, Readings, and Cases*. Blackwell Publishing, Oxford.

Laufer, A. and Tucker, R., 1987. *Is construction planning really doing its job? A critical examination of focus, role, and process*. Construction Management and Economics 5 (3), 243–266.

Meredith, J. R. and Mantel, S. J., 2006. *PM: A Managerial Approach*, Fourth edition. John Wiley and Sons, New Jersey.

Milosevic, D. Z., 1999, *Echoes of the silent language of PM*. Project Management Journal 30 (1), 27–39.

Turner, J. R., 1999, *The Handbook of Project-based Management*, Second Edition. McGraw Hill.

Zwikael, O. and Ahn, M., 2011, *The effectiveness of risk management*. Risk Analysis 31 (1), 25–37.

Interaction between Theory and Practice in Civil Engineering and Construction
Edited by Komurlu, R., Gurgun, A. P., Singh, A., and Yazdani, S.

AN AGENT-BASED APPROACH FOR MODELING THE EFFECT OF LEARNING CURVE ON LABOR PRODUCTIVITY

H. SHEHWARO, E. ZANKOUL, and H. KHOURY

Dept of Civil and Environmental Engineering, American University of Beirut, Beirut, Lebanon

The labor-intensive nature of construction projects requires proper management and efficient utilization of labor resources. Improvement of labor productivity can enhance project performance and thereby lead to substantial time and cost savings. Several studies focused on identifying the effect of different factors on labor productivity, whereby the learning curve factor proved of paramount importance. Although previous research efforts developed models to represent the learning curve effect using traditional simulation approaches such as System Dynamics (SD) and Discrete Event Simulation (DES), none of these studies used Agent-Based Modeling (ABM) techniques. This study takes the initial steps and presents work targeted at analyzing the effect of learning on labor productivity using ABM. Based on ABM, a construction site can be modeled as an active environment in which agents interact with each other and their surroundings thereby creating an adaptive environment open for learning and improvement. The solution to the problem is described in details using a simulation model developed in AnyLogic 7 (Educational Version). The components of the proposed model have been created and preliminary results highlighted the potential of using the agent-based modeling paradigm to simulate the effect of learning on labor productivity in the construction industry.

Keywords: Construction, Management, Simulation, Modeling, Proficiency, ABM.

1 INTRODUCTION AND PERTINENT LITERATURE

Among many factors that impact construction labor productivity (Gunduz 2004, Hanna *et al.* 2005, Lee 2007, Singh 2010, Hinze 2011, Hafez *et al.* 2014), the learning curve has long been proven to be of paramount importance (Wright 1936). The learning curve theory states that, under a repetitive process, whenever the production quantity of a product doubles, the unit required for production drops by a certain percentage of the previous unit referred to as the learning rate (Jarkas 2010). Many models have been developed to illustrate this learning curve phenomenon and show, for example, the relationship between the cycle number and the time per cycle. These models include: (1) The Straight-Line Model (Wright's Log-Linear Model) (2) The Stanford "B" Model (3) The Exponential Model (4) The De Jong's Model; and (5) The Cubic Power Model (The S-Curve Model) (Thomas *et al.* 1990, Hijazi *et al.* 1992, Naresh and Jahren 1998, Chen *et al.* 2009, Taylor *et al.* 2009, Jarkas, 2010, Shehata and El-Gohary 2012, Pellegrino *et al.* 2012, Panas and Pantouvakis 2014). On the other hand, using simulation for modeling the learning curve phenomenon has gained more and more attention over the last years. The most popular simulation techniques adopted were Discrete-Event Simulation (DES) (Hijazi *et al.* 1992, Lutz *et al.* 1994, Panas and Pantouvakis 2014) and System dynamics (SD) (Nasirzadeh and Nojedehi 2013). However, none of the previous studies used Agent-Based Modeling (ABM) to model the effect of learning on labor productivity. ABM can be defined as a computer simulation technique allowing the examination of how system patterns develop from the behaviors of individual agents. ABM creates virtual agents that have the ability to interact with each other and their environment and accordingly make autonomous decisions (Awwad *et al.* 2014). ABM was used to model the effect of congestion (Watkins *et al.* 2009, Marzouk and Ali 2013) and safety (Marzouk and Ali 2013) on labor productivity but the effect of learning development was not modeled. As a matter of fact, it was stated in prior ABM efforts that the limitation was the exclusion of the learning curve effect (Watkins *et al.* 2009). The objective of this paper is thereby to analyze the effect of learning on labor productivity by making use of the ABM paradigm and developing a model that incorporates different learning curve techniques into the simulation.

2 METHODOLOGY

2.1 Construction Process Description

In order to best illustrate the learning curve effect, a process of repetitive nature must be selected. Typically, high-rise buildings projects incorporate many tasks of this nature whereby the floors are almost identical and learning can be witnessed. For that reason, a case study of a multi-story building (50 stories) in the region of Beirut, Lebanon was adopted. The building consists mainly of a core wall, slabs and exterior walls. Only activities related to the structural construction, in particular erecting forms, installing steel rebars, and pouring concrete were modeled. Based on the RS Means Building Construction Cost Data book (RS Means Building Construction Cost Data 2014), the daily outputs of the aforementioned activities together with respective crews are shown in Table 1 for a slab and a wall. The corresponding volumes of concrete and steel needed are calculated and summarized in Table 2.

Table 1. Productivity rates and crews.

Task	Task Type	Daily Output	Crew	Number of Crews
Single Slab	Erect slab forms	470 S.F.[44 m^2]	1 Forman, 4 Carpenters	4
	Install steel rebars	2.9 Ton [2.6 Ton]	4 Rodmen	3
	Place concrete	160 C.Y. [122 m^3]	1 Forman, 5 Laborers	1
Single Wall	Install steel rebars	4 Ton [3.6]	4 Rodmen	3
	Erect Forms	280 S.F. [26 m^2]	1 Forman, 4 Carpenters	4
	Place concrete	95 C.Y. [73 m^3]	1 Forman, 5 Laborers	1

Table 2. Materials takeoff.

Task	Concrete (m^3)	Steel (T)	Area (m^2)
Single Wall	117	23	778
Single Slab	196	39	982

2.2 Agent-Based Model Development

The agent-based model is composed of agents whose behavior is defined by state-charts. State-charts define an agent's different states connected through a logical flow. In the case of the proposed model, five agent types were included, *Steel Crews, Formwork Crews, Concrete Crews, Slabs and Walls*, which can be divided into two categories; the crews and the constructed entities. All crews are assumed to have a similar behavior and state-charts and the same applies to constructed entities (Fig.1). To that end, only one of each was explained.

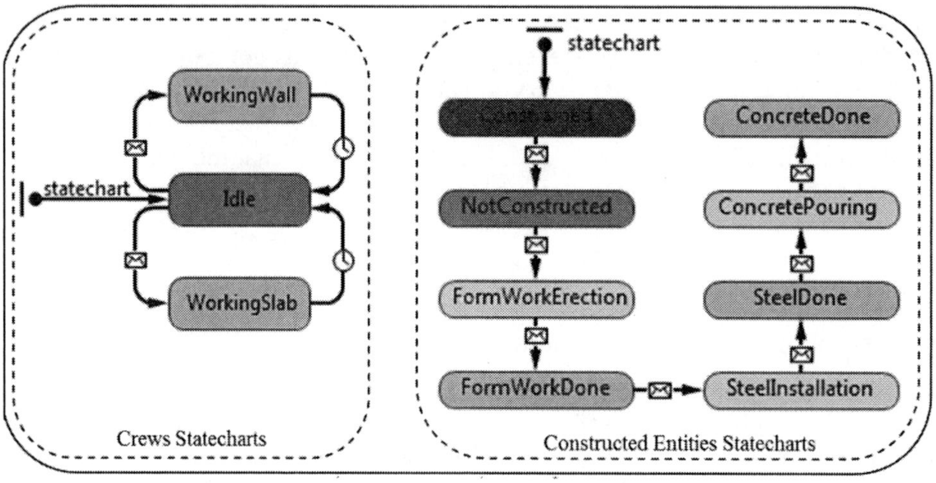

Figure 1. Statecharts of agents.

As shown in Fig. 1, the slab agent has eight states. The first one named *Constrained* is the initial state of every slab or wall whereby construction cannot begin due to predecessors or dependencies. The transition from *Constrained* to *NotConstructed* is triggered by the message "unconstrained", which is sent in the following three instances, (1) On start-up, the floor slab directly moves from *Constrained* to the state *NotConstructed* since it has no predecessors; (2) Each time a slab is constructed, the wall above becomes unconstrained; (3) Each time a wall is built, the slab above becomes unconstrained. More specifically, when a slab moves to the *NotConstructed* state, it sends "move to slab" message to the *Formwork Crew* agent. This allows the crew to move from its initial state *Idle* to the state *WorkingSlab* and accordingly, a message is sent to the slab to move from *NotConstructed* to *FormWorkErection* (Fig.1). The formwork crew becomes idle again once formwork erection is complete. As such, one of the most important transitions in the model happens when the crew moves from the working state to the idle state. In fact, this transition of type timeout represents the duration it takes to complete a certain task. In order to model the learning curve phenomenon, the timeout duration was assumed to vary each time the crew goes through its specific transition. In other words, using each of the aforementioned learning curve models (i.e., Straight-Line, Stanford "B", Exponential, De Jong, and Cubic Power) and assuming an 80% learning rate, the timeout duration was computed in such a way it is gradually decreasing with each task repetition. Once this specified time elapses, the crew moves back to the state *Idle* and the slab moves to the state *FormWorkDone*. The same process is repeated for other concrete activities related to slabs and the entire process is repeated for walls. The state-charts keep interacting until the whole building is completed. In addition to these charts, variables and functions were incorporated in the simulation model to represent factors affecting learning rate. These include shared worker experience, psychological effects and work interruptions.

3 RESULTS AND CONCLUSION

The proposed ABM model was run using each of the aforementioned methods (i.e. Straight-Line, Stanford "B", Exponential, De Jong, and Cubic Power). Fig.2 depicts the respective learning curves (time vs. floor number) for the slabs formwork activities under each method. On the other hand, Fig. 3 portrays the duration of all concrete activities under each method.

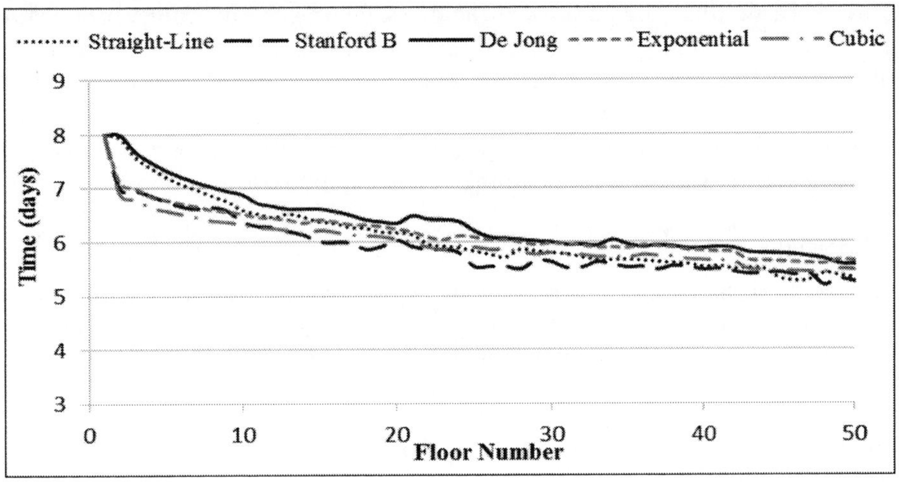

Figure 2. Learning curves for slabs formwork activities.

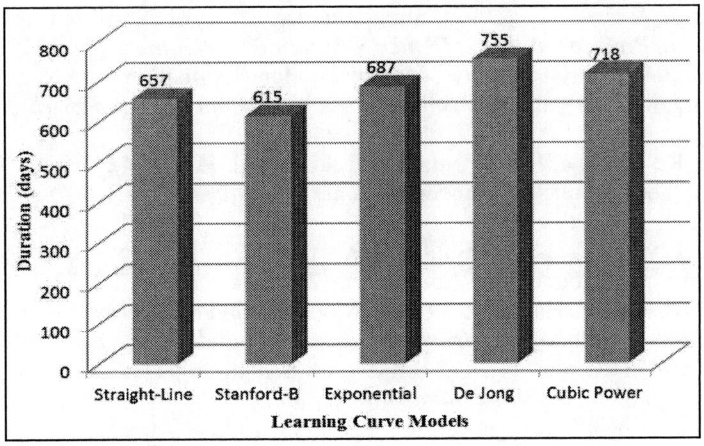

Figure 3. Total Duration of slabs and walls concrete activities (days).

Based on the results in Fig.2, the influence of learning under all five methods was clear in the first 18 floors then started to vanish. For instance, using the Straight-Line model, the estimated time at the beginning was 8 days/floor dropping to 3.2 days/floor when reaching the 18th floor. If the effect of learning and interaction was not considered, the 8 days/floor requirement would be constant and the formwork activities duration would be greater. The absence of learning development would thereby lead to an overall duration greater than 1000 days. Fig. 3 shows that the duration of all concrete activities, when the learning effect is taken into consideration, varies between 615 and 755 days whereby the Stanford-B model led to the shortest duration.

It can be concluded that ABM is a reliable technique that allows modeling the interaction of agents and the quick generation of different learning curves and activities overall durations. The proposed approach is flexible and user-friendly in that the user can try various scenarios with different learning rates and test different learning curve methods.

Future work will include additional factors affecting the learning curve such as the change in crew size, unforeseen conditions such as bad weather and accidents, etc. These factors will be added to the current simulation model and their effect on learning curve and thereby productivity will also be assessed. Moreover, the proposed agent-based model will be verified and validated by visualizing processes in 3D and graphically depicting variables and parameters as the model is running.

Acknowledgments

The presented work has been supported by AUB's University Research Board (URB). The authors gratefully acknowledge URB support. Any opinions, findings, conclusions, and recommendations expressed by the authors in this paper do not necessarily reflect the views of URB.

References

Awwad, R., Asgari, S., and Kandil, A. (2014). Developing a Virtual Laboratory for Construction Bidding Environment Using Agent-Based Modeling. *J. Comput. Civ. Eng.*, 10.1061/(ASCE) CP.1943-5487.0000440 , 04014105.

Chen, J., Taylor, J., & Unsal, H. (2009). Simulating the Effect of Learning Decay on Adaptation Performance in Project Networks. *Winter Simulation Conference.* 2712-2722

Gunduz, M. (2004). A Quantitative Approach for Evaluation of Negative Impact of Overmanning of Electrical and Mechanical Projects. *Building and Environment.* 39(5), 581-587.

Hafez, S., Aziz, R., Morgan, E., Abdullah, M., & Ahmed, E. (2014). Critical factors affecting construction labor productivity in Egypt. *American Journal of Civil Engineering*. 2(2), 35-40.

Hanna, A. S., Taylor, C. S., & Sullivan, K. (2005). Impact of Extended Overtime on Construction Labor Productivity. *Constr. Eng. Manage.* 131(6), 734–739

Hijazi, A., AbouRizk, S., & Halpin, D. (1992). Modeling and Simulating Learning Development in Construction. *Journal of Construction Engineering and Management*, 118(4),685-700.

Hinze, J. W. (2011). Chapter 11: The Impact of Scheduling Decisions on Productivity. In *Construction Planning and Scheduling, 4th Edition.*

Jarkas, A. (2010). Critical Investigation into the Applicability of the Learning Curve Theory to Rebar Fixing Labor Productivity. *Constr. Eng. Manage.*, 136(12), 1279-1288.

Lee, S. (2007). Understanding and Quantifying the Impact of Changes on Construction Labor Productivity: Integration of Productivity Factors and Quantifying Methods. *Doctoral Dissertation ,University of Berkeley.*

Lutz, J., Haplin, D., & Wilson, J. (1994). Simulation of Learning Development in Repetitive Construction. *Journal of Construction Engineering and Management*. 120(4), 753–773.

Marzouk, M., & Ali, H. (2013). Modeling safety considerations and space limitations in piling operations using agent based simulation. *Expert Systems with Applications*, 40(12), 4848–4857.

Naresh, A., & Jahren, C. (1998). Learning Ooutcomes from Construction Simulation modeling. Civil. Eng. and Env. Syst., 129-144.

Nasirzadeh, F., & Nojedehi, P. (2013). Dynamic modeling of labor productivity in construction projects. *International Journal of Project Management*, 31(6), 903-911.

Panas, A., & Pantouvakis, J. (2014). Simulation-Based and Statistical Analysis of the Learning Effect in Floating Caisson Construction Operations. *Constr. Eng. Manage.* 140(1), 04013033

Pellegrino, R., Costantino, N., Pietroforte, R., & Sancilio, S. (2012). Construction of multi-storey concrete structures in Italy: patterns of productivity and learning curves. *Construction Management and Economics*. 30(2),103-115.

RS Means Building Construction Cost Data. (2014).

Shehata, M., & El-Gohary, K. (2012). Towards improving construction labor productivity and projects' performance. *Alexandria Engineering Journal*. 50(4), 321-330.

Singh, S. P. (2010). Factors affecting the productivity of construction operations in the United Arab Emirates. *Master's Thesis. Heriot-Watt University.*

Taylor, J., Levitt, R., & Villarroel, J. (2009). Simulating Learning Dynamics in Project Networks. *Journal of Construction Engineering and Management*, 135(10), 1009–1015. 1009-1015.

Thomas, H., Maloney, W., W. Horner, R., Smith, G., Handa, V., & Sanders, S. (1990). Modeling Construction Labor Productivity. *Constr. Eng. Manage*. 116(4), 705-726.

Watkins, M., Mukherjee, A., Onder, N., & Mattila, K. (2009). Using Agent-Based Modeling to Study Construction Labor Productivity as an Emergent Property of Individual and Crew Interactions. *Journal of construction engineering and management*. 135(7), 657-667.

Wright, T. P. (1936). Factors affecting the cost of airplanes. Aeronaut.Sci. 3(4),122-128.

Interaction between Theory and Practice in Civil Engineering and Construction
Edited by Komurlu, R., Gurgun, A. P., Singh, A., and Yazdani, S.

DURABILITY STUDIES OF PQC MIX INCORPORATING RECYCLED CONCRETE AGGREGATE AND MINERAL ADMIXTURES

G. D. RANSINCHUNG R. N. and ABHISHEK JINDAL

Dept of Civil Engineering Indian Institute of Technology, Roorkee, India

Concrete as known globally is the most widely used construction material in all type of construction works. Recycled concrete aggregates have been eyed as a viable option for part replacement of natural coarse aggregates for concrete production by researchers for quite some time. However, their tendency to bring down the property parameters of new concrete makes it very important to investigate the extent of variation in durability properties of fresh concrete incorporating recycled concrete aggregates. Mineral admixtures are believed to improve the durability and mechanical aspects of concretes. This study makes an effort to understand the effects of incorporating recycled concrete aggregates and different mineral admixtures on durability aspects of pavement quality concrete mixes. In order to investigate the same, recycled concrete aggregates were used along with mineral admixtures in different proportions to study the variation in properties. Concrete mixes prepared with natural aggregates, recycled concrete aggregates and mixes containing recycled concrete aggregates and mineral admixtures were prepared and tested for variations in their durability properties. Laboratory investigations revealed that incorporating mineral admixtures does improves the durability aspects of pavement quality concrete having recycled concrete aggregates. Reduction in values for water absorption, sorptivity coefficients and chloride ion concentrations supported better durability aspects for mixes prepared with recycled concrete aggregates incorporating mineral admixtures.

Keywords: Concrete recycling, Rice husk ash, Bagasse ash.

1 INTRODUCTION

Recycled concrete aggregates as the name suggests are recycled aggregate obtained from concrete wastes. Demolished concrete waste can be procured from various sources such as construction and building waste, waste from demolished road slabs, waste from testing laboratories at construction sites etc. This waste could be processed for obtaining recycled concrete aggregates in requisite sizes as per the work requirements. Recycled concrete aggregates contains aggregates coated upon by the adhered mortar, this mortar is weak and porous in nature which tends to decrease the quality of concrete when incorporated in new works. Few surface treatment methods have been proposed by researchers for separating aggregate particles from adhered mortar. Tam and Le (2007) proposed an acid treatment technique for removing adhered mortar from recycled concrete aggregates. Tsujino *et al.* (2007) proposed a surface treatment technique to peel off adhered mortar from recycled aggregate particles by application of oil-type improving agent and silane-type improving agent. Akbarnezhad *et al.* (2011) proposed a new microwave assisted technique for removal of adhered mortar from recycled concrete aggregates in order to improve their quality. Larbi *et al.* (2000) tried a thermal treatment method to improve the quality of recycled aggregate.

Few researchers such as Hasaba *et al.* (1981), Hansen and Narud (1983) Katz (2003) investigated the content of old mortar and their effects in concrete. Studies carried out by Symonds (1999), Akash *et al.* (2007), Hendriks and Pietersen, Acker (1998), Ransinchung *et al.* (2012), Poon *et al.* (2002) and Farid Debieb (2005) discussed the inferiority and properties of concrete upon the incorporations of recycled concrete aggregates.

The motive of this study is to examine the improvements in durability properties of recycled aggregate concrete with the inclusions of mineral admixtures. Two different mineral admixtures were used as an addition to binder to understand the possible improvements in durability of recycled aggregate concrete due to mineral admixtures thereby suggesting the probable use of concrete for general construction works. This paper discusses a part of continuous study investigating the feasibility of incorporating recycled concrete aggregates in fresh concrete. In this part of study replacement of natural aggregates by recycled concrete aggregates is carried out for 30% as optimum percentage of using recycled concrete aggregates, investigated in previous part of study.

2 MATERIALS AND MIX

2.1 Materials

Recycled concrete aggregates used in the study were manufactured from processing of concrete waste in laboratory. Recycled concrete aggregates were surface treated by pre-soaking them in 0.1 molar acidic solution of H_2SO_4 followed by abrasion, leading to the removal of some adhered mortar in form of powder. In order to obtain best results, all the coarse aggregates were used in pre-saturated moisture state. Table 1 discusses the properties of all aggregates types used in study.

Rice husk ash and Bagasse ash were used as mineral admixtures in this study. Conplast SP430SRV obtained from Fosroc Chemicals India pvt. Ltd. was used in this study as superplasticizer to maintain required slump for the concrete mixes.

Table 1. Physical and mechanical properties of coarse aggregates.

Type of Aggregate	Specific Gravity	Water Absorption (%)	Aggregate Impact Value (%)	Crushing Value (%)
Natural Aggregate	2.677	0.274	13.88	17.775
Untreated RCA	2.417	3.18	22.23	19.42
Beneficiated RCA	2.660	1.88	15.67	17.91

2.2 Concrete Mixtures

A total of 8 mixes were prepared and investigated in this study. One concrete mix with natural aggregates alone was prepared to serve as control mix for conventional concrete (NAC mix), one mix was prepared with recycled concrete aggregates incorporated as 30% part replacement of natural coarse aggregates (RAC mix). Three concrete mixes were prepared for each mineral admixture, incorporating 30% recycled concrete aggregates and 5%, 10%, and 15% mineral admixtures by weight of cement.

Table 2. Mix proportions for materials.

Notation	Mineral Admixture %	Mineral Admixture Kg	Recycled Aggregate %	Recycled Aggregate Kg	Cement (Kg/m³)	Water (Kg)	Fine Aggregate (Kg/m³)	Coarse Aggregate (Kg/m³)	Super plasticizer (%)
NAC	0	0	0	0	400	173	660.22	1180	0
RAC	0	0	30	354	400	167	660.22	826	0
RACRA-1	5	20	30	354	400	167	660.22	826	0.5
RACRA-2	10	40	30	354	400	167	660.22	826	0.5
RACRA-3	15	60	30	354	400	167	660.22	826	0.5
RACBA-1	5	20	30	354	400	167	660.22	826	0.5
RACBA-2	10	40	30	354	400	167	660.22	826	0.5
RACBA-3	15	60	30	354	400	167	660.22	826	0.5

3 TESTING AND RESULTS

3.1 Compressive Strength

Incorporating recycled concrete aggregates reduced the compressive strength of concrete for RAC mix to around 11% in comparison to NAC mix at 28 days as visible from table 3, while the reductions at initial age of testing were not significant. Considerable increase in strength was noticed for concrete mixes with rice husk ash such as RACRA-2 showing a strength increase of around 9% with respect to RAC mix. Further increasing the content of mineral admixture to 15% brought the strength values of RACRA-3 mix at par with those of NAC mix. RACRA-3 mix observed a strength increment of approximately 13% and 2% w.r.t. RAC and NAC mixes respectively. Similar strength results were observed for mixes incorporating bagasse ash with RACBA-3 showing an increase of around 13% and 2.5% w.r.t. RAC and NAC mixes respectively.

Table 3. Properties of concrete mix under study.

Mix Designation	Compressive Strength (MPa) 3 days	Compressive Strength (MPa) 7 days	Compressive Strength (MPa) 28 days	W. A. (%) 28 days	Sorptivity (mm/min^0.5) 28 days	Cl Ion (mg/L) 28 days	Abrasion (gm) 28 days
NAC	18.162	27.169	39.573	2.117	0.116289	25	7.5
RAC	16.242	23.773	35.586	2.311	0.151176	31	9.5
RACRA-1	17.128	23.773	37.210	2.310	0.137221	28	9.5
RACRA-2	16.833	23.035	39.130	2.295	0.113964	25	8.5
RACRA-3	17.424	23.773	40.458	2.250	0.100009	22	7
RACBA-1	17.867	25.102	37.358	2.350	0.134896	24	10.5
RACBA-2	17.276	24.068	38.539	2.295	0.116289	22	9
RACBA-3	18.310	24.364	40.606	2.210	0.106986	18.5	8

3.2 Water Absorption

RAC mix reported water absorption values approximately 9% higher than that for NAC mix. For rice husk ash and bagasse ash admixed concrete mixes, RACRA-3 observed water absorption value only 6.09% higher than that of NAC mix while for RACBA-3 the increase was observed to be 4.29%. However, when compared to parameters recorded for RAC mix, the water absorption values were found to be decreasing for concrete upon admixing mineral admixtures.

Concrete mixes with bagasse ash reported highest reductions with RACBA-3 showing a reduction of around 4.5% w.r.t. RAC mix while rice husk ash admixed concrete showed a maximum reduction of 2.69% for mix RACRA-3. This reduction in water absorption values could be attributed to ultrafine particles of mineral admixtures rendering the mix thereby making it less permeable.

3.3 Chloride Ion Concentration

The increase of resistance to chloride concentration in concrete mixes was observed to be pronounced for concrete mixes containing bagasse ash. Mix RACBA-3 observed the concentration of 18.5 mg/L which was way lower than 25 mg/L as obtained for NAC mix. Addition of mineral admixtures up to 15% did reduced the concentrations of chloride ion in concrete mixes thereby bringing them within the permissible limits as specified by IS-456:2001 and ACI committee.

3.4 Sorptivity

The value of sorptivity coefficient was found to be increased when RCA were induced in the concrete mix making the concrete more water absorbing. Fine particles of rice husk ash when incorporated due to their pozollanic properties and higher surface area tends to reduce the pores thereby bringing down the porosity of concrete. This effect was observed to be enhanced with increase in percentage of mineral admixtures up to 15%. For mixes containing rice husk ash, RACRA-3 reported sorptivity values lower than that of NAC mix while for bagasse ash admixed concrete, RACBA-3 reported values comparable to those of NAC mix thereby indicating an improvement in durability of recycled aggregate concrete.

3.5 Abrasion

Concrete mixes prepared with rice husk ash and bagasse ash reported reductions in abrasion loss parameters. This reduction in abraded weight was observed to be highest for mixes with 15% mineral admixtures RACRA-3 followed by RACBA-3 having comparable parameters to that of NAC mix. Among different concrete mixes, increase in percentage of mineral admixtures led to increase in their resistance to abrasion. RAC mix observed abrasion loss approximately 23% higher than that for NAC mix. However, mixes admixed with rice husk ash recorded 11% and 33% reduction in abrasion loss for RACRA-2 and RACRA-3 mix respectively while RACBA-3 showed a reduction of approximately 17% from that of RAC mix.

3.6 SEM Analysis

SEM examinations were carried out on the fracture surfaces of the different concretes mixes. Figure 1 gives a view of SEM image of NAC mix virgin aggregate concrete-cement interface microstructure noticing formation of Calcium Silicate Hydrate (C-S-H) gels and hydrated compounds. Presence of hexagonal and fine bundle type structures indicate toward the abundance of hydroxide compounds and C-S-H gels as indicated in images below. While the SEM image of RAC mix Figure 2 shows abundance of needle type ettringite and hexagonal calcium hydroxide particles. Investigations of SEM images for mix containing rice husk ash showed increase in content of hydration products. This increase was well expected due to pozzolanic properties of rice husk ash which adds up to the hydration process.

Figure 1. SEM Image of NAC Mix

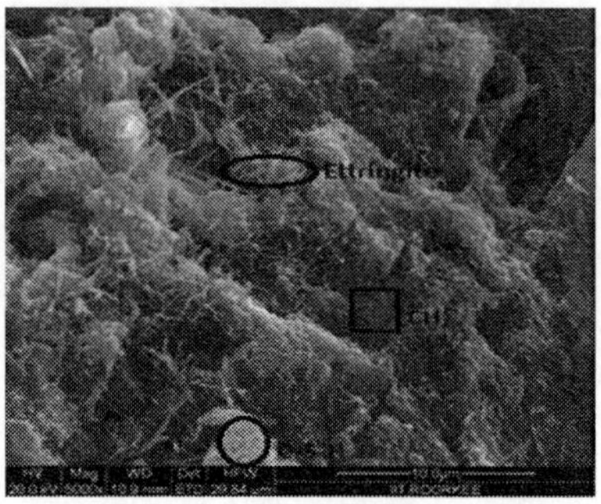

Figure 2. SEM Image of RAC Mix.

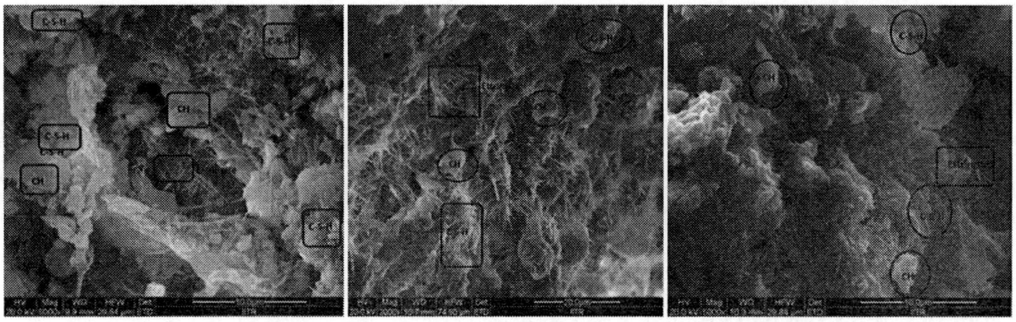

Figure 3, 4, 5. SEM Image of RACRA-1, RACRA-2 and RACRA-3 Mix.

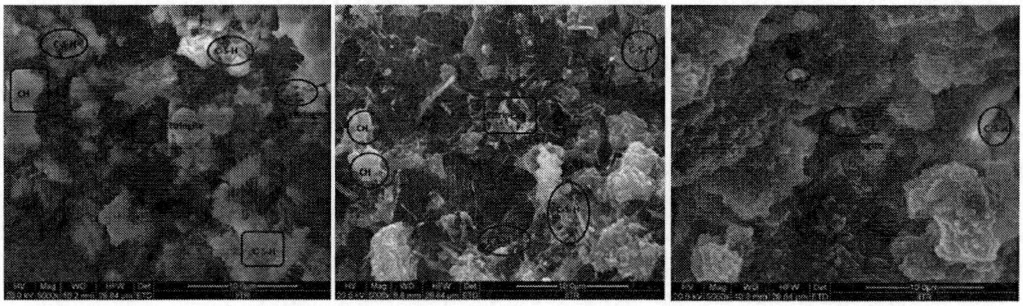

Figure 6, 7, 8. SEM Image of RACRA-1, RACBA-2 and RACBA-3 Mix.

SEM Figures 6 – 8 show that the particles of bagasse ash also added to the hydration products, thereby reducing the voids in concrete and improving the durability aspects. Aigbodion (2008) studied and concluded that the structure of bagasse ash particles could be classified in three main group's i.e. prismatic group, spherical and fibrous groups.

The increase in content of CSH due to pozzolanic properties of mineral admixtures led to better bonding between aggregate particle and mortar thus forming a strong interfacial transition zone (ITZ) among them. This strength of ITZ is visibly reflected as gain in strength parameters of compressive and betterment of durability aspects of concrete.

4 CONCLUSIONS

In the light of above discussed results it could be concluded that incorporating recycled concrete aggregates does leads to inferiority in properties of concrete however, inducing rice husk ash or bagasse ash improves the recycled aggregate concrete thereby depicting similar durability aspects as those of conventional concrete.

References

Akbarnezhad, A. *et al.*, "Microwave-assisted beneficiation of recycled concrete aggregates", Construction and Building Materials, 25(8), pp.3469–3479, 2011.

Acker A.V., "Recycling of concrete at a precast concrete plant. Sustainable construction: use of recycled concrete aggregate", Proceedings of the international symposium, London. 1998 and PP. 321–32.

Aigbodion. V. S, "Potential of using bagasse ash particle in Metal Matrix Composite", PhD work on-going, Department of Metallurgical and Materials Engineering, Ahmadu Bello University, Samaru, Zaria, Nigeria, 2008.

G. D. Ransinchung R. N., Chauhan A. K., Aditya Kumar Anupam, & Praveen Kumar, "Influence of Moisture Content on Use of Recycled Aggregates in PQC", International Journal of Global Technology Initiatives, ISSN 2277-6591, Issue 1, Page No. A-8 – A-15, March, 2012.

Hasaba S, Kawamura M, and Toriik K,"Drying shrinkage and durability of concrete made of recycled concrete aggregates", Japan Concrete Institute 1981; PP:55–60.

Hendricks, C. F., Pieterson H. S., and Fraay A. F. A., "Recycling of building and demolition waste, an integrated approach". In Proceedings of the International Symposium on 'Sustainable Construction: Use of Recycled Concrete Aggregate, London, UK, and PP.419–431.

Hansen T. C. and Narud H., "Strength of recycled concrete made from crushed concrete coarse aggregate", Concrete International (1983), PP. 79–83.

Kanai, S., Debieb, F., and Lakhdar, A., "Performance of concrete made with Recycled concrete and Brick Aggregate", Rio de Janeiro - Brazil, November 11 – 15th, 2005.

Katz, A., "Properties of concrete made with recycled aggregate from partially hydrated old concrete", Cement Concrete, Res 2003; 33:703–11.

Larbi, J.A., Heijnen W.M.M., Brouwerb J.P., E. Mulder, "Preliminary laboratory investigation of thermal treated recycled concrete aggregates for general use in concrete", Waste Materials in Construction-WASCON, Pergamon, England,2000, PP. 129– 139.

Rao, Akash, Kumar, N., Jhab, Mishra S., "Use of aggregates from recycled construction and demolition waste in concrete", V50, PP: 71-81, March 2007.

Symonds, "Construction and demolition waste management practices and their economic impacts", Report to DGXI, (1999), European Commission, PP.81-97.

Tam, V. W. Y., Tam, C. M., and Le K. N., "Removal of cement mortar remains from Recycled Aggregate using pre-soaking approaches"; PP. 82–101, 2007.

Tsujino, M. *et al.*., "Application of conventionally recycled coarse aggregate to concrete structure by surface modification treatment", Journal of Advanced Concrete Technology, 5(1), pp.13–25, 2007.

AUTHOR INDEX

AUTHOR INDEX

ISEC PRESS
PO Box 1334
Fargo, North Dakota 58107
USA

ISBN 978-1-7138-0002-6

9 781713 800026

CURRAN ASSOCIATES INC.
proceedings.com